AF725861

People and Wildlife in Northern North America

Essays in honor of R. Dale Guthrie

Edited by

S. Craig Gerlach and Maribeth S. Murray

Editorial Assistants

Meg L. Thornton, Tom Flanigan, Joshua Reuther and Mark C. Diab

with contributions from

P.M. Anderson, P.M. Bowers, J.W. Brink, L.B. Brubaker, A. Cannon, R. DeAngelo, A. Demma, M.C. Diab, J.C. Driver, A.S. Dyke, J. Fee, T.M. Friesen, D.M. Georgina, S.C. Gerlach, T.E. Gillispie, R.D. Guthrie, D. Hanson, G. Hare, C.R. Harington, J.L. Hofman, B. Kooyman, K.D. Kusmer, A.P. McCartney, A. Magoun, P. Matheus, R.O. Mills, M.L. Moss, M. Nagy, W.W. Oswalt, B. Saleeby, D.L. Sandgathe, R. Sattler, J.M. Savelle, A.V. Sher, R.O. Stephenson, M.L. Thornton, L.C. Todd, P. Valkenberg, D.M. Vinson and D.R. Yesner

BAR International Series 944
2001

Published in 2016 by
BAR Publishing, Oxford

BAR International Series 944

People and Wildlife in Northern North America

ISBN 978 1 84171 236 9

© The editors and contributors severally and the Publisher 2001

The authors' moral rights under the 1988 UK Copyright, Designs and Patents Act are hereby expressly asserted.

All rights reserved. No part of this work may be copied, reproduced, stored, sold, distributed, scanned, saved in any form of digital format or transmitted in any form digitally, without the written permission of the Publisher.

BAR Publishing is the trading name of British Archaeological Reports (Oxford) Ltd. British Archaeological Reports was first incorporated in 1974 to publish the BAR Series, International and British. In 1992 Hadrian Books Ltd became part of the BAR group. This volume was originally published by Archaeopress in conjunction with British Archaeological Reports (Oxford) Ltd / Hadrian Books Ltd, the Series principal publisher, in 2001. This present volume is published by BAR Publishing, 2016.

Printed in England

BAR
PUBLISHING

BAR titles are available from:

BAR Publishing
122 Banbury Rd, Oxford, OX2 7BP, UK
EMAIL info@barpublishing.com
PHONE +44 (0)1865 310431
FAX +44 (0)1865 316916
www.barpublishing.com

Dale Guthrie with Pete, Angel Creek Trail, Alaska, September 2000

CONTRIBUTORS

Patricia M. Anderson
Quaternary Research Center
University of Washington
Box 352100
Seattle, Washington
98195
USA

Peter M. Bowers
Northern Land Use Research
P.O. Box 83990
600 University Avenue, Suite 6E
Fairbanks, Alaska
99708
USA

Jack W. Brink
Archaeological Survey, Provincial Museum of Alberta
12854-102 Avenue
Edmonton, Alberta
T5N 0M6
Canada

Linda B. Brubaker
College of Forest Resources
University of Washington
Box 351360
Seattle, Washington
98915
USA

Aubrey Cannon
Department of Anthropology
McMaster University
1280 Main Street West
Hamilton, Ontario
L9H 1A3
Canada

Rebekka DeAngelo
4141 Autumn Court
Boulder, Colorado
80304
USA

Angela Demma
Minnesota Historical Society
St. Paul, Minnesota
55102
USA

Mark C. Diab
Department of Anthropology
P.O. Box 757720
University of Alaska Fairbanks
Fairbanks, Alaska
99775-7720
USA

Jonathon C. Driver
Department of Archaeology
Simon Fraser University
Burnaby, British Columbia
V5A 1S6
Canada

A.S. Dyke
Terrain Sciences Division
Geological Survey of Canada
601 Booth St.
Ottawa, Ontario
K1A 0E0
Canada

Jennifer Fee
Department of Anthropology
University of California, Davis
Davis, California
95616-8522
USA

T. Max Friesen
Department of Anthropology
University of Toronto
Toronto, Ontario
M5S 3G3
Canada

Dianna M. Georgina
Department of Anthropology
Washington State University
Pullman, Washington
99164-4910
USA

S. Craig Gerlach
Department of Anthropology
P.O. Box 757720
University of Alaska Fairbanks
Fairbanks, Alaska
99775-7720
USA

Thomas E. Gillispie
Department of Anthropology
P.O. Box 757720
University of Alaska Fairbanks
Fairbanks, Alaska
99775-7720
USA

R. Dale Guthrie
Department of Biology and Wildlife
University of Alaska Fairbanks
Fairbanks, Alaska
99705
USA

Diane Hanson
Office of History and Archaeology
Anchorage, Alaska
USA

Jack L. Hofman
Department of Anthropology
622 Fraser Hall, The University of Kansas
Lawrence, Kansas
66045-2110
USA

Gregory Hare
Yukon Heritage Branch, Department of Tourism
Box 2703
Government of Yukon
Whitehorse, Yukon
Y1A 2C6
Canada

C. Richard Harrington
Canadian Museum of Nature, Paleobiology
Ottawa, Ontario
K1P 6P4
Canada

Brian Kooyman
Department of Archaeology
2500 University Drive NW
University of Calgary
Calgary, Alberta
T2N 1N4
Canada

Karla D. Kusmer
Cheatgrass Lane
Sparks, Nevada
89436
USA

Allen P. McCartney
Department of Anthropology
University of Arkansas
Fayetteville, Arkansas
72701
USA

Audrey J. Magoun
Alaska Department of Fish and Game
1300 College Road
Fairbanks, Alaska
99701
USA

Paul Matheus
Alaska Quaternary Center
University of Alaska Fairbanks
Fairbanks, Alaska
99775
USA

Robin O. Mills
Bureau of Land Management
1150 University Avenue
Fairbanks, Alaska
99709
USA

Madonna L. Moss
Department of Anthropology
University of Oregon
Eugene, Oregon
97403-1218
USA

Maribeth S. Murray
Department of Anthropology
P.O. Box 757720
University of Alaska Fairbanks
Fairbanks, Alaska
99775-7720
USA

Murielle Nagy
GÉTIC
University of Laval
Quebec City, PQ
G1K 7P4
Canada

W. Wyatt Oswald
College of Forest Resources
University of Washington
Box 3521000
Seattle, Washington
98195

Becky Saleeby
National Park Service,
2525 Gambell St.
Anchorage, Alaska
99503-2892
USA

Dennis Sandgathe
Department of Archaeology
Simon Fraser University
Burnaby, British Columia
V5A 1S6
Canada

Robert Sattler
Alaska Quaternary Center
University of Alaska Fairbanks
Fairbanks, Alaska
99775
USA

James M. Savelle
Department of Anthropology
McGill University
855 Sherbrook St. West
Montreal, Quebec
H3A 2T7
Canada

Andrei V. Sher
Severtsov Institute of Evolutionary Animal Morphology and
Ecology, Russian Academy of Sciences

Robert O. Stephenson
Alaska Department of Fish and Game
1300 College Road
Fairbanks, Alaska
99701
USA

Meg L. Thornton
Department of Anthropology
P.O. Box 757720
University of Alaska Fairbanks
Fairbanks, Alaska
99775-7720
USA

Lawrence C. Todd
Department of Anthropology
Colorado State University
Fort Collins, Colorado
80523-1787
USA

Pat Valkenburg
Alaska Department of Fish and Game
1300 College Road
Fairbanks, Alaska
99701
USA

Dale M. Vinson
National Park Service
Alaska Regional Office
3636 Gambell Street
Anchorage, Alaska
99705
USA

David R. Yesner
Department of Anthropology
University of Alaska Anchorage
3211 Providence Drive
Anchorage, Alaska
99508
USA

TABLE OF CONTENTS

METHODS

PREFACE

S. Craig Gerlach
University of Alaska Fairbanks

The occasion was the retirement of Russell Dale Guthrie from the University of Alaska Fairbanks in the spring of 1996. Most of the contributors to this volume had gathered at the Westmark Hotel in Fairbanks for the 23rd Annual Meeting of the Alaska Anthropological Association to honor Dale's many contributions to archaeology, anthropology, paleontology, paleoecology, and taphonomy. As time passed and the completion date for the volume became more elusive, several new contributions were added, contributions by people who were unable to attend the original gathering, but who have been significantly influenced in one way or another by Dr. Guthrie and his far-reaching thoughts, theories and ideas. To the authors and contributors, we know that this volume has been a long time in the making; each paper was submitted, then reviewed by three or four independent reviewers, returned for revision, and in some cases re-consideration of style, data, content, and/or theoretical assumptions that were called into question. The editorial process continued here in Fairbanks as we attempted to standardize the volume. Dr. Murray's technical expertise helped bring the volume to completion, as did her familiarity with English as opposed to the American language that I speak, (having been raised in Oklahoma, English is my second language). I did, however, do much of the editing, so if you are unhappy with sentence and/or paragraph revisions and modifications, look to me as the culprit. We hope that everyone feels that the final product is worth the effort.

My first exposure to Dale Guthrie was through reading Body Hot Spots (1967) as an undergraduate. I was so impressed with the book that I presented an hour long seminar reviewing Dale's arguments about the biology of human behavior, human evolution, and the biological basis for sexual selection in *Homo Sapiens*. For reasons that you can draw your own conclusions about, I was immediately in trouble with a few of the cultural anthropologists, but then this has become an all too familiar position for me to be in. The cultural anthropologists were anticipating subsequent developments in post-modern critical theory (remember they didn't like Robin Fox and Lionel Tiger either), and Dale was experimenting with evolutionary theory and evolutionary ecology as applied to anthropology.

Dr. Guthrie came to the University of Alaska in 1963, the same year that he completed his Ph.D. at the University of Chicago. Dale's early interests were in art, biology, and anthropology. He received a B.S. in biology and anthropology at the University of Illinois in 1958 where he and Milford Wolpoff were undergraduates together. Although some might find such diverse interests incompatible, one medium informs the other in his scientific work. Dale continued to pursue his anthropological interests at the University of Chicago where he minored in the subject, and where he was exposed to F. Clark Howell and others who were working on problems in archaeology and human evolution. Dr Guthrie received his Ph.D. from the University of Chicago in 1963 in Geology and Geophysics, with an emphasis in evolutionary biology. Dale's first appointment at UAF was in the Department of Geology and Geophysics, with a primary obligation toward vertebrate paleontology. However, with his interests in evolutionary biology, he eventually gravitated to the Institute of Arctic Biology and the Biology Department. From 1963 to 1996 Dr. Guthrie taught at the University of Alaska in Fairbanks. Dale was a key participant in the landmark Beringian symposium organized by David Hopkins in 1979. Hopkins eventually came to UAF himself to direct the Alaska Quaternary Center, the primary interdisciplinary venue at UAF. Along with Robert Thorson, Jim Dixon, Carl Benson and others, including Dale, the AQC became the primary home for students and faculty who found themselves uncomfortable with the confines of a single academic discipline.

From 1965 to the present, Dale has published approximately 70 articles and two books on a wide variety of quaternary and evolutionary topics. While the number of publications alone is a significant achievement, what is most impressive perhaps is the scope of topics covered and the variety of disciplines represented. In 1992 the Quaternary Sciences Division of the Geological Society of America awarded Dale with the Kirk Bryan Award for his work on "Blue Babe," and the Frozen Fauna of the Mammoth Steppe (1993), research conducted in part with his wife, Mary Lee Guthrie. His current work in progress is a grand evolutionary synthesis of human evolution as viewed through the lens of Paleolithic art, a work that when published will raise eyebrows if not other things across the country. He continues to develop his considerable skills as a sculptor, graphic artist, scientific illustrator, and naturalist. His artwork is easily reviewed as he illustrates many of his own books and papers, and his sculptures of mammoth and bison are on display in some of the downtown Fairbanks art galleries.

I suppose that in some ways what I find most interesting about Dale Guthrie is his creative approach to research. Not only is he is an avid and active hunter, but he uses what he learns directly from his field experiences to better understand animal behavior and human hunting strategies. I have been told by reliable sources that biologists from the Alaska Department of Fish and Game used to track Dale to find out where the big rams were. He also kept a small flock of Dall Sheep in his backyard for several years, and has at various times dragged bison and other carcasses into his front yard to watch them dessicate. There are several areas of research where Dr. Guthrie's work paralleled or anticipated developments in zooarchaeology, taphonomy, and middle-range theory. In two papers written in 1968 Dale raised important questions about preservation bias and stratigraphic contexts for both large and small mammals recovered from Pleistocene Alaskan localities. In these papers he stressed the need to understand depositional contexts, site formation processes, and depositional preservation variables and the ways in which they bias the assemblages with which paleontologists work. These early papers were important because they reflected an awareness of collecting bias, sampling and sampling size, appropriate units of analysis, and other types of taphonomic factors that influence paleoecological reconstruction.

Dr. Guthrie's formal collaboration with archaeologists began in the early 1970s with Roger Powers (UAF) at the Dry Creek site in the Nenana Valley of central Alaska. Although only a relatively small assemblage of fauna was recovered, primarily teeth, Guthrie's efforts to understand the paleoecological context for early Alaskan hunters laid at least part of the foundation for what we all now recognize as the Mammoth Steppe concept. Site seasonality was also reconstructed and linked to the seasonal nutritional requirements (physiological specializations) and seasonal movements (migration patterns) of northern ungulates. Seasonality was used as a springboard to discuss late Pleistocene and early Holocene settlement and land use patterns. Finally, there was a clear recognition in Dale's early archaeological work of the importance of differential butchering practices in assemblage formation and assemblage composition, anticipating in some ways ideas that Lewis Binford later developed in greater detail in Nunamiut Ethnarchaeology (1978).

By the late 1970s the primary evidence for humans in Beringia at or before 12,000 BP consisted of controversial bone artifacts from Old Crow in the Yukon. These are arguments that we are all familiar with, but the spiral fractures, the crude mammoth bone "cores"and flakes, Pleistocene mammal bone with unusual wear and polished facets, and some with cutmarks caught Guthrie's and Robert Thorson's attention. Quite apart from the problem of a lack of site context for most of the specimens, Guthrie and Thorson experimented with alternative explanations to the idea that only humans could have broken bones in such a manner. In a paper entitled "River Ice as a taphonomic agent: an Alternative Hypothesis for Bone Artifacts," Guthrie and Thorson set up a series of experiments to create an experimental frame of reference for understanding what happens to bones when moved in a high energy system such as spring break-up of river ice. They were able to show that as bones were moved over long distances from primary to secondary contexts many of the very atrributes that were being attributed to humans were produced by transport in a frozen condition. This is not the place to review all of the experiments undertaken, but one is worth mentioning because it captures the spirit with which Guthrie approached his research. As the story goes, Guthrie and Thorson encased frozen bones in blocks of ice, tied them to the back of a pickup truck, and then careened through the streets of Fairbanks in an effort to simulate break-up by banging into whatever got in their way. I am sure that the experiment was more sophisticated than this, but this is the image that I prefer and will hold on to.

Guthrie may have retired from the University of Alaska, but he has not retired from active research. Apart from the new book that will soon be completed, he has two papers in the current volume. I am not sure if this is a first, but it is not common for the honoree of a volume like this to contribute two new research papers of his own. Dale Guthrie, Dave Hopkins, and a core of distinguished quaternary researchers left a real hole in what is still an active quaternary research program at UAF. I doubt is there is one quaternary researcher, student or otherwise, who was not mentored or inspired in one way or another by R. Dale Guthrie. So, Dale, to borrow from a bad commercial, "this Bud's for you."

STUDYING NORTHERN BONES: PERSPECTIVES ON PALEOBIOLOGY AND ZOOARCHAEOLOGY IN HIGH LATITUDE ENVIRONMENTS

David R. Yesner
University of Alaska Anchorage

Northern North America is an ideal place for studying the climatic, ecological, and evolutionary relationships affecting animals and the human hunter-gatherers that depend on them. There are a number of reasons why this is the case. First, faunal assemblages often preserve well and are highly visible in northern environments. This provides the opportunity for detailed studies of the ecology and distribution of animals and the subsistence patterns of human hunters. However, there are a number of taphonomic factors that are relatively unique to northern regions. Second, high latitude environments contain a number of unique features, including severe climatic conditions (especially in winter), extreme climatic and biotic seasonality, episodic fluctuations of animal populations, and animal population aggregations that are unmatched in low latitude environments and that form the basis of studying both animal and human adaptive responses to these conditions. Third, the original entry into and colonization of North America was, in the opinion of many archaeologists, through an arctic avenue. For this reason, Pleistocene and early Holocene faunal assemblages from northern regions are critical to deciphering the way of life of the earliest migrants to the New World, as well as their impacts on the native fauna. Furthermore, in a fashion unmatched in the Old World, where domestication of reindeer took place and southern peoples pushed strongly into northern habitats, hunter-gatherers in northern North America retained that niche until the time of European contact, with only minimal adoption of iron metallurgy after 1000 AD. In fact, it is a phenomenon matched only in areas such as Australia and perhaps some regions of Africa and western North America where continuity in hunter-gatherer lifestyles also took place. As such, northern North America is an excellent place for testing hypotheses about the impact of climatic and environmental changes on human societies, and the long-term evolution of technological and other adaptive responses by hunter-gatherers. In order to do so, it has been necessary to develop and test new theories and methodologies utilizing techniques appropriate to northern species, e.g., the biogeopgraphy of northern fauna, the development of animal utility indices for various northern terrestrial and marine taxa, the reconstruction of butchering patterns for northern animals (both extinct and extant), and the analysis of blood residues keyed to various northern species. All of these issues are explored in a more detailed fashion below.

PRESERVATION AND TAPHONOMY OF FAUNAL ASSEMBLAGES IN NORTHERN ENVIRONMENTS

First, faunal assemblages often preserve well in the cool conditions associated with arctic and subarctic environments. In the Far North, largely above the Arctic Circle, continuous permafrost helps to protect bone and other organic remains from decay. In some areas, ancient limestones have been exposed, rocks which not only provided stone materials such as chert for the weapons of early human occupants, but also caves in which animal bones, both paleontological and archaeological, have been well preserved.

In addition, as a part of adaptation to the conditions of the North in both the Pleistocene and Holocene, large bodied mammals have dominated faunal assemblages. It was the well-preserved evidence of Pleistocene megafauna in northern and interior Alaska that led Dale Guthrie in the 1960s to describe for the first time the "mammoth steppe," where large mammals dominated faunal assemblages in terms of the numbers of bones and individual animals represented.

Although the diversity and biomass of the steppe/tundra that supported these animals has been challenged, no one has seriously taken issue with the notion that, as in contemporary northern Eurasia, large mammals dominated these assemblages. Even with the disappearance of some of these taxa in the late Pleistocene (e.g., proboscideans, horses, sloths, camels, and large predators such as lions and short-faced bears), and the diminution in size of others in the Holocene (e.g., bison, wapiti, and moose), northern North America continued to harbor significant numbers of relatively large mammals. This is largely because, following Bergmann's Rule, among polytypic animal families (e.g., the Cervidae or deer family), larger body forms should be found in higher latitude environments, as a mechanism for conservation of heat energy through a reduced ratio of surface area to volume. As a result, the presence of such large animals, with their correspondingly larger bones, greatly increases the "visibility" of such taxa in both paleontological and archaeological assemblages.

Of course, small mammals are well represented in the Arctic as well, and are often prevalent in cave faunas and other paleontological assemblages. A somewhat different story is true of archaeological deposits, however. Northern hunters generally focus on large game whenever they can, especially those (like caribou) that aggregate in large numbers, both for reasons of availability and energetic efficiency. In addition, these animals represent large packages of fats and protein, important for adaptation to northern environments where higher basal metabolism offers a protective response to cold conditions by increasing thermogenesis. The presence of large animal bones, and often large assemblages of large animal bones, again increases archaeological site visibility, and thus an opportunity to study prehistoric human subsistence strategies. Small mammals present in some sites result from secondary deposition, through burrowing, (particularly in open-air sites), or deposition by large mammals or birds, (particularly in cave sites).

Further south, discontinuous permafrost continues into subarctic regions, sometimes counteracting the destructive effects of acidic soils normally associated with boreal forest environments. Where karst features exist in these environments, excellent faunal preservation may again be associated with cave deposits.

In coastal zones of southern Alaska and adjacent British Columbia, under slightly warmer conditions, the deposition of shellfish (mollusk and echinoderm) remains in archaeological sites additionally helps to neutralize soil acidity and make faunal assemblages there available for study. Fish and sea-mammal remains are often abundant in these sites, and offer opportunities for study of the ecology, distribution, and human utilization of these taxa.

Although many of these preservational factors are unique to the north, others apply to environments further south. On the northwestern Plains, for example, highly arid conditions, associated with shortgrass prairies, provide similar preservation of faunal assemblages. In addition, these environments also supported large game animals, in both Pleistocene and Holocene times, which are highly visible, both paleontologically and archaeologically. Aggregation of large herds of bison, wapiti, and antelope also promote high visibility of archaeological sites on the landscape, in a similar way to the Far North.

While cold environments tend to retard some of the biological decay operative in warmer environments, northern faunal assemblages are by no means insulated from all taphonomic effects. Permafrost itself has numerous impacts, along with cryoturbation, solifluction, and effects of freeze/thaw cycles. River ice may also impact bones redeposited along gravel bars, as Guthrie (1995) demonstrated. These factors have deeply impacted the analysis of paleontological and archaeofaunal assemblages (Thorson 1990), particularly in areas of shallow soils, steeper slopes, and north-facing exposures, or where exposure and redeposition has occurred. This raises some potentially significant difficulties in the interpretation of northern faunal assemblages.

Another major taphonomic factor operative in northern and arid environments is gnawing on bone both by rodents and carnivores. In these open environments, sources of calcium and phosphorus may have been few and far between, and large mammal bones and antlers provided excellent sources for these nutrients. One of the chief taphonomic problems for terrestrial faunal assemblages is distinguishing the Causes and nature of bone destruction associated with these agents.

Perhaps a more intractable problem is the attribution of sources of bone accumulations to human and non-human agencies. It is rarely a problem for open-air archaeological sites, although unquestionably both carnivores and birds, as well as burrowing small mammals, are making contributions to such sites. In such cases detailed study of bone part distributions and mortality curves is only partially helpful, since other carnivores can mimic human hunting behavior, and both attritional and catastrophic mortality profiles can be characteristic of both. It is a much bigger problem for cave sites, however, where it is sometimes difficult if not impossible to differentiate between carnivore dens and human camps. Distribution of gnawing patterns and animal toothmarks may be helpful in making some differentiations.

In coastal sites, the impact of these taphonomic agents tends to be less pronounced, but other factors intervene. While the problem for terrestrial northern environments is in distinguishing between the effects of human hunters and animal scavengers on faunal assemblages, there are relatively few archaeologists who believe that humans themselves were fundamentally scavengers, with the exception of scavenging bones and ivory for tool production (cf. Yesner 1995). However, whalebone found in northern archaeological sites may have been obtained through active hunting or scavenging of animal carcasses, and if scavenged it may be a whole or partial carcass, or only the bones for tool-making and/or house construction (Yesner 1993). This may lead to vastly disparate interpretations of the dietary importance of whales, e.g., in Thule period sites, depending on the assumptions one makes about how they were obtained and used. Because they are such large mammals, even a relatively modest contribution in numbers of individuals means a significant contribution to the overall diet (McCartney 1980)

LONG-TERM STUDIES OF HUNTER-GATHERER ADAPTATIONS

Hunter-gatherers have utilized landscapes in northern North America for at least 12,000 [radiocarbon] years (Yesner 1995, West 1996). How much longer is a matter of debate. Although some capability of adaptation to harsh northern Eurasian environments must have existed as early as Neanderthal times, and the Lake Baikal region may have been inhabited by the mid-Pleistocene, the paucity of archaeological sites in northeastern Russia (east of the Lena River basin) predating 14,000 years ago suggests that this area, possibly the original home of extremely cold-adapted species such as the woolly mammoth (Sher 1996), was not penetrated until after that time. Two sites, Berelekh and Ushki Lake in Kamchatka, have provided radiocarbon dates in that interval. A number of archaeologists believe that a coastal penetration of eastern Beringia (Alaska) could have taken place as early as 13-14,000 years ago. At this time, climatic conditions in Alaska had begun to ameliorate with the onset of the "Birch Period" (Ager 1975), deglaciation of the Cordilleran ice sheet was underway, and icebergs had begun to vacate the shores of southeastern Alaska (Ager 1999:pers. comm.; Mann and Hamilton 1995). However, viable terrestrial habitat may not have been available for human settlement in the intervening region of southwestern and southcentral Alaska until the early Holocene (Yesner 1995, 2000). Alternatively, human populations may have followed the course of the PaleoYukon River eastward from its mouth as late Pleistocene sea levels rose, eventually drowning the former land bridge. In that light, it is intriguing that the earliest well-dated sites in interior Alaska are found along the Tanana River and its tributary, the Nenana River. The Tanana itself is a tributary of the mighty Yukon where, one might predict, early sites will eventually be found in areas constrained by bedrock formations. A few early dates in northern Alaska suggest that it also may have been occupied by 11,700 years ago, but perhaps only scantily until the early Holocene. By 11,500 years ago, a corridor may have opened between the Cordilleran and Laurentide ice sheets sufficiently for southward movement (Duk-Rodkin 1997), although this may be irrelevant if earlier movement occurred along the southeastern Alaskan and British Columbia coasts (Dixon 1999). To date, however, no sites in that region predate the early Holocene. Clearly, the California and Peruvian coasts were occupied as early as 11,000 years ago, and the Chilean coast perhaps even earlier, but whether these sites can be linked to each other or to early coastal Alaskan sites, like beads on a chain, depends upon the demonstration of similarities of artifact assemblages which has not yet been demonstrated.

Few of the early open air sites in the North have produced well-preserved bone remains (or even organic artifacts) to use along side the traditional stone tools and site locations in reconstructing human subsistence and settlement patterns. One exception is the Broken Mammoth site in the Tanana Valley (Yesner 1995); a few fragmentary teeth were also recovered at the Dry Creek site (Guthrie 1983) and there are some bone remains from Onion Portage (Anderson 1988). Faunal remains are better preserved in cave sites, and early radiocarbon dates from these bones have been used to support notions of hypothetically early human occupation (i.e., predating 12,000 years ago). However, the analyses of these sites and faunal assemblages are beset by many of the taphonomic problems raised above. Issues of whether carnivores contributed significantly to late Pleistocene faunal assemblages in caves where human artifacts have also been found have been raised and accepted for Alaskan sites like the Porcupine Caves (Sattler et al. this volume) and Trail Creek Caves (Vinson 1993), but judged equivocal at the Lime Hills Caves (Georgina this volume) and rejected at the Bluefish Caves in the adjacent Yukon Territory. At Charlie Lake Caves in northern British Columbia (Driver, this volume), most of the faunal remains seem attributable to human occupants, although (as in the open-air sites) burrowing rodents also make a contribution.

The faunal remains that have been preserved in northern sites help to make a contribution to the understanding of human impacts on the extinction of late Pleistocene faunas, as well as to the overall picture of early human subsistence. In the few cases in which faunal remains have been well-preserved in clear late Pleistocene and early Holocene archaeological contexts, it appears that bison and wapiti were the major contributors to the large mammal inventory, and that they may have persisted in interior Alaska and the adjacent Yukon until well into the Holocene. The relationship of the late Pleistocene and early Holocene remains of *Bison priscus* to later Holocene remains of *Bison bison athabascae* is an issue that needs to be resolved, but is beset by small samples and a lack of diagnostic horn cores. The complexity of this issue has been raised by Guthrie (1980), Geist and Karsten (1977), McDonald (1981), and others, and needs to be pursued in conjunction with similar studies on the relationship between *Bison occidentalis* and *Bison bison* in the western Plains. In any case, although a few terminal Pleistocene dates for mammoth remains have been presented from paleontological and archaeological contexts, the bulk of data suggest either extinction or minimal contact of mammoths and humans until the latter reached northwestern North America south of the ice sheets. Perhaps the infamous Clovis fluted point was invented there as a mammoth hunting tool, eventually making its way back north through the now open corridor between the receding ice sheets. Precisely because faunal remains are relatively few and far between, even in northern sites, other types of data, particularly blood residue from stone tools, may offer our best hope of resolving these issues in sites where bones are absent.

The long-term presence of hunter-gatherers in interior Alaska yields a significant opportunity for the study of changing human hunting strategies and settlement patterns in response to climatic, vegetational, and faunal change. Long-term climatic and vegetational records in the form of studies of glacial expansion and contraction, snow line reconstructions, lake level fluctuations, and pollen sequences have provided the basic data used to test hypotheses about human demographic flux and shifting subsistence regimes. From these studies, it is generally clear that the late Pleistocene/early Holocene provided a significant ecological "window of opportunity" for early human immigrants, with abundant game associated with the open birch/poplar parkland environment. Evidence from the Broken Mammoth site (Yesner 1995) also suggests that a variety of small game, waterfowl, and even fish were equally important. With the advent of the spruce forest, and increasingly arid conditions after 8,000 years ago, some depopulation of the interior may have occurred, possibly driving some people toward the coast. The little faunal evidence we do have suggests that caribou may have become the mainstay for interior peoples, moose a distant second, and small game very important (Yesner 1989). With the establishment of modern salmon runs in mid-Holocene times, human populations undoubtedly rose, although we have much to learn about this process. Well-preserved faunal assemblages tend to date only from the late Holocene, and these are dominated again by caribou remains, although this is affected by the visibility issues raised above.

When did the coastal option for northern hunter-gatherers first become available? We don't know the answer but based on current data it is clear that there has been coastal settlement in some fashion for 10,000 years in the ice-free areas of southeastern Alaska and coastal British Columbia (the northern Northwest Coast). Dixon et al. (1997) have shown that early human remains from On-Your-Knees Cave reflect an individual consuming a significant amount of some type of sea food. Hidden Falls on Baranof Island and the Chuck Lake site on Heceta Island in southcentral Alaska, as well as Lawn Point on the Queen Charlotte Islands and Namu on the mainland coast of British Columbia, all provide evidence for significant use of fish, shellfish, and sea mammals during the early Holocene. However, they also make clear that terrestrial mammals were more important than at later time periods, and that use of salmon and shellfish intensified during the mid-to-late Holocene. Eventually, an intensive focus on these, particularly salmon, ushered in the sedentary lifestyles, large village populations, surplus food production, and patterns of social stratification and warfare so characteristic of the northern Northwest Coast.

Further to the west, in the contemporary Alutiiq (Pacific Eskimo) region, maritime adaptation appears to begin around 8,500 years ago, as evidenced on Anangula Island and Hog Island (eastern Aleutians) as well as Sitkalidak Island (Kodiak archipelago). There is tenuous evidence for relationships of both the early southeastern and southwestern cultures to early Holocene cultures in the interior, since all contain microblade assemblages. A few seal bones suggest that maritime hunting was being carried out, but most inferences rely on site locations. Systematic shellfish (mollusk and echinoderm) exploitation does not begin until after 4,000 years ago, in a similar fashion to the Northwest Coast. An exception is on Kodiak Island and the eastern Alaska Peninsula, where some middens date to the early Ocean Bay culture, perhaps as far back as 7,000 years ago. Preliminary evidence suggests (as in coastal California) attention early on to near-shore sea mammals such as sea otters, with larger sea mammals such as sea lions being taken in later prehistoric times. It also suggests more attention to cod and halibut, and a later shift to salmon (see Bowers and Moss, this volume). To some degree the development of these activities, both in southeastern and southwestern Alaska, are affected by Holocene climatic changes such as the Little Ice Age, which may have iced in protected bays, and possibly caused abandonment of some locales. Increased storminess associated with shifts in oceanic storm tracks may have caused increased die-offs of sea birds at different times. Changes in water temperature associated with El Nino-Southern Oscillation (ENSO) events may have episodically affected fish runs and sea mammal migration patterns, either directly or through changes of incidence in diseases such as paralytic shellfish poisoning (see Hanson and Kusmer, this volume).

In the regions to the north, where ice is present at least part of the year along the coast (from Bristol Bay, Alaska in the west across to Labrador in the east), the impact of such climatic change was probably more substantial. Shifting global climatic patterns affecting the Siberian high pressure and Aleutian low pressure air masses would have affected not only storminess but the latitudinal distribution of sea ice. Enigmatic protoEskimo cultures such as Old Whaling may best be explicable in those terms. Played out against this backdrop of climatic response was a longer-term signal of increased technological evolution outlined by Dumond (1975). An increased harvest of larger sea mammals, beginning with bearded seals and beluga whales, culminated in the development of full-scale hunting of baleen whales by around 500 B.C., not only in northernmost North America, but throughout the North Pacific Rim. The best of whaling times was associated with the early Thule villages in northern Alaska and high arctic Canada, where earlier Dorset non-whalers were apparently displaced (Maxwell 1985). The importance of the high fat yields from these animals for arctic survival cannot be overestimated. Nevertheless, hunting the largest baleen whales (e.g., bowhead whales) strains the capacity of even the most sophisticated skin boats and crew organizations (Gerlach *et al.* 1993), so that it is not surprising that two year old juvenile whales were favored, a point which probably simultaneously optimizes the nutrient yield and the energetics of pursuit.

As we get closer to the "ethnographic present" (i.e., the time of European contact) on both the coast and interior, several things become possible. First, faunal remains tend to preserve better, allowing greater site visibility and more data for paleoeconomic reconstructions. Second, it becomes possible to connect with greater confidence these societies with those reported ethnographically. It is impossible to overestimate the value of ethnographic data available for the North in helping to understand the archaeological record of hunter-gatherers, particularly for the late prehistoric period. A number of studies, such as Binford's (1978) on Nunamiut ethnoarchaeology have been designed with the archaeological record in mind, but numerous studies undertaken over the last two centuries contain useful elements in reconstructing everything from the butchering patterns to the resource preference schedules of northern peoples. Certainly, a more complete understanding of the response of human societies in the north to climatic change over time, as well as the capacity of northern peoples to adapt to harsh conditions including extreme climatic seasonality and episodic animal population fluctuations, requires a biocultural approach that integrates archaeological, biological, and ethnographic data in a rigorous fashion.

THE CONTRIBUTIONS OF R. DALE GUTHRIE TO PALEOBIOLOGY AND ARCHAEOLOGY AND ORGANIZATION OF THE VOLUME

Dale Guthrie has contributed immeasurably to all of the important issues concerning northern paleobiology and archaeology outlined above. These include the taxonomy, ecology, and behavior of both small and large mammals of northern North America during the Quaternary period; the reconstruction of the ecosystem of northern North America during that period based on faunal assemblages; and the hunting strategies and settlement patterns of early human occupants of the Far North, based on animal bones from archaeological sites. All of these studies are enriched because of Dale's contributions, and we dedicate this volume in his honor.

The current volume is largely comprised of a number of papers on various aspects of northern paleontology and zooarchaeology given at the Guthrie symposium in March 1996 at the annual meeting of the Alaska Anthropological Association in Fairbanks, Alaska. They also include a number of additional papers on similar subjects solicited from individuals who could not be present at the time of the original symposium. The papers focus on a number of broad issues of concern to Dale: northern animal taxonomy, ecology, and behavior; bone taphonomy; evolution of northern landscapes; human impact on northern animal communities, including the development of hunting technologies; and insights to be gained from the lifestyle of contemporary hunter-gatherers in northern regions. Together this set of papers represents an excellent cross-section of current topics in northern paleobiology and zooarchaeology that are of broad interest to all those interested in animal and human behavior in the north, and particularly to those interested in past human-animal relationships as recorded in the various proxy data sets with which paleoecologists work.

The volume is divided into three sections: one on paleoecology, one on archaeology, and one on methods. Papers within each section are organized alphabetically. In terms of subject matter, 13 of the papers deal essentially with terrestrial mammals and birds and their exploitation by humans, while seven deal with aquatic mammals, birds, and fish. Ten of the papers are based essentially on Alaskan data, nine on Canadian data, and one relies on data from the western US.

The papers in the first section, paleoecology, focus on paleoenvironmental reconstruction based on analysis of mammalian faunal assemblages from Alaska and Canada. Jonathan Driver's paper on "Paleoecological and Archaeological Implications of the Chalie Lake Cave Fauna," including a review of relevant paleontological and archaeological data from other sites in northern British Columbia, demonstrates that large mammal populations were absent during the last glacial maximum (between 21,000 and 11,600 years ago), reappearing at a time coincident with deglaciation of the region and colonization by open country vegetation (aspen, birch, and willow) similar to that of the "Birch Period" in interior Alaska. It was at this time (around 10,500 years ago) that the earliest archaeological sites in the region appear that contain well-preserved fauna (Charlie Lake Cave and Vermillion Lakes), and both suggest that (like interior Alaska) bison was the most important large game species. (No horn cores are present, but as in interior Alaska, this appears to be the larger *Bison priscus*).

Within a relatively short period of time, perhaps no more than 500 years, this open parkland was replaced by more closed boreal forest (by 10,500 years ago), once again analogous with the process occurring in interior Alaska, only on a more rapid time frame. The best mammalian indicators of this transition are shifts in lagomorphs (from Arctic hares or jackrabbits to snowshoe hares), the disappearance of ground squirrels and collared lemmings, and the appearance of species such as beaver. Again, this parallels what we see at the Broken Mammoth site in interior Alaska. However, this area was apparently occupied first by Laurentide ice, perhaps as late as 13,000 years ago, and then by a large glacial lake (Glacial Lake Peace) until perhaps 10,500 years ago, when the Charlie Lake Cave site was initially occupied. Driver argues, in fact, that the Charlie Lake Cave faunas were produced by some of the first hunters to be retreating northward along the eastern flanks of the Rocky Mountains as the environment ameliorated (and bison populations flourished). In doing so, he creates a strong contextual argument for a relatively late opening of the "western corridor," followed by a reverse migration of hunters toward Beringia. The main subsistence base for these northward-moving hunters was bison, which were butchered at or near the site. After deposition of the bones, carnivores created secondary alterations to the faunal assemblage. In this paper, then, Driver has contributed significant new information about the paleoenvironment and the exploitation patterns of early human hunters in the Peace River country of British Columbia.

The second paper in the volume was contributed by Dale himself, and both are vintage Guthrie material. In the first paper, entitled "Paleobehaviour in Alaskan Pleistocene Horses" Guthrie confronts an extremely male-biased distribution pattern in Alaskan horse fossils collected predominantly from the Fairbanks gold mining gravels by Otto Geist, the "father" of Alaskan paleontology, over a few decades. Offering several hypotheses to explain this male-bias in the horse fossils (misidentification, differential destruction, differential scavenging, collector bias, and differential animal behavior/habitat use), he systematically examines and rejects all of the hypotheses but the last: that male horse groups were differentially utilizing the upland hills, and their bones were eroded and recovered from these areas by gold mining operations. He offers this conclusion as a cautionary note to archaeologists, i.e., that not all skewed sex distributions in fossil assemblages are necessarily the result of deliberately skewed hunting strategies). Personally, however, I think that the collector bias hypothesis still has a lot to recommend it, particularly because, as Guthrie notes, other Pleistocene mammal collections are beset by the same male bias.

In his second contribution to the volume, Dale (along with Sher of the Russian Academy of Sciences and Harington of the Canadian Museum of Nature) tackles the paleoecological significance of new radiocarbon dates on saiga antelopes from Alaska, Canada, and Siberia. These dates suggest two clusters: one corresponding to Isotope stage 3 (25-40,000 BP), and the other to Isotope stages 1 and 2 (12-15,000 BP) correspondingly provides very few dates, suggesting the possibility that saiga antelope, while well adapted to dry, dusty steppe and semidesert environments throughout central Asia are really not well adapted to cold steppes, at least not the extremely cold version that developed in northeastern Asia and Beringia during the Late Glacial Maximum.

Saiga antelopes were never a major presence in the Beringian fauna, and never dispersed southward from Beringia to become a more general element of the Rancholabrean fauna of North America. The woolly rhinocerous and short-faced bear also seem to have been excluded from the Late Glacial Maximum, and the woolly rhinos never entered the record of eastern Beringia (Alaska and Yukon). In fact, the reappearance of saiga in the terminal Paleolithic (12-15,000 BP) suggests they were tolerant of the scrub growth (dwarf birch, willow, and alder parkland) vegetation of that period, and actually increased in numbers during this warmer period, until the much more mesic (warmer and wetter) conditions developed that were associated with the so-called "poplar rise."

Turning to questions of maritime paleoecology, Hanson and Kusmer ("Sea Otter Scarcity in the Strait of Georgia, British Columbia") present a problem of interest to archaeologists working in the entire North Pacific region, in fact from coastal California to the Aleutian Islands: describing the prehistoric (and modern) distribution of the sea otter, an animal utilized throughout that zone. Although the general association between sea otters and rocky, ice-free coasts has been known for a long time, the usual explanations for habitat preference involve such things as the presence of kelp beds for protection and as a habitat for sea-urchins, often a primary food source. Hanson and Kusmer, however, offer an innovative addition to this explanation: adaptations to high levels of paralytic shellfish poisoning (PSP). They note that, for example, the shallow, silty bays of the inner coast of the Strait of Georgia have relatively few sea otters, as do more northerly areas such as Prince William Sound, Alaska. For the Strait of Georgia, the fewer number of sea otters are linked to the much higher PSP levels in that region. The fact that this pattern was true of the past as well as the present is suggested both by ethnohistoric accounts, and by the absence of sea otter remains in Strait of Georgia archaeological sites going back at least 2500 to 3500 years, at the same time as such remains were present in outer coastal sites. There are no reasonable cultural hypotheses that could account for this differential absence, so their suggestion is a useful one.

Kooyman and Sandgathe, in their chapter on "Sexually Dimorphic Size Variation in Holocene Bison as Revealed by Carpals and Tarsals," return us to terrestrial paleoecology, specifically to the question of Holocene bison evolution. They provide an historical perspective on the question, noting the complexities of the taxonomic issue, and the fact that previous taxonomies (*Bison [bison] occidentalis*, *B. [b.] antiquus*, *Bison [b.] bison*, etc.) have been based almost exclusively on male horn core characters, with only minimal attention given to metapodials by Guthrie (1980) and MacDonald (1981). Kooyman and Sandgathe utilize a sample of bison carpal and tarsal bones from a series of archaeological sites in Alberta, Canada to redress this problem. They demonstrate that non-cranial bones can be used effectively to examine historic changes both in overall bison size and in relative proportion of body parts. Analysis of their sample of carpal bones clearly shows that a pattern of size diminution, differentially affecting male bison, continued throughout the mid-to-late Holocene period. At the same time, the tarsal bones show less size change, indicating that the forelimbs have reduced the most. Exactly what this means in terms of the taxonomic problem is difficult to say: it might suggest that in situ microevolution was a local (or at least northern Plains) phenomenon, so that at no point can one clearly differentiate between these taxa. In any case, it tends to support a point that Guthrie has been arguing for years: that most of the size diminution affected body parts related to male reproductive displays.

According to Guthrie's model, the factors involved with late Pleistocene size diminution would have involved a combination of reduction in the nutrient quality of forage and the shunting of energy away from biologically expensive intraspecific agonistic competition. Kooyman and Sandgathe have taken this process one step further by relating it to continued microevolutionary pressures for size diminution throughout the Holocene period. A very different argument for relating bison size diminution to human hunting pressures is presented elsewhere in this volume by Hofman and Todd.

In Matheus' paper "Pleistocene Carnivores and Humans in Beringia: Did Short-faced Bears Really Keep People Out of North America?, we turn from prey to predators in considering not only potential competition with human hunter-gatherers, but possible predation on human hunter-gatherers. Here Matheus takes up the hypothesis of Guthrie's mentor, Valerius Geist, who has hypothesized that humans were excluded from the New World prior to around 12,000 years ago by North American large carnivores, particularly the giant short-faced bear (*Arctodus simus*), against which they had few or no defenses. This, of course, assumes that humans were not present in the Americas prior to 12,000 years ago, which is itself a hotly debated issue. According to this hypothesis, furthermore, Ice Age hunters did not have adequate means to protect their kills from these predators. However, based in part on better dating of Pleistocene faunas, Matheus is able to show that the number of sympatric predators present at any one time in given regions of North America was no larger than in the Old World, nor did they have a larger body size or behave more aggressively. Focusing specifically on the short-faced bear, Matheus is able to show that while definitively a meat-eater (based on stable isotopes), its gracile body build indicates a scavenger that covers large territories, rather than an active hunter. Furthermore, based on new radiocarbon dates it appears to have become extinct by around 20,000 years ago, well before most archaeologists believe humans entered either northeastern Siberia or the New World.

The paper by Robert Sattler, Dale Vinson, and Thomas Gillispie on "Taphonomic History of Beringian Cave Faunas and the Disappearance of the Mammoth Steppe" returns us once again to the problem of how cave faunas might be used to elucidate the paleoecology of the late Pleistocene period. Sattler *et al.* focus on northern caves which have not definitively shown human presence in direct association with late Pleistocene faunal assemblages. Their data come primarily from excavations at the Lower Rampart Caves and Trail Creek Caves in Alaska, with comparisons made to other excavations at the Lime Hills Caves (see below), as well as Bluefish Caves in the Yukon Territory. Briefly, they describe the complex taphonomic history of these cave faunas, including carnivore gnawing, faunalturbation, weathering, and pedogenic processes. Based on skeletal part representations, bone breakage patterns, and gnaw and tooth marks, they conclude that carnivores were the major, if not sole, contributors to these faunal accumulations. Furthermore, they note that tool cutmarks on bones, or even human bone breakage, do not definitively indicate human contemporaneity with the Pleistocene faunal assemblages: humans may well have scavenged from such carnivore bone piles in order to obtain the raw materials needed for tool manufacture. In their model humans and carnivores are joint transformers of the assemblages, with each modifying the remains left by the others (a process that probably applies to European cave sites as well). Finally, using new radiocarbon dates, Sattler *et al.* are able to show convincingly that some of the key Beringian herbivores, especially mammoth and horse, appear to have become extinct by 13,000 years ago, and suggesting that their absence may have been a limiting factor for carnivores after that point in time. This collapse of the mammoth steppe, apparently associated with the rapid climatic warming of the "Birch Period," may have been a limiting factor on early human migrations as well, at least until bison and wapiti populations increased during the latter phase of that period (the "Poplar Rise"), as noted above.

The paper by Stephenson, Gerlach, Guthrie, Harington, Mills, and Hare, on "Wood Bison in Late Holocene Alaska and Adjacent Canada," is an exhaustive tour de force on the subject of late Pleistocene/Holocene survival of bison in extreme northwest North America. It is a textbook example of how best to integrate biological, paleontological, archaeological, and ethnohistoric data (including traditional place names) to produce a synthetic statement on the regional importance of Holocene bison. Armed with an impressive set of 50 radiocarbon dates and extensive oral histories, the authors have now definitively demonstrated that bison continued to inhabit the region continuously throughout the Holocene. Moreover, bison were present in the region until at least as 170 years ago, according to the radiocarbon dates (i.e., within 200 years of European contact), and probably later according to the oral histories (as the authors point out, paleontological specimens are only spottily available, particularly in more recent contexts). Although the sample sizes are small, the authors make a convincing case that the large-horned *Bison priscus* was present until the early Holocene, probably as late as 9300 years ago, according to our data from the Broken Mammoth site in the Tanana River valley of east-central Alaska (Yesner 1996), to be replaced by *Bison bison* (cf. *athabascae*) by as early as 8100 years ago. According to this framework, *Bison bison* then expanded northward in the early Holocene, with *B. [b.] athabascae* representing the variant at the northern end of its range. Clearly, the distribution of this subspecies was once much more extensive than at present. Although the set of factors underlying this expansion is not clear, it may be linked to the increasing aridity of interior Alaska and the Yukon during the mid-Holocene, as evidenced by palynological, sedimentological, and lake-level data. Thus, although spruce had expanded into the region by 9000 BP, it was not the closed boreal forest of today, but a mosaic spruce-birch-cottonwood parkland with significant understory resources that undoubtedly appealed to bison populations. Beginning about 2000 years ago, the modern boreal forest appears to have developed, associated with the formation of acidic podzols superimposed on the Holocene loess sedimentary record. How much of this was a function of cooler and moister Neoglacial climates is difficult to say, but it may have been linked to a downturn in the fortunes of bison and an increase in moose numbers (cf. Yesner 1989).

Stephenson *et al.* tend to downplay the importance of late Holocene climatic changes, including that of the Little Ice Age, in the expansion of moose and contraction of bison populations, but they do suggest that "the discontinuous nature of late Holocene habitats probably played an important, albeit indirect, role" in their causation. In any case, it appears that remnant bison herds may have become limited to isolated patches of upland and riparian habitat, where maintaining population numbers would have been difficult, and the likelihood of human extirpation may have been exacerbated or facilitated. Stephenson *et al.* believe that human hunting is "the most likely proximate factor that reduced [bison] numbers and prevented the recovery of subpopulations or recolonization of suitable habitat." After European contact, land clearance and more widespread fires probably further increased moose numbers at the expense of any remnant bison populations, and the advent of firearms may have finally done them in.

The paper by Oswald, Brubaker, Anderson, and Gerlach on "Late Holocene Environmental and Cultural Changes at Tukuto Lake, Northwestern Alaska" focuses on late Holocene (2000 BP to present) environmental changes in northern Alaska in considerably more detail. It attempts to link these changes with the expansion and contraction of caribou populations, and the changing fortunes of human hunters of caribou as reflected in site sizes, locations, and contents. The authors subdivide the late Holocene into five periods, alternating between conditions that are relatively cool and moist (Periods A, C, and E) and those that are relatively warm and dry (Periods B and D), with Periods D and E corresponding to the Medieval Optimum and the Little Ice Age, respectively. They are able to establish tentative correlates between some of these climatic changes and caribou numbers, particularly with an apparent caribou population crash during the relatively cool and moist Period B (1500 to 1200 BP), and a corresponding re-expansion during the relatively warm and dry Period C (1200 to 900 BP). However, caribou population changes seem to be less in evidence during the Medieval Optimum, and the effects of Little Ice Age climate changes seem to be mostly in the direction of increased snow depths that shifted both animal and human migration patterns, and necessitated some new hunting strategies such as the possible use of dog traction. The authors thus conclude that the relationship between climate, vegetation, caribou, and people is complex and "not isomorphic…[or] predictable," a worthwhile general caveat.

The second section of the volume deals more exclusively with zooarchaeology: using faunal remains from archaeological sites to reconstruct diet, subsistence, hunting strategies, and animal exploitation patterns characteristic of a broad cross-section of Northern societies.

In their paper on "The North Point Wet Site and The Subsistence Importance of Pacific Cod on the Northwest Coast," Bowers and Moss offer a challenge to the "salmonopia" (Monks 1987:119) of Northwest Coast specialists. They point out that cod predominate in many Northwest Coast assemblages, including those at Hidden Falls (Moss 1989) and Nash Harbor (Souders 1997). Perhaps the best case for cod use on the northern Northwest Coast can be made for the North Point Wet Site, which has produced nearly 3000 fish bones, two-thirds of which are from cod. Interestingly, this site is of the same age as Locarno Beach and Hoko River, contains similar artifact assemblages, and is likewise dominated by cod. In explaining this phenomenon, Bowers and Moss suggest a linkage to Neoglacial cooling which seems to have dominated the period from around 4200 to 2200 BP, based on data from glacial deposits as well as pollen records. There are, of course, alternative explanations for the larger amounts of cod recovered at North Point, including seasonality of site occupation and recovery methods (i.e., using ¼ inch screens), although their overall approach is an insightful one begging to be explored in much more detail on a regional basis.

Aubrey Cannon's paper "Was Salmon Important in Northwest Coast Prehistory" also offers a provocative challenge to the assumption that salmon were universally important in late prehistoric Northwest Coast economies, based on a model championed by Croes (1992) and Matson (1992) that suggests that salmon harvests were intensified over time in response to increased sedentism, population growth, and development of processing and storage technologies. He challenges their data, derived from sites excavated on the Olympic Peninsula of Washington state and the Fraser River estuary, respectively, as one that is based on small samples excavated from few sites, and suggests that differences in seasonality of occupation rather than age of these sites could well explain apparent differences in amounts of salmon bone recovered from them. He also challenges the purported increase in evidence for storage facilities, suggesting that many such facilities are probably perishable and relatively invisible in the archaeological record. Finally, he challenges their basic underlying philosophy of a "push" model in which salmon harvesting is viewed as a high cost, labor-intensive endeavor requiring elaboration of processing and storage technology. Cannon believes that such a position is refuted by evidence of early (i.e., 9000 yr BP) salmon utilization at the Namu site on the central coast of British Columbia, followed by a decline in salmon use there later in prehistory, while site use appears to have remained year-round throughout. Cannon's underlying assumption is based on a "pull" model in which salmon are viewed as attractive, easily harvested, and highly nutritious, particularly because of their high fat content, something particularly valuable to northern hunter-gatherers. It views salmon as a resource important on the Northwest Coast from early times, and evidence for intensification as weak and based on limited data. Of course, Cannon's data, like those of Croes and Matson, are based on a relatively restrictive sample, which he readily admits; both could easily be affected by local factors, both geological and climatic, that might affect salmon abundance and exploitation.

Is there room for compromise here between all of these disparate viewpoints on coastal adaptations? It may simply be the case that the Neoglacial cooling caused temporary deterioration of salmon runs, but that after around 2000 BP the process of intensification of salmon harvesting began to take hold across the region. For some reason, Namu remained outside that pattern, but what combination of regional paleoclimatic or biodiversity may have been responsible is difficult to say. As Cannon notes, we need more data before we can cast final judgement on these scenarios, but perhaps they are not mutually exclusive.

In her paper on "Animal Procurement and Seasonality: A Study of Faunal Remains from a Prehistoric Sod House Ruin in Northwest Alaska," Rebekah DeAngelo develops a detailed interpretation of caribou exploitation strategies along the Noatak River in northwestern Alaska around 400 years ago. The data on which the paper is based are from a residential site and associated midden, rather than from a kill/butchering site, and therefore allows some assessment of the transport of butchered game, as well as the nature of consumption patterns taking place on the site. DeAngelo analyzes 9000 bone fragments, or about a third of the total bone fragments from the house and midden, and presents detailed data from which she derived estimates of occupational seasonality (juvenile caribou molar eruption stages) as well as patterns of transport of carcass sections, bone fragmentation patterns, and nutritional yield (postcranial element and element fragment frequencies). On the basis of these data, as well as the patterning of birds and fish, and the ethnographic data, she is able to conclude that the season of site occupation extended from fall through spring, but that the major focus of caribou hunting and salmon fishing probably occurred in the fall, with a secondary focus of caribou hunting and both birding and fishing in the spring, and, one assumes, dependence on stored food through the winter. Ethnographic accounts of coastal trade appear verified by the recovery of a few seal bones at the site.

The Lime Hills caves form the basis of Dianna Georgina's paper on "The Small Mammals of Lime Hills Cave I." Here, Georgina reconstructs the paleoenvironmental setting for a possible late Pleistocene and Holocene human occupation of the caves, based on small mammal data (as well as the local pollen record). Her data show that an arctic tundra/steppe fauna was present during the Late Pleistocene (Stratum 5), that a shrub/dwarf tree fauna was present during the terminal Pleistocene (Stratum 4), and that a somewhat warmer and wetter fauna was present during the early Holocene (Stratum 3). The upper two strata date only to the last 600 years, and include modern boreal faunas. The data from Stratum 4 are potentially of greatest interest, since they are correlated with a well-defined Paleoarctic organic assemblage including microblade inserts and side-slotted antler arrowheads and spearheads (Ackerman 1996). It is also possible, as Georgina notes, that at least some of the small mammals were consumed by humans, although she favors the notion that most were brought to the cave by predators, including bears, wolves, and foxes; porcupines are also known to transport bones to cave dens.

Hofman and Todd's paper on the Lipscomb Site, "Tyranny in the Archaeological Record of Specialized Hunters," is the only paper in the collection not dealing with a data set from Alaska or Canada, but it contains both data and theory that are highly relevant to understanding faunal assemblages from northern archaeological sites. Here, Hofman and Todd attempt to dissect the concept of "specialized" hunting with reference to Paleoindian bison hunters on the northwestern Plains. Their particular focus is the Lipscomb site, a Folsom-age bison bonebed. Like other bison bonebeds, and further north, many caribou hunting sites, it reflects a short-term set of activities related to killing and butchering a species which is present in large aggregations that are restricted in time and space. Does it therefore represent "specialized" hunting, or only a localized phenomenon of relatively short duration? Hofman and Todd weigh in on this question, using the data from Lipscomb and other Folsom sites to suggest that it may be too early to cash in the specialized Paleoindian big game hunting model.

In spite of the problem of archaeological visibility affecting bison (or caribou) bonebeds, and the fairly large list of taxa taken to some degree by Folsom hunters, their reading of the data suggests that non-bison hunting activities were simply "embedded in an overall system geared to bison procurement." According to their view, "Folsom people were bison hunting specialists for some seasons of most years," and because of preservation problems we lack the evidence of even more extensive utilization of bison for non-food purposes (hides, sinew, bones, etc.) than can be observed archaeologically. This evidence derives from both the bonebeds and the associated lithic inventories. Hofman and Todd further argue that, in contrast to the Alaskan situation, where Pleistocene bison were apparently the focus of serious predation (Guthrie 1980), they were subjected to much less predation on the Plains, where they were largely limited by, and probably affected, the productivity of the grasslands. All this changed with human colonization of the regions; as top carnivores, humans redirected the limiting factors in local food webs, and began to impact bison populations. Thus, in contrast to the argument advanced earlier by Kooyman and Sandgathe, in which Plains bison size diminution was related to continued declines in the nutrient quality of forage, Hofman and Todd directly link it to the continued impact of human hunting.

Murielle Nagy continues the theme of the impact of human hunters on their prey, but in a coastal rather than a terrestrial setting. In her paper on "Seal Pups and their Moms: A Preferred Paleoeskimo Hunting Strategy in Ivujivik, Nunavik, Eastern Arctic," Nagy demonstrates a shift from hunting infant ringed seals and their mothers during the Pre-Dorset/Dorset Transition in northern Quebec, to taking more subadult seals in the Dorset period proper. Nagy suggests that this may result, in part, from changes in settlement pattern and seasonal occupation of coastal sites, such that the PD/DT groups were on the coast predominantly in spring and early summer, while the Dorset groups were there primarily in the late summer to winter. Female seals taken in spring/early summer would have been lean, but by taking their babies as well, the PD/DT folks would have had an instant source of dense marine mammal fat.

According to Nagy, the PD/DT population didn't take the larger male adult seals as well, based on small counts of bacula (although perhaps these could be missing for other reasons, such as use in tool-making or ritual activities). Why didn't they take the adult males? Citing modern biological data from ringed seals, she notes that up to 65% of infants could probably be harvested annually from a stable population. Other biological data suggest that sustained yield can be increased in harvesting of modern seal populations by a correspondingly increased emphasis on the taking of young individuals. Krupnik (1993) has suggested that this is because, after the death of infants, sexually mature females will ovulate again relatively quickly. This preference for hunting juvenile sea mammals is also the focus of the study in this volume by Savelle et al., which looks at whales rather than seals.

Becky Saleeby and Angela Demma focus on their interpretations of "The Kitluk River Site: A Protohistoric Faunal Assemblage from the Seward Peninsula." Here, relative numbers of animal bones from different taxa are converted into MAU ("Minimum Animal Unit") percentage data in order to compare the exploitation patterns at several sites at the mouth of the Kitluk River on the Seward Peninsula in northwestern Alaska, and to contrast these patterns with those found at Cape Espenberg to the south, as well as at Roger Harritt's excavations at Kuzitrin Lake in the interior of the Peninsula. Because these sites all date to the late prehistoric to protohistoric periods, Saleeby and Demma are able to take advantage of extensive ethnographic and historic data from the region, including the use of their own informants. Briefly, Saleeby and Demma determine that the focus at the Kitluk River mouth is on sea mammal hunting, with a major emphasis on ringed seals, and a secondary emphasis on spotted seals, walrus, and bearded seal. (Walrus are not available in the area today, but conditions may well have changed during the past century as the result of the impact of Yankee whalers hunting walruses for sea mammal oil). Birds and fish were also important, with the latter including sculpins and tomcod obtained by jigging through ice during the winter. In contrast, the Cape Espenberg folks hunted fewer caribou, either because they were not locally available, or because, as ethnographic accounts suggest, it was a site used by task groups focused specifically on seal hunting. (Some accounts also suggest that an ethnographic boundary separated these two nearby coastal regions.)

The Kuzitrin Lake site, in contrast, was apparently a specialized caribou kill/butchery site, although ethnographic data indicate that the Kuzitrin Lake people did fish and hunt seal and beluga at other times and places. Saleeby and Demma effectively use the MAU data to show that caribou sections were probably being transported out of the Kuzitrin Lake sites, and into the Kitluk River mouth sites. One possible explanation is that caribou populations were at a high during the mid-nineteenth century, when the Kitluk River mouth sites were occupied. The latter illustrates the advantage of excavating more recent sites, for which not only ethnographic but biological data may be available to facilitate explanation of faunal patterns. It also illustrates the kinds of interpretations we cannot make (without much greater speculation) at older sites!

Finally, the paper by James Savelle, Allen McCartney, and Arthur Dyke "Human Predators and Migratory Megafauna: The Case of Thule Inuit Bowhead Whaling" returns us to the question of the selective preference of sea mammal hunters for the age of their prey and how it is reflected in the mortality profiles that can be constructed from faunal assemblages from individual sites.

From a series of Thule Eskimo sites in the central archipelago of High Arctic Canada, Savelle et al.

successfully demonstrate (by comparing mortality profiles from Thule sites with those from naturally stranded whale populations) that first, Thule people were hunters rather than scavengers of bowhead whales; second, (based on measurement of Thule whalebones and comparison with controlled samples of known ages), that they greatly preferred to take yearling whales in the 7-9 m range; and third, that the Thule people were practicing an "intercept" strategy (based on the fact that the size of yearling whales increases with the position of sites along the annual bowhead migration route). The focus on yearling bowhead whales is explained by Savelle et al. as an optimization of three factors: (1) yield of meat, fat, and other nutrients, which, if all other things are equal, should be greater in larger mammals; (2) technological limits, which prevent Thule from being able to capture, retrieve, transport, and butcher (within 24-48 hours) animals much larger than yearlings; and (3) hunting danger, which affects any attempt to take infants who are still tied to mothers in their migration pattern. As such, this final paper in the volume is a clear statement of how optimization principles structure northern hunting adaptations, by affecting the decision-making of hunters "in the field."

The final section of the volume focuses on methods used by northern paleontologists and zooarchaeologists to understand the structure of faunal assemblages, and their implications for paleoecology as well as human diet and animal utilization. Three of the articles, Brink, Kooyman, and Friesen *et al.*, focus on the quantification of faunal assemblages and the construction of utility indices for northern taxa. These indices represent statistical descriptions of the relative nutritional "value" of animal carcass segments, in order to comprehend the body part distributions found in archaeological sites.

Brink's paper focuses on the use of bison utility indices to shed light on animal carcass treatment at a pair of sites (the Wardell sites) in Alberta, Canada, one of which is a kill site and the other a processing site. From these analyses, Brink concludes that the most relevant utility indices are constructed not through a "total products model" which includes all body parts (i.e., Binford's classic model), but by focusing on specific anatomic elements, and by considering the internal contents of the bones themselves (marrow and bone grease) as well as the fat and muscle in surrounding tissues. By utilizing this technique, Brink is able to show that bones associated with the greatest percentage of marrow and grease are differentially missing from the kill site, and are differentially represented at the processing site. This suggests the transport of such bones from the kill to the butchering area/ village campsite. Brink recognizes the potential contribution of non-human predators to the assemblages, as well as the non-food yield of bison body parts, but feels that both are minor players in the overall analysis of carcass utility at such sites. Brink's paper is a major contribution to the growing literature on utility indices, although he clearly carries the argument beyond simple assumptions of rational and maximizing behavior by hunter-gatherers.

The second paper in this section, by Max Friesen, James Savelle, and Mark Diab, "A Consideration of the Interspecific Application of Food Utility Indices, with Reference to Five Species in the Order Pinnipedia," presents new data on sea mammal utility indices, of interest particularly to zooarchaeologists working in coastal settings. Building on previous utility indices constructed for California sea lions (*Zalophus californianus*), hooded seals (*Cystophora cristata*), and harp seals (*Phoca groenlandica*), Friesen et al. calculate new indices for two other pinnipeds: harbor seals (*Phoca vitulina*) and ringed seals (*Phoca hispida*). Their data indicate strong similarities among all of the species within this order, implying that the utility indices are essentially interchangeable. They also demonstrate that indices developed for artiodactyls similarly show relatively interchangeable values within that order, while artiodactyls, pinnipeds, and cetacean values appear not surprizingly to be non-comparable. This suggests that utility indices should be applicable interspecifically within orders, a fact which will facilitate many analyses where species-specific data are unavailable.

In his second paper in the volume, on "The Roles of Bone Density and Cultural Behavior in Head-Smashed-In Bone Bed Taphonomy," Kooyman covers some of the same ground as Brink, but toward a different end. Head-Smashed-In, a mid-to-late Holocene bison jump in southern Alberta, is like Wardell in that there is both a kill site and a camp/ butchery area. Like Wardell, there tend to be few high "value" elements (in terms of food utility) left at the kill site, which is dominated by cranial, lower leg, and some upper leg bones. Conversely, there are many high "value" elements at the camp/butchery area. Kooyman also notes that there are two main areas at the kill site, which differ in terms of bone density: a "well-preserved" and a "poorly-preserved" area. In order to compensate for these differences, he multiplies MAU (Minimal Animal Unit) data from each area by the inverse of the bone density, and finds that the patterns are similar and "surprisingly linear." All of this suggests, that with proper manipulation even poorly-preserved samples can be made representative of the "target population." The remaining question, of course, is how much further the "well-preserved" MAU data need to be massaged in order to make them truly representative of the original butchered faunal assemblage.

The final two papers in this section deal with taphonomic methods, utilizing field data from interior Alaska. Magoun and Valkenburg, in their paper on "Caribou Remains at Kill Sites and the Role of Scavengers." provide important insights into carcass taphonomy that is useful to both paleontologists and archaeologists. According to the authors, five variables are most important in understanding carcass taphonomy: the nature of the carnivores present in the area; the length of time available for scavengers to feed on carcasses; the total number of carcasses available; the size of carcasses; and the season in which the carcass became available. Some carnivores (e.g., bears) are much more destructive in a short time, while others (e.g., wolves) lose interest after a few days and prefer to abandon carcasses in order to make new kills; still others (e.g., wolverines) will completely reduce a carcass, but only if given an entire season in which to do so.

In their paper on "Rodent Gnawing as a Taphonomic Agent," Thornton and Fee present a number of insights into this process, many of which are counter-intuitive. Rodents gnaw on bones to reduce the length of their incisors, and to obtain minerals (salt, calcium, phosphates). Large rodents such as porcupines, differentially gnaw on larger, denser bones. Smaller rodents such as mice appear to go for smaller bones or smaller projections on bones. They particularly like roughened ends or broken bone, which accelerates the destruction of already fragmentary specimens in favor of complete bones. Rodents frequently leave definitive striations on bone fragments, but in some cases they do not, and their presence may go unrecorded from a site report. Mammal bones were more frequently damaged than bird bones as well.

In combination, the papers in this volume illustrate the full range of modern techniques and theories being used by northern paleontologists and zooarchaeologists to understand the evolution of animal communities in the north, the response that humans have made to those evolutionary trends, and the impact that humans have had on northern animal communities. In keeping with the spirit of Dale Guthrie's contributions to the literature, they represent the best in innovative approaches to looking at people and animals in the North.

REFERENCES CITED

Ager, T.A.
1975 *Late Quaternary Environmental History of the Tanana Valley, Alaska.* Report No. 54, Institute of Polar Studies, Ohio State University, Columbus.

Anderson, D.D.
1988 *Onion Portage: The Archaeology of a Stratified Site from the Kobuk River, Northwest Alaska.* Vol. 22 of *Anthropological Papers of the University of Alaska*, University of Alaska Press, Fairbanks.

Binford, L.R.
1978 *Nunamiut Ethnoarchaeology*. Academic Press, NY.

Catto, N.R.
1996 Richardson Mountains, Yukon-Northwest Territories: The northern portal of the postulated "Ice-free Corridor." Pp. 3-20 in *The Ice-free Corridor Revisited*, edited by Carole A.S. Mandryk and Nathaniel Rutter. *Quaternary International*, Elsevier, Tarrytown, NY.

Croes, D.R.
1991 Exploring Prehistoric Subsistence Change on the Northwest Coast. Pp. 337-366 in *Long-term Subsistence Change in Prehistoric North America*, edited by Dale R. Croes, Rebecca A. Hawkins, and Barry L. Isaac. Supplement No. 6 of *Research in Economic Anthropology*, JAI Press, Greenwich, CT.

Croes, D.R., and S. Hackenberger
1988 The Hoko River Archaeological Complex: Modeling Prehistoric Northwest Coast Economic Evolution. *Research in Economic Anthropology* 3:19-86.

Dixon, E.J.
1999 *Bones, Boats, and Bison*. University of New Mexico Press, *Albuquerque.*

Dixon, E.J., T.H. Heaton, T.E. Fifield, T.D. Hamilton, D.E. Putnam, and F. Grady
1997 Late Quaternary Regional Geoarchaeology of Southeast Alaska Karst. *Geoarchaeology* 12:689-712.

Driver, J.C.
1996 The Significance of the Fauna from the Charlie Lake Cave Site. Pp. 21-28 in *Early Human Occupation in British Columbia*, edited by Roy L. Carlson and Luke Dalla Bonna. University of British Columbia Press, Vancouver.

Duk-Rodkin, A. and R.W. Barendregt
1997 Mid-Pliocene to Pleistocene Glaciations in the Tintina Trench, Dawson area, Yukon Territory. Pp. 50 in *Program and Abstracts, Beringian Paleoenvironments Workshop*, Florissant, CO.

Dumond, D.E.
1975 Coastal Adaptation and Cultural Change in Alaskan Eskimo Prehistory. Pp. 167-180 in *Prehistoric Maritime Adaptations of the Circumpolar Zone*, edited by William W. Fitzhugh. Mouton, The Hague.

Fladmark, K.R.
1975 *A Paleoecological Model for Northwest Coast Prehistory*. Mercury Series No. 43, Archaeological Survey of Canada Papers, National Museums of Canada, Ottawa.

Foronova, I.V., and A.N. Zudin
1999 Evolution of the Mammoth Lineage in Eurasia. Pp. 99-109 in *Quaternary of Siberia*, edited by Jiri Chlachula, Robert A. Kemp, and Jaroslav Tyracek. *Journal of Geological Sciences*, Czech Geological Survey, Prague.

Geist, V. and P. Karsten
1977 The Wood Bison in Relation to Hypothesis on the Origin of the American Bison. *Zeitschrift Saugertiek* 42:119-122.

Gerlach, S.C., C. George and R. Suydam
1993 Bowhead Whale (*Balaena mysticus*) Length Estimations Based on Scapula Measurements. *Arctic* 46(1):55-59.

Guthrie, R.D.
1980 Bison and Man in North America. Pp. 55-73 in *The Ice-free Corridor and the Peopling of the New World*, edited by Nathaniel W. Rutter and Charles E. Schweger. Vol. 1 of *Canadian Journal of Anthropology*, University of Alberta, Edmonton.

1983 Paleoecology of the Site and its Implications for Early Hunters. Pp. 209-287 in *Dry Creek*, edited by W. Roger

Powers, R. Dale Guthrie, and John F. Hoffecker. National Park Service, Washington, DC.

1984 The Evidence for Middle-Wisconsin Peopling of Beringia: An Evaluation. *Quaternary Research* 22:231-241.

Krupnik, I.
1993 *Arctic Adaptations: Native Whalers and Reindeer Herders of Northern Eurasia.* University Press of New England, Hanover, NH.

Mann, D.H., and T.D. Hamilton
1995 Late Pleistocene and Holocene Paleoenvironments of the North Pacific Coast. *Quaternary Science Reviews* 14:449-471.

Matson, R.G.
1992 The Evolution of Northwest Coast Subsistence. Pp. 367-430 in *Long-term Subsistence Change in Prehistoric North America*, edited by Dale R. Croes, Rebecca A. Hawkins, and Barry L. Isaac. Supplement 6 of *Research in Economic Anthropology*, JAI Press, Greenwich, CT.

McCartney, A.P.
1980 The Nature of Thule Eskimo Whale Use. *Arctic* 33:517-541.

McDonald, J.N.
1981 *North American Bison: Their Classification and Evolution.* University of California Press, Berkeley.

Monks, G.G.
1987 Prey as Bait. *Canadian Journal of Archaeology* 11:119-142.

Moss, M. L.
1989 Analysis of the Vertebrate Assemblage. Pp. 93-130 in *The Hidden Falls Site, Baranof Island, Alaska*, edited by Stanley D. Davis. *Aurora* Vol. 3, Alaska Anthropological Association, Anchorage.

Sher, A.V.
Antelopes and Mini-mammoths of Beringia: Strayed from the Southern Herds or Evolved in the Arctic? Where Have They Gone? Paper presented at the 47th Arctic Science Conference, Girdwood, AK.

Souders, P.
Ellikarrmiut Economy: Animal Resource Use at Nash Harbor, Nunivak Island, Alaska. M.S. Thesis, Department of Anthropology, University of Oregon, Eugene.

Thorson, R.M.
1990 Geologic Contexts of Archaeological Sites in Beringia. Pp. 399-420 in *Archaeological Geology of North America*, edited by Norman P. Lasca and Jack Donahue. Centennial Special Vol. 4, *Decade of North American Geology*, Geological Society of America, Boulder.

Thorson, R.M., and R.D. Guthrie
1984 River Ice as a Taphonomic Agent: An Alternative Hypothesis for Bone "Artifacts." *Quaternary Research* 22:172-188.

Vereschagin, N.K., and G.F. Baryshnikov
1984 Paleoecology of the Mammoth Fauna in the Eurasian Arctic. Pp. 267-280 in *Paleoecology of Beringia*, edited by David M. Hopkins, John V. Matthews, Jr., Charles E. Schweger, and Steven B. Young. Academic Press, NY.

Vinson, D.M.
1993 *Taphonomic Analysis of Faunal Remains from Trail Creek Caves, Seward Peninsula, Alaska.* M.A. Thesis, Dept of Anthropology, University of Alaska Anchorage.

West, F.H.
1996 *American Beginnings.* University of Chicago Press, Chicago.

Yesner, D.R.
1980 Caribou Exploitation in Interior Alaska: Evidence from Two Paxon Lake Sites. *Anthropological Papers of the University of Alaska* 19:15-32.

1989 Moose Hunters of the Boreal Forest? A Re-examination of Subsistence Patterns in the Western Subarctic. *Arctic* 42:97-108.

1995 Whales, Mammoths, and Other Big Beasts: Assessing Their Roles in Prehistoric Economies. Pp. 149-164 in *Hunting the Largest Animals: Native Whaling in the Western Arctic and Subarctic*, edited by Allen P. McCartney. Studies in Whaling No. 3, Occasional Pub. No. 36, Canadian Circumpolar Institute, University of Alberta, Edmonton.

1996 Human Adaptation at the Pleistocene-Holocene Boundary (circa 13,000 o 8,000 BP) in Eastern Beringia. Pp. 255 to 276 in *Humans at the End of the Ice Age: The Archaeology of the Pleistocene-Holocene Transition*, edited by Lawrence Guy Straus, Berit Valentin Eriksen, Jon M. Erlandson, and David R. Yesner. Plenum Press, New York.

PALEOECOLOGICAL AND ARCHAEOLOGICAL IMPLICATIONS OF THE CHARLIE LAKE CAVE FAUNA, BRITISH COLUMBIA,10,500 TO 9,500 B.P.

Jonathan C. Driver
Department of Archaeology, Simon Fraser University

Charlie Lake Cave is located near the city of Fort St. John in northeastern British Columbia (Figure 1). Fort St. John lies to the east of the Rocky Mountains, in the heart of the putative "ice-free corridor." The site is situated on an outcrop of sandstone on a steep slope above a small creek that drains via the Beatton River to the nearby Peace River. Although the site is named after a small cave, most excavations have taken place in front of the cave mouth where a deep gully has been infilling gradually with sediments for the past 10,500 years. Fladmark excavated the site in 1983 where he demonstrated a long cultural and paleoenvironmental sequence, beginning with a Paleoindian occupation containing a late fluted point component, and ending with late prehistoric components (Fladmark 1996; Fladmark *et al.*. 1988). Three studies of the fauna from the 1983 excavations have been published (Driver 1988, 1996; Driver and Hobson 1992). Further excavations were undertaken in 1990 and 1991, directed by Driver. These excavations increased sample sizes of artifacts and fauna from all periods, and identified another cultural component (Driver *et al.* 1996; Handly 1993).

Although the site contains a sequence of artifacts and fauna spanning the last 10,500 radiocarbon years, this paper deals with the first 1000 years of the site's formation and occupational history, documenting a transition from early post-glacial to Holocene faunas. This paper discusses the nature of the early environments, the timing of the transition from open to forested conditions, and the human use of the early environments.

STRATIGRAPHY AND DATING

Stratigraphy, culture history and radiocarbon chronology have been discussed and illustrated elsewhere (Fladmark 1996; Fladmark *et al.* 1988; Driver *et al.* 1996), so only a brief summary is presented here. During the last glaciation of the region a cave was formed in sandstone bedrock, possibly by sub-glacial water. At about 10,500 B.P. a large slab of sandstone (about 12 x 5 x 4 meters) was detached from the face of the small escarpment in which the cave was situated. The slab split away from the cliff face along a vertical joint in the sandstone. When the slab was detached from the bedrock it moved about three meters downslope from the bedrock face, but remained in a vertical position. This created a gully bounded to the north (upslope) by a newly exposed escarpment face containing a cave and to the south (downslope) by one side of the vertical slab. The east and west ends of the gully were open, so that one could enter the gully by walking along the hillside at the base of the escarpment. The floor of the gully sloped steeply south and was littered with sandstone boulders and crushed sandstone left behind as the slab moved downhill. The cave was about three meters above the floor of the gully.

Upslope from the gully the hillside was mantled in glaciolacustrine sediments. Over a period of about 1000 radiocarbon years these were redeposited in the gully, forming a layer of variable thickness over the boulders on the floor. Because the gully floor slopes to the south, the greatest thickness of redeposited glaciolacustrine sediments was against the upslope side of the detached slab on the south (or downslope) side of the gully. Up to a meter of sediment accumulated here. By about 9,500 B.P. the rate of deposition slowed sufficiently to allow soil horizons to form. The subsequent history of deposition is a mixture of allochthonous slopewash and autochthonous weathered sandstone, with numerous soil horizons.

Site stratigraphy is complex, and numerous layers in the lower portion of the site can be resolved into a series of stratigraphic zones. The lowest boulders form Zone I, which contains no artifacts or fauna. The redeposited glaciolacustrine sediments are in Zone II. After the 1983 excavations this was divided into two subzones (Fladmark *et al.*. 1988; Driver 1988); as a result of the 1991 field season, we now recognise four subzones, IIa through IId (Driver *et al.* 1996). The initiation of soil horizons marks the beginning of Zone III, divided into eight subzones. In this paper we are concerned only with the earliest, Subzone IIIa. Radiocarbon dating places Subzones IIa and IIb between 10,500 and 10,000 B.P., while IIc and IId date from 10,000 to slightly before 9,500 B.P. Zone IIIa has a single date of about 9,500 B.P. (Driver *et al.* 1996).

Twenty three one-by-one meter excavation units were placed in the gully and the mouth of the cave in three seasons of excavation. Of these, thirteen units reached Subzone IIIa and lower. Faunal remains from nine of these units have been studied. Faunal specimens were recovered in situ and in 3mm screens.

RECOVERY AND IDENTIFICATION OF FAUNA

Table 1 lists all mammal and bird specimens from the nine excavation units which have been identified to family, genus or species level. Specimens identified to order or class and unidentified specimens have been excluded. A few fish bones were recovered, but the fish assemblage from the site will be reported separately. Frog and snake bones were present in the strata reported here, but were not identified further. Identifications were made with the aid of comparative collections at Simon Fraser University, University of Puget Sound, University of Washington (Burke Museum), University of Toronto, Royal Ontario Museum, and Canadian Museum of Nature. All specimens were catalogued individually and are reposited at Simon Fraser University.

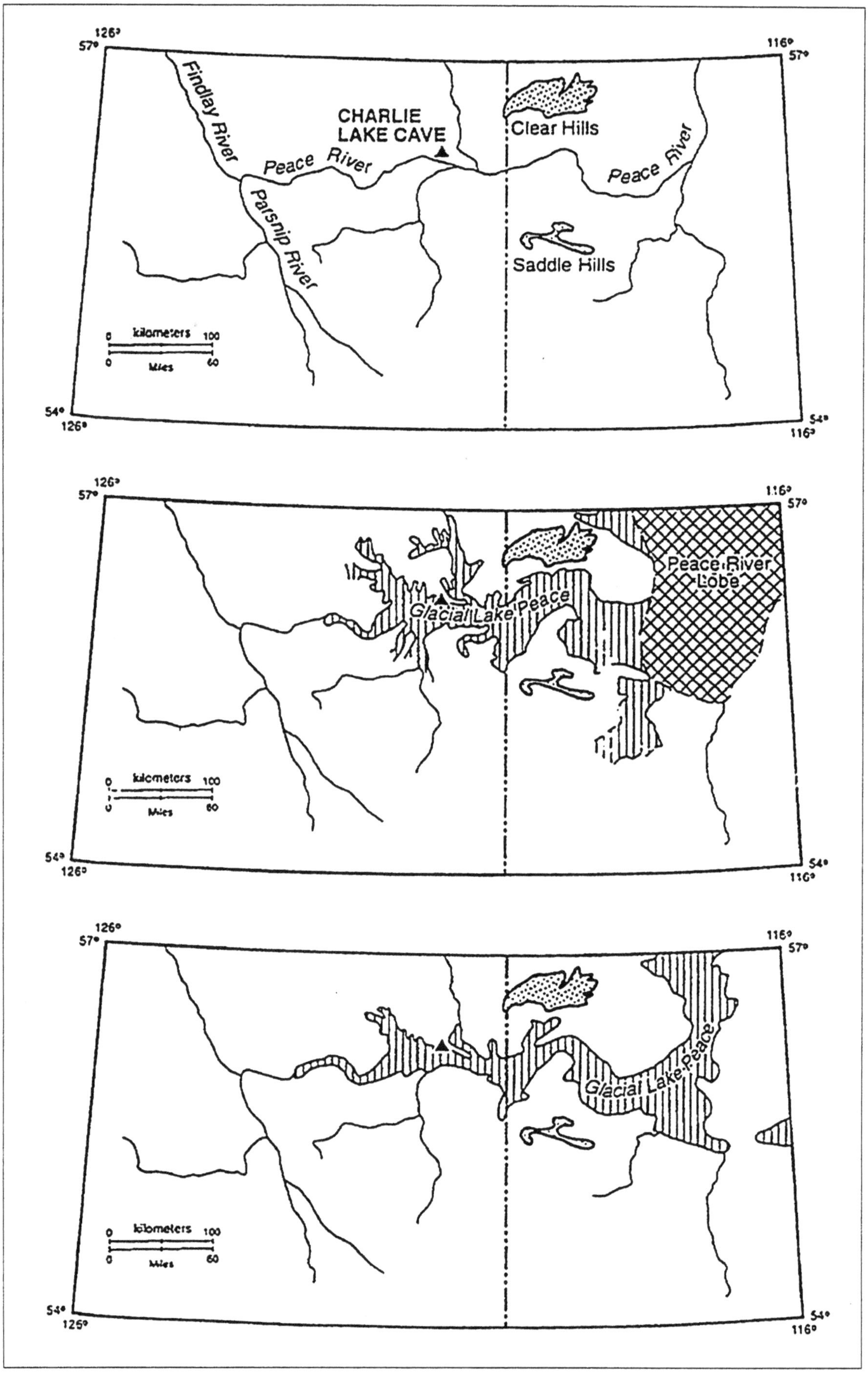

Figure 1. Location of Charlie Lake Cave in relation to modern topography (upper), Glacial Lake Peace, early Clayhurst Stage (middle), and Glacial Lake Peace, Indian Creek Stage (lower). Data from Mathews (1980).

A few identifications require further comment. Almost all lagomorphs at Charlie Lake Cave are identified as *Lepus americanus* (snowshoe hare), and this species dominates the mammal assemblage in most periods (Driver 1988). This species is distinguished from some other *Lepus* because it is much smaller, and differs from others of similar size *Sylvilagus* (cottontail rabbits) in cranial morphology. Measurements of selected post-cranial elements throughout the 10,500 year sequence show that a few early specimens are too large to be from *L. americanus*, and these are described as "large hare" in Table 1. In both size and morphology these may be from either arctic hare (*L. arcticus*) or one from one of the jackrabbits (*L. townsendii* or *L. californicus*).

As discussed elsewhere (Driver 1988) ground squirrels (*Spermophilus* sp.) are not identified to species, but most closely resemble *S. richardsonii* and *S. columbianus*. Collared lemming (*Dicrostonyx torquatus*) was identified from a single tooth in the identified sample. However, a complete mandible with associated teeth was found in a search of unanalyzed material from another excavation unit. Identification criteria are discussed elsewhere (Driver in press a).

FAUNA FROM SUBZONES IIA AND IIB

As there are very little fauna from Subzone IIa, and as the one radiocarbon date from this subzone is in the same range as the dates from Subzone IIb, the two subzones are considered together here. The earliest faunas from Charlie Lake Cave are notable for the presence of various animals which do not live in the Peace River region today, suggesting that environmental conditions were different from the boreal forest environments characterizing much of the Holocene in northern British Columbia. The only large mammal is *Bison*. The absence of bison cranial material precludes species identification, but limb bone measurements show it was significantly larger than modern species (Driver 1988). Although bison were present in the Peace River region until historic times, and while they are represented throughout the Charlie Lake Cave sequence, other artiodactyls such as deer, moose and elk are usually associated with them (see Stephenson et. al., this volume). The absence of cervids suggests a more open environment. Snowshoe hare (*Lepus americanus*) is poorly represented in the IIa/IIb fauna, co-occurring with larger specimens of *Lepus*, either a jackrabbit or Arctic hare. While snowshoe hare becomes very common in succeeding stratigraphic zones, the large hares disappear; only snowshoe hare is found in the region today. As all modern large *Lepus* species in North America prefer open habitats, it is likely that the large species from Charlie Lake Cave had similar habitat preferences. The dominant small mammal in IIa/IIb is *Spermophilus*. Although not all ground squirrels are associated with open environments, the Charlie Lake Cave specimens are closest to *S. richardsonii* or *S. columbianus*, both of which are found today in grasslands to the south of Charlie Lake Cave. *Dicrostonyx torquatus*, the collared lemming, is today confined to tundra habitats in the far north, although in Late Pleistocene times it was found to the south of the ice sheets (Mead and Mead 1989).

The bird assemblage is dominated by two species. Cliff swallow (*Hirundo pyrrhonota*) probably nested at the site; colonies are still found on sandstone cliffs around Charlie Lake today. Raven (*Corvus corax*) has a high NISP value (Table 1) because an almost complete skeleton was recovered from Subzone IIb. It is possible that this specimen was deposited at the site by people (Driver In press, b). In comparison with the later avian assemblages, water birds are notably lacking from the lowest fauna. The presence of woodpecker does suggest some tree cover in the area.

The combined fauna from IIa/IIb is not found as a living community today, because it contains species which are typical of areas to both the north and the south of the modern boreal forest. Ground squirrels are probably from a species found today in areas south of the boreal forest, while the lemming is now found to the north. The large lagomorph could be northern or southern, depending on whether it is a jackrabbit or Arctic hare. This mixture of "southern" and "northern" species has been reported many times for late glacial faunas (e.g. Lundelius *et al.*. 1983; Graham *et al.*. 1987; Graham 1992; FAUNMAP 1996). The absence of water birds is a notable contrast with later assemblages, and may suggest either that migratory routes were not well established, or that there were few productive local aquatic habitats.

FAUNA FROM SUBZONES IIC AND IID

The greater richness of the later Zone II assemblages is in part due to the increased size of the faunal sample. As can be seen from Table 1, larger assemblages have more identified taxa. Nevertheless, there are both quantitative and qualitative changes which suggest that an environmental change occurred at about 10,000 B.P. The variety of waterbirds suggests that productive aquatic habitats were established near the site. The birds represented are all found in the area today, and include western grebe, horned grebe, mallard, teal, ruddy duck, coot, and a rail. Among the upland birds, cliff swallow continues and grouse appears. A short-eared owl is represented by the bones of one foot. Ground squirrels remain an important part of the fauna, but snowshoe hare increases significantly, suggesting that forested environments were occupying increasingly more of the region around the site. Bison persists as the only large artiodactyl species.

FAUNA FROM SUBZONE IIIA

Trends begun in IIc and IId continue in IIIa. Aquatic mammals such as beaver and muskrat appear. Snowshoe hare dominates the small mammal assemblage, and ground squirrel disappears by the end of this subzone. An unusual occurrence is passenger pigeon, which appears by 9,000 B.P. and persists until historic times (Driver and Hobson 1992). By 9,000 B.P. the vertebrate fauna from the site consists entirely of species which could be found in the Peace River region in historic times, and suggests that a boreal forest with local wetlands was present.

TAXON		IIa, IIb	IIb/c	IIc, IId	IIIa	TOTAL
Aechmophorus sp.	Western or Clark's Grebe			7	7	14
Podiceps auritus	Horned grebe			10	4	14
Anas platyrhynchos	Mallard			1		1
Anas sp.	Tea'			1		1
Oxyura jamaicensis	Ruddy duck			1		1
Tetraonidae	Grouse		10	3	7	20
Fulica americana	Coot			11	1	12
Rallidae	Virginia or Sora rail			2		2
Ectopistes migratorius	Passenger Pigeon				1	1
Asio flammeus	Short-eared owl			8		8
Picidae	Woodpecker	1				1
Corvus corax	Raven	59		1		60
Hirundo pyrrhonota	Cliff swallow	5		20	1	26
Lepus sp.	Large hare	3		1		4
Lepus americanus	Snowshoe hare	2	44	201	167	414
Clethrionomys gapperi	Gapper's red-backed vole			3	2	5
Dicrostonyx torquatus	Collared lemming	1				1
Ondatra zibethicus	Muskrat				8	8
Peromyscus sp.	Deer mouse	1	2	1	1	5
Marmota sp.	Woodchuck/marmot	1		2	1	4
Eutamias sp.	Chipmunk				2	2
Spermophilus sp.	Ground squirrel	54		169	7	230
Sciuridae		5		3		8
Castor canadensis	Beaver				9	9
Canis sp.	Wolf-size canid			6		6
Canidae		2				2
Mustela nivalis	Least weasel				2	2
Mephitis mephitis	Striped skunk				1	1
Mustelidae	Weasel			1		1
Bison sp.	Bison	20	4	30	4	58
TOTAL		154	60	482	225	921
No. of discrete taxa		12	4	20	17	

Table 1. Identified fauna from Zone II and Subzone IIIa, Charlie Lake Cave, 1983 and 1990/1991 seasons.

MICROSTRATIGRAPHIC FAUNAL TRENDS

Zone II deposits are over one meter thick on the south side of the gully where vertebrate accumulations are relatively dense and bone preservation is quite good. The combination of abundant fauna and deep stratigraphy allows one to look at faunal changes in more detail than is revealed in the faunas organized by stratigraphic zones. Three excavation units have been selected for more detailed examination, because they combine abundant fauna, good stratigraphy, and a sequence of radiocarbon dates on animal bone. These are unit 3, excavated in 1983, and units 26 and 28, excavated in 1990 and 1991. The Zone II deposits in unit 3 were excavated in a series of arbitrary 10cm levels, and have not been divided into subzones. Unit 3 was close to the rock wall of the gully; because of the angle of the rock, the unit did not reach the lowest part of Zone II. Units 26 and 28 both contain a complete Zone II sequence, although no fauna was recovered from IIa in either unit. These two units were excavated stratigraphically, but some of the thicker layers were also subdivided and excavated in arbitrary levels. Depths below surface and radiocarbon dates allow Unit 3 levels to be correlated with Units 26 and 28.

In order to demonstrate the change from open to forested conditions, the absolute and relative frequency of ground squirrel and snowshoe hare remains are shown in Table 2. The radiocarbon dates are all from the excavation units studied, except the uppermost date which is from a unit adjacent to Unit 28. From about 10,500 to 10,000 B.P. ground squirrel was more common than snowshoe hare, conforming to the overall pattern for IIa/IIb (Table 1). Starting at about 10,000 B.P. there is an increase in the frequency of snowshoe hare relative to ground squirrel, but by about 9700 B.P. ground squirrel has almost completely disappeared and snowshoe hare is dominant. Tree ring calibration is not available for the full range of Zone II dates, but it appears that the transition from open to forested conditions is unlikely to have taken more than 500 calendar years. The following sections place this environmental change in a context of regional geology and palynology.

UNIT	SUBZONE	LAYERS	DATES	HARE NISP	SQUIRREL NISP	%HARE
26+28	IIIa	82-84,86,91	9490+/-140	59	0	100
3	IIIa	14		13	0	100
26+28	IId	93		22	0	100
3	II	15-1,-2		40	0	100
26+28	IIc	98 (upper)	9670+/-150	70	1	99
3	II	15-3,-4,-5		8	7	53
3	II	15-6	9990+/-150	14	35	29
26+28	IIc	98 (lower)		2	28	7
3	II	15-7,-8		1	23	4
3	II	15-9,-10	10100+/-210	0	36	0
26+28	IIb	104		0	18	0
26+28	IIb	105	10290+/-100 10560+/-80	0	14	0

Table 2. Absolute and relative frequencies of snowshoe hare and ground squirrel from excavation units 3, 26 and 28.

REGIONAL GEOLOGY

Mathews (1978) identified Cordilleran and Laurentide tills in the Peace River region, proposing that coalescence of eastern and western ice occurred at about 15,000 B.P. near Fort St. John. More recent work (Catto *et al.*. 1996; Bobrowsky and Rutter 1992) suggests that Cordilleran ice never extended as far east as Fort St. John during the late Pleistocene, and that there was a single late Pleistocene Laurentide advance dating later than 22,000 and earlier than 13,000 B.P. All authors agree that towards the end of the last glaciation Laurentide ice blocked drainage from the Rockies, creating extensive glacial lakes between the Rockies and the Laurentide ice. One of the largest of these lakes was Glacial Lake Peace. The chronology of the various stages of this lake is important for understanding the regional setting of Charlie Lake Cave.

Mathews (1980) has mapped the Glacial Lake Peace shorelines. Just below Charlie Lake Cave is a small remnant of a raised beach, at about 710 meters a.s.l. This elevation places it in Mathews' late Clayhurst stage, bearing in mind that on the Alberta/B.C. border beaches of this period are between 660 and 690 metres, and that there is a 0.4 metre per kilometer rise in elevation from east to west (Mathews 1978:12, 1980:17). All earlier stages of Glacial Lake Peace would have covered Charlie Lake Cave (see Figure 1).

In 1991 we excavated a stratigraphic section on the hill above the site and found a diamicton overlain by rythmites. This presumably reflects the Laurentide advance followed by a high glacial lake. The raised beach below the site is the result of a later, lower lake. Radiocarbon dates from the basal layers of Charlie Lake Cave average about 10,500 B.P., and are taken on terrestrial birds and mammals. Thus, all stages of Glacial Lake Peace prior to the late Clayhurst Stage must pre-date 10,500 B.P.

If Charlie Lake Cave was occupied when the raised beach below the site was being formed (late Clayhurst Stage), the site would have been situated at the end of a peninsula bounded to the west by a narrow arm of the lake, and to the east and south by extensive stretches of water. A similar situation, with less extensive water would have occurred during the Indian Creek stage which follows the late Clayhurst Stage (Figure 1, lower). Fladmark (1986:18-23) proposed this configuration for the earliest occupation at the site, suggesting that bison might have forded the shallow lake where it narrowed below the site, allowing hunters to ambush them as they emerged from the lake. However, new radiocarbon dates show that it is possible that Glacial Lake Peace had already drained prior to the occupation of Charlie Lake Cave, and that the Peace River was once again following its long-established drainage route.

Radiocarbon dates averaging 10,500 B.P. have been obtained on an articulated bison skeleton from terrace gravels within the Peace River valley near the Alberta/B.C. border (Apland and Harington 1994). The terrace gravels were deposited after the lowest stages of the glacial lake had drained. These dates are very similar to the radiocarbon dates from the base of Charlie Lake Cave, suggesting that the first occupation might have taken place after the drainage of the glacial lake. Based on available data, it appears that life did not return to the Peace River region until the drainage of Glacial Lake Peace, at about 10,500 B.P. However, there must have been a period of at least some centuries to develop an ecosystem capable of supporting bison herds and viable human social groups. There is currently no evidence for this from vertebrate fossils, but some palynological sequences provide relevant data.

REGIONAL PALYNOLOGY

Regardless of the status of the glacial lake, all elevations above Charlie Lake Cave must have supported terrestrial flora and fauna before the first human use of the site at 10,500 B.P. Although no pollen was preserved at Charlie Lake Cave, two palynological studies in nearby regions document a post-glacial sequence leading from open to forested environments. To the northeast, MacDonald's (1987) analysis of two lakes in the Clear Hills suggests a sparse post-glacial vegetation consisting of deciduous trees and shrubs such as aspen, birch and willow in association with herbs and grasses. Spruce arrived by 10,000 B.P. and was a dominant species by 9,800 B.P. in a boreal forest somewhat similar to today's northern boreal forest, although with a different species composition. To the southeast, a study of the Saddle Hills records similar early post-glacial vegetation, but both spruce and pine appear as early as 11,200 B.P. (White and Mathewes 1986). The Charlie Lake Cave dates and vertebrates suggest a somewhat later appearance of coniferous forest, with dates similar to those in the Clear Hills cores.

Most palynological sequences in previously glaciated areas situated along a 500 km wide area east of the Rockies show a similar post-glacial sequence of pioneering, colonizing non-coniferous vegetation succeeded by a coniferous forest (Driver 1998). It was into this early open environment that the first large mammals and humans migrated at the end of the Pleistocene. Although pollen deposition rates are low, for example, 1000 grains per cm2 per year in the early stages of the Clear Hills lakes, these rates are largely for non-arboreal taxa. As Guthrie (1985) pointed out, influx rates for pollen depend not only on biological productivity, but also on the types of plants represented. A comparison of influx rates for late Pleistocene/early Holocene samples in the Peace River region with modern Canadian prairie grasslands demonstrates similar rates for non-arboreal taxa, suggesting that the early vegetation communities may have been quite productive.

HUMAN ACTIVITIES AT CHARLIE LAKE CAVE COMPONENTS 1 AND 2

Three cultural components occur in Zone II and Subzone IIIa at Charlie Lake Cave (Driver *et al.* 1996; Handly 1993). Each component is defined by a relatively small number of artifacts and animal bones, and may represent only a few hours or days of human activity. The two lowest components, 1 and 2, are characterized by chert and quartzite stone tools, debitage from resharpening, and butchered bison bone. Component 1 dates to about 10,500 B.P. and Component 2 to about 10,000 B.P. The artifact inventory for both components includes carefully made chert artifacts, such as a projectile point and convergent scraper, and much larger quartzite artifacts which probably functioned as butchering tools, possibly for smashing bones to extract marrow (Driver at al. 1996). None of these artifacts were made on site, and the lack of debitage from tool manufacture suggests that hunters had "geared up" for a kill by preparing tools and implements elsewhere. Most of the quartzite artifacts could have been made expediently, and it is likely that they were discarded after a single use. Just over 100 specimens identified as either *Bison* or large artiodactyl were recovered from Zone II (Components 1 and 2) at Charlie Lake Cave (Table 3).

Determination of the precise number of specimens depends upon fragments that can be fitted together are counted. NISP figures in Table 3 count every specimen separately. MNE figures estimate the minimum number of complete elements required to account for all specimens recovered. As it is not known how many bison kills occurred at the site, the entire assemblage from Zone II is considered here. (Note that the specimens described here are from all Zone II deposits, not just those selected for detailed faunal analysis, and include specimens identified as large artiodactyl as well as bison. This is why the sample size is larger than the total for bison in Table 1).

The bison bones were recovered from fine grained sediments deposited in and around large boulders in Zone II. Preservation conditions were excellent in the southern half of the site where specimens seem to have been buried rapidly. Some indication of the excellent preservation is demonstrated by the recovery of fairly fragile bones such as a newborn calf's innominate or unfused epiphyses of

ELEMENT	NISP	MNE	CARNIVORE	CUTMARK
Cranium	9	1		
Teeth	3	3		
Cervical	1	1		
Thoracic	6	1	1	
Sacrum	1	1	1	
Vertebra	1	1		
Rib	3	1	2	1
Scapula	1	1		
Innominate	4	2	4	
Humerus	5	2	3	2
Radius	9	3	4	
Ulna	2	2	2	1
Carpal	11	11	3	
Metacarpus	3	2	1	
Metacarpus V	1	1		
Tibia	11	4	5	
Lat. mal/	2	2		
Tarsal	5	5	4	
Metatarsus	3	1	1	1
Metapodial	1	0		
Prox. phalanx	4	4	1	
Med. phalanx	9	9	4	
Dist. phalanx	8	8	3	
Sesamoid	4	4		
TOTAL	107	70	39	5

Table 3. Bison and large artiodactyl specimens from all Zone II deposits.

subadults. Another indicator of excellent preservation is the recovery of tracheal rings from the raven skeleton in Subzone IIb. Towards the northern side of the gully preservation was poor, although the number of bones recovered was conspicuously smaller. This is due to the greater amount of groundwater to the north, which seems to have leached bone from the sediments. The number of bison specimens per square metre in different areas of the gully gives some idea of the relative density of bones across the site. In the north the average is about 4 specimens per square metre, while in the centre it is 18, and in the south 12.

Because of good preservation conditions, surface modification of bone is easily seen and characterized. There is no evidence for significant abrasion resulting from transport by water. Bones are typically well preserved with sharply defined features. There is little evidence for weathering of bone surfaces, suggesting that the bones were buried rapidly. There is no evidence for differential weathering of upper and lower surfaces of the bones, which also suggests rapid burial.

Some bones were modified by people and animals prior to burial. Rodent activity is present, but very rare. However, just under 40% of the bison specimens exhibit traces of carnivore activity (Table 3), typically involving chewing of long bone ends, with punctures, furrows and crenulated edges (Binford 1981).

Cutmarks produced by stone tools attest to human interest in these bones (Fladmark *et al.* 1988). Cut marks have been found mainly on the forelimb - two examples on the humerus and one on the ulna suggest disarticulation of. the elbow. Cutmarks were also observed on a rib and a metatarsus Some bison bones were definitely transported to the gully as articulated sets of limb bones:

1) A complete ulna, radius and two carpals from an immature animal. It is likely that some immature phalanges are also from this specimen. This forelimb was scavenged by carnivores which chewed some of the carpals and phalanges and appear to have destroyed the metacarpus;
2) A largely complete ulna (with olecranon removed by carnivores), a radius smashed in the midshaft to obtain marrow, and four articulating carpals;
3) Four articulating carpals which cannot be fitted to any existing radius;
4) A complete forefoot, from the distal metacarpus down.
5) A proximal metatarsus and navicular-cuboid.
6) Various pairs of phalanges;
7) Complete tibia and lateral malleolus.

A significant proportion of bison bones arrived at the site as articulated portions of lower limbs, suggesting relatively little time between the death of the bison and deposition at the site. High frequency of carnivore damage might suggest bone transport by carnivores, but the close spatial association between stone artifacts and bison bones means that deposition by people is more likely. Because the gully was narrow, with a steep uneven floor, it is possible that people did not occupy the gully, but disposed of artifacts and bones into the gully from the hillside above. As discussed elsewhere (Driver, In press b) there may have been some ritual deposition taking place at the site, but much of the bison bone shows evidence for having been defleshed and broken for marrow before being deposited and before being damaged by carnivores.

Element frequencies are weighted towards the lower limbs. NISP and MNE figures show that the lower limbs are well represented when compared to the axial skeleton (Table 3). Element frequencies may be skewed from the expected distribution of elements by either cultural or natural processes (Lyman 1994). We must therefore ask whether the small sample of bison bones from the site owe their pattern of element representation to human behaviour or to natural processes. In other words, does the preponderance of lower limb bones reflect what was originally discarded by people, or have carnivores and other natural processes removed or destroyed other areas of the skeleton, leaving lower limbs as a residue?

One way of considering this is to look at bone density. Both carnivores and abiotic processes (such as weathering) typically destroy weaker and less dense bones first. Kreutzer (1992) has examined volume densities of modern bison bone in the belief that density measured in this way should be correlated with strength to withstand various mechanical and chemical weathering processes. If complete bison skeletons had been deposited at the site, it should be possible to see if low density bones occur in lower than predicted frequencies. However, because the original assemblage of bison bones at the site may not have consisted of all parts of the skeleton, this method cannot be used.

If we assume that the original sample of bison bones at the site did not consist of a representative sample of all elements of the skeleton, then we must develop predictions about how such an assemblage would be affected by processes which selectively removed weaker or less dense bones. In view of the good preservation of specimens and the lack of weathering, these predictions should concentrate on carnivore activity which is abundantly represented on the bones. The following predictions can be tested on the assemblage, but the small sample makes interpretation difficult. First, if low overall frequency of cranial and axial specimens is the result of carnivore destruction, then stronger and denser parts of the axial skeleton should be present.

Second, if carnivores have attacked lower limbs (as surface modification shows), there should be a preponderance of high density specimens and reduced numbers of lower density bones, and this pattern should be most obvious if carnivores have worked on the assemblage intensively. Third, for elements which are common in the assemblage, low density parts of the bones should be preferentially removed, leaving higher density parts intact.

The first prediction fails. The parts of the cranial area most likely to survive are teeth and parts of the mandible (Kreutzer 1992 Table 2). Teeth are underrepresented, and no portions of the mandible are present, even though volume densities of the mandible are similar to the tibia and radius, which are well represented. Furthermore, other areas of the axial skeleton, notably axis and atlas vertebrae, contain high density bone which should survive, but which is not found in the Charlie Lake Cave assemblages. Other parts of the axial skeleton with densities comparable to surviving limb bones include the scapula and parts of rib shafts, both of which are poorly represented.

The second prediction considers the limbs as a whole. There is some evidence to support the prediction that carnivores should selectively remove low density bones. The least dense of the major limb bones are the proximal humerus, and proximal and distal femur (Kreutzer 1992). No femur fragments were recovered, and only one fragment of proximal humerus (Table 4). These areas are high in bone grease (Brink 1997, see also Brink this volume) and therefore particularly likely to attract carnivores (e.g. Garvin 1987; Haynes 1982). It possible that upper limbs were deposited, but that carnivores selectively destroyed some specimens. However, the absence of any femur shaft fragments may mean that no femora were deposited in this area of the site.

ELEMENT	MN ENDS*	DENSITY**	FAT%***
Prox. humerus	1	.24	40.5
Dist. humerus	2	.38	22.0
Prox. radius	3	.38	(32.7)
Prox. ulna	2	.34	(32.7)
Dist. radius	3	.35	25.7
Prox. metacarpus	0	.59	8.9
Dist. metacarpus	1	.46	15.2
Prox. femur	0	.31	31.4
Dist. femur	0	.26	35.2
Prox. tibia	2	.41	33.5
Dist. tibia	4	.41	14.4
Prox. metatarsus	1	.52	12.4
Dist. metatarsus	1	.48	22.7

* minimum number of long bone ends, ** based on Kreutzer (1992) volume densitites, *** based on Brink (1997) combined value for proximal radius and ulna

Table 4. Element frequencies for bison and large artiodactyl limb bone ends.

The third prediction considers the relative representation of different parts of certain skeletal elements. Binford (1981) first suggested this method as a way of identifying assemblages affected by carnivores. Binford's work suggested that carnivores would preferentially attack less dense, more fatty long bone ends. Long bones with one end less dense than the other should show differential survival of the two ends. The two most common long bones at Charlie Lake Cave are tibia and radius. As measured by Kreutzer (1992), bone volume densities are similar for proximal and distal ends of both bones. However, fat content in these two long bones ranks them in the following order: proximal tibia, proximal radius/ulna, distal radius/ulna, distal tibia (Brink 1997 Table 1).

Following the same sequence of skeletal elements, figures for Charlie Lake Cave bison are 2, 2.5, 3 and 3 (Table 4). (Because Brink considered ulna and radius together, I have added Charlie Lake proximal radius and proximal ulna together and divided by two to reach the 2.5 value. If only proximal ulna is counted, this region is less well preserved). Thus, there is some evidence that the ends of bones most attractive to carnivores have been preferentially scavenged. However, we must also note that bones of comparable density but with a lower fat content (e.g. distal metacarpus and distal metatarsus) are more poorly represented than either tibia or radius. It may be that the assemblage is too small to discern clear patterns of carnivore activity at this level of detail, or that carnivore activity was not sufficiently intense to create a clearly defined pattern of damage

One cannot account for the differential representation of skeletal elements simply as a function of bone density or carnivore behavior. The disparity between the relatively well preserved limb elements, including some neonatal and juvenile material, and the almost complete absence of axial material cannot have been caused by natural processes such as carnivore activity and differential weathering, because high density axial elements are missing and some lower density limb bones are present. The high representation of limbs is probably the product of the original human discard activity, although carnivores may have selectively removed the weakest and fattiest limb bones - both ends of the femur and the proximal humerus. It should also be noted that assemblages which have been impacted heavily by carnivores contain high proportions of long bone diaphyses and splinters (Binford 1981; Haynes 1982), which is not the case at Charlie Lake Cave.

The cultural context in which the discard of articulated limb portions and artifacts occurred is difficult to establish on a small sample recovered from a restricted excavation area. From the end of the Pleistocene until the nineteenth century bison hunting was the primary way of life for the inhabitants of the grasslands of interior western Canada. The earliest human use of Charlie Lake Cave took place in an open environment in which, on the basis of paleontological evidence, bison were the most common large mammal (e.g. Churcher and Wilson 1979). It is not unexpected that bison hunting should have been important until boreal forest was established (see, however, Stephenson *et al.*. this volume). The bison assemblage and artifacts are consistent with material discarded at other bison kill sites.

Artifacts were brought to the site ready to use, and were discarded after use. There was little artifact manufacture taking place, although some resharpening occurred. Larger quartzite artifacts were probably used for smashing bones for marrow extraction, and impact points and spiral fractures on bones attest to this activity. Bone was not heavily processed, and there was patterned discard of skeletal elements, with this area of the site dominated by bones of the middle and lower limbs.

HUMAN ACTIVITIES AT CHARLIE LAKE CAVE COMPONENT 3

The third component is associated with Subzone IIIa, dating to about 9500 B.P. The main features of this component are (a) a large number of retouch flakes which suggest that at least two bifaces were resharpened, (b) another largely complete raven skeleton, and (c) a wedge-shaped microblade core associated with the skeleton. (Note that this raven skeleton does not appear in the totals in Table 1 because it was recovered from a unit in which the microfauna was not analysed in detail). The core is illustrated in Driver *et al.* (1996) and the raven is discussed in more detail elsewhere (Driver In press, b). By this time the site may have functioned as a lookout or game monitoring station where bifaces were prepared. The raven and microblade core may have been cached temporarily and abandoned, or they may have been deposited deliberately for ritual reasons.

CONCLUSION

The early deposits at Charlie Lake Cave contain some evidence about the people who first inhabited the post-glacial landscapes of western Canada, and some indications of the nature of the environment in which they lived. The early faunas contain a mixture of northern and southern species, but all seem to indicate a fairly open environment quite different from the boreal forest which developed between 10,000 and 9500 B.P. The lowest two components at Charlie Lake Cave were created during episodes of bison hunting and butchering. We do not have enough data to state whether bison were killed at the site, or whether the site contains refuse dumped into the gully from a nearby kill or processing area.

Archaeological sites dating earlier than 10,000 B.P. are very scarce in western Canada, and those with any quantity of faunal remains are rare (Driver 1998). Charlie Lake Cave and the Vermilion Lakes site near Banff (Fedje *et al.*. 1995) show that big game hunting was an important component of the early economy, and the short-lived open environment of the post-glacial period may have been very attractive to mobile hunter-gatherer bands who probably moved north along the eastern flanks of the Rocky Mountains between 11,500 and 10,500 B.P.

REFERENCES CITED

Apland, B. and C.R. Harington
1994 Pleistocene Bison Skeleton (*Bison bison cf. occidentalis*) from Clayhurst Crossing, British Columbia. *Geographie physique et Quaternaire* 48(2):213-223.

Binford, L.R.
1981 *Bones: Ancient Men and Modern Myths*. Academic Press, New York.

Bobrowsky, P. and N. Rutter
1992 The Quaternary Geologic History of the Canadian Rocky Mountains. *Geographie physique et Quaternaire* 46(1):5-50.

Brink, J.W.
1997 Fat Content in Leg Bones of *Bison bison*, and Applications to Archaeology. *Journal of Archaeological Science* 24: 259-274.

Catto, N., D.G.E. Liverman, P.T. Bobrowsky and N. Rutter
1996 Laurentide, Cordilleran and Montane Glaciation in the Western Peace River-Grande Prairie Region, Alberta and British Columbia, Canada. *Quaternary International* 32:21-32.

Churcher, C.S. and M. Wilson
1979 Quaternary Mammals from the Eastern Peace River District, Alberta. *Journal of Paleontology* 53(1):71-76.

Driver, J.C.
1988 Late Pleistocene and Holocene Vertebrates and Palaeoenvironments from Charlie Lake Cave, Northeast British Columbia. *Canadian Journal of Earth Sciences* 25:1545-1553.

1996. The Significance of the Fauna from Charlie Lake Cave. In *Early Human Occupation in British Columbia*, edited by R. Carlson and L. Dalla Bona, pp. 21-28. University of British Columbia Press, Vancouver.

1998 Human Adaptation at the Pleistocene /Holocene Boundary in Western Canada, 11,000 to 9000 BP. *Quaternary International* 49/50:141-150.

In press (a) Late Pleistocene Collared Lemming (*Dicrostonyx torquatus*) from Northeastern British Columbia, Canada. *Journal of Vertebrate Paleontology*

In press (b) Raven Skeletons from Paleoindian Contexts, Charlie Lake Cave, British Columbia. *American Antiquity*

Driver, J.C., M. Handly, K.R. Fladmark, D.E.Nelson, G.M. Sullivan and R. Preston
1996 Stratigraphy, Radiocarbon Dating, and Culture History of Charlie Lake Cave, British Columbia. *Arctic* 49(3):265-277.

Driver, J.C. and K.A. Hobson
1992 A 10,500 Year Sequence of Bird Remains from the Southern Boreal Forest Region of Western Canada. *Arctic* 45(2):105-110.

FAUNMAP Working Group
1996 Spatial Response of Mammals to Late Quaternary Environmental Fluctuations. *Science* 272:1601-1606.

Fedje, D.W., J.M. White, M.C. Wilson, D.E. Nelson, J.S. Vogel, and J.R. Southon
1995 Vermilion Lakes Site: Adaptations and Environments in the Canadian Rockies During the Latest Pleistocene and Early Holocene. *American Antiquity* 60:81-108.

Fladmark, K.R.
1986. *British Columbia Prehistory*. National Museums of Canada, Ottawa.

1996 The Prehistory of Charlie Lake Cave. In *Early Human Occupation in British Columbia*, edited by R. Carlson and L. Dalla Bona, pp. 11-20. University of British Columbia Press, Vancouver.

Fladmark, K.R., J.C. Driver and D. Alexander
1988 The Palaeoindian Component at Charlie Lake Cave (HbRf 39), British Columbia. *American Antiquity* 53(2):371-384.

Garvin, R.D.
1987 Research in Plains Taphonomy: the Manipulation of Faunal Assemblages by Carnivores. Unpublished M.A. Thesis, Department of Archaeology, University of Calgary.

Graham, M.A., M.C. Wilson and R.W. Graham
1987 Paleoenvironments and Mammalian Faunas of Montana, Southern Alberta and Southern Saskatchewan. In *Late Quaternary Mammalian Biogeography and Environments of the Great Plains and Prairies*, edited by R.W. Graham, H.A. Semken, Jr. and M.A. Graham, pp. 410-459. Illinois State Museum, Springfield.

Graham, R.W.
1992 Late Pleistocene Faunal Changes as a Guide to Understanding Effects of Greenhouse Warming on the Mammalian Fauna of North America. In *Global Warming and Biological Diversity*, edited by R.L. Peters and T.E. Lovejoy, pp. 76-87. Yale University Press, New Haven.

Guthrie, R.D.
1985 Woolly Arguments Against the Mammoth Steppe - A New Look at the Palynological Data. *Quarterly Review of Archaeology* 6(3): 9-16.

Handly, M.J.
1993 Lithic Assemblage Variability at Charlie Lake Cave (HbRf-39): A Stratified Rockshelter in Northeastern British Columbia. Unpublished M.A. thesis, Department of Anthropology, Trent University, Peterborough, Ontario.

Haynes, G.
1982 Utilization and Skeletal Disturbances of North American Prey Carcasses. *Arctic* 35(2):266-281.

Kreutzer, L.A.
1992 Bison and Deer Bone Mineral Densities: Comparisons and Implications for the Interpretation of Archaeological Faunas. *Journal of Archaeological Science* 19(3):271-294.

Lundelius, E.L., R.W. Graham, E. Anderson, J. Guilday, J.A. Holman, D.W. Steadman and D.S. Webb
1983 Terrestrial Vertebrate Faunas. In *Late-Quaternary Environments of the United States, Volume 1, the Late Pleistocene*, edited by S.C. Porter, pp. 311-352. University of Minnesota Press, Minneapolis.

Lyman, R.L.
1994 *Vertebrate Taphonomy*. Cambridge University Press, Cambridge.

MacDonald, G.M.
1987 Postglacial Development of the Subalpine-Boreal Transition Forest of Western Canada. *Journal of Ecology* 75:303-320.

Mathews, W.H.
1978 Quaternary Stratigraphy and Geomorphology of the Charlie Lake (94A) Map Area, British Columbia. *Geological Survey of Canada Paper* 76-20.

1980 Retreat of the Last Ice Sheets in Northeastern British Columbia and Adjacent Alberta. *Geological Survey of Canada Bulletin* 331.

Mead, E.M. and J.I. Mead.
1989 Quaternary Zoogeography of the Nearctic *Dicrostonyx* Lemmings. *Boreas* 18:323-332.

White, J.M. and R.W. Mathewes
1986 Postglacial Vegetation and Climatic Change in the Upper Peace River District, Alberta. *Canadian Journal of Botany* 64:2305-2318.

THE SMALL MAMMALS OF LIME HILLS CAVE I

Dianna M. Georgina
Department of Anthropology, Washington State University

Cave and rockshelters have often been used as habitations and resting places by humans and animals. Cave sediments frequently contain plant and animal fossils that are important indicators of regional paleoclimate (Farrand 1985). Lime Hills Cave I, in the Kuskokwim River drainage, near Stoney River (Figure 1), provides a rare opportunity for study of the paleoenvironment of interior southwest Alaska based on zooarchaeological data. Studies of paleoenvironments in this region have mainly focused on pollen; only a few faunal studies have been undertaken (Ager 1982; Ager and Brubaker 1985; Short *et al.* 1992). Elsewhere in northern North America, Beringian cave studies include Lower Rampart Cave in northeast Alaska (Dixon 1984; Sattler 1991, and this volume) the Trail Creek Caves on the Seward Peninsula (Larsen 1968; Schaaf 1988; West 1996; Vinson 1988, and this volume), Bluefish Caves in northwest Yukon Territory, Canada (Cinq-Mars 1979, 1990; Morlan and Cinq-Mars 1982) and Charlie Lake Cave in British Columbia (Driver, this volume).

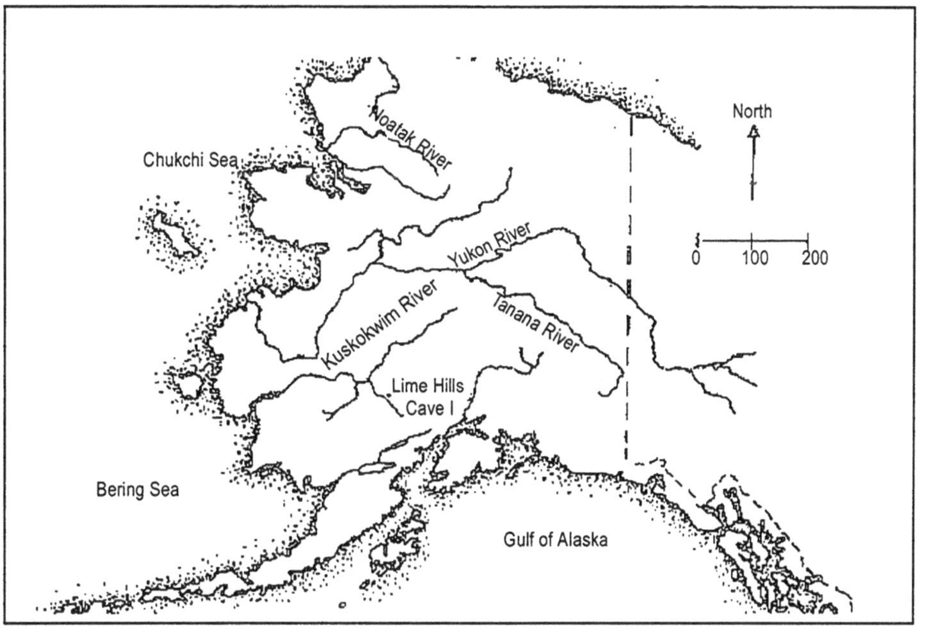

Figure 1. Map of Alaska showing Lime Hills Cave I.

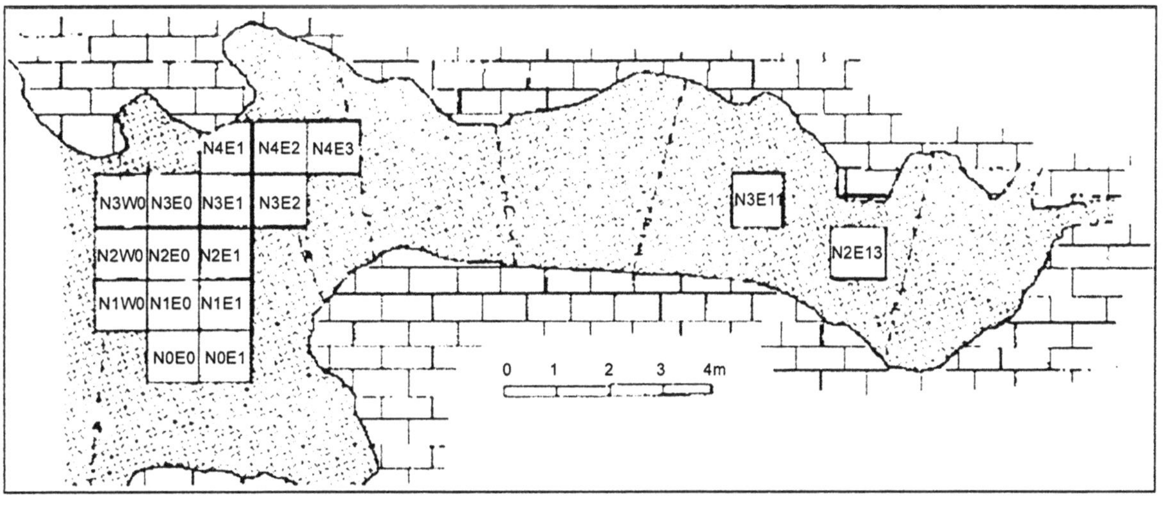

Figure 2. Planview of excavations.

This study presents data on the small mammal remains from Lime Hills Cave I. While large mammals have received more attention in the literature, small mammals are more useful as paleoenvironmental indicators due to their greater ecological sensitivity (Repenning *et al.* 1964; Guthrie 1968). The small mammal remains from Lime Hills Cave I reveal paleoenvironmental changes in the region beginning 38,000 years ago and continuing through the Holocene. These findings agree with pollen studies of the cave sediments (Ruter 1998, pers. comm.).

ENVIRONMENTAL SETTING

The environment within a few kilometers of the cave includes areas of moist tundra, low brush bog and muskeg, bottomland spruce-poplar forest, upland spruce-hardwood forest, and high brush (Selkregg 1976). Small mammals found in the area today include shrew (*Sorex* sp.), brown lemming (*Lemmus sibiricus*), red-backed vole (*Clethrionomys rutulus*), tundra vole (*Microtus oeconomus*), beaver (*Castor canadensis*), snowshoe hare (*Lepus americanus*), ermine (*Mustela erminea*), least weasel (*Mustela rixosa*), land otter (*Lontra canadensis*), red fox (*Vulpes vulpes*), and lynx (*Lynx canadensis*) (Manville and Young 1965; Smith 1979; Jarrel *et al.* 1998) (Table 1).

SITE TESTING

Before excavation, the cave entrance was 6.4m wide and 2.5m high. The cave extends backwards from the entrance as a narrow curving corridor about 17m long (Figure 2). Human activity was limited to the area of the cave entrance. The inner area of the cave, today accessible by crawling, appears to have been used only by animals, particularly procupine, and possibly bear. The floor sediments at the entrance to the cave were up to 1.5m deep, while those in units E11 and E13, at the back of the cave, were less than half a metre deep, much of which consisted of porcupine droppings and bedding material.

Lime Hills Cave I was excavated during the summers of 1993 and 1995 by a team of archaeologists from Washington State University. The cave was mapped in detail with control points for excavation established throughout. Thirteen 1x1m units were excavated in 1995, in addition to the four test squares excavated by Ackerman in 1993 (Ackerman 1996) for a total of 17 squares, 15 in the entrance area and two at 11 and 13 metres from the cave entrance. Excavations to bedrock were conducted in arbitrary five cm levels, and natural levels when applicable. Where possible each level was dry-screened through ¼ and 1/8 inch hardware cloth. Water screening was used for wet sediments that blocked the dry screens. In addition, samples were run through a flotation machine as a check on collection procedures; this check demonstrated that the screening procedures were nearly equally effective.

Cave stratigraphy is complex, and at least four major strata are observable. Strata are divided into five levels; the two lowest are inferred. Radiocarbon dates from the two lowest strata fall in three clusters: 38-28,000; 16-13,000; and 9,800-9,500. The hiatuses may result from the removal of cave sediments, or more likely, from different rates of sediment accumulation in the Pleistocene as compared to the Holocene (Ruter 1998, personal communication).

Frost wedges are evident in at least one profile, and several stratigraphically inverted dates may indicate downward movement of material and possible localized bioturbation (Ruter 1998, personal communication) Table 2 presents a general description of the major strata.

TAXON	STRATUM					PRESENT TODAY
INSECTIVORA	1	2	3	4	5	
Soricidae						
Sorex sp.			x			x
CARNIVORA						
Canidae						
Alopex lagopus (Arctic fox)			x		x	
Canis lupus (wolf)			x		x	x
Vulpes vulpes (red fox)			x		x	x
Mustelidae						
Lontra canadensis (river otter)		x				x
Martes americana (marten)		x	x	x		x
Mustela erminea (ermine)		x	x	x		x
RODENTIA						
Scuridae						
Marmota caligata (hoary marmot)			x	x		x
Spemophilus parryii (arctic ground squirrel)		x	x	x	x	x
Castoridae						
Castor canadensis (beaver)		x	x			x
Muridae						
Cleithionomys rutilus (northern red-backed vole)	x	x	x	x		X
Dicrostonyx groenlandicus (collared lemming)		x	x	x	x	
Lemmus trimucronatus (brown lemming)		x	x	x	x	x
Microtus miurus (singing vole)		x	x	x	x	
Microtus oeconomus (tundra vole)		x	x	x	x	x
Microtus pennsylvanicus (meadow vole)	x	x	x	x	x	x
Microtus xanthognathus (yellow cheeked vole)		x	x	x		
Ondatra zibethicus (muskrat)		x		x		x
Synaptomys borealis (northern bog lemming)		x				
Erethizontidae						
Erethizon dorsatum (porcupine)				x		x
LAGOMORPHA						
Leporidae						
Lepus sp. (tundra/snowshoe hare)		x	x	x	x	x

Table 1. Small mammals from Lime Hills Cave I.

STRATA	DESCRIPTION
1	220±70 *B.P.* (312 *B.P.* calibrated age) Organic histosol, very dark brown, up to 60% saturated fibrous organic material primarily derived from procupine dung. Approximately 15cm thick.
2	490±60 *B.P.* (623 *B.P.* calibrated age) Silty clay loam/clay loam, very dark greyish brown to dark brown, ranging from 30cm to nearly 50cm at the cave mouth, lensing out in the back of the cave.
3	9530±60 *B.P.* (10,360-8,385 *B.P.* calibrated age) Oxidized clay (silty clay/clay loam), yellowish brown ranging. Ranging in thickness from 20cm at the back of the cave where it is frozen and in direct contact with regolith, to the cave mouth were it is about 35cm.
4	9,080±50 *B.P.* (9,875-9540 *B.P.* calibrated age) Possibly 15,690±140 *B.P.* – 13,139±180 *B.P.* Silty clay loam, somewhat oxidized reddish brown, comprising a matrix with gravel and small clasts of eroded calcite and limestone.
5	27,950±560 *B.P.* – 38,860 *B.P.* Possibly 15,690±140 *B.P.* – 13,139±180 *B.P.* Sandy clay loam/clay loam, similar to stratum 4 with which it intergrades, somewhat oxidized reddish brown to a dark yellowish brown. IT comprises a matrix with gravel and small clasts of eroded calcite and limestone, as well as eroded regolith.

Table 2. Description of major strata, Lime Hills Cave I (Ruter 1998, personal communication).

FAUNAL ANALYSIS

Faunal material was recovered in the screens and *in situ* from all units and in all stratigraphic levels. Larger bones were mapped in place, while smaller remains were generally recovered during screening.

Faunal material was identified using the comparative collection at The Conner Museum, Washington State University. Additional reference sources included Bee and Hall (1956), Olson (1964), Maser and Storm (1970), Gilbert (1980), Smith (1979), and Zweifel (1994).

Postcranial elements of the various *Microtus species*, both in the Connor Museum collection and those recovered from the cave sediments, are identifiable only to Genus. However, *Microtus* can be identified to species on the basis of enamel patterns on dental occlusal surfaces (Bee and Hall 1956; Maser and Storm 1970; Gilbert 1980; Zweifel 1994). Due to the importance of *Microtus* sp. in this paleoenvironmental study, dental material was used exclusively.

DISCUSSION

The small mammal remains are in excellent condition; they are generally well-preserved, with little or no evidence of carnivore or rodent gnawing. There is little digestive etching or end rounding on most of the bones and the majority are whole, although some are missing epiphyses, apparently snapped off, and a number of mandibles have broken or missing teeth. Sharp broken ends, displaying no weathering, digestive rounding or etching, suggest that trampling and rockfall compression may account for some of the bone breakage. Very few crania were found, and except for atlases, no vertebrae. Ribs are abundant; the smaller bones of the foot were absent or not recovered.

The largest quantity of remains (40.7%) are from Stratum 4, followed by Stratum 3 (29.0%). Strata 5 and 2 each contain less than 20%, and Strata 1 less than 1% of the total sample. Table 3 lists by stratigraphic level the frequencies of identifiable small mammals found in the cave deposits. As Table 4 illustrates, the brown lemming (*Lemmus trimucronatus*, 28.8%) and the collared lemming (*Dicrostonyx groenlandicus* 19.2%) together comprise 50% of the total number of identified specimens. Yellow cheeked vole (*Microtus xanthognathus*, 9.4%), meadow vole (*M. pennsylivanicus*, 8.7%) and singing vole (*M. miurus*, 8.5%) round out the top 75%.

Small mammals present in the cave sediments but no longer found in the area include yellow cheeked vole, collared lemming and Arctic fox (Manville and Young 1965; Selkregg 1976). The presence of northern bog lemming in the region today is debated in the literature. Manville and Young (1965), and Selkregg (1976) view the species as absent, but Wilson and Reeder (1993) list it as present in the southern two thirds of Alaska.

TAPHONOMY OF LIME HILLS CAVE I

The means by which faunal remains are transported into caves strongly influences which species are represented. Guthrie (1968) points out that predators do not randomly sample fauna within their territory. Most predators preferentially choose a favored prey species, introducing a bias into the faunal assemblage in a diet. For many predators, the favored species is the most common rodent (Bee and Hall 1956; Andrews 1990; Andrews and Evans 1990; Sklepkovych and Montevecchi 1996).

Caves are used as den sites by both bears and porcupines (see Sattler *et al.*, this volume). Bears do not bring bones for gnawing to their den sites, but porcupines often do. In addition, bears may die during hibernation (Stiner *et al.* 1996), and some rodents will opportunistically live and caves and may also die there. Both events add to faunal assemblages in caves.

Limited human use of Lime Hills Cave I is evidenced by the presence, in some levels, of American Paleoartic tradition artifacts including microblade inserts, side-slotted antler arrowheads, and spearheads (Ackermann 1996, 1997).

Human Agents

Many researchers consider small mammals non-cultural in archaeological sites, ignoring their potential importance in prehistoric diet. However, in her analysis of human coprolites, Sobilik (1993) found that many small mammal species were consumed including voles, squirrels, and ground squirrels. Human digestion is a taphonomic process, and Crandall and Stall (1995) report that with regard to small mammal remains it results in extreme skeletal attrition, skull breakage, loss of distal elements, maxillary tooth loss, mandibular breakage, and a high proportion of isolated teeth. This pattern is consistent with some aspects of the Lime Hills faunal assemblage, and human predation on small mammals remains a possibility.

TAXA	STRATUM					TOTAL
	1	2	3	4	4	
Sorex sp.			1			1
Alopex lagopus			1		1	2
Vulpes vulpes			2		2	4
Canis lupus			2	1		3
Lontra candendsis		1				1
Martes americana		1	2	1		4
Mustela ermina		1	2	2		5
Marmota caligata			3	2		5
Spemophilus parryi		1	5	3	9	18
Castor canadensis		3	7			10
Cleithrionomys rutilus	2	10	18	4		34
Dicrostonyx groenlandicus		1	3	65	35	104
Lemmus trimuconatus		14	25	89	28	156
Microtus miurus		3	20	16	7	46
Microtus oeconomus		6	14	6	2	28
Microtus pennsylvanicus	1	12	16	15	3	47
Microtus xanthoghathur		12	27	12		51
Ondatra zibethicus		1		1		2
Synaotomys borealis		2				2
Erethizon dorsatum			2			2
Lepus sp.		4	7	3	2	16
TOTAL	3	72	157	220	89	541

Table 3. Frequencies of identifiable small mammal dentition.

TAXA	STRATUM					TOTAL
	1	2	3	4	5	
Sorex sp.			0.18			0.18
Alopex lagopus			0.18		0.18	0.37
Vulpes vulpes			0.37		0.37	0.74
Canis lupus			0.37	0.18		0.55
Lontra candendsis		0.18				0.18
Martes americana		0.18	0.37	0.18		0.74
Mustela ermina		0.18	0.37	0.37		0.92
Marmota caligata			0.55	0.37		0.92
Spemophilus parryi		0.18	0.92	0.55	1.66	3.33
Castor canadensis		0.55	1.29			1.85
Cleithrionomys rutilus	0.37	1.85	3.33	0.74		6.28
Dicrostonyx groenlandicus		0.18	0.55	12.01	6.47	19.22
Lemmus trimuconatus		2.59	4.62	16.45	5.18	28.84
Microtus miurus		0.55	3.70	2.96	1.29	8.50
Microtus oeconomus		1.11	2.59	1.11	0.37	5.18
Microtus pennsylvanicu s	0.18	2.22	2.96	2.77	0.55	8.69
Microtus xanthoghathur		2.22	4.99	2.22		9.43
Ondatra zibethicus		0.18		0.18		0.37
Synaotomys borealis		0.37				0.37
Erethizon dorsatum			0.37			0.37
Lepus sp.		0.74	1.29	0.55	0.37	2.96
TOTAL	0.55	13.31	29.02	40.67	16.45	100

Table 4. Percentage of taxa based on site total.

Non-human Agents

Non-human predatotrs that may have been active in the vicinity of Lime Hills Cave I include bear, canids, mustelids, wolverine, and raptors (Selkregg 1976). Carnivore remains identified in the cave include wolf, otter, marten, ermine, and possibly bear. Coyote is a recent migrant to Alaska (Mannville and Young 1965), and therefore could have made no contributions to the bone assemblages in the cave prior to about AD 1900.

McNamee (1984) found that bears do not habitually carry food into dens. Hibernating bears are particularly vulnerable to attack from wolves and other bears, and there is safety in cleanliness. Caves that are used as dens are kept clean of bones that might smell and attract other carnivores to the location. It is unlikely then that the bones of prey species in Lime Hills Cave I were transported there by bears.

Both red and arctic foxes are known to cache large amounts of food near their den sites (Sklepkovych and Montevecchi 1996). One type of larder hole is dug near the base of boulders, and the roof fall in Lime Hills Cave I provides many such locations. Most carcasses are buried intact, although sometimes heads and wings are removed before caching. Larder hoarding is associated with extreme seasonal shifts in food conditions; the fox hoards food during superbundance for use during lean times (Sklepkovych and Montevecchi 1996). Faunal remains from forgotton caches will show no signs of mastication or digestion. Arctic and red fox prey on small mammals, especially the brown and collared lemming (Bee and Hall 1956). Arctic fox has also been observed scavenging carcasses (Andrews and Evans 1983). Bone from the scat of both fox species exhibits similar characteristics, including extreme breakage and corrosion, and few cranial elements remain intact in fox scat. (Andrews and Evans 1983). The patterning of some of the faunal remains in Lime Hills Cave I is consistent with the faunal patterning from red and arctic fox caches.

Mustelids (weasels and martens) prey on small mammals. Very often the heads of prey are not eaten, and skulls are under-represented in mustelid produced faunal assemblages. The exception to this is assemblages produced by marten (*Martes martes*), where bone preservation is slightly better; mandibles and maxillae are broken, with rounded edges and nearly half the vole teeth are broken into lobes (Andrews and Evans 1983).

Prey assemblages accumulated by mustelids are not always directly representative of the local species representation. For example in Scotland, the pine marten's favored prey is *Microtus*, however 90% of the rodent fauna is *Cleithrionomys* and *Apodemus*; *Microtus* represents only 6% of the rodent fauna. At Lime Hills Cave I the faunal assemblage does not reflect mustelid prey choices, digestive or bone breakage pattens.

Eagles and kestrels and other diurnal raptors tear apart their prey as they eat it. This is reflected in the high proportion of broken bones in their prey remains. Owls tend to swallow their prey whole or in large chunks. Indigestible parts are regurgitated after a number of hours. Owl pellet assemblages exhibit large quantities of bones from a restricted number of species; a high relative frequency of complete femora, radii, mandibles, and humeri, and a low relative frequency of complete scapula and innominates; a large proportion of mandibles with damage to the ascending ramus; and a large proportion of bones in good condition, but with evidence of possible digestive erosion (Kusmer 1990).

The snowy owl (*Nyctea scandiaca*) preys primarily upon the most common rodent species in its territory. Snowy owls on Baffin Island prey mainly on voles and lemmings. The owls eat prey whole, head first, and pellets contain mostly broken bones, with the exception of skulls, mandibles and pelves (Watson 1957). In contrast, Andrews (1990) found a larger proportion of postcranial elements in the pellets that he examined.

Snowy owls nest in hollows in vegetation or sediments, in places where there is a good view; they have also been observed nesting on cliff edges (Watson 1957). The owls drop pellets in and around the nest site, and around perching sites (Watson 1957). There are few records of snowy owls entering caves, and there are few caves in the regions they now inhabit. However, an owl nest or perching site on the ridge above the cave could account for some of the small mammal bones found inside.

The most plausible explanation for the small mammals in Lime Hills Cave I involves the bone collecting activities of porcupines, the use of the cave as a larder/cache site by foxes, the deposition of scat (marten and possibly also human), the deposition of owl pellets and the natural death of some individuals.

ANALYSIS

Recent palynological studies in the Nushagak and Holitna lowlands of southwestern Alaska (Short *et al.* 1992) indicate a paleoenvironmental sequence comparable with the sequence described by Ager (1983). During the last glaciation, 25,000 to about 12,000 *B.P.*, the pollen indicates a cold, dry, herbaceous steppe-tundra. Birch appears in the area between 12,200 and 9,200 *B.P.*, indicating a mesic tundra and perhaps warmer, moister summers and possibly deeper winter snows. The transition from birch to alder seems to have occurred after 8,500 *B.P.* (Short *et al.* 1992).

In this study, the division of the stratigraphy below Stratum 3 into two strata is arbitrary; it is based on observed differences in the stratigraphy that may not be due to depositional variation. These inferred strata are analyzed separately.

Stratum 5

Collared lemming comprise more than one third (39.3%) of the total fauna from the level, followed by brown lemming (31.5%). These species are often found in adjacent communities, the brown lemming occupying wet lowlands. while the collared lemming inhabits the mesic and xeric upland tundra (Bee and Hall 1956; Batzli and Jung 1980; Rodgers and Lewis 1986; Batzli and Henttonen 1990). The two lemmings are non-competitive, eating foods that are inedible to the other (Rodgers and Lewis 1986). The collared lemming primarily eats *Dryas* (cottongrass) and some *Salix* (willow), while the brown lemming prefers grasses and moss. The collared lemming is the best-adapted microtine to freezing temperatures, snow and ice. It occupies a restricted habitat, preferring the tundra (Bee and Hall 1956), The brown lemming is also cold-weather adapted. It is known to burrow down to permafrost to stay cool in the summer, and to build snow nests in the winter (Bee and Hall 1956). The collared lemming is today found only in the northern coastal areas of Alaska, and in the Brooks Range; the brown lemming is found throughout most of the state (Bee and Hall 1956; Manville and Young 1965).

The singing vole (7.9%) is not found in the area today. It prefers tundra or grassy areas above the treeline in the Brooks and Alaska ranges (Manville and Young 1965). They are indicators of a windswept area with intermittent snow banks; while their habitat may today be restricted to windy alpine areas, during the Pleistocene they lived on rolling hills and in lowlands (Guthrie 1982).

Other muirid rodents in Stratum 5 at Line Hills include tundra vole (2.2%) and meadow vole (3.4%), both of which are present in the area today. The meadow vole tends to prefer grassy areas and spruce forest while the tundra vole occupies tundra, grass or sedge, tending more toward damp areas (Bee and Hall 1956; Manville and Young 1965; Batzli and Jung 1980; Douglass 1984; Batzli and Henttonen 1990).

The arctic ground squirrel found in this stratum (10.1%) was situated in the same square and depth below datum from which a radiocarbon date of 32,630±260 was obtained. Bee and Hall (1956) found that arctic ground squirrel is best adapted to areas where permafrost is several feet below the surface, most often on well-drained slopes. Arctic ground squirrel had a much wider distribution during the Pleistocene, requiring open country, plenty of seeds and forbs and a habitat of well-drained sites with irregular surfaces and cutbanks (Guthrie 1982).

Arctic fox (1.1%) is absent from the area today. Its range is limited o coastal areas, on open tundra or rocky beaches (Manville and young 1965). Red fox appears for the first time in this top level, near the transition into Stratum 4, Red fox are ubiquitous in Alaska today, adapted to most environments; Manville and Young (1965) indicate red fox prefer dry country at all elevations.

Fauna indicating woodland or forest is absent from this level. Large mammals included in this level are woolly mammoth, horse, and bison (Ackerman 1996), markers of the mammoth steppe (Guthrie 1982).

Stratum 4

Stratum 4 had the greatest quantity of identified teeth, more than 40% of the total from the site. The earliest evidence of wolf in the cave appears at this level. Wolves are ubiquitous and range over wide areas, and are therefore not good environmental indicators. However, their presence in the cave is significant as a possible taphonomic agent.

Hoary marmot live on talus slopes and excavate dens in boulder fields (Bee and Hall 1956), suggesting the possibility that marmots were denning in cave talus, This specimen, however, was found deep within the cave mouth, well back from the dripline, and may have been prey. The muskrat is a resident of streams and marshes throughout Alaska as far north as the Brooks Range (Manville and Young 1965).

The marten from the top of this level suggests the presence of dwarf trees or tall shrubs. Alder (*Alnus* sp.) pollen was found in all levels, while spruce does not appear until Stratum 2, after 4000 *B.P.* (Ruter 1998). Martens prefer mature conifers, from sea level to above timberline in the mountains (Manville and young 1965). Bee and Hall (1956) report finding marten tracks ten miles south of the crest of the Brooks range, on rock slopes above timberline.

Among the microfauna, brown and collared lemmings make up more that 70% of the total fauna for the level. The only areas where these two lemmings occur together today are on the arctic coast and adjacent tundra areas (Manville and Young 1965). While the ratio of brown to collared lemmings was more nearly even in Stratum 5, in this level the collared lemming numbers exceed those of brown lemmings by almost a third; this is the only level where the former outnumber the latter by a significant margin. This may indicate that areas such as ridges and the crests of hills, were still cold and dry enough to support xerophytic species like collared lemming and arctic fox.

The yellow-checked vole and red-backed vole make their first appearance in this level. The red-backed vole is present throughout Alaska today although, it is generally known to prefer areas where there is overhead protection. They are often found in rock fields and talus slopes, particularly the contact area between boulder fields and open country (Bee and Hall 1956). Plant communities commonly associated with the red-backed vole are *Cassiope*, dwarf willow and alder (Bee and Hall 1956). Douglass (1984) noted an association with lichens, birch, and Labrador tea; he also found more red-backed voles in the driest environments in his study area. The yellow cheeked vole is found today only in interior Alaska, mainly within the Yukon-Tanana drainages. It prefers dry soil, among trees and shrubs (Manville and Young 1965), and may be associated with boreal forest (Banfield 1974). Driver (1988) noted the appearance of these two rodents together in a similarly-aged level in Charlie Lake Cave, British Columbia, indicating a foresting of the area by 10,000 B.P. (see Driver, this volume).

Stratum 3

Twenty-nine percent of all identified specimens were recovered from this level. Beaver and porcupine first appear, indicating establishment of tree or tall shrub cover in the area by this time. Beavers use timber for dam and habitation construction and for food. Modern procupine climb trees to escape predators, and use trees for food.

Among the microtine fauna, collared lemming numbers drop in relation to brown lemming, approximately 2-16% respectively. This may indicate increased moisture and enlarging wet tundra areas, such as wet sedge meadows in the flats. Singing vole peaks in this level (12.7%), indicating the presence of dry tundra, probably on the alpine ridges. Yellow-cheeked vole also peaks (17.2%), evidencing a well-developed shrub zone providing cover.

Arctic fox last occurs in the top five cm of this level. It appears that for some time, beaver, collared and brown lemming, singing vole and arctic fox were simultaneously present in the area, a situation which today exists along Norton Sound and the Seward Peninsula (Manville and Young 1965).

Stratum 2

During this period of deposition, either fewer small mammals are being deposited in the cave, or preservation is compromised relative to the lower levels. Only 13.3% of the total of identified small mammals were recovered from this level. Higher acidic values due to organics would mitigate against preservation of bone, which is likely the case.

Among those recovered, river otter and northern bog lemming make their first appearance in the cave sediments. Northern bog lemming prefer cold bog or spring areas, and near rocky cliffs. They prefer somewhat warmer climates than other lemmings (Manville and Young 1965).

Hare, red-backed vole and beaver indicate forest; the tundra vole, damp lowlands; and the singing vole, a well-drained habitat with a deep permafrost level. This suggests the environment was more similar to that of the present. However, the presence of collared lemming and singing vole may indicate that the environment was still somewhat colder than today. Ruter's (1998, personal communication) pollen studies found that boreal forest was established in the Lime Hills area by 4000 *B.P.*, which appears consistent with the faunal material.

Stratum 1

The red-backed vole and meadow vole are the only two species whose teeth were found in this level. This stratum, the duff or modern cave floor, is composed mainly of porcupine dung and other organic material (Table 2). The lack of faunal remains in this level may be due to a combination of several factors, including increased acidity and the presence of porcupines, who are known to chew bones.

CONCLUSION

The small mammal remains in Lime Hills Cave I indicate environmental changes from the Boutellier Interstadial to the present. The findings in this study agree with the pollen data from the area.

Hiatuses and varying rates of deposition in the strata below Stratum 3 complicate our ability to differentiate between the

Boutellier interstade and the Late Glacial Maximum in the faunal record of the cave. However, we can make inferences based on the apparent differences in the stratigraphy between Strata 4 and 5.

During the Boutellier interstade (Stratum 5), the environment in the immediate area favored several types of arctic fauna. Areas of good drainage would have supported animals such as arctic ground squirrel, singing vole and collared lemming, while damper areas supported brown lemming and tundra vole. Grassy areas would have provided food for meadow and singing voles, and brown lemmings.

During the deposition of Stratum 4, marten, ermine, marmot and red-backed vole occupied the area in the vicinity of the cave. Marten and red-backed vole are commonly associated with dwarf tree and shrub forms, and colder, and dryer environments. The significant increase in collared lemmings during this period may indicate that some areas of the region, such as the crests of hills, were well-drained and cold enough to support xerophytic species, while muskrat provide evidence of swamp and bog environments.

In the fauna of Stratum 3, a sharp decrease in collared lemming and the presence of beaver and procupine may indicate a mixed environment with increasing moisture and enlargement of wet tundra areas.

During the deposition of Stratum 2 the local environment consisted of a mixture of forest, damp lowlands, dry upland tundra, and a deep permafrost layer. The ecosystem included northern bog lemming, beaver, and hare suggesting it was warmer than previous periods, but colder than present.

ACKNOWLEDGEMENTS

I am grateful to my committee chair, Robert E. Ackerman, for his patience and continued support and advice; to Carl Gustafson for his assistance in learning how to identify the Lime Hills faunal material; and Kevin Pullen of the Zoology Department for showing me around Conner Museum.

REFERENCES CITED

Ackerman, R.E.
1996 Cave One, Lime Hills. In *American Beginnings: The Prehistory and Paleoecology of Berengia*, edited by F.H. West, pp. 470-477. University of Chicago Press, Chicago.
1997 Report on the 1995 Field Season, Lime Hills Archaeological Project, Southwestern Alaska. Report to the Office of History and Archaeology, State of Alaska and Calista Corporation, Washington State University.

Ager, T.A.
1982 Vegetation History of Western Alaskan Wisconsin Glacial Interval and the Holocene. In *Paleoecology of Beringia*, edited by D.M. Hopkins, J.V. Mathews, Jr., C.E. Schweger, and S.B. Young,. Academic Press, Inc., New York.

Ager, T.A. and L. Brubaker
1983 Quaternary Paleoecology and Vegetational History of Alaska. In *Pollen Record of Late Quaternary North American Sediments*, edited by V.M. Bryant, Jr., and R.G. Holloway. American Association of Stratigraphic Palynologists Foundation.

Andrews, P. and Evans, E.
1990 Small Mammal Bone Accumulations Produced by Mammalian Carnivores. *Paleobiology* 9:289-307.

Andrews, P.
1990 *Owls, Caves and Fossils: Predation, Preservation, and Accumulation of Small Mammal Bones in Caves*. University of Chicago Press, Chicago.

Banfield, A.W.F.
1974 *The Mammals of Canada*. University of Toronto Press, Toronto.

Batzli, G.O. and H-J G. Jung
1980 Nutritional Ecology of Microtine Rodents: Resource Utilization Near Atkasook, Alaska. *Arctic and Alpine Research*. 12(4):483-499.

Batzli, G.O. and H. Henttonen
1990 Demography and Resource Used by Microtine Rodents Near Toolik Lake, Alaska, U.S.A. *Arctic and Alpine Research* 22(1):51-64.

Bee, J.W. and E.R. Hall
1956 *Mammals of Northern Alaska*. University of Kansas Museum of Natural History, Lawrence.

Cinq-Mars, J.
1979 Bluefish Cave 1: A Late Pleistocene Eastern Beringian Cave Site in the Northern Yukon. *Canadian Journal of Archaeology* 3:1-32.

1990 La Place des Grottes du Poisson-Bleu dans la Prehistorie Beringienne. *Revista de Arquelog 'a Americana* 1:9-32.

Crandall, B.D. and P.W. Stahl
1995 Human Digestive Effects on a Micromammalian Skeleton. *Journal of Archaeological Science* 22:789-797.

Dixon, James E.
1984 Context and Environment in Taphonomic Analysis: Examples from Alaska's Porcupine River Caves. *Quaternary Research* 22:201-215.

Douglass, R.J.
1984 Ecological Distribution of Small Mammals in the De Long Mountains of Northwestern Alaska. *Arctic* 37(2):148-154.

Driver, J.C
1988 Late Pleistocene and Holocene Vertebrates and Paleoenvironments from Charlie Lake Cave, Northeast British Columbia. *Canadian Journal of Earth Science* 25:1545-1553.

1996 The Significance of the Fauna from the Charlie Lake Cave Site. In *Early Human Occupation in British Columbia.* R.L. Carlson and L. Dalla Bona, eds. pp 21-28. UBC Press, Vancouver.

Farrand, W.R.
1985 Rockshelter and Cave Deposits. In *Archaeological Sediments in Context.* J.K. Stein and W.R. Farrand, eds. Center for the Study of Early Man.

Gilbert, M.B.
1980 *Mammalian Osteology.* Modern Printing Company, Laramie.

Guthrie, R.D.
1968 Paleoecology of a Late Pleistocene Small Mammal Community from Interior Alaska. *Arctic* 21(4):223-244.

1982 Mammals of the Mammoth Steppe as Plaeoenvironmental Indicators. In *Paleoecology of Beringia* edited by D.M. Hopkins, J.V. Mathews, Jr., C.E. Schweger, and S.B. Young, pp. 307-326. Academic Press, Inc., New York.

D.M. Hopkins, J.V. Mathews, Jr., C.E. Schweger, and S.B. Young, (editors)
1982 *Paleoecology of Beringia.* Academic Press, Inc., New York.

Jarrel, G.H., S.O. MacDonald, and J.A. Cook
1998 Checklist to the Mammals of Alaska. Electronic Document.
http://zorba.uafdm.alaska.edu/museum/AK_Mammals/Checklist.html.

Kusmer, K.D.
1990 Taphonomy of Owl Pellet Deposition. *Paleobiology* 64(4):629-637.

Larsen, H.
1968 Trail Creek: Final Report on the Excavations of Two Caves in Seward Peninsula, Alaska. *Acta Arctica* 15:7-79.

Manville, R.H. and S.P. Young
1965 *Distribution of Alaskan Mammals.* U.S. Government Printing Office.

Maser, C. and R.M. Storm
1970 *A Key to the Microenvironment of the Pacific Northwest (Oregon, Washington, Idaho).* Oregon State University Bookstores, Inc., Oregon.

McNamee, T.
1984 *The Grizzly Bear.* Alfred A. Knopf, New York.

Morlan, R.E. and J. Cinq-Mars
1982 Ancient Beringians: Human Occupation in the Late Pleistocene of Alaska and the Yukon Territory. In *Paleoecology of Beringia* edited by D.M. Hopkins, J.V. Matthews, Jr., C.E. Schweger, and S.B. Young, pp. 353-381. Academic Press, Inc., New York.

Olsen, S.J.
1964 *Mammal Remains from Archaeological Sites. Part I: Southeastern and Southwestern United Stated.* Peabody Museum, Cambridge, Mass.

Repenning, C.A., D.M. Hopkins and M. Rubin
1964 Tundra Rodents in a Late Pleistocene Fauna from the Tofty Placer District, Central Alaska. *Arctic* 17(3):177-197.

Rogers, A.R. and M.C. Lewis
1986 Diet Selection in Arctic Lemmings (*Lemmus sibiricus* and *Dicrostonyx groenlandicus*): Forage Availability and Natural Diets. *Canadian Journal of Zoology* 64:1684-1689.

Sattler, R.A.
1991 Paleoecology of a Late Quaternary Cave Deposit in Northeast Alaska. Unpublished MA thesis, University of Alaska, Fairbanks.

Schaaf, J.
1988 Trail Creek Caves, In *Bering Land Bridge National Preserve, An Archaeological Survey.* Volume 1. J, Schaaf, ed. Alaska Region National Park Service, Anchorage.

Selkregg, L. ed.
1976 *Alaska Regional Profiles, Southwest Region.* Arctic Environmental Information and Data Center, University of Alaska, Fairbanks.

Short, S.K., S.A. Elias, C.F. Waythomas, and N.E. Williams
1992 Fossil Pollen and Insect Evidence for Postglacial Environmental Conditions, Nushagak and Holitna Lowland Regions, Southwest Alaska. *Arctic* 45(4):381-392.

Sklepkovych, B.O. and W.A. Montevecchi
1996 Food Availability and Food Hoarding Behaviour by Red and Arctic Foxes. *Arctic* 49(3):228-234.

Smith, G.S.
1979 *Mammalian Zooarchaeology, Alaska: A Manual for Identifying and Analyzing Mammal Bones from Archaeological Sites in Alaska.* Anthropology and Historic Preservation Cooperative Park Studies Unit, University of Alaska, Fairbanks.

Sobolik, K.D.
1993 Direct Evidence for the Importance of Small Animals to Prehistoric Diet: A Review of Coprolite Studies. *North American Archaeologist* 14(3):227-244.

Stiner, M.C., G. Arsebuk, and F.C. Howell
1996 Cave Bears and Paleolithic Artifacts in Yarimburgaz Cave, Turkey; Dissecting a Palimpsest. *Geoarchaeology: An International Journal* 11(4):279-327.

Vinson, D.
1988 Preliminary Report on Faunal Identifications from Trail Creek Caves. In *Bering Land Bridge National Preserve: Archaeological Survey.* Volume 1. Jeanne Schaff, ed., Alaska Region, National Park Service, Anchorage.

Wilson, D.E. and D.M. Reeder
1993 *Mammal Species of the World*. Smithsonian Institution Press, Washington D.C.

Watson, Adam
1957 The Behavior, Breeding, and Food-Ecology of the Snowy Owl, *Nyctea Scandiaca*. *Ibis* 99:419-461.

West, F.H.
1996 Trail Creek Caves, Seward Peninsula. In *American Beginnings: The Prehistory and Paleoecology of Beringia* edited by F.H. West. University of Chicago Press, Chicago.

Zweifel, M.K.
1994 *A Guide to the Identification of the Molariform Teeth of Rodents and Lagomorphs of the Columbia Basin*. Unpublished Masters thesis, Washington State University, Pullman.

PALEOBEHAVIOR IN ALASKAN PLEISTOCENE HORSES: SOCIAL STRUCTURE, MATURATION DATES, USES OF THE LANDSCAPE, AND MORTALITY PATTERNS

R. Dale Guthrie
Biology Department, University of Alaska, Fairbanks

When one digs a stained bone out of the mud, washes it clean in the stream, and recognizes its identity, it is not easy to imagine it as part of a once-living animal, a living lever for tendon and muscle. But it is an even greater stretch to imagine that animal as part of a complex herd of its own kind, and part of an ever-changing community of other species. Furthermore, it is very difficult to see it in the context of a local landscape. Although the stained bone is a geological specimen, it is part of an animal that may have experienced lust, jealousy, deep hunger, delight, status, fear, nurture, a bond of attraction, senility, and much more. We can seldom get close to reconstructing these kinds of past complexities from an old bone, but sometimes the rare opportunity does come. Let me present a brief for such an argument—in this case, fossil horses from the Fairbanks area.

The focus of this study is the behavior and ecology of late Pleistocene Alaskan horses. These little caballoids were apparently a relatively common and important part of the extensive Mammoth Steppe landscape during the late Pleistocene. Their fossils are ubiquitous wherever muck deposits of late Pleistocene age are found in Alaska. In such deposits bones of horse and bison are about equal in number, and are usually more abundant than all the other large mammal fossils combined (Guthrie 1967). In this paper I try to relate information obtained from a sample of Pleistocene horses to modern equid behavior, and in combination these a fascinating story is exposed.

While deep historical context is important in understanding modern ecology, such depth is not always easy to come by. Our record of the past is always somewhat obscured, incomplete, and biased. Fossils are often found one at a time, they are usually fragmentary, and are difficult to date exactly. These deficiencies normally discourage any broad paleobiological interpretation. But there are cases where we can turn such normal deficiencies to our advantage. The fossil-rich Alaskan muck deposits from the Fairbanks area provide such an opportunity. These frozen sediments mostly date from the latest Pleistocene - less than 50 thousand years (Péwé 1975; 1989), and over two-thirds of the fossils fall within the finite range of Radiocarbon dating (less than 38-40ka). Scattered through these organic-rich sediments are hundreds of fragments of horse bones and skulls. As such, they do not represent one population, or any one specific slice of time.

This is not a paleontologist's dream sample. But for certain questions a broad chronological sample actually offers an analytical advantage because it presents a more general picture. This averaging over many generations avoids the problem of extreme interannual population fluctuations that might cause frequent demographic imbalances in instantaneous samples (Murie 1944). So, this time-transgressive sample should provide an even better general picture of a species biology than the statistical ideal of a controlled instantaneous snap-shot sample. At least this is my proposal in this paper.

Likewise, one normally assumes that taphonomic forces, which usually distort preservation ratios of sex and age, offer unwanted hazards to biological reconstruction. But again, in this case I want to take advantage of these forces, to turn them around, and exploit their biases to try to enlighten us about paleobehavior. The mortality patterns of this aggregate sample will be the key, because demographic data from a death assemblage are loaded with more information than simple life-expectancy.

During the last 50 years, paleontologists have reconstructed mortality patterns for a variety of fossil mammals. Their research has greatly enriched our understanding of these species' lives and surroundings (e.g., Kurtén 1953, 1958, 1964, 1976; Van Valen 1964; Voorhies 1969; Saunders 1980; Klein 1982; Conybeare and Haynes 1984; Barnowsky 1985; Lyman 1987; Stiner 1991, 1994; Haynes 1991;). For example, paleodemographic patterns are used to evaluate whether a particular fossil assemblage was a product of human or non-human accumulation and, if the assemblage was directly within an archaeological context, whether it was of pre-domestic or post-domestic origin (Klein 1982; Levine 1983). Additionally, my orientation in this paper builds on a long tradition of studying details of modern animal demographics by using mortality patterns (see Caughley 1966).

ALASKAN HORSES IN THE LATE PLEISTOCENE: EVOLUTIONARY BACKGROUND OF MY SAMPLE

Systematics of late Pleistocene horses has been in an abysmal state for decades. The closely related caballoid or caballoid-like horses have been given regionally-specific names: *Equus latipes, E. spelaeus, E. transilvanicus, E. gmelini, E. occidentalis, E. complicatus, E. fraternus, E. scotti, E. lambei, E. alaskae, E. conversidens, E. excelsus, E. leninensis*, and many more. The species diversity among these caballoid horses does not seem to match reality, but is built on a topological approach from the last century which does not acknowledge intraspecific and intraspecific variability, especially with regard to dental anatomy, body size, leg length and hoof size. Eisenmann (1991), using skull and dental morphology, prefers to separate late Pleistocene and modern caballoids into three species: *Equus caballus,* domestic horses; *E. przewalskii,* a geographic variant of the living Asian wild horses, and *E. ferus,* wild ancestors of domestic horses.

In comparison to caballoid horses, species like reindeer, *Rangifer tarandus*, which occupied much the same range as Pleistocene horses, have (and had) considerable local morphological diversity in body size, dental patterns, leg length, and hoof spread (Flerov 1952), but are still thought of as a single diverse species. This pattern of wide distribution and local specializations is a common pattern among the broadly-dispersed species of the Mammoth Steppe fauna: eg., *Ovibos moschatus, Bison priscus, Mammuthus primigenius, Coelodonta antiquitatis, Ursus arctos, Panthera leo, Canis lupus*, etc. While the technical status of these horses from Alaska is still unclear, researchers seem to agree that they are closely related to northern caballoids. Hopefully, mitochondrial studies will provide some clarification of Pleistocene equid systematics in the near future.

There is another factor, besides this matter of local diversity and broad distribution which frustrates equid systematics (combined, of course, with their lack of horns or antlers); this is the fact that equids exhibit an unusual combination of varying superficially in details, yet have a consistently conservative ecological adaptation. Which is to say, that despite dramatic differences in their respective appearances in life, asses, hemionids, caballoids, and the three African zebra species vary only subtly when their crania and postcranial skeleton elements are examined. This osteological conservatism is unusual and is different than bovids. The bones as well as the flesh-and-blood body shape of African buffalo, Asian buffalo, cattle, and bison are strikingly different. As a result, each has been given separate generic status despite their rather recent radiation in Pleistocene times, paralleling that of *Equus* both in time and space. Living equid species do differ in things like physiology, social structure, and vocalization patterns, but only subtly in other areas such as dietary preferences. Definite osteological and dental patterns separate the extant species, but they are subtle enough that multiple metric characters are preferred when making taxonomic identifications.

Harington and Clulow (1973), and Harington (1989), followed earlier researchers, in arguing that there was one fossil horse species in Pleistocene Beringian sediments, *Equus lambei*. However, Harington has also recovered stilt-legged kiang-like metapodials from the Yukon Territory (Harington, pers. comm). Winans (1989) agreed, proposing that all fossil equids consisted of only one single polymorphic species, but referred these to *E. alaskae*. Likewise, Uerpmann (1989) described some of the slim metapodials found along with those of caballoids from Alaskan-Yukon Territory, as hemiones, or hemionid-like.

While useful, metapodials are not the best skeletal element for systematic purposes. Skulls are much better. But, in the case of the hemionid or hemionid-like metapodials, it is difficult to connect skulls to this postcranial material. Unfortunately, no complete specimens have been found in Alaska or the Yukon Territory which are clearly hemionid in character; rather, all of the skulls which have been studied seem to be caballoids (e. g., Eisenmann 1984, 1991; Forsten 1986). Radiocarbon dates show that hemiones were contemporaries of the Alaskan caballoid horses only during the late Pleistocene. However, my count of the metapodials in the American Museum of Natural History and the University of Alaska Museum shows that the characteristic hemionid (quite narrow and elongated) metapodials constitute only around 6% of the collection. So, it is possible that a very small number of incisors I have included in this study may be those of hemiones, as I was unable to separate incisors of the two groups (if indeed hemiones incisors were present). However judging from the number of complete skulls used in this study, which all belonged to caballoids, they suggest that, if they were present, hemiones or hemionid-related forms in the sample were a minor element.

As part of a recent radiocarbon dating program to date over 100 horse metacarpals from the Fairbanks area, I dated both the caballoid and hemionid patterns (Guthrie, in prep). Unlike the horse teeth used for the following demographic study these metacarpals were not chosen at random but spread more evenly across the size ranges to insure that I had a good coverage of the size-shape spectrum (and to include a disproportionate number of the less common, stilt-legged, hemionid-like metapodials). As part of that study I took length and width measurements. When these are plotted on a bivariate plot (Figure 1) there is a good separation of two clusters. This plot clearly delineates the two main groups, with no overlap; one of caballoid-like individuals and the other of hemionid-like individuals.

There is another complicating factor to northern equid evolution. Some deposits date from early to mid Pleistocene age. The Alaskan sites of Gold Hill and Cripple Creek Sump, for example, are quite old. That is, they are dated as being considerably older than the typical Interior Alaskan muck deposit sites by using tephra dates (Péwé 1975, 1989; Preece *et al.* 1999). The fossil mammals from these two sites which are not from the upper parts of the deposits (the late Pleistocene portions) are heavily stained, and some are partially mineralized. A few are even contained in globular calcareous concretions. The chronology of specimens from these early sites is rough, but the trend is clear - the older fossils and the more recent horse fossils are not of the same size as are the horses from later Pleistocene deposits.

By comparing site locations with metapodial length and robusticity it is apparent that older (outside the radiocarbon dating range) Pleistocene Alaskan caballoid horses were quite large (metacarpal length roughly in the 225-250mm range). Recall, there is virtually no sexual dimorphism in equid metacarpal shape or size, which simplifies the assessment of size patterns. Dated horse metacarpals, which fall into the early parts of the radiocarbon range (during Isotope stage 3) were medium sized (metacarpals roughly in the 210-225mm range). This size decline continued, culminating in the small horses of the latest Pleistocene (metacarpals roughly in the 190 to 210mm range). Burke and Cinq-Mars' (1997) plot of the caballoid horse teeth from Blue Fish Cave in the Yukon Territory, Canada shows those horses to be smaller than living equids (their samples were from the last full Glacial). This same general directional

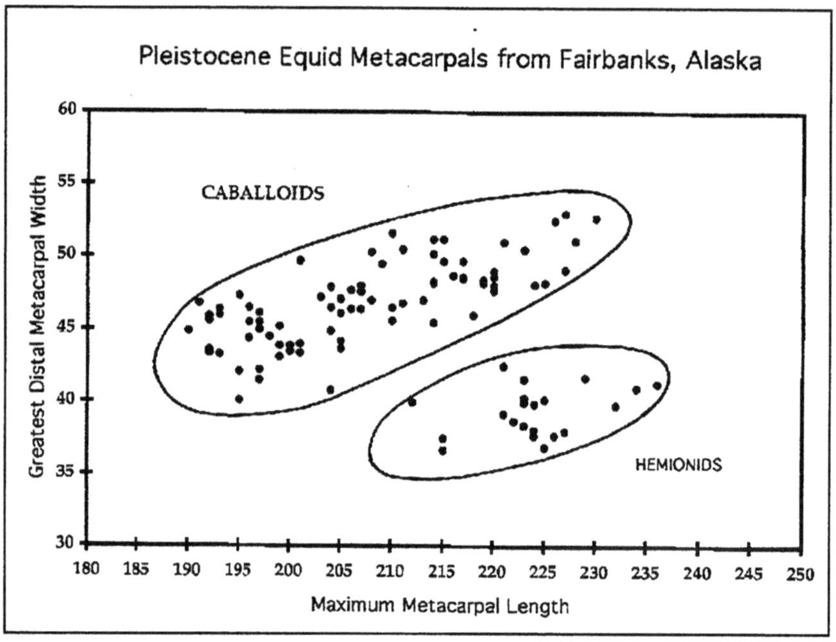

Figure 1. Plot of Pleistocene Equid Metacarpals from the Fairbanks Mining District, Fairbanks, Alaska. Measurements of these specimens were taken from collections mostly assembled by Otto Geist of the University of Alaska and sent to a private collector, Childs Frick. The Frick collection is now at the American Museum of Natural History. The fossil metacarpals can be separated into main clusters. I suggest that the stout metacarpals are caballoid and that the more stilt-legged cluster represents kiang-like animals of the hemionid group.

trend of size decrease is seen throughout much of Eurasia (Azzaroli 1990; Forsten 1991, 1993; Spassov and Iliev 1997), a Pleistocene decline in horse weight by a factor of two. The long diagonal axis of the caballoid metacarpal plot shown in Figure 1 is indicative of that time-transgressive shift in body size.

Although these Alaskan caballoid or caballoid-like metapodials differ greatly in size, they are of the same basic shape. This preliminary evidence points to the possibility that at least in Alaska the decline in body size occurred without significant breaks along the size gradient. However, it is probable that this gradient of size reduction may not have been continuous and smooth. The sample of incisors that I use for aging were almost totally from the small late Pleistocene variants, but the phenomena I derive from them probably extended well back through the Pleistocene.

The Alaskan and Siberian variants of this common caballoid horse seem to have been part of a wide-spread panmictic group (Groves 1986), though local populations may have varied in external appearance more so than they did osteologically. From mummies of this species found in Siberia (Vereschchagin and Lazarev 1977) we can say that the specimen was similar in appearance to the extant wild Przewalski's horse: golden dun color in summer and rich ochre in winter, with bay points of black mane, tail, and legs (there may have been gray population variants, as one finds with the tarpan in western Europe). Like all wild equids it had an upright mane. Some variants have a dark stripe down the back, ending in a tail that always seems to have been well "feathered" at the dock.

The one horse mummy from the Yukon Territory described by Harington and Eggleston-Stott (1966) had a lightly colored non-erect floppy mane. This may be evidence to support the idea of a separate group in Alaska-Yukon Territory. However, the unique features of this specimen (no other wild equid is known to have a lightly colored or non-erect mane) may be due to post-depositional diagenetic forces which bleached the melanin and reduced the hair stiffness. Some of the partial Alaskan mummies of mammoth, caribou, bison, moose, and horses, do have the hair bleached in this manner. Frequently, when the black melanin is leached away, the reddish eumelanin remains (and perhaps some iron-staining as well), leaving a peculiar translucent, rusty-reddish color (similar to the hair-braids from Neolithic women found in Danish peat bogs, on display at the Danish National Museum, Copenhagen). This bleaching phenomenon gave rise to the earlier idea that Siberian woolly mammoths were red. Some of the poorly preserved specimens do have red hair. But the mammoths which are well preserved show a richly complex dark pelage with a black ground combined with ruffous undertones. These, better-preserved mummies do not have much diagenetic melanin alteration. Several partial mummies of caballoid horses have been found in Alaska (e.g., Guthrie 1990, Guthrie and Stoker 1990), but these are rare and are only partial specimens. They do not have sufficiently preserved remnants of pelage to make any contributions to the question of pelage coloration in life.

Using images of Pleistocene horses drawn in the Paleolithic art of western Europe, we can also say that striping on legs, neck, and face were common, especially in southwestern Europe (Guthrie, in press). Such geographic and temporal continuity in appearance further supports the idea of a widely dispersed species with local variants.

DENTAL AGING

Because teeth wear with time, there is a good correlation between tooth height and age. At least for equids, teeth are not only good indicators of age, the presence of large canines in males, and their absence in females (a few adult females have a small vestigial canine), allows one to separate the sample by sex. The dentition of fossil horses is thickly enameled, dense and rugged, which probably explains why they are so numerous and unusually well preserved in the record, allowing reasonable assessment of mortality patterns.

Cheek-tooth height can be used to assess age, but in most equids this measurement is not straight forward because most of their tall hypsodont molars and premolars are concealed by dense jaw bone. Levine (1983) used individual cheek-teeth from several European sites to effectively approximate the age of horses. She used this technique mainly because the teeth occurred as isolated specimens, and thus there was no way to improve on this strategy with her sample. But these isolated teeth present a potential problem. Each adult horse has 24 cheek teeth, 12 incisors, and 4 canines. If the samples are small, as indeed they were at some of Levine's localities, an over-representation of most of the teeth from one or two individuals can greatly affect the pattern, unless one corrects for minimum number of individuals. Burke and Cinq-Mars (1997) using a similar sample of isolated cheek teeth, also studied demographics using cheek-tooth height.

Because of the large collection of specimens available for this study, I chose to use incisor eruption and wear rate to estimate age. And, since canines (in males) are just a few centimeters away from the incisors it is easy to sex each specimen. I only used those specimens with both age and sex data. The scale for aging incisors has been checked with horses of known-age for over 200 years among experienced horse breeders and is apparently reliable. I employed the guide used by veterinarians: the AAEP Official Guide for Determining the Age of the Horse (Anonymous 1981), Sack and Sadler (1982), Pasquini *et al.* (1983), and Gaughan and DeBowes (1998). All of these are similar but use slightly different wear features (Figures. 2, 3 and 4 and Table 1). These guides are almost identical to, Joubert's (1972) technique developed with Hartmann zebra, *Equus zebra hartmannae*. Incisor wear has been used on feral horses on natural range which seem to wear their incisors at about the same rates as domestic horses (Berger 1986) and asses (Johnson et al 1987). Theoretically, there could be considerable variability in wear rate due to differences in environment and diet. But, in fact, other equids show very similar incisor emergence patterns and wear rates (Klingle and Klingle 1966; Penzhorn 1982).

It seems possible that most of the dietary differences in wear rate affect cheek teeth used to masticate rather than incisors used mainly to nip stems. Using incisors in this nipping fashion rather than the mortar-pestle grinding process of the cheek teeth may not be so affected by dust in the food (horses also use their lips to pull or nip soft leafy material). From a histological analysis of plant samples removed from the incisor infundibular pits (Figure 2) of living equids, I was able to show a much larger percentage of woody plant fragments than existed in the actual diet (Guthrie, in press). I showed that this was also true of Alaskan Pleistocene horse incisors. this is curious, and may indicate a much higher use of incisors on shrubs than on grasses. Horses do eat some shrubs, particularly willows in spring (the fraction of woody plants in the incisors of fossil Alaskan horses was mostly willows).

	I1	I2	I3
Eruption	2.5	3.5	4.5
In wear	3.0	4.0	5.0
Level	5.0	6.0	7.0
Open pit gone	6.0	7.0	8.0
Dental star	8.0	9.0	10-11
Oval shape	9.0	10	11-12
Infindibulum gone	13-16	14	15-16
Triangular shape	14	15	16-17
Rectangular	18	19	20-21

Table 1. Tabulation of the timing of incisor eruption and chronological sequence of different crown features. These show timing for first (nearest the midline) second and third (most lateral) permanent incisors (applies to both uppers and lowers). This is compiled from several publications referred to within the text.

Fortunately, there are several ways to check whether the incisors of Alaskan Pleistocene horses were worn at an unusual rate. I compared the wear rate among Pleistocene Alaskan horses to that of other equids by examining wear at the time deciduous incisors are replaced by permanent ones; sequentially the first, second, and third incisors are shed and replaced from beneath. At their respective times of emergence, the Alaskan samples showed wear rates identical to the guides mentioned above for domestic horses. This preliminary evidence is so unequivocal with respect to the proposition that the incisor wear rate of Alaskan wild horses and domestic horses is roughly the same, I did not pursue a separate study of histological corroboration, though Burke and Castanet (1997) did so in their study of horse teeth in archaeological sites (which included samples of fossil northern horses). Aging beyond 20 is probably difficult, so I listed that age class as 20+, meaning 20 years or older. A few individuals in this class had wear comparable to domestic horses of 30 years of age.

Equids are adapted to be able to resort to diets of coarse grass, and as part of that adaptation, their incisor and molar batteries are extremely hypsodont, greater than any bovid. This complexity compensates for the combined abrasive wear of opaline phytoliths in grasses, and attrition of the lengthy mastication that this diet requires. Ruminants can presoak their coarse food in the rumen prior to mastication, but horses must process a mouthful of dry grasses as is. Opposable rows of sharp-edged incisors allow equids to clip tough grass stems and leaves of grass which the bovid-cervid occlusion of upper incisors against the premaxilary pad cannot manage on a sustained basis.

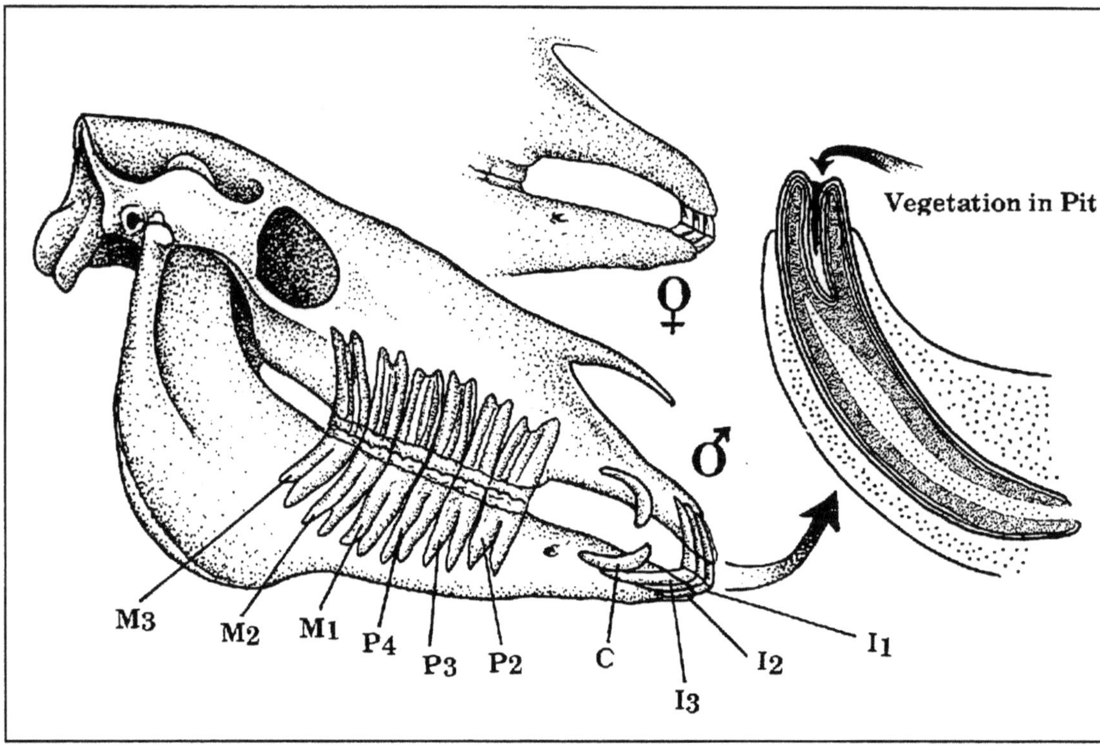

Figure 2. A diagrammatic "x-ray" view of horse teeth. This diagram illustrates hypsodont cheek teeth (three premolars and three molars on each quadrant), canines (one in each quadrant) and incisors (three in each quadrant). Male and female's are shown for comparison. To the far right is a longitudinal section of a single incisor showing how deeply it is imbedded in the mandible. On young animals there is an infundibular pit which among living equids (and in well-preserved Alaskan fossils) is usually packed with vegetation.

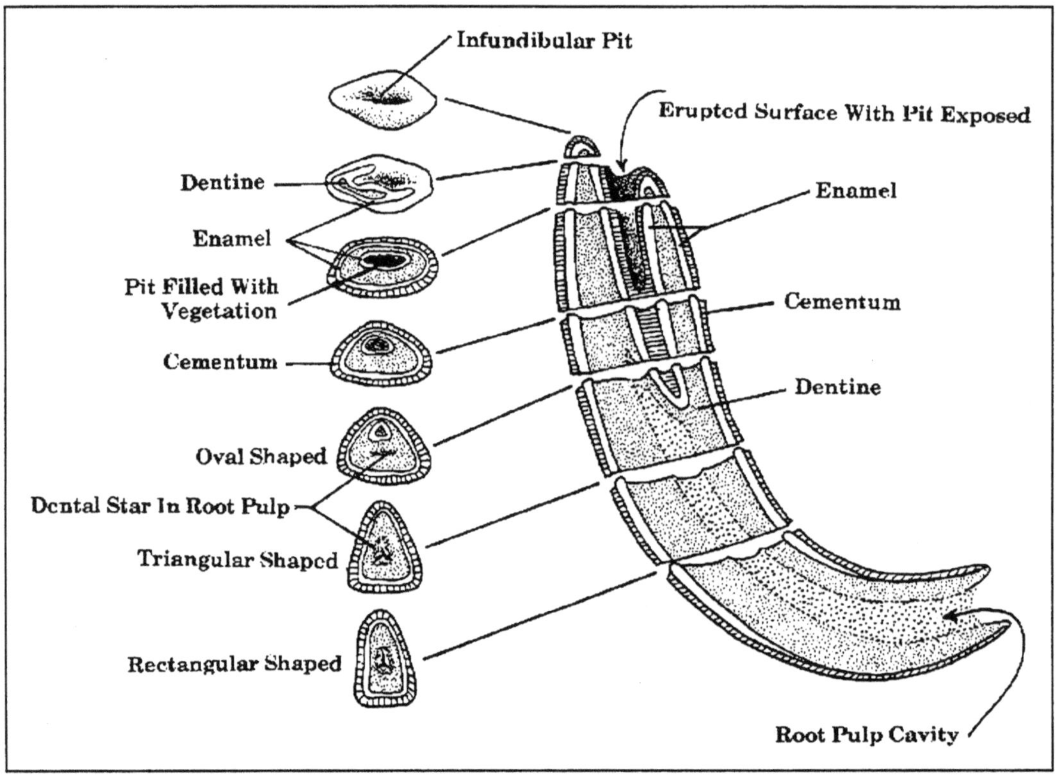

Figure 3. Longitudinal section of an incisor showing histological parts in detail. As the tooth wears the occlusal surface forms a different pattern which changes with age. Not only does the interior of the worn crown vary with age, the shape of the outer perimeter also changes from oval shaped, to triangular and finally to rectangular in very old animals.

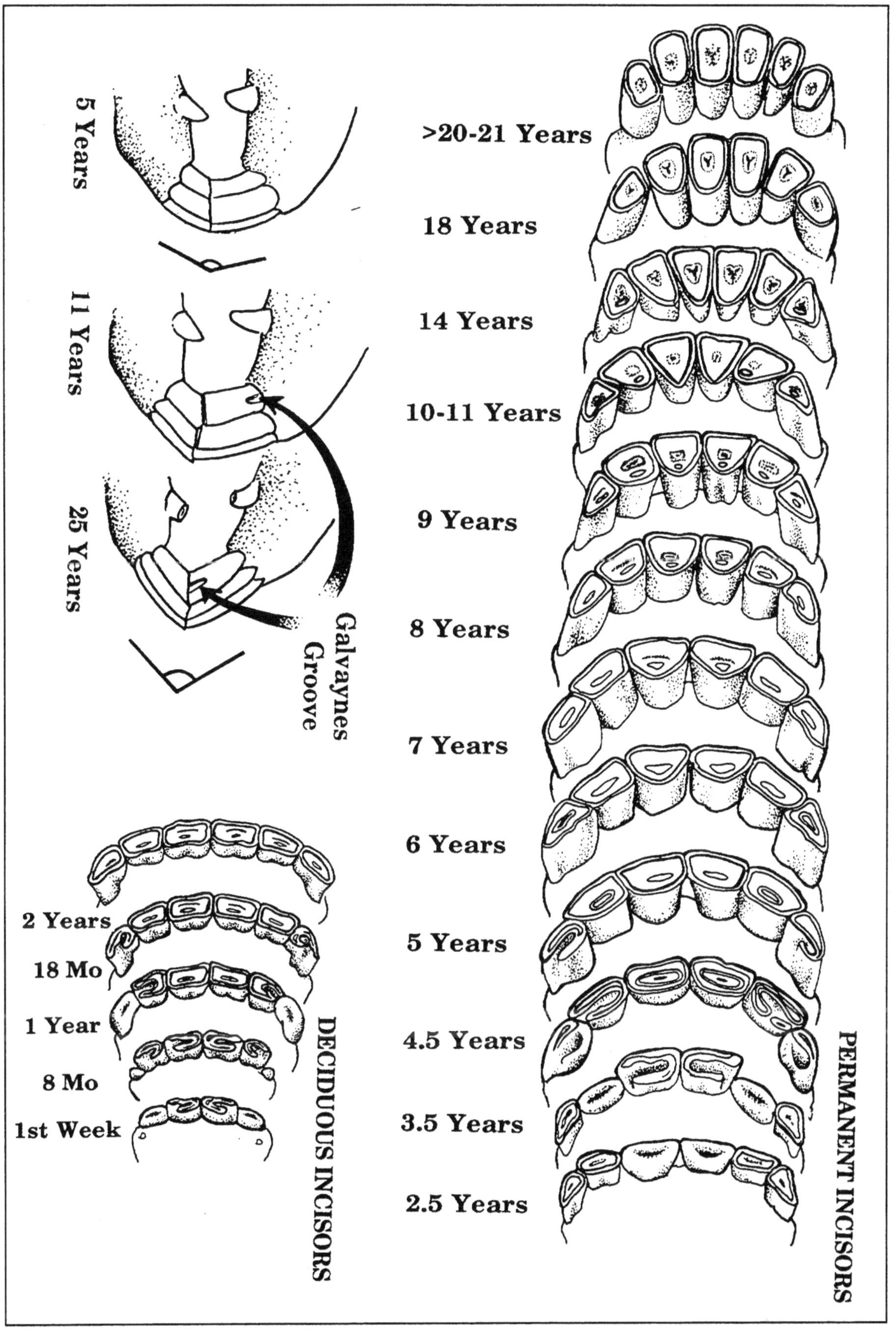

Figure 4. Occlusal view of incisor wear pattern as it changes with age, which helped in assigning age to Alaskan Pleistocene horses. These are derived from a composite of several publications listed in the text. Horses have three deciduous and three permanent incisors in each quadrant. Only males have large canines, both above and below. At lower left is a lateral view of male incisors showing how the angle changes with age and exposure of Galvaynes groove.

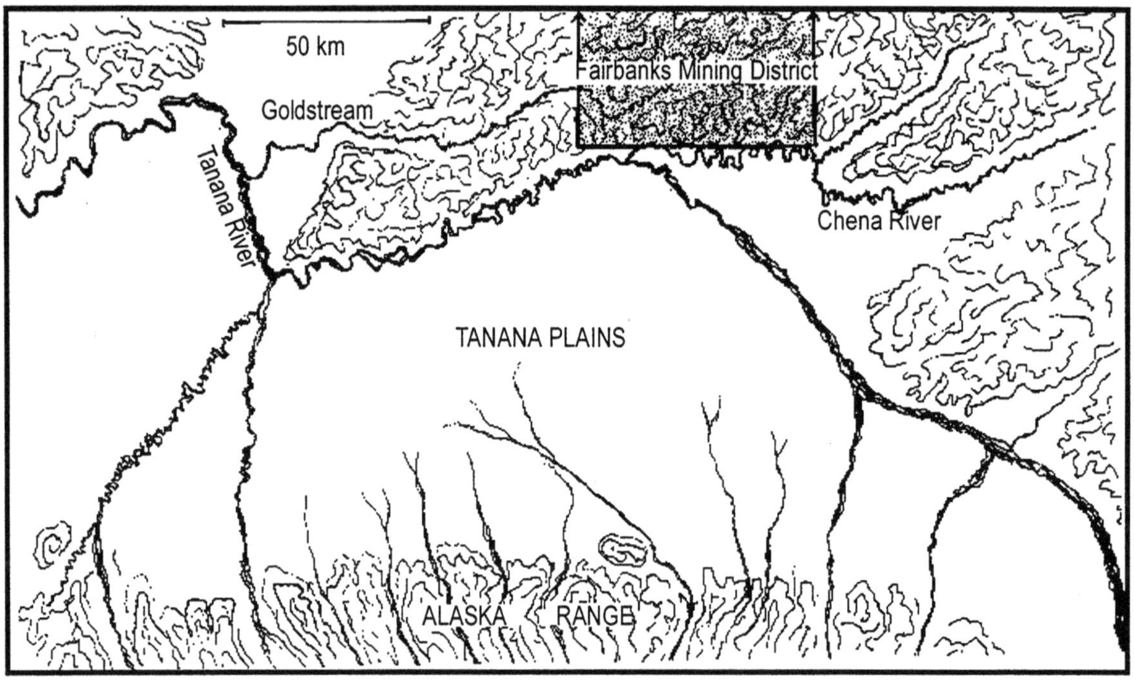

Figure 5. Generalized map of the Fairbanks Mining District. This shows how the low rolling Yukon-Tanana Uplands lie in relation to the Tanana Plains. It is this mining district around Fairbanks produced most of the late Pleistocene mammalian fossils found in interior Alaska. Glaciation in the Tanana Uplands area involved only small mountain glaciers.

In areas where silt is common in the form of airborne dust it may have increased the rate of dental wear. Certainly evidence shows that Alaska during Pleistocene times was a more arid and dustier climate than today. Other grazers from those areas (there are large collections of bison teeth for example) reveal no special pattern or unusual rate of tooth wear (see Guthrie 1990). As dust falls on all plants, this would be expected to affect all large ungulates equally. It would have even affected species with very low-crowned dentition, species such as caribou and muskoxen from the same localities in which the horses are recovered. But there is no evidence of any unusual dietary high-abrasive effect in any of the fossils from the Pleistocene north.

There is necessarily some edge of subjectivity to this technique, e.g. whether to call one tooth a 10 or an 11 year old individual. I did not split-weight into half points, but put each individual fossil into the age group which seemed closest. But, as with any relativistic aging technique, natural variations in wear may eclipse any of the measuring variables.

THE ECOLOGICAL SETTING FOR THE ALASKAN DEATH SITES

Though equid fossils come from all over Alaska, the large collection used for this study was mainly recovered from around Fairbanks Alaska by Otto Geist. These were produced in conjunction with the large-scale placer gold mining operations that operated in interior Alaska during the first half of the 20th century. Thousands of Pleistocene fossils were unearthed and Geist collected material for several decades. Select items from his collections were shipped to Childs Frick's laboratory, associated with the American Museum of Natural History in New York. Radiocarbon dates on fossil horses from interior Alaska (Guthrie 1990) cluster during the last glacial (Isotope stage 2), a few scatter back into the previous Interstade (Isotope stage 3) and even fewer fall beyond the limits of radiocarbon dating. The most recent dates thus far are around 12-13 ka.

We know from other studies (Péwé 1975) that the reason so many specimens date within the radiocarbon range is that sediments around Fairbanks date mostly from the late Pleistocene (Péwé 1975). Apparently, older loess deposits were removed by episodes of major erosion during Interglacial (Isotope stage 5), and to a lesser degree during Interstadial times (Isotope stage 3). Thus, horses were not necessarily more abundant during the late Pleistocene of Alaska than during earlier parts of the Pleistocene. Data from late Pliocene-early Pleistocene localities in both Alaska and Siberia (Guthrie and Matthews 1971; Sher 1986) show that horses have a long Pleistocene history in the North.

The mines from which most of the horse fossils were collected are located among the rolling foothills of the Yukon Tanana Upland (Figure 5). During the Pleistocene, eolian silt, or loess, which cloaks the uplands, was washed downslope where it accumulated in the valleys as dark, organically-rich muck containing many fossils. This reworked loess covered the auriferous gravels and thus had to be removed by miners. The muck which contained the fossils was permanently frozen (however, permafrost in the Fairbanks area is not continuous), so miners devised large scale thawing and washing operations using high pressure jets of running water. These operations recontoured whole drainage systems.

The creeks and valleys which drain this upland flow into the broad flood plain of the Tanana River. The dimensions of this plain, while not vast, are still substantial. Its width, from Fairbanks southward to the Alaskan Range, is over 100 km, and in the other, east-west direction, it is over 400 km across; that is, over twice the size of the Serengeti Plain. I think the juxtaposition of this Tanana flood plain with the south-facing foothills of the Yukon Upland offers a significant taphonomic bias which will figure importantly in this story of horse mortality patterns.

MORTALITY PATTERNS

Implications

A census of modern mammal populations normally finds many young and few very old individuals. This age structure is the kind of curve one might expect in a catastrophic paleontological sample, where large numbers were killed without age-sex bias (Kurtén 1953, 1964). This seldom happens, except in such cataclysms as volcanic eruptions. The other extreme of mortality pattern is an attritional curve, which is more bimodal in shape; the young and the elderly are over represented, as per graveyard tombstone dates, while prime aged animals are rare (Kurtén 1953, 1964). But this extreme is also not frequent because of other taphonomic biases. So paleodemographers are usually left with an oddly shaped bumpy curve, where the distorting taphonomic agents remain to be accounted for.

Mortality Patterns for the Alaskan Horses

For the sake of clarity and comparability I express the mortality patterns of fossil Alaskan horses in simple bar graphs of the numbers of specimens in each age and sex class (c (x)= age specific mortality). This is essentially the raw data and as I will propose later, an attempt to convert this into survivorship curves is difficult. The mortality patterns for Alaskan stallions and mares which appear here fit neither the attritional nor catastrophic ideals (Figure 6). Four unusual things stand out as important about these plots: (1) few young show up in the sample. (2) there are many more males than females. (3) there are two peaks of mortality among the males, one around four to five years old and one in the 10 to 13 range. (4) mortality seems comparatively low between ages six and nine in both sexes. Only a modest number from either sex survived beyond age 13, but, among those that do, they may live a rather long time.

We can examine some possible hypotheses as to this unusual pattern and its deviation from theoretical survivorship curves outlined above.

Missing Foals and the Fossil Record

Young mammals are seldom preserved in the fossil record; not only their lives but their deaths disappear. The major factor must be the fragile nature of poorly ossified young bones. The thin cortex of the skull and long bones allows carnivores to eat these bones of young animals almost complete, or at least to break and disperse them. Carnivores would not be expected to eat the teeth. However, when the fragile mandible and maxilla of a young animal are broken to obtain the digestible tissue inside, the teeth do not remain in association as a rugged block. This makes teeth from young animals less likely to preserve and/or less likely to be retrieved by the collectors. In general, there are few young animals in most assemblages of fossil mammals. Levine (1983) found the same pattern in several European paleontological and archaeological fossil horse sites.

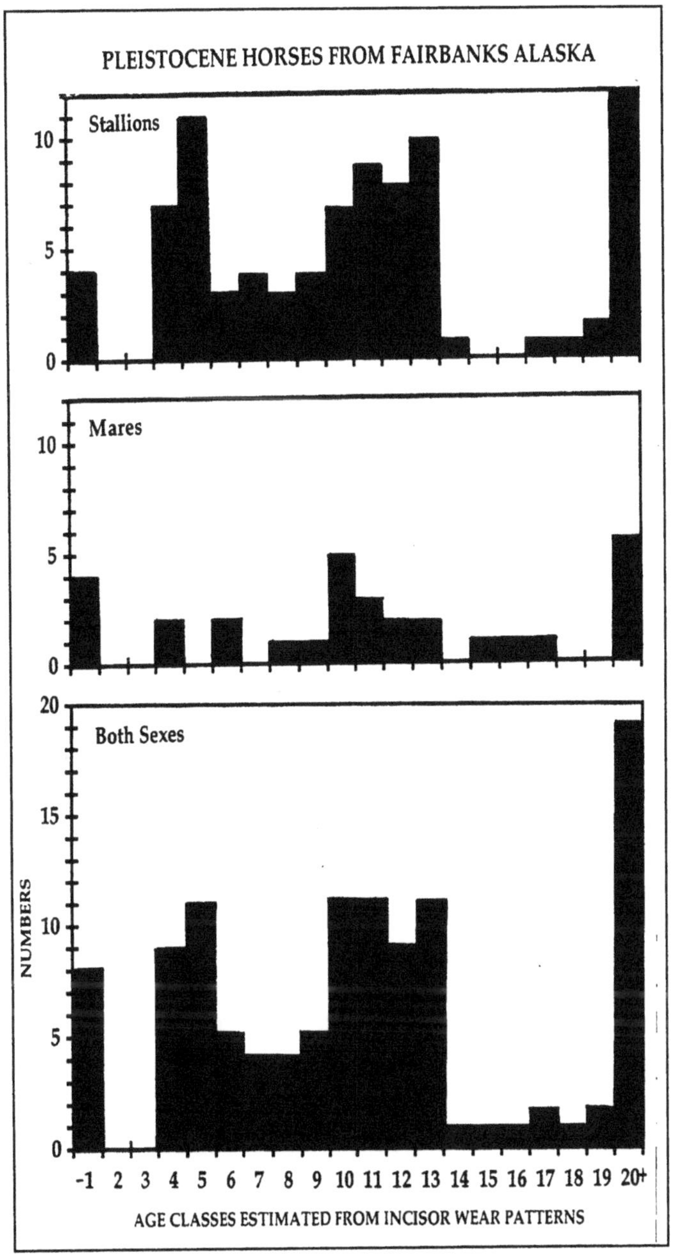

Figure 6. Bar graph of the numbers of Alaskan fossil horses falling within each sex and age class. There are many more stallions than mares, and most of the stallions fall within the 4 to 5 and 10 to 13 year age classes. The bar to the left (-1) include all specimens prior to 2 years old and younger, bar 2 includes 2 year olds up until age 3, and so on. The bar to the right (20+) includes all of those animals estimated to be 20 years and over. These are based on estimates of age from degree of incisor wear. Males were separated on the basis of large canines.

Van Valen (1964) found no young of the fossil equid genus *Merychippus* under the age of one. In this Alaskan sample of 114 individuals there are only eight under two years of age. Young animals are, by far, the most common member of the death assemblage, but the death assemblage is not the fossil assemblage.

Schaller (1972) observed a 33% decline in the numbers of live Burchell's zebra (*Equus burchelli*) foals in the first few months of life, but could only account for a small number of carcasses. Wildlife managers of wild ungulate populations generally plan on the loss of 50% of the first year's age class, as a mean estimate assuming stable densities and the presence of native predators. And since this age class is by far more abundant than any other age class, the year-after-year contribution of this many young animals would, theoretically, heavily bias their representation in the death assemblage. However, the low preservability is so potent it acts not only to obliterate this mortality bias, but to virtually eliminate the young from the fossil assemblage. Levine (1983), using a complex compensatory equation, tried to reconstruct the early part of the mortality curve, but I do not think this exercise is appropriate in the Alaskan sample for reasons I discuss below.

Berger (1986), working with an expanding population of wild mustangs in the American Great Basin, found a very high foal survival rate (averaging around 90% over the 5 year study), and an even higher yearling survival rate (averaging around 95%) which at face value corresponds to the Alaskan mortality data. But his data seem to not only be atypical of wild horses living with predators, but of those living without predators as well. In another study of a more stable population of feral horses, Welsh (1975) showed that 25% of the foals died before the age of four months. Most wild horse foal mortality is related to winter stress - the debilitating combination of hypothermia and nutritional deficits (Berger 1986), and that combined with (and contributing to) predation was probably similar to the moderately high foal mortality seen among wild equids, and which probably existed in Pleistocene Alaska.

There are some special situations which result in a more balanced age representation. Kurtén's (1958, 1976) study of cave bears showed a preponderance of young animals and very old adults. Many mammals, especially young, die in hibernation, so location of death is already taphomomically biased. Hibernating deep in a limestone cave, on a floor of terra rosa, seems to have been conducive to preserving young cave bear bones and teeth. The alkaline substrate and the scavenging carnivores going deep into caves, greatly increased the chance for preservation. Likewise, Conybeare and Haynes (1984) showed the bias toward preservation of the remains of young African elephants that died near water holes in arid regions.

Protected Childhood

We may use fragility to explain the absence of very young animals from the Alaskan sample, but what about the absence of two and three year old animals? A three year old horse is almost adult size and has a moderately rugged skeleton, even though its adult teeth are still emerging and the skull sutures have not yet quite fused. Fragility cannot explain the absence of fossils in this age group. It seems likely that more individuals of this age would have shown up in the fossil record if preservability were the only bias. I think it is more likely that specimens of this age class are simply rare because individuals did not often die at this age, or did not die in these depositional localities.

Among many mammals, the time after babyhood to early maturity is a period of grace, when one is under the full tutelage and protection of an experienced adult. Among wild horses one's adult protectors are experienced survivors, and the band is a protective social unit. Not by chance, this is the peak time of mammalian play. It is during this time of tutelage that the young must learn what situations of predatory risk to avoid, how to find good forage, what is right to eat, how to fight, how to behave socially, and on down a very long list. While still young, equids are able runners, built like thoroughbreds, faster than the adults—that is the special nature of "Derby" races of very young animals, who have not yet become loaded down with thick fighting muscles and reproductive structures. Among social species, such as equids, this is a golden time; one has the grace of an adult's life, but not the demands of territorial or harem protection or nurturing; one can rely on experienced eyes for early warning, rapid escape, and knowledgeable defense from potential predators. Juveniles are exempt from the demands of reproduction, gestation, lactation, courtship, etc. Even the defense of its social position is mainly taken care of by the parent. The equid grace period has its analog among humans. While human mortality has been traditionally high from birth to five, mortality is rather low between five to late teens, even in pre-industrial societies.

The Wild Horses's Fourth Year – Life in the Adult Lane

For Alaskan male horses, the age classes four and five show a sharp increase in mortality if we take the data at face value. Berger (1986) found that emigration from the harem occurred between one and two years of age on high-quality ranges, but nearer three years of age on low quality range. Burros (*Equus asinus*) seem also to mature early, with 14% of animals pregnant at one and half years of age in Death Valley (Johnson *et al.* 1987). For most mammals, the rate of maturity is related to diet. This graduation from the harem would correspond to the young male's onset of puberty. Berger calculated female emigration at about the same time, varying in a corresponding way between high and low-quality ranges. Berber's evidence of age at female puberty was corroborated by his data of age at first copulation and the age of first parturition. The filly leaves the parental band when she is sexually attracted to another male outside the harem.

It seems probable that this peak of mortality at age four to five in Alaskan fossil horses may not represent emigration timing, but onset of psychological or social maturity. Smuts (1976a, 1976b) found that while zebra begin to show signs of puberty at three years of age they do not reach full testes size until four and half to five years. At five and six they are as common as harem masters as are older stallions. I would suggest then that this increased mortality of Alaskan stallions at four to five years of age corresponds to their first entry into adult status and their consequent inexperienced participation in serious harem battles, with the accompanying risks.

Young mares of feral mustangs and zebras leave the harem around two to three years of age (where sometime they have a year's grace in the new stallion's band before they are bred), and they drop their first foal around four years of age. Though the evidence is not as clear as it is for the stallion curve because of the low absolute numbers of females in the Alaskan sample, there is a suggestion that females may also have experienced some increase in mortality at this age (a slight hump in the mortality plot). The drain of pregnancy and lactation during this first reproductive experience can be significant mortality forces.

It is interesting that other published mortality curves for fossil equids show this same rise in mortality, peaking at about four years of age, slightly earlier than for Alaska, but not much. Levine's (1983) data from Jaurens, Combe Grenal levels 14, 22, 23, Gönnersdorf, Arlay, Solutré square P 16, and Solutré square L 13, all show this peak. And it appears in her averages of pooled European assemblages and of individual samples. Van Valen (1964) found twice as many *Merychippus* fossils in the four year age class as either in age classes three or five, using first molar height as an indicator of age.

This spike in mortality among many mammalian species at the onset of full sexual maturity is common, so common in fact that it is curious that it goes unnamed. But it is not universal. In some animals, like Dall sheep, *Ovis dalli*, males enter the ranks of social reproduction slowly instead of precipitously. This is reflected in age-specific mortality slowly increasing about five years of age, and at its extreme beginning about age eight, when rams enter into full contests for dominance (Murie 1944). Presumably these mortality peaks at early maturity among the female mammals are related to first reproduction, and among males, to the high-risk behavior in assuming mature social roles. (Automobile insurers recognize a comparable high-risk peak among young men and increase rates accordingly).

Prime of Life and Adult Well-Being

The mortality pattern for the Alaskan horses shows a strikingly low plateau from the sixth through the ninth year. This is true for both males and females. Again, this is not unexpected as it is a feature common to many mammalian mortality patterns, including humans. Levine's pattern for Jauren's cave, a non-archaeological site, shows a similar plateau beginning one year earlier, at five rather than six. And, it is interesting that there is no mid-life plateau in her horses from archaeological sites. In fact, it was just the opposite; more animals of prime age appear in the archaeological sites. This suggests human hunting was age-selective, or at least did not discriminate against prime-aged individuals.

A prime-aged adult in the wild is less likely to meet with death threatening experiences than at most other times in full-grown life. A grown mare is protected within a harem, gaining experience every year. A harem master stallion is at its peak in strength, in ability to avoid predation, and in ability to resist debilitating environmental forces such as tolerating cold, making efficient use of forage, etc. During this prime of life, stallions are most likely to have a harem, which means living with the alert mare's watchful eyes for predators, and prime of life is associated with the greatest success in deterring aggression from other males.

The Onset of Senility

Murie (1944), in his classic study of sheep mortality in McKinley (Denali) National Park, showed a rapid acceleration in male mortality between ages eight and nine. This can be attributed to the onset of senility, but the senility part of mortality statistics are complicated in sheep because of the postponement of participation in the rut until late in life. It is only the older rams who become severely debilitated in competitions over the privileges of tending and defending the large numbers of ewes. They go without eating for the several weeks of rut, during late November and December, leaving them without reserves in the worst part of the winter.

Though both male and female equids still appear to be prime animals, in many respects, at nine or ten years of age most are, in fact, beginning to show some difficulties of age. For example, this is past time to retire racehorses from the track. Johnson *et al.* (1987) showed a visibly marked decline in body condition of feral burrows after the age of 10.5 years of age. Humans in their late 40s and early 50s are still relatively prime, physically and mentally, yet few of that age participate in demanding, active, competitive sports. It is this time in the human mortality curve that a tilt begins. Among equids subject to the full demands of living in the wild, the dip is steeper. Smut's (1976a, 1976b) data on both sexes of Burchell's zebra shot at random show this same decline in numbers after age 12. The Alaskan horse data indicate an earlier age for an essentially similar phenomenon. Mourer-Chauviré (1980) and Levine (1983) showed the sharpest mortality in the nine year age class, a year earlier than in Pleistocene Alaskan horses. Van Valen's (1964) analysis of *Merychippus* showed a rise in mortality around ages 7 and 8, and for *Griphippus* around age 8. Unlike promiscuous Dall sheep there would not necessarily be a wild episode of violent rut battles at mating time, but fighting would certainly have been on the increase because during the month leading up to the mares' estrus a harem stallion would be at his worst physical condition during the year, especially if he was starting to experience some disability from senility. However, if a harem loser had not experienced broken bones but was only worn down from being displaced, he would have the green of spring to recover along with the bachelor band.

As far as the mares are concerned, spring would also have been a time of difficulty for all, but particularly those mares bearing their first young and the more aged mares. Even mild debilitation from senility makes it more difficult to carry a foal and to undergo the increased load of lactation, which doubles the demand for her consumption of digestible energy, protein, calcium, phosphorous, and vitamin A (National Research Council 1978).

As ovulation is controlled by body condition, a mare of advancing years is less and less likely to have sufficient fatty acid levels to trigger ovulation after a long winter. Johnson *et al.* (1987) found that body condition among feral female burros was directly related to reproductive state. Pregnancy and lactation were associated with poorer body condition. And, since estrous occurs after +11 months (the gestation period for caballoids), that is, in early spring, Alaskan mares would have to conceive during the worst time of year, at the end of a long winter. Theoretically, at least, the aging Alaskan mares would be less and less likely to conceive and foal. Smuts (1976a, 1976b) found no reproductive senescence in zebra mares, but the rich Low Veldt of Kruger Park does not normally have severe environmental conditions preceding ovulation.

Smuts (1976a, 1976b) also found that few zebra live beyond age 16. Lion predation may be unusually high in Kruger because of the high lion numbers that are sustained there by optimum predation situations during the dry season, created by natural and artificial water holes. Among his sample of 271 mares, none had reached the age of 18 years. In quite another death profile, of 631 burros exterminated in the Mojave Desert, the oldest animals lived to 15.5 years for females and 20.5 for males. The Alaskan caballoids in my sample, not only appear to have had a longer life expectancy, but also appear to have lived much longer in absolute terms.

The debilitating effects of senescence can be seen clearly in populations of male equids, especially in the age of stallions who control the harems. Smuts (1976a, 1976b) found that few Burchell's stallions possess a harem after age 13. Fights between harem masters and the contenders are serious battles, and well they should be for the stakes in evolutionary fitness between being an adult bachelor and having the social license to fertilize an entire band of mares is significantly different. I have seen heated fights among zebra stallions in Etosha Park, Namibia. Not only is there biting and head throwing, but kicks are common. Legs are one of the main attack zones in a heated equid battle, and one stallion I observed at Etosha had suffered a broken tibia, making him totally vulnerable to large predators. Injury to the legs during a fight is serious. Their large temporal muscles give equids a powerful bite. The diastema, the long gap between incisors and cheek teeth, of stallions fits neatly around an opponent's leg allowing the sharp canine to bore into tendon and muscle. African wild dogs, wolves, and hyenas, and to some extent lions, select their targets from a zebra herd by watching for the slightest sign of limps or hesitations. Berger (1986) found that in any one year, 96% of adult stallions showed bite-related wounds. Of stallions four years and older 13% showed some degree of lameness. One can see by this that the strategy of harem take-over or defense is not a game for young males, the weak at heart, or more senile individuals.

The fossil record accompanying the Alaskan equid material allows us to ascertain that lions, wolves, brown bears, and probably short-faced bears were contemporaries with these horses (Guthrie 1990). All of these carnivores would have been formidable large mammal predators, and one would think that the presence of this predatory array would have affected the equid mortality curve, yet the Alaskan horses seem to have regularly lived until old age.

Retirement

At least among male ungulates, and probably females too, there can be a life after reproductive age. For males, retirement from reproduction is due to social reasons, not impotence. Once past the likelihood of a male ever winning a harem battle, the incidence of a serious fight seems to decline and individuals depend either on not reproducing or on unusual reproductive events, like finding an estrous immigrant female. This non-aggressive strategy for an old stallion would increase life expectancy considerably and may account for the significant percentage of Alaskan stallions reaching 20 and over. There are several large elephants in Kruger Park which are quite old. While these individual bulls have enormous tusks, they seem not to be actively participating in musth. I noticed most of these bulls are slack in muscle, show deep wounds from the past, and probably have molars so worn that their efficiency of processing coarse roughage is poor. Yet, in bypassing the stresses of frequent participation in musth they are able to live on for some years. Mammoth teeth with virtually exhausted molar crowns do occur in the fossil record and are seen occasionally among fossil specimens of Alaskan mammoths and horses as well.

It is interesting that in the Alaskan sample of fossil equids, about 25% of both male and female adults (that is, aged 4 and over) were senior animals (14 years and older). This is atypical for equids, and if these data are representative, even with the caveat of a powerful taphonomic distortion, a significant fraction of Alaskan Pleistocene horses, both mares and stallions, were able to live to old age. This suggests they were well-adapted to rigors of late Pleistocene climate at high-latitudes, a climate, so harsh during full Glacial as to exclude virtually all trees (Hopkins *et al.* 1981).

Distortion of the Ratio of Stallions to Mares

The most surprising thing which comes from the plots of age and sex for the Alaskan horses is the great disparity between numbers of males and females. The preponderance of adults are males, a total of 74% of the sample, with a resulting sex ratio of 3:1. At face value, this is not unusual, because it is common to have more males than females in the mammalian fossil record. For example, among bison skulls (*Bison priscus*) from interior Alaska the ratio of males to females is about 10:1, and it is the same in the late Pleistocene Rhine sand deposits in Germany (Guthrie and von Koenigswald 1996). And I know that skulls of fossil Dall sheep (*Ovis dalli*) show an even more extreme disparity, though I have collected no figures for this. One can see why this is true by examining the modern death assemblage. In Dall sheep country one finds a number of ram carcasses and skulls, a testimony to past mortality on high alpine winter range. But it is rare to find winter-killed ewes and much rarer to find a complete skull of a ewe. Much the same is true for muskoxen skulls, both in the fossil record and on wild muskoxen ranges. Again most of the skulls are from males.

The obvious explanation for this ungulate sex bias, both in theory and from direct observation, is that wolves and other predators and scavengers, like wolves, can and do chew through the more fragile female skulls to reach brain, nasal epithelium, etc., inside the skull. In so doing they consume virtually all of the bones or leave only fragments behind. Male skulls are simply more hard-headed; the rugged skulls of males are adapted for head-to-head clashing to determine status disputes. This construction is comparatively impenetrable by carnivores. But horse skulls, except for the presence or absence of the large canine, are not sexually dimorphic.

How can we explain this striking distortion in the fossil equid assemblage around Fairbanks Alaska? I have formulated several hypotheses which might explain this difference in male and female numbers:

1. They were misidentified, male and female fossils are at parity.
2. Physical processes destroyed more female skeletons than males.
3. Scavengers destroyed more female than male skeletons.
4. There actually were three times more males than females, in life.
5. Male skulls were selectively chosen by fossil collectors.
6. There were male-female sex differences in habitat use.
7. Males were selectively chosen by predators in the hills.

Though they had a small sample, Burke and Cinq-Mars (1997) also found predominantly males at Blue Fish Caves I, II and III. They questioned if these Pleistocene populations of caballoids were sexually dimorphic with regard to canines. But, other Alaskan and Siberian samples show a clear dimorphism, without overlap. So, it is very unlikely that there was any mistake in sexing the skulls. Sexes in all other species and subspecies of equids are quite distinctive due to the presence or absence of the large male canine, or "wolf tooth" in horseman's parlance. One of the main criteria for including fossils in my sample was that each could be clearly sexed; thus, I only included specimens sufficiently complete that this part of the premaxilla or dentary allowed me to sex each individual. I should note here that three of the adult mares from Fairbanks did show a small vestigial permanent canine, like those pictured by Penzhorn (1982) for mountain zebra (*Equus zebra*), and discussed by Smuts (1996b) for Burchell's zebra. However, there were no intermediate sized canines which made sex uncertain. So in can eliminating misidentification as an explanation, we can reject hypothesis 1.

Horses are unusual among northern large mammals, in that there is virtually no sexual dimorphism in skull size or robusticity between stallions and mares. Thus, one would assume that the physical processes which destroy or preserve skulls would have treated males and females similarly; that is, that the likelihood of preservation for both sexes would be close to parity. Indeed, for the late Pleistocene skulls from the Kolyma Lowlands in the collection of the Russian Academy of Science, I found 12 males and 13 females. So fossil horses from this lowland sample do not show a greater abundance of males. So we can also confidently reject hypothesis number 2. Likewise, hypothesis number 3 does not seem to account for the sex bias. Once killed, there is no reason to imagine that female carcasses would be more thoroughly eaten than those of males. I know of no modern equid predator that leaves behind male skulls and not those of females.

Sex ratios of mammals in the wild can become quite disparate, but this is generally not the case with equids. Living wild equids show adult sex ratios close to parity, but there are some exceptions. Schaller (1972) found a very high mortality of Burchell's zebra mares in their mid-years, which he attributed to disease. It was particularly apparent in age classes 9-13 years of age. In those year classes Serengeti mares were twice as likely to die as males. His life tables thus showed twice as many stallions as mares in the age classes 12 or older. But this seems anomalous. And of course the total adult skeletons from dead animals on the landscape would still be close to parity, as every animal must eventually die somewhere. Rudnai (1973) found a male:female ratio of 0.37:1 near Nairobi, Kenya. Johnson *et al.* (1987) found the sex ratio of 0.6:1 among burros. From his studies in Kruger Park, Smuts (1976b) found that the ratios average around 0.7:1. He found that lions in Kruger selectively killed male zebras over females. This same selectivity was also found by Rudnai (1973). This is probably because harem masters approach water holes first, taking the risk, and hang behind to defend the harem when chased by predators.

At the other extreme, Berger (1986) found a preponderance of stallions (1.30:1) in an expanding population, more males than females. The average for all the equids in his review was around 0.7:1, a more typical ungulate ratio of slightly fewer adult males than females. Mourer-Chauviré (1980) using subtle size differences among fossil horses proposed the proportion of males to females in Jaurens cave to be 1:2.6. But there are potential problems in using metric dimensions to sex equids. Haynes (pers. comm. 1993) found a distorted sample of feral mustang carcasses favoring males at a water hole in Australia, which did not correspond to the sex ratio in the living wild herds. Generally then, from this summary we can expect a population of wild equids to be near parity or perhaps with slightly fewer males, and so reject hypothesis 4. It appears highly unlikely that there were three times as many male as female horses reaching adulthood on the Alaskan Pleistocene landscape. Some other taphonomic bias seems to have been at work to distort the sample in this manner.

Collector preference can easily bias fossil samples, for example the enormous antlered skulls of male Irish elk, *Megaloceros*, were highly sought after by foreign buyers who wanted the spectacular male skulls. Collectors used a long metal rod to probe down through the bog peat to locate the antlered male skulls. The likelihood of hitting the three meter wide antlers was itself a tremendous bias in determining which sex was ultimately located and excavated.

And, in fact, in collections outside the British Isles one sees only the exported-for-sale antlered male skulls of this species. There are several skulls of females and young *Megaloceros* from Ireland, all found in Irish and British collections, probably because of their more limited sale or trade value.

Collection of Alaskan Quaternary fossils in the first half of the 20th century was funded by an independently wealthy New York collector, Childs Frick. He preferred that his collectors send only skulls, teeth, and complete limb bones, not vertebrae, ribs, broken bones, horn sheaths, etc. Alaskan fossil horses were collected with that charge, so they do come from a biased collection, but it is difficult to imagine how it could have been sex-biased for equids. The collection bias for fossil stallions over mares in this case is not credible, because it made little difference to the miners, the collectors, or to Frick as to whether an equid skull was male or female, and the miners at least could probably not have determined the difference, allowing us to reject hypothesis 5. So how can we account for the skewed ratio?

Do other upland sites in Alaska and the Yukon Territory contain equid fossils which are mostly male? As I indicated above, an equid sample from the uplands of the Yukon Territory examined by Burke and Cinq-Mars (1997) showed a similar distortion favoring males in three cave deposits (Blue Fish I, II, and III). These were among a large assemblage of bones that predators had dragged into caves, probably feeding young in dens, located on the nose of a very high ridge. The accumulation spanned a 10,000-15,000 year period. Burke and Cinq-Mars (1997) also observed the collections from Alaskan sites near the Canadian border accessioned to the Canadian Museum of Nature. These also showed a preponderance of males. From a sample of 18 adult jaws, 15 were male (though the sample is small, the skew on this is statistically robust).

I think hypothesis 6 is getting close; that male and female differences in habitat use may have been a factor. For example, if more males frequented the hills around Fairbanks than females, more males would also have died there in the hills. This was one of the three hypotheses Burke and Cinq-Mars (1997) suggested to explain the preponderance of males in their sample from the hills in northern Yukon Territory. Burke and Cinq-Mars rejected this hypothesis, but I propose there is sufficient data to make us reconsider the point. But, there are enough data to suggest that this may have been the case. First, we must consider caballoid behavior as well as the fossil record, to reconstruct how and why the landscape might have been used differently by male and female horses of different ages.

Relevant Caballoid Social Behavior

Early in their evolution caballoids probably broke with a more ancient equid territoriality. By abandoning territory-based social organization, these caballoids could use resources which occurred in broad ecological and seasonal gradients (Forsten 1986). This continual adjustment of local habitat and density made different demands on the caballoid social system (Kingdon 1979). In this new system permanent attachment of mares to males became the norm. Mares normally deferred directly to stallion-stallion combat to settle who is the most confident and strongest—the best stud with whom to attach their immediate survival and genetic future. Mares benefit in this system by having a stallion to take risks in exploring, out ahead, new ranges or water holes, and to attack predators in defense of his offspring. Males benefit in this system of exclusive reproductive rights by being certain of paternity. Independently, Burchell's zebra (*Equus burchelli*) converted to this same system, undoubtedly in response to the similar environmental pressures and opportunities.

Unlike asses (*E. asinus*) and Grevy's zebra (*E. grevyi*) in which males defend a breeding territory, the caballoids and Burchells zebras are semi-nomadic, organized into small harems of mares and young with a single lead stallion. Unattached males often collect together in bachelor bands (Klingle 1974). At puberty, young males may remain in the family group or take up residence in one of the bachelor bands. As they continue to grow and gain experience, young bachelors can obtain females either by luring a young filly away from her maternal harem, or by challenging a stallion for his harem. In either event they have to fight the lead stallion. These fights are far from ritualized. They are drawn out serious battles with considerable kicking and biting with apparent intent to cause debilitating damage.

Starting a new harem with fillies involves accumulating them one at a time, fighting for each one. This normally continues until three or more mares have been won, producing a harem with attached young of about 5-8 animals, and can include, as many as 20 (Berger 1986, Klingle 1974). Displacing the herd stallion is a sweepstakes route to acquiring a mature harem, and is usually attempted only when he has been killed by a predator, or is debilitated from accident, infirmity, or senescence.

A bachelor band thus contains young untried stallions, unsuccessful young combatants, and older males that have been displaced, as well as a few stallions in their prime who do not have a harem. Most prime stallions are absent from the bachelor group because they are the ones that have harems. Thus the ages found in a bachelor band are precisely the same pattern represented by the "fossil" stallions from my Alaskan Tanana Hills sample. I think this coincidence is significant.

Sex-Age Specific Use of the Hills versus the Plain?

Studies have shown that among most ungulate species, male and female animals do not use the landscape in identical ways. There is some disagreement about the reasons for this, which relate to reducing intersex competition, dominance-subordination behavior, different travel energetics, variation in food preference, susceptibility to predation, and others. But whatever the reasons, some degree of sexual separation for much of the year is common. Could it be that in some cases this produces a distortion as to which sex predominates in the fossil record?

Studies of living equids show that harems and bachelor bands use the landscape in different ways. There is no reason to assume these socially organized patterns would be different in late Pleistocene Alaska; such behavior is conservative. Let me propose that the bachelor bands of Alaskan horses tended to used the Tanana hills around Fairbanks more than the harems. Berger (1986) found that in the American West, bachelor mustangs (especially the old bachelors) preferred the uplands compared to the harem bands. Haynes (1991, pers. comm.) found indirect evidence of similar sex-specific mortality patterns at "mass-death" sites around different water holes in central Australia following severe drought. The water holes he examined had mostly male carcasses - which he interpreted as being members of bachelor bands. Whereas, at other water holes females and young predominated.

I showed in an earlier study of fossil large mammals (Guthrie 1968) in this same region that the proportions of horses in the large mammal community increase as one goes from uplands to the lowlands. Since this is a proportional change it is difficult to know if it represents a meaningful separation, i. e., a preference of horses for the flatlands, but it is suggestive. Equids use flight as a means of escape and are therefore less vulnerable in the open level country. Berger (1986) showed that harems preferred the flatlands, and that harems required smaller home ranges. Berger's bachelor bands had almost twice the home range of harem bands, and bachelors moved up to twice as much as harems. Bachelors move considerably at night, while harems move very little at night. All of this suggests that uplands are a less desirable habitat. Although caballoids are not territorial, they have harem-band specific home ranges, which are generally better in quality than those occupied by bachelors. This probably accounts for the harem's smaller home-range size. Though harem stallions do not technically defend territories they do not allow bachelors near their feeding harem, and constantly harass unrelated bachelors in their vicinity.

In general, mares are more cautious than stallions, preferring places where predation is less likely. From my own experiences watching zebra approach water holes in Namibia, Botswana, and South Africa, it is apparent that females were by far the warier. In every instance the harem stallion was the first to venture near water holes frequented by lions. As lions will chase any zebra which comes into their attack zone, stallions, in suitable lion hunting terrain, will be the ones most likely to be caught first.

I propose the most favored rangeland of the harem masters in the Fairbanks region was out on the flat lands of the Tanana Plain. Today, these flats are cloaked in a thick tangle of shrub cover, interlaced with muskeg, and dotted with thaw-lakes. Away from waterways the flats are almost impenetrable in summer. However, pollen records (Ager 1975) and other evidence (Guthrie 1990) indicate that during the cool and dry episodes of the Pleistocene the flats were more arid and more open, with braided glacial streams coming out of the Alaska Range to the south (Guthrie 1990). Such lowland soils would have been nutrient rich. The wide plain was then an arid grassland with scattered sage, *Artemisia,* and nutritious forbs, with no trees or lakes. Evidence from the now-vegetated dune field patterns suggests these plains were raked year-around with strong winds, carrying the glacial dust aloft and redepositing it as loess onto the Yukon-Tanana Upland. This wind would cause snows to drift, thus allowing easy access of grazers to their winter forage. Though the south facing hill country would have been good rangeland in spring, wind probably occluded the shallow valley bottoms with snow drifts, leaving only ridge tops and bare south faces of the hills snow free.

According to this hypothesis the greater use of the hill-country by bachelor bands should have produced a non-random frequency in the sex of equids dying there in winter (the season of major mortality). The death assemblage in the hills would thus have favored the young bachelors and the older males most likely to have been debilitated from social battles. Comparatively few stallions of prime ages, between 6 and 9, would have died there, as they would have been on the plains with their harems. Berger (1986) showed that stallions between 7-10 years of age leave more foals than any other age class and have a greater chance of winning predator battles.

The harem has vested genetic interest in one another and look out for each other's interests, give early warning signals in case of trouble, and flee and fight as a group. Bachelors on the other hand share few genes with compatriots and it behooves each to act in their own interests - to act independently in case of trouble. While they are social company, they are also genetic competitors for the next open harem master slot. Co-operative warning and defense behaviors against predation are different in the two groups.

I initially assumed the rarity of young animals in my sample was due to poor preservation, but in this scenario, equid harems would have spent most time in the lowlands. So the age bias of the sample could also be an aspect of this behavioral partitioning of the landscape, harems with young animals being much less likely to utilize hill habitats which offered more cover to lions.

Predation as the Main Mortality Force

Hypothesis 7 may not explain the sex-distortion, but it may be the reason the more conservative harems avoided the hills in favor of the plains. Circumstantial evidence suggests that lions were more efficient predators in the hills than on the plains. Lions have a moderately short attack zone and depend on cover and stealth to put them within killing distance before the potential prey is alerted (Schaller 1972). Horses are fast, have good stamina and can outrun a lion if given a modest lead. Hill country, with its irregular terrain, would offer much more cover than the flats, and irregular snow drifts, erosion gullies, and scattered shrubs along the streams would also provide cover for lions. The plains lack relief, and allow a predator to be seen from considerable distance. The plains also offer long stretches of good footing for a fleet hoofed animal. African lions rely heavily on darkness, but at 65° north there is no darkness for much of the year and snow reflection, starlight, moonlight and auroral haze makes complete darkness rare during winter.

Lions were present in the Fairbanks hills at the same time as the horses. Museums have over 45 fossil lion skulls which have been collected from this area, usually from the same gold mines where the fossil horses were recovered. And the few ^{14}C dates we have for lions show they shared the habitat with horses during the same time period. We also know that lions preyed on large mammals in these hills and valleys. A frozen bison mummy, found at one of these mines, has evidence on soft tissue showing how it was killed, partially skinned, and eaten by lions (Guthrie 1990).

It was the search for gold weathering out of rock in these Fairbanks hills which led miners to move whole valleys of muck and turn up the remains of horses that died in the surrounding hills. Though the many of the fossil remains of Pleistocene horses that died on the Tanana River flood plain may indeed be out on those flats, there are no Pleistocene horse fossils available from the area. Such fossils may be deeply buried, and lacking the incentive from miners to open that area they have not been unearthed. The lack of gold in the flats, or road-building or nearly any other large scale human-made earth moving means the mortality pattern of horses on the plains will remain a mystery. If the opportunity ever arises to collect such a sample, one could predict a more harem-like mortality pattern skew on the plains.

I think this "bachelor-bands in the hills" model best accounts for the sex and age bias of the fossil horse assemblage. This is, of course, a stochastic explanatory model. Harem stallions, their mares and young would probably have used these hills to some extent, perhaps seasonally, as our sample shows, just not as often as bachelors.

To help us envision the proximate mortality forces, we can look at Burchell's zebra for a rough modern analog of predator-equid interactions. Today, in most areas where lion and zebra exist together, the latter is a preferred prey of the former. Zebra normally comprise 15-20% of lion diets in East Africa (Schaller 1972). Furthermore, Schaller found that lions were almost three times as successful in stalks of single zebra (22%) than for 2-75 zebra (6%). Unattached bachelors have only a loose bond to fellow bachelors, and are more likely to be physically dispersed while feeding than the tightly defended harem. Hyenas and wild dogs are a threat to sick or debilitated zebra, but adult zebra are a much smaller part in the diet of other carnivores than of lion. Presumably, Alaskan wolves (one of the other major predators on the Mammoth Steppe) would have acted much like African coursers, hyena and wild dog, which generally prey selectively on the weaker animals. So, some Alaskan fossil horses could have been wolf-kills. But, as coursers, wolves would be as likely to hunt in the plains as the hills, while, the non-coursing lions may have been at a greater handicap in the plain which probably had a low grass sward and generally lacked trees and physical relief that lions typically employ in their stalks.

Sex Bias of Habitat Use by Other Alaskan Fossil Species

Skulls from other fossil species, like bison and muskoxen, are highly biased toward males, but this seems to be a factor of differential preservation due to physical robustness. In fact, there is such sexual dimorphism in skull and skeleton ruggedness of male muskoxen, sheep, bison, moose, caribou, mammoth, saiga and others that it would be difficult to make any comparable assessment as has been done with fossil equids. However, it is suggestive that for steppe bison metapodials, males outnumber females 3050 to 1788, a ratio of almost 2:1 (Skinner and Kaisen 1947). These metapodials, are rugged, the greatest cortex thickness (about 1.0 cm) of any bison limb bone. The marrow cavity of bison metapodials is quite small hence the cortex is seldom broken by scavengers—both chew marks and cracking for marrow extraction are almost unknown on fossil bison metapodials from Alaska. Yet there is a definite sexual dimorphism in robusticity, so it is not as clear as among the horse skulls that physical destruction can be discredited as an explanation. Bison bulls would have also been more susceptible to lion attack in these hills, as older bison bulls are more likely to be solitary, less wary, and probably make use of higher-risk irregular terrain than cows. The occurrence of a bull bison mummy (Guthrie 1990) killed by lion(s), is poignant bit of evidence for this image, but it is of course insufficient to allow us to unconditionally accept the hypothesis of disproportionate male bison mortality in these hills.

CONCLUSIONS

The bias in sex and age-specific mortality in this Alaskan sample of fossil equids seems to be best accounted for by differential range use. The rarity of young equids may in part be explained by that factor, but may also be a product of the more fragile bones of young animals being less likely to preserve. The most abundant age groups in the fossil assemblage are young adults, 4 and 5 year olds, and older animals, over the age of 10. The best explanations for this assemblage lies in the equid pattern of bachelor-bands. These bachelor bands would have been composed of stallions not yet old enough to take harems, and individuals too old to retain them. Studies of other extant equids, show the age of socio-sexual maturity peaks at about 4-5 years of age. Our statistics from fossil Alaskan stallions suggest they may have matured, at least socially, a little later than extant caballoids. It is interesting that the rigors of the northern climate did not curtail maximum life expectancy, as a significant number of these horses lived to a very old age.

One important cautionary note for archaeologists arises from this study. Identifying a site as a product of human refuse based only on the presence of highly distorted age or sex distributions of a given large mammal species is tricky, because such distortions can occur naturally due not only to different predator patterns in prey selection, but to the common biological pattern of differential range use by different ages of male and female large mammals.

ACKNOWLEDGMENTS

I am especially grateful to the American Museum of Natural History for permission to work on their collections off and on over the last 30 years and to sample material for radiocarbon analysis. Dick Tedford has been especially helpful and welcoming. Also, thanks to Dick Harington, Lee

Lyman and Mary Lee Guthrie for their careful reviews of the manuscript. The University of Alaska Institute of Arctic Biology provided salary for this study, and the National Science Foundation assisted in the expenses of travel, dating, and some summer salary during the project. This paper was a product of a long-term dating study of Alaskan Pleistocene Mammals. Dating has been performed at the United States National AMS Laboratory in Tucson, Arizona, where Warren Beck, Rosemary Maddock and Doug Donahue have done a very commendable job.

REFERENCES CITED

Ager, T.A.
1975. *Late Quaternary Environmental History of the Tanana Valley, Alaska.* Institute of Polar Studies, Report No. 4. Ohio State University, Columbus.

Anonymous.
1981 *American Association of Equine Practitioners (AAEP) Official Guide For Determining the Age of the Horse.* AAEP Golden, Colorado.

Azzaroli, A.
1990 The genus *Equus* in Europe. pp. 339-356. In European Neogene Mammal Chronology. E.H. Lindsay editor. Plenum Press, New York.

Barnowsky, A.D.
1985 Taphonomy and Herd Structure in the Extinct Irish Elk, *Megaloceros giganteus. Science* 228:340-344.

Berger, J.
1986 *Wild Horses of the Great Basin.* University of Chicago Press, Chicago.

Burke, A. and J. Cinq-Mars.
1997 Late Pleistocene Equids in Eastern Beringia: Dental Characteristics and Mortality Profiles of *Equus lambei* from the Bluefish Caves, Yukon Territories, Canada. *Anthropozoologica* 25-26:217-224.

Caughley, G.
1966 Mortality Patterns in Mammals. *Ecology* 47: 906-918.

Conybeare A. and G. Haynes
1984 Observations on Elephant Mortality and Bones in Water Holes. *Quaternary Research* 22:189-200.

Eisenmann, V.
1984. Sur Quelques Caractères Adaptitifs du Squelette d'*Equus* et Leur Implications Paleoecologiques. *Bulletin Museé National Histoire Naturel*, Paris. 6:185:194.
1991 Les Chevaux Quaternaires Européens (Mammalia, Perissodactyla). taille, typologie, biostratigraphie et taxonomie. *Geobios* 6:747-759.

Flerov, K. K.
1952 *Fauna of the USSR Mammals.* Vol. 1. Musk Deer and Deer. Academy of Sciences USSR, Moscow.

Forsten, A.
1986 Middle Pleistocene Replacement of Stenonid Horses by Caballoid Horses-Ecological Replacement. *Paleogeography, Paleoclimatology, and Paleoecology* 65:23-33.

1991 Size Decrease in Pleistocene-Holocene True or Caballoid horses in Europe. *Mammalia* 55:407-420.

1993 Size-decrease in Late Pleistocene Caballoid Horses (Genus *Equus*) Intra or Interspecific Evolution? A Discussion of Alternatives. *Quaternary International*, 19: 71-75.

Gaughan, E. and R.M. DeBowes.
1998 The Veterinary Clinics of North America. *Equine Practice, Dentistry* 14: 1-437.

Groves, C.
1986 The Taxonomy, Distribution, and Adaptations of Recent Equids. In *Equids in the Ancient World.* R.H. Meadow and H.P. Uerpmann (Eds.) pp. 11-66. Dr. Ludwig Reichert Verlag, Wiesbaden.

Guthrie, R.D.
1967. Differential Preservation and Recovery of Pleistocene Large Mammal Remains in Alaska. *Journal of Paleontology* 41(1):243-262.

1968 Paleoecology of the Large Mammal Community in Interior Alaska During the late Pleistocene. *American Midland Naturalist* 79: 346-363.

1990. *Frozen Fauna of the Mammoth Steppe.* University of Chicago Press, Chicago.

In press Origin and Causes of the Mammoth Steppe: A Story of Cloud Dover, Woolly Mammals, Buckles and Inside-Out Berengia.

Guthrie, R.D. and J.V. Matthews.
1971 The Cape Deceit Fauna - Early Pleistocene Mammalian Assemblage from the Alaskan Arctic. *Quaternary Research* 1:474-510.

Guthrie, R.D. and S. Stoker.
1990 Paleoecological Significance of Mummified Remains of Pleistocene Horses from the North Slope of the Brooks Range, Alaska. *Arctic.* 42: 267-274.

Guthrie, R.D. and W. von Koenigswald.
1996 Puncture Wounds in Bison Skulls: A Case of Archaeology or Paleontology? pp. 415-425. In *Spuren der Jagtd: Die Jagd nach Spuren, Festschrift für Hansjürgen Müller-Beck.* H. Ingo Campen, J. Hahn, and M. Uerpmann (eds.) Mo Vince Verlag, Tübingen.

Harington, C.R. and F.V. Clulow.
1973 Pleistocene Mammals from Gold Run Creek, Yukon Territory. *Canadian Journal of Earth Science.* 10: 697-759.

Harington, C. R.
1989 Pleistocene Vertebrate Localities in the Yukon. In: L.D. Carter, T. Hamilton and J.P. Galloway (eds.) *Late Cenozoic History of the Interior Basins of Alaska and Yukon.* United States Geological Survey Circular 1026: 93-98.

Haynes, G.
1991. *Mammoths, mastodonts and elephants.* Cambridge University Press.

Hopkins, D.M., P.A. Smith, and J.V. Matthews Jr.
1981 Dated Wood from Alaska and the Yukon: Implications for Forest Refugia in Beringia. *Quaternary Research* 15: 217-249.

Johnson, R.A., S.W. Carothers, and T.J. McGill.
1987 Demography of Feral Burros in the Mojave Desert. *Journal of Wildlife Management.* 51: 916-920.

Joubert, E.
1972 Tooth Development and Age Determination in the Hartmann Zebra, *Equus zebra hartmannae. Madoqua, Ser.* 6: 5-16.

Kingdon, J.
1979 *East African Mammals, Vol. III Part B.* pp. 1-210. University of Chicago Press, Chicago.

Klien, R.G.
1982 Age (mortality) Profiles as a Means of Distinguishing Hunted Species from Scavenged Ones in Stone Age Archeological Sites. *Paleobiology* 8:151-158.

Klingle, H. and U. Klingle.
1966 Tooth Development and Age Determination in the Plains Zebra. *Zoologisch Garten* 33:34-54.

Klingle H.
1974 A Comparison of the Social Behavior of the Equidae. In *The Behavior of Ungulates and its Relation to Management.* V. Geist and F. Walther (Eds.) pp. 124-132. IUCN Publication No. 24.

Kurtén, B.
1953 On the Variation and Population Dynamics of Fossil and Recent Mammal Populations. *Acta Zoologica Fennica.* 76:1-122.

1958 Life and Death of the Pleistocene Cave Bear. *Acta Zoollogica Fennica.* 95: 1-59.

1964 Population Structure and Paleoecology. In *Approaches to Paleoecology,* J. Imbrie and N. Newell (Eds.) pp. 123-144. New York, Wiley.

1976 *The Cave Bear Story.* Columbia, New York.

Levine, M.
1983 Mortality Models and the Interpretation of Horse Population Structure. In *Hunter-gatherer Economy in Prehistory: A European Perspective.* G. Bailey (ed.) pp. 23-46. Cambridge University Press, Cambridge.

Lyman, R.L.
1987 On the Analysis of Vertebrate Mortality Profiles: Sample Size, Mortality Type, and Hunting Pressure. *American Antiquity* 52:125-142.

Mourer-Chauviré, C.
1980 *Le Gisement Pléistocène Supérieur de la Grotte De Jaurens, A Nespouls, Corrèze, France: Les Équidés (Mammalia, Perissodactyla).* Nouvelle Archaeology. Museé Histoire Naturel. Lyon 18: 17-56.

Murie, A.
1944. *The Wolves of Mount McKinley,* U.S. Dept. Int. Nat. Park Service, Fauna Ser. 5.

NA
1978 *Horses.*National Research Council, Animal Care Publication,.

Pasquine, C., V.K. Reddy, M.H. Ratzlaff.
1983. Atlas of Equine Anatomy. Sudz Publishing, Eureka, California. pp. 339.

Penzhorn, B.L. 1982 Age Determination in Cape Mountain Zebra, *Equus zebra,* in the Mountain Zebra National Park. *Koedoe* 25: 89-102.

Péwé, T. L.
1975 Quaternary Geology of Alaska. *United States Geological Survey Professional Paper* 835.

1989 Quaternary Stratigraphy of the Fairbanks area, Alaska. In: Carter, L. D., T.D. Hamilton, and J.P. Galloway, (eds.) Late Cenozoic History of Interior Basins of Alaska and the Yukon. *United States Geological Society Circular* 1026:72-77.

Preece, S. J., J.A. Westgate, B.A. Stemper, and T.L. Péwé
1999 Tephrochronology of Late Cenozoic Loess at Fairbanks, Central Alaskas. *United States Geological Society Bulletin* 111: 71-90.

Rudnai, J.A.
1973 The Pattern of Lion Predation in Nairobi National Park. *(photocopy report).*

Sack, W.O. and L.L. Sadler
1982 *Rooney's Guide to the Dissection of the Horse.* Veterinary Textbooks, Ithica, New York. pp. 350.

Saunders, J.J.
1980 A Model for Man-mammoth Relationships in Late Pleistocene North America. *Canadian Journal of Anthropology* 1:87-98.

Schaller, G.B.
1972 *The Serengeti Lion.* University of Chicago Press, Chicago.

Sher, A.V.
1986 On the History of Mammal Fauna of Beringia. *Quartärpaleontologie.* 8:185-193.

Skinner, M.F. and O.C. Kaisen.
1947 The Fossil Bison of Alaska and Preliminary Revision of the Genus. *Bulletin of the American Museum of Natural History.* 89:131-144.

Smuts, G.L.
1976a Reproduction in the Zebra Stallion (*Equus burchelli antiquorum*) from Kruger National Park. *Zoologica Africana* 11:207-220.

1976b Population Characteristics of Burchell's Zebra *(Equus burchelli antiquorum*) from Kruger National Park. *Koedoe* 18:139-146.

Spassov, N. and N. Iliev
1997 The Wild Horses of Eastern Europe and the Phylogenetic Origin of Domestic Horses. *Anthropozoologica* 25-26: 753-761.

Stiner, M.C.
1991 (editor) *Human Predators and Prey Mortality.* Westview Press, Boulder Colorado.

1994 The Use of Mortality Patterns in Archaeological Studies of Hominid Predatory Adaptations. *Journal of Anthropological Archaeology* 9: 305-352.

Uerpmann, H.P.
1989 Altweltiche Faunenelemente im Spätpleistozän Alaskas und die Frühe Besiedlungsgeschichte Nordamerikas. *Beiträge zur Archäozoologie, Archaologie, Anthropologie, Geologie, und Paläontologie.* (Festschrift für Hans R. Stampfli. pp. 303-307. Helbing und Lichtenhahn, Tübingen.

Van Valen, L.
1964 Age in Two Fossil Horse Populations. *Acta Zoologica* 45:93-106.

Vereshchagin, N.K. and P.A. Lazarev
1977 Description of Soft Parts and Skeleton of the Selerikan Horse. In *Fauni i flori anthropogena Severo-Vostoka Sibiri.* O. A. Skarlato (Ed.) pp. 85-185. Akademiya Nauk, Trudy Zoologischeskologo Instituta, Vol. 63.

Voorhies, M.R.
1969 Taphonomy and Population Dynamics of an Early Pliocene Vertebrate Fauna. Knox County, Nebraska. *University of Wyoming Special Contributions to Geology, Special Paper* 1:1-69.

Welsh, D.A.
1975 *Population, Behavioral, and Grazing Ecology of the Horse of Sable Island, Nova Scotia.* Unpublished PhD. Dissertation, Dalhousie University, Halifax.

Winans, M.C.
1989 A Quantitative Study of North American Fossil Species of the Genus *Equus* in *The Evolution of Perissodactyls*, edited by D.R. Prothero and R.M. Schoch. Oxford University Press, New York.

NEW RADIOCARBON DATES ON SAIGA ANTELOPES (*SAIGA TATARICA*) FROM ALASKA, CANADA AND SIBERIA: THEIR PALEOECOLOGICAL SIGNIFICANCE

R. Dale Guthrie
Institute of Arctic Biology, University of Alaska

Andrei V. Sher
Severtsov Institute of Evolutionary Animal Morphology and Ecology, Russian Academy of Sciences.

C. Richard Harington
Canadian Museum of Nature, Ottawa

Saiga antelopes (*Saiga tatarica*) (Figure 1) now live on the dry steppes and semi-desert in southern Russia and northern China, but during the late Pleistocene, they roamed across most of Europe and northern Asia (Bannikov 1958; H.D. Kahlke 1975; R.D. Kahlke 1991; Baryshnikov and Tikhonov 1994). Saiga fossils are scattered as far west as France and England, and eastward toward Manchuria and the Russian Far East. But the most remarkable aspect of saiga distribution are fossils found north of the Arctic Circle in Siberia, Alaska, Yukon and Northwest Territories (Sher 1968; Harington 1980, 1981; Guthrie 1982), 10,000 km from their present range. These latter areas are now particularly unsuitable for saiga (Sher 1968; Guthrie 1968; Harington 1978).

Though saiga fossils are widely distributed in Beringia (northeastern Russia, Alaska, and northwestern Canada), they are uncommon compared to the tens of thousands of bones from other large mammal species: bison (*Bison*), horse (*Equus*), mammoth (*Mammuthus*), caribou (*Rangifer*), muskox (*Ovibos*), etc. Saiga fossils constitute only a small fraction of one percent of the total assemblage (Guthrie 1968; Sher 1968; Harington 1978). We do not attribute this solely to preservational differences because saiga remains are uncommon even when compared to species of roughly the same size, like mountain sheep (*Ovis*) or caribou. Beringian Pleistocene saigas simply do not seem to have reached the densities one finds today on the Asian steppes, where scattered herds of thousands occur. Nevertheless, Beringian fossils do indicate that saigas were able to penetrate and exploit the far north, which is important paleoecological information. Furthermore, saigas may have done this several times, starting early in the Pleistocene (Sher 1986). Indeed, they existed as a characteristic arctic species, at least for episodes of tens of thousands of years, if not longer. This fact is informative both about saiga adaptations and about northern climates and environments in the Pleistocene.

Harington (1981:213) first noted the slightly large size of Pleistocene saigas, but includes them with the modern species under *Saiga tatarica* (Harington 1981; Harington and Cinq-Mars 1995). Baryshnikov and Tikhonov (1994) found them to be roughly 10-15% larger than extant saigas in inferred body size (so much larger that they referred saigas from the Khazarian Fauna of the Volga and Mammoth Fauna of Europe and Siberia to a separate specific rank, *Saiga borealis* Tschersky 1876. It is an axiom among range managers that body size in a wild ungulate population gives one a rough assessment of range quality, and available forage per individual during the growing season (Klein 1965; Guthrie 1984b). On this basis, Pleistocene saigas had better forage than living ones. Larger body size is a general phenomenon among bovids on the Mammoth Steppe (Harington 1978; Guthrie 1984b). It suggests that there was little competition for food and better than adequate forage quality and volume during the growth season.

SAIGA SPECIALIZATIONS

Saigas are more specialized than most large mammals. They are particularly adapted to hard-packed flat substrates. Saigas are pacers; they have evolved a suite of anatomical traits which allow them to use a rapid pacing or ambling gait; this means that both legs on the same side move in step. Saigas are the only bovid to specialize in this gait. Although pacing is one of the more efficient forms of quadrupedal locomotion (a trot has limits in the arc of foreleg leg swing), its chief drawback is instability. Pacing causes the body mass to be shifted left and right over the changing center of gravity, and thus is unsuited to rough, broken ground.

Because of their anatomical specializations for pacing, saigas are poor jumpers, and they prefer to walk around even the slightest irregularity (Heptner *et al*., 1988). To diminish the instability of pacing, saigas run with their head low, thereby lowering the center of gravity. The long proboscis seems to be one of several adaptations to lowering the head into the dust generated by the herd. Chronic aridity combined with incomplete ground cover of steppe vegetation results in particularly dusty landscapes. The enlarged nasal chamber forms a moist trap for dust (Bannikov *et al*., 1967; Vereshchagin and Baryshnikov 1982), minimizing dust inhalation. Pleistocene saigas seem to have had disproportionately longer nasal bones, suggesting a proboscis smaller than the enlarged one of living saigas. A smaller proboscis suggests either not-so-dusty conditions, smaller herds to stir up dust, or both during the late Pleistocene.

Perhaps this preference for flat surfaces prevented saigas from dispersing across the rugged terrain south of the Central Siberian Plateau and also through Canada to the American Great Plains. Every other large mammal, except saigas, that penetrated the lowlands of northern Siberia also spread southward to Manchuria. Likewise, all, except the extinct muskox *Praeovibos*, the other Asian large mammal that reached Alaska, were able to disperse southward. Saigas' unique limitations, their precise steppe adaptations, make

Figure 1. Saiga antelope. Males are horned and females not. This illustration shows a male in rut. With thick neck, beard, enlarged proboscis, and a face stained black from preorbital gland secretions. Low head carriage and short legs compared to the relatively long body are probably adaptations to an efficient pacing gait.

saiga fossils informative resources for paleoecological assessment (Harington 1981:216-219; and see Harington and Cinq-Mars 1995 for special adaptations of Eastern Beringian saigas). How did saigas once live in areas which now are so unsuitable?

SAIGAS AND THE MAMMOTH STEPPE

Eurasia was different in the Pleistocene. Late Pliocene Himalayan uplift impeded the northward flow of southern moisture (Ruddiman and Kutzbach 1989) and intensified the dryness of Eurasian steppes. Due to radical shifts in atmospheric flow during the Pleistocene, these steppes experienced episodic expansion, exaggerated during cyclic troughs in solar insolation. Pleistocene cold steppes stretched from England to Japan. In addition to this lateral expansion, the steppes extended northward beyond the Arctic Circle, creating a vast cold-arid biome, designated the Mammoth Steppe (Guthrie 1982, 1990). A complex fauna adapted to low-growing grasses and forbs and cold temperatures evolved within these steppes. Russian paleontologists refer to these animals as the mammoth fauna (woolly mammoth, horses, bison, reindeer/caribou, etc.). Most of these grassland mammals originally entered Beringia from the west, including ancestors of extant forms such as: bison (*Bison bison*), mountain sheep (*Ovis canadensis* and *O. dalli*), wapiti (*Cervus elaphus*), caribou (*Rangifer tarandus*), moose (*Alces alces*), brown bears (*Ursus arctos*), and wolves (*Canis lupus*). Extinct genera: mammoth (*Mammuthus*), muskoxen (*Soergelia*), and lion (*Panthera*) also followed this route to Beringia and midcontinent North America.

Because saigas' adaptations to certain steppe vegetation and landscapes have been well studied (Bannikov *et al.*, 1967; Heptner, *et al.* 1988), they are useful as a general proxy indicator of paleoclimatic conditions during these poorly understood (and controversial) steppe expansions. But chronology is critical. Saigas may not have lived continuously in Beringia during the most extreme Pleistocene climatic swings; their paleoecological utility is dependent on good dating. Unfortunately, few saiga fossils in museum collections have been radiocarbon dated. Most saiga specimens lack a geochronologically-controlled stratigraphic provenience, so each fossil must itself be dated. Dating of individual specimens was seldom done because museums and private collectors were understandably reluctant to sacrifice large parts of such rare fossils to obtain sufficient carbon for traditional, radiometric dating. Furthermore, radiocarbon dating is very costly. Prior to this study only three Beringian saiga dates were published (Harington 1980, 1981; Harington and Cinq-Mars 1995).

GAP IN THE SAIGA DATES

Fortunately, the technique of accelerator mass spectrometry (AMS) radiocarbon dating requires a very small bone sample which can be taken from places resulting in minimal damage to a skull's scientific or display value. We obtained specimens from our own field work, museums and private collections; and AMS-dated most specimens that can be clearly identified as saiga – mainly skull caps (Table 1) with attached horncores. All of the latter were males. We attribute this gender skew to the fact that the fragile female skulls can be entirely eaten by predators such as wolves or bears and are thus much less likely to be preserved. Only the collagen fraction of the bone samples was dated, and all had sufficient protein (1.07-3.51 %) to indicate good preservation.

Saiga dates (Figure 2) show a bimodal distribution, so far lacking points in the period between 15,000 to 25,000 yr. excepting the marginal date of 15,460 ± 130 yr BP (AA-3892) from Kolyma River. Our sample is too small to interpret this lack as total absence, and it is quite possible that future dates from Beringia will fall within this now empty span. Still, the two modes (student t = 9.0 probability) indicate comparatively higher densities during these two time periods (40,000 to 25,000 and 15,000 to 12,000 yr), although some part of the bimodality may be caused by depositional bias, discussed later. Continued search for more Beringian saiga fossils and further dating is important to clarify this pattern.

The gap in Beringian saiga dates corresponds approximately to the peak full-glacial (marine isotope stage 2), and even more closely to Hopkins' (1982) Beringian Duvanny Yar Interval 25,000 to 14,000 yr. The Duvanny Yar pollen record shows only herbaceous pollen and some low shrubs (Livingston 1955; Colinvaux 1964; Ager 1975; Brubaker *et al.*, 1983; Ritchie 1984; Anderson 1988).

The usual outer limit of radiocarbon dating is approximately 40,000 ka, so these dates have no resolution beyond that point. The cluster of older dates from that outer limit to 25,000 yr, apparently corresponds to the rather long interstade (marine isotope stage 3), Hopkins' (1982) Boutellier Interval. The Beringian climate of this interval is not clearly understood, but at least small amounts of arboreal pollen were produced which indicate partial tree cover, suggesting slightly warmer/wetter conditions, but not nearly to the Holocene, or present extent (Brubaker *et al.*, 1983).

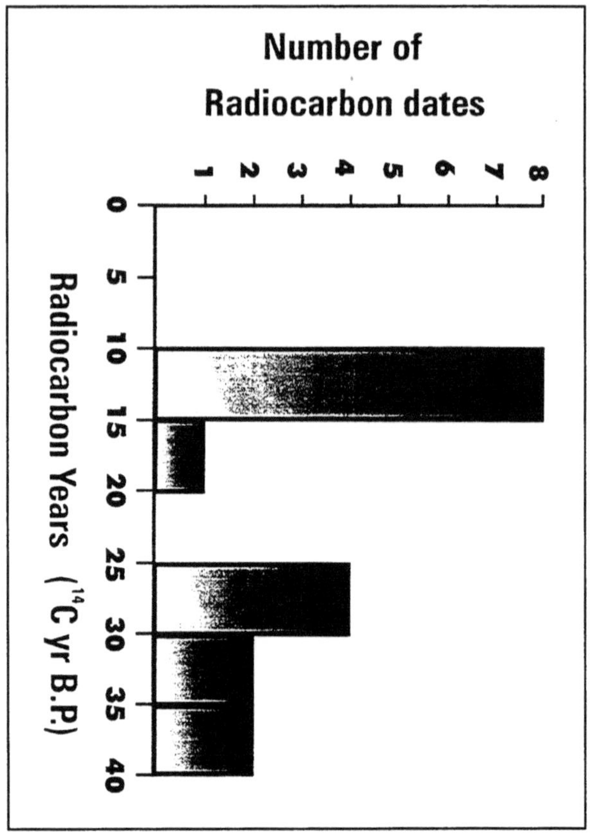

Figure 2. Bar graph showing the distribution, over 5000-yr intervals, of Beringian radiocarbon dated saiga fossils. Note the bimodal curve suggesting population peaks before and after the maximum cold period of the last glaciation (about 25–15,000 yr.).

DATE	LABORATORY #	LOCALITY AND MAP NUMBER (FIGURE 3)	SPECIMEN	REFERENCES
12,220 ± 130	AA-3077	unknown locality, Alaska	Skull, (private collection)	This paper
12,390 ± 120	AA-3075	Lost Chicken Creek, Alaska (1)	Skull, UA-V-54-196	This paper
12,620 ± 110	AA-3074	Lost Chicken Creek, Alaska (2)	Skull, UA-V-54-970	This paper
12,780 ± 140	AA-3215	Fairbanks, Alaska (3)	Skull, (type of Saiga ricei), FAM 30495	This paper
13,220 ± 70	(Beta-79853, CAMS-18416)	Old Crow River Loc. 149, Yukon, Canada (4)	Right radius	This paper
13,260 ± 250	AA-3895	Kolyma River, Russia (5)	(?) (S-PIN 761-1)	This paper
13,390 ± 180	(RIDDL-279)	Bluefish Cave III, Yukon, Canada (6)	Cranial fragment	Harington and Cinq-Mars 1995
14,920 ± 160	(Beta-25119, ETH-3898)	Baillie Islands, Northwest Territories, Canada (7)	Cranial fragment (NMC 12090)	Harington and Cinq-Mars 1995
15,460 ± 130	AA-3892	Kolyma River, Russia (8)	Radius, (S-3914-21)	This paper
25,750 ± 450	AA-3214	Gold Hill, Alaska (9)	Skull, UA/FAM 30407	This paper
26,070 ± 280	AA-3894	Kolyma River, Russia (10)	(?) (S-3914-64)	This paper
27,740 ± 330	AA-3891	Kolyma River, Russia (11)	Skull, (S-3915-232)	This paper
28,930 ± 560	AA-3073	Braux Creek, Alaska (12)	Skull, (private collection)	This paper
31,390 ± 780	AA-3076	Lost Chicken Creek, Alaska (13)	Skull, USNM 374407	Porter 1988
34,400 ± 750	AA-3893	Kolyma River, Russia (14)	(?) (S-3914-36)	This paper
36,650 ± 850	AA-3216	Cripple Creek, Alaska (15)	Skull, UA/FAM 30408	This paper
37,000 ± 910	GSC-3050	Usuktuk River, Alaska (16)	Cranial fragment (USGS M1422)	Harington 1980

Table 1. List of Radiocarbon Dates on Saiga (Saiga tartarica) Fossils from Beringia.
AA- Arizona Accelerator Laboratory; GSC - Geological Survey of Canada Radiocarbon Laboratory, Ottawa; AU - University of Alaska Museum; FAM - Frick Collection, American Museum of Natural History, New York; NMC - Canadian Museum of Nature, Ottawa; USNM - United States National Museum, Washington; RIDDL - Department of Archaeology, Simon Fraser University, Burnaby; S - field number, Paleontological Institute, Russian Academy of Sciences, Moscow. USGS - United States Geological Survey collections.

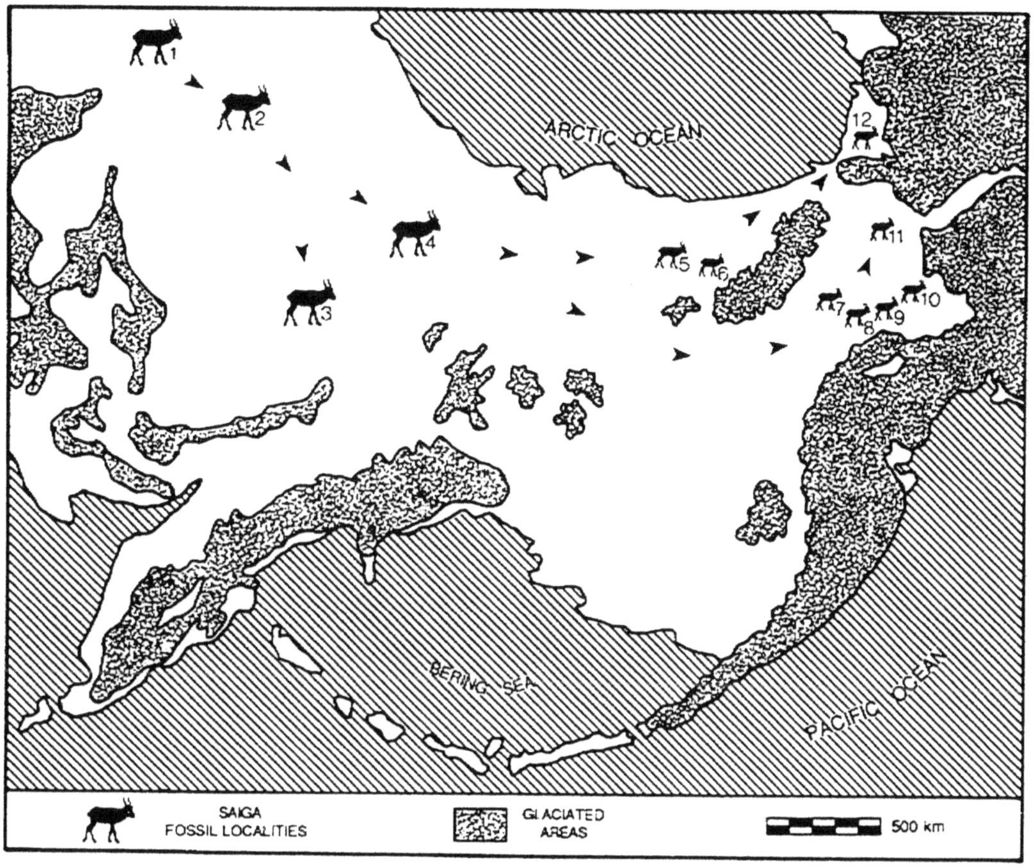

Figure 3. Map showing localities of radiocarbon dated Beringian saiga fossils (see Table 1 for specimen and locality data).

Further, a small spruce (?) stump collected in place at Sixtymile, Yukon yielded a radiocarbon date of about 26,000 yr (Harington 1989). The Boutellier Interval (marine isotope stage 3), during which one cluster of saiga dates fall, was a glacial interstade, not just a local climatic amelioration but part of a global interlude between the two extremes of glacial climate (marine isotope stages 2 and 4).

The other saiga dates cluster between approximately 15,000 and 12,000 yr. This was a time of rapid glacial recession, during which the Beringian climate was irregularly shifting toward the warm/wet Holocene pattern which today supports wet tundra and boreal forest landscapes (Ager 1975; Anderson 1988). Boreal treeline (spruce, *Picea;* larch, *Larix;* aspen, *Populus;* pine, *Pinus;* and alder, *Alnus*) was moving northward, but had not reached the regions where these saiga fossils were collected. Hopkins (1982) calls the time from 14,000 to 10,000 yr, the Birch Period, as birch (*Betula*) pollen is significant in the assemblage, probably representing dwarf birch shrubs (Brubaker *et al.,* 1983).

Habitat Windows for Saiga in the North?

If we assume that the two clusters of saiga dates represent real population density differences over time (and that may be an incorrect assumption, as we will discuss later), then we must try to address both sides of the problem: approximating conditions necessary for the presence of saigas during these periods, and an accounting for their absence in between and afterwards. The ecology of living saigas is vital to these explanations, but one always has to be cautious when using analogues (Sher 1990). However, because many of the features which characterize saigas are so fundamental to their body form, and involve evolutionarily conservative traits, we can expect this suite of characteristics was shared by late Pleistocene saigas. Late Pleistocene fossils give no indication of remarkable differences from modern saiga. We propose therefore, that the ecological limits of modern saigas do give us relevant information regarding constraints on Pleistocene saigas (Harington 1981).

The primary natural limit to saiga expansion in historic times is the extent of the arid steppe. This includes a firm substrate and limited winter snow cover, both of which allow saigas to use their efficient pace. This gait is an essential part of the species' long-range migratory behavior (which maximizes range use) and its speed in escaping predators (Heptner *et al.*, 1988). We propose that the presence of saiga fossils indicates such steppe-like habitat was present or nearby. Thus, hard dry substrates and steppe-like habitats likely existed in Beringia during the periods in which the saiga dates cluster.

Conversely, the steppe adaptations outlined above also clarify why saigas evidently were not present in the far north after 12,000 yr. Saigas simply cannot now, and probably could not then, withstand interglacial moisture extremes when tundra-taiga conditions predominated.

In summer, cushion plants and tussocks created a rough rubbery surface which precludes the use of saigas' special locomotion adaptations. Cold, water-logged muskeg and tundra soils do not support nutritious forbs like those which grow on dry steppe soils (Guthrie 1990). The eastward retreat of the steppes from Europe, at this same time, also resulted in an elimination of European saigas (Currant 1987; Delpech 1983).

Saigas' small size and specialized anatomy means that their maximum snow tolerance is close to 25 cm (Heptner *et al.*, 1988). Snow more than 20 cm deep makes access to food and locomotion extremely difficult for these small ungulates. Heavy snow years result in widescale die-offs among modern saiga herds in central Asia (Bannikov *et al.*, 1967; Heptner *et al.*, 1988). Deep or crusted snow, now typical of much of the north, would make food access and efficient or rapid locomotion impossible for saigas. The larger body size of Pleistocene saigas would both increase their cold and maximum snow-depth tolerance, but not dramatically. The present ubiquity of deep or packed snow makes it easy to understand why saigas could not live in Beringia today (Sher, 1968; Harington, 1978; Guthrie, 1982). The real puzzle posed by these new saiga dates is why saigas were apparently scarce or absent during the peak of the last glaciation.

Summer aridity and winter wind are two other significant ecological forces that affect saiga distribution today, and would have in the Pleistocene. Even though they are adapted to arid steppes, modern saigas are water-limited; local summer drought induces migration, and wide-scale drought causes mass die-offs (Heptner *et al.*, 1988). Beringia was very dry (Hopkins 1982), and perhaps aridity was a direct factor in decreased saiga presence during peak glacial times. However, circumstantial evidence makes this doubtful. The abundant highlands and mountains discharged networks of streams across the lowlands which constituted saigas' main habitat. Also, meltwater from snowdrifts probably continued to produce small amounts of water in the foothills throughout the summer.

Beringia during Duvanny Yar times was quite arid, but it could hardly have been as arid as saiga habitat today. Lower summer temperatures and lower sun angles in the far north would moderate extremes of evapotranspiration-induced aridity experienced in lower latitude deserts. Duvanny Yar fossils of horses and bison are common in Beringia (Matthews 1982, Figure 2), and our knowledge of their analogous physiological potentials (travel efficiency to and from water, thermoregulation, fecal water loss, etc.) makes it difficult to argue that Beringian bison and horses were somehow able to live in radically lower moisture conditions than saigas.

In some cases aridity affects ungulates more indirectly by reducing appropriate forage. What about the effects of low summer moisture on vegetation? Despite feeding very selectively on particular plant parts and growth stages, saigas are unusually eclectic in their dietary breadth (Heptner *et al.*, 1988). Glacial climates reduced the more toxic browse species, like spruce (*Picea*), Labrador tea (*Ledum*), and favored high-nutrient, energy-rich, herbs (Klein 1970; Guthrie 1984b), the very sort of plants to which saigas were adapted. And there is ample evidence that other grazers were in Beringia during the Duvanny Yar Interval. Perhaps something other than food resources restricted saigas? Not that dietary resources sufficient to one grazing species necessarily implies food appropriate for another, but in this particular context, the presence of large grazers, like horses and bison, does suggest that adequate range existed for saigas. Saigas are adapted to operating on rangeland which can be marginal for other large grazers, and are known to tolerate species of plants which other large bovids cannot detoxify (Heptner *et al.*, 1988).

Recolonization of Europe by saigas during the last glaciation may provide clues about conditions that limit their distribution in Beringia. Saigas became abundant in archaeological sites in western Europe during the peak of the last glacial phase (isotope stage 2), and can be specifically dated at about 18,300 yr in southwestern France (Delpech 1983). This is a time when aridity and cold were most extreme in Europe, and steppe conditions were at their maximum (Woillard, 1978). Moderate wind, aridity, and maybe even cold, seem to result in habitats which are optimum for saigas. But the arctic extremes of these forces in Beringia may have eventually restricted saigas.

Winters are quite cold in the subarctic and arctic today, but evidence indicates that Duvanny Yar Beringian winter temperatures were even colder than today's, ultimately due to less net insolation (Berger 1978). For example, there were more extensive permafrost conditions (Péwé 1975). And these colder temperatures were combined with much stronger winds (Begét 1990). While winter winds in central Asia today benefit saigas by redistributing snow and exposing forage, there also must have been a point of detrimental wind intensity. Even at today's southern latitudes, saigas are very sensitive to wind accompanying cold weather (Bannikov *et al.*, 1967; Heptner *et al.*, 1988). Wind is a major factor precipitating long migrations to more congenial pastures (Bannikov *et al.*, 1967). Also, there are many instances of mass die-offs due to large-scale winter storms (Heptner *et al.*, 1988).

Unlike the southern parts of the Mammoth Steppe, stronger winds and lower temperatures in Beringia would have resulted in long episodes of temperatures near the -56.6°C (-70°F) range in exposed areas. Saigas tend to stick to open terrain because they are more vulnerable to predators in hilly or broken ground. Even if they did seek shelter away from the plains, persistent wind, lasting days, would have ultimately required a return to the open country to find food. Perhaps temperatures in these open areas were simply too often below saigas' metabolic thermoneutral zone. Extreme wind-chill on the small saiga bodies could produce rapid metabolic imbalance, especially in lean-weight winter condition.

Though all of the large mammal species of the Mammoth Steppe existed contemporaneously and overlapped greatly in their distribution, they certainly did not share exact distributions in time nor in space. That is, while each was part of the same fauna, they reacted individually to specific environmental forces (FAUNMAP Working Group, 1996). Some clusters of species show more distributional correspondence than others (Matthews 1982). The characteristic members of the mammoth fauna: woolly mammoth, steppe bison, horse, and reindeer/caribou, seem to occur as a cohesive fauna. Others, such as moose (*Alces*, including *A. alces* and *A. latifrons*) and wapiti or red deer (*Cervus*) are more variable in time and space. As with saiga, no Beringian moose (*Alces* alces) fossils have been dated to the Duvanny Yar Interval.

Moose are a browsing specialist and apparently this was a time with little browse (Guthrie 1984b, 1990). The story with sheep is the opposite. But habitat is not a simple matter of food. As grazers, mountain sheep (*Ovis*), rely on broken ground for escape terrain, (opposite that of saigas) and today these sheep are severely restricted in their distributions to grassy alpine habitat. Duvanny Yar conditions which spread grasses and forbs at lower altitudes greatly expanded mountain sheep habitat, judging from the presence of their fossils far from their occurrence in recent times. Woolly rhinoceros (*Coelodonta*) distributions, overall, correspond closely to those of the mammoth fauna in Eurasia, but unlike most other members of the mammoth fauna, woolly rhinos have yet to be found in Alaska and Yukon, for reasons we simply do not understand. Faunas are an amalgam of species which more-or-less hang together but seldom conform exactly, and these unconformities are a normal part of widespread faunas.

The issue of the existence of steppe vegetation and steppe faunas in Alaska and Yukon has a long history of debate (Cwynar 1982; Ritchie and Cwynar 1982; Ritchie 1984; Colinvaux and West 1984; Guthrie 1985, 1990; Sher 1990) particularly with regard to the use of mammals as paleoecological indicators. Our arguments propose an arid steppe-like grassland which supported a diverse community of large mammals. This proposal relies on fossils of the mammoth fauna which date to glacial times, marine isotope stage 2. Saigas were one of the key species we used to develop the Mammoth Steppe concept: positing a cold and arid Pleistocene landscape, occupied by a diverse grazing fauna, even during the climatic extremes. The rarity of saiga fossils during isotope stage 2, means our image may need modification.

Fine-tuning of this broad model of the Mammoth Steppe is a process we are just beginning. Vereshchagin and Baryshnikov's (1982) dates for a variety of species in the Siberian mammoth fauna on careful examination reveal a 2,000-yr interval, from 16,000 to 18,000 yr into which no dates seem to fall. New Siberian mammoth dates also show a bimodality, harmonically consistent with, but not so extreme as the saiga dates (Sher and Sulerzhitsky, in press). There are, however, many dates on mammoths and other key members of the mammoth fauna during the Duvanny Yar Interval from 15,000 to 25,000 yr. These dates clearly indicate the presence of the mammoth fauna during Duvanny Yar, but this bimodality around the peak glacial is new. At this point the reasons for the bimodality are unclear. It may indicate a reduced population density during the full glacial or it may be due to reduced deposition because of restricted moisture, and thus may be an artifact of preservation (Sher and Sulerzhitsky, in press). Significantly, the plot of plant radiocarbon samples taken to date strata, shows similar biomodality as is seen in the mammoth and saiga dates (Sher and Sulerzhitsky, in press). The period of climatic changes occurring during global glacial recession seems to have resulted in active silt retransport and redeposition perhaps providing optimum preservation of fossils and producing the peak of Beringian mammal dates in the 12,000 to 14,000 yr range. A similar episode of active silt reworking (Carter 1981; Nelson 1982) seems to have occurred during parts of the interstade, marine isotope stage 3, the other peak of Beringian mammalian radiocarbon dates. At this point it is not easy to tease apart the components of taphonomic preservation and relative population density in order to evaluate their contributions to these apparent chronological patterns, but both may be involved.

These radiocarbon dates alter our picture of paleoecological changes in Beringia through the late Pleistocene. Understanding the paleoecology of large mammals is important: top-of-the-pyramid mammals are key proxy monitors of entire ecosystem changes. The pattern emerging from dated mammalian fossils shows the potential for a more complete picture of changing faunal complexity in Beringia, and perhaps a clearer image of Quaternary climates and landscapes in the far north.

REFERENCES CITED

Ager, T.A.
1975 *Late Quaternary Environmental History of the Tanana Valley, Alaska*. Ohio State University Institute of Polar Studies, Report.

Anderson, P.M.
1988 Late Quaternary Pollen Records from the Kobuk and Noatak River Drainages, Northwestern Alaska. *Quaternary Research* 29:263-276.

Bannikov, A.G.
1958 Distribution Géographique Actuelle et Biologique de la Saïga en Europe. *Mammalia* 22: 208-225.

Bannikov, A.G., Zhirnov, L.V., Lebedeva, L.S., and Fandeev, A.A.
1967 "Biology of the Saiga." (Translation from Russian) Israel Program for Scientific Translation. IPST Catalogue Number 1833.

Baryshnikov, G.F. and Tikhonov, A.N.
1994 Notes on Skulls of Pleistocene Saiga of Northern Eurasia. *Historical Biology* 8:209-234.

Béget, J.E.
1990 Middle Wisconsin Climate Fluctuations Recorded in Central Alaskan Loess. *Géographie physique et Quaternaire* 44:3-13.

Berger, A.L.
1978 Long-term Variations of Caloric Insolation Resulting from the Earth's Orbital Elements. *Quaternary Research* 9:139-167.

Brubaker, L.B., Garfinkel, H.L., and Edwards, M.E.
1983 Late-Wisconsin and Holocene Vegetation History of the Upper Koyukuk Region, Alaska: Implications for Alaskan Paleoecology. *Quaternary Research* 20:194-214.

Carter, L.D.
1981 Middle Wisconsinan through Holocene Climate in the Ikpikpuk River Region, Alaska. Pp. 5-9. Abstracts. 10th Arctic Workshop, Boulder.

Cinq-Mars, J., and Nelson, D.E.
Unpublished data. AMS Bone Dates from the Bluefish Caves, Northern Yukon, Canada.

Colinvaux, P.A.
1964 The Environment of the Bering Land Bridge. *Ecological Monographs* 34:297-329.

Colinvaux, P.A., and West, F.H.
1984 The Beringian Ecosystem. *Quarterly Review of Archaeology* 5:10-16.

Currant, A.P.
1987 Late Pleistocene Saiga Antelope *Saiga tatarica* on Mendip. *Proceedings, University of Bristol Spelaeological Society* 18:74-80.

Cwynar, L.C.
1982 A Late-Quaternary Vegetation History from Hanging Lake, Northern Yukon. *Ecological Monographs* 52:1-24.

Delpech, F.
1983 Les Faunes du Paleolithique Supérieur dans le Sud-Ouest de la France. *Cahiers du Quaternaire* 6:1-453.

FAUNMAP Working Group
1996 Spatial Response of Mammals to Late Quaternary Environmental Fluctuations. *Science* 272:1601-1606.

Guthrie, R.D.
1968 Paleoecology of the Large Mammal Community in Interior Alaska during the Late Pleistocene. *American Midland Naturalist* 79:346-363.

1982 Mammals of the Mammoth Steppe as Paleoecological Indicators. In *Paleoecology of Beringia* edited by D.M. Hopkins, J.V. Matthews, Jr., C.E. Schweger, and S.B. Young, pp. 307-326. Academic Press, Inc., New York.

1984a Mosaics, Allelochemics, and Nutrients: An Ecological Theory of Late Pleistocene Extinctions. In *Quaternary Extinctions* edited by P.S. Martin, and R.G. Klein, pp. 259-298. University of Arizona Press, Tuscon.

1984b Alaskan Megabucks, Megabulls, and Megarams: The Issue of Pleistocene Gigantism. In *Contributions in Quaternary Vertebrate Paleontology* edited by H.H. Genoways, and M.R. Dawson, pp. 242-510. Carnegie Museum of Natural History Special Publication No. 8.

1985. Woolly Arguments Against the Mammoth Steppe - A New Look at the Palynological Data. *Quarterly Review of Archaeology* 6:9-14.

1990 *Frozen Fauna of the Mammoth Steppe: The Story of Blue Babe.* University of Chicago Press, Chicago.

Harington, C.R.
1978 Quaternary Vertebrate Faunas of Canada and Alaska and their Suggested Chronological Sequence. *Syllogeus* 15:1-105. Ottawa.

1980 Radiocarbon Dates on Some Quaternary Mammals and Artifacts from northern North America. *Arctic* 33:815-832.

1981 Pleistocene Saiga Antelopes in North America and their Paleoenvironmental Implications. In *Quaternary Paleoclimates.* edited by W.C. Mahaney, pp. 193-225. Geoabstracts, Norwich.

1989 Pleistocene Vertebrate Localities in the Yukon. In *Late Cenozoic History of the Interior Basins of Alaska and the Yukon.* edited by L.D. Carter, T.D. Hamilton, and J.P. Galloway, pp. 93-98. U.S. Geological Survey Circular 1026.

Harington, C.R., and Cinq-Mars, J.
1995 Radiocarbon Dates on Saiga Antelope (*Saiga tatarica*) Fossils from Yukon and Northwest Territories. *Arctic* 48:1-7.

Heptner, V.G., Nasimovich, A.A., and Bannikov, A.G.
1988 *Mammals of the Soviet Union.* (Translation from Russian) Amerind Publishing Co. Pvt. Ltd., New Delhi.

Hopkins, D.M.
1982. Aspects of the Paleogeography of Beringia during the Late Pleistocene. In *Paleoecology of Beringia.* edited by D.M. Hopkins, J.V. Matthews, Jr., C.E. Schweger, and S.B. Young, pp. 3-28. Academic Press, New York.

Kahlke, H.D.
1975 Der Saiga-Fund von Bottrop/Westfalen. *Quatärpaleontologie* 26:135-146.

1991 Pleistocene Distributional and Evolutionary History of the genus *Saiga* Gray, 1843 (Mammalia, Artiodactyla, Bovidae) in the Palaearctic. *Vertebrata PalAsiatica* 29:314-322.

Klein, D.R.
1965 Ecology of Deer Range in Alaska. *Ecological Monographs* 35:259-284.

1970 Tundra Ranges North of the Boreal Forest. *Journal of Range Management* 23:8-14.

Livingston, D.A.
1955 Some Pollen Profiles from Arctic Alaska. *Ecology* 36:587-600.

Matthews, J.V., Jr.
1982 East Beringia During late Wisconsin Time: A Review of the Biotic Evidence. In *Paleoecology of Beringia* edited by D.M. Hopkins, J.V. Matthews, Jr., C.E. Schweger, and S.B. Young, pp. 127-150. Academic Press, Inc., New York.

Nelson, R.E.
1982 *Late Quaternary Environments of the Western Arctic Slope, Alaska*. Ph.D. dissertation, University of Washington, Seattle. 149 pp.

Péwé, T.L.
1975 *Quaternary Geology of Alaska*. U.S. Geological Survey Professional Paper 835.

Porter, L.
1988 Late Pleistocene Fauna of Lost Chicken Creek, Alaska. *Arctic* 41:303-313.

Ritchie, J.C.
1984 *Past and Present Vegetation of the far Northwest of Canada*. University of Toronto Press, Toronto.

Ritchie, J.C., and Cwynar, L.C.
1982 The Late Quaternary Vegetation of the north Yukon. In Paleoecology of Beringia edited by D.M. Hopkins, J.V. Matthews, Jr., C.E. Schweger, and S.B. Young, pp. 113-126. Academic Press, Inc., New York.

Ruddiman, W.F., and Kutzbach, J.E.
1989 Forcing of late Cenozoic Northern Hemisphere Climate by Plateau Uplift in Southern Asia and the American West. *Journal of Geophysical Research* 94:18.409-18.427.

Sher, A.V.
1968 Fossil Saiga in Northeastern Siberia and Alaska. *International Geology Review* 10:1247-1260.

1986 On the History of the Mammal Fauna of Beringia. *Quärtarpaleontologie* 8:185-193.

1990 Actualism and Disconformism in Studies of Ecology of Pleistocene Mammals. *Journal of General Biology* 51:163-177. (In Russian).

Sher, A.V., and Sulerzhitzky, L.D.
In press Radiocarbon dating of fossil mammals in Siberian Arctic: Implications on paleoecology of mammoth fauna. *Boreas*.

Vereshchagin, N.K., and Baryshnikov, G.F.
1982 Paleoecology of the Mammoth Fauna in the Eurasian Arctic. In *Paleoecology of Beringia* D.M. Hopkins, J.V. Matthews, Jr., C.E. Schweger, and S.B. Young, pp. 267-278. Academic Press, Inc., New York.

Woillard, G.
1978 Grande Pile Peat Bog: A Continuous Pollen Record for the Last 140,000 Years. *Quaternary Research* 9:1-21.

SEA OTTER *(Enhydra lutris)* SCARCITY IN THE STRAIT OF GEORGIA, BRITISH COLUMBIA

Diane K. Hanson
Alaska Office of History and Archaeology, Anchorage

Karla D. Kusmer
Sparks, Nevada

Archaeological data are used to establish the distribution of sea otters before historic contact to determine if sea otters inhabited the region prehistorically, or if they had been present but over-hunted by the Coast Salish participating in the fur trade on the outer coast of British Columbia. Biological studies are used in an attempt to identify causes for the biogeographical limits to sea otter distribution, and to understand the relation of these to prehistoric environments.

The first European explorers to venture to the Northwest Coast reported that sea otters *(Enhydra lutris)* were common along the rest of the coast, but they were rare or absent in the Strait of Georgia (Cook 1973). During his 1792 voyage with Captain Vancouver, Menzies noted that few sea otter skins were found among the inhabitants of the Strait of Georgia and the Puget Sound:

We saw but few Sea Otter Skins amongst them which shews that these Animals do not much frequent the interior Channels & perhaps only straggling ones at particular Seasons, for the Fur of the few pieces we saw was of a very inferior quality to those found along the exterior edge of the Coast (Menzies in: Newcombe 1923:82-83).

Ethnographic and other historic accounts provide similar reports. Barnett (1955:92) stated that sea otters were extremely rare and were not sought by the Coast Salish. Mayne (1969) was probably referring to marine dwelling river otters (*Lutra canadensis*), a common animal in the Strait of Georgia, when he said that he observed sea otters being eaten in the late 1850's.

Historic accounts about fauna in the Strait of Georgia post-date the 1790's, by which time sea otters were rare or near extinction along much of the west coast of Vancouver Island and presumably the outer coast of Washington (Cook 1973; Mozino 1970). Sea otter populations in southern British Columbia and northern Washington suffered from intense exploitation by participants in the fur industry relatively early in the historic period. By 1792, sea otters were rare on the west coast of Vancouver Island and the sea otter fur trade on the Northwest Coast had collapsed by 1825, although some hunting continued (Cook 1973:283; Lensink 1960; Mozino 1970:48; Riedman and Estes 1988). Sea otters were nearly extinct from Prince William Sound to central California by 1911. A small colony survived in the Queen Charlotte Islands, but even these animals disappeared before 1919 (Banfield 1974; Lensink 1960; Riedman and Estes 1988). Modern populations of sea otters on the outer coast of British Columbia and Washington are from animals transplanted from Amchitka and Prince William Sound during the late 1960's and early 1970's (Bigg and MacAskie 1978; Jameson *et al.* 1982).

SEA OTTER DISTRIBUTION ON THE CENTRAL NORTHWEST COAST

The primary sources for information about past distributions of animals are paleontological and archaeological records. Archaeological sites on the Northwest Coast are particularly useful because humans tend to concentrate animals from a wide range of local environments, and the shell-rich sediments are conducive to bone preservation. Archaeological sites are also normally well excavated with relatively tight chronological controls.

Faunal data from archaeological sites in the Strait of Georgia and from the outer coast of Washington and British Columbia are used for this study. The cultural context of sea otter bones was recorded to determine the geographic distributions of sea otters through time. The prehistoric cultures in the Strait of Georgia are:

Developed Coast Salish horizon	1500 BP - Contact
(San Juan Phase, Gulf of Georgia Culture Type, Stselax Phase)	
Marpole Phase	2500 - 1500 BP
Locarno Beach Phase	3500 - 2500 BP
Charles Culture	4500 - 3500 BP
(Mayne or St. Mungo Phase)	

Sea otter remains from Developed Coast Salish horizon assemblages are found in archaeological sites along the outer coast (Figure 1). A species list from Fossil Bay on Sucia Island (45 WH 105) includes the tentative identification of a sea otter bone from the Developed Coast Salish component (Kidd 1969), but as the identification has not been verified it should be reexamined. Kusmer identified a sea otter femur at Crescent Beach. Unfortunately, the bone was found in the back dirt of a backhoe digging sewer lines through the town, so there is no contextual information. Because historic materials were mixed with the prehistoric, we do not know with which cultural phase this bone was associated. These sites appear on the distribution map as open circles (Figure 1). With those two exceptions, there are no other sea otters identified from the 27 Developed Coast Salish sites with faunal assemblages in the Strait of Georgia. The distribution of sea otter remains from Marpole Phase sites (2500-1500 BP) was compared to other coastal sites to determine if this pattern characterized earlier periods, Again, sea otters are absent from archaeological assemblages in the Strait of Georgia but are present in assemblages from the outer coast (Figure 2). The pattern also holds for the Locarno Beach Phase (Figure 3). There are sea otter elements at the Beach Grove Site (DgRs 1) in the older St. Mungo components (3800-3900BP) (Kusmer 1996:3), but there is little to compare this assemblage to in the Strait of Georgia to determine if sea otters were more abundant during this early prehistoric period.

Factors Potentially Limiting Sea Otter Distribution

An obvious explanation for the lack of sea otters in Strait of Georgia archaeological sites is that they were not hunted. This is unlikely since all other marine and medium to large sized terrestrial mammals are represented in the archaeological sites. This includes their similarly adapted relative, the river otter, which was actively trapped historically. Smaller mustelids including mink (*Mustela vison*), ermine (*M. ermina*), long-tailed weasel (*M. frenata*), fishers or marten (*Martes* sp.), and skunks (*Spilogale gracilis* or *Mephitis mephitis*) also appear in Strait of Georgia sites (Hanson 1991). These animals were trapped in dead falls (Suttles 1974: 96). Sea otters would most likely have been hunted in the water as they were elsewhere along the eastern Pacific coast. The Coast Salish were familiar with hunting seals and sea lions with harpoons, clubs, and nets. Delphinids (porpoises, killer whales) were hunted using the same harpoons used for seals (Suttles 1974). Therefore, neither animal size nor technology should have prevented sea otters from being hunted. The ethnographic literature contains no references to avoidance of sea otters for spiritual reasons; even so, extending religious beliefs back 3500 years is tenuous. Sea otters do appear in archaeological deposits in the outer Juan de Fuca Strait and along the outer coast where they were observed historically. Sea otters evolved in the North Pacific during the early Pleistocene (Riedman and Estes 1988). Their remains appear in early archaeological sites in California (ca. 7000 - 5000 BP; Erlandson 1994), Oregon (ca. 3000 BP, Lyman 1991; Bennett and Lyman 1991), Washington and British Columbia (Bear Cove ca. 8000 BP, Carlson 1979; Namu ca. 6300 BP, Cannon 1996). Concerns that sea otters may have been recent inhabitants in the region and did not have time to colonize the Strait of Georgia are unwarranted.

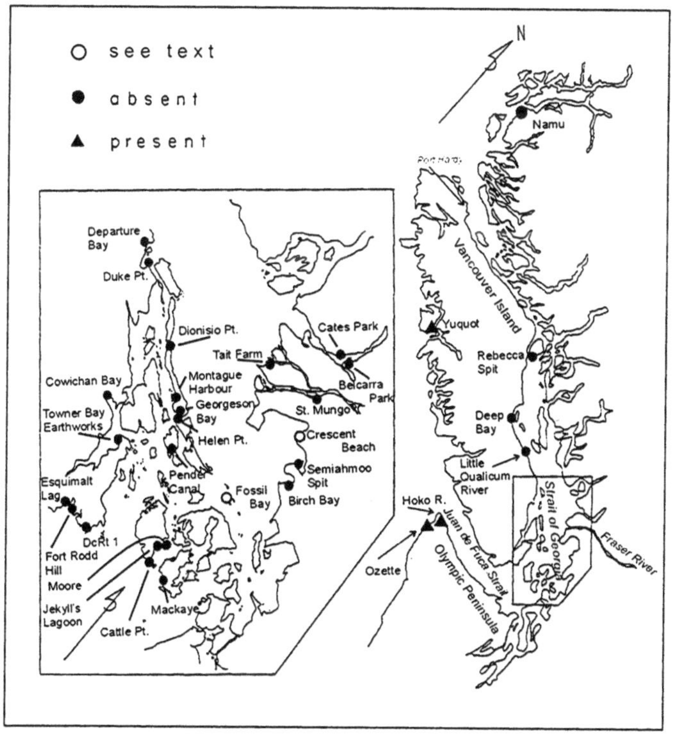

Figure 1. Distribution of sea otter remains from Developed Coast Salish sites.

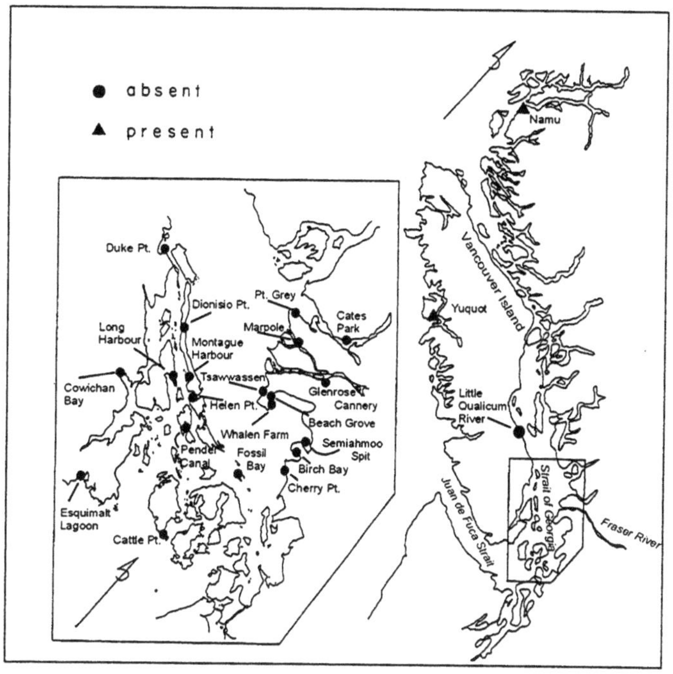

Figure 2. Distribution of sea otter remains from Marpole Phase sites.

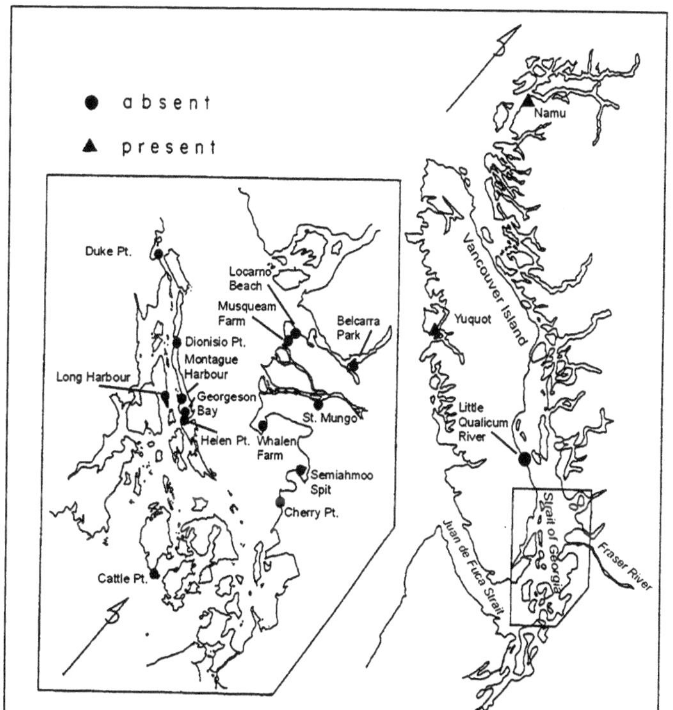

Figure 3. Distribution of Sea Otter Remains from Locarno Beach Phase sites.

Sea otter populations were decimated during the fur trade in the 1700 and 1800's, leading to the extinction of the animals from small remnant populations in Queen Charlotte Islands, Central Prince William Sound to Baja California, with the exception of California and Islas San Benitos in Mexico. Since the 1960's there have been successful reintroductions of sea otters along the British Columbia and Washington coasts. These sea otters have not colonized the Strait of Georgia area.

Sea otters transplanted from 1969 through 1972 in the northwestern part of Vancouver Island and have spread along the outer coast of Vancouver Island and central British Columbia to the Queen Charlotte Islands (Bigg and MacAskie 1978; Watson 1997: pers. comm.). They have not moved into the protected waters of Johnstone Strait north of the Strait of Georgia (Bigg and MacAskie 1978; Watson 1997: pers. comm.). Otters released on the outer coast of Washington in 1969 have moved just within the entrance of the Juan de Fuca Strait to southern Vancouver Island (Osborne *et al.* 1988; Watson 1997: pers. comm.). The occasional animal has been observed in Neah Bay, Washington (Osborne *et al.* 1988), and solitary otters have been observed near Saltspring Island and San Juan Island in the Strait of Georgia. There has also been an unconfirmed sighting at Dungeness Spit on the south side of the Juan de Fuca Strait (Watson 1997:pers. comm.). No otter colonies have yet been observed in these areas.

The absence of sea otters in the Strait of Georgia is puzzling because river otters are common in marine waters of the Strait of Georgia as are other sea mammals. One of the principal differences between the marine and sea otters is their diet. Marine dwelling river otters are primarily fish eaters. Fish were identified in 99.4% of the scats and in 86.9% of the stomachs of river otters living along the east coast of Vancouver Island in the Strait of Georgia (Stenson *et al.* 1984). In the northern Strait of Georgia region, crustaceans comprise only 7.2% and birds 4.2% of the diet, based on scat analysis (Stenson *et al.* 1984).

In contrast, sea otters subsist primarily on shellfish, although they will incorporate a large proportion of fish into their diet when shellfish are scarce (Estes *et al.* 1978:831; 1982). On rocky coasts, sea otter diets focus on mussels *(Mytilus* sp.), urchins *(Strongylocentrotus* sp.) and abalones *(Haliotis* sp.), but along coasts with soft substrates they prefer butter clams *(Saxidomus giganteus)*. Kvitek and Oliver (1988:23) report that clams make up 68-90% of the prey captured by volume. Butter clams dominate the diet of sea otters living around Kodiak Island except where the butter clams had been over-exploited, and littleneck clams *(Protothaca staminea)* were more commonly eaten (Kvitek *et al.* 1992). Soft shelled clams *(Macoma* sp., *Mya* sp.) and cockles *(Clinocardium* sp.) were eaten more frequently only where sea otters were forced to exploit deeper waters or use alternative foods (Kvitek *et al.* 1992:422). A similar pattern was observed on the west coast of Vancouver Island where otters were seen eating butter clams most often, but other foods include urchins, sea stars *(Pisaster* sp.), crabs, California mussels *(Mytilus californianus)*, gumboot chitons *(Cryptochiton stelleri)*, abalone *(Haliotis kamtschatkana)* gastropods and octopus (Morris *et al.* 1981:294). Sea otters have different requirements than river otters, so the presence of river otters should not be used as an indication of habitat suitability for the larger shell fish-eating mustelid.

Lower benthic biomass has been cited by Irons *et al.* (1988:21) to account for lower sea otter density in the bays and fjords of Prince William Sound. They assume that this is related to the lower benthic biomass as compared to the biomass of the outer coastal waters. The "primary productivity" is the production of unicellular marine algae (Hobson 1980), and therefore the production of food for larger organisms. The Strait of Georgia has a primary productivity over twice as high as the open ocean (Hobson 1980; Harrison *et al.* 1983). Therefore, the absence of sea otters should not be associated with a lower benthic biomass than the outer coasts of Washington and British Columbia.

Niether is scarcity of butter clams or other suitable clams or prey species in the Strait of Georgia a likely cause for sea otter absence. Butter clams and blue mussels *(Mytilus edulis)* are abundant, and occur in large numbers in prehistoric sites throughout the Strait of Georgia as do other bivalves, chitons, urchins, and gastropods. Commercial butter clam canneries were located at Sidney on southern Vancouver Island, Nanaimo east Vancouver Island, and at Friday Harbour on San Juan Island in the early part of the twentieth century (Quayle and Bourne 1972).

Physically limiting forces for northern sea otter populations, such as sea ice in the Bering Sea, (e.g. Schneider and Faro 1975) are not likely to have affected sea otter distributions in the temperate waters of the Strait of Georgia. While sea ice will form along bays and protected waters, habitats most favorable to the otters are not likely to become ice bound, and if they did become frozen over it would be a temporary condition. Open water at the mouths of the larger rivers and the center of the strait would also provide refuges for the otters. Ice is, therefore, an unlikely deterrent to otters in the Strait of Georgia.

Kelp beds are not necessary for sea otter protection or survival and otter populations are often found well away from kelp beds in open waters (e.g. Lensink 1960). Nevertheless, kelp beds are present in the Strait of Georgia. Shallow waters, however, are important to sea otters since they must dive to the sea floor to forage for their preferred foods. Thus water depth affects otter distributions and abundance (Lensink 1960; Irons *et al.* 1988). Sea otters prefer waters fewer than 50 fathoms (300 feet) deep inthe Aleutian Islands (Lensink 1960:181), and in Prince William Sound most bivalve foraging occurs at 10 fathoms (60 feet) or less (Irons *et al.* 1988:3). Sea otter density in Prince William Sound was greatest around islands and the mouths of fiords and bays and shallow exposed areas, while the fewest animals were observed in deep or exposed waters (Irons *et al.* (1988:12). Much of the Strait of Georgia is less than 50 fathoms deep (Figure 4), and the numerous islands appear to provide ideal habitat for foraging otters, yet the otters appear to have been scarce historically and prehistorically.

One explanation for the absence of sea otters in the Strait of Georgia may be their ability to taste high doses of paralytic shellfish poison (PSP) toxins in their food (Kvitek *et al.* 1991). PSP toxins are produced by dinoflagellates associated with "red tide" blooms and consist of Saxitoxin and other toxins. The butter clam is unique in its ability to store PSP for more than two years after exposure to the red tide (Quayle and Bourne 1972). If red tide blooms are frequent, butter clams can potentially be chronically toxic. Captive animals appear to be able to taste or detect PSP in their prey and will avoid or discard clams with toxic levels of Saxitoxin (Kvitek *et al.* 1991). The otters also discarded those portions containing the highest levels of the toxins when no other foods were available. Kvitek *et al.* (1991) concluded that sea otters did not occur in southeastern Alaska where butter clams contained greater than 149 µg of Saxitoxin/100g, suggesting that this is a variable restricting sea otters to the outer coast. Historic records also indicated that they were virtually absent from the inside passage, and transplanted populations have not migrated from the outer coast to the inner waters of southeast Alaska. Sea otters have been observed, however, in areas with exceptionally high levels of PSP in the fjords of southeastern Alaska (Bodkin 1996: pers. comm.) and feeding in areas with high PSP levels on the central coast of British Columbia (Watson 1997: pers. comm.). These events are taking place in areas that are newly colonized by transplanted populations, so it still needs to be determined if these animals will remain in these regions. Nor do we know if the otters are exploiting rocky shore invertebrates or fishes rather than the butter clams, which would sequester high levels of Saxitoxins.

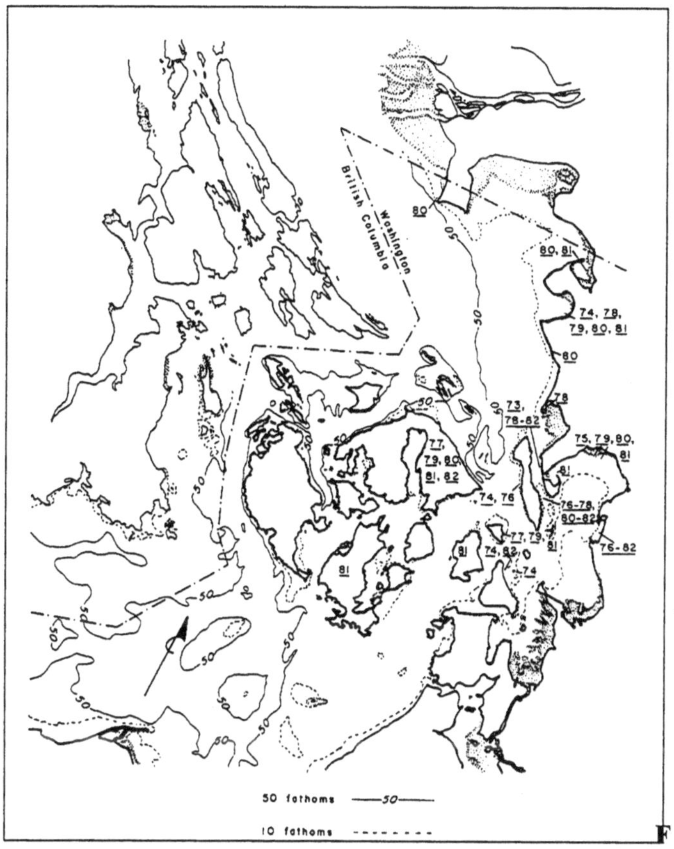

Figure 4. Locations with high PSP levels in San Juan, Whatcom and Skagit Counties, Washington. These locations are indentified by the years in which the PSP levels were at or greater than 49µg/100g shellfish meat.

FREQUENCY OF PSP IN THE SOUTHERN STRAIT OF GEORGIA

The PSP monitoring program in Washington began after high PSP levels were recorded in British Columbia in 1957. Beaches are closed to shellfish harvesting when PSP levels reach 80 µg/100g shellfish meat (Nishitani and Chew 1988:654). The waters of the Juan de Fuca Strait often had PSP levels high enough to justify the closure of associated bays to shellfish harvesting, while this occurred only infrequently on the outer coast (Nishitani and Chew 1988:660). In the early 1970's, beaches in the San Juan Islands and near Belllingham were closed to shellfish harvesting, and Samish Bay was closed because of high toxin levels in 1974 and nearly every year since then (Nishitani and Chew 1988). In contrast, the protected waters of the southern Strait of Georgia are considered one of the safest areas for shellfish harvesting in British Columbian coastal waters (Schallie 1996: pers. comm.). PSP toxins in shellfish meat occur when clams ingest the *Alexandrium* sp. dinoflagellates. These dinoflagellates are found in waters of the outer coast as well as in the Strait of Georgia and Puget Sound (Harrison *et al.* 1983). On the outer coast, toxic levels of PSP are associated with dinoflagellate blooms drifting in from the open ocean, while PSP toxins in the protected waters are the result of local blooms associated with warm surface waters (Nishitani and Chew 1988). Dinoflagellate cycts resting within the substrate sediments may be the source of blooms in the more protected waters. Protected waters such as bays are more likely to have toxic levels of PSP than are turbulent waters. Some *Alexandrium* sp. also prefer areas with low salinity such as estuaries. These dinoflagellates are associated with the eastern side of the Strait of Georgia near the Fraser River outlet (Harrison *et al.* 1983).

Data on PSP levels recorded from 1967 to 1982 in San Juan, Whatcom and Skagit Counties of the state of Washington were provided by Linda Hanson of the Office of Shellfish Programs, Department of Health from the State of Washington. We have plotted the years in which PSP levels in butter clams have been greater than 149 µg/100g of shellfish meat (Figure 4). These are the PSP levels used by Kvitek *et al.* (1991) to illustrate the limits of sea otter distributions in southeast Alaska. All areas were not sampled each year, but it appears from these data that the eastern portion of the strait had higher levels of PSP than the western side and that most of the eastern shoreline had PSP levels above 149 µg/100g between 1978 or 1979 and 1982. The apparent increase in PSP toxins in butter clams in the Strait of Georgia may be a recent phenomenon (Nishitani and Chew 1988:661). Alternatively, it may be the result of increased awareness of the problem by residents in the area and the greater number of biologists working on the problem.

PARALYTIC SHELLSH POISONING AND HUMANS

Humans are considerably more sensitive to the toxins stored by clams than are sea otters. The LD_{50} is the dose at which 50% of a tested population dies. The LD_{50} for Saxitoxin (one of 8 toxins found in the dinoflagellates associated with red tides) for mammals are provided in Table 1. The LD_{50} for sea otters is estimated to be within the range of the other medium sized mammals (Kvitek *et al.* 1991:399). Humans are extremely sensitive to the toxins, but because sea otters normally have a diet of 68 - 90% clams (Kvitek and Oliver 1988:23), they may be particularly vulnerable to the effects of PSP.

Unlike the butter clam, which sequesters the poison in the siphon for as long as two years, other clams that accumulate toxins store the PSP toxins in the gut, and the toxins usually disappear within weeks after exposure to red tide (Kvitek 1990:113). The littleneck clam, for example, converts PSP toxins into less toxic forms which disappear in fewer than five weeks (in: Kvitek 1990:126). Fish are considerably less toxic than clams upon exposure to red tide, and they may avoid foods containing the toxins (White 1981; Kvitek 1991). Therefore, river otters and the other fish-eating sea mammals may be less likely to suffer from poisoning than sea otters.

ANIMAL	SAXITOXIN µg/kg BODY WEIGHT
Guinea pig	135
Rabbit	181
Dog	181
Rat	192
Cat	254
Mouse	382
Monkey	364-727
Human	10-20 (Kao 1966 in Baslow 1969:151)
Sea Otter	est. 130-400 (Kvitek *et al.* 1991:399)

Table 1. Saxitoxin LD_{50} for mammals.

The abundant shell deposits in Strait of Georgia archaeological sites, do not indicate that humans avoided butter clams in the past. Seventeen of eighteen Developed Coast Salish sites with analyzed shellfish assemblages had butter clams (Hanson 1991). The one exception was a fishing camp on the Fraser River Delta (Tait Farm DhRt-36; Ham 1987). Butter clams may still have been used because of the way they were processed. Steaming butter clams for ten minutes reduces paralytic shellfish poison levels by 90%, although cooking the meat longer does not provide any additional benefits (Quayle and Bourne 1972:17). Removing the whole siphon reduces toxin levels by 60% or more while removing just the tip reduces toxicity by 24% (Quayle and Bourne 1972:16). Clams with 500 µg of toxins fall within safe levels when the black siphon tips are removed, and clams with 1000 µg of Saxitoxin fall to levels of about 200 µg by removing the entire siphon (Quayle and Bourne 1972:18).

The Manhousat, living in protected inlets on western Vancouver Island which are also subject to red tide blooms, cut off the black tip of the butter clam siphon, an act which was reportedly "done only for etiquette" (Ellis and Swan 1981:50). From the information presented by Quayle and Bourne (1972), it is apparent that cutting off the black tip may not simply be a case of good manners but also of safety. Butter clams were not usually eaten raw because, according to the Manhousat, "they have a part like a worm that can get caught in your throat and doesn't feel good" (Ellis and Swan 1981:50). This may refer to the initial stages of paralytic shellfish poisoning, which begins with a tingling and numbness in the lips and tongue, followed by similar sensations in the fingertips and toes, and a final loss of control of voluntary movements. There may also be difficulty in breathing" (Quayle and Bourne 1972:15). These symptoms can appear within 30 minutes of eating the clams (Dymsza *et al.* 1980).

Like the Manhousat, the Coast Salish of the Strait of Georgia treated butter clams differently than other clam species. Although red tide blooms are more frequent in the summer, clams were still collected in large numbers during the summer for winter storage (Suttles 1974). Horse clams might be dried by roasting the raw meat and littleneck clams were sometimes simply sun dried, but butter clams were always steamed for approximately 15 minutes before being roasted over a fire and dried (Suttles 1974:65-67). The steaming time is nearly the same as the time recommended by Quayle and Bourne (1972) to reduce the toxins to safe levels. Differences in the way the Coast Salish treated the various clam species suggests that they were aware of a problem with the meat of butter clams, a problem consistent with paralytic shellfish poisoning.

CONCLUSIONS

Faunal data from archaeological sites support the suggestion that sea otters were not present, or were rare prehistorically in the southern Strait of Georgia. In contrast, archaeological sites on the outer coast frequently contained sea otter remains. Historic records also note this disparity in sea otter distributions. While the sea otter trade was a lucrative business on the outer coast, the absence of sea otters in the Strait of Georgia caused disinterest among the European and America/Canadian traders (Cook 1973; Menzies in: Newcombe 1923).

Biologists have several explainations for the biogeographical limits of sea otters. These include waters too deep for foraging, sea ice, an inadequate food supply, and red tide blooms or paralytic shellfish poisoning. High levels of PSP and the tendency for the primary food of sea otters to sequester the poisons long after the red tide blooms are gone may have caused sea otters to avoid the Strait of Georgia (Kvitek *et al.* 1991; Kvitek 1990). Sea otters generally appear to avoid areas with high PSP levels, although there are recent accounts of otters forayint into regions with high PSP. Humans deal with PSP by removing shellfish parts in which the poisons are stored, and by cooking clam meat to further reduce the toxicity. If red tide was a recent phenomenon, the Central Coast Salish would not be expected to have methods for reducing toxicity in the one bivalve that sequesters toxins. Nor would one expect differential treatment of bivalve species. These observations do not provide direct evidence that PSP toxins caused the absence of sea otters in the southern Strait of Georgia, but otter scarcity in the archaeological record may indicate that red tide was an environmental variable with which both sea otters and humans had to contend.

ACKNOWLEDGMENTS

A number of people have provided us with information and discussed this paper with us. Braulio Jimenez of the University of Puerto Rico Medical School, F.J.R. Max Taylor of the Department of Oceanography at the University of British Columbia, and Klaus Schallie were kind enough to discuss the problems of Paralytic Shellfish Poisoning. James Bodkin and Jane Watson generously took the time to answer questions about sea otter distributions. We appreciate their help. Linda Hanson also sent information about PSP levels in northern Washington. Lydia Black, William Prentiss, Jane Watson and one anonymous reviewer's suggestions improved the paper from the original versions. The primary author is happy to take this opportunity to thank Dr. R. Dale Guthrie for starting her out in the career of zooarchaeology.

REFERENCES CITED

Abbott, D.N.
n.d. *Preliminary report on the Beach Grove site. Permit 1961-3*. Report to the Heritage Conservation Branch, Victoria, British Columbia.

Banfield, A.W.F.
1974 *The Mammals of Canada*. University of Toronto Press, Toronto.

Barnett, H.G.
1955 *The Coast Salish of British Columbia*. University of Oregon, Eugene

Bernick, K.
1983 A Site Catchment Analysis of the Little Qualicum River Site, DiSc 1: A Wet Site on the East Coast of Vancouver Island, B.C. *Archaeological Survey of Canada, Mercury Series Paper No. 118*. National Museum of Man, Ottawa.

Bennett, A.C., and R.L. Lyman
1991 Archaeology of Whale Cove (35LNC60). In: *Prehistory of the Oregon Coast: The Effects of Excavation Strategies and Assemblage Size on Archaeological Inquiry*, edited by R.L. Lyman, pp. 241-277. Academic Press, Inc., San Diego

Bigg, M.A., and I.B. MacAskie
1978 Sea Otters Reestablished in British Columbia. *Journal of Mammalogy* 59(4):874-876.

Blacklaws, R.W.
1979 *Excavations at Esquimalt Lagoon: A Contribution to Straits Salish prehistory*. Unpublished M.A. thesis, Department of Archaeology, Simon Fraser University, Burnaby, British Columbia.

Bodkin, J.L.
1996 Biologist, National Biological Service, United States Fish and Wildlife Service, Anchorage.

Boehm, S.G.C.
1973 *Cultural and Non-cultural Variation in the Artifact and Faunal Samples from the St. Mungo Cannery Site, B.C. DgRr 2*. Unpublished M.A. thesis, Department of Anthropology and Sociology, University of Victoria, Victoria.

Boucher, N.D.
1976 *Prehistoric Subsistence at the Helen Point Site*. Unpublished M.A. thesis, Department of Archaeology, Simon Fraser University, Burnaby.

Cannon, A.
1996 The Early Namu Archaeofauna. In *Early Human Occupation in British Columbia*, edited by R.L. Carlson and L. Dalla Bona, pp. 103-110. University of British Columbia Press, Vancouver.

1991 *The Economic Prehistory of Namu*. Department of Archaeology, Simon Fraser University, Burnaby.

Carlson, C.
1979 The Early Component at Bear Cove. *Canadian Journal of Archaeology* 3:177-194.

Carlson, R.L.
1954 *Archaeological Investigations in the San Juan Islands*. Unpublished M.A. thesis, University of Washington, Seattle.

Charlton, A.S.
1977 *The Archaeology of the Belcarra Park Site: A Contribution to Strait of Georgia Prehistory*. Unpublished M.A. Thesis, Department of Archaeology, Simon Fraser University, Burnaby.

n.d. *Archaeological Investigations of the Cates Park Site*. Permit 1974-24 CHA. Report on file Archaeological Sites Advisory Board, Victoria.

Cook, W.L.
1973 Flood Tide of the Empire: Spain and the Pacific Northwest, 1543-1819. Yale University Press, New Haven.

Crockford, S.
1986 In *Excavations at the Whalen Farm Site (Canadian) DgRs 14 Boundary Bay, B.C.*, by D. Hammon. The Delta Museum and Archives, Delta B.C.

Dewhirst, J.
1979 *An Archaeological Pattern of Faunal Resource Utilization at Yuquot, a Nootkan Outside Village: 1000 BC - AD 1966*. Paper Presented at the 49th Annual Meeting of the SAA 23-25 April. Vancouver, B.C.

Dymsza, H., Y. Shimizu, F.E. Russel, and H.D. Graham
1980 Poisonous Marine Animals. In: *The Safety of Foods*, 2nd ed. Edited by Horace D. Graham pp. 625-651 Publishing Company, Inc., Westport.

Ellis, D.W. and L. Swan
1981 *Teaching of the Tides: Uses of Marine Invertebrates by the Manhousat People. Nanaimo, British Columbia*: Theytus Books.

Erlandson, J.M.
1994 *Early Hunter-gatherers of the California Coast*. Plenum Press, New York.

Estes, J.A., N.S. Smith and J.F. Palmisano
1978 Sea Otter Predation and Community Organization in the Western Aleutian Islands, Alaska. *Ecology* 59(4): 822-833.

Estes, J.A., R.J. Jamieson, and E.B. Rhodes
1982 Activity and Prey Selection in the Sea Otter: Influence of Population Status on Community Structure. *The American Naturalist* 120(2):242-258.

Gaston, J. and G.F. Grabert
1975 *Salvage Archaeology at Birch Bay Washington: A Report*

of Investigations Conducted in 1975. Prepared for URS/Hill, Ingman, Chase and Co. Department of Sociology and Anthropology: Western Washington State College, Bellingham.

Grabert, G.F., J.A. Cressman and A. Wolverton
1978 *Prehistoric Archaeology at Semiahmoo Spit, Washington*. Department of Anthropology, Western Washington University, Bellingham.

Haggarty, J.C. and J.H. Sendey
1976 *Test Excavation at the Georgeson Bay Site, Gulf of Georgia Region, British Columbia*. Province of British Columbia Victoria.

Ham, L.
1987 *An Archaeological Heritage Resource Overview of Richmond, B.C.* (Permits 1986-19 and 1987-1). Report to Richmond Heritage Advisory Committee for the Richmond Museum, Richmond.

1982 *Seasonality, Shell Midden Layers, and Coast Salish Subsistence Activities at the Crescent Beach Site, DgRr 1*. Unpublished Ph.D. Dissertation, University of British Columbia, Vancouver.

n.d. *A Charles Culture Fishing Village: The 1982/1983 Archaeological excavations at the St. Mungo Site (DgRr 2), North Delta, British Columbia*. (Permit 01982-21) Report to the Heritage Conservation Branch, Victoria.

Hanson, D.K.
1995 Subsistence During the Late Prehistoric Occupation of Pender Canal, British Columbia. *Canadian Journal of Archaeology* 19:29-48.

1991 *Late Prehistoric Subsistence in the Strait of Georgia Region of the Northwest Coast*. Unpublished Ph.D. Dissertation, Department of Archaeology, Simon Fraser University, Burnaby.

Hanson, D.K. and H.A. van Gaalen
in press. Subsistence at Cherry Point, Washington (45WH1), in *I Heard the Owl Call My Name: Papers in Archaeology in Honor of Dr. G.F. Grabert*, edited by M. McCormick and B.R. Roulette, Jr., Harrison, P.J., J.D. Fulton, and F.J.R. Taylor,

Harrison, P.J., J.D. Fulton, F.J.R. Taylor and T.R. Parsons
1983 Review of the Biological Oceanography of the Strait of Georgia: Pelagic Environment. *Canadian Journal of Fisheries and Aquatic Sciences* 40 (7): 1064-1094.

Hobson, L.A.
1980 Primary Productivity of the North Pacific Ocean-A Review. In: *Salmonid Ecosystems of the North Pacific*, edited by William J. McNeil and Daniel C. Himsworth. Oregon State University Press and Oregon State University Sea Grant College Program, Corvallis.

Huelsbeck, D.R.
1994 Mammals and Fish in the Subsistence Economy of Ozette. In: *Ozette Archaeological Project Research Reports, Vol II: Fauna*, edited by S.R. Samuels. Reports of Investigations 66, Washington State University, Pullman, and National Park Service, Pacific Northwest Regional Office, Seattle.

Imamoto, S.S.
1974 *Analysis and Interpretation of Faunal Remains from a Complex Site in the Fraser Delta Region of British Columbia: Glenrose Cannery DgRr 6*. Unpublished M.A. thesis, Department of Anthropology and Sociology, University of British Columbia, Vancouver.

1976 An Analysis of the Glenrose Faunal Remains. In: The Glenrose Cannery Site, edited by R.G. Matson pp. 21-41. *Archaeological Survey of Canada, Mercury Series Paper No. 52*. National Museums of Man, Ottawa.

Irons, D.B., D.R. Nysewander and J.L. Trapp
1988 *Prince William Sound Sea Otter Distribution in Relation to Population Growth and Habitat Type*. US Fish and Wildlife Service, Anchorage.

Jameson, R.S., K.W. Kenyon, A.M. Johnson and H.M. Wight
1982 History and Status of Translocated Sea Otter Populations in North America. *Wildlife Society Bulletin* 10:100-107.

Johnstone, D.
1989 *Long Harbour Interim Report: 1988 Excavations*.(Permit 88-42). Simon Fraser University. Report Prepared for Ministry of Municipal Affairs, Recreation and Culture, Victoria.

Kidd, R.S.
1969 The Archaeology of the Fossil Bay Site. Sucia Island, Northwestern Washington State, in Relation to the Fraser Delta Sequence. Pp. 32-67. *National Museums of Canada Bulletin 232, Contributions to Anthropology VII: Archaeology*. National Museums of Canada, Ottawa.

King, A.R.
1950 Cattle Point: A Stratified Site in the Southern Northwest Coast Region. *Memoirs of the Society for American Archaeology No. 7*. Society for American Archaeology and the Tulane University of Louisiana.

Kusmer, K.D.
1996 *Beach Grove (DgRs 1), Faunal Analysis*. Unpublished Ms. Report to Arcas Consulting Archeologists Ltd., Coquitlam, B.C.

1994 Changes in Subsistence Strategies at the Tsawassen Site, A Southwestern British Columbia Shell Midden. *Northwest Anthropological Research Notes* 28(2):189-210.

1988 *Faunal Remains from the Marpole Site (DhRs 1)*. Unpublished Ms. Report to Arcas Consulting Archaeologists Ltd., Coquitlam, British Columbia.

Kvitek, R.G.
1991 Paralytic Shellfish Toxins Sequestered by Bivalves as a Defense Against Siphon-nipping Fish. *Marine Biology* 11(3):369-374.

1990 *Sea Otters and Paralytic Shellfish Poisoning Toxins in Soft-Bottom Communities: Influence and Interactions*. Ph.D. Dissertation, University of Washington, Seattle.

Kvitek, R.G., A.R. DeGange and M.K. Beitler
1991 Paralytic shellfish poisoning toxins mediate feeding behavior of sea otters. *Limnology and Oceanography* 36(2):393-404.

Kvitek, R.G., J.S. Oliver, A.R. DeGange and B.S. Anderson
1992 Changes in Alaskan Soft-Bottom Prey Communities Along a Gradient in Sea Otter Predation. *Ecology* 73(2):413-428.

Kvitek, R.G. and J.S. Oliver
1988 Sea Otter Foraging Habits and Effects on Prey Populations and Communities in Soft-Bottom Environments. In *The Community Ecology of Sea* Otters (Ecological Series: v. 65), edited by G.R. VanBlaricom and J.A. Estes, pp. 22-47.: Springer-Verlag, Berlin.

Lensink C.J.
1960 Status and Distribution of Sea Otters in Alaska. *Journal of Mammalogy*. 41(2):172-182.

Lyman, R.L
1991 Archaeology of Umpqua/Eden (35DO83). In *Prehistory of the Oregon Coast: The Effects of Excavation Strategies and Assemblage Size on Archaeological Inquiry*, edited by R. L. Lyman, pp. 102-174. Academic Press, Inc., San Diego.

Matson, R.G., D. Ludlowicz and W. Boyd
1980 *Excavations at Beach Grove in 1980. Permit 1980-14 MAT*. Report to the Heritage Conservation Branch, Victoria.

Maxwell, D.B.S. and P.M.K. Cunneyworth
1989 *Change Through Time? A Preliminary Analysis of the Faunal Remains from Long Harbour (DfRu 44)*. Paper presented at the 42nd Northwest Anthropology Conference.

Mayne, R.C.
1969 *Four Years in British Columbia and Vancouver Island*. S.R. Publishers Limited.

McMurdo, J.
1974 *Archaeology of Helen Point*. Unpublished M.A. Thesis, Dept. of Archaeology, Simon Fraser University, Burnaby.
Mitchell, D.H.
1968 Excavations at Two Trench Embankments in the Gulf of Georgia Region. *Syesis* 1:29-46.

1971 (a) Archaeology of the Gulf of Georgia Area, A Natural Region and its Culture Types. *Syesis* 4(supp. 1):1-228.

1971 (b) Dionisio Point Site and Gulf Island Culture History. *Syesis* 4:145-168.

1980 DcRt-1: A Salvaged Excavation from Southern Vancouver Island. *Syesis* 13:37-51.

1981 DcRu 78: A Prehistoric Occupation of Fort Rodd Hill National Historic Park. *Syesis* 14:131-150

Monks, G.G.
1987 Prey as Bait: The Deep Bay Example. *Canadian Journal of Archaeology* 11:119-142.

Montgomery, J.A.
1979 *Prehistoric Subsistence at Semiahmoo Spit 45 WH 17*. Unpublished M.A. thesis, Department of Sociology and Anthropology, Western Washington University, Bellingham.

Morris, R., D.V. Ellis and B.P. Emerson
1981 The British Columbia Transplant of Sea Otters *Enhydra lutris. Biol. Conserv*. 20:291-295.

Mozino, J.M.
1970 *Noticias de Nutka*: An Account of Nootka Sound 1792. Translated and edited by Iris Higbie Wilson. University of Washington Press, Seattle.

Murray, R.A.
1982 Analysis of Artifacts From Four Duke Point Area Sites, Near Nanaimo, B.C.: An Example of a Cultural Continuity in the Southern Gulf of Georgia Region. *Archaeological Survey of Canada, Mercury Series Paper No. 113*. National Museum of Canada, Ottawa.

Newcombe, C.F.
1923 Menzies' Journal of Vancouver's Voyage: April to October, 1792. *Archives of British Columbia, Memoir No. 5*. William H. Cullin Printer, Victoria.

Nishitani, L. and K. Chew
1988 PSP Toxins in the Pacific Coast States: Monitoring Programs and Effects on Bivalve Industries. *Journal of Shellfish Research*. 7(4): 653-669.

Osborne, R., J. Calambokidis and E.M. Dorsey
1988 *A Guide to Marine Mammals of Greater Puget Sound*. Island Publishers, Anacortes, Washington

Quayle, D.B. and N. Bourne
1972 The Clam Fisheries of British Columbia. *Fisheries Research Board of Canada, Bulletin 179*. Fisheries Research Board of Canada, Ottawa.

Riedman, M.L. and J.A. Estes
1988 A Review of the History, Distribution and Foraging Ecology of Sea Otters. In: *The Community Ecology of Sea Otters (Ecological Series: v. 65)*, edited by G.R. VanBlaricom and J.A. Estes, pp. 4-21. Springer-Verlag, Berlin.

Schneider, K.B. and J.B. Faro
1975 Effects of Sea Ice on Sea Otters *(Enhydra lutris)*. *Journal of Mammalogy* 56(1): 91-101.

Schallie, K.
1996 Personal Communication. Fisheries Inspection Branch of British Columbia

Seymour, B.
1976 1972 Salvage Excavations at DfRs 3, The Whalen Farm Site. In *Current Research Reports*, edited by R.L. Carlson. Publication No. 3, Department of Archaeology. Simon Fraser University, Burnaby.

Stenson, G.B., G.A. Badgero and H.D. Fisher
1984 Food Habits of the River Otter *Lutra canadensis* in the Marine Environment of British Columbia. *Canadian Journal of Zoology* 62(1):88-91.

Stiefel, S.K.
1985 *The Subsistence Economy of the Locarno Beach Culture (3300-2400 BP)*. Unpublished M.A. thesis, Department of Anthropology and Sociology, University of British Columbia, Vancouver.

Suttles, W.P.
1974 *The Economic Life of the Coast Salish of Haro and Rosario Straits*. New York: Garland Publishing Inc.

Unfreed, W.J.
1989 *Faunal Analysis of the Point Grey (DhRt 5) Site, Vancouver, B.C.* Unpublished M.A. thesis, Department of Archaeology, University of Calgary, Calgary.

Watson, J.
1997 Personal Communication. Malaspina University-College, Nanaimo, British Columbia.

White. A.W.
1981 Sensitivity of Marine Fishes to Toxins from the Red-tide Dinoflagellate *Gonyaulax excavata* and Implications for Fish Kills. *Marine Biology* 65:255-260.

Wigen, R.J. and B.R. Stucki
1988 Taphonomy and Stratigraphy in the Interpretations of Economic Patterns at Hoko River Rock Shelter. In *Prehistoric Economies of the Pacific Northwest Coast*, edited by B.L. Isaac, pp. 87-146. Research in Economic Anthropology, Supplement 3. JAI Press Inc., Greenwich

Williams, J.
n.d. *Faunal Remains from the Cates Park Site (DgRr 8) North Vancouver, British Columbia*. Ms on file at the Zooarchaeology Laboratory, Department of Archaeology, Simon Fraser University, Burnaby.

Wilson, I.R.
1990 (a) *Archaeological Evaluation and Impact Assessment, DhRx 16, Departure Bay*. Prepared for Ministry of Municipal Affairs, Recreation and Culture, Victoria.

1990 (b) *Archaeological Evaluation and Impact Assessment, DhRx 16, Departure Bay*. Prepared for Ministry of Municipal Affairs, Recreation and Culture, Victoria.

1991 *Evaluative Archaeological Testing, Departure Bay Site, DhRx 16*. Prepared for Ministry of Municipal Affairs, Recreation and Culture, Victoria.

Yip, A.J.
1982 *Archaeological Investigations at the Cowichan Bay Site, DeRv 107, Duncan, B.C.* Report Submitted to the Heritage Conservation Branch, Victoria.

SEXUALLY DIMORPHIC SIZE VARIATION IN HOLOCENE *BISON* AS REVEALED BY CARPALS AND TARSALS

Brian Kooyman
Department of Archaeology, University of Calgary, Calgary

Dennis Sandgathe
Department of Archaeology, Simon FraserUniversity, Burnaby

PREVIOUS INTERPRETATIONS OF CHANGES IN BISON SIZE

The diminished size of modern North American *Bison* relative to those of earlier Holocene and pre-Holocene times has been noted and discussed since the mid-nineteenth century (Roe 1970:18-25; Wilson 1975, 1978, 1992:2-10; Bedord 1978:43; Peterson and Hughes 1980; McDonald 1981:39-41; Zeimens 1982; Todd 1987:163-164; Morlan 1992). There has been considerable debate centered on whether these various forms, as well as modern forms, are distinct species or distinct geographic or chronological subspecies (McDonald 1981:250-263; Geist 1991b; Wilson 1992:4-10). Recent work (Kooyman 1994) at the Spring Kill site (EgPs-51) (Figure 1) shows that even comparatively recent *Bison*, radiocarbon dated to between 5300 and 5000 B.P., appear larger than modern *Bison*. The timing of the transition to modern-sized *Bison* form continues to be an issue of debate (e.g. Walker 1992). The present study was initiated to trace *Bison* size changes through the Holocene.

Species and subspecies designation for extinct North American *Bison* taxa is based on cranial features, particularly the size and form of male horn cores (McDonald 1981:54). Little attempt has been made to use postcranial elements, although metapodial size change has been documented (McDonald 1981:257, 259). McDonald studied a number of limb elements and concluded that they showed insufficient variation to be taxonomically useful (1981:54-5), although his bivariate plots (1981:143-146) do show some variation. McDonald also showed consistent patterns of sexual dimorphism in limb bones and cranial traits of the five species of *Bison* he accepted (1981:149-152). Sexual size dimorphism is well studied in the genus *Bison* (Duffield 1973; Bedord 1974, 1978; Shackleton et al.1975; Peterson and Hughes 1980; Rehr and Frison 1980; Roberts 1982; Speth 1983; Walde 1985; van Zyll de Jong 1986). McDonald also noted that not all traits studied showed Holocene diminution in size, and that those that did show the size change did not all do so at the same rate (1981:257-8, 265). The present study lends support to this observation.

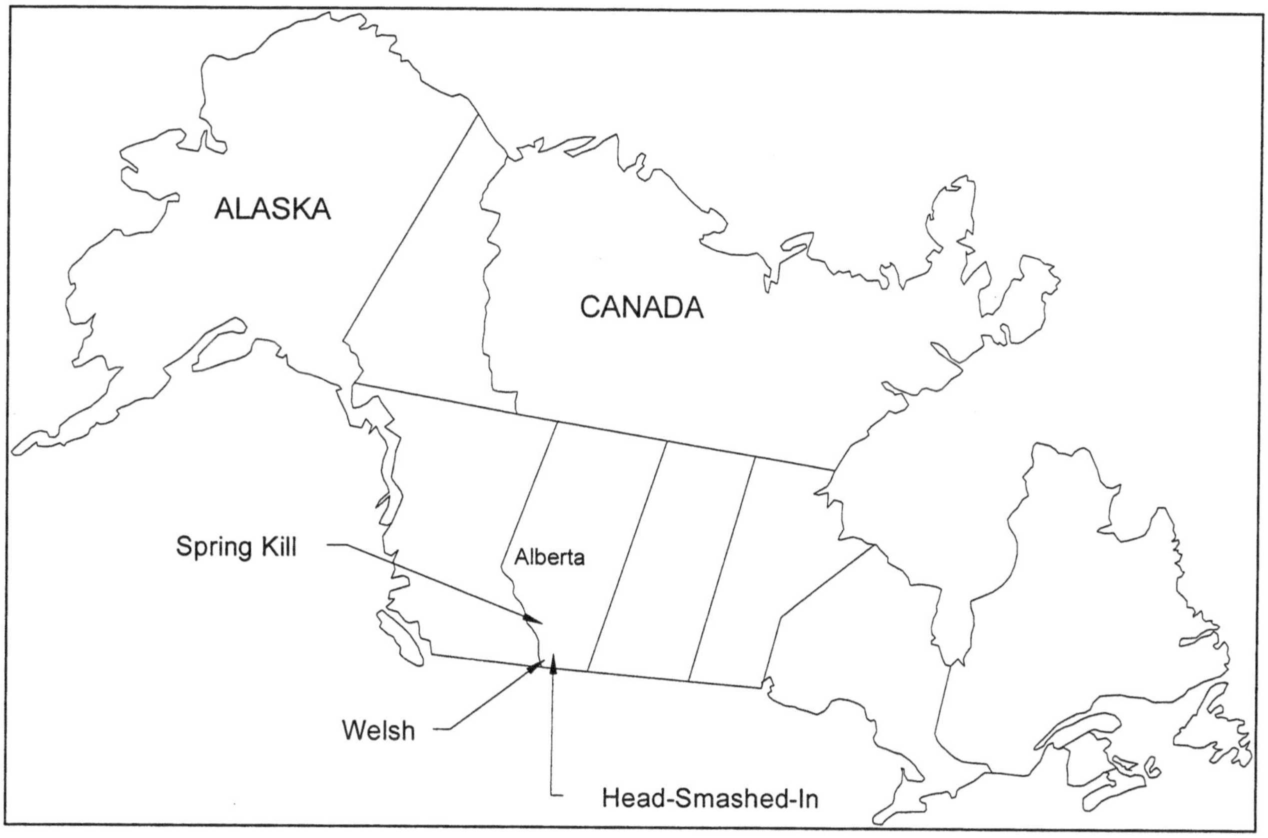

Figure 1. Location of sites used in analysis.

The causes and implications of changing *Bison* size, particularly during the Pleistocene and early Holocene, have engendered considerable debate. Guthrie has contributed much to this discussion (e.g. 1970, 1980, 1984a, 1984b), as have others (e.g. Geist 1974, 1978, 1989, 1991a, 1991b; Wilson 1975, 1978, 1980; Peterson and Hughes 1980; McDonald 1981, 1984; Bamforth 1988:145-148 Walker 1992). Guthrie (1984b:260-262) traces the variations in large ungulate size back to the early Tertiary. Climatically, from this point in time forwards there has been a variable, gradual worldwide trend towards increased seasonality, particularly in terms of seasonal distribution of moisture, as well as cooler temperatures. These trends have resulted in a change to more open savanna and steppe habitats from an earlier dominance of closed forests. These changes have favored ungulate species. There has also been a decrease in annual growing season, which initially produced vegetational mosaics increasingly dominated by non-woody plants. These were rich, diverse habitats for ungulates and the animals responded to this food potential with a more diverse range of species and an increase in size. An optimum period for ungulates was reached in the late Miocene and Pliocene after which the shortened growing season caused the vegetational mosaics to fragment into vegetational zones. These zones each had fewer plant species and a lesser diversity of growth habitats for any one plant species, resulting in a decreased ungulate diversity as a response to the decreased plant diversity. Decreased plant diversity also resulted from plant competition in these less favorable environments and this intensified the homogenizing trend as each plant species became increasingly restricted to habitats closest to its climatic optimum.

The more homogeneous vegetational zones produced increased ungulate competition for the same food species and some ungulates became extinct. This situation resulted in a relative increase in plant food biomass and quality for the fewer successful ungulate species, including *Bison*, and these species experienced an increase in body size. For many species, again including *Bison*, these circumstances occurred in the mid-Pleistocene and resulted in these few species dominating new ecological conditions. However, the continuing trend towards shorter growing seasons reduced plant biomass and diversity, resulting in a decrease in the size of these ungulates in the later Pleistocene. Late Pleistocene and early Holocene extinction of species (mammoth, horse, etc.) that had used primarily low quality plant foods, removing these plants and allowing more vigorous growth of the species and plant forms *Bison* used, may have strengthened the trend toward smaller *Bison* (Guthrie 1984a:494).

The warming accompanying deglaciation at the end of the Pleistocene may have had another effect on the generalized decrease in *Bison* size in the Holocene. Within mammals, larger animal size allows better control of body temperature (e.g. McNab 1990:12). A larger animal has a smaller surface area for a given volume compared to a smaller animal, and hence loses less heat proportionally. This relationship is the basis of Bergmann's rule and explains why homiothermic animals in colder climates tend to be larger than those in warmer climates (Wells 1978:109; Lundelius et al. 1983:346). The terminal Pleistocene and early Holocene warming may have allowed somewhat smaller *Bison* to be as competitive as larger individuals in this regard. Since larger size does require extra energy to maintain, smaller individuals may have become more common. There is a limit to this trend, however, since even in the comparatively warm modern climate small ungulates are not present in colder environments, probably because they are too small to maintain body temperature, but too large to burrow to escape the cold (Maiorana 1990:83).

The late Pleistocene environments of North America south of the ice sheets in many cases included examples of co-existing plant and animals species that are currently allopatric. Such distributions are often interpreted as indications of a more equable climate at the end of the Pleistocene, compared to the current situation, with cooler summers allowing lower elevation and southern range extensions for species that currently have a more "boreal" distribution (Lundelius et al. 1983:346; Graham et al. 1987:427; Walker 1987:359; Wendland et al. 1987:464-466). Walker (1992:77-78) has suggested that the decreased seasonality (compared to the modern climate) and longer growing season (compared to the full glacial climate) of the late Pleistocene may have allowed smaller *Bison* to be more successful. However, the changes were not of sufficient magnitude to alter the established vegetational zonation. As well, the onset of the Holocene saw seasonality again become more accentuated (Lundelius et al. 1983:346, Semken and Falk 1987:184). In view of these circumstances, Walker's hypothesis cannot explain the continued Holocene decrease in *Bison* size through to ca. 5000 B.P.

Geist (1978, 1989:163-165) characterized phenotypic responses to changing resource availability as occurring along a continuum between what he referred to as "dispersal" and "maintenance" phenotypes. The "dispersal" phenotype arises in resource rich circumstances and is characterized by rapid growth and a realization of maximum phenotypic size. This phenotype is also characterized by well developed tissues and organs, even if these structures have low growth priority. "Maintenance" phenotypes arise in response to resource shortages, and use resources more efficiently. This phenotype does not attain maximum phenotypic size and low growth priority structures are not maximally developed. "Maintenance" phenotypes are in a sense the norm, since resource rich circumstances generally "fill" over time, and increased competition for resources results in a need for a more efficient exploitation of resources. The mid-Pleistocene optimum for *Bison* would have promoted a dispersal phenotype, including a larger size. The "trigger" for development of this phenotype was likely a higher protein content in the maternal diet and hence in the fetal and neonatal diets (Geist 1989:164).

Guthrie views the duration of the seasonal growth peak, during which there is abundant forage of high nutritional quality, as the key factor in the onset of dwarfing in ungulates (1984b:267-269; 1989a:483-485,506). As the growth season progresses, the nutritional quality of forage decreases and antiherbivore devices such as lignin increase, decreasing protein content. Dietary protein plays a major role in determining growth of the ungulate body frame. Most large mammal herbivores maximize the potential of the growth season by various behaviors that ensure use of a variety of plant species, exploiting their variation in nutrients and in the timing of the availability of those nutrients (e.g. altitudinal movements between local resource patches that mature at different times). As environments became more homogenized and zoned, this scheduling became less possible; the growth season could not be extended or maximized to the same degree by "piecing together" a diet. The result, especially for species adapted to a longer growing season, was a reduction in the energy available for increased body frame size and social organ size, the latter including structures such as horns and antlers.

Geist (1991b:294-296) suggested that a restricted ability to schedule use of varying seasonal food resources after the initial spring growth flush was a basic factor in body size and antler/pelage display characteristic differences in wapiti (*Cervus elephus*), moose (*Alces alces*), and *Bison* populations in Elk Island National Park and Banff National Park (Alberta). Wapiti, moose, and *Bison* have a larger body size at Elk Island compared to Banff, but the Elk Island wapiti and moose have smaller antlers and the *Bison* have Plains style pelage. These "social organs" are developed mainly later in the growth season, at which time the Banff animals shift their forage ranges (e.g. altitudinally) to take advantage of new plant growth, but this is not possible for the fenced animals in the uniform elevation Elk Island area. The Elk Island populations have lesser resources for social organ growth and these are "sacrificed" for maintenance of general growth.

Guthrie followed Geist (e.g. 1978:192) in referring to features of animal "dress" such as antlers and horns as "organs of low growth priority," whose growth is sacrificed during times of low food availability or low nutritional value. These structures are used only for early or expanded reproduction in a single season and are therefore expendable. A male animal that expends too much of its limited energy on reproduction will not accumulate large stores of fat for the winter season (in temperate and arctic climates), and hence may not survive the winter season. Such an animal will not contribute as much to the succeeding generation's gene pool (Geist 1974:240). As a result, when environmental conditions were less ideal for growth, growth in these non-essential "social" structures was differentially sacrificed. Guthrie's experimental work with Dall sheep (1984a, 1984b) showed that although the growth rate of horns could not be altered with a high protein diet, the length of the growth season could be increased by maintenance of a high protein diet. Guthrie also noted that life curves for Pleistocene mammals did not show a shorter life span associated with dwarfing (1984b:271), indicating that some other factor, such as growth season length, was fundamental. This suggestion is supported by the observation that in modern, closely related artiodactyls, larger species and subspecies are larger at each age stage compared to smaller species and subspecies. This pattern shows that increased size is not just a result of growing through a greater number of growth seasons (Guthrie 1984a:485).

Many authors have noted that the change in the size of the horns and the form of the skull in *Bison* have been more dramatic than other aspects of Pleistocene and Holocene change. However, the factors involved are more complex than those affecting general body size due to the roles of social/reproductive behavior and defense (e.g. Guthrie 1970, 1980, 1984a, 1984b; Geist 1974, 1978; McDonald 1981; Wilson 1992). For example, Geist noted (1974:240) that ungulates can lessen rutting energy output during difficult climatic periods by simply using displays to assert dominance. This might include greater use of vocalization or physical structures, such as large antlers, to accomplish the end. He also noted that during periods of deglaciation ungulates moved into new and unexploited habitats where they were able to achieve maximum growth due to the abundance of food. However, as Guthrie (1984a:491) has also noted, the colonization phenomenon, and any other exploited range phenomenon, can only continue to promote larger size as long as non-growth season mortality increases so that the range remains understocked. Guthrie saw severe winter mortality during the Pleistocene as the key factor (1984a:506-507). Geist (1974:241) also saw that this maximization potential would no longer be available once these new habitats were filled, but he suggested that increased body and horn/antler growth might be selected for because these would enhance reproductive success due to greater success in male dominance confrontations. Geist (1978:189) suggested that these circumstances arose a number of times during the Pleistocene, allowing a generalized trend to larger size to build upon itself. Only factors that somehow favored greater reproductive success for smaller, non-dominant males would reverse such a trend, a situation not seen in modern studies (Geist 1974:241).

We believe that Guthrie's discussion of the shortening of the growth season peak, in combination with Geist's observations about compromised dominant male success if they did not survive the winter, might provide a situation where smaller, non-dominant males could achieve greater reproductive success.

Modern *Bison* horns are deflected outwards or backwards and are more blunt than cattle horns. The eye orbits are also "telescoped" to bring the eyes out somewhat from the skull (Guthrie 1980:59-61). Guthrie interpreted the horn deflection and orbit telescoping as indicative of a change in male/male rutting confrontation, from the "clashing" impacts of *Bison* to the interlocking and pushing of cattle. This situation necessitated moving the horns out of the clash line to avoid damage to them. It also resulted in a thickening of the skull skin and hair to absorb clash impacts, hence the need to move the eyes outwards so that vision was not impared. The skull also developed an enhanced pneumatic cavity to aid in protection of the brain tissue. Sharp horn tips in bovids are used in defense against predators and the blunting of these in modern *Bison* probably reflects the disappearance of the large predators that pursued steppe bison (*Bison priscus*).

Substantial amounts of energy are expended in male dominance displays. Geist (1974:239-241) observed that there are a number of behaviors that can be employed to minimize this energy output, such as having a restricted rut season, and separate male herds so that territory and females do not need to be defended during the entire year. These behaviors would reduce the need for energetically expensive display mechanisms such as larger body and larger horn size. This situation might particularly apply to horn size if clashing became a major dominance display and the horns became reduced to allow this to occur. Guthrie (1980:70-71) also cited Lott's work showing that modern male *Bison* mimic female features during the winter. Neutral sexual appearance allows males to graze peacefully without wasting energy on rut style confrontation with other males. Guthrie (1980:62) also cited Estes' research on African bovids showing a sexually monomorphic tendency among smaller savanna grazers such as the wildebeest, suggesting that this reduces combative behavior. Geist (1991b:284) has also drawn attention to research indicating that this reduction in male display varies seasonally and with age. We hypothesize that given this tendency, male form and size in the smaller late Holocene *Bison* might also more closely approach that of females.

MATERIALS AND METHODS

It was in the context of these various studies that the current research was undertaken. As was previously noted, work at the Spring Kill site had shown that these 5000 year old *Bison* appeared larger than modern *Bison*. Placing them in a broader chronological framework seemed appropriate. Recent archaeological studies of *Bison* have examined sexual dimorphism in many different postcranial elements with the objective of better understanding variation in seasonal hunting patterns and animal exploitation (Speth 1983; Walde 1985; Brink *et al.* 1986; Todd 1987; Morlan 1991). The analytical techniques developed have been useful in examining these aspects of human behavior, but caution has been expressed about applying these techniques to earlier, larger *Bison* taxa (Walde 1985:31-34). Research by Todd (1987) and Morlan (1992) has clearly shown that the measurements can be used, but these authors have emphasized that there is a clear shift in the size ranges for male and female animals in more ancient taxa.

Morlan's results from the Gowen sites suggested that the use of carpals or tarsals might be fruitful. Although the sample of measurable bones from the Spring Kill site was not large (23 carpals and 10 tarsals used in this study), the carpals were best represented and so these were chosen for study. A desire to evaluate size changes in non-cranial elements also contributed to this decision. If animal science research into environmental impacts on growth (Geist 1989:156-157) can be applied to *Bison*, lower limb elements such as carpals and tarsals may be particularly good indicators of non-environmental change since they develop well even under poor nutrition. A good sample (80 carpals and 22 tarsals used here) of relatively recent material, dating to 1000 to 1500 B.P., from Head-Smashed-In Buffalo Jump (DkPj-1) (Kooyman 1990, 1992) was available for comparative use. Excavations at the Welsh site (DjPm-36) (Van Dyke *et al.* 1991) in the Old Man River Dam project provided material (101 carpals and 46 tarsals used) from an earlier period, *ca.* 8000 B.P. The objectives of our study were to determine changes in male and female *Bison* through time, and to determine whether changes were consistent between sexes, and determine if size changes in different traits varied at the same rate and time.

The reasons for the current study have been indicated but some comments on the specific approach are warranted. Sexual dimorphism in body size is a characteristic of most mammal species. Todd (1986:110) has shown that in *Bison* these differences are more easily seen in proximal limb bones that in distal limb bones. However, studies of distal limb elements of *Bison* and related species have clearly shown sexual dimorphism in metapodials (Higham 1969; Bedord 1974, 1978; Peterson and Hughes 1980; Speth 1983; Walde 1985; Berteaux and Guintard 1995), phalanges (Roberts 1982), and carpals and tarsals (Higham 1969; Zeimens and Zeimens 1974; Vigne 1988; Zeimens 1982; Morlan 1991). Morlan's paper employed a relatively small sample of known sex specimens, but his more recent work (personal communication, 1998) has significantly expanded the sample size with continued good separation of the sexes. Jack Brink (personal communication, 1998) has measured additional known sex individuals and has found that certain of Morlan's measurements work well for separating the sexes in adult *Bison*. Both Morlan and Brink have found that some of the measurements provide better separation than do others. Male and female measurement size differences in carpals and tarsals are generally only a few millimeters, but are consistent and document a real difference (see Maiorana's (1990:71-72) comments on scale and perception).

Walker (1995, 1998:pers. comm.) used various combinations of carpal and tarsal measurements, including Morlan's, but found that separating known male and female *Bison* in the University of Wyoming comparative collection was problematic. His sample included as many as four, two year old males, depending on the bone, and one or more of these often plotted in his female range on scatter graphs. He found that at times some mature males plotted in the female range while some mature females plotted in the male range. The specimens used were from animals that had all or most epiphyses of all bones fused, although at least one of the two year old males did have some unfused elements. Walker found that separation did sometimes occur, but that it was a separation of a few millimeters at best. His main conclusion was that there was considerable overlap in the measurements between male and female animals.

Various authors note that only skeletally mature individuals can be used to isolate sexual dimorphism (e.g. Todd 1987:163-164; Fandos and Vigal 1993), or that only certain measurements can be used to isolate the sexes when immature males are present (Shackleton et al. 1975:881). This is particularly important when measurement differences of only a few millimeters are involved, as is the case with carpals and tarsals. Sexual maturity cannot be the basis for defining material as "mature" for these purposes. Sexual maturity in *Bison* can occur as early as one or two years of age (Roe 1970:506; Banfield 1974:406), even if active participation in the rut usually occurs much later in the wild (ca. two to four years for females and two to six and one-half years for males) (Banfield 1974:406; Shackleton et al. 1975:881). Peterson and Hughes (1980:172-174) show that, based on the work of Halloran, female *Bison* cease gaining weight by their third year, but that male *Bison* may continue weight gain well into their eighth or ninth year. Clearly full sexual dimorphism, only achieved as full size is attained, will not be expressed in many younger male animals. Morlan's ongoing research shows some younger males to be problematic. In this regard we urge caution in using modern ranch-raised *Bison* as a comparative base since the males in particular may be generally killed at less than two years of age (Zeimens 1982:214). Feeding and husbandry practices used with ranch animals could also influence results (cf. Reitz and Ruff 1994). Although Walker's work clearly indicates that more research is needed to better document variation in carpal and tarsal form, most evidence supports the veracity of detectable size differences in carpals and tarsals between male and female animals. Morlan's carpal and tarsal sexing measurements are employed in the current study.

RESULTS

The carpals from both the 5000 year old Spring Kill site and the 8000 year old Welsh site are larger than the carpals from 1500 year old Head-Smashed-In Buffalo Jump. The Head-Smashed-In sample is intended to serve as a modern comparative baseline. Visual inspection of the Head-Smashed-In sample in the plots (Figures 2-9) suggest male/female divisions as follows: scaphoid (Figure 2), 3 females and 5 males; cuneiform (Figure 3), 11 females and 8 males; lunate (Figure 4), 6 females and 4 males; unciform (Figure 5), 9 females and 10 males; magnum (Figure 6), 16 females and 8 males; navicular cuboid (Figure 7), 5 females and 0 males; calcaneum (Figure 8), 3 females and 1 male; cuneiform pes (Figure 9), 10 females and 3 males.

Dick Morlan kindly provided us with his known sex comparative data to test our hypothesized division of the Head-Smashed-In sample and our use of it as representative of modern *Bison*. When compared to Morlan's mature, healthy modern sample, with the exception of the calcaneum (Figure 8) all Head-Smashed-In hypothesized females are within his female range and all males within his female range. Only one Head-Smashed-In calcaneum falls within Morlan's female range, showing the difficulty of defining divisions with samples that are too small. Despite the inadequate calcaneum sample, the Head-Smashed-In sample divisions into hypothesized male and female animals are consistent with the size range of modern specimens of known sex.

The Spring Kill carpals (Figures 2-6) are generally intermediate in size as compared to the Head-Smashed-In and Welsh sites, although they are larger than the recent Head-Smashed-In material. The tarsal sample sizes are not generally as good (Figures 7-9) as the carpal samples, but the calcaneum (Figure 8) and cuneiform pes (Figure 9) plots do indicate that the size decrease through time also seems to hold for some of the tarsals. The pattern is weaker in tarsals, particularly for the astragalus (not shown) and the navicular-cuboid (Figure 7). Higham (1969:64-65) found domestic cattle carpals to show greater sexual dimorphism tarsals. He also found the astragalus to exhibit less sexual dimorphism than the calcaneum or navicular-cuboid.

A few of the smallest Welsh site specimens from each sex of each carpal usually extended into the upper end of the range for the corresponding Head-Smashed-In specimens, hence also generally encompassing the Spring Kill specimens. However the majority of the Welsh site specimens clearly fell beyond the size range of the same sex material from the other two sites. This general decrease in size of both male and female *Bison* probably resulted from the circumstances outlined above - a reduced length of seasonal growth peak and a reduction in overall quantity of forage. It is significant that these materials clearly document the size change in elements that are not low growth priority social structures. As already noted, carpals and tarsals grow well even under poor nutrition, indicating that causal factors were long term and relatively severe. Environmental conditions must have changed to a degree such that sacrificing growth of social structure alone was not adequate to maintain body size.

Our results also suggest that not all measurements changed at the same rate. On the scaphoid graph (Figure 2), for example, both variables measured show considerable shift between the Head-Smashed-In specimens and the other two, but between the Spring Kill and Welsh material there is quite a shift in the y-axis variable (Depth), but relatively little in the x-axis variable (Length). The lunate (Figure 4), unciform (Figure 5), and magnum (Figure 6) graphs show a similar pattern of much greater shift in the depth measurement compared to the other measurements when the Welsh and Spring Kill results are compared. Again, there is a shift in both variables that is clearly seen when these two sites are compared to the Head-Smashed-In material for these three carpals.

The cause of the greater change in the depth measurement, which is a cranial to caudal dimension in each case, between the Welsh and Spring Kill sites is not clear but it is consistent. Morlan (personal communication, 1998) has also noted more marked difference in depth measurements. Higham (1969:64) found breadth measurements to give better separation of the sexes in cattle than did length measurements, and he cited similar results in an earlier study by Zalkin. Significantly, Scott (1990:312) has found that non-length dimensions of ungulate long bones are better correlated with body mass than are length measurements. It is probable that depth measurements are more responsive to the body mass changes seen in Pleistocene to Holocene *Bison* dwarfing.

What is additionally interesting is that generally the male specimens from the Spring Kill site, and particularly the Welsh site, fall well outside the range of the Head-Smashed-In specimens. There was much greater overlap between the female distributions from the three sites than there was between the male distributions. This pattern may indicate that diminished size in Holocene bison was more marked in male animals than female. This situation can also be seen in the percentage size change in males versus females seen in Table 1. The greater decrease in male size compared to female size probably reflects male *Bison* increasingly mimicing female form and size (Guthrie 1980), reflecting perhaps a tendency to reduce obvious male features seasonally and with greater age (Geist 1991b). Presumably decrease in forage (due to reduced growth season length and/or reduced quantity of forage) would have affected male and female animals to the same extent proportionally. The foraging strategy would remain the same for each gender

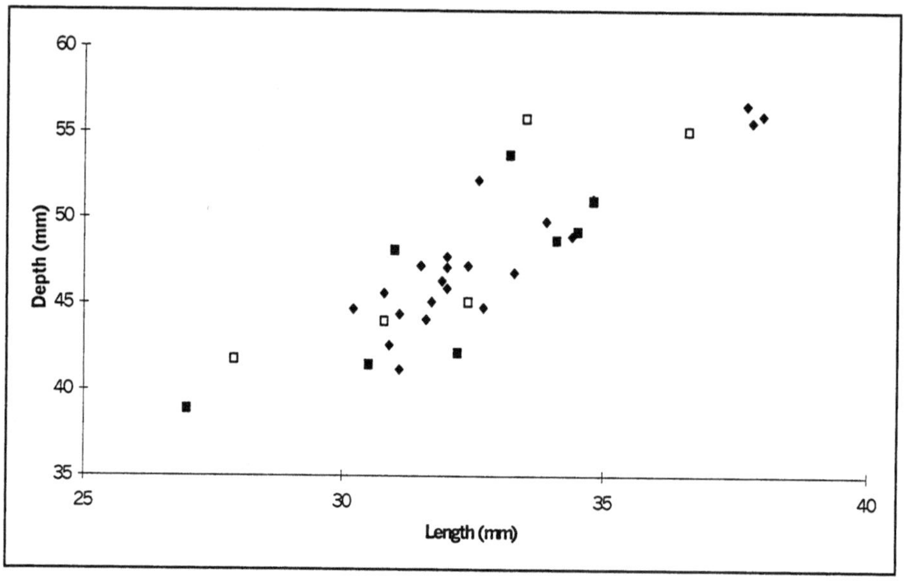

Figure 2. Scaphoid size (solid squares = HSI, open squares = Spring Kill, solid diamonds = Welsh)

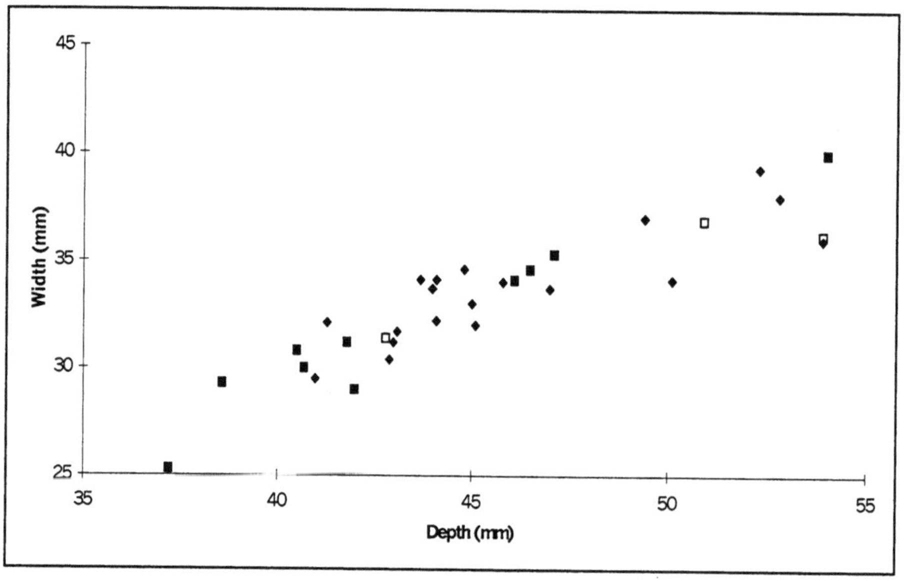

Figure 3. Cuneiform size (solid squares = HSI, open squares = Spring Kill, solid diamonds = Welsh).

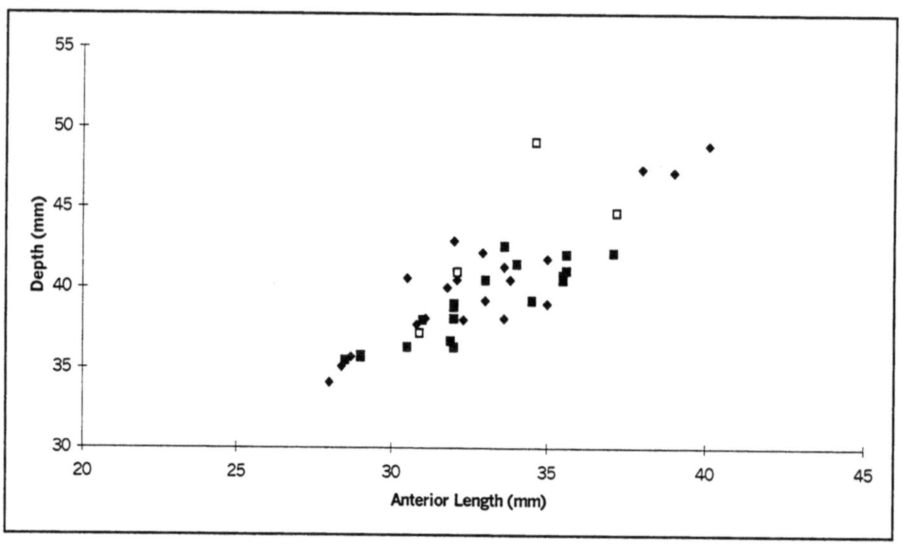

Figure 4. Lunate size (solid squares = HSI, open squares = Spring Kill, solid diamonds = Welsh).

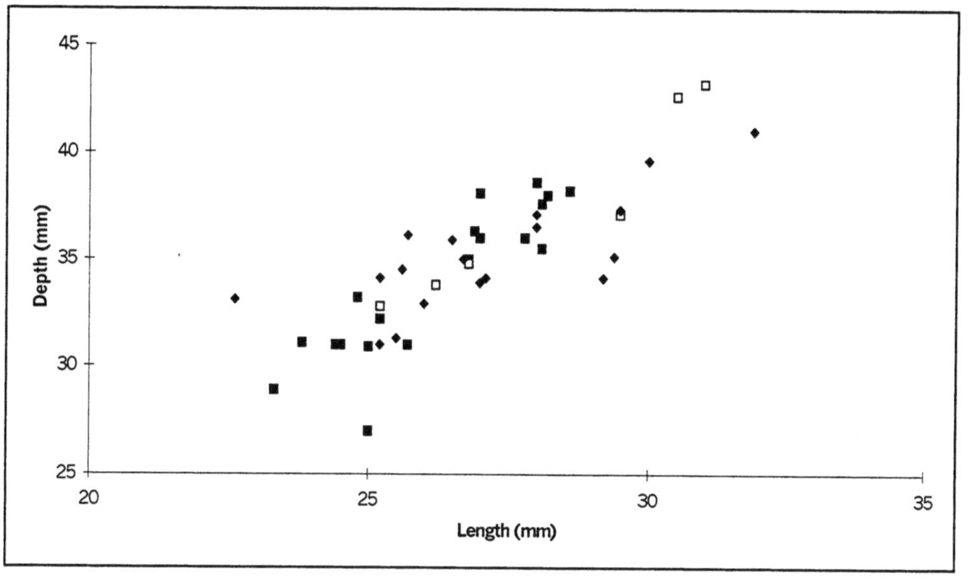

Figure 5. Unciform size (solid squares = HSI, open squares = Spring Kill, solid diamonds = Welsh).

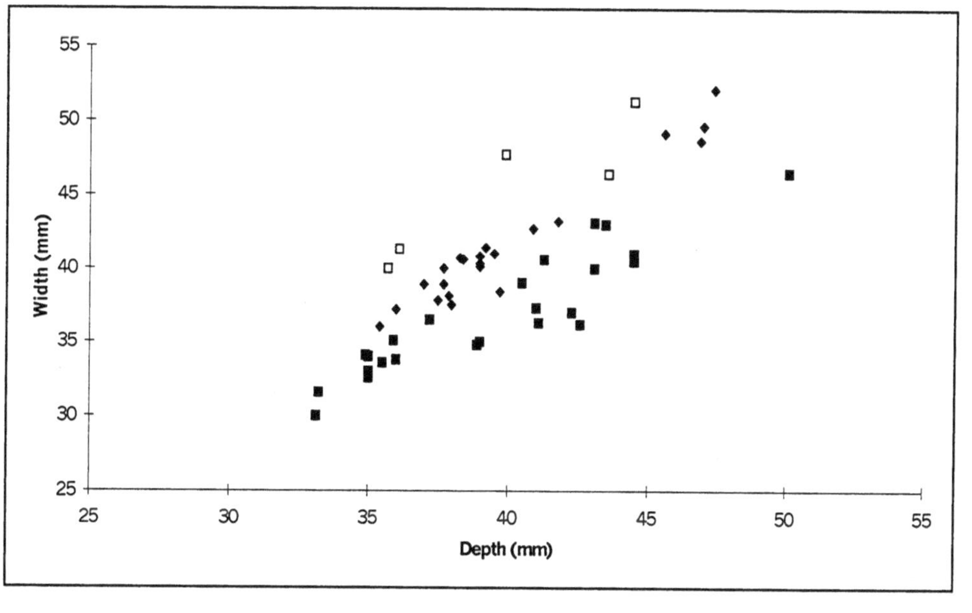

Figure 6. Magnum size (solid squares = HSI, open squares = Spring Kill, solid diamonds = Welsh).

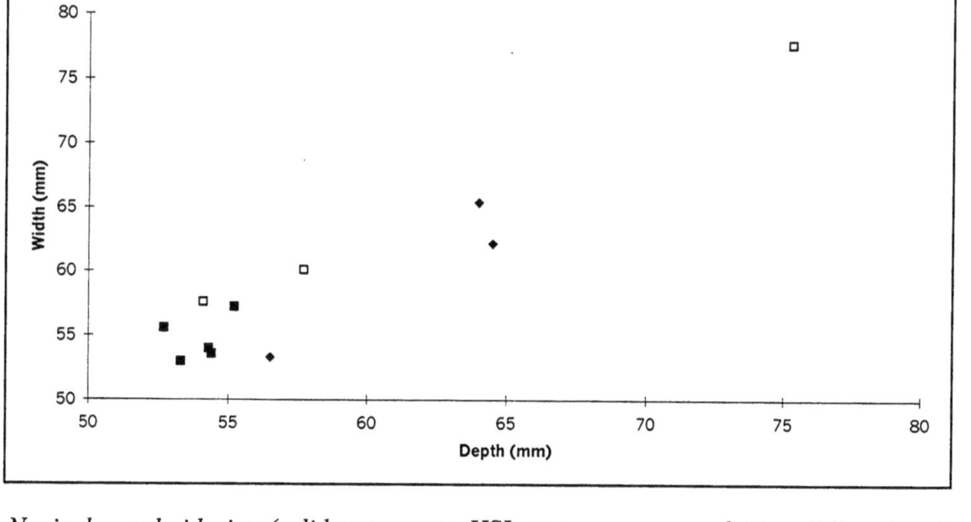

Figure 7. Navicular cuboid size (solid squares = HSI, open squares = Spring Kill, solid diamonds = Welsh).

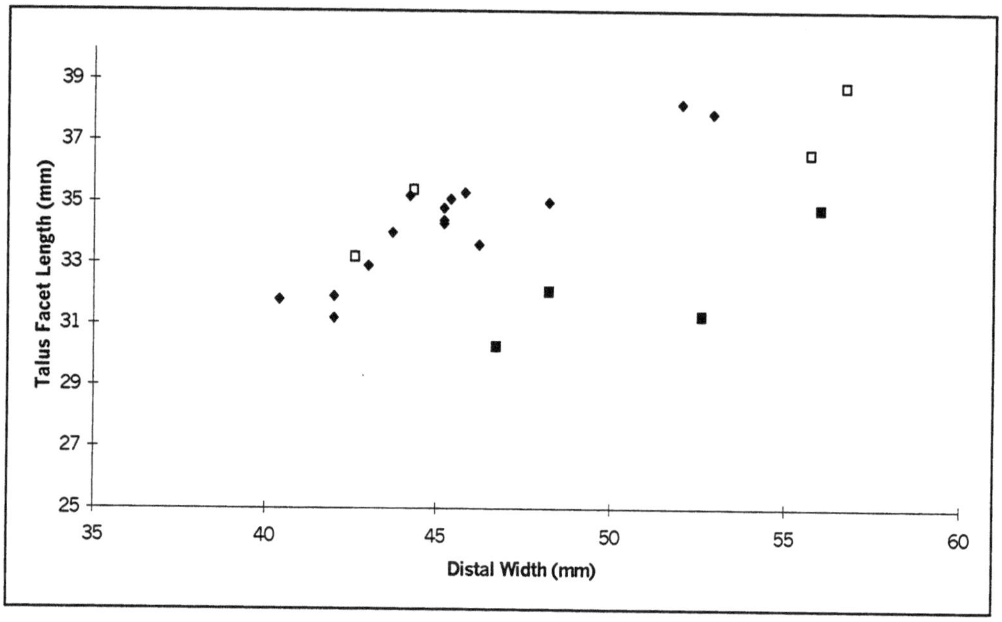

Figure 8. Calcaneum size (solid squares = HSI, open squares = Spring Kill, solid diamonds = Welsh).

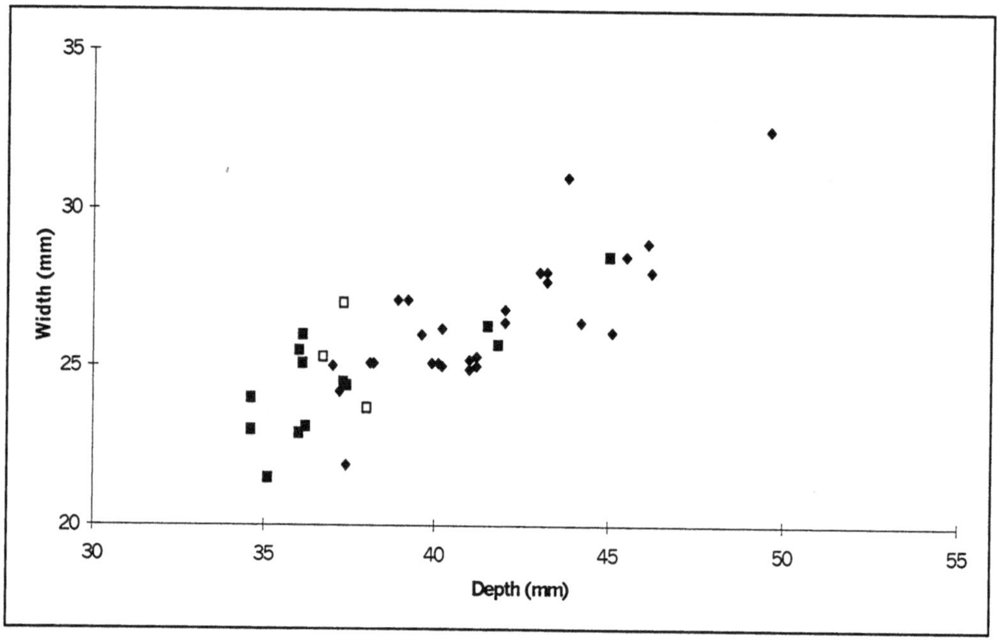

Figure 9. Cuneiform pes size (solid squares = HSI, open squares = Spring Kill, solid diamonds = Welsh).

and the success rate for search time would decrease at the same rate as a consequence of the length of the search season and/or the quantity of forage to be found. The greater decrease in male size indicates that male *Bison* increasingly approached the female form.

The proportionally greater change in the size of carpals, as compared to tarsals, may relate to other changes in the front limb. Guthrie (1980:61) argues that the longer and more muscled front limb in *Bison*, compared to cattle, allows *Bison* to have a rocking canter that is a very efficient gait for travelling large distances over the open plains (see also Smiley 1979; Bamforth 1988). In this gait the front limb provides the main power stroke, causing it to be more heavily muscled. The front limb and carpals might be more able to accommodate a decrease in size since this limb is proportionally larger in the first place as a result of these developments. A carpal depth reduction would reduce unnecessary bulk without compromising the leverage advantage the long limb provids for gait. The small absolute size of the length dimensions (proximal to distal) may put a practical absolute limit on how much reduction in that dimension was possible. Although the mechanisms are not clear, the fact that there is a patterning to carpal change, mainly in depth (at least during the earlier Holocene), suggests that there is an underlying functional explanation. In contrast, McDonald (1981:191) did not observe an appreciable difference in the reduction in length of front and hind limb elements in his comparison of *Bison bison bison* with *B.b.occidentalis* and *B. b. antiquus*.

BONE	DIMENSION	SITE	FEMALE % WELSH	MALE % WELSH
Scaphoid	Depth	Spring Kill	94	99
Scaphoid	Depth	HIS	88	90
Scaphoid	Length	Spring Kill	94	93
Scaphoid	Length	HIS	93	89
Cuneiform	Depth	Spring Kill	100	98
Cuneiform	Depth	HIS	95	87
Cuneiform Anterior	Length	Spring Kill	99	92
Cuneiform Anterior	Length	HIS	98	90
Unciform	Depth	Spring Kill	99	102
Unciform	Depth	HIS	90	92
Unciform	Length	Spring Kill	98	98
Unciform	Length	HIS	93	89
Magnum	Depth	Spring Kill	93	91
Magnum	Depth	HIS	97	95
Magnum	Width	Spring Kill	103	97
Magnum	Width	HIS	88	84

Table 1 Element size as percentage of Welsh site element size.

CONCLUSION

Carpals and tarsals from the Welsh, Spring Kill, and Head-Smashed-In Buffalo Jump sites show a general decrease in size. The results support the general pattern of Holocene *Bison* dwarfing documented in studies of other elements, including skulls. This trend is probably related to increasing seasonality, shorter growth season, and changing vegetation patterns beginning in the late Pleistocene and extending into the Holocene. The carpals particularly also show a more marked decrease in the size of male elements, indicating that during the Holocene there was a tendency for the size of male *Bison* to approach the female size. This change probably reflects the same trend seen in *Bison* pelage characteristics and some aspects of *Bison* behavior such as seasonal herd composition. This aspect of size change in males probably involved decreased male energy expenditure on reproductive displays. The greater change in carpals compared to tarsals, and especially the greater change in carpal depth measurements during the earlier Holocene, may reflect a need to reduce bulk while maintaining front limb length.

ACKNOWLEDGMENTS

Major funding for the research at Head-Smashed-In Buffalo Jump was provided by the Social Sciences and Humanities Research Council of Canada. Additional funding was provided by the Alberta Historical Resources Foundation, the Friends of Head-Smashed-In Buffalo Jump Society, the Archaeological Survey of Alberta, and the Department of Archaeology at the University of Calgary. The senior author is much indebted to Joe Crowshoe and Leo Pard for sharing Piikáni traditional knowledge during the excavations at Head-Smashed-In, and for Leo's continued advice during the excavations at the Spring Kill site. Excavations at the Spring Kill site were funded by Alberta Transportation and Utilities. We also wish to thank the Archaeological Survey of Alberta for permission to use the Welsh site material. Dick Morlan, Jack Brink, and Danny Walker shared ideas and comments about the research. Their inputs have contributed significantly to the work, although the conclusions and any errors remain the responsibility of the authors.

REFERENCES CITED

Bamforth, Douglas B.
1988 *Ecology and Human Organization on the Great Plains*. New York: Plenum.

Banfield, A.W.F.
1974 *The Mammals of Canada*. Toronto: University of Toronto Press.

Bedord, Jean Newman
1974 Morphological Variation in Bison Metacarpals and Metatarsals. In *The Casper Site: A Hell Gap Bison Kill on the High Plains*. George C. Frison,ed. Pp.199-240. New York: Academic Press.

1978 A Technnique for Sex Determination of Mature Bison Metapodials. In *Bison Procurement and Utilization: A Symposium*. Leslie B, Davis and Michael C. Wilson, eds. Pp.40-43. Plains Anthropologist Memoir 14, Lincoln: Plains Anthropological Society.

Berteaux, Dominique, and Claude Guintard
1995 Osteometric Study of the Metapodials of Amsterdam Island Feral Cattle. *Acta Theriologica* 40:97-110.

Brink, Jack, Milt Wright, Bob Dawe, and Doug Glaum
1986 *The Final Report of the 1984 Season at Head-Smashed-In Buffalo Jump Alberta*. Archaeological Survey of Alberta Manuscript Report Series No.9. Edmonton: Alberta Culture.

Duffield, Lathel F.
1973 Aging and Sexing the Post-cranial Skeleton of Bison. *Plains Anthropologist* 18:132-139.

Fandos, Paulinno, and Carlos R. Vigal
1993 Sexual Dimorphism in the Size of the Skull of Spanish Ibex *Capra pyrenaica*. *Acta Theriologica* 38:103-111.

Geist, Valerius
1974 On the Relationship of Ecology and Behaviour in the Evolution of Ungulates: Theoretical Considerations. In *The Behaviour of Ungulates and Its Relation to Management*. V. Geist and F. Walther, eds. Pp.235-246. Morges: International Union for the Conservation of Nature and Natural Resources New Series 24.

1978 *Life Strategies, Human Evolution, Environmental Design: Toward a Biological Theory of Health*. New York: Springer-Verlag.

1989 Environmentally Guided Phenotype Plasticity in Mammals and Some of its Consequences to Theoretical and Applied Biology. In *Alternative Life-History Styles of Animals. M. N. Bruton*, ed. Pp.153-176. Dordrecht: Kluwer Academic Publishers.

1991a Bones of Contention Revisited: Did Antlers Enlarge with Sexual Selection as a Consequence of Neonatal Security Strategies? *Applied Animal Behaviour Science* 29:453-469.

1991b Phantom Subspecies: The Wood Bison *Bison bison "athabascae"* Rhoads 1897 is Not a Valid Taxon, but an Ecotype. *Arctic* 44:283-300.

Graham, Mary Ann, Michael C. Wilson, and Russell W. Graham
1987 Paleoenvironments and Mammalian Faunas of Montana, Southern Alberta, and Southern Saskatchewan. In *Late Quaternary Mammalian Biogeography and Environments of the Great Plains and Prairies*. Russell W. Graham, Homes A. Semken, and Mary Ann Graham, eds. Pp.410-459. Springfield: Illinois State Museum Scientific Papers Volume 22, Illinois State Museum.

Guthrie, Dale
1970 Bison Evolution and Zoogeography in North America During the Pleistocene. *The Quarterly Review of Biology* 45(1):1-15.

1980 Bison and Man in North America. *Canadian Journal of Anthropology* 1:55-73.

1984a Alaskan Megabucks, Megabulls, Megarams: The Issue of Pleistocene Gigantism. In *Contributions in Quaternary Vertebrate Paleontology: A Volume in Memorial to John E. Guilday*. H.G. Genoways and M.R. Dawson, eds. Pp.482-510. Pittsburgh: Special Publication of Carnegie Museum of Natural History.

1984b Mosaics, Allelochemics and Nutrients: An Ecological Theory of Late Pleistocene Megafaunal Extinctions. In *Quaternary Extinctions: A Prehistoric Revolution*. P.S. Martin and R.G. Klein, eds. Pp.259-298. Tucson: The University of Arizona Press.

Higham, Charles F.W.
1969 The Metrical Attributes of Two Samples of Bovine Limb Bones. *Journal of Zoology* 157:63-74.

Kooyman, Brian P.
1990 *Final Report of the 1989 Excavation in the North Kill at Head-Smashed-In Buffalo Jump* (DkPj-1). Submitted to the Archaeological Survey of Alberta, Provincial Museum of Alberta, Edmonton.

1992 *Preliminary Report of the 1991 Excavations in the North Kill at Head-Smashed-In Buffalo Jump* (DkPj-1). Submitted to the Archaeological Survey of Alberta, Provincial Museum of Alberta, Edmonton.

1994/1993 *Excavations at the Spring Kill Site (EgPs-51) Near Exshaw, Alberta*. Submitted to the Archaeological Survey of Alberta, Provincial Museum of Alberta, Edmonton.

Lundelius, Ernest L. Russell W. Graham, Elaine Anderson, John Guilday, J. Alan Holman, David W. Steadman, and David S. Webb
1983 Terrestrial Vertebrate Faunas. In *Late Quaternary Environments of the United States Volume 1 The Late Pleistocene*. Stephen C. Porter,ed. Pp.311-353. Minneapolis: University of Minnesota Press.

Maiorana, Virginia C.
1990. Evolutinary Strategies and Body Size in a Guild of Mammals. In *Body Size in Mammalian Paleobiology: Estimation and Biological Implications*. John Damuth and Bruce J. MacFadden, eds. Pp.69-102. Cambridge: Cambridge University Press.

McDonald, Jerry N.
1981 *North American Bison: Their Classification and Evolution*. Berkeley: University of California Press.

1984 The Reordered North American Selection Regime and Late Quaternary Megafaunal Extinctions. In *Quaternary Extinctions: A Prehistoric Revolution*. Paul S. Martin and Richard G. Klein, eds. Pp.404-439. Tucson: The University of Arizona Press.

McNab, Brian K.
1990 The Physiological Significance of Body Size. *In* Body Size In *Mammalian Paleobiology: Estimation and Biological Implications*. John Damuth and Bruce J. MacFadden, eds. Pp.11-23. Cambridge: Cambridge University Press.

Morlan, Richard E.
1991 Bison Carpal and Tarsal Measurements: Bulls Versus Cows and Calves. *Plains Anthropologist* 36:215-227.

1992 Appendix 2: Bison Size and Gender at the Gowan Sites. In *The Gowen Sites: Cultural Responses to Climatic Warming on the Northern Plains* (7500-5000 B.P.). E.G. Walker, ed. Pp.203-208. Archaeological Survey of Canada Mercury Series Paper 145, Ottawa: Canadian Museum of Civilization.

Peterson, Robert R., and Susan S. Hughes
1980 Appendix 2: Continuing Research in Bison Morphology and Herd Composition Using Chronological Variation in Metapodials. In *The Vore Site, 48CK302, A Stratified Buffalo Jump in the Wyoming Black Hills. Charles A. Reher and George C. Frison*, eds. Pp.170-190. Plains Anthropologist Memoir 16, Lincoln: Plains Anthropological Society.

Reher, Charles A., and George C. Frison
1980 *The Vore Site, 48CK302, A Stratified Buffalo Jump in the Wyoming Black Hills*. Plains Anthropologist Memoir 16, Lincoln: Plains Anthropological Society.

Reitz, Elizabeth J., and Barbara Ruff
1994 Morphometric Data for Cattle from North America and the Caribbean Prior to the 1850s. *Journal of Archaeological Science* 21:699-713.

Roberts, Linda J.
1982 *The Formulation and Application of a Technique, Based on Phalanges, for Discriminating the Sex of Plains Bison (Bison bison bison)*. Unpublished M.A. thesis, Anthropology Department, University of Manitoba.

Roe, F.G.
1970 *The North American Buffalo: A Critical Study of the Species in Its Wild State* (Second Edition). Toronto: The University of Toronto Press.

Scott, Kathleen M.
1990 Postcranial Dimensions of Ungulates as Predictors of Body Mass. In *Body Size in Mammalian Paleobiology: Estimation and Biological Implications*. John Damuth and Bruce J. MacFadden, eds. Pp.301-335. Cambridge: Cambridge University Press.

Semken, Holmes A., and Carl R. Falk
1987 Late Pleistocene/Holocene Mammalian Faunas and Environmental Changes on the Northern Plains of the United States. In *Late Quaternary Mammalian Biogeography and Environments of the Great Plains and Prairies*. Russell W. Graham, Homes A. Semken, and Mary Ann Graham, eds. Pp.176-313. Illinois State Museum Scientific Papers Volume 22, Springfield: Illinois State Museum.

Shackleton, D.M., L.V. Hills, and D.A. Hutton
1975 Aseccts of Variation in Cranial Characters of Plains Bison (*Bison bison bison* Linnaeus) from Elk Island National Park, Alberta. *Journal of Mammalogy* 56:871-887.

Smiley, F.E.
1979 *An Analysis of Cursorial Aspects of the Biomechanics of the Forelimb in Wyoming Bison*. Unpublished M.A. thesis, Anthropology Department, University of Wyoming.

Speth, J.
1983 *Bison Kills and Bone Counts: Decision Making by Ancient Hunters*. Chicago: University of Chicago Press.

Todd, Lawrence W.
1986 *Determination of Sex of Bison Upper Forelimb Bones: The Humerus and Radius*. Wyoming Archaeologist 29(1/2):109-123.

1987 *Taphonomy of the Horner II Bone Bed. In The Horner Site: The Type Site of the Cody Complex*. George C. Frison and Lawrence C. Todd, eds. Pp.107-198. Orlando: Academic Press.

Van Dyke, S., S. Hanna, W. Unfreed, and B. Neal
1991 That Dam Archaeology: Campsites Archaeology in the Oldman River Reservoir. In *Archaeology in Alberta, 1988 and 1989*. M. Magne, ed. Pp.25-65. Archaeological Survey of Alberta Occasional Paper No.33, Edmonton: Alberta Culture.

van Zyll de Jong, C.G.
1986 *A Systematic Study of Recent Bison, with Particular Consideration of the Wood Bison (Bison bison athabascae* Rhoads 1898). Ottawa: National Museums of Canada Publications in Natural Sciences No.6.

Vigne, Jean-Denis
1988 *Les Mammifères de Corse: Etude Archéozoologique* 26. Paris: Supplement to Gallia Préhistoire.

Walde, D.
1985 *On Assigning Gender to Post-cranial Bison Bones*. Unpublished Master's thesis, Department of Anthropology, University of Victoria.

Walker, Danny N.
1987 Late Pleistocene/Holocene Environmental Changes in Wyoming: The Mammalian Record. *In Late Quaternary Mammalian Biogeography and Environments of the Great Plains and Prairies*. Russell W. Graham, Homes A. Semken, and Mary Ann Graham, eds. Pp.335-392. Illinois State Museum Scientific Papers Volume 22, Springfield: Illinois State Museum.

1992 A *Bison bison bison* Skull from 48CR4897, Seminoe Reservoir, Carbon County Wyoming. *Wyoming Archaeologist* 36:71-84.

1995 Body Size in Bison: A Review of the Field and Presentation of New Evidence. Paper presented at the 53 Plains Conference, October 1995, Laramie Wyoming.

Wells, R.T.
1978 Fossil Mammals in the Reconstruction of Quaternary Environments with Examples from the Australian Fauna. In *Biology and Quaternary Environments*. D. Walker and J.C. Guppy, eds. Pp.103-124. Canberra: Australian Academy of Science.

Wendland, Wayne M., Ann Benn, and Holmes A. Semken
1987 Evaluation of Climatic Changes on the North American Great Plains Determined from Faunal Evidence. In *Late Quaternary Mammalian Biogeography and Environments of the Great Plains and Prairies*. Russell W. Graham, Homes A. Semken, and Mary Ann Graham, eds. Pp.460-472. Illinois State Museum Scientific Papers Volume 22, Springfield: Illinois State Museum.

Wilson, M. C.
1975 *Holocene Fossil Bison from Wyoming and Adjacent Areas*. Unpublished M.A. thesis, Anthropology Department, University of Wyoming.

1978 Archaeological Kill Site Populations and the Holocene Evolution of the Genus *Bison*. In *Bison Procurement and Utilization: A Symposium*. Leslie B, Davis and Michael C. Wilson, eds. Pp.9-22. Plains Anthropologist Memoir 14, Lincoln: Plains Anthropological Society.

1980 Morphological Dating of Late Quaternary Bison on the North American Plains. *Canadian Journal of Anthropology* 1:81-85.

1992 Bison in Alberta: Paleontology, Evolution, and Relations with Humans. In *Buffalo*. J. Foster, D. Harrison, and I. MacLaren, eds. Pp.1-17. Edmonton: The University of Alberta Press.

Zeimens, George M.
1982 Analysis of Postcranial Remains. In *The Agate Basin Site: A Record of the Paleoindian Occupation of the Northwestern High Plains*. George C. Frison and Dennis J. Stanford, eds. Pp.213-240. New York: Academic Press.

Zeimens, George M. and S. Zeimens
1974 Volumes of Bison Astragali. In *The Casper Site: A Hell Gap Bison Kill on the High Plains*. George C. Frison, ed. Pp.245-246. New York: Academic Press.

PLEISTOCENE PREDATORS AND PEOPLE IN EASTERN BERINGIA: DID SHORT-FACED BEARS REALLY KEEP HUMANS OUT OF NORTH AMERICA?

Paul E. Matheus
Alaska Quaternary Center, University of Alaska Museum
and The Institute of Arctic Biology, University of Alaska Fairbanks

Geist (1989, 1994) developed an ecological model in which he hypothesizes that aggressive large carnivores in Pleistocene North America prevented humans from crossing the Bering Land Bridge and gaining a permanent foothold in eastern Beringia (unglaciated Alaska and the western Yukon Territory) prior to around 12,000 B.P. According to Geist, large carnivores in Pleistocene North America were more diverse and more aggressive than were those in the Old World – a scenario punctuated by the presence of the giant short-faced bear (*Arctodus simus*) in North America. Geist further contends that humans had limited technologies for dealing with the threat presented by North America's large predators, especially a super-huge bear, the likes of which people had presumably never before encountered (Figure 1). Only with the extinction of these "super-predators" at the end of the Pleistocene could humans safely enter North America. This intriguing hypothesis has been largely ignored by anthropologists concerned with the peopling of the New World. But Geist has presented enough compelling data and arguments in support of his hypothesis that make this notion about human-predator interactions worthy of further consideration.

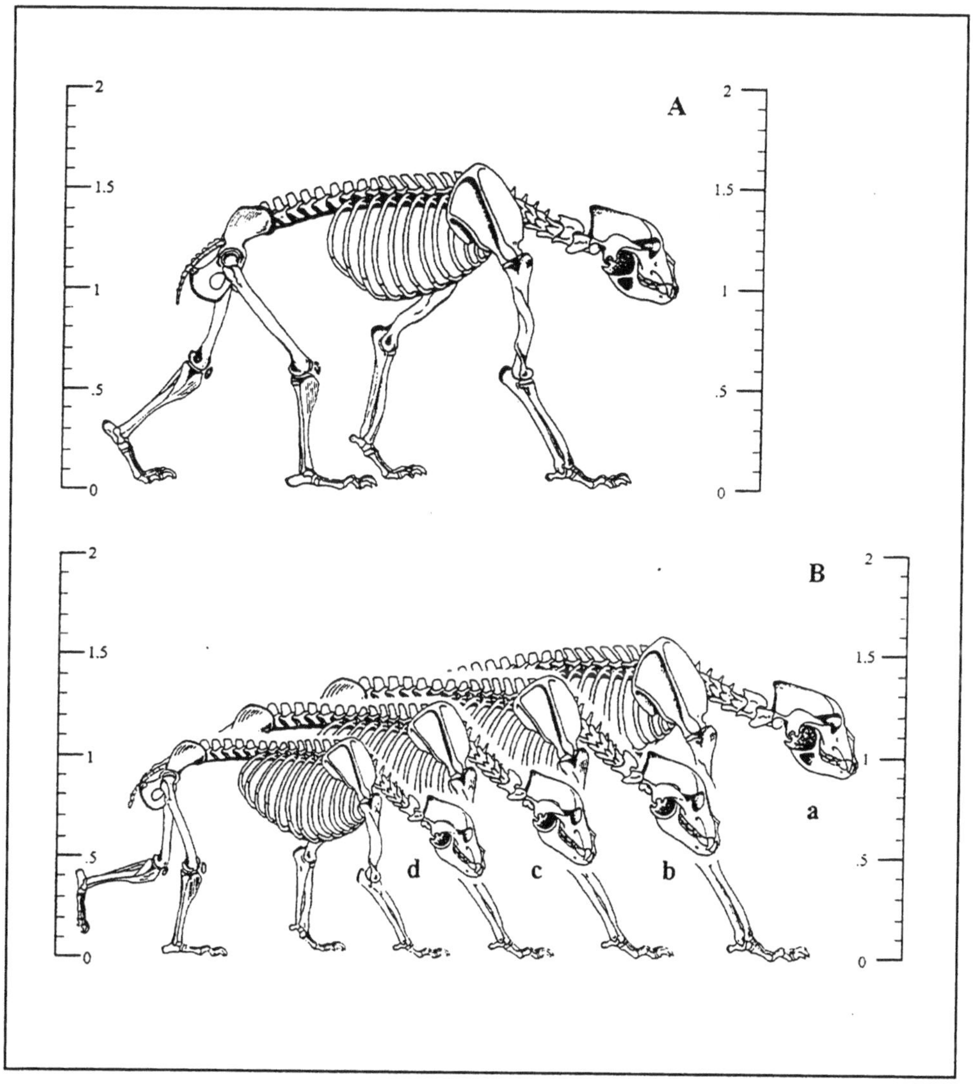

Figure 1. A) Reconstruction of the short-faced bear's skeletal conformation. B) Size comparison between short-faced bears and various brown bear morphs: a = large male short-faced bear, b = largest size attained by coastal brown bears today (could also represent the largest modern polar bear), c = large male brown bear typical for eastern Beringia during the Pleistocene (also approximate size of a large male modern coastal brown bear), d = large male grizzly from modern interior Alaska (units in meters) (original illustrations by author).

In this paper, I examine the biological assumptions and fossil data upon which Geist's hypothesis rests, and I present new information about *Arctodus*, including behavioral, ecological, and chronological data that help us to more accurately infer the potential interactions between humans and short-faced bears during the late Pleistocene. While Geist's hypothesis is biologically sound, the fossil record does not unequivocally support his notions about Pleistocene large carnivores. Moreover, there is a large chronological gap between the apparent timing of *Arctodus*' extinction in Beringia, and the time that humans were poised to enter Alaska.

SUMMARY OF GEIST'S HYPOTHESIS

Mammalian Faunas of North America and Eurasia

In order to follow Geist's arguments, it is important to understand his vision of Pleistocene North America. He proposed that North American Rancholabrean[1] large mammals south of Beringia evolved under an intense co-evolutionary predator-prey regime. According to Geist, North American (Rancholabrean) ungulates were larger and developed more elaborate horns and antlers compared to their Eurasian relatives. This, Geist asserted, occurred as an anti-predatory response in North America where predators were supposedly more numerous, larger bodied, and more aggressive. Geist portrayed North America's predators and ungulates as engaged in a sort of co-evolutionary arms race. He also argued that both ungulate and carnivore diversity, and thus competition, was higher in North America than in Eurasia, providing further selection for increased body size.

As evidence for his model, Geist contends that fossils of late Pleistocene ungulates in North America were larger than ungulates in Europe or Asia. Moreover, Geist argued, when Eurasian species migrated to the New World they quickly evolved larger body sizes and enlarged horns/antlers in response to competition. As examples, he cites the observation that *Bison*, *Ovis*, and *Cervalces* all became larger and developed more elaborate horn/antler structures after migrating from Eurasia to North America in the mid to late Pleistocene. Thus, Geist used the morphology of North American ungulates as indicators of an intensified predatory regime.

In Geist's opinion, North America's highly co-evolved Rancholabrean fauna, excluding eastern Beringia, was composed primarily of ecological specialists such as long-horned bison (*Bison longifrons*) and various species of equines and antilocaprids. The Siberian fauna, from which a considerable number of North America's modern mammals are derived, was composed of species that were relatively less specialized. These "cold climate generalists" (Geist 1989:282) from Siberia included such animals as brown bear (*Ursus arctos*), wapiti (*Cervus elaphus*), caribou (*Rangifer tarandus*), wolverine (*Gulo gulo*), and steppe bison (*Bison priscus*).

According to Geist, North American and Siberian faunas met and mixed in eastern Beringia during the Wisconsinan glacial (~115,000-12,000 B.P.). The purportedly smaller and less specialized Siberian mammals found it difficult to make inroads into competitive communities in North American ecosystems south of Beringia, the only successful ones being those that adapted to the more intense predatory regime via increased body and/or antler/horn size. As examples, Geist asserts that herbivores like *Bison*, *Ovis*, and *Cervalces*, increased body size and horn/antler size, while carnivores, such as the lion (*Panthera leo atrox*) became larger after migrating to areas of North America south of Beringia. Geist was arguing that Rancholabrean specialists and super-predators formed an ecological barrier to the spread of "ill-adapted" Siberian mammals including humans. In the closing millennia of the Pleistocene, Rancholabrean ecosystems began to collapse, culminating in the extinction of many large mammals, including its super-predators and super-competitors. Only then, Geist contends, could humans and other Siberian species that were bottled up in Beringia spread south into the rest of North America.

Human Biogeography

According to Geist, humans inhabited northeastern Asia and were poised to enter North America in mid-Wisconsinan times (~40,000-30,000 B.P.). He believes that there is archaeological evidence showing that humans even made a few mid-Wisconsinan forays into eastern Beringia (Geist 1989:283). Geist's theory is that these early occupations of Alaska and the Yukon were ultimately untenable because some of North America's super-aggressive large Rancholabrean predators also lived in the region, especially the putative king of super-predators, the giant short-faced bear. He argues that there was no room for an additional predator of large game in the "super-competitive" and "super-predatory" world of Rancholabrean communities south of Beringia, particularly a small bipedal predator equipped only with light weapons. Indeed, late Paleolithic weapons seem to have been mostly designed to kill small, thin-skinned game at long range, primarily reindeer (Guthrie 1983). Moreover, Geist suggests that predatory pressures of Rancholabrean North America would have turned human hunters into prey. To this effect, Geist (1994) made the following statement summarizing his position on the role of *Arctodus* in human expansion into the New World:

> "Perhaps only with the disappearance of the short-faced bear - humankind's single most ferocious predator - could New World hunters live off ungulates like the proverbial mice in cheese."

Geist argues that without firearms, humans have difficulty contending with large, aggressive predators, particularly bears (Geist 1989:290-291). Geist tried to show, for instance, that brown bears approach the upper limit of carnivore size and aggressiveness that human hunters can safely deal with while either hunting bears directly, or when defending themselves and their possessions. According to Geist, large bodied subspecies of brown bears and polar bears probably exceed safe limits (without the use of dogs), and short-faced bears would have far surpassed these limits. Furthermore, special weaponry is needed to hunt bears, including lances with extraordinarily long, stout shafts tipped with broad, sturdy points; according to Geist, there is no evidence for these types of weapons in the late Paleolithic on either side of Beringia. Even with lances and dogs, bear hunting and defending oneself against bears is risky business - too risky if the bear is larger than an average grizzly bear (around 300 kg) (Geist 1989:290). Escaping bears by climbing trees was not a practical option, either, on Beringia's mostly treeless landscape.

The main thrust of Geist's hypothesis, then, is that expanding populations of late paleolithic hunters in northeastern Asia had no means by which to protect themselves from predation by short-faced bears, the dominant aggressive predator in North American ecosystems. Furthermore, when hunters killed game they did not have adequate means to protect their kills from North America's "single most ferocious predator" (Geist 1989).

TESTING GEIST'S HYPOTHESIS WITH THE FOSSIL RECORD

Mammalian Faunas of Western Beringia, Eastern Siberia, and Central North America

Geist placed considerable emphasis on perceived differences among Pleistocene faunas in western Beringia (Siberia), eastern Beringia (Alaska and the western Yukon), and areas of North America south of Beringia. He argued that 1) North American faunas south of Beringia were characterized by large "super-aggressive," "super-competitive," and "super-specialized" mammals, 2) Siberian mammals were less aggressive and less competitive, ecological generalists, and 3) eastern Beringia contained a mix of Siberian and North American species. Two critical aspects of this scenario play into Geist's hypothesis regarding human migration. First, super-aggressive predators in North America south of Beringia absolutely excluded humans from that region. Second, mammalian faunas in eastern and western Beringia differed significantly from each other. In particular, eastern Beringia had enough aggressive "North American" predators to make life difficult, but not impossible, for humans in Alaska and the Yukon.

I would argue that there was not a large difference among the mammalian fauna of eastern and western Beringia. Beringia's biotic history, especially the faunal connections between Beringia and North America, is better explained by paleogeography, namely the coupled effects of sea level changes and the size of the Laurentide and Cordilleran ice sheets, than it is by competitive exclusion or super-predators.

During glacial maxima like the late Wisconsinan maximum, 21,000-16,000 B.C., global sea levels were lower, and the continental shelf in the region of the Bering Straits became dry, connecting eastern and western Beringia via the Bering Land Bridge. However, according to our best evidence, the Laurentide and Cordilleran ice sheets coalesced in central Canada during glacial maxima, effectively blocking passage between eastern Beringia and the rest of North America (Levson and Rutter 1996).[2] Siberian mammals were able to enter Alaska and the western Yukon Territory via the Bering Land Bridge, but could not proceed any further east or south into North America (Guthrie 1990a, 1990b). During periods of climatic amelioration, warmer and wetter phases such as that of the last interstadial, (65,000-28,000 B.C.), glaciers retreated somewhat and an ice-free corridor re-connected eastern Beringia with areas of the North American continent further south. For most species in eastern Beringia, such as muskox (*Ovibos moschatus*), caribou (*Rangifer tarandus*), woolly ponies (*Equus caballus*), saiga antelope (*Saiga tatarica*), and woolly mammoth (*Mammuthus primigenius*), the open corridor did not provide much opportunity to spread very far south because these species were essentially adapted to cold and/or arid environments. However, mammals from central North America would have been able to expand north during interglacials and interstadials, both because of the open corridor (retreated glaciers) and because of the return of more temperate, mesic conditions in eastern Beringia, to which these species were adapted. During warmer/wetter climatic periods, however, sea levels were high enough to inundate the Bering Land Bridge and prevent North American mammals from entering Siberia via Alaska. The distribution of mammals in eastern Beringia, western Beringia, and North America may thus be explained effectively by the paleogeography of these regions without invoking competition and predation.

It is Geist's contention that more intense predatory pressure in North America led ungulates living there to evolve larger body size and larger, more elaborate antlers/horns, which in turn, provides selective pressure for North American predators to become even larger and more aggressive. Geist also argues that there simply were more species of large predators in Pleistocene North America during the mid-to-late Pleistocene. While there were some obvious differences in the identity and nature of large predators in North America and Eurasia, these reflect distinct biogeographic legacies, such as those outlined above. The notion that there were significant differences in the body sizes, diversity, and levels of aggression of predators in these regions is challenged below.

CARNIVORE DIVERSITY IN NORTH AMERICA, BERINGIA, AND EURASIA

North America

Were there more large predators in North America, and were they bigger than their Old World counterparts? Figure 2 lists the species that Geist considered to be part of the large predator fauna in North America, excluding Beringia, during Wisconsinan times. The diversity of predators in Geist's list is misleading, however, since a number of the species are not even significantly carnivorous, much less predatory or aggressive. Carnivorans in this latter group include species such as the Florida cave bear (*Tremarctos floridanus*), the black bear (*Ursus americanus*), and probably the obscure and primitive lesser short-faced bear *(Arctodus pristinus)* (Kurtén and Anderson 1980).

More importantly, the fossil record of carnivores in North America (i.e., their chronology and distribution) provides little reason to believe that more than about half of the species on Geist's list were sympatric at any given time

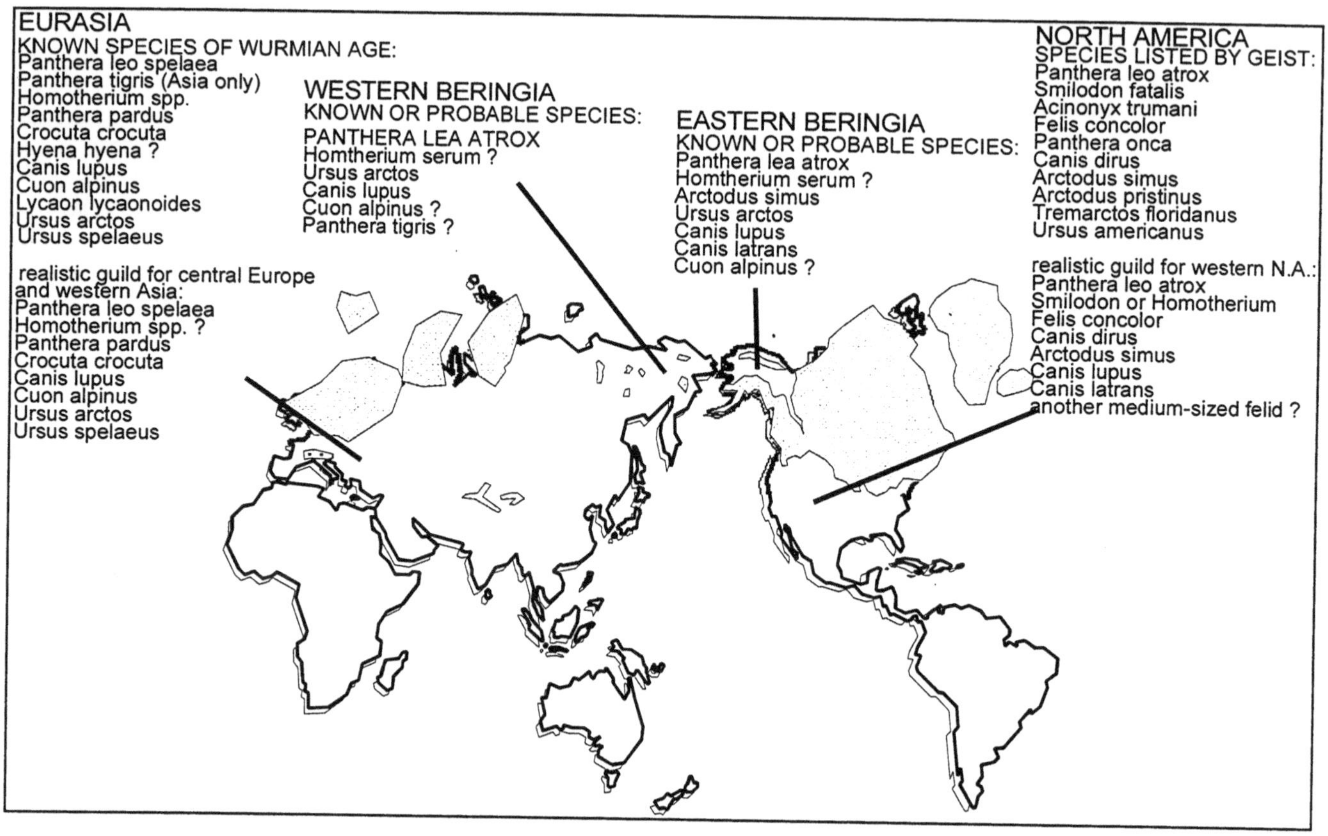

Figure 2. Disparate reconstructions of large carnivore guilds in Eurasia and North America during the late Pleistocene (Würm/Wisconsinan).

(Kurtén and Anderson 1980; Harris 1985; Graham and Lundelius 1995). Lesser short-faced bears, for example, mainly inhabited North America's eastern seaboard and the south-central United States (Kurtén and Anderson 1980; Richards *et al.* 1996), and jaguar (*Panthera onca*) only inhabited tropical, subtropical, and mild temperate habitats, mostly in the southern and western United States. Jaguars expanded up into the northern United States only during warm periods prior to the Wisconsinan Glaciation (Kurtén 1973a; Seymour 1993). The most diverse, yet plausible Rancholabrean large carnivore guild conceivable for North America might include: short-faced bears, lions, a single sabertooth species (*Smilodon* or *Homotherium*), pumas, perhaps another medium sized felid (cheetah on the plains, jaguar in warm, forested environments), dire wolves, coyotes, and perhaps grey wolves in the north (Figure 2). Most large carnivore guilds in North America, however, probably exhibited only about half this much diversity.

Geist did not include grey wolves in his list of North American predators presumably because he followed the commonly-held belief that grey wolves did not occur in substantial numbers in most of Pleistocene North America, other than in eastern Beringia and periglacial regions in the mid-continent (northern U.S. and southern Canada) (Harington 1977; Kurtén and Anderson 1980). Proponents of this idea argue that it was not until about 12,000 B.P. that grey wolves expanded to become North America's dominant predator. Seldom do grey wolves and dire wolves co-occur at fossil localities, and it has been suggested that competition between the two species kept grey wolves out of central and southern North America (Kurtén and Anderson 1980). The majority of fossil grey wolves south of the ice sheets are Wisconsinan age, mainly late Wisconsinan, and most are from northern states. However, some Illinoisan-age fossils have been found in Nebraska and Arkansas (Nowak 1979; Graham and Lundelius 1995). Contrary to Geist, I think there are enough late Pleistocene records of grey wolves in the lower 48 states and southern Canada to consider them to be a significant Rancholabrean predator, at least during the Wisconsinan Glaciation (Nowak 1979, Harris 1985, Graham and Lundelius 1995).

The coyote, while not on Geist's list, was a fairly common Rancholabrean canid throughout most of North America. Coyotes can kill large, healthy, male whitetail deer (*Odocoileus virginianus*) (Hilton 1992), and considering that the Pleistocene coyote was larger than the present form, it was probably also a significant predator of large game (Nowak 1979; Kurtén and Anderson 1980; Harris 1985, Stock and Harris 1992; Graham and Lundelius 1995). The coyote's threat to humans may have been minimal, however, and it is probably for this reason that Geist did not give it any consideration.

Beringia

Since the earliest humans entered the New World through Beringia, it is important to review the record of Beringia's late Pleistocene large carnivores (Figure 2). During the

Wisconsinan glaciation, the following large carnivores were present in eastern Beringia: lions (*Panthera leo atrox*), grey wolves (*Canis lupus*), short-faced bears (*Arctodus simus*), brown bears (*Ursus arctos*), and probably scimitar cats (*Homotherium serum*). Dholes (*Cuon alpinus*) and perhaps pumas (*Felis concolor*) also may have been present at times.[3] Wolverines (*Gulo gulo*) played a role in this guild, too, but the wolverine is not discussed further here because it poses no immediate threat to humans (Péwé and Hopkins 1967; Péwé 1975; Guthrie 1968, 1976, 1982, 1990a; Harington 1977, 1978; Kurtén and Anderson 1980; Matheus 1997). Radiocarbon dates on carnivore fossils from eastern Beringia are quite rare, so it is difficult to equivocally establish the contemporaneity of all the carnivores listed above, or even whether they all survived into the late Wisconsinan.

Guthrie (1996:pers. comm.) suspects that Scimitar cats died out in Beringia before the Wisconsinan, based on their dark staining and general rarity among Beringian fossils. But Harington (1977) reports at least one specimen from near Dawson, Yukon Territory, that he assigns to the late Wisconsinan because it appears "very fresh" (Harington 1977:531).

Evidence for the presence of dhole in east Beringia comes from only two fossils, one from Cripple Creek near Fairbanks and the other from Old Crow Locality 14N in the Yukon Territory (Harington 1977; Kurtén and Anderson 1980). Both Beringian specimens are suspected of being Illinoisan age, and a fossil dhole of probable Wisconsinan age has been found at San Josecito Cave in Mexico (Kurtén and Anderson 1980). In Asia, where the dhole still lives today, its fossils occur in early and mid Pleistocene deposits as far north as Choukoutien, China (Pei 1934), but dhole fossils are yet to be found in western Beringia. Since dholes primarily inhabit mountainous woodlands, the dhole most likely was present mainly in Alaska and the Yukon Territory during the last interstadial and interglacial periods, but like other species from these warm periods, the dhole left few fossils. Dholes still live in woodlands of southern Siberia today. Geist's reconstructions probably overestimate carnivore diversity, so alternative, realistic guilds are presented.

The Rancholabrean large carnivore guild of Eastern Beringia did not differ significantly from large carnivore guilds further south in North America. Most notably, eastern Beringia lacked: 1) *Smilodon*, although *Homotherium* was present, 2) another medium-sized felid, particularly puma (see footnote 4), and 3) the dire wolf.[4] Furthermore, all carnivores that were present in eastern Beringia also were present elsewhere in North America, with the exception of the brown bear[5] and probably the dhole. Therefore, eastern Beringia's carnivore fauna was essentially a less diverse, northern version of the Rancholabrean fauna with species particularly adapted for open habitats.

The difference between carnivore guilds in eastern Beringia and the rest of North America is not as critical to Geist's model as is the difference between guilds in eastern and western Beringia. A key test of Geist's hypothesis is whether large carnivore guilds differed significantly between western Beringia, where humans co-existed with the carnivore guild, and eastern Beringia, where they apparently did not. According to Geist's reasoning, eastern Beringia should have had more aggressive large carnivores. We know little about the large carnivores of late Pleistocene western Beringia because few carnivore fossils have been recovered there. Those found suggest that carnivores inhabiting western Beringia did not differ substantially from those in the east. They consist mostly of wolf and lion, with a few brown bear and a rare scimitar cat (Sher 1974, 1986, 1987) (Figure 2).[6]

The only large carnivore in eastern Beringia that did not also inhabit western Beringia was the short-faced bear; Alaska was the northern and western limit of its range (Kurtén and Anderson 1980; Richards *et al.* 1996). Athough it was prevalent in Alaska, *Arctodus* apparently did not cross into Siberia. This is significant because Geist's hypothesis points to *Arctodus* as a conspicuous species that could have excluded humans from eastern Beringia.

My interpretation of the fossil record is that, except for the presence of *Arctodus* in Alaska and the Yukon, the carnivore guilds of eastern and western Beringia did not differ substantially, especially qualitatively. It is true, however, that not much farther west and south of Beringia (i.e., northern China), typical Eurasian elements start to appear, like spotted hyenas (*Crocuta crocuta*), leopards (*Panthera onca*), and tigers (*Panthera tigris*) (Pei 1934; Colbert and Hooijer 1953; Kurtén 1968; Vereshchagin and Baryshnikov 1982, 1984). Many of these species would seem to be aggressive carnivores with whom humans had learned to live before entering either western or eastern Beringia.

Eurasia

A comparison of North American carnivores with those of Europe during the last (Wisconsinan/Würm) glaciation shows that it is difficult to substantiate Geist's claim of a more intense predatory regime in North America. Europe provides a good comparison because its fossil record is excellent, and because predators did not exclude humans from Europe during the late Würm.

European faunas have been highly diverse in the past because, in addition to indigenous species, the region has functioned as a crossroad for migrating species originating in Asia and Africa (Kurtén 1960, 1968; Biberson 1970; Azzaroli and Ambrosetti 1970; Azzoroli 1990; Masini and Torre 1990; Qiu 1990). Figure 2 lists large the carnivore fossils known from the Würm period of Europe, and presents a summary of those species that would be found in a typical large carnivore guild of central Europe (for general reference see Kurtén 1968, 1976b; Vereshchagin and Baryshnikov 1982, 1984; Anderson 1984). This guild includes: lions (*Panthera leo spelaea*), spotted hyenas (*Crocuta crocuta*), leopards (Panthera pardus), grey wolves (*Canis lupus*), brown bears (*Ursus arctos*), cave bears (*Ursus spelaeus*) (see qualifications below), and at times scimitar cats (*Homotherium latidens*), dholes (*Cuon alpinus*), hunting dogs (*Lycaon lycaonoides*), and/or striped hyenas (*Hyaena hyaena*).

As with Geist's list of North American carnivores, perhaps not all of the species on this European list would have been sympatric, and thus is not a true guild. Lions, spotted hyenas, wolves, and cave bears are the most common fossil carnivores from late Pleistocene Europe, while others like the hunting dog and scimitar cat apparently were uncommon (Kurtén 1968). Striped hyenas, too, may have been extinct in Europe before the Würm (Thenius 1980). Some of the more renowned European giants, like the hunting hyena (*Chasmaporthetes spp.*), the short-faced hyena (*Hyena brevirostris*), and the sabertooth cat (*Megantereon spp.*) also were extinct in Europe by the end of the late middle Pleistocene (Riss Glaciation, ~130,000 B.P.) (Kurtén 1968; Kurtén and Garevski 1989; Anderson 1984).

There is compelling morphological and stable isotope data demonstrating that cave bears were primarily herbivorous (Kurtén 1976b, Bocherens *et al.* 1990). Still, it is reasonable to infer that cave bears were quite dangerous to humans. It is believed that bears, in general, have evolved their large body size and belligerent temperament, at least in part, as a means of defending themselves and their young (Kurtén 1976b; Herrero 1978; Stirling and Derocher 1990). Indeed, all species of bears can behave aggressively when provoked. Geist invoked the cave bear in his hypothesis, arguing that cave bears are a good example of a bear that was probably too big, and thus too dangerous, for humans to safely defend themselves against using pre-firearm technology. Geist bases this argument partly on the lack of archaeological evidence for cave bear hunting in paleolithic sites. But whereas Geist used these inferences about human-cave bear interactions to argue that humans are vulnerable to large Carnivorans, it seems equally reasonable to use these inferences to argue that humans have been able to co-exist with aggressive Carnivorans.

Carnivore faunas certainly had a different character in late glacial Europe than in North America, but the European fauna does not appear to have been any less diverse, and it would seem difficult to argue that overall predatory pressures were any less intense in Europe. Ungulates and humans in Europe minimally had to deal with various giant felids, huge hyenas, packs of predatory canids of one type or another, big brown bears, plus the cave bear. The existence of Paleolithic archaeological sites indicates that humans managed to coexist with these species and even successfully hunt most of them.

Body Size: Were North American Predators Larger and More Aggressive than in Europe and Asia?

As further evidence of more intense predation in North America, Geist contends that North America's carnivores were larger bodied compared to conspecifics in the Palearctic (north-temperate Europe and Asia). I would argue that evidence for this is to the contrary, or at least equivocal. Geist, for example, argued that lions in North American were the largest-bodied subspecies (*Panthera leo atrox*) of Pleistocene lions, and that the largest individuals of this subspecies were found in eastern Beringia. Guthrie (1976) made a similar contention based on his personal observation of lion skulls in Europe and North America, but neither Geist nor Guthrie have published morphometric data to support their statements. While published morphometric data sets (Merriam and Stock 1932; Harington 1969, 1977; Vereshchagin 1971; Kurtén 1968, 1985; Kurtén and Anderson 1980) demonstrate that Pleistocene lions were larger than modern African lions, the data categorically show that Pleistocene North American lions were not significantly larger than those were from Europe or Asia. In fact, craniometric data in Kurtén (1985) specifically show that not only was the European subspecies (*Panthera leo spelaea*) bigger than the North American subspecies *(Panthera leo atrox)*, but that Beringian lions were smaller than either of those subspecies. Guthrie (1997:pers. comm.) disagrees in part with this assessment. He thinks that large European lion fossils are too rare to draw a conclusion on this point, and that the size of lions in Europe is overstated.

Very large lion skulls have been found in eastern Beringia, but this probably results from the fact that an unusually large number of fossils have been collected there, providing a better sample of eastern Beringia's overall lion population, including more samples from the extreme ends of the size spectrum. The largest specimens have garnered the most attention, and in the past collectors had a decided bias for large specimens. Beringia's reputation for large-bodied individuals seems to extend to other species, including short-faced bears. It is becoming apparent, however, that many short-faced bear fossils from areas south of Beringia are in the same size class as the famous large specimens from Alaska and the Yukon Territory (Richards and Turnbull 1995, Richards *et al.* 1996, Matheus 1997). I believe that Beringia's reputation for supporting some of the largest-bodied mammals of the Pleistocene is unjustified.[7]

Data on fossil jaguars also dispute Geist's notion that late Pleistocene carnivores in North America necessarily evolved larger body sizes. Jaguars evolved in the New World from a larger, archaic species of *Panthera* that migrated from Eurasia during the late Blancan/early Irvingtonian, about two million years ago. This ancestor, which probably also gave rise to lions and tigers in the Palearctic, was much larger than modern jaguars; relative to its ancestors, jaguars have grown steadily smaller throughout the Pleistocene (Simpson 1941; Kurtén 1965a, 1973a; Kurtén and Anderson 1980, Seymour 1993).

Leopards, hunting dogs, spotted hyenas, dholes, brown bears and wolves of Würmian Europe all were larger than modern conspecifics (see Kurtén 1968, and references cited below for each species). Of these species, leopards, hunting dogs (*Lycaon lycaonoides*), and spotted hyenas never migrated to North America, but they were highly successful in Europe and Asia (Kurtén 1968, 1976b). Compared with modern specimens and with Pleistocene fossils from Africa, these three species all were much larger during the Würm in Europe.[8]

Hunting dogs are canids adapted for very swift, high-endurance locomotion using highly co-ordinated group hunting tactics to capture small to medium-sized ungulates in open terrain (Kruuk and Turner 1967; Kruuk 1972, Bertram 1979). They originated in Africa, but apparently inhabited

Europe during full glacial times when steppes and periglacial environments dominated (Kurtén 1968). The dhole also is a social canid with high endurance, but it specializes in large woodland prey (Davidar 1975). Therefore, at any given time in the Pleistocene, perhaps only one of these canids was present in Europe. Dhole fossils are more common by far, and have a much longer fossil record in Europe than do hunting dogs, but interglacial and interstadial deposits are better preserved in Europe, which undoubtedly biases the fossil record toward preservation of a higher number of dhole bones of (Kurtén 1968, 1976b; Anderson 1984).

Spotted hyenas reached as far north as 56° in Siberia during the middle to late Pleistocene (Kurtén 1968, Vereshchagin and Baryshnikov 1984), but they never entered North America. Apparently paleolithic humans rarely killed this giant subspecies of spotted hyena (*Crocuta crocuta spelaea*), since its modified remains are rare in paleolithic archaeological sites (Vereshchagin and Baryshnikov 1984). Scimitar cats formerly inhabited parts of Europe, Asia, and North America, and over much of this range they survived until the end of the Pleistocene. A little smaller than a lion, but built with sloping hind quarters and a stouter frame, the scimitar cat did not differ substantially in size from Texas to Great Britain, although late Pliocene to mid Pleistocene scimitar cats in China were smaller than elsewhere in the Holarctic (Turner 1997). Confusion regarding size comparisons in scimitar cats may result from the high degree of sexual dimorphism exhibited by the genus *Homotherium* and felids in general (Martin and Schultz 1975; Kurtén and Anderson 1980; Kurtén 1985; Martin *et al.* 1988, Marean 1989; Kurtén and Werdelin 1990; Turner 1997).

Brown bears, seem to have entered eastern Beringia from northeast Asia sometime in the early or mid Wisconsinan (perhaps earlier), but did not colonize areas farther south until the ice sheets receded, about 12,000-13,000 B.P. (Kurtén 1960, 1963, 1966a, 1968, 1973b, 1976a; Guilday 1968; Kurtén and Anderson 1974, 1980; Harris 1985). As with other mammals, late Pleistocene brown bears in both unglaciated Beringia and Europe appear to have been larger when compared to brown bears inhabiting those regions today (Erdbrink 1953; Kurtén 1959, 1968, 1973b), with the exception of the large bears on Kodiak Island and the Kamchatka Peninsula.[9] Geist's hypothesis specifically predicts that species like the brown bear should have become larger as it entered eastern Beringia, in response to supposedly more intense competition and predation. However, craniometric and dental data (*Ibid.*) show that this is not the case.

Wolves do not conform to Geist's predictions either. The European wolf of the last interglacial (Eem) and glacial (Würm) were somewhat larger than present (Kurtén 1965b, 1968). The picture is less clear for Wisconsinan-age wolves from North America south of Beringia, as very large and very small specimens have been found there, sometimes at the same localities. However, it is not clear if these two morphs were contemporaneous (Nowak 1979). Guthrie (1995: pers. comm.) believes that Beringian wolves were smaller than present day wolves. Based on limited available data, Harington (1977) suggests that late Illinoisan and Sangamonian wolves in the Yukon Territory may have been smaller than modern wolves, but he did not draw conclusions about Wisconsinan wolves. Harington (1977, Table 26) lists mandibular dimensions for two moderate-sized wolves of probable Wisconsinan age, but this is too few to make a firm conclusion about individual body sizes within the population as a whole. Up to now, this widely-held belief that late Pleistocene wolves in eastern Beringia were smaller than the present day wolves has not been systematically tested or documented.

To shed some light on the issue of Pleistocene wolf size, I measured condylobasal lengths on fossil wolf skulls (n = 22) of presumed Wisconsinan-age from the Fairbanks area and compared them to 38 modern skulls (20 males and 18 females) from the same region (Tables 1 and 2 and Fig. 3).[10] The size distribution and statistical analyses of these data support the contention that east Beringian wolves were indeed significantly smaller than present, but only by about 4% (one-tailed two sample t-test, assuming unequal variances: t Critical = 2.0223, T = 3.5127, df = 39; p ($T \le t$) = 0.00057).

Sampling biases, however, may be overestimating the difference between modern and fossil wolves because my measurements on modern wolves were made on specimens in the mammalogy collection in the University of Alaska Museum. Many of these wolves were collected by fur-trappers, and it is unclear what sort of sampling bias this may introduce. On the one hand, trappers typically configure their equipment and operation to capture the largest wolves possible, which would bias the museum's collection towards large specimens. On the other hand, young dispersing wolves are probably most vulnerable to being caught in traps, which could also potentially bias the collection for smaller wolves. However, for both modern and fossil wolves, I intentionally excluded specimens that exhibited juvenile traits, such as clearly open cranial sutures or teeth that were not fully erupted. Even with this precaution, it is easier to exclude juveniles from a collection of modern, complete skulls than from incomplete and damaged fossil skulls.

The distribution of data in Figure 3 suggests that there was a small sampling bias for smaller (younger?) individuals among modern male wolves, based on the data's leftward skew (skewness, g_1 = -0.62337), the skewness in modern females is slight and to the right (g_1 = +0.13531).[11] Among modern wolves, neither males nor females show much variation in skull length (S.D.=7.1 and 7.2, respectively). The combined standard deviation for all modern wolves was 9.5, compared to 10.3 in Pleistocene wolves. Because the sample's standard deviation is similar to the combined standard deviation of modern males and females, I conclude that the size distribution for Pleistocene wolves in Figure 3 is a fairly representative sampling of the population as a whole, and contains roughly equal numbers of males and females.

A slight leftward skew in the Pleistocene distribution (g_1 = 0.3339) suggests there may have been some taphonomic bias favoring the preservation of younger wolves. It is not uncommon for natural attritional death assemblages to preserve a high percentage of young (small) individuals, due

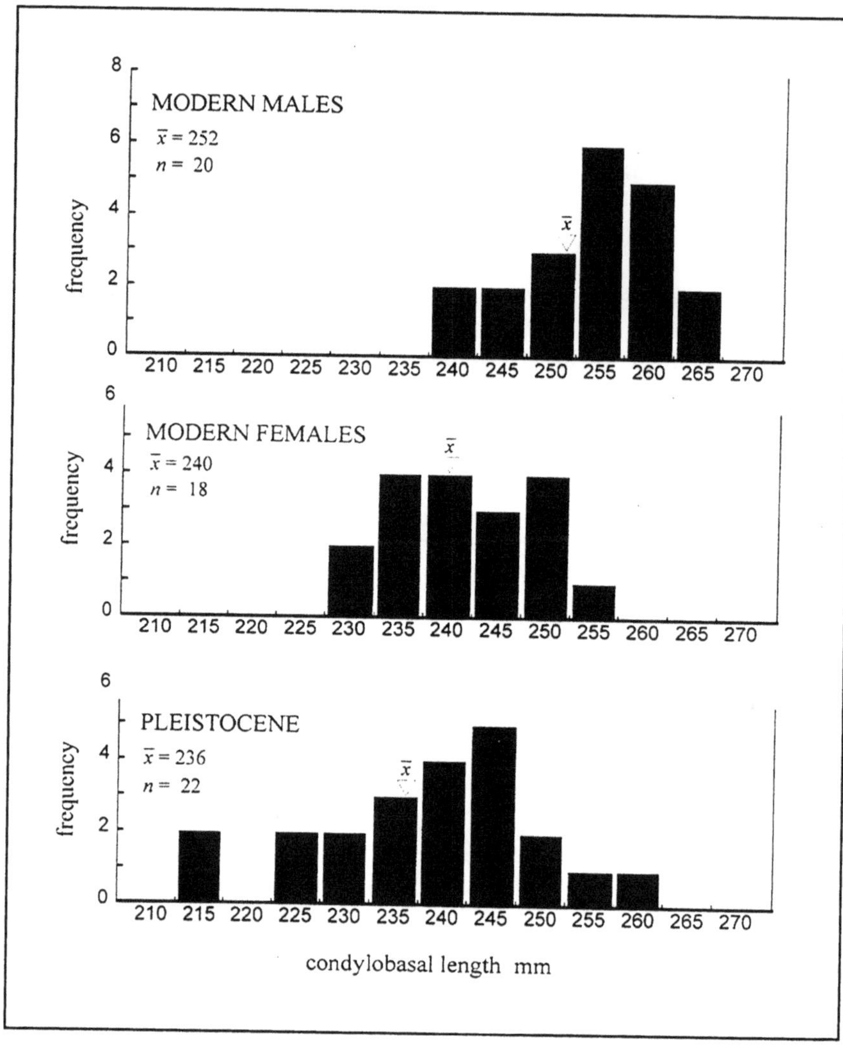

Figure 3. Condylobasal (skull) lengths of late Pleistocene and modern wolves (Canis lupus) from interior Alaska. Unlike most large mammals, interior wolves were actually smaller bodied during the Wisconsinan compared to today, but it is not known whether this reflects an ecophenotypic size change within a continuous population, as opposed to a Holocene replacement by a larger-bodied population. All samples were collected within 200 km of Fairbanks.

to their higher mortality rates compared to pre-senescent adults (Klein and Cruz-Uribe 1984; Haynes 1985, 1990; Lyman 1994). Indeed, among modern wolves, it is the young, dispersing individuals that have the highest mortality rates (Wasser 1996). Juvenile male and juvenile female wolves disperse from natal territories with equal frequency in modern populations, and both experience high mortality during dispersal (Waser 1996), so the size distribution of Pleistocene wolves cannot be used to indicate any discernable difference in mortality among males and females.

Kurtén (1968) published a conceptual diagram, but no data, to show that European wolves were about two percent larger during the last glacial and interglacial, compared to Holocene wolves in Europe. My data show that Alaskan wolves from the last glacial were actually smaller than modern Alaskan wolves by about four percent. Even though body sizes of modern and Pleistocene wolves are significantly different in the two regions, the differences are small, and considering potential taphonomic and sampling biases, it probably is best to conclude that late Pleistocene wolves were *about* the same size as present, both in Europe and eastern Beringia. This is an interesting pattern because it differs from that of most other Pleistocene large mammals.

I can offer two preliminary explanations for this. First, *C. lupus* has been slowing evolving a larger body size ever since its divergence from *C. etruscus* in the Villafranchian/Irvingtonian; thus, *C. lupus* did not experience any full glacial gigantism so commonly exhibited by other large mammals (Pei 1934; Kurtén 1968). Relative to other mammals, grey wolves did not experience any substantial post-Pleistocene diminution, either, because they essentially did not have a "pre-gigantic" size to which to return. Instead, wolves seem to be displaying a continuation of their slow trend toward larger body size, as illustrated by the Alaskan wolf data presented above. Alternatively, we might predict that a higher diversity of carnivores and a more intense competitive regime in the Pleistocene (throughout the Holarctic, not just North America) forced some carnivores to become larger and others to become smaller. This prediction follows from principles of resource partitioning and competitive exclusion, the main mechanisms driving the structure of consumer guilds. During the Pleistocene, wolves had to compete with more and larger sympatric carnivores, most notably lions, than during the Holocene. Because of the lion's larger size and abilities to hunt socially, it is reasonable to assume that lions killed larger prey than wolves. A basic prediction derived from principles of resource partitioning and competitive exclusion is that wolves shifted to smaller prey where they were sympatric with lions, in turn, this placed selective pressure on wolves to evolve smaller body size. With the large-scale range contraction of lions at the end of the Pleistocene/early Holocene, competitive release facilitated the expansion of the wolf's niche to include larger prey, thus shifting the selective regime back to one favoring larger body size.

This hypothesis is consistent with morphometric data on wolves in Europe and Alaska. The morphometric data for Pleistocene wolves are inconsistent with a model that predicts this prevalent carnivore should have been larger in Beringia and the rest of North America than it was in Eurasia. Nor have I seen any data demonstrating that wolves were universally larger in the rest of North America (Nowak's 1979 data is equivocal). Prey diversity was greater in Pleistocene Europe and wolves could have shifted to other prey, even large-bodied prey, to avoid competition with lions. There were fewer prey among larger-bodied herbivores in eastern Beringia; and there, in the face of competition, wolves probably would shift to smaller prey.[12].

In this model of competitive dynamics, increased competition among carnivores should cause some carnivores to evolve smaller body sizes. This is contrary Geist's competitive model, in which increased competition leads to larger body sizes in all competitors. In our fascination with Pleistocene gigantism, and the ecological/evolutionary forces driving it, we may fail to consider that the evolution of smaller, not larger, body sizes may have been selectively advantageous.

The pattern of Pleistocene large carnivores in North America and Asia is similar to the pattern of large carnivores in North America and Europe but in Asia there were even more giant predators. For example, tigers (*Panthera tigris*) grew to a huge size in the Pleistocene and were found in woodlands which ranged far into northern Asia, (Pei 1934; Colbert and Hooijer 1953). The presence of giant lions, hunting dogs, cheetahs, wolves, brown bears, and tigers in Asia must have placed its large carnivore community at least on par with those in eastern Beringia and the rest of North America with respect to carnivore diversity, body size, and aggressiveness.

It is difficult to argue that overall Pleistocene predatory pressures were escalated in North America as compared to in Europe and Asia. Nearly all Pleistocene carnivores were huge throughout their range, not just in North America. While it may be true that their large size was a response to larger bodied prey, this phenomenon was not unique to North America. The significant dimension of the carnivore community of North America missing in Europe and Asia was a super-huge bear.

NEW DATA ON SHORT-FACED BEARS AND THEIR RELEVANCE TO HUMAN HISTORY

Previous Models of Short-Faced Bear Paleoecology

Since the earliest discoveries of short-faced bear fossils in the late nineteenth and early twentieth centuries (Cope 1879, 1891; Lambe 1911; Merriam 1911; Merriam and Stock 1925), the creature has been a biological enigma to paleontologists. *Arctodus* was larger than the largest modern brown bear, yet built with long, gracile limbs and a slender body capped by a large, wide, and robust skull (Fig. 1). It was recognized early that short-faced bears belonged to the subfamily of New World bears referred to as the Tremarctinae, not the Old World bears of the subfamily Ursinae (Cope 1879, 1891). Tremarctine bears share a number of common diagnostic traits, including a masseteric fossa that is split by a bony ridge into an anterior and posterior portion, the presence of an epicondylar foramen in the distal humerus, an anteriorly-posteriorly shortened rostrum, and a host of lesser dental and skeletal traits (Merriam and Stock 1925; Kurtén 1966b, 1967). More significant, however, is the fact that Tremarctine and Ursine bears have not shared a common ancestor for at least five million years, meaning there may have been significant ecological and behavioral differences between short-faced bears and the more familiar modern (Ursine) bears.[13]

Kurtén (1967) published the first thorough overview and interpretation of *Arctodus'* morphology and distribution, culminating in his suggestion that *Arctodus* was a swift predator specialized for hunting Pleistocene megaherbivores. Kurtén's model of *Arctodus*' ecology, re-iterated in Kurtén and Anderson (1980), became the widely accepted paradigm for the species, and Kurtén's 1967 paper is still regarded as the most authoritative work on *Arctodus*. Kurtén's reputation as an expert on fossil bears and the scope of his 1967 publication has perhaps prevented a serious challenge to his model. Today, short-faced bears are sometimes referred to as "running bears."

By way of exception, Emslie and Czaplewski (1985) attempted to refute Kurtén's predatory model by pointing out logical and empirical weaknesses in his morphological assessments and ecological conclusions. They especially challenged the notion that *Arctodus*' limbs were significantly elongated or cursorily adapted, and they argued that its dentition lacked the sectorial modifications expected in a carnivorous bear. They also chose to emphasize similarities between *Arctodus* and its closest living relative, the extant spectacled bear (*Tremarctos ornatus*), which is almost exclusively herbivorous. Citing body size patterns in Eisenberg (1980), these authors further assert that *Arctodus* was well beyond the size range of any known terrestrial carnivore, necessarily relegating it to a life of herbivory (based on energetic constraints). While Emslie and Czaplewski point out some shortcomings of Kurtén's thesis and argumentation, they do not provide support for any particular alternative model, including the herbivore model they favor.

Emslie and Czaplewski's herbivore model has received little acceptance among paleontologists, and over the past 30 years, most authors invariably make reference to *Arctodus* as a carnivore, and specifically a predator (e.g., Harington and Clulow 1973; Harington 1977, 1991; Voorhies and Corner 1982, 1986; Nelson and Madsen 1983; Agenbroad and Mead 1986; Guthrie 1988, 1990a; Agenbroad 1990; Richards and Turnbull 1995; Richards *et al.* 1996). Following this convention, Geist expounded on the implications of the presence of such a large and swift predator on Pleistocene landscapes for humans and ungulates. However, there has been little evidence put forward to support the hypothesis that short-faced bears were carnivores, much less predatory. Without firm evidence of *Arctodus*' diet, ecology, behavior, or niche, any suggestion as to *Arctodus*' influence on human expansion is simply speculative.

A New Model of *Arctodus*' Ecology and Niche

Recently I proposed a new model of *Arctodus'* ecology and niche based on stable isotope data and by applying information gained from recent studies of locomotor mechanics in large mammals (Matheus 1994, 1995, 1997). These data are summarized below.

Stable Isotopes

Stable isotope ratios of carbon (^{13}C:^{12}C) and nitrogen (^{15}N:^{14}N) recorded in the tissues of modern and fossil animals can be used to elucidate certain aspects of animal diets. For instance, stable isotopes have been used to determine the percentage of flesh versus vegetation in an animal's diet, the amount of marine versus terrestrial foods being consumed, and, in some cases, the specific type of vegetation on which a herbivore has fed. The ratio of ^{13}C:^{12}C in terrestrial food webs and consumers is determined primarily by the mechanism of carbon fixation employed by terrestrial plants at the base of food webs (Figure 4). Plants fixing carbon by C_3 photosynthesis generally have $\delta^{13}C$ values in the range of -22 to -31 ‰, while the δ13C values of C_4 plants are in the range of -10 to -16 ‰ (Ehleringer and Rundel 1988).[14] The range of $\delta^{13}C$ values is large in both C_3 and C_4 plants because factors such as the level of water stress, position in a forest canopy (high versus low), and growth form (herb versus shrub versus tree) influence carbon isotope ratios in plants. Herbivores building tissues from a diet of plants have $\delta^{13}C$ values consistent with those of the plants they ate. There is a predictable shift (fractionation) in $\delta^{13}C$ values in the tissues of a herbivore versus those of the plants it eats, but the degree of fractionation is specific to each type of animal tissue; fat, muscle, bone, etc. Bone collagen, the material most often examined in paleodietary studies involving fossil animals, is enriched by about four to five ‰ over dietary plant material in herbivores. However, the tissue-specific $\delta^{13}C$ values of carnivores differ little from those of the animals they consumed. Carbon isotope values in consumers therefore provide some information about the types of plants on the landscape and about environmental conditions. But because carbon isotopes fractionate very little as they pass from one trophic level to another, they provide almost no information about trophic cascades (who is eating who) unless used in conjunction with other information about a consumer's trophic position. If a consumer's trophic position is known, carbon isotopes sometimes can be used to determine the plants or animals on which a consumer fed (Figure 4).

Unlike carbon isotopes, the ratio of nitrogen isotopes in consumers is strongly influenced by trophic dynamics (Figure 4). Food web studies that employ both carbon and nitrogen isotopes tend to be more informative than those that use only one or the other. The "starting point" of nitrogen ratios in terrestrial primary producers is basically determined by whether nitrogen is taken up in mineral form by plants or assimilated as diatomic nitrogen via symbiotic nitrogen-fixing bacteria. In the former case, soil processes that influence the state of nitrogen (i.e., degree of mineralization and reduction) have the greatest influence on $\delta^{15}N$ values in plants, because soil-forming processing fractionate nitrogen isotopes. Thus, environmental factors such as soil moisture content, temperature, and rates of decomposition can influence $\delta^{15}N$ values of soils and plants that derive their nitrogen from them. Symbiotic fixation of diatomic nitrogen results in plant $\delta^{15}N$ values that are not significantly different than that of atmospheric nitrogen, which by definition is 0‰. Nitrogen values ($\delta^{15}N$) in proteinaceous animal tissues are enriched by about 3 ‰ over the animal's diet because nitrogen becomes fractionated during protein metabolism. The lighter ^{14}N isotope is preferentially excreted in urea, leaving the body's pool of nitrogen enriched in ^{15}N relative to an animal's diet (Figure 4).

In order to tackle the fundamental question of whether short-faced bears were carnivores or herbivores I examined stable isotope ratios in collagen extracted from fossil bones of short-faced bears and modern bears, the latter for comparison

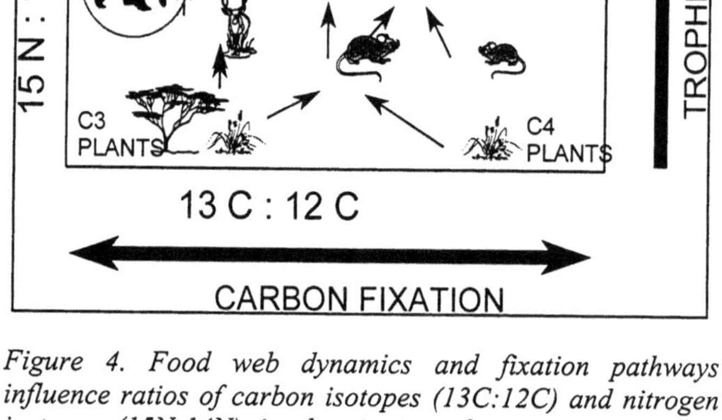

Figure 4. Food web dynamics and fixation pathways influence ratios of carbon isotopes (13C:12C) and nitrogen isotopes (15N:14N) in the tissues of consumers. Carbon isotopes primarily reflect pathways of carbon fixation by primary producers, while nitrogen isotopes primarily reflect a consumer's trophic level.

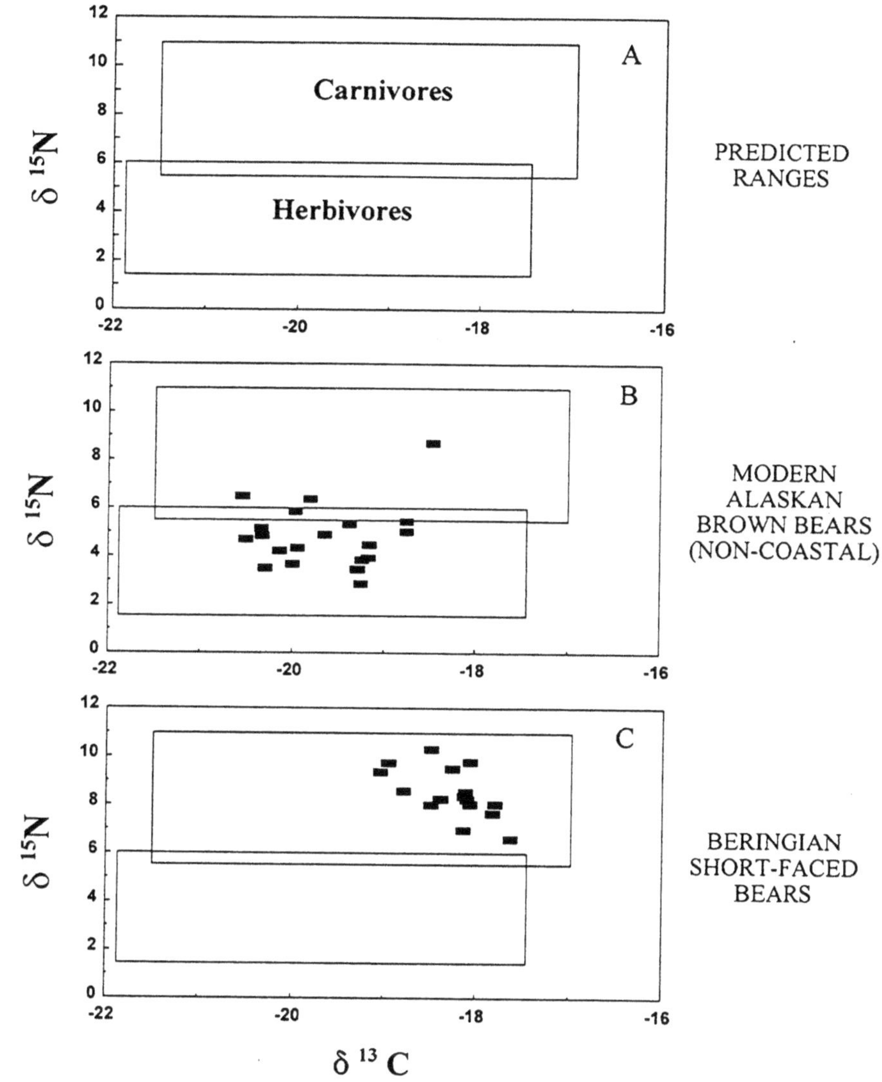

Figure 5. Stable carbon and nitrogen isotope values in Beringian short-faced bears (C), compared to values in known carnivores and herbivores (A) and data from modern brown bears of interior Alaska (B). The modern brown bear data demonstrate how stable isotopes accurately reflect this species' omnivorous, but mostly vegetarian, diet. In comparison, the data for short-faced bears reveals that it was highly carnivorous. Isotopes were measured in purified bone collagen extracted from well-preserved fossils.

and interpretation (Matheus 1995, 1997). Figure 5a shows the range of isotopic signatures predicted for carnivores and herbivores in a terrestrial ecosystem (compiled from isotope ratios in modern terrestrial consumers). Figure 5b shows the actual isotope values in the bone collagen of 24 modern brown bears from interior Alaska, and Figure 5c shows the values recovered from purified bone collagen of 16 short-faced bear fossils (late Wisconsinan age) from Alaska and the Yukon Territory.

The high $\delta^{15}N$ values found in *Arctodus* reveal that it was a highly carnivorous bear. Moreover, the low variation among $\delta^{15}N$ values suggests that there was no significant trophic variation in the diets of individual bears. It may be significant, however, that increasing $\delta^{15}N$ values are commensurate with a trend towards decreasing $\delta^{13}C$ values. In some cases, researchers studying modern consumers have noted a tendency for $\delta^{13}C$ to become *enriched* up to about one part per mil per trophic step. The reasons for this are unclear, but a trophic *depletion* of ^{13}C is not known (DeNiro and Epstein 1978; Chisholm *et al.* 1982; Schoeninger and DeNiro 1984). If there was some trophic variation in the sampled bears, then rising $\delta^{15}N$ values should be accompanied by decreasing $\delta^{13}C$ values. As this does not occur, it suggests that rising $\delta^{15}N$ values in *Arctodus* do not reflect increasing trophic position, but instead indicate that the meat being consumed came from more than one source, that is, a variety of herbivores were eaten. Alternatively, the variation in $\delta^{13}C$ may reflect climatic variation over time, which in turn caused slight changes in the physiology and/or type of primary producers. Testing this hypothesis will require radiocarbon dating all the samples graphed in Figure 5C to determine if $\delta^{13}C$ values (or $\delta^{15}N$) vary in accordance with known climatic intervals, such as the last interstadial versus the last glacial maximum.

Reconsidering *Arctodus*' Functional Morphology:

Stable isotope data refute Emslie and Czaplewski's (1985) herbivore/omnivore model for Arctodus, but examination of *Arctodus*' post-cranial morphology suggests that this was not a bear that had evolved as a predator (Matheus 1997).

Kurtén (1967) argued that *Arctodus'* gracile build was evidence that it had evolved towards a high-speed predatory niche, and he interpreted Arctodus' large size as an indication that it directed its predatory efforts towards large prey. Nearly all large predators rely on a combination of high top speed (pursuit), rapid acceleration (ambush), and/or maneuvering at high speeds to capture prey. However, the appendicular skeleton of *Arctodus*, while decidedly cursorial, appears *not* to be adapted for generating high top speeds or for acceleration, nor could this bear make sharp turns at high speed (Matheus 1997). Its' legs were far too gracile to handle the dynamic forces that are generated when such a large animal manoeuvres at high speeds or tries to accelerate rapidly.

Moreover, from the viewpoint of a functional morphologist, increased predation in a group of a large-bodied mammals (i.e. bears) should be accompanied by a *decrease*, not an increase, in body size. This is because bears are already so much larger than the ideal, or "target," size for a fast predator, which is around one kg (Garland 1983). Absolute running speed increases with body size in mammals, but only up to about 119 kg (Garland 1983), after which the dynamic stresses of running increase at a greater rate than the ability of bone to handle those stresses (Garland 1983; Biewener and Taylor 1986; see summary in Matheus 1997). This is one of the main reasons why there are no "super-large" (>250 kg) predators (Figures 6 and 7). One would predict that a "super-predatory" bear would evolve to be leaner and smaller than a typical bear, unless it had evolved to hunt very large and slower prey. If that was the case for *Arctodus*, then how do we explain its very lean build and long-leggedness?

Recent studies in locomotor mechanics indicate that the greatest advantage of larger body size in terrestrial mammals is increased locomotor efficiency. This is the ability to travel uninterrupted over long distances at moderate speeds (Heglund et al. 1982; Heglund and Cavagna 1985; Kram and Taylor 1990; summarized in Matheus 1997). In an animal as large as *Arctodus* (about 700 kg), light, gracile legs are good only for running in a relatively straight line or for highly efficient sustained locomotion or endurance. *Arctodus*' large size and gracile build are best explained as the result of selective pressure to increase locomotor efficiency. The most plausible source of such pressure in a carnivore probably stems from its need to repeatedly and efficiently traverse a large home range for the purpose of seeking out widely dispersed carrion resources and I propose that *Arctodus*' was a specialized scavenger (Matheus1997).[15].

The home range size and energetic requirements of a short-faced bear functioning as a scavenger, can be explored using calculations regarding the productivity of the east Beringian landscape, the carcass biomass it could have produced, and the dietary requirements of a carnivore the size of *Arctodus* (Matheus 1997). First, by developing models of natural mortality rates and predation rates amongst large mammals on the Beringian landscape, I have calculated that east Beringian ecosystems were capable of producing about 27.1 kg/km^2/year of carrion for short-faced bears to scavenge. Using allometric equations that scale body size to dietary requirements in carnivores, I also calculated that an average (700 kg) short-faced bear needed to consume 5853 kg of flesh annually to meet its energetic needs for maintenance and reproduction. This means that the east Beringian landscape could have supported a minimum density of 0.00463 short-faced bears/km^2. Maps in Hopkins *et al.* (1982) suggest that the size of unglaciated eastern Beringia was about 2,000,000 km^2, so a density of 0.00463 individuals/km^2 means that about 9260 short-faced bear would have inhabited eastern Beringia. To assess the biological plausibility of this population level, I calculated the minimum viable population size necessary for a 700 kg carnivore with conservative reproductive habits to avoid chance extinction. This exercise indicates that about 2000bears would have constituted a minimal viable population. For comparison, there are about 35,000-45,000 brown bears in eastern Beringia (Alaska and the Yukon) today, not [2]including Kodiak Island (Brown 1993).

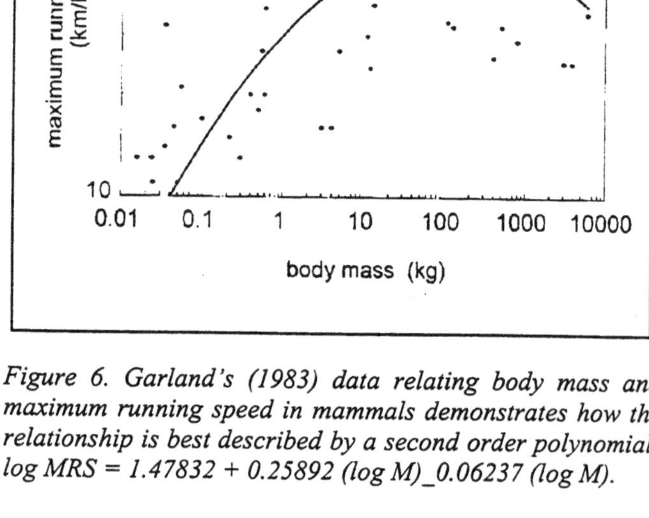

Figure 6. Garland's (1983) data relating body mass and maximum running speed in mammals demonstrates how the relationship is best described by a second order polynomial: log MRS = 1.47832 + 0.25892 (log M)_0.06237 (log M).

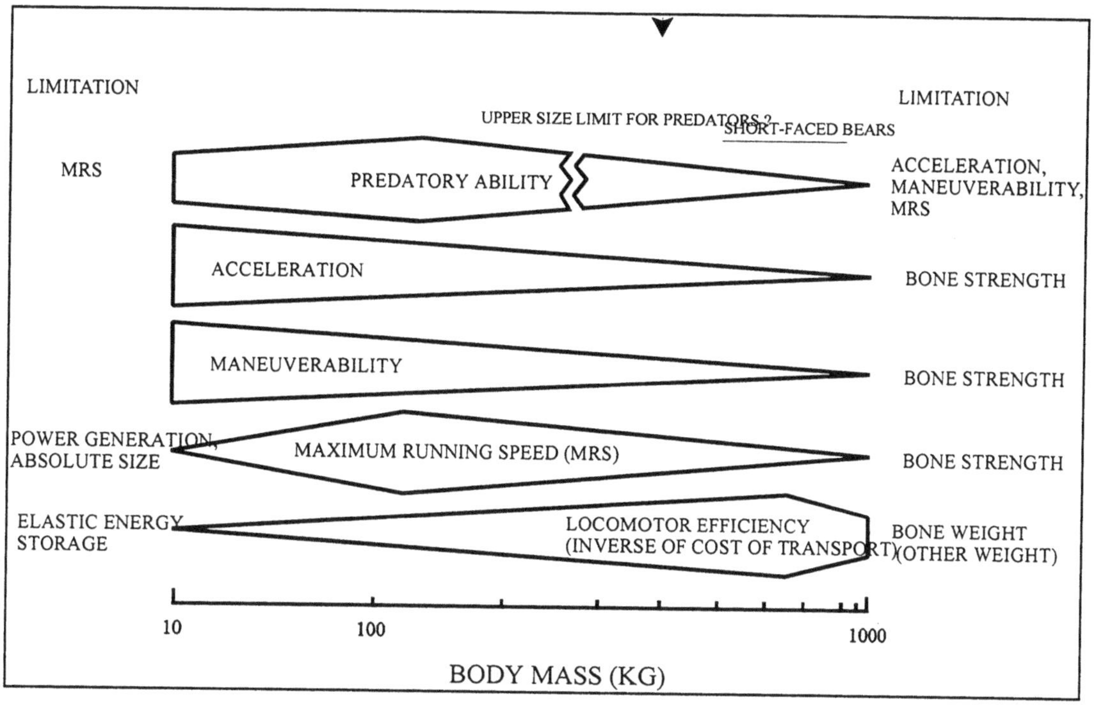

Figure 7. Spectrum of locomotor skills in mammals as they relate to body size. Predatory ability is a function of acceleration, manoeuvrability, and maximum running speed (MRS). Acceleration and manoeuvrability decrease with increasing body size, while MRS increases up to ~119 kg, then decreases as body size increases. These patterns place an upper limit on the body size of an effective predator, which is around 250 kg. Conversely, locomotor efficiency increases with an increasing body size, up to about 700-800 kg. Larger mammals are more efficient at locomotion because their size allows them to have large stretch tendons that store elastic strain energy. The size of short-faced bears (~700 kg) would have meant that they probably were poor predators, but highly efficient locomotors – an ability which is adaptive for a wide-ranging scavenger.

These calculations conservatively suggest that there may have been 2000 to 10,000 short-faced bears in the 2,000,0000 km^2 of eastern Beringia. That equates to an average exclusive home range size of 200 to 1000 km^2. Normal annual home range sizes for modern brown bears, a much smaller bear, are around 100-300 km^2 (Nowak 1991; see discussion in Matheus 1997). It is logical that there would have been considerable selective pressure for increased locomotor efficiency in an animal that repeatedly traversed a large home range, perhaps as large as 1000 km^2, searching for carrion. It is my argument that short-faced bears responded to such selective pressure by evolving a larger but more gracile and long-legged morphology. Body size, and leg length are the two most important determinants of locomotor efficiency in large mammals.

Arctodus' large size is consistent with a scavenging model for reasons beyond locomotion. In large carnivore guilds with high diversity, there is considerable competition among species for control of carrion. Many carnivores that forage in open terrain, including lions, spotted hyenas, and African hunting dogs, have evolved social units (packs, prides, etc.) at least in part to defend and even steal large mammal carcasses from other carnivores. As body size increases in these species, pack size decreases, because it requires fewer individuals to defend a carcass. A lone male lion, for example, is able to defend a carcass from up to a dozen hyenas (Eaton 1979). Bears do not form cohesive social foraging units, and it seems reasonable that the large short-faced bear was a solitary animal. We know from the stable isotope data that it was carnivorous and the fossil record of eastern Beringia indicates that short-faced bears were sympatric at least with lions, wolves, and brown bears, and probably with coyotes and scimitar cats (*Homotherium*). Large size would have been advantageous to a specialized scavenger that defended and even procured carcasses from other large carnivores.

The scavenging model also is consistent with *Arctodus*' cranial morphology, which indicates a very strong masticatory apparatus and a mechanical configuration that imparts strong occlusal forces to the distal teeth, especially during a half-gape. The high amount of wear present on the lingual occlusal surface of the P^4 of mature short-faced bears indicates that this tooth was used as a bone cracking hammer, in conjunction with the talonid of M_1 (Matheus 1995, 1997). While it may seem heretical to suggest that the carnassials were used in this manner, *Arctodus* inherited typical unspecialized, ursid carnassials, which, in my opinion, did not form a functional shear surface, and thus were of limited use for cutting flesh. However, the distal position and conical shape of *Arctodus*' carnassials suggests they were excellently pre-adapted for use in cracking cortical bone. The short rostrum in *Arctodus* brought its distal teeth closer to the jaw articulation (fulcrum), increasing the mechanical advantage to these teeth and thereby imparting more force to them. Bite forces would have been focused towards the apices of the conically-shaped P^4 and the trigonid of M_1; such a configuration that concentrates bite forces towards two opposing points is necessary for initiating the cracking process in a dense, crystalline material like cortical bone. Bone-cracking is a recurrent theme in the evolution of scavengers, because it enables them to expose marrow cavities and to extract the maximum nutritional content of a carcass (i.e., accessing lipids in marrow and eating whole bone). This is an important asset for scavengers since any carcass they find may already be cleaned of flesh by other carnivores.

Finally, it is informative to consider how and why *Arctodus'* dentition differs so much from polar bears (*Ursus maritimus*), the only other bear which has evolved towards exclusive carnivory. Polar bear cheek teeth are unique among bears. In their brief evolution (probably <700,000 years), polar bear teeth have evolved rapidly towards increased slicing duties to match the polar bear's carnivorous niche (Kurtén 1964). Polar bear teeth are more trenchant and reduced in size, particularly in their width, compared to all other ursids, forming a molar battery of taller, narrower teeth with increased sectorial capabilities. The highly carnivorous short-faced bears had a slightly longer evolutionary history (about one million years) (Kurtén 1967), yet they show trends in their dentition opposite to those in polar bears; rather than evolving narrow, trenchant molars for cutting flesh, *Arctodus* retained broad, robust molars with minimal sectorial qualities. For a carnivorous animal, this provides little advantage unless the animal was processing very durable material such as bone

Whereas the predator and herbivore models for *Arctodus* have inconsistencies, the scavenger model most congruously links all the data for this unusual bear, from stable isotopes to cranial features to total post-cranial design. Information from stable isotopes eliminates the herbivore model, and it is difficult to demonstrate how *Arctodus* was specifically adapted, morphologically, to be a predator, since it could not catch particularly fast prey, nor did it have the acceleration necessary to be an ambush predator. If *Arctodus* preyed on large, slow pachyderms, it seems unlikely that selection would favor such a gracile build.

Implications of *Arctodus*' Scavenging to Human Occupation

My goal here is not to recapitulate all the evidence indicating that *Arctodus* was a scavenger, but rather to point out the implications of the scavenging model for humans. Geist contended that the Pleistocene large carnivore communities of the Old World and New World differed significantly in species composition and the body size of its member species. In this regard, that the presence of *Arctodus* in North America was just one aspect that distinguished carnivore communities in the New World from those in the Old World. I tried to show that the overall carnivore community did not differ substantially between the Old World and New World and that carnivores were not larger in North America.

Despite my refutation of Geist's argument that the overall carnivore communities differed, the presence of *Arctodus* in North America is a key feature distinguishing Nearctic and Palearctic carnivore faunas in the late Pleistocene. I agree with Geist that *Arctodus* was a unique large carnivore in the New World, but I disagree with his assumption that short-faced bears were predatory. At this point it is reasonable to ask whether it mattered to humans if this carnivorous bear was primarily a predator or a scavenger? Probably not. If *Arctodus* was a predator, this meant that a large, lethal carnivore was prowling the landscape. However, the scavenging model also predicts that *Arctodus* was aggressive, sought out meat resources, and keyed in on the activities of other predators, potentially including humans. In both cases, humans and their meat resources (kills and caches) might have been under constant threat by short-faced bears in eastern Beringia. Below I will show that this may be a moot point because short-faced bears probably became extinct in most areas of North America prior to the arrival of humans.

New Radiocarbon Dates for *Arctodus*

Geist's thesis rests critically on two temporal considerations: 1) the timing of the earliest human occupation of far Northeast Asia and Alaska, and 2) the period that *Arctodus* inhabited Alaska, in particular, the timing of extinction.

There is no unequivocal archaeological evidence for human occupation of far northeast Siberia (especially the Chuckchi Peninsula) prior to about 14,000 to 18,000 B.P. (Dikov 1978; Yi and Clark 1985; Clark 1988; Grayson 1988; Powers 1990, 1996; Kelly 1996). Archaeological sites appear farther west and south in Siberia perhaps as early as 35,000 B.P. (Mochanov 1978; Müller-Beck 1982) although some suggest that even these are likely less than 20,000 years old (Yi and Clark 1985; Clark 1988). In fact many authors (e.g., Haynes 1982; Powers 1990; Guthrie 1996), doubt that humans were poised to enter Alaska much before 12,000 B.P.

Guthrie (1996) expounds the notion that the ancestors of the first Americans (neomongoloids) were biologically and technologically adapted to the mammoth steppe environment and he contends that neomongoloids diverged from a paleomongoloid stock during the last interstadial and expanded north, but did not reach Alaska during that early expansion. Their range contracted southward during the glacial maximum and expanded again in conjunction with woody environments during the subsequent Birch Period (ca. 14,000 B.P.), finally culminating in their habitation of the New World by 12,000 B.P. By contrast, Geist did not reject the notion of a much earlier human occupation in eastern Beringia. This notion was based on controversial, and subsequently rejected, evidence for an interstadial (or older) human occupation of the Yukon Territory's Old Crow Basin. Geist used this evidence to argue that neomongoloids (Guthrie's term) made forays into North America during their interstadial expansion in northern Asia, but had failed to gain permanent footings in North America until about 12,000 B.P. He further suggested that the reason for this failure was primarily the presence of *Arctodus* and other large, aggressive carnivores in North America.

Geist's scenario presumes that short-faced bears were in eastern Beringia throughout the late Pleistocene and it predicts that their extinction should slightly precede the earliest unequivocal date of *permanent* human occupation.

That date, accepted by Geist, is around 11,500 to 12,000 B.P. (Powers 1990; Hoffecker *et al.* 1993). This is also the terminal date for *Arctodus* in the rest of North America (Kurtén and Anderson 1980; Nelson and Madsen 1983; Gillette and Madsen 1992), and Geist assumes that this is a valid terminal date for *Arctodus* in Beringia. Based on their position in Pleistocene muck deposits, most *Arctodus* fossils in Beringia likely date to the Wisconsinan period, although a few are probably Illinoisan-age, and some may be older. But up to now, only three radiocarbon dates have been published on Beringian *Arctodus*. Table 3 lists these three dates and eight new dates, five from this study, and three unpublished dates communicated by C.R. Harington. These 11 dates, depicted graphically in Figure 8, suggest that *Arctodus* was present in eastern Beringia during the interstadial as well as early stages of the glacial maximum, but there are no dates younger than about 20,000 B.P. (uncalibrated). One must be cautious about making conclusions based on a lack of dates from a certain time period, and eleven dates are by no means exhaustive. Even with these considerations, however, the radiocarbon data preliminarily suggest that *Arctodus* did not survive the glacial maximum in Beringia.

There are a number of important paleoecological implications to be drawn from the pattern of radiocarbon dates, but in terms of this discussion, it suggests that *Arctodus* was extinct in Alaska before humans crossed the Bering Land Bridge. Geist's notion that *Arctodus* may have put an end to interstadial forays of humans into the New World also is not supported by the radiocarbon data, otherwise one would expect the earliest permanent archaeological sites in Alaska to be just under 20,000 B.P. That is to say, if *Arctodus* was precluding early human attempts to permanently inhabit eastern Beringia, then as soon as *Arctodus* became extinct (about 20,000 B.P.) humans should have appeared. I am not vehement about this point, however, because it rests on the assumption that Paleolithic people had the capability to survive in the Arctic during the peak of the last glacial maximum. Indeed, the lack of full-glacial archaeological sites from northern Europe and Asia suggests that extreme northern environments were uninhabited by humans, and that was related to technological limitations (Soffer and Gamble 1990).

Any future discovery of Beringian *Arctodus* fossils substantially younger than 20,000 B.P., or the discovery of considerably older archaeological sites in eastern Beringia, will nullify most of the above. Older archaeological sites have been sought for quite some time, but continue to elude researchers. However, it would not be so surprising for a younger date on *Arctodus* to turn up. At this point, it seems that the timing of key events does not support Geist's hypothesis.

These conclusions suggest that no human ever laid eyes on a short-faced bear. However, that may not be the case. An *Arctodus* specimen, dated to 12,650 ± 350 B.P., from the Lubbock Lake site in Texas, is reported to show signs of butchering (Johnson 1987). If this report holds up to scrutiny, then paleoindians south of Beringia apparently encountered this giant bear and survived. To some, this evidence alone may disprove Geist's hypothesis. I think it could also mean that the appearance of humans in southern North America coincided with the last years of *Arctodus'* existence.

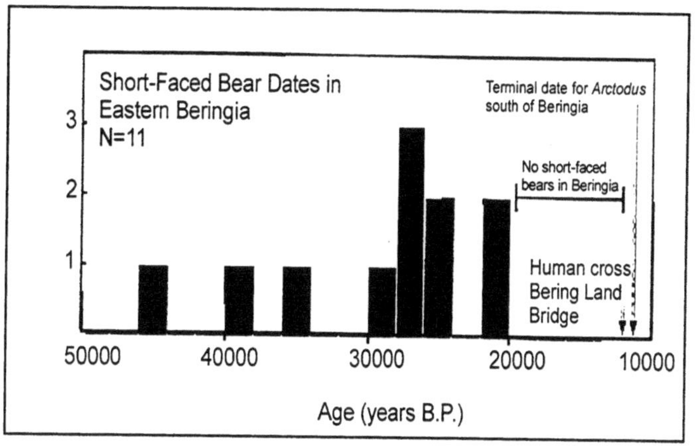

Figure 8. Distribution of all known radiocarbon dates for Beringian short-faced bears (data from Table 1). Arctodus was present in eastern Beringia through the last interstade and the beginning of the last glacial maximum, but the data tentatively suggest that Arctodus died out in Beringia shortly after 20,000 B.P. The gap of nearly 8000 years between Arctodus' apparent extinction and the oldest verifiable archaeological sites in eastern Beringia (11,800 B.P.) is inconsistent with Geist's contention that short-faced bears were excluding humans from North America.

Summary

Geist (1989, 1994) presented a biologically insightful and creative hypothesis that short-faced bears and other North American large carnivores presented the main obstacle to humans trying to inhabit the New World at the end of the Pleistocene. Geist's hypothesis centered on two key arguments: 1) that compared to Pleistocene Eurasia, North America's Rancholabrean fauna was characterized by more intense predation imparted by larger, more aggressive predators, and 2) that North America had a unique and very large predator in the short-faced bear, a predator against which humans were not equipped to defend themselves. The premises of Geist's hypothesis are biologically sound, but several underlying assumptions are flawed because they are difficult to reconcile with the fossil record of large carnivores.

First, faunal assemblages and body size data do not support Geist's argument that North American predators were more diverse or larger than in Europe and Asia. Throughout the Holarctic, carnivore diversity was roughly equal and Pleistocene carnivores displayed nearly equal levels of gigantism (and probably aggression). Second, the assumption that *Arctodus* was an active predator can be refuted. Recent stable isotope data and mechanical analyses of *Arctodus'* morphology more consistently support the alternative hypothesis that *Arctodus* was a specialized scavenger. The scavenging hypothesis specifically predicts that short-faced bears were aggressive and skilled in locating and procuring meat resources from other predators, which would have included humans. Theoretically, this would have made *Arctodus* even more of a threat than if it was strictly predatory. Thus, Geist's hypothesis is compelling because the presence of *Arctodus* in North America indeed may have been the most critical factor distinguishing carnivore guilds in the Palearctic and Nearctic during the Pleistocene.

However, current data on the chronology of short-faced and bears in Beringia and humans in northeast Siberia and Alaska are inconsistent with Geist's predictions about the timing of human migration events. Humans probably were not poised to enter Alaska until about 12,000 to 14,000 years ago, and all eleven radiocarbon dates on *Arctodus* indicate that it may have become extinct in Beringia shortly after 20,000 B.P. When Geist presented his hypothesis, these facts were not readily apparent.

Geist's hypothesis was virtually ignored by anthropologists. perhaps because it seems so incredible, but possibly also because it may at first appear impossible to test or refute. I have tried to show that this is not the case, as the hypothesis has specific predictions and assumptions that can be tested. It is important to use this exercise not only to remind ourselves that difficult questions can be tested in the paleo-record, but also that the biological and evolutionary limitations of humans should not be ignored. Indeed, if new dates on humans or short-faced bears in Beringia fill the chronological gap between the two, then anthropologists will have to seriously reconsider the question posed by Geist: did "humankind's single most ferocious predator" foil early attempts by Asians to inhabit North America?

SPECIMAN	COLLECTION LOCALITY	CONDYLOBASAL LENGTH	ZYGOMATIC WIDTH
AMNH 30433	Fairbanks vicinity	214	126
AMNH 30458	Goldstream	215	118
AMNH 30457	Cleary Creek	223	132
AMNH 67157	lower Goldstream at Fox	225	131
ANMH 67158	Engineer Creek, Dawson cut	230	138
AMNH 67164	lower Goldstream at Fox	230	130
AMNH 30432	Fairbanks vicinity	232	135
AMNH 67159	Cripple Creek	234	140
AMNH 67167	Lower Goldstream	234	148
AMNH 67160	Little El Dorado Creek	236	142
AMNH 67166	upper Cleary Creek	236	142
AMNH 30452	Manley Hot Springs	238	140
AMNH 67163	Ester Creek	238	146
AMNH 30431	Fairbanks vicinity	241	140
AMNH 67170	Ester Creek	241	152
AMNH 67165	Glimore deadline	242	146
AMNH 30451	Gravel drift at upper Cleary	244	144
AMNH 30450	lower Goldstream	244	154
AMNH 30453	Goldstream	245	145
AMNH 67169	Fairbanks Creek	247	145
AMNH 97079	Wiseman	255	152
AMNH 30430	Goldstream	256	-

Table 1. Skull measurements of fossil wolves collected from central Alaska. All except one from Manley Hot Springs and one from Wiseman come from the Fairbanks vicinity.

SPECIMAN	COLLECTION LOCALITY	SEX	CONDYLOBASAL LENGTH	ZYGOMATIC WIDTH
UAM 28899	Fairbanks	M	237	145
UAM 17341	Fairbanks	M	240	140
UAM 17265	Fairbanks	M	241	150
UAM 17090	Fairbanks	M	245	144
UAM 17154	Fairbanks	M	248	146
UAM 17348	Fairbanks	M	249	149
UAM 28906	Healy	M	249	144
UAM 2188	Big Delta	M	252	147
UAM 17349	Fairbanks	M	253	145
UAM 18391	Fairbanks	M	253	149
UAM 17261	Fairbanks	M	254	148
UAM 18249	Fairbanks	M	255	150
UAM 17345	Fairbanks	M	255	157
UAM 17242	Fairbanks	M	256	150
UAM 17351	Fairbanks	M	256	146
UAM 17353	Fairbanks	M	257	154
UAM 3416	Tanacross	M	258	156
UAM 28909	Fairbanks	M	259	148
UAM 31718	Minchumina	M	261	160
UAM 28907	Healy	M	264	160
UAM 28898	Fairbanks	F	228	133
UAM 2187	Big Delta	F	230	123
UAM 28904	Fairbanks	F	232	133
UAM 28908	Fairbanks	F	233	140

SPECIMAN	COLLECTION LOCALITY	SEX	CONDYLOBASAL LENGTH	ZYGOMATIC WIDTH
UAM 17257	Fairbanks	F	234	136
UAM 28910	Fairbanks	F	235	132
UAM 17356	Fairbanks	F	236	140
UAM 28901	Fairbanks	F	237	128
UAM 17174	Fairbanks	F	238	137
UAM 28902	Fairbanks	F	239	137
UAM 17318	Fairbanks	F	242	146
UAM 17369	Fairbanks	F	242	138
UAM 28903	Fairbanks	F	243	140
UAM 28900	Fairbanks	F	247	141
UAM 28905	Healy	F	249	142
UAM 18045	Fairbanks	F	249	147
UAM 17317	Fairbanks	F	249	142
UAM 28911	Fairbanks	F	251	146

Table 2. Skull measurements of modern wolves collected from central Alaska.

ENDNOTES

[1]The Rancholabrean is a land mammal age (i.e., a time period) in North America spanning from approximately 400,000-10,000 B.C. The initiation of the Rancholabrean is marked by the first appearance of *Bison* in North America (Kurtén and Anderson 1980). In Geist's publications, he uses of the term more in reference to the fauna itself, rather than the time period, and his use of the term "Rancholabrean fauna" excludes Beringia.

[2]The nature and timing of a mid-continental "Ice-Free Corridor" are still uncertain. There is little consensus, as to whether the corridor was blocked (closed) only during the late Wisconsinan or if ice sheets also coalesced during previous glacial maxima. New geologic evidence suggests that the Laurentide and Cordilleran ice sheets may have advanced and retreated out of phase with each other, such that the corridor may never have been closed. Still others argue that the corridor, when present, would have been uninhabitable to most animals, including humans, except perhaps for very short periods (see papers in Mandryk and Rutter 1996).

[3] Only a single fossil puma has been recovered in eastern Beringia, at Blue Fish Cave in the Old Crow Basin of the central Yukon Territory. It was recovered from cave strata that have been dated to between 25,000 - 12,000 B.C. (Cinq-Mars 1979). Of the thousands of late Pleistocene mammal bones recovered from interior Alaska, not a single puma fossil has been identified.

[4]Yesner (1994, 1996) reported a canid humerus from the Broken Mammoth Site in Alaska's Tanana Valley that he identified as a dire wolf because of its large size, but this identification is questionable because humeri of *C. dirus* are nearly impossible to separate from *C. lupus*.

[5] Brown bears probably migrated into eastern Beringia during the early or mid-Wisconsinan, but they did not spread south until about 12,000-13,000 B.P., after the Canadian ice-free corridor opened (Kurtén 1960, 1963, 1966a, 1968, 1973b, 1976a; Guilday 1968; Kurtén and Anderson 1974, 1980; Harris 1985). Brown bears seem to be the only Siberian large carnivore that immigrated to eastern Beringia during the Wisconsinan.

[6] It is possible that dholes ranged into western Beringia, too, as they live in southern Siberia today and Pleistocene-age fossils occur in Northern China (Pei 1934).

[7] Today many mammals in eastern Beringia are larger-bodied than conspecifics in other regions of the Holarctic. However, while many Pleistocene mammals were even larger than the largest individuals today, Beringia's Pleistocene mammals were not any more gigantic than those in other regions. In fact, some were smaller. For example, the woolly mammoth (*Mammuthus primigenius*) was the smallest of mammoths, and Beringian horse faunas are dominated by small-bodied forms (Guthrie 1968, 1982, 1990a; Harington and Clulow 1973; Harington 1977; Sher 1974, 1986, 1987; Kurtén and Anderson 1980).

[8] It is significant that while late Pleistocene leopards in Europe were huge, those in Africa were not much larger than the present form (Kurtén 1968). Following Geist, this might be taken to mean that predation levels were higher in Europe than in Africa.

[9] Brown bears from Kodiak Island and the Kamchatka Peninsula are as large as the biggest brown bear fossils recovered anywhere. There are no fossils of brown bears from Kodiak Island or the Kamchatka Peninsula, so it is not possible to compare the size of Pleistocene versus Holocene bears in these regions. Both of these regions were extensively covered by massive glacial ice during full glacial periods, however, so it is almost certain that all bears there descended from post glacial, Holocene immigrants.

[10] The fossil wolf skulls analyzed in this study are housed in the Department of Vertebrate Paleontology of the American Museum of Natural History. The modern wolf skulls are from the collection of the Mammalogy Department of the University of Alaska Museum.

[11] I only measured modern specimens for which there was information on gender

[12] These mechanisms also may explain why Pleistocene jaguars show a trend towards smaller size (i.e., they made a niche out of hunting smaller prey).

[13] The only extant Tremarctine bear is the endangered spectacled bear (*Tremarctos ornatus*), sometimes referred to as the Andean bear, of central South America. Spectacled bears are small, reclusive, mainly herbivorous, forest-dwelling bears that retain many characteristics of ancestral Tremarctine bears.

[14] The symbols "$\delta^{13}C$ ‰" and "δ ^{15}N ‰" refer to the difference ("del") between the ratio of ^{13}C:^{12}C and ^{15}N:^{14}N in a sample versus a standard measured in parts per mil (‰). For carbon, the standard is PDB belemnite, a fossil cephalopod; for nitrogen, the standard is atmospheric nitrogen.

[15] Some predators, particularly large canids, use long range pursuit to capture prey, and they too, are built for endurance. However, this

seems an unlikely strategy for *Arctodus* since most pursuit predators are much smaller and almost always hunt socially. There is no evidence or compelling reason to suspect that short-faced bears were social.

ACKNOWLEDGMENTS

I would like to thank Val Geist for providing such a provocative hypothesis to test, and one which compelled me to present this paper at the 1996 Alaska Anthropological Association Meeting. While I disagree with his conclusions regarding *Arctodus*, I respect Geist as a leading theoretician in mammalian evolutionary ecology. Likewise, many thanks are extended to Dale Guthrie, with whom I was fortunate enough to study at UAF. Any abilities I may have to read and interpret the fossil record or understand mammalian ecology are owed to him. Environmental forces have played a major role in the expansion of human populations, and indeed, physical and biological environments have shaped human adaptations. Few people have contributed more to our understanding of this than Dale Guthrie. Dale's long tenure in the far north and his biological approach to archaeological issues give him a powerful insight that has influenced many of us through his creative papers and frank discussions about adapting to life in Beringia. It is to this inspiring mentor that I dedicate this paper.

C.R. Harington continues to provide critical data and ideas regarding Pleistocene chronologies in Beringia, and I thank him for generously allowing me to present three of his unpublished radiocarbon dates. Any reference to those three dates should credit him. We are fortunate to have creative practitioners like Geist, Guthrie, and Harington who not only formulate innovative hypotheses, but also test them with the grit of real-life experiences. I thank the NSF-University of Arizona AMS lab for providing five radiocarbon dates, and especially Rosemary Maddock for her kind assistance. The Department of Mammalogy, University of Alaska Museum, facilitated my study of modern wolf skulls and kindly allowed me to sample their collection of Alaskan bears for isotopic analysis. The Department of Vertebrate Paleontology at the American Museum of Natural History generously allowed me to study and sample fossil carnivores in its Alaskan collection. Thanks to Craig Gerlach and David Yesner for a superb job organizing this session on zooarchaeology. Thanks especially to Craig for remaining upright and in his chair during my presentation. Finally, a big thanks to Toos, for her help with this manuscript and especially for all her patience.

REFERENCES CITED

Agenbroad, L.D.
1990 The Mammoth Population of the Hot Springs Site and Associated Megafauna. In *Megafauna and Man: Discovery of America's Heartland*, edited by L.D. Agenbroad, J.I. Mead, and L.W. Nelson, pp. 32-39. The Mammoth Site of Hot Springs, South Dakota, Inc. Scientific Papers, Volume 1 and Northern Arizona University Press, Flagstaff.

Agenbroad, L. D., and Mead, J. I.
1986 Large Carnivores from Hot Springs Mammoth Site, South Dakota. *National Geographic Research* 24:508-516.

Anderson, E.
1984 Who's Who in the Pleistocene: A Mammalian Bestiary. In *Quaternary Extinctions: a Prehistoric Revolution*, edited by P.S. Martin and R.G. Klein, pp. 40-89. University of Arizona Press, Tucson.

Azzaroli, A. and Ambrosetti, P.
1970 Late Villafranchian and Early Mid-Pleistocene Faunas in Italy. *Palaeogeography, Palaeoclimatology, Palaeoecology* 8:107-111.

Azzaroli, A.
1990 Quaternary Mammals and the "End-Villafranchian" Dispersal Event - A Turning Point in the History of Eurasia. In *European Neogene Mammal Chronology*, edited by E. H. Lindsay, V. Fahlbusch, and P. Mein. Plenum Press.

Bertram, B. C. R.
1979 Serengeti Predators and Their Social Systems. In *Serengeti: Dynamics of an Ecosystem,* edited by A.R.E. Sinclair and M. Norton-Griffiths. University of Chicago Press.

Biberson, P.
1970 The Problem of Correlation Between South Europe and North Africa During the Pleistocene. *Palaeogeography, Palaeoclimatology, Palaeoecology* 8:113-128.

Biewener, A. A. and Taylor, C. R.
1986 Bone Strain: A Determinant of Gait and Speed? *Journal of Experimental Biology* 123:383-400.

Bocherens, H., Fizet, M., and Mariotti, A.
1990 Mise en Évidence du Régime Alimentaire Végétarien de l'Ours des Cavernes (*Ursus spelaeus*) par la Biogéochimie Isotopique (^{13}C, ^{15}N) du Collagène des Vertébrés Fossiles. *Comptes Rendus de l'Académie des Sciences de Paris* 311:1279-1284.

Brown, G.
1993 The *Great Bear Almanac*, Lyons and Burford Publishers.

Chisholm, B.S., Nelson, D.E., and Schwarcz, H.P.
1982 Stable-Carbon Isotope Ratios as a Measure of Marine versus Terrestrial Protein in Ancient Diets. *Science* 216:1131-1132.

Clark, G.A.
1988 The Upper Paleolithic of Northeast Asia and its Relevance to the First Americans: a Personal View. *Current Research in the Pleistocene* 5:3-6.

Colbert, E.H. and Hooijer, D.A.
1953 Pleistocene Mammals from the Limestone Fissures of Szechwan, China. *Bulletin of the American Museum of Natural History* 102:1-134.

Cope, E.D.
1879 The Cave Bear of California. *American Naturalist* 13:791-800.

1891 The California Cave Bear. *American Naturalist* 25:997-999.

Davidar, E. R. C.
1975 Ecology and Behavior of the Dhole or Indian Wild Dog *Cuon alpinus* (Pallas). In *The Wild Canids: Their Systematics, Behavioral Ecology, and Evolution,* edited by M. W. Fox. Von Nostrand Reinhold Company.

DeNiro, M.J. and Epstein, S.
1978 Influence of Diet on the Distribution of Carbon Isotopes in Animals. *Geochimica et Cosmochimica Acta* 42:495-506.

Dikov, N.N.
1978 Ancestors of Paleo-Indians and Proto-Eskimo-Aleuts in the Paleolithic of Kamchatka. In *Early Man in America from a Circum-Pacific Perspective*, edited by A.L. Bryan, pp 68. Department of Anthropology, University of Alberta Occasional Papers 1. Alberta.

Eaton, R. L.
1979 Interference Competition among Carnivores: A Model for the Evolution of Social Behavior. *Carnivore* II 1:9-16.

Ehleringer, J.R. and Rundel, P.W.
1988 Stable Isotopes: History, Units, and Instrumentation. In *Stable Isotopes in Ecological Research*, edited by P.W. Rundel, J.R. Ehleringer, and K.A. Nagy. Springer-Verlag.

Eisenberg, J.F.
1980 *The Mammalian Radiations*. University of Chicago Press, Chicago.

Emslie, S.D. and Czaplewski, N.J.
1985 A New Record of Giant Short-Faced Bear, *Arctodus* simus, from Western North America with a Re-evaluation of its Paleobiology. *Natural History Museum of Los Angeles County Contributions in Science* 371:1-12.

Erdbrink, D.P.
1953 *A Review of Fossil and Recent Bears of the Old World.* Lange Publishers, Deventer.

Garland, T., Jr.
1983 The Relationship Between Maximum Running Speed and Body Mass in Terrestrial Mammals. *Journal of Zoology, London* 199:157-170.

Geist, V.
1989 Did Large Predators Keep Humans out of North America? In *The Walking Larder: Patterns of Domestication, Pastoralism, and Predation*, edited by J. Clutton-Brock. Unwin Hyman, London.

1994 Why Antlers Branched Out. *Natural History* 94(4):66-69.

Gillette, D. D., and Madsen, D. B.
1992 The Short-faced Bear *Arctodus simus* from the Late Quaternary in the Wasatch Mountains of Central Utah. *Journal of Vertebrate Paleontology* 121:107-112.

Graham, R.W., and Lundelius, E.L.
1995 FAUNMAP. Internet-accessible and Searchable Database Cataloging the Distribution of Pleistocene Mammals in the United States. http://www.museum.state.il.us/research/faunmap/aboutfaunmap.html.

Grayson, D.K.
1988 Perspectives on the Archaeology of the First Americans. In *Americans before Columbus: Ice-Age Origins*, edited by R.C. Carlisle, pp. 107-123.

Guilday, J. E.
1968 Grizzly Bears from Eastern North America. *American Midland Naturalist* 791:247-250.

Guthrie, R.D.
1968 Paleoecology of the Large-mammal Community in Interior Alaska during the Late-Pleistocene. *American Midland Naturalist* 79(2):346-363.

1976 Environmental Influences on Body Size, Social Organs, Population Parameters and Extinction of Beringian Mammals. In *Beringia in the Cenozoic Era*, edited by V.L. Kontrimavichus, translated by R. Chakravarty, Amerind Publishing Company, New Delhi, for the United States Department of the Interior and the National Science Foundation (1984).

1982 Mammals of the Mammoth Steppe as Paleoenvironmental Indicators. In *Paleoecology of Beringia*, edited by D. M. Hopkins, J. V. Matthews, C. E. Schweger, and S. B. Young. Academic Press.

1983 Osseous Projectile Points: Biological Considerations Affecting Raw Material Selection and Design Among Paleolithic and Paleoindian Peoples. In *Animals and Their Prey*, edited by J. Clutton-Brock and C. Grigson. BAR International Series 163. British Archaeological Reports, Oxford.

1988 Bone Litter from an Alaskan Pleistocene Carnivore Den. *Current Research in the Pleistocene* 5:69-71.
1990a *Frozen Fauna of the Mammoth Steppe: The Story of Blue Babe*. University of Chicago Press, Chicago.

1990b Late Pleistocene Faunal Revolution: A Perspective on the Extinction Debate. In *Megafauna and Man: Discovery of America's Heartland*, edited by L.D. Agenbroad, J.I. Mead, and L.W. Nelson., pp. 42-53. The Mammoth Site of Hot Springs, South Dakota, Inc. Scientific Papers, Volume 1 and Northern Arizona University
Press, Flagstaff.

1996 The Mammoth Steppe and the Origin of Mongoloids and Their Dispersal. In *Prehistoric Mongoloid Dispersals,* edited by T. Akazawa and E.J.E. Szathmáry, pp. 172-186.Oxford University Press, Oxford.

Harington, C.R.
1969 Pleistocene Remains of the Lion-Like Cat (*Panthera atrox*) from the Yukon Territory and Northern Alaska. *Canadian Journal of Earth Sciences* 6:1277-1288.
1977 Pleistocene Mammals of the Yukon Territory. Unpublished Ph.D. dissertation, University of Alberta, Edmonton.

1978 Quaternary Vertebrate Faunas of Canada and Alaska and Their Suggested Chronological Sequence, *Syllogeus* no. 15, National Museum of Canada.

Harington, C. R., and Clulow, F. V.
1973. Pleistocene Mammals from Gold Run Creek, Yukon Territory. *Canadian Journal of Earth Sciences* 10:697-759.

Harris, A.H.
1985 *Late Pleistocene Vertebrate Paleoecology of the West.* University of Texas Press, Austin.

Haynes, C.V.
1982 Were Clovis Progenitors in Beringia? In *Paleoecology of Beringia*, edited by D.M. Hopkins, J.V. Matthews, C.E. Schweger, and S.B. Young, pp. 383-398. Academic Press, New York.

Haynes, G.
1985 Age Profiles in Elephant and Mammoth Bone Assemblages. *Quaternary Research* 24:333-345.

1990 The Mountains that Fell Down: Life and Death of the Heartland Mammoths. In *Megafauna and Man: Discovery of America's Heartland*, edited by L.D. Agenbroad, J.I. Mead, and L.W. Nelson., pp. 22-31. The Mammoth Site of Hot Springs, South Dakota, Inc. Scientific Papers, Volume 1. Northern Arizona University Press,
Flagstaff.

Heglund, N. C., Fedak, M. A., Taylor, C. R., and Cavagna, G. A.
1982 Energetics and Mechanics of Terrestrial Locomotion Part IV: Total Mechanical Energy Changes as a Function of Speed and Body Size in Birds and Mammals. *Journal of Experimental Biology* 97:57-66.

Heglund, N.C. and G.A. Cavagna
1985 Efficiency of Vertebrate Locomotor Muscles. *Journal of Experimental Biology* 115:283-292.

Herrero, S.M.
1978 A Comparison of Some Features of the Evolution, Ecology, and Behavior of Black and Grizzly/Brown Bears. *Carnivore* 1:7-17.

Hilton, H.
1992 Coyotes in Maine: A Case Study. In *Ecology and Management of the Eastern Coyote*, edited by A.H. Boer.

Hoffecker, J.F., Powers, W.R., and Goebel, T.
1993 The Colonization of Beringia and the Peopling of the New World. *Science* 259:46-53.

Hopkins, D. M., Matthews, J. V., Schweger, C. E., and Young, S. B. (editors)
1982 *Paleoecology of Beringia*, Academic Press.

Johnson, E.
1987 *Lubbock Lake: Late Quaternary Studies on the Southern High Plains*. Texas A&M Press, College Station.

Kelly, R.L.
1996 Ethnographic Analogy and Migration to the Western Hemisphere. In *Prehistoric Mongoloid Dispersals,* edited by T. Akazawa and E.J.E. Szathmáry, pp. 228-240. Oxford University Press, Oxford.

Klein, R.G., and Cruz-Uribe, K.
1984 *The Analysis of Animal Bones from Archeological Sites*. Prehistoric Archeology and Ecology Series, edited by K.W. Butzer and L.G. Freeman. University of Chicago Press, Chicago.

Kram, R., and Taylor, C. R.
1990 Energetics of Running: A New Perspective. *Nature* 346:265-267.

Kruuk, H.
1972 *The Spotted Hyaena*, University of Chicago Press.

Kruuk, H., and Turner, M.
1967 Comparative Notes on Predation by Lion, Leopard, Cheetah and Wild Dog in the Serengeti Area, East Africa. *Mammalia* 31:1-27.

Kurtén, B.
1959 On the Bears of the Holsteinian Interglacial. *Stockholm Contributions in Geology* 2:73-102.

1960 An Attempted Parallelization of the Quaternary Mammalian Faunas of China and Europe. *Commentationes Biologicae Societas* 23:1-12.

1963 Notes on Some Pleistocene Mammal Migrations from the Palaearctic to the Neararctic. *Eiszeitalter u. Gegenwart* 14:96-103.

1964 The Evolution of the Polar Bear, *Ursus maritimus* Phipps. *Acta Zoologica Fennica* 108:1-30.

1965a The Pleistocene Felidae of Florida. *Bulletin of the Florida State Museum* 9(6):215-273.
1965b The Carnivora of the Palestine Caves. *Acta Zoologica Fennica* 107:1-74.

1966a Pleistocene Mammals and the Bering Bridge. *Commentationes Biologica* 298:1-7.

1966b Pleistocene Bears of North America I: Genus *Tremarctos*, Short-faced Bears. *Acta Zoologica Fennica* 117:1-60.

1967 Pleistocene Bears of North America, II: Genus *Arctodus*, Short-Faced Bears. *Acta Zoologica Fennica* 117:1-60.

1968 *Pleistocene Mammals of Europe*. Weidenfield and Nicolson Publishers, London.

1973a Pleistocene Jaguars in North America. *Commentationes Biologicae* 62:1-23.

1973b Transberingian Relationships of Ursus arctos Linné (brown and grizzly bears). *Commentationes Biologicae* 65:3-10.

1976a Trans-Beringian Carnivore Populations in the Pleistocene. In *Beringia in the Cenozoic Era*, edited by V. L. Kontrimavichus. Translated from Russian by Amerind Publishing Company, New Delhi (1984).

1976b *The Cave Bear Story*. Columbia University Press, New York.

1985 The Pleistocene Lion of Beringia. *Ann. Zoologica Fennici* 22:117-121.

Kurtén, B. and Anderson, E.
1974 Association of *Ursus arctos* and *Arctodus simus* Mammalia: Ursidae in the Late Pleistocene of Wyoming. *Breviora*, 426, 1-6.

1980 *Pleistocene Mammals of North America*. Columbia University Press, New York.

Kurtén, B. and Garevski, R.
1989 Giant Hyena, *Hyaena brevirostris* Aymard (Mammalia, Carnivora), from the Middle Pleistocene of Manastirec, Yugoslavia. *Ann. Zoologica Fennici* 26:145-147.

Kurtén, B. and Werdelin, L.
1990 Relationships between North and South American *Smilodon*. *Journal of Vertebrate Paleontology*, 10(2):158-169.

Lambe, L.M.
1911 On Arctotherium from the Pleistocene of Yukon. *The Ottawa Naturalist* 25(2):21-31.

Levson, V.M. and Rutter, N.W.
1996 Evidence of Cordilleran late Wisconsinan glaciers in the "Ice-Free Corridor." *Quaternary International* 32:35-51.

Lyman, R.L.
1994 Vertebrate Taphonomy. Cambridge University Press.
Mandryk, C.A.S. and Rutter, N. (editors)
1996 The Ice-Free Corridor Revisited. *Quaternary International* 32:1-126.

Marean, C. W.
1989 Sabertooth Cats and Their Relevance for Early Hominid Diet and Evolution. *Journal of Human Evolution* 18:559-582.

Martin, L.D. and Schultz, C.B.
1975 Cenozoic Mammals from the Central Great Plains. Part 5, Scimitar-toothed Cats, *Machairodus* and *Nimravides*, from the Pliocene of Kansas and Nebraska. *Bulletin of the University of Nebraska State Museum* 10:55-63.

Martin, L.D., Schultz, C.B., and Schultz, M.R.
1988 Saber-Toothed Cats from the Plio-Pleistocene of Nebraska. *Transactions of the Nebraska Academy of Sciences* 16:153-163.

Masini, F. and Torre, D.
1990 Large Mammal Dispersal Events at the Beginning of the late Villafranchian. In *European Neogene Mammal Chronology*, edited by E. H. Lindsay, V. Fahlbusch, and P. Mein. Plenum Press.

Matheus, P. E.
1994 The Late Pleistocene Paleoecology of Short-Faced Bears (*Arctodus simus*) and Brown Bears (*Ursus arctos*) in Eastern Beringia (*Abstract*). In *Bridges of Science between North America and the Russian Far East, Abstract Book 2*. 45th Arctic Science Conference of the AAAS (Arctic and Pacific Division), and the Far Eastern Branch of the Russian Academy of Sciences. August 29-September 2, 1994. Vladivostok, Russia.

1995 Diet and Co-Ecology of Pleistocene Short-Faced Bears and Brown Bears in Eastern Beringia. *Quaternary Research* 44:447-453.

1997 Paleoecology and Ecomorphology of the Giant Short-Faced Bear in Eastern Beringia. Unpublished Ph.D. dissertation, University of Alaska Fairbanks.

Merriam, J.C.
1911 Note on a Giant Bear from the Pleistocene of Rancho La Brea. *Bulletin of the Department of Geology, University of California Publications* 6:163-166.

Merriam, J.C. and Stock, C.
1925 Relationships and Structures of the Short-Faced Bear, *Arctotherium*, from the Pleistocene of California. *Carnegie Institute of Washington Publications* 347(1):1-35.

1932 The Felidae of Rancho La Brea. *Carnegie Institute of Washington Publications* 422(1):1-132.

Mochanov, I.A.
1978 Stratigraphy and Absolute Chronology of the Paleolithic of Northeast Asia According to the Work of 1963-1973. In *Early Man in America from a Circum-Pacific Perspective*, edited by A.L. Bryan, pp 68. Department of Anthropology, University of Alberta Occasional Papers 1. Alberta.

Müller-Beck, H.
1982 Late Pleistocene Man in Northern Alaska and the Mammoth-Steppe Biome. In *Paleoecology of Beringia*, edited by D.M. Hopkins, J.V. Matthews, C.E. Schweger, and S.B. Young, pp. 329-352. Academic Press, Inc., New York.

Nelson, M. E., and Madsen, J. H.
1983 A Giant Short-Faced Bear (*Arctodus simus*) from the Pleistocene of Northern Utah. *Transactions of the Kansas Academy of Science* 86:1-9.

Nowak, R.M.
1979 North American Quaternary *Canis*. *The Museum of Natural History Monograph* No. 6, University of Kansas, Lawrence.

1991 *Walker's Mammals of the World*, 5th edition, Johns Hopkins University Press.

Pei, W.C.
1934 On the Carnivora from Locality 1 of Choukoutien. *Palaeontologia Sinica, Series C* 8(1):1-217.

Péwé, T. L.
1975 *Quaternary geology of Alaska*, U.S. Geological Survey Professional Paper 835, Washington D.C.

Péwé, T. L., and Hopkins, D. M.
1967 Mammal Remains of Pre-Wisconsin Age in Alaska. In *The Bering Land Bridge*, edited by D. M. Hopkins. Stanford University Press.

Powers, W.R.
1990 The Peoples of Eastern Beringia. In *Prehistoric Mongoloid Dispersals*, No. 7, Special Issue, pp. 53-74. The University Museum, University of Tokyo.

1996 Siberia in the Late Glacial and Early Postglacial. In *Humans at the End of the Ice Age: The Archaeology of the Pleistocene-Holocene Transition,* edited by L.G. Straus, B.V. Eriksen, J. M. Erlandson, and D. R. Yesner. Plenum Press, New York.

Qiu, Z.
1990 The Chinese Neogene Mammalian Biochronology - Its Correlation with the European Neogene. In *European Neogene Mammal Chronology*, edited by E. H. Lindsay, V. Fahlbusch, and P. Mein. Plenum Press.

Richards, R.L. and Turnbull, W.D.
1995 Giant Short-Faced Bear (*Arctodus simus yukonensis*) Remains from Fulton County, Northern Indiana. *Fieldiana: Geology New Series* 30:1-34.

Richards, R.L., Churcher, C.S., and Turnbull, W.D.
1996 Distribution and Size Variation in North American Short-Faced Bears, *Arctodus simus*. In *Paleoecology and Paleoenvironments of Late Cenozoic Mammals: Tributes to the Career of C.S. (Rufus) Churcher*, edited by K.M. Stewart and K.L. Seymour, pp. 191-246. University of Toronto Press, Toronto.

Schoeninger, M.J. and DeNiro, M.J.
1984 Nitrogen and Carbon Isotopic Composition of Bone Collagen from Marine and Terrestrial Animals. *Geochimica et Cosmochimica Acta* 48:625-639.

Seymour, K.
1993 Size Change in North American Quaternary Jaguars. In *Morphological Change in Quaternary Mammals of North America*, edited by R.A. Martin and A.D. Barnosky. Cambridge University Press, Cambridge.

Sher, A. V.
1974 Pleistocene Mammals and Stratigraphy of the Far Northeast USSR and North America. *International Geological Review* 16:1-284.

1986 On the History of Mammal Fauna of Beringia. *Quartärpaläontologie* 6:185-193.

1987 Olyorian Land Mammal Age of Northeastern Siberia. *Palaeontographia Italica* 74:97-112.

Simpson, G.G.
1941 Large Pleistocene Felines of North America. *American Museum Novitates*, 1131:1-12.

Soffer, O. and Gamble, C. (editors)
1990 *The World at 18,000 BP: Volume 1, High Latitudes*, Unwin Hyman.

Stirling, I., and Derocher, A. E.
1990 Factors Affecting the Evolution and Behavioral Ecology of the Modern Bears. In *Bears - Their Biology and Management*, edited by L. M. Darling and W. R. Archibald. *Eighth International Conference on Bear Research and Management* 8:189-204.

Stock, C., and Harris, J.M.
1992 Rancho LaBrea: a Record of Pleistocene Life in California. 7th edition. *Natural History Museum of Los Angeles County, Science Series* No. 37.

Thenius, E.
1980 *Grundzüge der Faunen und Verbreitungsgeschichte des Säugetiere: Eine Historische Tiergeographie.* Gustav Fischer Verlag, Jena.

Turner, A.
1997 *The Big Cats and Their Fossil Relatives: An Illustrated Guide to Their Evolution and Natural History*, Columbia University Press.

Vereshchagin, N.K.
1971 The Cave Lion and its History in the Holarctic and in the Limits of the USSR (in Russian). *Trudy Zoologichekogo Instituta, Akademiya Nauk SSSR* 69:123-199.

Vereshchagin, N.K. and Baryshnikov, G.F.
1982 Paleoecology of the Mammoth Fauna in the Eurasian Arctic. In *Paleoecology of Beringia*, edited by D.M. Hopkins, J.V. Matthews, C.E. Schweger, and S.B. Young, pp. 267-280. Academic Press, Inc., New York.

1984 Quaternary Mammalian Extinctions in Northern Eurasia. In *Quaternary Extinctions: a Prehistoric Revolution*, edited by P.S. Martin and R.G. Klein, pp. 483-516. University of Arizona Press, Tucson.

Voorhies, M. R., and Corner, R. G.
1982 Ice Age Superpredators. *Museum Notes* no. 70, The University of Nebraska State Museum and Planetarium.

1986 The Giant Bear *Arctodus* as a Potential Breaker and Flaker of Late Pleistocene Megafaunal Remains. *Current Research in the Pleistocene* 3:49-51.

Waser, P. M.
1996 Patterns and Consequences of Dispersal in Gregarious Carnivores. In *Carnivore Behavior, Ecology, and Evolution, vol. 2*, edited by J. L. Gittleman. Comstock Publishing Associates.

Yesner, D.R.
1994 Subsistence Diversity and Hunter-Gatherer Strategies in Late Pleistocene/Early Holocene Beringia: Evidence from the Broken Mammoth Site, Big Delta, Alaska. *Current Research in the Pleistocene* 11:154-156.

1996 Human Adaptation at the Pleistocene-Holocene Boundary (circa 13,000 to 8,000 BP) in Eastern Beringia. In *Humans at the End of the Ice Age: The Archaeology of the Pleistocene-Holocene Transition*, edited by L.G. Straus, B.V. Eriksen, J.M. Erlandson, and D.R. Yesner, pp. 255-276. Plenum Press, New York.

Yi, S. and Clark, G.A.
1985 The "Dyuktai Culture" and New World Origin. *Current Anthropology* 26(1):1-20.

LATE HOLOCENE ENVIRONMENTAL AND CULTURAL CHANGES AT TUKUTO LAKE, NORTHWESTERN ALASKA

W. Wyatt Oswald
College of Forest Resources, University of Washington

Linda B. Brubaker
College of Forest Resources, University of Washington

Patricia M. Anderson
Quaternary Research Center, University of Washington

S. Craig Gerlach
Department of Anthropology, University of Alaska Fairbanks

INTRODUCTION

Archaeological research in northern Alaska indicates that many changes in settlement patterns, economic orientation, and technology occurred throughout the region during the last 2000 years (Giddings 1966; Bockstoce 1979; Gerlach and Hall 1988; Sheppard 1988; Gerlach and Mason 1992; Mason and Gerlach 1995). Although few detailed faunal analyses are available, prehistoric subsistence strategies in inland northern Alaska (the Arctic Foothills and Brooks Range) appear to have been based on the harvesting of caribou (*Rangifer tarandus*), mountain sheep (*Ovis dalli*) and fish (Bockstoce 1979; Gerlach and Hall 1988; see also D'angelo, this volume). Because of their reliance on subsistence resources, the people of the region were likely affected by environmental variations that either enhanced or depleted the availability of key species. Past changes in climate and vegetation may have had important consequences for arctic people (e.g., Gerlach and Hall 1988; Minc and Smith 1989), although paleoecological confirmation of this from archaeological sites in northern Alaska is limited (Bockstoce 1973; 1979, Lutz 1982; Mason and Gerlach 1995). At this point, the interpretation of variations in late Holocene subsistence and settlement patterns is hindered in part by the inability to place cultural changes in a paleoenvironmental context.

In this study archaeological and paleoecological records from Tukuto Lake are used to infer past relationships among human activities and changes in climate and ecosystems in northern Alaska. Interest in late Holocene environmental variability has grown in recent years (e.g., Jones *et al.* 1996; Overpeck *et al.* 1997), but paleoenvironmental studies in northern Alaska have not examined past changes at a spatial and temporal scale that is relevant to ecosystems and hunter-gatherer land use patterns. The data from Tukuto Lake suggest that 1) resources, in this case caribou, are sensitive to relatively minor environmental perturbations, and 2) that human subsistence activities are significantly influenced by shifts in the distribution and abundance of important resources, especially in areas such as interior northern Alaska where secondary or 'backup' resources are rare or few in number.

BACKGROUND

Study Site

Tukuto Lake (68° 30' N, 157° 03' W, elevation 541 m) is located in the northern foothills of the western Brooks Range (Figure 1). The lake basin sits in rolling hills ca. 25 km northwest of Howard Pass. The lake is fed by two small inlet streams with a single outlet that flows into Tukuto Creek, a tributary of the Etivluk River. The surrounding vegetation is mixed shrub-Cyperaceae tussock tundra, composed primarily of *Eriophorum vaginatum*, *Carex bigelowii*, *Betula nana*, *Ledum decumbens*, *Vaccinium vitis-idaea*, and *Vaccinium uliginosum*. *Salix* species and *Alnus crispa* occur in protected sites and along lake shores and streams (Viereck *et al.* 1992).

Late Holocene Climatic Variability

Evidence of glacial activity in the Brooks Range provides some information on regional climatic changes during the last 2000 years (Calkin and Ellis 1980, 1981; Ellis *et al.* 1981; Ellis and Calkin 1984; Haworth 1988). Although lichenometric dates are relatively imprecise, the timing of glacial advances in the Survey Pass, Arrigetch Peaks, Killik River, and Atigun Pass areas (Figure 1) appears to correspond to periods of alpine glacial advance throughout the northern hemisphere (Porter 1986). Glacial advances in the central and western Brooks Range date to ca. 1750, 1050, and 400 cal yr B.P. (Table 1)[1]. Expansion of glaciers in the Brooks Range probably reflects increased winter precipitation and/or decreased summer temperatures.

Tukuto Lake Archaeological Research

At least twelve archaeological sites have been identified in the immediate vicinity of Tukuto Lake. Intensive investigations have been undertaken at the Sikoruk site at the northern end of the lake, and at the Croxton site on the southeastern lake shore (Figure 1, Hall 1976; Gerlach and Hall 1988; Gerlach 1989). The earliest occupations at Tukuto

[1] All ^{14}C and lichenometric dates presented in this paper have been transformed to calibrated years using CALIB Method A (Stuiver and Reimer 1993). We report the range of age values for ^{14}C dates with multiple intercepts.

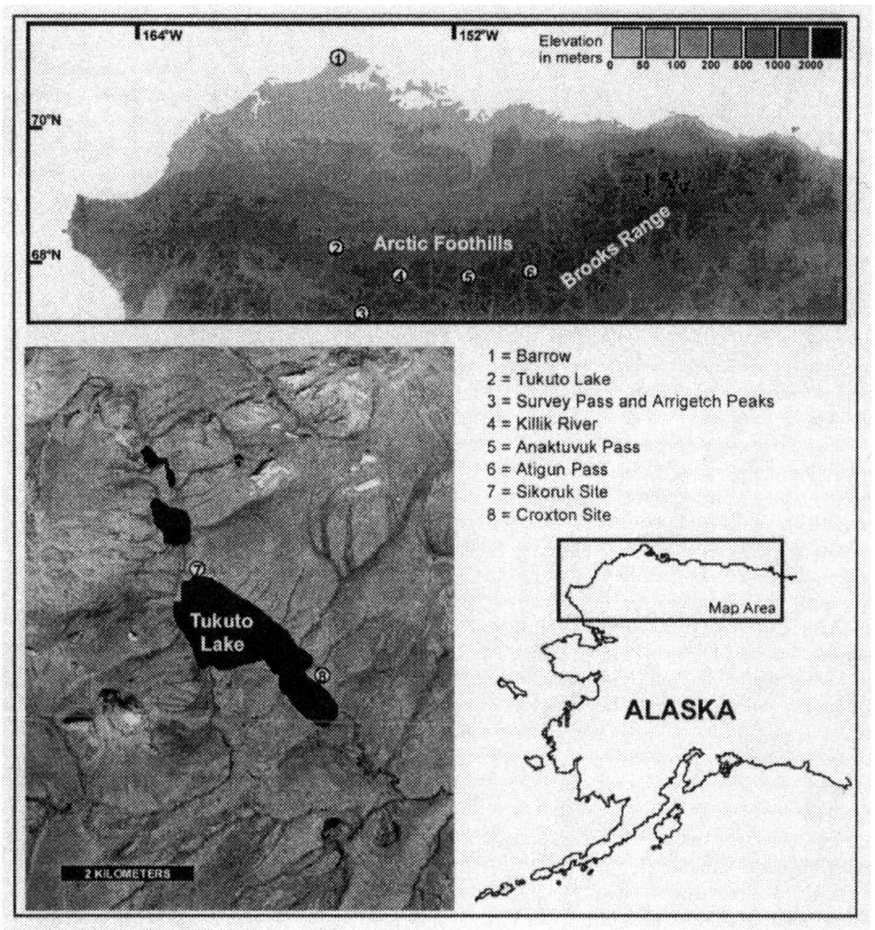

Figure 1. Locations discussed in the text.

Lake are dated at the Croxton site to 4990-3570 cal yr B.P. and 2300-1550 cal yr B.P.[2] and are attributed to an early phase of the Arctic Small Tool tradition (Gerlach and Hall 1988; Gerlach 1989). Use of Croxton during these intervals is limited compared to later occupations. Late Arctic Small Tool tradition people occupied the site occasionally between 1280 and 540 cal yr B.P.[2] Most of the archaeological materials from this period are associated with summer seasonal activities, specifically caribou hunting, butchering, and processing (Gerlach and Hall 1988; Gerlach 1989).

The Sikoruk site was first occupied by Arctic Woodland (possibly Ekseavik) peoples at 550 cal yr B.P., more or less coincident with the end of the late Arctic Small Tool tradition occupation at Tukuto Lake (Hall 1976; Gerlach and Hall 1988). The earliest settlement at Sikoruk probably involved a small group of people who utilized tent sites with slightly depressed floors. Human activities at Tukuto Lake intensified between 400 and 200 cal yr B.P., when subterranean houses with interior hearths were constructed at the Sikoruk site. This period of intense use was followed by three brief occupations during the 19th and early 20th centuries (Hall 1976, Gerlach and Hall 1988; Mason and Gerlach 1995).

Tukuto Lake Sediment Record

A 389-cm-long sediment core was recovered from Tukuto Lake in April, 1996. Analyses of the core yielded a late Quaternary pollen record more than 30,000 years in age (Oswald *et al.* in press). Similar to other pollen records from, northern Alaska (e.g., Livingstone 1955), percentages of *Alnus* pollen increase at ca. 7000 cal yr. B.P.. Following the *Alnus* rise, percentages of the major taxa, including *Betula* Cyperaceae Ericales, *Sphagnum*, Salix, and Poaceae, remain fairly stable. Here we focus in the 2000 years of the Tukuto Lake pollen record, paying particular attention to small changes in pollen and spore percentages during this interval.

Chronological control for this portion of the record is provided by four AMS ^{14}C dates in the upper 125 cm of the core (Table 1). Two minor dating reversals complicate interpretation of late Holocene changes, but a second-order polynomial curve fit to the four calibrated dates provides a reasonable chronology (Figure 2).

---

[2] Controversy continues to surround the validity of the Croxton site radiocarbon dates provided by DICARB (Gerlach 1989:152; Minc and Smith 1989). In 1999 Beta Analytic Inc. provided the following five new dates from the upper component in one area of the site (see Carrigan n.d.:17). The new uncalibrated dates are: 1320±40 B.P. (B-129943), 1400±40 B.P. (B-129944), 1330±40 B.P. (B-129945), 1270±40 B.P. (B-129946), and1250±50 B.P. (B-130192).

SITE AND SAMPLE	LAB NUMBER	^{14}C AGE B.P.	CALIBRATED AGE B.P. -1 SE [CAL DATE] +1 SE	REFERENCES
Glacial Advances				
Survey Pass glacial expansion	L	270 ± 60	430 [300] 0	Haworth 1988
Atigun Pass wood beneath moraine	BSG-522	320 ± 100	500 [420, 410, 320] 280	Calkin and Ellis 1980
Killik River glacial expansion	L	370 ± 80	510 [460, 340] 310	Haworth 1988
Atigun Pass glacial expansion	L	390 ± 80	510 [470] 310	Ellis and Calkin 1984
Arrigetch Peaks glacial expansion	L	390 ± 90	520 [470] 310	Ellis *et al.* 1981
Atigun Pass moss beneath glacier	BGS-614	1120 ± 180	1260 [1050, 1040, 990] 800	Calkin and Ellis 1981
Arrigetch Peaks glacial expansion	L	1120 ± 300	1300 [1050, 1040, 990] 720	Ellis *et al.* 1981
Atigun Pass glacial expansion	L	1150 ± 230	1290 [1060] 790	Ellis and Calkin 1984
Killik River glacial expansion	L	1200 ± 240	1320 [1080] 910	Haworth 1988
Survey Pass glacial expansion	L	1200 ± 240	1320 [1080] 910	Haworth 1988
Survey Pass glacial expansion	L	1800 ± 360	2140 [1710] 1320	Haworth 1988
Arrigetch Peaks glacial expansion	L	1800 ± 400	2290 [1710] 1300	Ellis *et al.* 1981
Killik River glacial expansion	L	1900 ± 380	2330 [1830] 1400	Haworth 1988
Archaeological Dates				
Sikoruk Site initial occupation	D		550	Hall 1976, Gerlach and Hall 1988
Croxton Site charcoal and peat	DIC-2203	360 ± 175*	540 [440, 350, 330] 0	Gal 1982, Gerlach 1989
Croxton Site wood, charcoal	DIC-2461	550 ± 125	650 [540] 500	Gal 1982, Gerlach 1989
Croxton Site wood, charcoal	DIC-2459	570 ± 45	630 [550] 530	Gal 1982, Gerlach 1989
Croxton Site wood, charcoal	DIC-2468	580 ± 100	650 [550] 510	Gal 1982, Gerlach 1989
Croxton Site charcoal	DIC-2205	670 ± 170	730 [650] 520	Gal 1982, Gerlach 1989
Croxton Site charcoal	DIC-2206	710 ± 215	890 [660] 520	Gal 1982, Gerlach 1989
Croxton Site charcoal	DIC-2207	910 ± 65	920 [790] 730	Gal 1982, Gerlach 1989
Croxton Site wood, hearth charcoal	DIC-2463	1040 ± 50	970 [940] 920	Gal 1982, Gerlach 1989
Croxton Site wood, charcoal	GX-8635	1075 ± 120	1070 [960] 910	Gal 1982, Gerlach 1989
Croxton Site wood, hearth charcoal	GX-8634	1135 ± 135	1220 [1060, 1030, 1000] 930	Gal 1982, Gerlach 1989
Croxton Site charcoal	DIC-2020	1260 ± 65	1270 [1170] 1070	Gal 1982, Gerlach 1989
Croxton Site wood, charcoal	GX-8633	1350 ± 140	1350 [1280] 1090	Gal 1982, Gerlach 1989
Croxton Site charcoal	GX-8636	1670 ± 160	1730 [1550] 1360	Gal 1982, Gerlach 1989
Croxton Site wood, charcoal	DIC-2465	2210 ± 155	2350 [2270, 2270, 2150] 1990	Gal 1982, Gerlach 1989
Croxton Site charcoal	DIC-2469	3350 ± 60	3680 [3570] 3470	Gal 1982, Gerlach 1989
Croxton Site charcoal	GX-8637	3680 ±205	4350 [3980] 3700	Gal 1982, Gerlach 1989
Croxton Site charcoal	DIC-2204	4420 ± 420	5590 [4990] 4440	Gal 1982, Gerlach 1989
Lake Sediment Record				
Tukuto Lake core 34-40 cm	CAMS-44172	2240 ± 80	2340 [2310, 2230, 2206] 2140	Oswald *et al.* in press
Tukuto Lake core 51-55 cm	CAMS-42612	2030 ± 240	2320 [1980, 1970, 1960] 1710	Oswald *et al.* in press
Tukuto Lake core 102-107 cm	CAMS-44173	4350 ± 90	5040 [4870] 4840	Oswald *et al.* in press
Tukuto Lake core 122-124 cm	CAMS-42613	4250 ± 140	4970 [4830] 4570	Oswald *et al.* in press

* possibly contaminated

Table 1. Calibrated radiocarbon, lichenometric (L), and dendrochronological (D) dates.

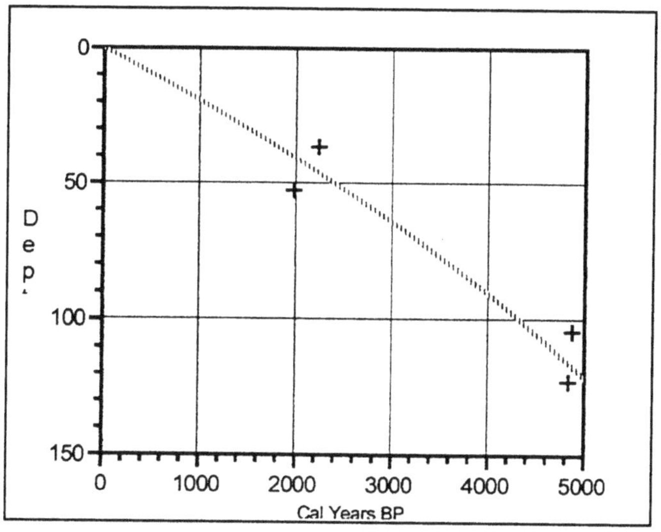

Figure 2. Tukuto Lake Late Holocene chronology.

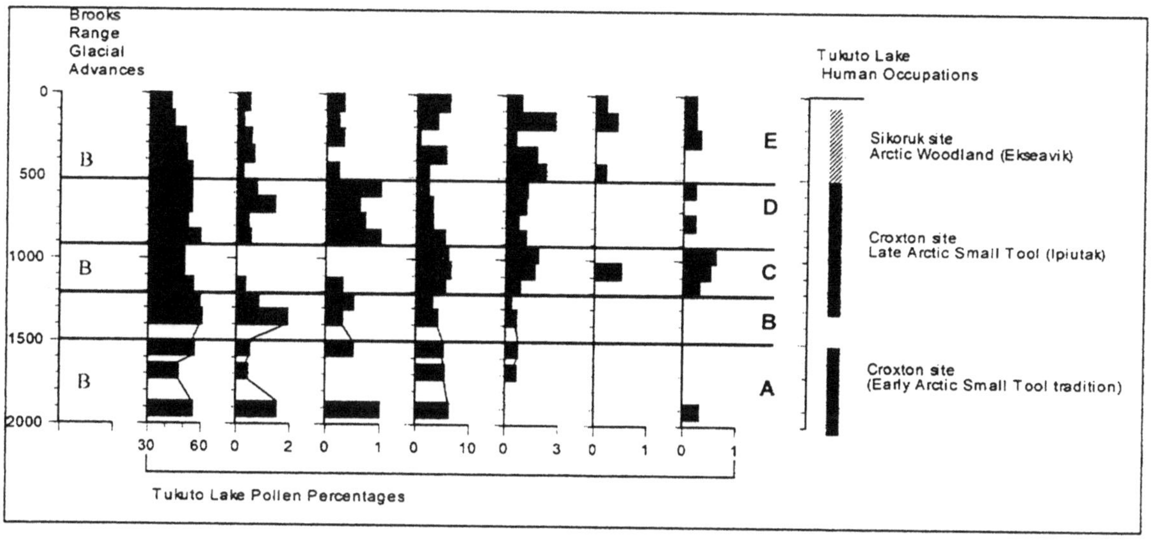

Figure 3. Tukuto Late pollen record.

ENVIRONMENTAL AND CULTURAL CHANGES

Small, but distinctive variations in pollen and spore percentages in the Tukuto Lake record appear to be contemporaneous with Brooks Range glacial advances and intervening periods of glacial retreat (Figure 3). Despite the potential difficulties in interpreting minor changes (i.e., <5%) in palynological data, inferences about climatic and ecological change during the late Holocene can be made from the response of key taxa. Attention to those taxa that have specific habitat requirements, indicator taxa, enhances previous interpretations of arctic paleoenvironments (e.g., Fredskild 1973; Cwynar 1982; Anderson *et al.* 1994; Oswald *et al.* in press). Based on pollen assemblage changes, the last 2000 years of the Tukuto Lake sediment record is divided into five intervals of relatively moist or dry conditions, with the Tukuto lake archaeological record examined in terms of inferred paleoenvironmental changes.

Period A (2000-1500 cal yr B.P.)

This interval is characterized by glaciation in the Brooks Range and mesic conditions near Tukuto Lake. For example, a large glacial advance is dated to 1710 cal yr B.P. in the Arrigetch Peaks area (Ellis *et al.* 1981). Although major ice advances are not as well defined in the Survey Pass and Killik River study areas, some moraines equate with the glacial advance at Arrigetch Peaks. Glacial deposits date to 1710 cal yr B.P. at Survey Pass, and to 1830 cal yr B.P. at Killik River according to lichenometric research by Haworth (1988).

Tukuto Lake pollen spectra from this interval have relatively high percentages of *Salix* and relatively low percentages of *Betula* pollen compared with other late Holocene periods. The high percentages of *Salix* pollen may indicate moist soils and perhaps deep snow, as many common *Salix* species occur in marshy tundra, riparian areas, and snow beds (Wiggins and Thomas 1962; Hultén 1968; Murray 1974; Cwynar 1982). The reduced percentages of *Betula* pollen between 2000 and 1500 cal yr B.P. may represent the decline of this taxon during an interval of cool summer temperatures.

Greenhouse experiments with *Betula nana* show this species to be temperature dependent, with increased growth under conditions that are warmer than present (Chapin *et al.* 1995, Chapin and Shaver 1996). Cooler temperatures in the spring and summer may have reduced the abundance of *Betula nana*, increased the number and duration of snow beds, and resulted in a generally more mesic landscape.

During this relatively cool and moist period the Croxton site at Tukuto Lake appears to have been occupied occasionally during the spring, summer, and fall by people of the Arctic Small Tool tradition (Gerlach and Hall 1988). There is no evidence of permanent settlement or winter utilization of the site between 2300 and 1550 cal yr B.P. Occupations were likely small, ephemeral, and discontinuous in time (Gerlach 1989). People who camped in the vicinity of Tukuto Lake would have been well situated to procure pregnant female caribou migrating north from the mountains in spring, smaller groups wandering through the Brooks Range in July and August, and larger aggregations of the Western Arctic caribou herd headed south in the fall (Lent 1966; Speiss 1979).

Period B (1500-1200 cal yr B.P.)

Both glacial and palynological evidence suggest that warmer and drier conditions occurred between 1500 and 1200 cal yr B.P.. No major glacial advances date to this period (Ellis *et al.* 1981; Ellis and Calkin 1984; Haworth 1988), and changes in the Tukuto Lake pollen record indicate that the tundra vegetation responded to increased aridity. For example, the relatively high percentages of *Encalypta* cf. *rhaptocarpa* spores suggest a greater presence of sparsely vegetated, xeric substrates near Tukuto Lake (LaFarge-England 1988; Vitt *et al.* 1988; Brubaker *et al.* in press). The peak in Brassicaceae pollen percentages may also represent drier conditions, as many species within this pollen type occur in xeric sites in the Arctic (e.g., *Lesquerella arctica*, various *Draba* spp., and *Erysimum inconspicuum*; Wiggins and Thomas 1962, Hultén 1968, Murray 1974).

Higher percentages of *Betula* pollen suggest that summer temperatures were warmer than during the previous interval. These inferred changes in climate and vegetation coincide with the end of the early Arctic Small Tool tradition occupations of the Croxton site (Gerlach and Hall 1988). This pattern is similar to that from other sites in northern Alaska where human activities were interrupted between 1450 and 950 cal yr B.P. (Giddings 1952, 1966; Irving 1964; Anderson 1968, 1972, 1988; Hall 1974). Although more archaeological research and radiocarbon dating needs to be done to confirm that this is more than a sampling problem, this widespread hiatus has been attributed to a decline of caribou populations in northern Alaska (Giddings 1964; Bockstoce 1979; Mason and Gerlach 1995). The coincidence of these cultural and environmental changes suggests that human subsistence was influenced by a change in climate, vegetation, and caribou population during this time period.

Period C (1200-900 cal yr B.P.)

The period between 1200 and 900 cal yr B.P. was characterized by glaciation in the Brooks Range and a shift towards more mesic tundra near Tukuto Lake. Moraines at Atigun Pass were dated to 1060 cal yr B.P. using lichenometry, and to 990-1050 cal yr B.P. using ^{14}C dating (Calkin and Ellis 1981; Ellis and Calkin 1984). This glacial advance dates to 990-1050 cal yr B.P. in the Arrigetch Peaks (Ellis *et al.* 1981), and to 1080 cal yr B.P. at both the Survey Pass and Killik River study areas (Haworth 1988). At this time, percentages of taxa indicative of warmer and drier conditions (e.g., *Betula*, Brassicaceae, and *Encalypta*) decrease, and percentages of mesic taxa (e.g., *Salix*, *Rubus chamaemorus*, and *Lycopodium annotinum*) increase. Higher percentages of *Salix* pollen probably represent the increased presence of moist soils and snow beds. Similarly, increased percentages of both *Rubus chamaemorus* and *Lycopodium annotinum* support interpretation of Period C as an interval of higher effective moisture in northern Alaska, as these species require moist, organic soils. In contrast, the elevated percentages of *Artemisia* pollen between 1200 and 900 cal yr B.P. are more difficult to interpret. This taxon is often inferred as indicative of cold, dry conditions, but species within this genus actually occur in a wide range of habitats (Cwynar 1982). Some species of *Artemisia* are commonly found in mesic habitats in northern Alaska (e.g., *Artemisia arctica*; Hultén 1968; Murray 1974), and the peak in *Artemisia* percentages in this interval may represent the positive response of these species to cooler and moister conditions.

The Croxton site was used by late Arctic Small Tool tradition people during this period of glacial advance and cool, moist climatic conditions. Occupations during this interval were short-lived and restricted to late spring, summer, and fall months, as evidenced by caribou remains and the lack of obvious dwellings and storage facilities (Gerlach and Hall 1988; Gerlach 1989). People camping at the Croxton site probably hunted caribou in the vicinity of Tukuto Lake with small-scale water drives and ambushes predominating (Gerlach 1989; Mason and Gerlach 1995).

Period D (900-600 cal yr B.P.)

The interval between 900 and 600 cal yr B.P. corresponds to the Medieval Optimum, a period of warm temperatures throughout the northern hemisphere (Porter 1986), and glacial retreat in the Brooks Range (Ellis *et al.* 1981; Ellis and Calkin 1984; Haworth 1988). The Tukuto Lake pollen assemblage from this period resembles that from 1500-1200 cal yr B.P. (see above), with relatively high percentages of *Betula* pollen, peaks in percentages of *Encalypta* spores and Brassicaceae pollen, and reduced percentages of *Salix*, *Artemisia*, and *Rubus chamaemorus* pollen and *Lycopodium annotinum* spores. These changes indicate warmer and more arid conditions, although the presence of *Lycopodium annotinum* suggests that the Medieval Optimum in northern Alaska may have been somewhat less dry than the interval between 1500 and 1200 cal yr B.P.. The archaeological record indicates no alteration of human subsistence activities near Tukuto Lake in response to these environmental changes. In fact, archaeological data suggest continuity in subsistence activities across these intervals (Gerlach 1989).

Period E (600-100 cal yr B.P.)

Glaciers in the Brooks Range advanced again during the Little Ice Age, and the Tukuto Lake pollen spectra suggest a return to the mesic conditions that occurred during Period C (1200-900 cal yr B.P.). A glacial advance at Atigun Pass has a lichenometric date of 470 cal yr B.P. and a ^{14}C date of 320-420 cal yr B.P. (Calkin and Ellis 1980; Ellis and Calkin 1984). Moraines of the Little Ice Age are dated with lichenometry to 470 cal yr B.P. in the Arrigetch Peaks area (Ellis *et al.* 1981), 300 cal yr B.P. at Survey Pass, and 340-460 cal yr B.P. in the Killik River area (Haworth 1988).

This period is characterized by decreased percentages of taxa indicative of aridity and warm temperatures (e.g., *Betula*, Brassicaceae, and *Encalypta*), and higher percentages of mesic taxa, namely *Salix*, *Artemisia*, and *Rubus chamaemorus*. However, climatic and ecological conditions during the Little Ice Age may have differed somewhat from those of Period C, because percentages of Brassicaceae pollen and *Encalypta* spores are slightly higher, and the percentages of *Lycopodium annotinum* spores are lower during the Little Ice Age than in the previous cool interval.

A significant cultural change occurred at Tukuto Lake at the same time as the transition from the relatively warm and dry Medieval Optimum to the cooler, moister Little Ice Age. At 540 cal yr B.P., the Croxton site was apparently abandoned by the Ipiutak people, prior to the first occupations of the Sikoruk site by Arctic Woodland (Ekseavik) peoples (Giddings 1952; Gerlach and Hall 1988). The Sikoruk settlement consists of hundreds of cache pits and many deep, semi-subterranean houses that were inhabited during winter months. During the fall migrations, the first occupants of the Sikoruk site hunted caribou by communally driving them into corrals or into the lake itself, where the animals were pursued and killed from kayaks (Speiss 1979; Gerlach and Hall 1988). The analysis of the caribou remains from the Sikoruk site suggests that caribou were hunted during their major spring and fall migrations, and that adult females were also hunted during the winter months (Speiss 1979). Artifacts found at the Sikoruk site indicate that the Arctic Woodland groups had probably developed dog sled technology which may have enabled hunting over a larger area (Giddings 1952; Gerlach and Hall 1988; Mason and Gerlach 1995).

ENVIRONMENT-HUMAN INTERACTIONS

The comparison of archaeological and paleoecological records from Tukuto Lake indicates coincident cultural and environmental changes during the late Holocene. We suggest that the following two linkages help to explain the complex relationships between climate, vegetation, caribou, and human populations in the different records from Tukuto Lake.

1. Subtle changes in climate and vegetation during the late Holocene affected the size and distribution of the caribou population in inland northern Alaska. For example, the Tukuto Lake pollen record suggests that relatively warm and dry conditions prevailed between 1500-1200 cal yr B.P., approximately the same period when a caribou population crash is proposed for northern Alaska (Giddings 1964; Bockstoce 1979; Spiess 1979; Dumond 1979). Warm and dry weather may have adversely affected caribou by changing the type and reducing the quality of their forage (Mason and Gerlach 1995:113). Warm conditions increase the growth of woody shrubs, such as *Betula nana* and *Salix* species, but do not favor *Eriophorum vaginatum* (Chapin *et al.* 1995; Chapin and Shaver 1996), the most important food source for caribou during the spring calving season (Lent 1966; Kuropat and Bryant 1980; White 1983). Moreover, during times of warm temperatures and moisture stress many tundra plants, including those that serve as summer caribou forage (e.g., *Betula nana* and *Salix* species), produce secondary compounds that deter foraging by herbivores (Bryant and Kuropat 1980; Kuropat and Bryant 1980; Bryant *et al.* 1983; Jonasson *et al.* 1986; Bryant and Reichardt 1992).

During modern times, prolonged periods of warm temperatures are characterized by declines in caribou body condition, which can lead to lower calf birth weights and reduced winter survivorship (Cameron *et al.* 1993; Mason and Gerlach 1995; Valkenburg unpublished). Increased aridity in northern Alaska between 1500 and 1200 cal yr B.P. may have caused changes in plant chemistry and community composition that led to similarly disastrous consequences for caribou populations.

Conversely, it appears that caribou populations may have recovered during the cooler and moister period between 1200 and 900 cal yr B.P.. Increased moisture and lower summer temperatures may have improved forage quality and thus initiated the end of the caribou population decline. As another example, the transition from the Medieval Optimum to the Little Ice Age may have had an important impact on caribou in inland northern Alaska. Colder and moister conditions between 600 and 100 cal yr B.P. probably led to increased snow depth in the Brooks Range. This change may have induced some female-juvenile bands of caribou to winter in the Arctic Foothills in the vicinity of Tukuto Lake, where snowfall tends to be lighter than in the southwestern drainages of the Brooks Range (Speiss 1979; Lent 1980). Indeed, archaeological analysis of caribou remains from the Sikoruk site suggests that caribou killed during the Little Ice Age winters were primarily adult females (Spiess 1979:160). Snow conditions are thought to be largely responsible for the winter distribution of caribou (Pruitt 1959; Henshaw 1968; Lent 1966; Skoog 1968; Jakimchuk and McCourt 1975; Stardom 1975; Roby 1980; Whitten and Cameron 1980; Adams *et al.* 1988), and groups of caribou have wintered north of the Brooks Range during recent years (Lent 1966; Calef and Heard 1980; White and Trudell 1980).

2. Late Holocene human settlement and subsistence patterns in inland northern Alaska were strongly influenced by the size and distribution of the caribou populations. The data

from Tukuto Lake suggest that the main cultural changes of the past 2000 years are likely related to changes in the size, location, and distribution of the caribou herds. For example, the apparent absence, or at least reduction, of the human population at Tukuto Lake between 1550 and 1280 cal yr B.P. may be attributed to warmer, drier climatic conditions during this interval, and a resulting reduction in the size of the caribou population that served as the primary food source. As caribou numbers increased during the cooler and moister period from 1200 to 900 cal yr B.P., people were once again motivated to resume seasonal occupations of the Croxton site at Tukuto Lake. Moreover, environmental changes during the Little Ice Age may have altered caribou migration patterns, and thus enabled people to persist at Tukuto Lake throughout the year. The presence of scattered individuals and small bands of caribou wintering north of the Brooks Range during this cold, snowy interval may have provided Arctic Woodland peoples with an adequate supply of food throughout the winters. Prime adult females, which appear to have been selectively killed by the Sikoruk residents during winter months, would have been a particularly important part of the fat-starved diet of these inland hunting people (Spiess 1979; Speth 1983). Such a highly selective foraging strategy may also indicate either the abundance of and/or easy access to large numbers of animals.

In addition to year-round access to the caribou resource, successful subsistence at Tukuto Lake by Arctic Woodland peoples was also probably related to their refinement of dog sled technology. Dog teams would have improved winter hunting success and facilitated interactions with other settlements in the foothills of the Brooks Range and along the coast (Sheppard 1988; Mason and Gerlach 1995).

CONCLUSIONS AND FUTURE RESEARCH

The magnitude of environmental variations that occurred in the Tukuto Lake area during the last 2000 years is small when compared with the major climatic and ecological changes of the past 30,000 years (Oswald *et al.* in press). Nonetheless, the coincidence of these late Holocene variations with cultural changes at Tukuto Lake suggests that environmental changes were strong enough to affect the people who survived primarily by hunting caribou, and by utilizing other albeit limited natural resources in the area. Subsistence activities were likely altered when the ecosystems around the lake responded to even slight changes in climate. However, the Tukuto Lake data suggest that the relationships between past variations in climate, vegetation, caribou, and human activities in northern Alaska are not isomorphic (Minc and Smith 1989). A shift to either warmer or cooler climatic conditions does not necessarily have a predictable impact on caribou demography and, in turn, human subsistence patterns. These environmental and cultural changes deserve further study, and the framework that we propose for late Holocene environment-human interactions at Tukuto Lake must be examined in terms of future research. The relationships illustrated by the Tukuto Lake records are broadly consistent with knowledge of centennial-scale climatic variations, and of the ecological characteristics of plants and animals used as paleoenvironmental indicators, and patterns of prehistoric subsistence in northern Alaska.

The interpretation of late Holocene human and environmental history in inland northern Alaska stands to be improved in several ways. The chronology of cultural changes at Tukuto Lake and other archaeological sites in this region should be further defined with more ^{14}C dates, and lake sediment pollen records should have ^{210}Pb profiles and additional ^{14}C dates to improve chronological control (Gerlach and Mason 1992). Furthermore, the source area and dispersal of pollen in tundra environments and the significance of relatively minor changes in pollen percentages are not well understood. Thus, studies of the relationship between modern pollen assemblages and vegetation are needed to improve the interpretation of past ecological changes. Nevertheless, the provisional framework of environmental and cultural interactions presented here can be used in evaluating other late Holocene archaeological and paleoecological records from northern Alaska.

ACKNOWLEDGEMENTS

We thank Charlie Warbalow for field assistance, and Dan Gavin, Kendra McLauchlan, and an anonymous reviewer for their careful reviews of this manuscript. The research was supported by National Science Foundation grant ATM-9423392. This paper is contribution #135 from the Paleoclimates from Arctic Lakes and Estuaries (PALE) program.

REFERENCES CITED

Adams, L.G., Dale, B.W., and Singer, F.J.
1988 Neonatal Mortality in the Denali Caribou Herd. In: *Proceedings of the Third North American Caribou Workshop*, edited by R.D. Cameron and J.L. Davis, pp. 33-34. Alaska Department of Fish and Game, Wildlife Technical Bulletin No. 8, Chena Hot Springs .

Anderson, D.D.
1968 A Stone-age Campsite on the Gateway to America. *Scientific American* 218:24-33.

1972 An Archaeological Survey of the Noatak Drainage, Alaska. *Arctic Anthropology* 9:66-117.

1988 Onion Portage: The Archaeology of a Stratified Site from the Kobuk River, Northwest Alaska. *Anthropological Papers of the University of Alaska* 22:1-163.

Anderson, P.M., Bartlein, P.J., and Brubaker, L.B.
1994 Late Quaternary History of Tundra Vegetation in Northwestern Alaska. *Quaternary Research* 41:306-315.

Bockstoce, J.R.
1973 A Prehistoric Population Change in the Bering Strait Region. *Polar Record* 16:793-803.

1979 *The Archaeology of Cape Nome*. University Museum, University of Pennsylvania, Philadelphia.

Brubaker, L.B., Anderson, P.M., Murray, B., and Koon, D.
In press. A Palynological Investigation of True Moss (Bryidae) Spores: Morphology and Occurrence in Modern and Late-Quaternary Lake Sediments of Alaska. *Canadian Journal of Botany*.

Bryant, J.P., and Kuropat, P.
1980 Subarctic Browsing Vertebrate Winter Forage Selection: The Role of Plant Chemistry. *Annual Review of Ecological Systematics* 11:261-285.

Bryant, J.P., and Reichardt, P.B.
1992 Controls Over Secondary Metabolite Production by Arctic Woody Plants. In: *Arctic Ecosystems in a Changing Climate: An Ecophysiological Perspective*, edited by F.S. Chapin III, R.L. Jefferies, J.F. Reynolds, G.R. Shaver, and J. Svoboda, pp. 377-390. Academic Press, New York.

Bryant, J.P., Chapin, F.S., III, and Klein, D.R.
1983 Carbon Nutrient Balance of Boreal Plants in Relation to Herbivory. *Oikos* 40:357-368.

Calef, G.W., and Heard, D.C.
1980 The Status of Three Tundra Wintering Herds in Northeastern Mainland Northwest Territories. In: *Proceedings of the Second International Reindeer/Caribou Symposium*, edited by E. Reimers, E. Gaare, and S. Skjenneberg, pp. 582-594. Direktoratet for vilt og ferskvannsfisk, Trondheim.

Calkin, P.E., and Ellis, J.M.
1980 A Lichenometric Dating Curve and its Application to Holocene Glacial Studies in the Central Brooks Range, Alaska. *Arctic and Alpine Research* 12:245-264.

Calkin, P.E., and Ellis, J.M.
1981 A Cirque Glacier Chronology Based on Emergent Lichens and Mosses. *Journal of Glaciology* 27:511-515.

Cameron, R.D., Smith, W.T., Fancy, S.G., Gerhart, K.L., and White, R.G.
1993 Calving Success of Female Caribou in Relation to Body Weight. *Canadian Journal of Zoology* 71:480-486.

Carrigan, K.J.
n.d. The Croxton Site: Analysis of Lithic Tools and Debitage. Unpublished Masters Paper on File, Department of Anthropology, University of Alaska Fairbanks, Fairbanks, Alaska.

Chapin, F.S., III, Shaver, G.S., Giblin, A.E., Nadelhoffer, K.J., and Laundre, J.M.
1995 Responses of Arctic Tundra to Experimental and Observed Changes in Climate. *Ecology* 76:694-711.

Chapin, F.S., III, and Shaver, G.S.
1996 Physiological and Growth Responses of Arctic Plants to a Field Experiment Simulating Climatic Change. *Ecology* 77:822-840.

Cwynar, L.C.
1982. A Late Quaternary Vegetation History from Hanging Lake, Northern Yukon. *Ecological Monographs* 52:1-24.

Dumond, D.E.
1979 Eskimo-Indian Relationships: A View from Prehistory. *Arctic Anthropology* 16:3-22.

Ellis, J.M., Hamilton, T.D., and Calkin, P.E.
1981 Holocene Glaciation of the Arrigetch Peaks, Brooks Range, Alaska. *Arctic* 34:158-168.

Ellis, J.M., and Calkin, P.E.
1984 Chronology of Holocene Glaciation, Central Brooks Range, Alaska. *Geological Society of America Bulletin* 95:897-912.

Fredskild, B.
1973 Studies in the Vegetational History of Greenland: Palaeobotanical Investigations of Some Holocene Lake and Bog Deposits. *Meddelelser om Grønland* 178:1-245.

Gal, R.
1982 An Annotated and Indexed Roster of Archaeological Radiocarbon Dates from Alaska, North of 68° Latitude. *Anthropological Papers of the University of Alaska* 20:159-180.

Gerlach, S.C.
1989 *Models of Caribou Exploitation, Butchery and Processing at the Croxton Site, Tukuto Lake, Alaska*. Ph.D. dissertation, Department of Anthropology, Brown University, Providence.

Gerlach, S.C., and Hall, E.S., Jr.
1988 The Later Prehistory of Northern Alaska: The View from Tukuto Lake. In: *The Late Prehistoric Development of Alaska's Native People*, edited by R.D. Shaw, R.K. Harritt, and D.E. Dumond, pp. 107-135. Aurora, Alaska Anthropological Association Monograph Series No. 4.

Gerlach, S.C., and Mason, O.K.
1992 Calibrated Radiocarbon Dates and Cultural Interaction in the Western Arctic. *Arctic Anthropology* 29:54-81.

Giddings, J.L.
1952 *The Arctic Woodland Culture of the Kobuk River*. University Museum Monograph, University of Pennsylvania, Philadelphia, PA.

1964 *The Archaeology of Cape Denbigh*. Brown University Press, Providence.

1966 Cross-dating the Archaeology of Northwestern Alaska. *Science* 153: 127-135.

Hall, E.S., Jr.
1974 Archaeological Investigations in the Noatak River Valley, Summer 1973. In *The Environment of the Noatak River Basin, Alaska*, edited by S.B. Young, Contributions from the Center for Northern Studies No. 1, Wolcott.

1976 A Preliminary Analysis of House Types at Tukuto Lake, Northern Alaska. In: *Contributions to Anthropology: The Interior Peoples of Northern Alaska*, pp. 98-134. National Museum of Man, Mercury Series, Archaeological Survey of Canada Paper No. 49, Ottawa.

Haworth, L.A.
1988 *Holocene Glacial Chronologies of the Brooks Range, Alaska and their Relationship to Climate Change*. Ph.D. dissertation, State University of New York, Buffalo.

Henshaw, J.
1968 The Activities of Wintering Caribou in Northwestern Alaska in Relation to Weather and Snow Conditions. *International Journal of Biometeorology* 12:21-27.

Hultén, E.
1968 *Flora of Alaska and Neighboring Territories*. Stanford University Press, Stanford.

Irving, W.N.
1964 *Punyik Point and the Arctic Small-Tool Tradition*. Ph.D. dissertation, University of Wisconsin, Madison.

Jakimchuk, R.D., and McCourt, K.H.
1975 Distribution and Movements of the Porcupine Caribou Herd in the Northern Yukon. In: *Proceedings of the First International Reindeer and Caribou Symposium*, edited by J.R. Luick, P.C. Lent, D.R. Klein, and R.G. White, pp. 140-154. Biological Papers of the University of Alaska, Special Report No. 1, Fairbanks.

Jonasson, J., Bryant, J.P., Chapin, F.S., , and Andersson, M.
1986 Plant Phenols and Nutrients in Relation to Variations in Climate and Rodent Grazing. *American Naturalist* 128: 394-408.

Jones, P.D., Bradley, R.S., and Jouzel, J.
1996 Climatic Variations and Forcing Mechanisms of the Last 2000 Years. *NATO ASI Series I: Global Environmental Change, Vol. 41*, Springer-Verlag, Berlin.

Kuropat, P., and Bryant, J.P.
1980 Foraging Behavior of Cow Caribou on the Utukok Calving Grounds in Northwestern Alaska. In: *Proceedings of the Second International Reindeer/Caribou Symposium*, edited by E. Reimers, E. Gaare, and S. Skjenneberg, pp. 64-70. Direktoratet for vilt og ferskvannsfisk, Trondheim.

LaFarge-England, C.
1988 The Contemporary Moss Assemblages of a High Arctic Upland, Northern Ellesmere Island, N.W.T., Canada. *Canadian Journal of Botany* 67:491-504.

Lent, P.C.
1966 The Caribou of Northwestern Alaska. In: *Environment of the Cape Thompson Region, Alaska*, edited by N.J. Wilimovsky, pp. 481-517. U. S. Atomic Energy Comission, Oak Ridge.

Lent, P.C.
1980 Synoptic Snowmelt Patterns in Arctic Alaska in Relation to Caribou Habitat Use. In: *Proceedings of the Second International Reindeer/Caribou Symposium*, edited by E. Reimers, E. Gaare, and S. Skjenneberg, pp. 71-77. Direktoratet for vilt og ferskvannsfisk, Trondheim.

Livingstone, D.A.
1955 Some Pollen Profiles from Arctic Alaska. *Ecology* 36: 587-600.

Lutz, B.J.
1982 Populations Pressure and Climate as Dynamics within the Arctic Small Tool Tradition of Alaska. *Arctic Anthropology* 19:143-149.

Mason, O.K., and Gerlach, S.C.
1995 Chukchi Hot Spots, Paleo-polynas, and Caribou Crashes: Climatic and Ecological Dimensions of North Alaska Prehistory. *Arctic Anthropology* 32:101-130.

Minc, L., and Smith, K.P.
1989 The Spirit of Survival: Cultural Responses to Resource Variability in North Alaska. In: *Bad Year Economics: Cultural Responses to Risk and Uncertainty*, edited by P. Halstead and J. O'Shea, pp. 8-39. Cambridge University Press, Cambridge.

Murray, D.F.
1974 *Notes on the Botany at Selected Localities in the Alatna and Killik River Valleys, Central Brooks Range, Alaska*. U.S. National Parks Service, Anchorage.

Oswald, W.W., Brubaker, L.B., and Anderson, P.M.
In press. Late Quaternary Vegetational History of the Howard Pass Area, Northwestern. Alaska.*Canadian Journal of Botany*.

Overpeck, J., Hughen, K., Hardy, D., Bradley, R., Case, R., Douglas, M., Finney, B., Gajewski, K., Jacoby, G., Jennings, A., Lamoureaux, S., Lasca, A., MacDonald, G., Moore, J., Retelle, M., Smith, S., Wolfe, A., and Zielinski, G.
1997 Arctic Environmental Change of the Last Four Centuries. *Science* 278:1251-1256.

Porter, S.C.
1986 Pattern and Forcing of Northern Hemisphere Glacier Variations During the Last Millennium. *Quaternary Research* 26:27-48.

Pruitt, W.O.
1959 Snow as a Factor in the Winter Ecology of the Barren Ground Caribou. *Arctic* 12: 159-179.

Roby, D.D.
1980 Winter Activity of Caribou on Two Arctic Ranges. In: *Proceedings of the Second International Reindeer/Caribou Symposium*, edited by E. Reimers, E. Gaare, and S. Skjenneberg, pp. 537-544. Direktoratet for vilt og ferskvannsfisk, Trondheim.

Sheppard, W.L.
1988 The Later Prehistory of Norton Sound as Viewed from Norton Bay. In: *The Late Prehistoric Development of Alaska's Native People*, edited by R.D. Shaw, R.K. Harritt, and D.E. Dumond, pp. 137-147. Alaska Anthropological Association Monograph Series No. 4.

Skoog, R.O.
1968 *Ecology of the Caribou (Rangifer tarandus granti) in Alaska*. Ph.D. dissertation, University of California, Berkeley.

Spiess, A.E.
1979. *Reindeer and Caribou Hunters*. Academic Press, New York.

Speth, J.D.
1983 *Bison Kills and Bone Counts*. University of Chicago Press, Chicago.

Stardom, R.R.P.
1975 Woodland Caribou and Snow Conditions in Southeast Manitoba. In: *Proceedings of the First International Reindeer and Caribou Symposium*, edited by J.R. Luick, P.C. Lent, D.R. Klein, and R.G. White, pp. 324-334. Biological Papers of the University of Alaska, Special Report No. 1, Fairbanks

Stuiver, M., and Reimer, P.J.
1993 Extended ^{14}C Database and Revised CALIB Radiocarbon Calibration Program. *Radiocarbon* 35:215-230.

Viereck, L.A., Dyrness, C.T., Batten, A.R., and Wenzlick, K.J.
1992 *The Alaska Vegetation Classification*. U.S. Department of Agriculture, Portland.

Vitt, D.H., Marsh, J.E., and Bovey, R.B.
1988 *Mosses Lichens and Ferns of Northwest North America*. Lone Pine Publishing, Edmonton.

Wiggins, I.L., and Thomas, J.H.
1962 *A Flora of the Alaskan Arctic Slope*. University of Toronto Press, Toronto.

White, R.G.
1983 Foraging Patterns and Their Multiplier Effects on Productivity of Northern Ungulates. *Oikos* 40:377-384.

White, R.G., and Trudell, J.
1980 Habitat Preference and Forage Consumption by Reindeer and Caribou near Atkasook, Alaska. *Arctic and Alpine Research* 12:511-529.

Whitten, K.R., and Cameron, R.D.
1980 Nutrient Dynamics of Caribou Forage on Alaska's Arctic Slope. In: *Proceedings of the Second International Reindeer/Caribou Symposium*, edited by E. Reimers, E. Gaare, and S. Skjenneberg, pp. 159-166. Direktoratet for vilt og ferskvannsfisk, Trondheim.

CALIBRATED RADIOCARBON AGES AND TAPHONOMIC FACTORS IN BERINGIAN CAVE FAUNAS AT THE END OF THE PLEISTOCENE

Robert A. Sattler
Alaska Quaternary Center, University of Alaska Fairbanks

Dale M. Vinson
National Park Service, Anchorage

Thomas E. Gillispie
Department of Anthropology, University of Alaska Fairbanks

INTRODUCTION

Several small caves and rockshelters in Eastern Beringia have produced heterogenous fill deposits consisting of large mammal faunas, human manufactured artifacts, and features representing cultural and carnivore activities (Larsen 1968; Cinq-Mars 1979; Dixon 1984; Dixon and Smith 1986; Thorson 1990; Vinson 1993; Ackerman 1994; 1996; Sattler 1997). Initially tested for archaeological remains, these caves produced late Pleistocene mammalian faunas including bones of extinct megaherbivores. Both humans and carnivores used these sites.

Humans appear to have utilized these localities for camp or hunting sites, while carnivores used the same places as dens. Herbivore limb-bones dominate these late Pleistocene faunal assemblages. Each cave or rockshelter includes megafaunal elements dated and calibrated to between c. 15,500 to 17,000 cal. B.P.. (13,000 to 14,000 B.P., uncalibrated), although no megafaunal elements are dated and calibrated to younger than around 15,500 cal. B.P. (13,000 B.P., uncalibrated). These radiocarbon dates provide new insight into the chronology of late Pleistocene paleoecological changes linked to the break-up of the "mammoth steppe" in Eastern Beringia (cf. Guthrie 1984, 1990a, b, 1995).

Here we focus on the late Pleistocene and Holocene faunas and radiocarbon ages from three small caves and compare them to other cave and shelter faunas in the region (Figure 1). The large mammal elements in these caves are presented in order to explore inter-assemblage variation, and to identify taphonomic agents that acted on these assemblages. The fauna from Trail Creek Caves A and B (Vinson 1993), and Lower Rampart Cave 1 (Sattler 1997) are compared to Trail Creek Caves 2 and 9 (Larsen 1968), the Bluefish River Caves (Cinq-Mars 1979; Morlan and Cinq-Mars 1982; Burke and Cinq-Mars 1998), the Lime Hills Cave 1 (Ackerman 1996; Georgina this volume), and Dyuktai Cave (Mochanov and Fedoseeva 1996). These localities are all small caves or rockshelters, either frost vugs or short segments of pre-Pleistocene karsts. As they have been described as caves in previously published reports, they are synonymously referred to as caves and rockshelters, and the animal bones in these sites are called cave faunas.

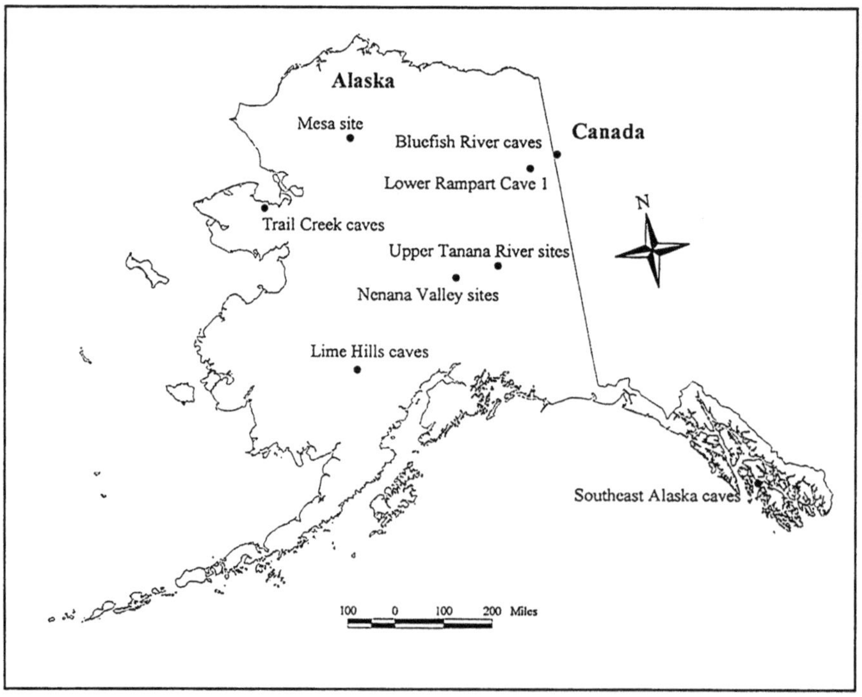

Figure 1. Map of Alaska with place names used in the text.

A summary of the taphonomic histories among these cave faunas addresses questions about late Pleistocene paleoecology, the timing of megafaunal extinctions, and the peopling of Eastern Beringia. Radiocarbon ages on bone collagen and charcoal are calibrated and compared to the earliest charcoal ages from open-air archaeological sites in Eastern Beringia. Comparison among the different series of calibrated radiocarbon ages indicate that the late Pleistocene use and abandonment of caves by "mammoth steppe" carnivores and the initial occupation of Eastern Beringia by late Pleistocene people did not overlap in time. Further, there is a period of nearly a millennium during which large mammal bones are absent in both the paleontological and archaeological records, revealing a pattern that may indicate a transitional phase in post-glacial Beringian large mammal faunas.

PROCEDURE

Cave sediments were systematically removed from Lower Ramparts Cave 1 and dry screened through 1/4 inch mesh or a series of smaller standardized mesh sizes (Dixon 1984; Sattler 1997). Testing at the Trail Creek Caves A and B included meticulous hand troweling, and the systematic collection of faunal remains (Vinson 1993). Stratigraphic units in Lower Rampart Cave 1 were designated alphabetically from top to bottom, while stratigraphic designations in the Trail Creek caves were assigned labels from bottom to top. Faunal remains and cultural artifacts were collected by sample units based on arbitrary levels in undifferentiated breccia, but wherever possible these were collected by natural stratigraphic units based on observable sediment boundaries. The excavation at Lower Rampart Cave 1 included approximately eight square meters, while one square meter test units were excavated at Trail Creek Caves A and B.

Observations made on the bone assemblages include counts of anatomical parts, taxonomic richness, and general patterns of bone destruction. We use the number of identified specimens (NISP) as the number of bone fragments identified to the level of genera. We utilize the minimum number of elements (MNE) as a measure of complete skeletal parts, or elements, as represented by the number of partial or incomplete skeletal elements in the cave faunal samples. Definitions and descriptions of punctures, striations, scores, pitting, and furrowing patterns of carnivore bone destruction follow Binford (1981:44-49) and Haynes (1983).

Based on the available literature for the other Beringian caves, we present calibrated radiocarbon ages, a comparison of the relative quantities of animal bones among caves, and species identified as the basis for discussing taphonomic histories and site formation processes. Radiocarbon ages are calibrated using the Seattle decadal 14C atmospheric, marine coral and varve calibration data set (Stuvier and Polach 1977; Stuvier and Reimer 1993). Radiocarbon ages are presented in terms of an uncalibrated time scale where noted by B.P. and as a calibrated time-scale where noted by cal. B.P.

EARLY-MAN ARCHAEOLOGY IN EASTERN BERINGIA

The earliest widely accepted cultural deposits in Eastern Beringia date to approximately 11,700 B.P. and the earliest cultural complexes with associated faunal remains are represented by several sites in Nenana River valley (Powers *et al.* 1983; Guthrie 1983; Powers and Hoffecker 1989; Hoffecker *et al.* 1993), the upper Tanana River valley (Yesner 1996; Holmes 1996; Holmes *et al.* 1996; Holmes 1998). One site without faunal remains is the Mesa Site in northern Alaska (Kunz and Reanier 1994). This phase of human occupation of the region encompasses the Nenana, Chindadn, and Mesa Complexes (Cook 1968; Powers and Hoffecker 1989; Kunz and Reanier 1994; Yesner 1996). The Nenana Complex sites may represent the earliest culture in Eastern Beringia (Dixon 1985, 1993; Meltzer 1995; West 1996).

Identifiable large mammal bones at the Alaskan archaeological sites indicate that bison (*Bison* sp.), wapiti (*Cervus canadensis*), moose (*Alces alces*), and mountain sheep (*Ovis dalli*) were prey to the earliest confirmed residents of Interior Alaska (Guthrie 1983:211, 1995; Yesner 1996). The proxy evidence in blood residue on Eastern Beringian fluted projectile points suggests that mammoth, caribou, bear, and musk-ox may also be included (Loy and Dixon 1998:38). This late Pleistocene archaeofauna may represent a transitional large-mammal community that occupied the region between the classic "mammoth steppe" fauna (cf. Guthrie 1968, 1982, 1990a,b) consisting of woolly mammoth (*Mammuthus primigenius*), bison, horse (*Equus* sp.), caribou (*Rangifer tarandus*) and mountain sheep, and the earliest Holocene fauna consisting of bison, moose, wapiti, caribou, mountain sheep and black bear (*Ursus americanus*) (Guthrie 1995; Yesner 1996; Sattler and Dixon 1997).

On the western side of the Bering Straits, the microlithic industries found in Dyuktai Cave are considered by some to be ancestral to the earliest core and blade industries in Alaska (Powers 1973; Yi and Clark 1985; Clark 1988; Mochanov and Fedoseeva 1996; Holmes 1998). The faunal assemblage at Dyuktai Cave includes abundant mammoth bones along with horse, moose, and bison that are thought to be associated with multiple episodes of human utilization of the cave near the close of the Pleistocene (Mochanov and Fedoseeva 1996). These faunal remains are in stratigraphic association with bifacial and microblade industries in three cultural horizons with uncalibrated ^{14}C ages between c. 12,000 to 14,000 B.P. While all of the associated faunal material is inferred to be archaeological, some of the remains almost certainly resulted from bone-collecting carnivores that were sympatric with humans. This assertion is based on observations resulting from taphonomic analyses on cave faunas demonstrating that bone beds in caves are largely products of natural mortality and bone-collecting carnivore activities (Brain 1981; Gamble 1983; Binford 1984; Stiner 1993, 1994).

Taphonomic issues fundamental to discussion of the latest Pleistocene Beringian cave sites and open-air early archaeological sites include differential transport of animal parts by carnivores and humans. Corroborating evidence in central Alaska indicates that the earliest Alaskans did scavenge mammoth ivory from older geological contexts, and transport it as raw material to overlook campsites for the manufacture of tools (Yesner 1994; Holmes 1996).

Given such background on the timing of late Pleistocene extinctions, Eastern Beringia cave taphonomy, and the peopling of the region we evaluate three propositions. First, the calibrated radiocarbon record between cave faunas and open-air cultural sites provides insight into the relationship between the timing of late Pleistocene extinctions and the peopling of Eastern Beringia. Second, the late Pleistocene cave faunas have similar large mammal assemblages and were affected by common taphonomic agents. Third, taphonomic histories of the cave faunas yield a new perspective on carnivore use of northern caves, and about paleoecological change in large mammal community structure during the latest Pleistocene in Eastern Beringia.

RADIOCARBON CHRONOLOGY

The published ^{14}C ages among nine Beringian caves are all within the range of conventional radiocarbon dating (cf. Taylor 1987, 1991), but not all are within the range of high-precision calibration. These 21 collagen dates consist of faunal elements ranging in age from approximately 12,300 to 38,300 B.P. on an uncalibrated time scale (Table 1). Fourteen dates are within the range of extended calibration made possible by recent advances in high-precision Uranium-Thorium (U-Th) dating (Bard *et al.* 1990; Stuvier and Reimer 1993). Further, these ages include a mix of conventional and AMS dates on charcoal and bone taken over a period of three decades by different laboratories using variable pretreatment methods. Given this variability, the small sample can be qualified in general because the ages from gas counting (conventional) are probably not as reliable as ages measured by direct counting methods (accelerator mass spectrometry, or AMS). Another criterion to consider in evaluating the ages are extended count estimates, but there is only one such date: Trail Creek Cave B produced an extended count age on mammoth (11,360+/-100 B.P., Beta-13811), but a resubmitted sample of the same bone produced an older, more concordant age of 13,570+/-120 B.P. (Beta-35841) in a normal AMS count. We are treating all ages as valid dates for the purpose of making general inferences about the cave versus archaeological age series.

Figure 2 illustrates in a cumulative frequency plot the series of archaeological ages from Alaskan open-air sites (left side), and the radiocarbon ages from caves and rockshelters (right side). Table 2 presents the dates used here from open-air archaeological sites. Intermingled between the two series and within the range of cave ages are the anthropogenic charcoal dates from Dyuktai Cave located in the Aldan River valley of the Russian Far East. Figure 2 also illustrates the relationship and temporal ranges of paleoenvironmental events in the region. These include the inundation of the Bering Straits, the two major climatic episodes at the end of the Pleistocene recognized in the Greenland ice-cores (Stuiver, Grootes, and Braziunas 1995), and the regional paleoenvironmental intervals recognized in Beringia by Hopkins (1982:9-11). The latter are the Duvanny Yar interval (Isotope Stage 2, Last Glacial Maximum) that was characterized by colder and drier conditions when the vegetation consisted of a treeless flora across most of Beringia. The following Birch Interval was marked by warmer and moister climates that initiated widespread establishment of dwarf birch shrublands in the latest Pleistocene, followed by forested conditions established by at least the early Holocene. Several authors (Hopkins 1982; Ager and Brubaker 1985; Barnosky *et al.* 1987; Anderson and Brubaker 1994) have reported the timing of the vegetation change at around 14,000 B.P., but we calibrated slightly younger ages (12,500 B.P.) based on AMS dating (Bigelow 1997).

The date series illustrates a relatively narrow range from archaeological sources, while the cave ages encompass a broader time period. In effect, the cultural dates indicate a more narrowly defined, widespread period of occupation, while the cave ages indicate an episodic pattern of occupation and suggest that there may be two periods of cave use by carnivores within the range of high-precision calibration. However, since the number of ages constitutes a statistically small sample, and the caves are spread out over a large geographic area, it is not clear whether these groupings reflect paleoecological change or whether they reflect biases in sampling, dating, or differential preservation. It is clear that nearly all the ages exceed c. 15,000 cal. B.P. (13,000 B.P.), and that each cave is represented with an age between c. 15,000 to 17,000 cal B.P. (13,000 to 14,000 B.P.). The radiocarbon dates from cave faunas on a regional scale possibly identify a young age limit not previously recognized, and a temporal relationship between ages taken on large mammals from cultural versus non-cultural assemblages that has not previously been illustrated. Most significant is the temporal gap of approximately a millennium between the cave and the archaeological series from Eastern Beringian sources. Significant paleoecological change may be responsible for the temporal gap.

Only the ages within the calibration range are presented in Figure 2. The older ages reported from the caves include multiple dates between c. 21,000 to 22,700 B.P., a date at 28,000 B.P. and one at 38,300 B.P. All of these ages should be considered minimum estimates. The temporal gaps between these older dates do not appear to reflect similarly scaled paleoecological change during the latest Pleistocene. These dates constitute a relatively small sample and may be biased due to sampling or analytical constraints (cf. Freisen 1997; Burke and Cinq-Mars 1998). In fact, taphonomic factors such as diagenesis, density-mediated attrition, and severe carnivore ravaging have not been adequately assessed in any of the northern caves. Additional analysis and further dating will be needed to more thoroughly explore the taphonomic histories among these cave and possible carnivore den deposits during the Last Glacial Maximum (Isotope Stage 2) and the last interstadial (Isotope Stage 3). Recent work in Southeast Alaska cave deposits documents a significant change in large mammal faunas in coastal refugia during Isotope Stages 2 and 3 (Heaton *et al.* 1996; Dixon *et al.* 1997; Guthrie 1988).

CAVE	TAXA	TYPE	STRATUM	MATERIAL	^{14}C YR B.P.	CAL ^{14}C YR B.P.	LAB NUMBER
TCCB	*Mammuthus*	C	A[1]	Collagen	11,360±100	13,410-13,150	Beta-13811
DYK			VIIa	Wood	12,100±120	14,320-13,920	LE-907
BFC3[2]	*Equus lambei*			Collagen	12,290±440	15,350-13,850	NMC-1236
DYK	*Mammuthus*		VIIa	Collagen	12.530±250	15,100-14,300	IM-462
DYK			VIIb	Charcoal	12,690±120	15,170-14,690	LE-860
BFC1	*Equus*			Collagen	12,900±100	15,480-15,050	GSC-28881
TCC9	*Bison*	C	IV	Collagen	13,010±350	15,050-14,850	K-1327
DYK			VIIb	Charcoal[3]	13,070±90	15,730-15,350	LE-784
LHC	*Rangifer*			Collagen	13,130±180	15,950-15,300	Beta-67671
DYK			VIIa	Charcoal	13,200±250	16,150-15,300	GIN-405
LRC1	*Equus*	AMS	D	Collagen	13,350-120	16,160-15,750	Beta-37056
BFC3	*Saiga*	AMS		Collagen	13,390±180	16,300-15,700	RIDDL-279
TCCB	*Mammuthus*	AMS	A[1]	Collagen	13,570±120	16,440-16,060	Beta-35841
LRC1	*Mammuthus*	AMS	C	Collagen	13,725±110	16,620-16,280	Beta-29166
DYK			VIIb	Charcoal[3]	14,000±100	16,940-16,640	GIN-404
TCCB	*Mammuthus*	C	A	Collagen	14,270±950	18,250-15,850	Beta-20027
BFC	*Mammuthus*			Collagen	15,500±130	18,560-18,260	GSC-3053
LHC	*Rangifer*			Collagen	15,690±140	18,750-18,420	Beta-67669
TCC9	*Equus*	C	IV	Collagen	15,750±350	19,050-18,250	K-1210
TCCA	*Rangifer*	AMC	A	Collagen	16,3000±140	19,400-19,000	Beta-35839
BFC1[2]	*Equus lambei*			Collagen	17,440±220	21,200-20,300	RIDDL-278
LRC1	*Bison*	C	C	Collagen	21,050+320/-340	Age out of range	DIC-1333
LRC1	*Equus*	C	C	Collagen	21,780±310	Age out of range	DIC-1334
LRC1	*Ovis*	AMS	C	Collagen	22,480±200	Age out of range	Beta-29165
BFC	*Equus lambei*			Collagen	22,680±530	Age out of range	NMC-1237
LHC	*Bison*			Collagen	27,950±560	Age out of range	Beta-67670
LRC1	*Ovis*	AMS	D	Collagen	38,260±830	Age out of range	Beta-37057

[1]Split sample of the same bone. [2]Does not appear in Figure 2.[3]Charcoal from the same hearth.

Table 1. Radiocarbon ages on large mammal elements from nine Beringian caves. Acronyms are: TCCA=Trail Creek Cave A; TCCB=Trail Creek Cave B; DYK=Dyuktai Cave; BFC=Bluefish River Cave; TCC2=Trail Creek Cave 2; TCC9=Trail Creek Cave 9; LHC=Lime Hills Cave; LRC1=Lower Rampart Cave 1 (after Larsen, 1968; Cinq-Mars, 1979; Dixon, 1984; Vinson, 1993; Ackerman, 1994; Harington and Cinq-Mars, 1995; Mochanov and Fedoseeva, 1996; Sattler, 1997; and Burke and Cinq-Mars, 1998).

SITE	STRATUM	PHASE OR COMPLEX	^{14}C YR B.P.	CAL ^{14}C YR B.P.[4]	LAB NUMBER
Healy Lake Village[5]	Level 6	Earliest Chindadn	11,100±60	13,090-12,940	Beta-76067, Cams-1591
Dry Creek	Component I	Nenana	11,120±85	13,130-12,940	SI-2880
Walker Road	Component I	Nenena	11,010±230	13,170-12,700	AA-1683
Mesa	Locality B	Mesa	11,190±70	13,190-13,010	Beta-57430
Broken Mammoth	Cultural Zone IV	Earliest Chindadn	11,040±260	13,230-12,690	Uga-6257D
Walker Road	Component I	Nenana	11,170±180	13,290-12,910	AA-1681
Broken Mammoth	Cultural Zone IV	Earliest Chindadn	11,280±190	13,420-13,010	WSU-4265
Walker Raod	Component I	Nenana	11,300±120	13,370-13,090	AA-2264
Broken Mammoth	Cultural Zone IV	Earliest Chindadn	11,280±190	13,420-13,010	WSU-4265
Owl Ridge	Component I	Nenana	11,340-150	13,440-13,100	Beta-110209
Healy Lake Village	Level 6	Earliest Chindadn	11,410±60	13,440-13,250	Beta-76060, Cams 15914
Broken Mammoth	Cultural Zone IV	Earliest Chindadn	11,420±70	13,460-14,250	Cams-5358
Broken Mammoth	Cultural Zone IV	Earliest Chindadn	11,500±80	13,560-13,320	Cams-8261
Swan Point	SP-7	Earliest Chindadn	11,660±60	13,730-13,500	Cams-12389
Mesa	Locality B	Mesa	11,660±80	13,740-13,470	Beta-55286, Cams-3572
Swan Point	SP-7	Earliest Chindadn	11,660±70	13,740-13,490	Cams-4252
Moose Creek	Component I	Nenana	11,730±250	14,020-13,410	GX-6281
Broken Mammoth	Cultural Zone IV	Earliest Chindadn	11,770±210	14,020-13,490	WSU-4351
Broken Mammoth	Cultural Zone IV	Earliest Chindadn	11,770±220	14,030-13,480	WSU-4364
Walker Road	Component I	Nenana	11,820±200	14,070-13,560	Beta-11254

[4]Calibrated age range, 68.2% confidance interval, per Stuiver and Reimer 1993. [5]Sources of dates are: Kunz and Reanier 1994; West 1996.

Table 2. Radiocarbon ages of earliest unequivocal archaeological cultures in East-Central Beringia. Discordant dates, as identified in the primary publications, have been disregarded.

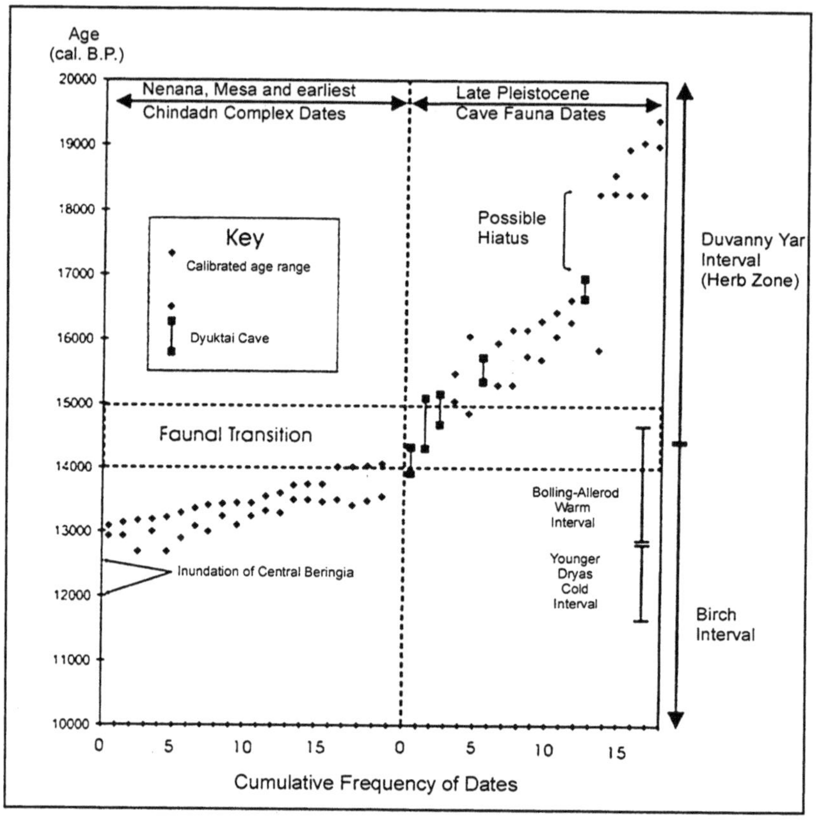

Figure 2. Calibrated radiocarbon ages of Late Pleistocene Cave fauna, earliest certain archaeological cultures, and key paleoenvironmental event in east-central Berengia. Bold line dates are on anthropogenioc charcoal from Dyuktai Cave in the Aldan River Valley. Environmental event dates after Bigelow (1997), and Elias et al. (1993). (Figure after Sattler and Gillispie 1998).

TAXONOMIC RICHNESS

The Lower Rampart Cave 1 and Trail Creek Caves A and B late Pleistocene faunal assemblages are taxonomically diverse and the Holocene assemblages are numerically impoverished (Table 3). Pleistocene levels among the three caves yield bison, mammoth horse, caribou, mountain sheep, red fox (*Vulpes vulpes*), and arctic fox (*Alopex lagopus*), while Holocene levels include black bear, caribou, and mountain sheep. The most common taxon in late Pleistocene and Holocene levels from these caves is *Rangifer*. Bear bones at the Trail Creek Caves are in Holocene levels, similar to their occurrence at Lower Rampart Cave 1, and may be in Pleistocene levels. Although the lowest bear bones in the Trail Creek caves could not be identified to species, they are either black or brown bear (*Ursus arctos*). These bear remains are in stratigraphic association with red fox in the early Holocene levels at the Trail Creek caves A and B, while arctic fox (*Alopex lagopus*) is identified only in the late Pleistocene levels. At Lower Rampart Cave 1, red fox and arctic fox are identified only in late Pleistocene levels. The fauna from Lime Hills Cave 1 resembles assemblages from the Trail Creek Caves and Lower Rampart Cave 1 in the presence of bison, horse, caribou, and bear (Ackerman 1994; Georgina this volume). Wolf may be represented by a single element from Trail Creek Cave A, and is reported at the Bluefish River caves (Harington 1989) and Dyuktai Cave (Mochanov and Fedoseeva 1996).

Though Trail Creek Caves A and B, Rampart Cave 1, and Lime Hills Cave 1 have similar faunas, the Bluefish River (Cinq-Mars 1979; Harington 1989; Burke and Cinq-Mars 1998) and Dyuktai caves (Mochanov and Fedoseeva 1996) are characterized by larger counts of skeletal elements and by taxonomically more diverse assemblages. The latter two assemblages contain not only a larger number of herbivore and carnivore taxa, but a higher density of bones and bone fragments. In the Bluefish River caves, the herbivores consist of mammoth, horse, bison, saiga antelope (*Saiga tatarica*), musk-ox (*Ovibos* sp.), moose, and wapiti (Cinq-Mars 1979; Harington 1989; Harington and Cinq-Mars 1995). The latest Pleistocene levels (c. 12,000 to 16,000 B.P.) at Dyuktai cave contain mammoth, bison, horse, musk-ox, moose, sheep, and caribou (Mochanov and Fedoseeva, 1996). Carnivores identified in the Bluefish River caves include the Beringian lion (*Panthera leo)*, mountain lion (*Puma concolor*), wolf (*Canis lupus*), bear, red fox, and arctic fox (Haringto n 1989). The latest Pleistocene levels (c. 12,000 to 14,000 B.P.) of Dyuktai Cave include *Panthera*, wolf, red fox, and arctic fox. The faunal assemblages from Dyuktai and Bluefish River caves consist of much larger and higher density "bone-beds" with more taxonomic-richness. These taphonomic patterns represent a larger scale of utilization by bone collecting carnivores.

TAXON	LOWER RAMPART CAVE 1		TRAIL CREEK CAVE A		TRAIL CREEK CAVE B	
Equus sp. (horse)	-	X	-	X	-	-
Mammuthus primigenius (woolly mammoth)	-	X	-	-	-	x
Bison *sp.* (bison)	-	X	-	X	X	X
Ovis dalli (mountain sheep)	-	X	-	-	X	X
Ursus americanus (black bear)						
deciduous teeth	X	-	-	-	-	-
post cranial bones	X	-	-	-	-	-
Ursus sp. (black or brown bear)						
deciduous teeth	-	-	-	x	-	-
post cranial bones	-	-	-	x	-	x
Rangifer tarandus (caribou)	X	-	X	X	X	X
Vulpes vulpes (red fox)	-	X	-	X	X	X
Alopex lupus (arctic fox)	-	-	-	X	-	X
Canis sp. (wolf?)	-	-	-	X	-	-

Table 3. Taxonomic representation for Pleistocene and Holocene levels in the three caves.

TAXA	LOWER RAMPART CAVE 1		TRAIL CREEK CAVE A			TRAIL CREEK CAVE B		
	UU	LU	C	B	A	C	B	A
Mammuthus	-	2	-	-	-	-	-	2
Equus	-	4	-	-	1	-	-	-
Bison	-	2	-	-	3	2	-	1
Ovis	-	10	-	-	-	3	-	7
Rangifer	2	6	12	-	23	33	8	41
Ursus	34[6]	-	1	-	3	-	-	3[7]
Canis	-	-	-	-	1	-	-	-
Alopex	-	1	-	-	6	1	-	7
Vulpes	-	1	-	-	6	-	-	22[8]
MNE	36	26	12	0	43	37	8	60
NISP	162	28	24	0	50	104	19	83

Table 4. Minimum number of identified skeletal elements (MNE) of herbivores and carnivores in the three caves: Lower Rampart Cave 1 - UU is upper units (Holocene) and LU is Lower units (Late Pleistocene); Trail Creek Cave A - C is Holocene, B is transitional, and A is Late Pleistocene (after Vinson 1993; Sattler 1997).

SURVIVORSHIP OF SKELETAL PARTS AND MNE

In the late Pleistocene levels of Lower Rampart Cave 1 and Trail Creek Caves A and B, anatomical part survivorship is biased toward the head and legs, especially the lower portions of legs including the feet (Vinson 1993; Sattler 1997). The most abundant elements are the metacarpals and metatarsals, phalanges, carpals, tarsals, and sesamoids. A tibia and a radio-ulna, cranio-facial fragments, isolated teeth, and fragments of shed and unshed antlers complement these. The articular ends of most caribou limb bones from all caves are destroyed or damaged by carnivore chewing. Wavy corrosion of the surface on some of the small diaphyseal fragments suggests that they passed through the digestive system of a carnivore. Survivorship patterns of all large mammal skeletal elements are not available for the Lime Hills, Bluefish River, or Dyuktai Caves. However, minimum numbers of elements and specimens of *Equus* mandibles and teeth are provided by Burke and Cinq-Mars (1998) for the Bluefish River caves.

Table 4 presents the MNE for large mammal and carnivore taxa in Trail Creek Caves A and B, and Lower Rampart Cave 1. The Holocene and Pleistocene levels at each cave reveal distinctively different patterns of element representation by taxa. At the Trail Creek Caves, the Holocene and Pleistocene levels include several times more caribou elements than at Lower Rampart Cave 1.

Disparities still exist despite the fact that four times the square meter area was excavated at Lower Rampart Cave 1 (8 meters square) than was excavated at either of the Trail Creek caves (2m square). *Rangifer* is the most abundantly represented large mammal from any of the caves. Further, there are fewer than ten elements per any taxon among the three caves with the exception of sheep and bear in Lower Rampart Cave 1, and red fox in the intermediate level at Trail Creek Cave B. The numbers of identified elements in Lower Rampart Cave 1 and Trail Creek Caves A and B are extremely low for all taxa. This is particularly significant when compared to the hundreds and thousands of bones and bone fragments in the Bluefish River and Dyuktai Caves.

[6] Includes 23 deciduous teeth.

[7] All are elements of a bear cub.

[8] Elements are deciduous and rooted teeth.

The MNE for ursid remains in Lower Rampart Cave 1 far exceeds the MNE of bear in Trail Creek Caves A and B. The skeletal parts in Lower Rampart Cave 1 consist of deciduous canines and third incisors representing several individuals, limb and foot elements representing one individual, and rare vertebrae fragments. There are numerous small, highly fragmented specimens of ribs and probable scapula pieces.

Collectively, these elements represent the most complete bear skeleton among all the caves discussed here, although complete skeletons of bears are known from karst caves in southeast Alaska (Heaton *et al.* 1996). Bear elements at the Trail Creek Caves include rare bones of two individuals: an adult lower incisor and carpal of one individual in the lower levels of Trail Creek Cave A, and a deciduous canine of a yearling or cub. Skeletal remains of bear in lower levels at Trail Creek Cave B consists of three elements identified to a separate young individual based on its provenance in an adjacent cave. Although the bones of a bear cub in Trail Creek Cave B are believed to be late Pleistocene in age, none of the ursid elements are carbon dated.

Bone destruction patterns in the three shelters indicate that the soft, spongy bone at the ends of limb elements were largely consumed by carnivores, and that marrow was obtained from cylinder shafts by successively removing bone flakes from the ends instead of breaking the cylinders mid-shaft. These same patterns of carnivore processing of ungulate limb-bones are described in varying degrees for carnivores in Eastern Beringia (Haynes 1980a,b, 1982, 1983; Binford 1981) and for other regions (Blumenshine and Marean 1993; Stiner 1991a,b, 1993, 1994). An expanded description about carnivore processing of the medium to large mammal long bone assemblages in Trail Creek Caves A and B, and Lower Rampart Cave 1 is reported elsewhere (Vinson 1993; Sattler 1997).

ARCHAEOLOGY OF EASTERN BERINGIAN CAVES: HUMANS AND BEARS

The taphonomic histories of Trail Creek Caves A and B, and Lower Rampart Cave 1 provide a basis to evaluate cultural assemblages in other caves in the region. Previous arguments made for cultural association in late Pleistocene caves are not widely accepted or appreciated. For example, Larsen (1968) argued for culturally modified bones based exclusively on broken megafaunal elements in Trail Creek Cave 2, although analysis of latest Pleistocene faunas in the adjacent Trail Creek Caves A and B provides an alternative hypothesis that refutes Larsen's observations (Vinson 1993). Cinq-Mars (1979), Morlan and Cinq-Mars (1982), and Burke and Cinq-Mars (1998) argue for a pre-12,000 B.P. component in the Bluefish River caves based on microdebitage, extinct faunal elements with cut marks, stone artifacts intermingled with extinct megafauna, and bone flakes representing a bone-core technology. However, these purported archaeological materials were not found in association with Late Pleistocene hearths such as the archaeological context for Late Pleistocene levels at Dyuktai Cave. Though direct association of these finds with hearths may be viewed by some as a too high standard, such evidence is difficult to refute. As it stands the available evidence is ambiguous.

Though assertions of human involvement in the breakage patterns observed on late Pleistocene faunal elements in Eastern Beringian caves continue to be made, only Holocene cultural deposits are clearly demonstrated in any of the caves discussed here. Irrefutable evidence from Holocene archaeological deposits consists of stone tool assemblages, well-preserved faunal remains, and anthropogenic charcoal or cultural hearths. The most extensive lithic assemblages are in Trail Creek Cave 2 where multiple components represent episodic cultural occupation spanning most of the Holocene culture history of Northwest Alaska (Larsen 1968; Vinson 1993).

Lime Hills Cave 1 has produced a well-defined cultural level with an assemblage of microblades, bone projectiles, an adze blade, and possible associated faunal material along an isolated stratigraphic boundary from what appears to be a hearth with multiple radiocarbon ages between 8,150 and 9,530 B.P. (Ackerman 1994, 1996; Georgina, this volume). Cinq-Mars (1979) and Morlan and Cinq-Mars (1982:371) describe a burin in the humus-rich upper levels at the Bluefish River Cave 1 that typologically resembles the widely recognized Denali Complex dated to the early Holocene (cf. West 1967; Dixon 1985). Sattler (1997) describes a late-Holocene archaeological component consisting of microblades, flakes, a biface fragment, and bear bones all associated along a stratigraphic boundary containing a hearth dated to 3,500 B.P.

The most perplexing and enigmatic evidence of human use in Eastern Beringian cave taphonomy consists of the microflakes, modified bones and flaked bone specimens in the Bluefish River caves. Bones modified by cutting tools in Bluefish Cave 1 are identified from both inside and outside the dripline (Cinq-Mars 1979; Morlan and Cinq-Mars 1982). The largest portion of the late Pleistocene aged bone bed is outside the shelter and consists of thousands of bones and bone fragments. Though the bones are unequivocally modified by sharp-edged tools, photographs of some faunal elements show that the cut marks are superimposed on bones that were previously gnawed by carnivores (Morlan and Cinq-Mars 1982: Figure 10, photographs 4, 10, and 11). This suggests that cultural modification on the bones lagged during the period in which bones were initially deposited at the cave. Consequently, people may have scavenged bones from the diverse bone bed, perhaps millennia after carnivores formed the assemblages. Further, the microdebitage found amongst the extinct fauna may have originated from above and been mixed downward by animal or other forms of post-depositional turbation.

One important taphonomic effect that must be considered in any cave analysis is the potential turbation of fill deposits resulting from the interspecific use of caves, especially that of canids and ursids. The effect of animals that dig into cave deposits, which in our case involves overwintering bears and summer denning canids, will yield disturbances to the primary integrity of the fill deposits. For example, stratigraphic evidence of overwinter denning of bears has been recognized at Lower Rampart Cave 1 and Trail Creek Caves A and B, and Caves 2 and 9 (Dixon and Smith 1986; Vinson 1993; Sattler 1997). The faunal signature for denning bears in these caves consists of anterior teeth and post-cranial skeletal elements from ontogenically young individuals. Stratigraphic patterns supporting the evidence for turbation consists of displaced bones and refitted fragments across lithostratigraphic boundaries defined primarily by textural differences.

Stratigraphic and faunal analysis at Lower Rampart Cave 1 demonstrates at least two events in which primary sediment from the back of the cave was displaced onto spoil piles below and outside the dripline (Sattler 1997). Meticulous stratigraphic analysis delineated the turbated levels from sediments that had not been turbated. Observations on displaced bones and refitting of artifacts in Trail Creek Caves A and B indicate that animal turbation probably accounts for the displaced artifactual materials described for Trail Creek Caves 2 and 9. Larsen (1968) described the anomalous placement of refit artifacts in those caves resulting from possible ursid or other cave-dwelling carnivore bedding activities. Turbation is not reported in the Lime Hills, Dyuktai, and Bluefish River caves, but skeletal elements of bear and canids are reported in all faunas (Harington 1989; Ackerman 1994; Mochanov and Fedoseeva 1996).

LATE PLEISTOCENE CAVE FAUNAS, ECOLOGICAL CHANGE, AND EARLIEST HUMANS

The broad age ranges indicated by latest Pleistocene collagen dates suggest that carnivores episodically used these caves as den sites for a period of several millennia. The accumulation of bones in the caves ended abruptly about a millennium before humans arrived. The widespread use of caves by carnivores during the latest Pleistocene (15,500 to 17,000 cal B.P.; 13,000 to 14,000 B.P.) was coincidental with major vegetation change from a herbaceous dominated flora to a dwarf birch shrubland (Ager and Brubaker 1985; Barnosky, Anderson, and Bartlein 1987; Anderson and Brubaker 1994; Bigelow 1997). Though the exact mechanisms for the pattern of carnivore use of caves and for the bone accumulations during a phase of major vegetation change are still unclear, the subsequent abandonment of caves by bone-collecting carnivores around 13,000 B.P. is clear (15,500 cal. B.P.). The dramatic cessation of bone recruitment into Eastern Beringian cave and rockshelter sites also coincides with the initial phase of heightened selection pressures that eventually resulted in extinctions and/or rapid evolution between c. 13,000 to 9,000 B.P. (Guthrie 1990a,b, 1995). The end to the use of these small shelters by the "mammoth-steppe" carnivores must surely signal a threshold of paleoecological change in the timing of terminal Pleistocene megafaunal extinctions such as for mammoth and horse.

The temporal gap of approximately one millennia in the radiocarbon age series between carnivore-derived cave assemblages and culturally-derived open-air assemblages illustrates a dichotomy in late Pleistocene faunas. Curiously, the ages on anthropogenic charcoal at Dyuktai cave fall between the Eastern Beringian cave and the cultural age series (Figure 2, bold symbols), suggesting that the late Pleistocene human occupations at Dyuktai Cave were contemporaneous with a transitional period in Eastern Beringia. This transitional period between break-up of the mammoth steppe (Guthrie, 1995) and the earliest arrival of a well-defined archaeological culture provides chronological support for the hypothesis that Dyuktai may be ancestral to the earliest microblade horizon in Eastern Beringia.

Accumulations of megafaunal skeletal parts during the several millennia period in the late Pleistocene reflects a period when large carnivores (i.e., wolves) transported bones to shelters when utilized as dens or rendezvous sites. The representation of limb and foot bones, along with head parts is a pattern suggesting that assemblages were differentially transported from kill sites (cf. Binford 1981:197-237; Haynes 1980a,b).

The most striking observation in the chronology of this process is that the shelters were abandoned nearly simultaneously across a broad geographic area. The pervasive abandonment of caves by "mammoth steppe" carnivores may reflect migration, adaptation, or extinction. We hypothesize that the pattern of abandonment reveals a changed scavenging niche in which large predators responded to declining populations of herbivore prey. Should the proxy data derived from the cave ages prove to be a threshold point in the timing of Pleistocene extinctions, large herbivores that either survived major environmental change (bison and caribou) or arrived after the extirpation of some "mammoth steppe" taxa (elk and black bear) may all have been a limiting resource for human colonization of Eastern Beringia between 11,700 to 13,000 B.P.

Differences in the faunal assemblages among the Beringian caves do suggest that different site formation processes acted on the shelter deposits. This paper offers new understanding of taphonomic processes acting on cave and rockshelter sites during a period of major environmental change at the end of the Pleistocene. Although there are shared taphonomic histories regarding the differential transport of large mammalian herbivore bones to the caves and shelters by bone-collecting carnivores, the more diverse faunas at Dyuktai and Bluefish caves greatly contrast to the smaller and less diverse faunas at Trail Creek Caves A and B, Lower Rampart Cave 1, and Lime Hills Cave 1. This variability may be a result of differences in local habitats. local fauna, carnivores species that used the caves, and duration of cave use by those carnivores.

In the context of archaeology, the anthropogenic charcoal ages associated with bifacial and blade industries at Dyuktai Cave may be viewed as the precursor of early sites in Eastern Beringia. Dyuktai Cave has multiple cultural levels greater than 12,000 B.P. with bifacial tools, a microlithic core and blade industry, and associated anthropogenic deposits. There is still a need to know the taphonomic history of carnivore and cultural overprinting on the Dyuktai Cave fauna as a comparative tool for interpreting Eastern Beringian cave faunas. In our view, caves and shelters in Eastern Beringia have yet to produce unequivocal anthropogenic deposits, namely hearth features intermingled with lithic tool assemblages, culturally modified faunal remains, and radiocarbon ages between c. 12,000 to 14,000 B.P.

CONCLUSION

The "mammoth steppe" faunal remains that accumulated in Eastern Beringian caves and rockshelters in the latest Pleistocene were derived from carnivores, but this taphonomic process ended abruptly approximately a millennia (13,000 B.P.) before humans widely occupied the region (11,700 B.P.). The observation that bone-collecting carnivores abandoned all caves across the region nearly simultaneously intimates a significant phase of paleoecological change. The question of whether this change was due to extinction, migration, or adaptation is crucial for understanding the place of humans in the late Pleistocene community structure. We hypothesize that the temporal gap between the cessation of bone collecting activities by shelter dwelling carnivores and the widespread appearance of humans in open-air sites represents a period of intensified ecological or environmental stress on prey species. The period between 11,700 to 13,000 B.P. may represent an interval of large mammal scarcity that posed a limiting resource for mammalian predators, including humans.

Finally, the variability in the faunas are characterized by two generally dissimilar types of bone assemblages that reflect different site formation processes in these caves and rockshelters: 1) slow or episodic accretion of faunal remains whereby a relatively small number of animal bones accumulated in faunas characterized by relatively low taxonomic richness, and 2) faster or punctuated accretion of faunal remains whereby a relatively large number of animal bones accumulated with a relatively high taxonomic richness. The site formation processes contributing to these patterns and their paleoecological significance needs to be further explored as more detailed taphonomic analyses become available from Beringian cave and rockshelter assemblages.

ACKNOWLEDGMENTS

This article began as two separate papers presented at the Alaska Anthropological Association meeting in honor of R. Dale Guthrie by RAS and DMV. The papers were combined as a result of the perceived need to begin synthesizing research on caves in Beringia as a new data set to add to discussions on the paleoecology of the region. We thank Craig Gerlach and two anonymous reviewers for constructive advice on this paper. Dr. Guthrie was an inspiration for the three of us who were students at UAF during the 1980's. During his 30 or so years at UAF, Dale was able to make some sense of the extraordinary collection of Pleistocene faunas collected during an equivalent period of time by former students and faculty (c. 1930-1960) throughout the heyday of industrial placer mining in Interior Alaska. We would like to acknowledge several persons and entities for support of the original research leading to this paper: Dr. E. James Dixon and Dr. David M. Hopkins, the Geist Fund of the University of Alaska Museum, the Alaska Quaternary Center, the National Park Service, and the National Science Foundation (DPP-8800011).

REFERENCES CITED

Ackerman, R. E.
1994 Early Cultural Complexes in Southwestern Alaska. *Current Research in the Pleistocene* 11:109-111.

1996 Bluefish Caves. In *American Beginnings: The Prehistory and Paleoecology of Beringa*, edited by F. H. West, pp. 511-513. University of Chicago Press, Chicago.

Ager, T. A., and L. Brubaker
1985 Quaternary Palynology and Vegetational History of Alaska. In *Pollen Records of Late Quaternary North American Sediments,* edited by V.M. Bryant and R.G. Holloway, pp. 353-384. American Association of Stratigraphic Palynologists, Dallas.

Anderson, P.M., and L.B. Brubaker
1994 Vegetation History of Northcentral Alaska: A Mapped Summary of Late-Quaternary Pollen Data. *Quaternary Science Reviews* 13: 71-93.

Bard, E., B. Hamelin, R.G. Fairbanks, and A. Zindler
1990 Calibration of the 14C Timescale Over the Past 30,000 Years Using Mass Spectrometric U-Th Ages from Barbados Corals. *Nature* 345:405-9.

Barnosky, C.W., P.M. Anderson, and P.J. Bartlein
1987 The Northwestern U.S. During Deglaciation: Vegetational History and Paleoclimatic Implications. In *North America and Adjacent Oceans During the Last Deglaciation*, edited by W.F. Ruddiman and H.E. Wright, Jr., pp. 289-321. Geological Society of America, Boulder.

Bigelow, N.H.
1997 *Late Quaternary Vegetation and Lake Level Changes in Central Alaska*. Unpublished PhD. dissertation. Department of Anthropology, University of Alaska Fairbanks, Fairbanks.

Binford, L.R.
1981 *Bones: Ancient Men and Modern Myths*. Academic Press, New York

1984 *Faunal Remains from Klasies River Mouth.* Academic Press, New York.

Brain, C.K.
1981 *The Hunters or the Hunted? An Introduction to African Cave Taphonomy*. The University of Chicago Press, Chicago.

Blumenshine, R. J. and C.W. Marean
1993 A Carnivore's View of Archaeological Bone Assemblages. In *From Bones to Behavior: Ethnoarchaeology and Experimental Contributions to the Interpretation of Faunal Remains*, edited by J. Hudson, pp. 273-300. Center for Archaeological Investigations, Occasional Paper No. 21. Southern Illinois University, Carbondale.

Burke, A. and Cinq-Mars, J.
1998 Paleoethological Reconstruction and Taphonomy of Equus lambei from the Bluefish Caves, Yukon Territory, Canada. *Arctic* 51(2):105-115.

Cook, J. P.
1968 *Prehistory of Healy Lake*. Unpublished PhD. dissertation. Department of Anthropology, University of Wisconsin, Madison.

Cinq-Mars, J.
1979 Bluefish Cave 1: A Late Pleistocene Eastern Beringian Cave Deposit in the Northern Yukon. *Canadian Journal of Archaeology* 3:1-32.

Clark, G. A.
1988 The Upper Paleolithic of Northeast Asia and its Relevance to the First Americans: A Personal View. *Current Research in the Pleistocene* 5:3-7.

Dixon, E.J.
1984 Context and Environment in Taphonomic Analysis: Examples from Alaska's Porcupine River Caves. *Quaternary Research* 22:201-215

1985 Cultural Chronology of Central Interior Alaska. *Arctic Anthropology* 22:47-66.

1993 *Quest for the Origins of the First Americans*. University of New Mexico Press, Albuquerque.

Dixon E.J., and Smith, G.
1986 Broken Canines from Alaskan Cave Deposits: Re-evaluating Evidence for Domesticated Dog and Early Humans in Alaska. *American Antiquity* 51:341-351.

Dixon, E.J., T.H. Heaton, F.E. Fifield, T.D. Hamilton, D.E. Putnam, and F. Grady
1997 Late Quaternary Regional Geoarchaeology of Southeast Alaska Karst: A Progress Report. *Geoarchaeology* 12(6):689-712.

Friesen, M.T.
1997 Effects of Collector Bias on Pleistocene Bone Breakage Frequencies from Old Crow, Northern Yukon Territory. *Current Research in the Pleistocene* 14:113-115.

Gamble, C.
1983 Caves and Faunas from Last Glacial Europe. In *Animals and Archaeology, Volume : Hunters and Their Prey*, edited by J. Clutton-Brock and C. Grigson, pp. 163-172. British Archaeological Reports S163, Oxford.

Guthrie, R. D.
1968 Paleoecology of the Large Mammal Community in Interior Alaska During the Late Pleistocene. *American Midland Naturalist* 79:346-363.

1982 Mammals of the Mammoth Steppe as Paleoenvironmental Indicators. In *Paleoecology of Beringia*, edited by D. M. Hopkins, J. V. Mathews, Jr., C. E. Schweger, S. B. Young, pp. 307-326. Academic Press Inc., New York.

1983 Paleoecology of the Site and its Implications for Early Hunters. In *Dry Creek*, edited by W. R. Powers, R. D. Guthrie, and J. F. Hoffecker, pp. 211-287. Washington: National Park Service.

1984 Mosaics, Allelochemics, and Nutrients. In *Quaternary Extinctions*, edited by P. S. Martin and R. G. Klein, pp. 259-298. University of Arizona Press, Tucson.

1988 Bone Litter from an Alaskan Pleistocene Carnivore Den. *Current Research in the Pleistocene* 5:69-70.

1990a Late Pleistocene Faunal Revolution--A New Perspective on the Extinction Debate. In *Megafauna and Man: Discovery of American's Heartland*, edited by L. D. Agenbroad, J. I. Mead, L. and W. Nelson, pp. 42-53. The Mammoth Site of Hot Springs, South Dakota, Inc., Scientific Papers, vol. 1, Hot Springs.

1990b Frozen Fauna of the Mammoth Steppe. University of Chicago Press, Chicago.

1995 Mammalian Evolution in Response to the Pleistocene-Holocene Transition and the Break-up of the Mammoth Steppe: Two Case Studies. *Acta Zool. cracov.* 38:139-154.

Harington, C.R.
1989 Pleistocene Vertebrate Localities in the Yukon. In *Late Cenozoic History of the Interior Basins of Alaska and the Yukon*, edited by L. D. Carter, T. D. Hamilton, and J. P. Galloway, pp. 93-98. U.S. Geological Survey Circular 1026 Government Printing Office, Washington.

Harington, C. R. and J. Cinq-Mars
1995 Radiocarbon Dates on Saiga Antelope (Saiga tatarica) Fossils from Yukon and the Northwest Territories. *Arctic* 48:1-7.

Haynes, G.
1980a Evidence of Carnivore Gnawing on Pleistocene and Recent0 Mammalian Bones. *Paleobiology* 6:341-351.

1980b Prey Bones and Predators: Potential Ecologic Information from Analysis of Bone Sites. *OSSA* 7:75-97.

1982 Utilization and skeletal disturbances of North American prey carcasses. *Arctic* 35:266-281.

1983 A Guide for Differentiating Mammalian Carnivores Responsible for Gnaw Damage to Herbivore Limb Bones. *Paleobiology* 9:164-172.

Heaton, T.H., S.L. Talbot, and G.F. Shields
1996 An Ice Age Refugium for Large Mammals in the Alexander Archipelago, Southeastern Alaska. *Quaternary Research* 46:186-192.

Hoffecker, J. F., W.R. Powers, and T. Goebel
1993 The Colonization of Beringia and the Peopling of the New World. *Science* 259:46-53.

Holmes, C. E.
1996 Broken Mammoth. In *American Beginnings: The Prehistory and Paleoecology of Beringia*, edited by F. H. West, pp. 312-318. University of Chicago Press, Chicago.

1998 New Data Pertaining to Swan Point, the Oldest Microblade Site Known in Alaska. *Current Research in the Pleistocene* 15:21-22.

Holmes, C. E., R. VanderHoek, and T. E. Dilley
1996 Swan Point. In *American Beginnings: The Prehistory and Paleoecology of Beringia*, edited by F. H. West, pp 319-323. University of Chicago Press, Chicago.

Hopkins, D. M.
1982 Aspects of the Paleogeography of Beringia During the Late Pleistocene. In *Paleoecology of Beringia*, edited by D. M. Hopkins, J. V. Matthews, Jr., C. E. Schweger, and S. B. Young, pp. 3-28. Academic Press, Inc., New York.

Kunz, M.L. and R. E. Rainer
1994 Paleoindians in Beringia: Evidence from Arctic Alaska. *Science* 263:660-662.

Larsen, H.
1968 Trail Creek: Final Report on the Excavation of Two Caves on Seward Peninsula, Alaska. *Acta Arctica* 15:7-79.

Loy, T.H. and E.J. Dixon
1998 Blood Residues on Fluted Points From Eastern Beringia. *American Antiquity* 63(1):21-46.

Meltzer, D.J.
1995 Clocking the First Americans. *Annual Review of Anthropology* 24:21-45

Mochanov, Y. A. and S. A. Fedoseeva
1996 Dyuktai Cave. In *American Beginnings: The Prehistory and Paleoecology of Beringia*, edited by F.H. West, pp. 164-174. University of Chicago Press, Chicago.

Morlan, R.E. and J. Cinq-Mars
1982 Ancient Beringians: Human Occupation on the Late Pleistocene of Alaska and the Yukon Territory. *In* Paleoecology of Beringia, edited by D. M. Hopkins, J. V. Matthews, Jr., C. E. Schweger, S. B. Young, pp. 307-326. New York: Academic Press.

Powers, W. R.
1973 Palaeolithic Man in Northeast Asia. *Arctic Anthropology* 10:1-106.

Powers, W. R., R. D. Guthrie, and J. F. Hoffecker
1983 *Dry Creek: Archeology and Paleoecology of a Late Pleistocene Alaskan Hunting Camp*. Washington: National Park Service.

Powers, W.R., and J. F. Hoffecker
1989 Late Pleistocene Settlement in the Nenana Valley, Central Alaska. *American Antiquity* 54:263-287.

Sattler, R. A.
1997 Large Mammals in Lower Rampart Cave 1, Alaska: Interspecific Utilization of an Eastern Beringian Cave. *Geoarchaeology* 12:657-688.

Sattler, R.A. and T.E. Gillispie
1998 Paleoenvironments, Cave Faunas, and Human Migration in Late-Pleistocene Beringia: A Comparison of Calibrated Age Ranges. *Current Research in the Pleistocene* 15: 124-126.

Sattler, R.A. and E. J. Dixon
1997 Timing the Arrival of Black Bear (Ursus americanus) in eastern Beringia. *Current Research in the Pleistocene* 14:155-157.

Stiner, M.C.
1991a An Interspecific Perspective on the Emergence of the Modern Human Predatory Niche. In *Human Predators and Prey Mortality*, edited by M.C. Stiner, pp. 149-185. Westview Press, Boulder.

1991b Food Procurement and Transport by Human and Non-human Predators. *Journal of Archaeological Science* 18:455-482.

1993 The Place of Hominids among Predators: Interspecific Comparisons of Food Procurement and Transport. In *From Bones to Behavior: Ethnoarchaeology and Experimental Contributions to the Interpretation of Faunal Remains*, edited by J. Hudson, pp. 38-61. Center for Archaeological Investigations, Occasional Paper No. 21, Southern Illinois University, Carbondale.

1994 *Honor Among Thieves A Zooarchaeological Study of Neanderthal Ecology*. Princeton University Press, Princeton.

Stuiver, M., and H. A. Polach
1977 Discussion: Reporting of 14C Data. *Radiocarbon* 19:355-363.

Stuiver, M., and P. J. Reimer
1993 Extended 14C Data Base and Revised CALIB 3.0 ^{14}C Age Calibration Program. *Radiocarbon* 35(1): 215-230.

Stuiver, M., P.M. Grootes, and T.F. Braziunas
1995 The GISP ^{18}O Climate Record of the Past 16,500 Years and the Role of the Sun, Ocean, and Volcanoes. *Quaternary Research* 44:341-354.

Taylor, R.E.
1987 *Radiocarbon Dating: An Archaeological Perspective*. Academic Press, London.

1991 Radioisotope Dating by Accelerator Mass Spectrometry: Archaeological and Paleoanthropological Perspectives. In *Scientific Dating Methods*, edited by Goksu,

H.Y, M. Oberhoffer, and D. Regulla, pp. 37-54. Netherlands: Kluwer Academic Publishers.

Thorson, R. M.
1990 Geologic Context of Archaeological Sites in Beringia. In *Archaeological Geology of North America, Centennial Special Volume 4*, edited by N.P. Lasca and J. Donahue, pp. 399-420. Geological Society of America, Boulder.

Vinson, D. V.
1993 *Taphonomy and Archeology of the Trail Creek Caves, Western Alaska*. Unpublished Master's thesis. University of Alaska Fairbanks.

West, F. H.
1967 The Donnelly Ridge Site and the Definition of an Early Core and Blade Complex in Central Alaska. *American Antiquity* 23(3):360-382.

1996 *American Beginnings: The Prehistory and Paleoecology of Beringia*. Chicago: University of Chicago Press.

Yi, S. and Clark, G.A.
1985 The "Dyuktai Culture" and New World Origins. *Current Anthropology* 16:1-20.

Yesner, D.
1994 Subsistence Diversity and Hunter-Gatherer Strategies in Late Pleistocene/Early Holocene Beringia: Evidence from the Broken Mammoth Site, Big Delta, Alaska. *Current Research in the Pleistocene* 11:154-156.

1996 Human Adaptation at the Pleistocene-Holocene Boundary (circa 13,000 to 8,000 B.P.) in Eastern Beringia. In Humans at the End of the Ice Age: The Archaeology of the Pleistocene-Holocene Transition, edited by L.G. Straus, B.V. Ericksen, J.M. Erlandson, and D.R. Yesner, pp. 255-272. Plenum Press, New York.

WOOD BISON IN LATE HOLOCENE ALASKA AND ADJACENT CANADA: PALEONTOLOGICAL, ARCHAEOLOGICAL AND HISTORICAL RECORDS

Robert O. Stephenson
Alaska Department of Fish and Game, Fairbanks

S. Craig Gerlach
Department of Anthropology, University of Alaska Fairbanks

R. Dale Guthrie
Department of Biology and Wildlife, University of Alaska Fairbanks

C. Richard Harington
Quaternary Zoology, Canadian Museum of Nature, Ottawa

Robin O. Mills
Bureau of Land Management, Fairbanks

Gregory Hare
Yukon Heritage Branch, Whitehorse

INTRODUCTION

Bison originated in Eurasia, dispersing to North America by the Bering Isthmus several hundred thousand years ago, possibly during a transition from Irvingtonian to Rancholebrean fauna (Harington 1984). Large-horned forms similar to steppe bison (*Bison* priscus McDonald 1981, Guthrie 1990; Bison *bison priscus;* van Zyll de Jong 1986) prevailed until near the end of the Pleistocene when a number of large Pleistocene mammals became extinct or shifted their ranges southward. Large-horned bison in Siberia, Alaska and northwestern Canada evolved into modern, small-horned bison during the late Pleistocene and early Holocene, culminating in the wood bison (*B. b. athabascae*). Bison were abundant on the Alaskan landscape during most of the last 100,000 years (Guthrie 1968; 1990), and wood bison were the last subspecies to occupy Alaska and adjacent regions (Skinner and Kaisen 1947; Harington 1977; van Zyll de Jong 1986).

Bison have long been thought to have been a major food resource for humans in Beringia and elsewhere. Evidence from archaeological sites in Alaska and a large body of archaeological evidence demonstrates late Pleistocene and early Holocene associations of hunting implements with bison remains (cf. Guthrie 1980; Powers et al. 1983; Frison 1991, 1996; Yesner 1994; Loy and Dixon 1998; Brink, this volume; Kooyman, this volume). However, with few exceptions (Gordon and Savage 1973; Holmes and Bacon 1982; West 1982; Guthrie 1990; Gates et al. 1992), ethnographers, archaeologists and biologists have given little consideration to the possibility that bison played a role in the late Holocene subsistence system of Athabascan people in interior Alaska and adjacent parts of northwestern Canada. On the basis of paleontological and zooarchaeological evidence, Skinner and Kaisen (1947), Holmes and Bacon (1982), Guthrie (1990), and others have suggested that northern bison populations disappeared shortly before the arrival of EuroAmericans. Van Zyll de Jong (1986) referred to the "prehistoric" range of wood bison as extending "north and west into Alaska and the Yukon," noting that "the exact boundaries of the historical distributional range of *B.b. athabascae* cannot be determined at present on the basis of the available information." He also described the occurrence of small-horned bison similar to wood bison in eastern Siberia during the Holocene.

In this study we review new and existing paleontological and archaeological data and oral accounts from Native elders in Alaska and Canada, focusing on the late Holocene history and human use of wood bison in Alaska, Yukon and the western Northwest Territories (N.T.). These data include oral narratives provided by Athabascan elders in several interior Alaskan villages, similar accounts obtained from First Nation elders in Yukon (Lotenberg 1996) and N.T., and radiocarbon dates from bison remains collected at various locations throughout the study area (Figure 1). We also explore cultural and ecological factors that may have caused the demise of bison in part of their original range in Alaska and Canada. In the course of our discussions with Athabascan elders we were also provided with information regarding the occurrence of muskoxen (*Ovibos moschatus)* prior to their extirpation. Information pertaining to muskoxen is included where it clarifies issues pertaining to the late Holocene range of wood bison and muskoxen.

WOOD BISON ECOLOGY AND BEHAVIOR

Wood bison are adapted to boreal regions, having a highly efficient digestive system and an ability to forage on a variety of common grasses and sedges found in meadows and early successional habitats (Reynolds and Hawley 1987; Reynolds et al. 1978; Larter and Gates 1991). They are highly mobile, use a variety of open and forested habitats, and maximize seasonal foraging efficiency by selecting habitats where crude protein is most available (Larter and Gates 1991). Bison are adapted to low temperatures (Fuller 1962; Peters and Slen 1964; Christopherson et al. 1978, 1979) and snow conditions typical of northern latitudes (Larter and Gates 1991; Carbyn et al. 1993). Unlike caribou,

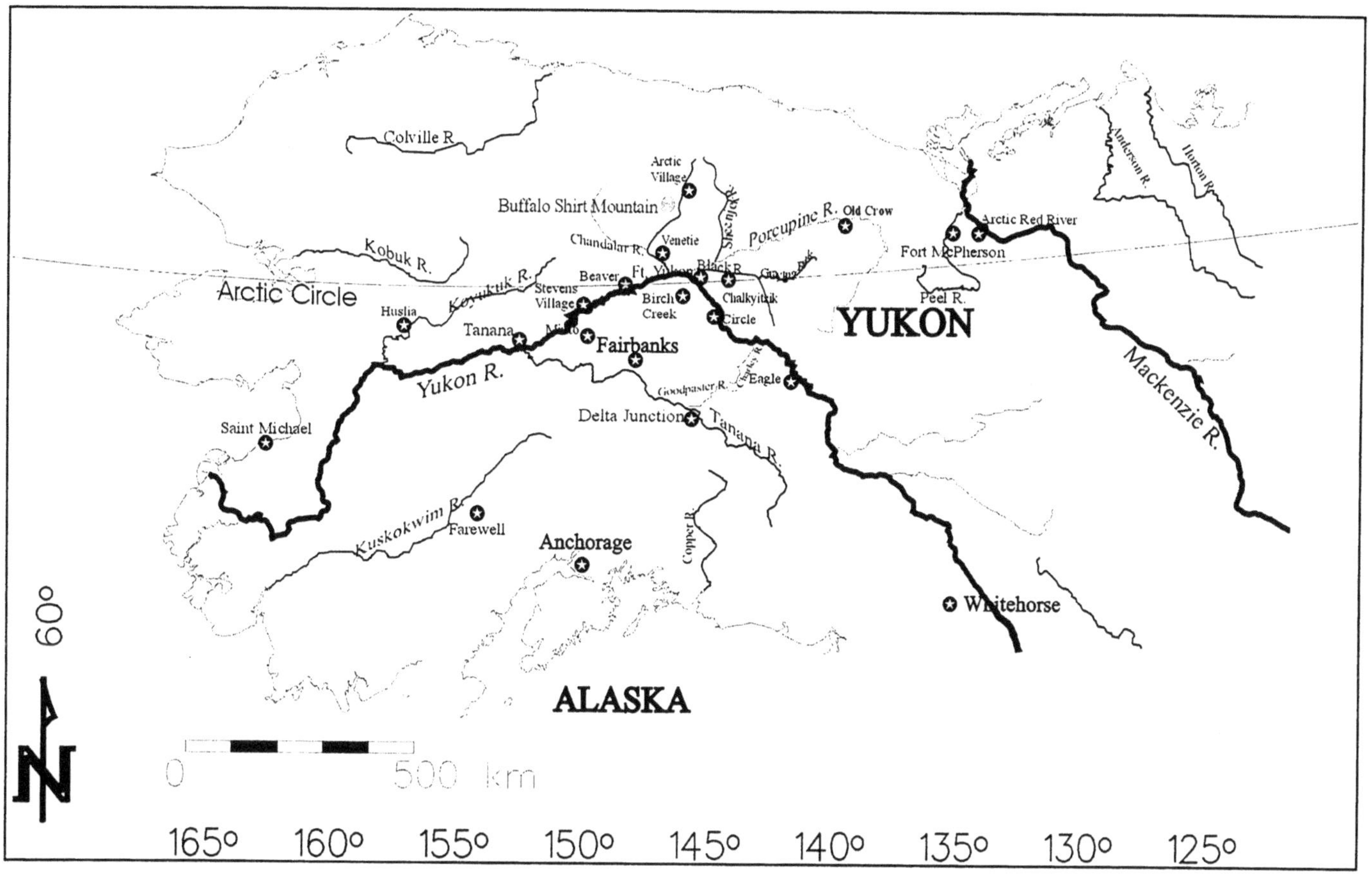

Figure 1. Map of study area showing major geographic features and locations mentioned in the text.

which crater snow with their hooves, bison expose forage by sweeping aside snow with their heads (Guthrie 1990).

The social behavior of wood bison is adapted to northern forests where forage occurs in scattered meadows separated by woodlands. Wood bison are gregarious animals that exist in several types of groups during the course of the year, although group size is generally smaller than among plains bison (*B.b. bison*) (Komers et al 1993). Cows, calves and yearlings are usually found in matriarchal groups ranging up to a few dozen animals or so. Maternity groups consisting of cows with young calves also occur. Mature bulls seldom form groups of more than a few animals, and solitary bulls are common (Komers et al 1992). Male wood bison show a linear, age-related, dominance hierarchy similar to that observed in plains bison (Komers et al 1994). Wood bison occur in small, mixed herds during the rut, apparently as an adjustment to resource availability in forested terrain (Melton et al. 1989; Komers et al 1993), but adult bulls generally remain separate during other seasons (Reynolds et al. 1978). Adult female and yearling bison occur in larger groups and have larger home ranges than do adult male bison (Larter and Gates 1994).

Soper (1941) estimates a total population of 168,000 wood bison in North America in 1800. However, the fate of wood bison was similar to that of plains bison. By the end of 19th century they had nearly vanished as a result of over-exploitation following the fur trade and westward expansion of European settlement (Gates et al. 1992). Subsequent conservation efforts improved their status in Canada (Reynolds and Gates 1991). There are currently about 2800 disease-free wood bison in six, free-ranging herds, and an additional 700 in captive herds in Canada. Prior to population declines during the 19th century, wood and plains bison occupied different habitats. During winter their movements may have overlapped in a limited area along the southern edge of wood bison range, but different migratory patterns suggest they did not share a common range during the breeding season. Differing habitat preferences and seasonal behavior maintained reproductive isolation (van Zyll de Jong 1986).

The evolutionary history of the North American bison is described in detail elsewhere (Skinner and Kaisen 1947; Harington 1977; van Zyll de Jong 1986, 1993; McDonald 1981; Guthrie 1990). These studies are in general agreement regarding the pattern of evolutionary divergence and geographic discontinuities lead to the existence of two morphologically distinct bison types during the Holocene. Modern forms evolved relatively recently, resulting in minor genetic, morphological and behavioral differences between subspecies. Studies of bison systematics show primarily clinal variation among contiguous populations, with a phenotypic discontinuity between boreal (*B. b. athabascae*) and plains (*B. b. bison*) populations supporting the designation of only two late Holocene subspecies in North America (van Zyll de Jong 1986, 1993).

A typological approach to systematics during the late 1800s and early 1900s led to a bewildering array of species designations within the genus *Bison* (Guthrie 1980; van Zyll de Jong 1986). Subsequent recognition of the clinal variation in certain characteristics, and the relatively labile nature of social organs such as horn cores (Guthrie 1966, 1990), resulted in a trend toward synonomy in bison taxonomy (Skinner and Kaisen 1947; Guthrie 1980; van Zyll de Jong 1986).

The evolution of northern bison reflects the clinal variation and gradual reduction in body and horn size that has occurred in *Bison* over time (Guthrie 1990). Steppe bison evolved into an intermediate form, *B.b. occidentalis*, during the late Pleistocene-Holocene transition, which in turn gave rise to *B.b. athabascae* (van Zyll de Jong 1986). It is clear that wood bison are the most recent northern variant of *B. bison*, just as plains bison are the most recent southern variant. It is also known that wood bison were the last type of bison to inhabit Alaska and northwestern Canada, and were once widely distributed in this region (Skinner and Kaisen 1947; Harington 1977; Guthrie 1980; van Zyll de Jong 1986; 1993).

Wood bison are recognized as a subspecies based on morphology (Figure 2), blood characteristics and DNA. Geist (1991) suggested the subspecific status is not warranted and that observed differences are environmentally induced. However, morphological studies of plains bison and wood bison show differences in cranial and skeletal characteristics (van Zyll de Jong 1986), as well as in the anterior slope of the hump, location of the highest point on the hump, angle of the hump, cape variegation and demarcation, upper front leg hair, frontal display hair, ventral neck mane and beard, indicating that phenotypic differences are genetically controlled (van Zyll de Jong et al. 1995).

There are similarities as well as differences in mtDNA, genomic DNA (Bork et al. 1991; Strobeck et al. 1993), and erythrocyte antigens and blood proteins in plains bison and wood bison (Zamora 1983; Peden and Kraay 1979). Wilson and Strobeck (1999) investigated variability in 11 microsatellite loci of genomic DNA in 11 North American bison herds. Genetic distances between wood bison and plains bison subpopulations were usually larger than those within either subspecies, indicating that wood bison continue to function as genetic entities separate from plains bison, despite the introduction of plains bison into wood bison range in the 1920s (van Zyll de Jong et al. 1995). Strobeck et al. (1993) compared sequence divergence in a section of D-

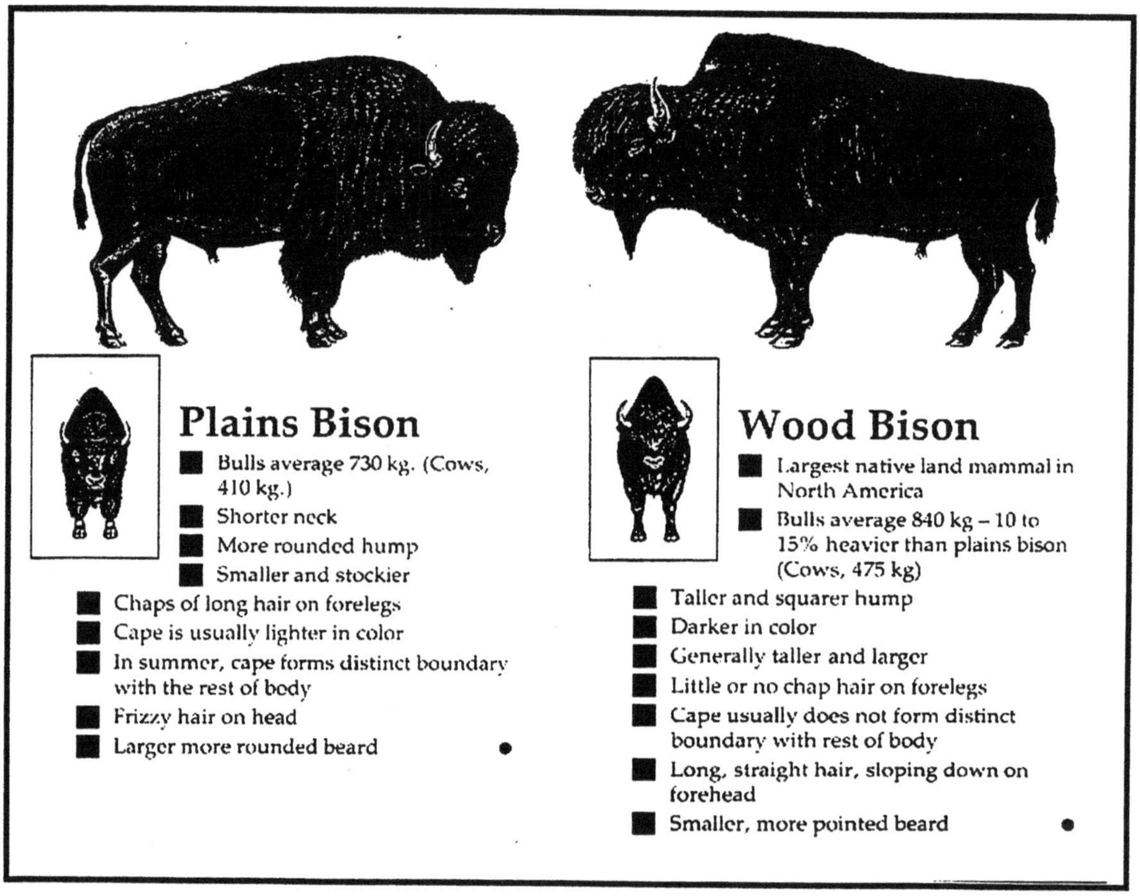

Figure 2. A comparison of the morphological characteristics of plains bison and wood bison. Reproduced with permission of Environment Canada, Elk Island National Park.

loop in the mtDNA of a small number of wood and plains bison and found that differences between the two subspecies are approximately the same or less than within plains bison. The rate of sequence divergence in mtDNA is on the order of 1% to 2% per million years (Wilson et al. 1985). These findings are consistent with the view that wood bison and plains bison existed as reproductively isolated populations during the last 5-10,000 years, a relatively short time in evolutionary terms, and probably also reflect some introgression of plains bison mtDNA into remnant wood bison populations during the 1920s (Van Zyll de Jong et al. 1995).

PROCEDURES

We first became aware of the existence of traditional knowledge about late Holocene bison in Alaska in the fall of 1991, when a resident of Fort Yukon informed the senior author that his mother had told stories describing how bison had once lived in the area around Fort Yukon. This prompted efforts to determine whether historic information might be present in the oral traditions of other Athabascan people in the region. We subsequently learned that elders in several villages in interior Alaska could recount credible and substantive stories about both the presence and the hunting of bison. We interviewed Athabascan elders in nine interior Alaska communities including Fort Yukon, Birch Creek, Beaver, Chalkyitsik, Venetie, Arctic Village, Minto, Nenana and Tanana (Figure 1) between November 1991 and July 2000. In June 1998, residents of several communities in northern Yukon and NT traveled to Alaska to attend the Sixth Biannual Gwich'in "Giikhii" Gathering at Fort Yukon, where two residents of Arctic Red River, NT contributed information regarding the occurrence of bison within recent times near that community. Individuals providing information in this study are identified in connection with their accounts, which are detailed below. Most informants ranged in age from 60 to over 90 years.

Oral accounts were summarized in writing following each interview based on notes taken during the conversation. Interviews were generally conducted in English, but in some cases in *Gwich'in* with the aid of a translator. Photographs of wood bison and muskoxen were used to help confirm the identity of the animal in question. Individuals were specifically asked to identify the animals represented and to provide, if possible, Athabascan or English names for them. Efforts were made to collect information about the abundance and spatial and temporal distribution of bison, their physical and behavioral characteristics, other details relating to their occurrence and how they were hunted and used, and if and where bison remains had been found.

Several accounts were substantiated and elaborated on through second and third discussions with elders. In subsequent discussions we reviewed and clarified details, particularly if there appeared to be confusion about the literal or interpretive meaning of an account (see Burch 1975). Information from two informants was independently reviewed and verified by researchers not directly involved in the study. William Schneider, a long time associate of Moses Cruikshank, and Curator of Oral History at the University of Alaska Fairbanks (UAF), conducted additional interviews with Mr. Cruikshank, independently confirming our results in the process. Historic accounts provided by Reverend David Salmon were compiled and reviewed by UAF anthropologist Thomas O'Brien who, in 1994, began a long-term collaborative project to record Rev. Salmon's life history (e.g. geneology, land use, historic travel narratives), and to collect narratives pertaining to a variety of other ethnographic and historic topics (O'Brien 1997). O'Brien conducted a line-by-line review with Rev. Salmon to evaluate with him the accuracy of accounts recorded previously by Stephenson.

We recognize the potential for bias in oral accounts, just as there is a bias in written and other documentary sources. We also recognize their utility for historical reconstruction and value in understanding natural history (cf. Garraghan 1940; Dundes 1965; Cruikshank 1981; Vansina 1986; Burch 1991; Beaudry 1994; Stevenson 1996; Wenzel 1999). We find no reason to discount or dismiss the following accounts pertaining to the late Holocene occurrence of bison in Alaska. While individual narratives vary in level of detail, they provide a relatively robust set of historical data with which to reevaluate the biogeography of wood bison and some aspects of late Holocene Athabascan subsistence, economy and ecology.

The emergence of oral accounts describing the presence of bison in recent times led to a more detailed examination of archaeological, paleontological and previously collected ethnohistorical data. We also obtained radiocarbon dates from bison skeletal material recovered from the Yukon Flats and other parts of Alaska and Canada, dating those remains that, based on morphology and preservation, appeared to represent Holocene specimens. In addition, we worked with other investigators to compile new and existing radiometric data for late Pleistocene and Holocene bison discoveries (Table 1). We relied on published literature and skeletal measurements to distinguish specimens of male and female bison where possible. In 1993 Gerlach and Mills conducted a 10 day archaeological survey in the vicinity of Smoke Creek, approximately 45 km southwest of Arctic Village, Alaska to search for evidence of bison hunting in the form of kill-, drive- or jump-sites. The survey occurred near "Buffalo Mountain" (Figure 1), a place name reported by T.E. Taylor in 1956 (Orth 1971). This survey was brief and logistically constrained, and no positive evidence was found (Mills 1993).

ATHABASCAN ORAL NARRATIVES ABOUT LATE HOLOCENE BISON ON THE YUKON FLATS AND IN OTHER AREAS OF INTERIOR ALASKA

Athabascan elders residing in the upper Yukon and Tanana River drainages in interior Alaska provided oral accounts referring specifically to late Holocene bison. Consistent oral narratives provided by multiple elders indicate bison were present and were hunted in parts of interior Alaska within the last few hundred years. There are a number of persistent themes in the oral narratives that provide insight into the late Holocene distribution, human use, and disappearance of wood bison in Alaska. Although the narratives are substantive by current historical and ethnographic standards, there is little doubt that more detailed information would have been available to earlier ethnographers.

The accounts indicate that wood bison were widely distributed in the upper Yukon drainage, and apparently in the Tanana drainage as well, until late in the Holocene. Specific areas referred to as being inhabited by bison include the flats along the Yukon River from the vicinity of Beaver east to the lower Chandalar, Porcupine and Black River drainages in the vicinity of Fort Yukon, the Hodzana River drainage northwest of Beaver, the Grayling Fork area in the upper reaches of the Black River, and the flats adjacent to the Tanana River.

Discussions with several elders in Alaska also revealed an awareness of the fact that bison were still hunted in the Mackenzie River-Lake Athabasca region in Canada many years after significant populations had disappeared from the upper Yukon basin in Alaska. This knowledge is based on accounts from relatives who, on at least a few occasions, traveled to and from this region via the Porcupine River in connection with the fur trade during the late 1800s and early 1900s (McKennan 1965).

Oral narratives provided by thirteen Athabascan elders in the Yukon Flats communities of Beaver, Birch Creek, Chalkyitsik, Fort Yukon, and Venetie, and two elders from the communities of Minto and Nenana (Figure 1) indicate bison were sufficiently abundant to be a resource of some importance as recently as 200 to 300 years ago. One elder stated that a group of bison was seen near Eagle in 1916 or 1917, and two others report that small herds were sighted near Circle and Fort Yukon during the same period. Two elders describe instances in which single bison were killed in Alaska in the early 1900s. A few elders reported observations of tracks or other signs of bison in northeastern Alaska during the mid 1900s.

Details in these accounts are consistent with existing knowledge of bison ecology, morphology and behavior. Informants characterize the animals as being of unusually large size with small horns, a large hump, a large head, short legs, long hair and a long tail. These features are typical for wood bison (Figure 2), whose hump is especially prominent and whose tail is substantially longer than that of other ungulates present in Alaska during the Holocene (Guthrie 1990, van Zyll de Jong 1993). The elders further distinguish between bison and muskoxen, and include information that agrees with paleontological, archaeological and other historical evidence regarding their Holocene distribution (Lent 1998).[1]

Bison are said to provide high quality food and raw materials. Hides were tanned and used to make robes, pillows, and clothing, and sewing thread was made from bison hair. Bison were hunted with bow and arrow, spears, dogs, and with snowshoes to gain advantage in deep snow, and possibly with drives. The accounts obtained indicate that resident populations of wood bison declined or disappeared from Alaska by the early or mid 1800s, with the occurrence of small numbers in the eastern interior into the early 1900's.

Several elders were familiar with one or more *Gwich'in* terms for bison. Some informants indicated that one *Gwich'in* name for bison is *Dachantèe aak'ii*, which they translate as "cow in the forest." Another provided the name *Dachantèe qwaak'ii,* indicating it had a similar meaning. One informant described how the above terms were applied to either bison or muskoxen, and were always accompanied by a hand sign that identified the species being referred to.

These terms appear to be the most well known *Gwich'in* terms for bison; the first translates as "muskox among timber" and the second "the hefty one among timber" (J. Kari, pers. comm.). The *Gwich'in* term for muskoxen is *aak'ii*, which appears to originate from the verb theme "to be hefty, meaty" and is literally "the one that is hefty" (J. Kari, pers. comm.). Another *Gwich'in* name for bison is said to be *Ch'itthay dighan*, which describes a large, humped animal or "humped game" (R. Mueller, pers. comm.), or "humped meat" (J. Kari, pers. comm.). A similar term, *Ch'atthaii daghan choo,* was provided by another elder (O'Brien pers. comm.). One elder provided the name *Nan'aak'ii choo*, indicating it meant simply, "big animal." *Nan'aak'ii choo* translates as "big hefty one on the land" (J. Kari, pers. comm.).

Gwich'in terms for bison appear to be a lexical innovation perhaps coined at a time when the animal was already scarce, and after a more ancient term had become obsolete. The term for bison is not as widely distributed in northern Athabascan languages, and is not as ancient as, for example, terms for moose or caribou, which are cognate terms. The apparently limited contemporary linguistic terminology related to bison in Alaska may reflect the fact that wood bison have been absent from the region, and culturally obsolete, for some time. (J. Kari, pers. comm.).

Different linguists transcribe and gloss *Gwich'in* Athabascan words in different ways. For example, one may transcribe bison as "*Dachantèe aak'ìi*", another as "*Dachantèe akìi.*" The observed variation may reflect lexical, phonemic, dialectical, geographic, or idiosyncratic differences. We are unable to clarify all linguistic issues, but simply present them here as they have been interpreted for us.

Oral Narratives about the Upper Yukon Basin

Mrs. Annie James of Fort Yukon recognized a picture of bison and said that her mother, grandfather, and other "old-timers" told stories about bison in the Fort Yukon area, and in the upper Mackenzie River area in Canada. These elders described how bison were hunted in the upper Yukon area in Alaska during the "skin-clothes days," and were aware of the fact that bison still occurred and were hunted to the east in Canada after they had disappeared from the Fort Yukon area.

Mrs. James said she was told there were once "lots" of bison in the Fort Yukon area, that they were hunted with bows and arrows and spears, that the meat was sometimes dried, and that some "old-timers" thought the meat was tough. Mrs. James mentioned a number of other elders who should have information about the history of bison in the region, indicating that her knowledge of bison was not unusual.

Among these knowledgeable elders was her brother, the Reverend David Salmon of Chalkyitsik.

Rev. Salmon is acknowledged for his extensive and detailed knowledge of *Gwich'in* Athabascan history and traditions. He recounted bison stories that were related to him by his father, mother, maternal father-in-law, and additional stories based upon accounts provided by his paternal grandfather and other elders. Prior to their disappearance, bison were an important source of food for the *Gwich'in* on the Yukon Flats. Rev. Salmon emphasized that "they lived on it," especially before moose (*Alces alces*) became more common, adding that mosquitoes did not bother bison because of their long hair.[i] Bison were said to be a "good animal," providing valuable food and material for people. In discussing the history of bison in the region, Rev. Salmon often commented that the Yukon Flats "is their country…they belong to it."

According to stories passed on to Rev. Salmon, the bison that lived in this area had a large hump. During the 1970s his grandson found the skeleton of a bison in a cutbank along the Black River, about 20 miles west of Chalkyitsik (Figure 1). The remains were brought to Chalkyitsik where people noted the long spines on the cervical and thoracic vertebrae. Rev. Salmon pointed out that this was consistent with the large hump described in *Gwich'in* oral narratives. We obtained a radiocarbon date of 1730±60 BP for this specimen (Table 1, Map No. 37). Rev. Salmon has often found bison horns and other skeletal remains on the Yukon Flats, and observes that many of the bones have a "fresh" appearance and appear to be of recent origin. The Grayling River flats in the upper Black River drainage was one area known as bison country. Rev. Salmon mentioned that a bison skull had been found there and that his father found a frozen bison head along the Black River upstream from Chalkyitsik in 1926. It was recovered from permafrost and looked "fresh." Both horn sheaths were attached, and Rev. Salmon described it as being a small-horned bison.

According to Rev. Salmon, *Gwich'in* names for bison include *Dachantèe aak'ii* which he translated as, "cow in the forest", and *Ch'atthaii daghan choo*, which he translated as "large animal with a hump." He indicates these terms could be applied to either muskoxen or bison, emphasizing that only bison occurred on the Yukon Flats. Rev. Salmon explained that these two animals were distinguished in conversation on the basis of the distinctive shape of their horns, and clarified any ambiguity regarding the name *Dachantèe aak'ii*. He describes the characteristic downswept horns which "cover the head" of a muskoxen, contrasting them with the upwardly curving horns of bison. When referring to muskoxen the speaker would typically place an open palmed right hand above the ear, move it gently downward over the ear and then out and up in an arc, indicating the downswept curve and upturned end of the horn. When referring to wood bison a hunter would extend and slightly curve the first and second fingers, placing them against the upper temple, indicating the outward and upward curve of wood bison horns. The phrase "*Dachantèe aak'ii vijì vikì' noiinjik*" means "cow in the forest with the horn that sweeps downward to protect the head." In contrast, the phrase "Dachantèe *aak'ii vijì neekwaii gwanlii*" means "cow in the forest with two short horns turning upward." Rev. Salmon stated that even during the early 1900's hunters discussing bison or muskoxen would qualify the term with hand sign, adding that *Gwich'in* people in a large region extending from Fort McPherson to Fort Yukon commonly used the term *Dachantèe aak'ii*, accompanied by hand sign, to denote wood bison. It is interesting that this hand sign for bison is identical to that used in the traditional hand sign language of Plains Indians, as well as other indigenous peoples in North America (T. O'Brien, pers. comm.).[ii]

Rev. Salmon believes that Athabascan stories about bison hunting refer to a time no longer than 300-500 years ago, in part because he doubts that stories of this nature could persist for a longer period of time. Moreover, he indicates that bison were still being hunted at the time that moose first became known to the *Gwich'in*. Rev. Salmon said the first moose (*Alces alces*) was seen near Birch Creek about 400 years ago. People were initially afraid, but as moose became more common, they began to rely on them for food (Salmon, n.d.). McKennan (1965) describes a similar oral tradition among the Chandalar *Gwich'in* regarding the historic occurrence of moose in the vicinity of Arctic Village. Paleontological and archaeological evidence suggests that moose were present in Interior Alaska throughout the Holocene, but may have been locally absent or rare for extended periods in certain areas (Yesner 1989).

Rev. Salmon clearly distinguished bison from muskoxen and notes that bison bones are commonly found on the Yukon Flats, where the remains of muskoxen are unknown. He indicates that while oral tradition describes how wood bison were once common on the Yukon Flats, there is no such tradition regarding muskoxen, stating "…all they had was wood bison down this way." Rev. Salmon states that muskoxen were once an important resource for people to the northeast in the upper Porcupine-Peel river region, and describes hunting techniques used to take muskoxen with bow and arrow, and with muzzle loaders after they became available during the mid 1800s. Rev. Salmon also refers to a story originating with the *Dihaii Gwich'in* describing how muskoxen were sometimes killed by being driven over a cliff in the upper Chandalar River country. He indicates this technique may have also been used somewhere north of the Porcupine River, north and east of the Yukon Flats, in the "Old Crow country."

Athabascan people relied extensively on their highly developed archery to take bison and other animals. Rev. Salmon describes hunting strategies and archery technology that were effective in harvesting large game animals such as moose and grizzly bears (*Ursus arctos*), and believes they would have been adequate for bison hunting as well (O'Brien 1997). Hunters used finely crafted "self-bows" made from selected birch wood, along with a wide array of arrow types designed for specific purposes. Archery tackle was designed for long range shooting, with the draw weight of bows used by Athabascan men estimated to have ranged from 60 to 80 pounds (O'Brien pers. comm.). Arrows used to kill large game animals were generally tipped with one of a variety of bone or antler broadheads (O'Brien 1997).

Rev. Salmon states that archery training was extensive and thorough, and was intended to produce highly competent archers with years of experience, knowing that the survival of people from one generation to the next depended heavily on their skill. Hunting strategies were adapted to environmental conditions and to the behavior of animals, and often involved the efforts and expertise of several hunters. For example, during winter animals were driven into deep snowdrifts, where hunters using large hunting snowshoes could shoot at close range. Big game animals were often killed as they crossed rivers or lakes in spring and summer (O'Brien 1997).

Rev. Salmon describes a special arrowhead used to kill large animals which includes a detachable head that was serrated and barbed in such a way that it worked its way through tissue as the animal moved (O'Brien 1997). The hunter had only to penetrate the hide with this arrowhead in order to kill an animal. Although Rev. Salmon believes snares could have been used to take bison, just as they were used to take moose and caribou (*Rangifer tarandus*), he is unaware of specific stories describing their use, stating instead that the bow and arrow was the most effective weapon for this species.

Based on his knowledge of the skill and purpose of earlier generations of hunters, Rev. Salmon believes that hunting played a role in the disappearance of wood bison from the Yukon Flats, emphasizing the ease with which Athabascan archers killed moose, bears and other large game. Stressing the cyclic pattern of human populations increase and subsequent decline to very low levels due to shortages of big game and other resources, Rev. Salmon described how during hard times people hunted intensively, and may have "hunted out" bison (Salmon/O'Brien, taped interview, June 3, 1997; see also Campbell 1978; Mishler 1995). He points out that "mountain people" were generally "wiped out" during periods of starvation, but that a small number of people would typically survive along the Yukon River where salmon and other fish provided a more reliable supply of food. Human populations would subsequently increase and disperse from these areas. Rev. Salmon further said that the more open nature of the terrain in the past would have made bison vulnerable to hunters, and also believes that intensive hunting might have caused the last bison to leave the country and go to the "Canadian side."

Rev. Salmon recalled stories told to him by his maternal grandfather, John Chitleii (1840-1923), a noted hunter from Old Crow, Yukon (O'Brien 1997), regarding the disappearance of bison in the region. These accounts described the occurrence of bison in the country lying between Old Crow and Arctic Red River, NT, including the Peel River area, during a time when habitat was more open than at present. Bison were once hunted in this area but populations eventually declined, first in the western portion of this range and later to the east as well. Rev. Salmon said the stories told by John Chitleii indicated that bison disappeared from the region east of Old Crow after they had disappeared from the Yukon Flats.

Rev. Salmon and other elders emphasize geographical and temporal differences between oral traditions relating to bison in Canada, some of which were obtained during periodic long-distance trading forays in the mid 19th century. Rev. Salmon is aware that bison occurred in the area near Lake Athabasca in Canada during the mid 19th century, after they had largely disappeared from the Yukon Flats, based on information passed on by his father, William Salmon, reflecting the experiences of his paternal grandfather, King Salmon. During the period from approximately 1850-1868 King Salmon made several journeys into Canada via the Porcupine and Mackenzie Rivers in connection with the Hudson Bay Company. He traveled as far as Lake Athabasca and the Wabasca River, a tributary of the Peace River (Salmon n.d.).

Mr. Moses Cruikshank of Beaver said there were many *Gwich'in* stories describing how bison inhabited the Yukon Flats in the old days when, "big herds" of these animals occurred in the area. Large numbers of bison were sometimes killed in the fall when much of the meat was dried and "used all winter long." A story attributed to Chief Christian[iii] relates to "a mountain up in the Sheenjek River country" called "Buffalo Shirt Mountain." Mr. Cruikshank said "a large herd of bison came through and covered the mountain like a shirt" at this location. Bison were hunted there for several years, and were guided by fences down on the flats and driven over a cliff. Mr. Cruikshank said many bison were killed during this period. He noted that the taking of large numbers of bison by people at "Buffalo Mountain" occurred sometime before Chief Christian was born, and before firearms were generally available. Rev. Salmon estimates that Chief Christian was born about 1855 and was 93 years old when he died (T. O'Brien, pers. comm.). Campisi (1993) indicates Chief Christian was born in Arctic Village in 1866. Mr. Cruikshank said bison lingered in the area for a few years, but eventually disappeared. He remembered the "old-timers" saying there were two types of bison, one living on the flats and the other in the mountains, and that one was larger than the other. He thought the "mountain buffalo" were larger, but was not certain about this point. Mr. Cruikshank further states that a strong bow was required to kill bison, and that bison hunting was sometimes dangerous.

Mary Sam was born in 1914 and lived in the Black River area near Chalkyitsik until 1947. She presently lives in Beaver, Alaska. Mrs. Sam recalls that when she was about nine years old her grandparents, Mariah and Henry Gwats'oo, told stories originating from her great-grandfather that described how bison were once abundant in the Black River country. Mrs. Sam believes that her great-grandfather might have seen bison in that area. She was told these animals had small horns, a big hump, short legs, long hair, and were hunted with bow and arrow. They were said to have been a major source of food, were a "counterpart" (literal translation from *Gwich'in*) to moose in the diet, and were valued for their high quality meat.

With regard to the disappearance of wood bison, Mrs. Sam said that "maybe they ate it up," suggesting that hunting might have contributed to the disappearance of bison in the Black River area. She remembers her grandfather saying that moose were also likely to become scarce in the future because there would be more people in the area. She also described how on one occasion her grandparents pointed to another young girl, saying, "...When this young girl grows up, and her children grow up, then the bison will come back." She recalled a story about a hunter who returned to a village in the area near Beaver after unexpectedly encountering a bison north of the Yukon River, describing how "his eyes were really big." Although her grandparents had told her the *Gwich'in* name for wood bison, she could not recall it. Mary Sam clearly differentiated muskoxen from bison, noting that a muskox had recently been seen near Arctic Village or Venetie. Muskoxen were reintroduced to northern Alaska in 1969-70 and have expanded their range in recent years, with some males dispersing south of the Brooks Range (Reynolds 1998).

Ms. Ann Fisher describes how her mother, the late Charlotte Adams of Beaver, often talked about the former occurrence of bison in the Hodzana River area northwest of Beaver. These accounts originated from Charlotte's father, "Old Adam" who, with his family, lived and traveled in the Hodzana drainage (Cruikshank 1986). He often talked about seeing bison in this area when he was a young man, later referring to them as "white mans moose." Old Adam hunted with bow and arrow and spears in the early days and is said to be one of the last people to have killed a bear with a spear. The inscription at his gravesite indicates he was born in 1848 and died in 1944, at the age of 96. This account indicates that resident bison populations existed in the vicinity of Beaver during or after the mid-1800s.

Mrs. Virginia Titus provided stories regarding the presence of bison in interior Alaska that were conveyed to her by her father and grandfather. Her family originally lived in the Koyukon Athabascan community of Tanana. However, her father spent most of his life in Fort Yukon and travelled extensively between the two communities. Her father heard many stories about bison and their value to people in the "skin clothes days," when animal skins were the only materials available for clothing. She was told that bison were once common and widespread in Alaska, although they were found mostly at low elevation and were scarce in the mountains. The flats in the Tanana and Yukon drainages were said to support bison in the early days. According to Mrs. Titus, bison were second only to moose as a source of food, and were an important source of material for clothing and shelter as well. Bison had a "big head," and the hides were tanned with the hair on to make warm robes and clothing. The hair was soft, and bison hides were preferred for clothing because they did not cause allergic reactions in people.[iv]

Mrs. Titus said these stories described the presence of bison in the 1700s and into the early 1800s, as well as earlier. Her grandfather said bison were hunted with bow and arrow, with spears, with the aid of snowshoes in winter, and with the aid of dogs. Mrs. Titus said the dogs used by people in the early days were larger than the sled dogs generally used today. She adds that there was more snow in the early days, noting this may have increased the vulnerability of bison to hunters. She indicated that the disappearance of bison in this region was due to hunting, stating "...they were easy to kill, that's why they are not here," Mrs. Titus recalled the name for bison as *"nan' aak'ii choo,"* which she translated as "big animal."[v] Mrs. Titus stressed that stories relating to bison were not recorded earlier because of the generally poor communication between interior Alaska Natives and early settlers, missionaries and explorers.

Mr. Elliot Johnson, age 96, of Fort Yukon, also states that bison once lived on the Yukon Flats and were hunted in the early days. He had heard other "old-timers" talk about the former occurrence and use of bison on the Yukon Flats, specifically mentioning Chief Christian and Chief Robert, two prominent leaders in the Fort Yukon area during the late 1800s and early 1900s (T. O'Brien, pers. comm.). In addition to being an important source of food, bison provided raw materials, with hides making good blankets and pillows and the hair being used to make thread for sewing. Bison hides were left outside for a time to rid them of "bugs" before using them because they harbored parasites.[vi] According to Mr. Johnson, the decline of bison populations was followed by a period of food scarcity, which continued into the 1900's when he was a young man. He indicates that people referred to this period with the phrase "no buffalo, no power." Big game was extremely scarce in the Fort Yukon area during this period, which was characterized by social instability and extensive movements in search of bison and other game. Mr. Johnson states that he saw four bison along the Porcupine River a few miles from Fort Yukon on one occasion in the early 1900s. He describes them as having black, upward curving horns, long hair, and being about five feet high at the shoulder. He did not shoot these animals because it was the first time he had seen "buffalo", and later he returned to the site with other people in an unsuccessful effort to relocate them. Mr. Johnson also recounts a story told by his father-in-law, Frank Alexander, relating to an animal that he encountered as he was travelling from Tanana to Fort Yukon. He shot the animal with his rifle, thinking it was a moose, but on approaching it found it had a long tail and was actually a bison.

The late Julia Tritt of Venetie recounted stories told by her grandfather and other elders about how "buffalo" were hunted on the Yukon Flats. She referred to bison as the "big animal," stating that elders often remarked on the animals' "big head," long tail and large size. She said hunters often found their large round tracks in the snow. They were said to be fairly easy to hunt and to kill with bow and arrow or spears, and dogs were also used to help bring them down. Bison were sometimes caught in, and often ruined, snares set for moose or caribou. These snares were often not strong enough to hold them. She said bison were "good eating" and provided high quality food for people. Sewing thread was made by plaiting together several of the longest hairs, and a single bison hair was used to suture cuts on people. Mrs. Tritt said bison hides were hard to tan compared to those of other animals, and were sometimes only partially tanned and used to cover the floor in a dwelling. She said that bison eventually disappeared or left the country. Mrs. Tritt indicated these accounts pertained to the early 1800s and earlier.

Mr. Earl Erick of Venetie recalls stories told by his grandmother, the late Myra Roberts of Venetie. As a young woman, Mrs. Roberts lived at White Eye, located about 20 miles northeast of Beaver along the Yukon River. Her grandfather, father, and mother, described how small numbers of bison still roamed the flats adjacent to the Yukon River near White Eye in the recent past, with populations extending to the east at least as far as the area around the lower Chandalar River, the Porcupine River and Fort Yukon. Mrs. Roberts said these animals had small horns. Mr. Erick said Mrs. Roberts indicated that people in her grandfather's generation knew others who had seen bison in this area, and believes the stories about bison pertain to a period ending about 200 years ago.

Other elders in the Fort Yukon area also know that bison inhabited the area in the past, but provided fewer details about them or their relationship to people. Mr. Steven Henry of Chalkyitsik knew of their former presence, and states that bison were once abundant on the Yukon Flats. Based on stories told by his father, Paul Henry, Steven Henry believes that bison disappeared about 200 years ago. Mr. Richard James of Birch Creek said his grandfather, Birch Creek Jimmy, told him that bison once lived on the Yukon Flats and were a source of food for people in the area. Birch Creek Jimmy died in 1979 at the age of 111. Prior to his death in 1997, Mr. Steven Peter of Arctic Village (then residing in Fort Yukon) stated that bison lived on the Yukon Flats and were called *Dachan aak'ii*. He was also aware that bison existed to the east in Canada in more recent times. One of their characteristics was that they were difficult for wolves (*Canis lupus*) to kill. Mr. Peter said bison were known to kill wolves by kicking them. Mr. Daniel Flitt of Fort Yukon also said bison were once abundant on the Yukon Flats.

Oral Narratives from the Tanana Drainage

Mr. Peter John of Minto recognized an illustration of bison and noted that the animal had once lived in the Minto area. He stated that long ago "we used to hunt that animal," and described a place on the Chatanika River, near Minto, where old bones are often found. The remains of Pleistocene fauna, including bison, have been recovered from the Chatanika River and are deposited in the University of Alaska Museum (Gerlach lab notes, June 19, 1997).

Mr. Howard Luke, originally from Nenana and now residing near Fairbanks (see Luke 1998), stated that as a young boy he was told that bison once occurred "on the Fort Yukon side." He describes how people traveled from the Tanana drainage up the Goodpaster River and into the Yukon drainage in the early days, probably by way of the Charley River. He said these animals were wood bison and existed well before the introduction of plains bison to Alaska (Luke 1998). Plains bison were introduced near Delta Junction, Alaska, in 1928 (Dubois and Stephenson 1998). Although he did not know exactly how long ago bison were present, he was clear about the identity of the animal and certain that they once lived in the Fort Yukon area. Mr. Luke is not aware of stories regarding the occurrence of bison in the Tanana drainage. Mrs. Catherine Attla of Huslia, Alaska was present during the discussion with Howard Luke, and states that she too is unaware of any oral tradition regarding bison in the Koyukuk River area near Huslia.

Other 20th Century Oral and Documentary Accounts from Alaska

In addition to the accounts describing the presence of wood bison 200 years ago and earlier, a few accounts suggest that a few bison occurred in the upper Yukon area in the early 1900s. Mr. Bill Goebel, a resident of Eagle, relates an account provided to him in 1992 by his aunt, the late Sarah Malcolm of Eagle, shortly before her death at the age of 87. Mrs. Malcolm was a Han Athabascan who spent her life in Eagle. She stated that several bison were seen near Eagle in 1916 or 1917, when she was 11 or 12 years old. The bison became entangled in, and broke, snares set for moose or caribou. She said people were afraid of the animals because of their large size. Mr. Goebel questioned her closely as to the identity of these animals, but Mrs. Malcolm was certain they were "buffalo," and that this was the only time they were seen near Eagle in recent times. Lotenberg (1996) notes that a Vuntut *Gwich'in* First Nation Elder said there was a place near Eagle called Buffalo Hill.

Mr. William Joseph of Chalkyitsik, originally from the village of Circle, recalls a story told by his grandfather, Zias Joseph, who spent his life in the Circle area. Zias Joseph described how he once saw a group of 5-7 bison on the hills near Circle as they traveled through the area. Zias Joseph died in the 1930's when he was about 60 years old, suggesting these bison were observed in the late 1800s or early 1900s.

Mary Sam recalled an incident that occurred in 1922, when she was 8 years old, when her grandfather was hunting moose during winter near the south end of Ohtig Lake, about 6 miles south of Chalkyitsik. He killed a moose, and after returning to their camp near Chalkyitsik, described how he had found large, round tracks of a big animal he had never seen before. He thought a bison, an animal he said previously occurred in the area, made the tracks.

Mr. Joe Herbert of Chalkyitsik describes a similar incident that occurred in January 1941, when his family was living at Shuman House, on the Porcupine River about 25 miles north of Ohtig Lake. His grandfather (also named Joe Herbert) found large, round tracks of what he thought was a bison, and droppings in large mushy piles he described as looking like "cow patties" a short distance north of Shuman House. Drag marks showed that the animal's chest touched the snow, which was no more than 20 inches deep. They returned to the area the following day but the tracks had been largely obscured by drifting snow during the night. The droppings described are characteristic of bison, which unlike muskoxen or moose, do not produce fecal pellets.

Virginia Titus recalls that her father, Robert Albert, described to her how he and his adopted father, Pretty Albert, encountered a bison near Tanana, probably in the winter of 1918. This occurred when they were on their trapline and when her father, who was born in 1904, was fourteen years old. Her father remembered being scared when they encountered a large animal in the brush. His father shot the animal with a lever-action rifle, the first cartridge rifle they had obtained. The animal was a large bison. After butchering the animal they stored the meat in an underground cellar insulated with grass. The hide was given to their Chief, which he used in their "talking house" as a place to sit. The carcass provided food for their dogs for a long time. Mrs. Titus said this was the last known occurrence of bison in the Tanana area. Recall that Elliot Johnson reports seeing four bison near Fort Yukon, and described how his father-in-law killed a bison in the upper Yukon area, probably in the early 1900s. Mr. Arthur Adams of Beaver said that Jack VanHatten, who with his family lived in the upper Black River area above Grayling Fork during the early 1900s, recounted the story of a trapper who shot a bison in that area. However, the date of this occurrence is unknown. A number of residents of the upper Yukon basin in Alaska provided additional accounts describing tracks or other sign suggesting that a few bison may have occurred in this region as late as the mid 20th century.

The journal of James Geoghegan (James Geoghegan Collection, University of Alaska Archives), who lived near Delta Junction, Alaska in the early 1900s, includes a reference to the presence of bison near Donnelly, Alaska in 1918 or 1920. Unfortunately, it is not possible to determine whether the author is referring to plains bison introduced to this area in 1928 or to the presence of a few indigenous wood bison prior to that time.

Oral Narratives From and About Adjacent Canada

Mr. Moses Sam of Arctic Village said his uncle, Henry John, told him stories about "Indians over Old Crow way" who hunted buffalo for food, although he is not aware of their presence in the upper Chandalar River-Arctic Village area. The late Mr. Isaac Tritt, Sr. of Arctic Village knew that bison occurred in Canada, specifically mentioning the Mayo area in Yukon. However, he was not aware that they occurred in eastern Alaska. He said the *Gwich'in* name for bison was *Dachantèe qwaak'ii*, which he translated as "in the timber, cow."

Mr. Hyacinthe Andre, age 86, and his younger brother Gabe Andre, of Arctic Red River, NT provided information regarding the occurrence of bison in this area (Figure 1). They are most familiar with the lower Mackenzie River area near Arctic Red River. Hyacinthe Andre spent much of his life in the Anderson River country to the east, between the Mackenzie and Horton rivers. The Andre brothers referred to oral traditions describing the former occurrence of bison in this area. Hyacinthe Andre said he had heard "old stories" referring specifically to the Travaillant Lake area east of Arctic Red River as having once supported bison, that bison were once hunted to the north near the arctic coast, and that bison once occurred on the "barren-grounds" adjacent to the Anderson River, mentioning that bison skulls had been found there. He does not recall the *Gwich'in* name for bison, but notes they disappeared a long time ago and that it has been at least 200 hundred years since bison were hunted in the region north of Arctic Red River. Mr. Andre clearly differentiates bison from muskoxen, saying the bison were much larger than muskoxen, and that these animals "were powerful and could run through four feet of snow with no problem."[vii]

Gabe Andre adds that bison also occurred adjacent to the Mackenzie River near Arctic Red River, including the area near the Ontaratue River south of the Mackenzie. He recalls one story that describes how the bison disappeared after a bison head had been sent south to be examined. An account presented by Lotenberg (1996) indicates that bison were hunted near Fort MacPherson, about 30 miles west of Arctic Red River, in the 1820s. These accounts appear to be consistent with the presence of late Holocene bison remains along the arctic coast near Cape Bathurst, on the Old Horton River Channel, and Harowby Bay, about 250 miles northeast of Arctic Red River (Figure 3). As described below, one of these specimens is radiocarbon dated at 420±65 years BP.

PALEONTOLOGICAL, RADIOMETRIC, ARCHAEOLOGICAL AND ETHNOGRAPHIC INFORMATION

Paleontological and Radiometric Data

Paleontological discoveries often result from incidental encounters, are influenced by the vagaries of preservation, and in Alaska and Canada often represent isolated rather than mass mortality events (Lyman 1994). Radiometric data for Holocene bison are the product of the chance discovery of specimens and the selection of samples for dating by various investigators during the last few decades. The oral narratives presented above reflect the accumulated knowledge of people whose accounts are chronologically ordered, but sometimes do not contain precise temporal referents.

Table 1 presents available radiometric data for bison remains collected in Alaska and northwestern Canada that date to either late in the Pleistocene-Holocene transition (Guthrie 1995) or to the Holocene. Except for specimens dated by the sedimentary matrix within which they were found (Nos. 4, 9, 17, 18, 23, 25, 26, and 41), dates for Holocene bison are derived directly from bone. The sample of dated specimens includes some from known geological and sedimentary contexts, others found in direct association with archaeological and cultural material, and some without established associations.

The sample of Holocene bison includes 47 specimens from 44 locations, including 19 from Alaska and 28 from Canada (Table 1, Map Nos. 1-44). Conventional radiocarbon ages provided by the laboratory conducting each assay, or as reported in published literature are listed in Table 1. Males (n=15) predominate over females (n=4) in the sample of Holocene remains of known sex, probably because the remains of male bison are more robust and tend to persist longer than the remains of females. The ten earliest radiocarbon dates represent temporally late, eastern Beringian representatives of the Mammoth-Steppe Fauna, a widespread faunal community associated with steppe-tundra and found throughout Beringia (Guthrie 1982,1985, 1990; Matthews 1982; Vereschagin and Barynishnikov 1982; Burke and Cinq-Mars 1998).

MAP NO.	LOCATION	CONVENTIONAL 14C AGE	REFERENCE	LAB NO.	COMMENTS
-	Old Crow, (Loc. 11-1) Y.T.	11,990±180	Harington 1978	I-7765	Bison scapula
-	Fairbanks Creek, AK	11,980±135	Harington 1978	ST-1633	Bison bone
-	Birch Creek, Yukon Flats, AK	11,900±70	ADF&G, this study	Beta-67494	Female bison skull with horn sheaths
-	Cleary Creek, AK	11,735±130	Péwé 1975	ST-1631	Bison bone
-	Old Crow Flats, Y.T.	11,530±200	Harington 1977	QU-780	Bison humerus
-	Broken Mammoth Site, Delta Junction, AK Cultural Zone IV	11,510±120 11,420±70	Holmes 1996	WSU-4262 CAMS 5358	Bison bones with processing marks; associated hearth charcoal dates
-	Dry Creek, AK	10,715±225	Guthrie 1985	ST-1561	Bison bone
-	Lost Chicken Creek, AK	10,370±160	Harington 1978	I-8582	Distal portion of bison tibia
-	Broken Mammoth Site, Delta Junction, AK Cultural Zone III	10,290±70 10,270±110	Holmes 1996	CAMS-5357 WSU-4263	Bison bones with processing marks; associated hearth charcoal dates
-	Bluefish Cave II (MgVo-2) Y.T.	10,230±140	Burke & Cinq-Mars 1998	RIDDLE-561	Bison metacarpal at archaeological site
1	Engigstciak, Y.T.	9870±180 9770±180 9400±230	Cinq-Mars et al. 1991	RIDDL-362 RIDDL-281 RIDDL-319	Bison bone (tibia, metacarpal, and metatarsal) showing processing marks
2	Muskeg River, N.T.	9645±190	Harington, this study	I-9997	Bison cranial fragment
3	Cape Bathurst, N.T.	9560±60	Harington, this study	Beta-79861 CAMS-18424	Left bison scapula
4	Gerstle River Quarry Site, AK	9970±60 8280±60	Holmes 1998	Beta-98432 Beta-98434	Bracketing dates from charcoal for 6 bison bones
5	Porcupine River, AK	9000±250	UAF Museum, unpublished	Beta-18552	Bison bone
6	Victoria Island, Minto Inlet, Kuujjua River, N.T.	8080±60	Harington, this study	TO-3709	Partial male *B. bison* skeleton with cranium and horn cores
7	Broken Mammoth Site, Delta Junction, AK. Cultural Zone II	7600±140	Holmes 1996	WSU-4264	Bison bones with processing marks; associated hearth charcoal date
	Cultural Zone IA	2260±40	D. Yesner, pers. comm.	Beta-128716	Bison naviculo-cuboid; associated charcoal date
8	Mt. Granger, Whitehorse, Y.T.	7510±90	M. Hoefs, pers. comm.	Beta-135361	Female *B. bison* horn sheath from alpine ice patch
9	Canyon Site, Aishihik River, Y.T.	7195±100	Workman 1978 Harington 1978	SI-1117	Bison bone fragments around buried hearth; associated charcoal dates
10	Sullivan Pit, AK	6730±260	Repenning et al. 1964	W-1108	Bison bone
11	McIntyre Creek, Y.T.	ca 5840±70	Hare, this study	Beta-70100 CAMS-11243	Bison bone associated with cultural material
12	Goldstream Creek, AK	5340±110	Péwé 1975	SI-845	Bison horn sheath
13	Harrowby Bay, Beaufort Coast, N.T.	5230±110	Cinq-Mars 1991 Harington 1990	RIDDL-321	Metacarpal at archaeological site
14	Fort Yukon, AK	5045±45	Guthrie, this study	AA4379 VP4157	Male *B. bison* skull
15	Carmacks, Y.T.	4880±80	Harington, this study	Beta-25120	*B. bison* skull from terrace
16	Julian Site (JcRw-13) Fisherman Lake, N.T.	4800±160	Morlan 1999	S-0906	Bison bone at archaeological site
17	Canyon Site (JfVg-1) Y.T.	4730±320	MacNeish 1964	W-1125	Bison bone; associated charcoal date
18	Kusawa Bluffs Site (JdVa-2), Y.T.	4490±130	Greer 1986	Beta-14402	Date from elk bone located below bison bones in archaeological site
19	Black River, Yukon Flats, AK	4495±60	ADF&G, this study	Beta-65662	Male *B. bison* horn core and cranium
20	Black River, Yukon Flats, AK	4390±70	ADF&G, this study	Beta-136731	Male *B. bison* skull with horn sheaths
21	Fort Yukon, AK	3710±70	ADF&G	Beta-74344	Female *B. bison* horn core and part of cranium
22	Fort Yukon, AK	3520±40	Gerlach, this study	Beta-104823	Male *B. bison* skull with both horn sheaths
23	Delta River Overlook Site (XMH-297), Delta Jct., AK	3980±150 2285±145	Holmes and Bacon 1982	GX-6752 GX-6750	Bison tibia fragment; associated charcoal dates
24	Ruby Range, Kluane District, Y.T.	3470±70	M.Hoefs, pers. comm.	Beta-136362	Bison tibia at archaeological site
25	Pelly Farms Site (KfVd-2) Y.T.	3160±70	MacNeish 1964	S-193	*B. bison*; associated charcoal date
26	Pelly Farm Site (KfVD-2) Y.T.	2920±140	MacNeish 1964	GSC-127	*B. bison*; associated charcoal date
27	Fairbanks, AK (railroad terminal)	2900±80	Guthrie, this study	AA3320, AMNH A-501-5331	Male *B. bison* skull
28	Montague House, Y.T.	2800±60	Hare, this study	Beta-70101	Bison ribs

MAP NO.	LOCATION	CONVENTIONAL 14C AGE	REFERENCE	LAB NO.	COMMENTS
29	¾ mile downstream from Circle, Yukon Flats, AK	2545±80	Guthrie, this study	AA3217, AMNH A-479-4783	Male *B. bison* skull
30	Lower Tanana River, AK	2460±70	Guthrie, this study	unknown	Male *B. bison* skull
31	Braeburn, Y.T.	2460±40	M.Hoefs, pers. comm.	Beta-137731	*B. bison* skeleton in dry lake bed
32	Killik River (Site KIR 275), AK	2330±50	C. Martin, T. Birkedal, pers. comm.	Beta-130571	Bison metatarsal near archaeological site
33	Kluane Lake, (Congdon Creek) Y.T.	2180±30	M.Hoefs, pers. comm.	Beta-91755	Male *B. bison* cranium
34	Takhini River, Y.T.	2150±40	M. Hoefs, pers comm.	Beta-91756	Male *B. bison* frontal
35	Finlayson River, Y.T.	2200±60	Harington, this study	Beta-79854	Young male *B. bison* cranium
36	Baillie Islands, N.T.	1890±90	Harington 1980	I-5407	Bison horn sheath
37	Black River, Yukon Flats, AK	1730±60	ADF&G, this study	Beta-62999	Male *B. bison* skeleton
38	Dawson (Loc. 16), Y.T.	1565±85	Harington, this study	I-11051	Bison tibia, apparently fractured by humans
39	Quartz Creek, Dawson, Y.T.	1430±95	Harington 1977	I-5405	*B. bison* horn core
40	Tetlin-Tanacross area, AK	1270±55	Guthrie, this study	AA3218, AMNH A-393-1013	Male *B. bison* skull
41	Frenchman Lake Site (KaTx-6), Y.T.	<1250	J. Hunston, pers. comm.		Bison bone above White River Ash
42	Cowley Lake, Y.T.	940±90	Harington, this study	Beta-69762	Female *B. bison* skull
43	Old Horton River mouth, N.T.	420±65	Harington 1990 Morrison 1997	Beta-28765	Small adult male *B. bisons* skull showing cut marks
44	Anchorage, AK	170±30	Gerlach, this study	Beta 136732	Male *B. bison* skull with horn sheath

Table 1. Location and radiocarbon dates for bison specimens representing the end of the Pleistocene-Holocene transition or the Holocene in Alaska and adjacent Canada. Map numbers for specimen dates within the last 10,000 years correspond to those in Figure 3. Most radiocarbon ages are corrected for isotopic fractionation.

Several dated and undated specimens of northern small horned bison have been taxonomically referred to *B.b. athabascae* (cf. Skinner and Kaisen 1947; Harington 1977; van Zyll de Jong 1986). Measurements from these and other Holocene specimens for which cranial and postcranial material are sufficiently intact to allow morphometric analysis (specimens designated as *B. bison* in Table 1) indicate they are appropriately regarded as *B.b. athabascae*. This conclusion is based on comparisons with morphological data published by van Zyll de Jong (1986). The temporal and geographic distribution of these remains (Table 1, Figure 3) is consistent with earlier conclusions regarding the widespread occurrence of relatively small-horned bison, representing *B.b. athabascae*, in Alaska and northern Canada during the mid and late Holocene.

Radiometric data for a large number of Pleistocene and some Holocene bison have been previously published by Guthrie (1990), Harington (1977, 1978, 1980a, 1980b, 1989, and 1990) and others. The geographic location of dated Holocene specimens, and four undated specimens of small-horned bison, is shown in Figure 3. The distribution of wood bison during the mid and late Holocene as indicated by currently available radiometric data is also shown, as is the region in Alaska and northwestern Canada where oral and/or written accounts describe the presence of bison during this period (van Zyll de Jong 1986; Gates et al. 1992; Lotenberg 1996; this study).

The temporal distribution of available radiocarbon dates representing Pleistocene and Holocene bison in Alaska and adjacent Canada, based on data compiled in this study and by Guthrie (1990), C.R. Harington (radiocarbon table in Annotated Bibliography of Quaternary Vertebrates in Northern North America) and others, is shown in Figure 4. These data are the cumulative result of a variety of paleontological and archaeological studies over many years. While this sample reflects some degree of randomness, the compilation also reflects an early emphasis on dating relatively large horned Pleistocene bison, as well as a more recent emphasis on dating small horned Holocene specimens and bison remains found in archaeological contexts. Additional late Pleistocene dates were recently obtained from bison remains found on Alaska's North Slope (Matheus et al. 1999), but are not represented in Figure 4.

Radiometric and zooarchaeological data demonstrate that bison occurred in this region during the late Pleistocene and Holocene, and that the late Holocene distribution of wood bison included much of eastern Alaska, southern Yukon, the western NT, and possibly western Alaska as well (Skinner and Kaisen 1947). Well preserved bones indicate their presence along the arctic coast of the Yukon and NT, and on Victoria Island (Figure 3); oral accounts, as well as the archaeological evidence reviewed below (Morrison 1997), indicate the distribution of wood bison overlapped to some extent with that of muskoxen and caribou in parts of this region. Radiometric evidence discussed below also indicates that bison occurred near the arctic coast of Canada and in the northern Brooks Range in Alaska during the late Holocene.

Radiocarbon data indicate that bison occurred in Alaska at least as recently as 170±30 BP, based on a skull found at Anchorage (Map No. 44).[viii] Although there are currently no historic accounts regarding the presence of bison south of the Alaska Range, this specimen indicates they were present in

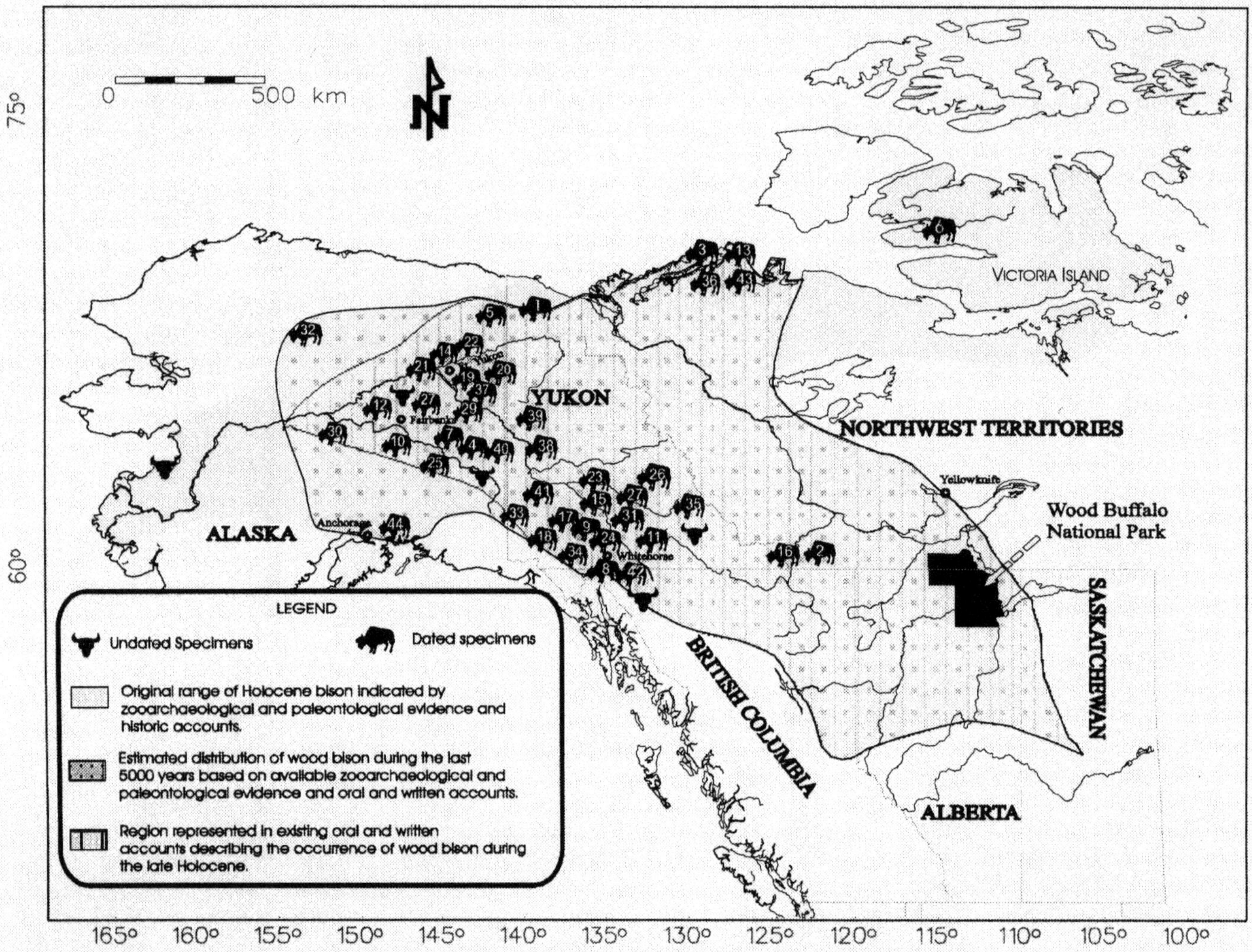

Figure 3. Approximate location of Holocene bison remains in Alaska and adjacent Canada, and estimated original and late Holocene range of wood bison in North American based on available zooarchaeological, paleontological, oral and written historical documentation (Skinner and Kaisen 1947; McClellan 1975, 1981; van Zyll de Jong 1986; Gates et al. 1992; Lotenberg 1996; this study). Location numbers correspond to map numbers in Table 1.

this area during the late Holocene. The lower Matanuska and Susitna river drainages near Anchorage contain extensive meadows that may have supported bison populations in the past. Radiocarbon dates for specimens collected in the upper Yukon and Tanana river lowlands, as well as oral accounts, indicate that bison also inhabited those areas during the late Holocene, further suggesting that low-elevation meadows were a component of wood bison habitat in Alaska.

Bison remains are distributed geographically from west to east across Alaska and into Canada (Figure 3), and temporally throughout the Holocene (Figure 4). The area represented by oral narratives encompasses the area where late Holocene bison remains appear to be most common. The available chronometric data and historic accounts suggest that wood bison were more common in eastern Alaska and the southern Yukon during the late Holocene than in areas to the north and west. However, the two specimens yielding the most recent radiometric dates occur some distance outside (Anchorage), or near the edge (arctic coast, Northwest Territories) of the area referred to in the oral accounts, indicating that wood bison persisted into the last millennia over a broader area than is indicated by currently available historical accounts. The area south of the Alaska Range in Alaska is the only extensive region where physical evidence for late Holocene bison is not presently corroborated by oral and/or written historical documentation (Figure 3).

ARCHAEOLOGY

Alaska

Archaeological investigations in interior Alaska are limited, consisting largely of site surveys (e.g., Dixon et al. 1985; Libby and Medlock 1979; Roseneau et al. 1975) with limited systematic excavation. Alaskan archaeological sites at which bison remains have been positively identified include the Dry Creek site near Healy (Powers et al. 1983), the Broken Mammoth (Holmes 1996), XMH-297 (Holmes 1979; Holmes and Bacon 1982) and Gerstle River (Holmes 1998) sites near Delta Junction, and the KIR-275 site in the north-central Brooks Range.

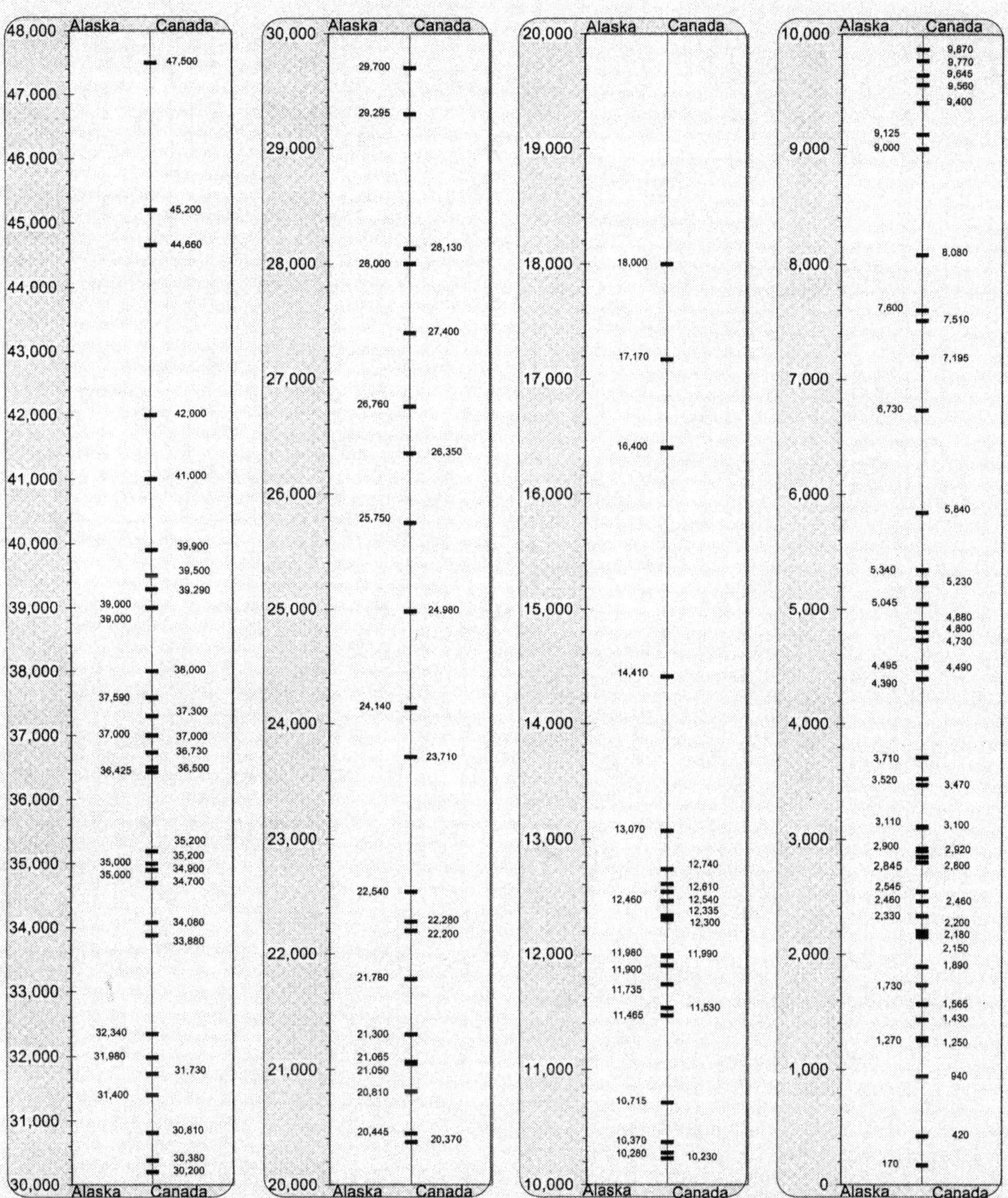

Figure 4. Temporal distribution of uncalibrated radiocarbon dates (years before present) for late Pleistocene and Holocene bison in Alaska and adjacent Canada. The median date was used in two cases where bracketing dates were available for Holocene bison. Dates later than 12,000 B.P. are detailed in Table 1. Pleistocene dates are based on published and unpublished data from Larsen (1968); McDonald (1981); Guthrie (1990); Bureau of Land Management, Fairbanks, open files; S. Dickson (pers. comm.); and the Annotated Bibliography of Quaternary Vertebrates of Northern North America, C.R. Harington, manuscript.

The Broken Mammoth site yielded a variety of bison bones in four "cultural zones." Five radiocarbon dates for bison remains range from 7600±140 to 11,420±70 BP, and a sixth bison bone was recently dated stratigraphically at 2260±40 BP (D. Yesner, pers. comm.). Part of a bison tibia found at the XMH-297 site was stratigraphically dated to between 2285±145 BP and approximately 3980±150 BP, shortly after the fall of the Jarvis Ash (Holmes and Bacon 1982). Bison remains at the Gerstle River site include a phalanx, metatarsal, naviculo-cuboid, calcaneum, and fragments of a scapula and sacrum in strata bracketed by charcoal dates of 8280±60 to 9970±60 BP (Holmes 1998). A bison metatarsal found near the KIR-275 archaeological site in the Killik River drainage has been dated at 2330±50 BP (C. Martin and T. Birkedal, pers comm.). The late Holocene date was subsequently corroborated by dates of 2470±50 and 2500±50 BP (P. Matheus, pers comm.). Isolated bison remains have also been found in late Holocene flood-plain deposits in interior Alaskan streams, but most are not dated or are from ambiguous stratigraphic contexts (Péwé 1975a, 1975b).

Yukon and Northwest Territories

Archaeological surveys and excavations conducted in northwest Canada, primarily in Yukon, provide additional insight into the occurrence of bison during the Holocene. Radiocarbon dates for bison remains spanning much of the past 10,000 years have been obtained from a number of archaeological sites, including those in Table 1 (Map Nos. 1, 9, 11, 13, 16, 17, 18, 24, 25, 41 and 43). The Engigstciak site in northern Yukon represents the earliest known association of bison remains with cultural artifacts in this region (Cinq-Mars et al. 1991). The majority of dates in the sample of Holocene sites occur between about 3,000 and 6,000 years BP. The only archaeological site with bison remains dated to within the past 1,200 years is the Bison Skull site (OaRw-2). In 1987 a bison skull with attached horn cores was discovered at this site, located along the banks of the Old Horton River Channel (Le Blanc 1988), near the extreme northwestern arctic coast of the Northwest Territories (Map No. 43, Table 1). The specimen yields a radiocarbon date of 420±65 BP (Harington 1990). The skull was found in close association with the skulls of two muskoxen at a site that had a long history of human use, primarily for caribou hunting and butchering (Le Blanc 1991; Morrison 1997). Cut marks on the upper surface of the bison skull suggest it was used as a cutting platform.

Late Holocene sites in the Old Crow area include two extensive spring caribou hunting sites: Klo-kut MjVl-1 (Morlan 1973) and Rat Indian Creek MjVg-1 (Le Blanc 1984). Both sites contained large, well-preserved faunal assemblages with abundant caribou remains and a diverse array of other species, including moose. No bison bones were reported at these sites. Other large-scale excavations include the Old Chief site MjVk-7 (Cinq-Mars 1974; Donahue 1974), NbVk-1 (Morlan 1972a), Cadzow Lake (Morlan 1972b) and the Whirl Lake site MjTp-1 located near the MacKenzie River Delta (Gordon and Savage 1973, 1974). These sites range from mid-to-late Holocene in age and several contain extensive faunal assemblages. The late Holocene component at the Whirl Lake site (dating to uncalibrated AD 1740±90 and AD 1730±90) contained remains of caribou and moose, 15 elements from an unidentified artiodactyl and 17 unidentified large-sized mammal bones (Gordon and Savage 1974). Gordon and Savage (1973) discuss the possibility that bison are represented at this site, although the reasons for this suggestion are not explicit.

Cultural practices and/or poor preservation may obscure the significance of bison as a food resource during the last millennium, at least in southern Yukon. Bone preservation appears to be less problematic at certain sites in northern Yukon where permafrost and rapid soil deposition improve preservation. Faunal remains of large mammals such as moose and bison are well represented at various mid-Holocene sites in southern Yukon (cf. MacNeish 1964; Workman 1978), but there appears to be a paucity of identifiable large mammal bones at late Holocene sites.

Greer (*in prep*) examined faunal assemblages from late Holocene components at 38 archaeological sites in southern Yukon, representing the past 1200 years. These include more than 125,000 bone fragments, most of which could not be specifically identified. The large mammal component is determined for only 16 bones or bone fragments. Caribou occurred with the highest frequency, followed by Dall sheep (*Ovis dalli*). Moose were identified at only one site (JcUj-12), and no bison remains were identified. Greer (ibid.) indicates that several unidentified bone samples were large enough to represent moose, although the possibility exists that some may represent bison.

The intensive processing of large mammal bones during the late Holocene may also be related to changes in cooking and preparation techniques. Late Holocene archaeological sites in southern and central Yukon are characterized by an increase in the occurrence of fire-cracked rock, which Charlie and Clark (1993) indicate may be related to the increased use of stone boiling techniques (Clark and Gotthardt 1997). Intensive processing of large mammal bones to produce bone grease is well documented in the ethnographic record (cf. Leechman 1954; Binford 1978, 1981). Greer (*in prep*) suggests large mammal bones were smashed into small fragments to facilitate the extraction of bone grease, often making it difficult or impossible to determine the species represented at late Holocene sites. Other cultural practices can further complicate the interpretation of faunal remains at archaeological sites, including the practice of disposing of the bones of some species in rivers, lakes or in fires, placing remains in trees and using bones to manufacture tools (McClellan 1975; O'Brien 1997).

Paleontological, archaeological (Table 1) and ethnographic evidence (Lotenberg 1996) shows that bison occurred in Yukon and were harvested by First Nation hunters during the late Holocene, although bison are not highly visible in the archaeological record. The same appears to be true for moose, which, on the basis of Greer's sample, are nearly invisible in the late Holocene archaeological record. The relatively large size of bison and moose probably increases the probability that carcasses were processed extensively at kill sites, further reducing the likelihood that largely intact bones would accumulate at archaeological sites (Binford 1978, 1981, 1983). The fact that high-ranked resources (Smith 1983; Simms 1984; Smith and Winterholder 1992; Broughton 1997; Janetski 1997) such as moose and bison are most susceptible to overexploitation could also account for the low frequency of their remains in archaeological sites (Broughton 1997; Kay 1994a, 1994b, 1997).

ETHNOHISTORY AND ETHNOGRAPHY

There are numerous ethnohistorical studies describing *Gwich'in* Athabascan lifestyles during the early contact period (Franklin 1970, 1971; Dall 1898; Hardisty 1872; Jones 1872; Kirkby 1865; Mackenzie 1801; Murray 1910; Petitot 1970; Richardson 1851, Vol 2; Whymper 1868), but these accounts include no mention of bison. The same is true of ethnohistorical research into the *Gwich'in* fur trade era during the mid- to late-nineteenth century (Krech 1976, 1978, 1979, 1981, 1987), focused loosely as it is on historical context, economics and the response of Native societies to Euroamerican influence. Although written records place bison in southern Yukon and the western NT during and after the mid 1700s (Gates et al. 1992; Lotenberg 1996), we have yet to find written documents placing bison in Alaska at the same time. Similarly, traditional knowledge regarding the late Holocene occurrence of bison in the upper Porcupine-Peel River area is not reflected in the journals of Archdeacon Robert McDonald (Archives of the Ecclesiastical Province of Rupert's Land, Winnipeg), who traveled in the region from 1862 until the early 1900s, or in other written accounts from the region (R. Le Blanc, pers. comm.).

Formal ethnographic studies in northeastern Alaska began in the early 1930s when Osgood (1936) studied the Peel River and Crow Flats *Gwich'in*, and McKennan (1965) began systematic studies with the Chandalar *Gwich'in*. Subsequent research includes studies by Balikci (1963a, 1963b), West (1963), Leechman (1954) and Slobodin (1962, 1981). Except for one intriguing clue, these studies contain no specific references to bison procurement or use.

McKennan (1965:18) notes: "Muskoxen, now extinct in the area, were said to have frequented the Chandalar territory in former days, and a small mountain near the forks of Smoke Creek [Figure 1] is known to the natives as [*ch'itthay ik*; James Kari pers. comm.) which they translate as Muskox Shirt Mountain." However, McKennan's original field notes, and information supplied by *Gwich'in* elder Moses Cruikshank and by Thomas E. Taylor, suggest McKennan may have been provided with a name reflecting the occurrence of "buffalo," but later concluded his informants were referring to muskoxen.

McKennan's original field notes,[ix] reflecting information obtained in 1933, primarily from Chief Christian, read as follows:

<u>Musk Ox</u> [*ch'itthay*]
[*ch'itthay ik*] = Buffalo (sic) shirt.
Little Mt. on Smoke Creek.
Musk ox formerly driven
over cliff on top.

McKennan apparently assumed the informant's reference to "Buffalo" was incorrect (as indicated by his insertion of "sic"), whereupon he changed it to "Musk ox."[x] Moses Cruikshank indicates that a mountain known as "Buffalo Shirt Mountain" was located in the "Sheenjek River country," in the next large drainage east of the Chandalar River. This geographic discrepancy may simply reflect Mr. Cruikshank's reference to a general geographic region in the foothills of the Brooks Range. United States Geological Survey (USGS) topographic maps identify a mountain between the lower reaches of Smoke Creek and the Wind River as "Buffalo Mountain" (Figure 1), based on information provided to USGS engineer T. E. Taylor by local residents (Orth 1971).[xi] His records, as cited by Orth (1971), indicate the name "Buffalo Mountain" was provided to him by residents of Venetie when he visited that community in November 1956. Taylor reviewed local place names with Venetie elders, using aerial photographs and 1:250,000 quadrangles to identify topographic features and inquire as to their names. Mr. Taylor recorded phonetic spellings of *Gwich'in* place names in notes and on maps. Many *Gwich'in* also spoke English and provided English equivalents for place names, as shown on current USGS topographic maps.

Venetie elders informed Taylor that the English name for the mountain was "Buffalo Shirt," and that the name arose from stories indicating that "buffalo" were formerly hunted there. Oral tradition described how buffalo were driven over an escarpment at this site. Taylor's informants stated this animal no longer occurred in the region and that they themselves had not seen or hunted buffalo. They did not speculate regarding the reasons for their disappearance, but were resigned to the fact that the buffalo were gone. Taylor said his informants stated "emphatically" that these animals were "buffalo," and were aware of the difference between bison and muskoxen. He indicates further that he did not question these statements, in part because there seemed to be no doubt as to their meaning, and because he was aware of the existence of wood bison in Canada.

The fact that a few muskoxen apparently persisted in the eastern Brooks Range until the late 1800s makes it difficult to clarify the ambiguity reflected in McKennan's field notes. Lent (1998) reviews historical data regarding the presence of tundra muskoxen in Alaska, and notes there is no subfossil evidence of Holocene muskoxen in the eastern Brooks Range. Oral and written accounts cited by Lent show that small numbers were present in this area in the late 1800s, and that several muskoxen were killed in the upper reaches of the Sheenjek and Chandalar drainages on at least two occasions in the 1890s. However, knowledge of group sizes characteristic of muskoxen and wood bison (Tener 1965; Komers et al 1993), as well as the apparent scarcity of muskoxen in the area, suggest that a herd of the size implied in oral accounts about "Buffalo Shirt Mountain" would be more typical of bison.

DISCUSSION

Zooarchaeological, paleontological, oral and written historical data demonstrate that wood bison populations persisted in a large part of their original range in Alaska and Canada during the late Holocene, and were a component in the economies of Athabascan people in central and eastern Alaska during this period. Their corresponding significance in adjacent parts of Yukon and NT is also indicated. The zooarchaeological record and oral accounts from a wide geographic area indicate that wood bison occurred on more than a sporadic basis in the late Holocene, and were sufficiently abundant to persist for a period of several thousand years. Oral narratives from Athabascan elders in Alaska and Yukon indicate bison were procured on more than an occasional or intermittent basis, with some historic accounts emphasizing the importance of bison as a resource. Although multiple oral accounts indicate that bison were indeed a source of food and materials, the precise nature of their seasonal, annual, and long term significance relative to other plant and animal resources is unspecified.

Extensive paleontological and limited archaeological data, in combination with oral accounts, indicate that wood bison were hunted by humans until their disappearance from Alaska during the last few hundred years. Wood bison had apparently become scarce shortly before early Euroamerican explorers, naturalists, and entrepreneurs entered Alaska from the east, and before firearms became widely available (Jennings 1968, Holmes and Bacon 1982, Guthrie 1990). In terms of historical context this decline coincided with the period during which bison were largely extirpated from the woodlands east of the Mississippi River (Dary 1989; Belue 1996), the period between the landing of the Mayflower in 1620 and the Lewis and Clark Expedition in 1804-1806, and with the Russian exploration of Alaska's western coast (Bancroft 1886). Wood bison disappeared from substantial portions of their original range in Alaska and Canada by the early 19th century.

Lotenberg (1996) reevaluated the history of wood bison in Yukon primarily on basis of traditional knowledge obtained through interviews with Native elders representing six First Nations in Yukon and northern British Columbia, as well as from previously published records. These accounts indicate wood bison were relatively abundant in a large portion of this region approximately 400 years ago, and continued to inhabit parts of southern Yukon during most of the last 400 years. Based on reports from Native elders, McClellan (1975) concluded that bison populations were declining in parts of Yukon during the eighteenth century, and that Inland Tlingit shot bison with arrows, although their importance in the subsistence economy was not addressed (McClellan 1981). Other traditional stories from southern Yukon also mention the former presence of bison (Sidney et al. 1977). The occurrence of bison in Yukon during the 1800s and early 1900s is corroborated by a limited number of documentary records (Clarke 1945; Jeckell 1933; Gates et al. 1992). The reported occurrence of bison near the Canadian border at Eagle, near Circle, and near Fort Yukon, Alaska, and two reports of bison being killed in interior Alaska during the early 1900s (this study), further suggest that wood bison persisted into the 1900s in Yukon, and that small numbers of bison also occurred, at least on a transient basis, in eastern Alaska adjacent to the Canadian border during this period.

Oral accounts detailed above indicate substantial bison populations were present in Alaska as recently as 200 years ago. Radiometric data corroborate their presence during the late Holocene, with one radiocarbon date indicating bison were still present in Alaska during the early 19th century, although in an area where there are no known historic accounts. The likelihood of finding and dating remains of the most recent wood bison is influenced by many factors, including the rate and duration of population decline prior to the disappearance of the species. Considering the improbability of finding or obtaining samples from the last wood bison in Alaska, the temporal gap between oral accounts and radiometric data in interior Alaska and southern Yukon is not unexpected.

The ability of humans to observe environmental phenomena, and to record and transmit historical knowledge, provides more comprehensive insight into the recent past than can be obtained from zooarchaeological data alone. This is especially true for phenomena that are difficult or impossible to reconstruct through paleontological or other proxies.

A substantial but not exhaustive search for written documents has been completed. The apparent absence of written documents referring to bison in Alaska may indicate that bison had become scarce by the early 1800s, before there was a significant Euroamerican presence in eastern Alaska. Although small numbers apparently persisted until the early 1900s, the handful of early travelers who produced journals or other written records may have remained unaware of, or failed to record, their presence. That early 20th century ethnographers failed to collect information about bison hunting in the north may reflect the limitations of a restrictive working paradigm that focused largely on the contemporary landscape, as well as a sometimes limited sample of informants (Osgood 1936; McKennan 1965; Burch and Mishler 1995). Knowledge of Alaska's faunal history was comparatively limited when the early ethnographic studies were undertaken, with little or no published information indicating that bison might have persisted into the late Holocene. Information about an animal that was no longer present may not have been elicited nor, in the absence of a direct question, offered. Alternatively, as often occurs with other early records, pertinent information may have been available but was simply not included in written accounts (Lyman 1998).

There are similarities in the oral narratives provided by Yukon (Lotenberg 1996) and Alaska Native elders, as well as some elements that are found only in one or the other oral tradition. The diverse content of these accounts is typical of the varying nature of detail encountered in oral narratives passed from one generation to another, representing events and observations occurring from one to several generations earlier. Similarities include: (1) descriptions of the presence of substantial bison populations at one time; (2) indications that bison were hunted extensively, were a major source of food, and provided raw material for clothing and other uses prior to their decline and disappearance during the last few hundred years; (3) indications that late Holocene bison populations were found primarily in low elevation habitats; (4) reports that bison skulls and other remains continue to be found in areas historically known as bison range; (5) observations that moose were absent or scarce at the time bison were abundant; (6) consistent narratives indicating that bison were hunted with bow and arrow; (7) indications that trees and shrubs expanded during and after the period when bison declined and disappeared, and moose populations expanded; and (8) that several elders from different areas remembered traditional names for bison.

Information found only in Yukon accounts includes descriptions of the use of snares to take bison, and the use of bison horn to make spoons. One account describes the yellow fat that is peculiar to bison, and a few elders indicated that their parents or grandparents hunted bison (Lotenberg 1996). Information peculiar to the Alaskan accounts includes: 1) observations that meat was dried for winter use; 2) that there may have been two types of bison; 3) that hides were used for blankets and pillows and hair was used as thread; 4) that the animal had a long tail and a large hump and head; 5) a reference to the non-allergenic nature of bison hides; 6) indications that bison and moose contributed jointly to the Athabascan subsistence economy at one time; and 7) that dogs were used to hunt bison. There are also indications that drives were used to kill bison in the mountains north of the Yukon Flats in Alaska. However, other accounts mention that drives may have been used to take muskoxen in the upper Chandalar drainage and in the upper Porcupine River region in Canada. The degree to which drives were used to procure bison and/or muskoxen remains unclear.

Accounts from both Alaska and Yukon include a few observations of bison or their sign indicating that small numbers of bison occurred, at least on a transient basis, after the decline of larger and geographically more widespread populations. Both sets of accounts include a few possible reasons for the decline or disappearance of bison populations. Yukon elders mention the expansion of forests, deep or crusted snow, cold winters and volcanic ash. Narratives from Alaska suggest that hunting by humans acted in combination with other environmental factors to cause the disappearance of wood bison, and that the less forested nature of the terrain during the late Holocene made bison especially vulnerable to hunters.

The hunting technology and strategies described in Athabascan oral accounts are consistent with existing knowledge of big game hunting techniques used by early Athabascan hunters. However, the reported use of dogs as an aid in hunting bison warrants discussion. Dogs were widely used by the *Gwich'in* to hunt moose and caribou, assisting hunters on snowshoes, especially in late winter when snow was crusted, and to pursue wounded animals (Osgood 1936; McKennan 1965; Nelson 1973).

Behavioral studies of interactions between wolves (*Canis lupus*) and wood bison indicate that dogs may have been effective in holding bison in a given area, and diverting their attention while hunters approached. Carbyn et al. (1993:178-210) describe hunting strategies of wolves in Wood Buffalo National Park, Canada, noting that bison herds often remain stationary for extended periods in the presence of wolves, relying on joint defensive behavior rather than flight. Wolves may spend hours or days in proximity to bison herds before causing a stampede and attempting a kill, or isolating a vulnerable individual. Bison often travel considerable distances after being stampeded by wolves. These researchers also found that a careful and patient approach allowed human observers to follow and observe wood bison at close range, with little disturbance to the animals. The behavioral response of bison to domestic dogs may be much the same as their response to wolves.

Radiocarbon dates for relatively recent bison specimens from Alaska, Yukon and NT range from 170 to 11,900 years before present (Table 1). A few additional specimens from this region have been taxonomically referred to *B.b. athabascae*, but are not yet dated. In combination with archaeological and paleontological data, the oral narratives suggest that bison also occurred along the western arctic coast of Canada during the Holocene, where they likely coexisted with caribou and muskoxen (McGhee 1996:85-93). Archaeological evidence representing early (Cinq-Mars 1991) and late (Harington 1990; Morrison 1997) Holocene sites, as well as oral accounts, indicate humans in this region hunted bison. The distribution of wood bison remains and other historic information suggest that Holocene bison populations were most prevalent in interior Alaska and in northwestern Canada (Figure 3). If the late Holocene climate was similar to the present, the distribution of late Holocene bison remains may reflect relatively low snow accumulation, and less icing and wind-packed snow in this region than in areas to the west (National Oceanic and Atmospheric Administration 1986).

The distribution of late Pleistocene fauna was strongly affected by changes in the Cordilleran and Laurentide ice sheets, which covered much of western Canada and southern Alaska during this period (cf. Matthews 1982; Pielou 1991). Radiocarbon dates for late Pleistocene bison in western Canada (Figure 4) were obtained from localities in west central Yukon which, along with interior Alaska, remained largely ice-free during this period. The apparent hiatus in radiocarbon dates between about 20,000 and 13,000 BP for bison in northwestern Canada may reflect glacial advances and associated changes in environmental conditions which occurred between about 24,000 and 17,000 BP (Hollin and Schilling 1981; Dyke and Prest 1987). The subsequent glacial recession and concomitant expansion of bison populations eastward is reflected in the increased number and wider distribution of dated bison remains between about 13,000 BP and the late Holocene (Figures 3 and 4, Table 1). Bison also colonized south central Alaska after the last major glacial recession, although there is currently only one specimen recorded in this region. The available data suggest that bison colonized new habitat as glaciers retreated, reaching areas as far east as Victoria Island and southern NT during the early Holocene (Figure 3). In contrast, paleontological and zooarchaeological data indicate that bison inhabited the ice-free refugia in interior Alaska continuously during the 48,000-year period represented in Figure 4.

A number of authors have already suggested that bison persisted in Alaska into the late Holocene, and were an important resource for humans during this period (Skinner and Kaisen 1947; Jennings 1968; Harington 1977; Holmes and Bacon 1982; Peek et al. 1987; Guthrie, 1990:286). Holmes and Bacon (1982), and West (1982) suggest that increasingly limited habitat and deep snow, combined with hunting by "Beringian hunters" likely caused the extirpation of bison. The reasons for the late Holocene disappearance of wood bison in Alaska, Yukon, and parts of Northwest Territories and British Columbia will likely never be understood in the detail that would accompany a similar event in contemporary times. However, published literature and historic accounts suggest a combination of factors that must be considered in explaining the decline and disappearance of wood bison, including predation by wolves and bears, hunting by humans, environmental factors that altered the amount and distribution of suitable habitat, and meteorological or climatic factors.

Predation

Studies of extant northern bison herds suggest it is unlikely that predation by wolves or bears, alone or in combination, would have led to the extirpation of healthy wood bison herds. Introduced or reintroduced herds at several locations in Alaska and northern Canada have achieved, and maintain, viable population levels without significant predator control. Mortality from predation is rare in most herds, even where wolves, black bears (*Ursus americanus*) and grizzly bears are common. Several northern bison herds are limited primarily by regulated hunting, with predation being only a secondary cause of mortality (Gates and Larter 1990; DuBois and Stephenson 1998; Whitman and Stephenson 1998; B. Hayes, pers. comm.). However, other bison populations recently infected with cattle diseases (bovine tuberculosis and brucellosis) may be more susceptible to the demographically limiting effects of predation (Joly et al. 1998).

Several studies have applied metapopulation modeling and knowledge of island biogeography, habitat fragmentation, and population ecology to the question of large mammal extinction (Armbruster and Lande 1993; Beck 1996; Ward 1997). These studies show that even modest levels of exploitation by predators, including humans, have substantial effects on the survival of isolated large mammal populations. Various studies show that isolated populations of birds and mammals are more vulnerable to extinction than larger, contiguous populations (MacArthur and Wilson 1967; Brown 1978; 1986: Grayson 1991). Lande (1988) emphasized the importance of stochastic demographic factors and habitat distribution in determining extinction threshholds for wildlife populations.

It has been suggested that hunting by humans affected the distribution and abundance of a number of large mammal species in North America during the late Pleistocene and Holocene (cf. Martin and Klein 1984; Martin and Wright 1989; McGhee 1996:85: Ward 1997; Whitney-Smith 1998; Martin and Szuter 1999). In this model the high return per unit effort in hunting large mammals resulted in their being preferred prey for early humans (Winterhalder 1981a, 1981b; Feit 1987; Smith 1983; Stephens and Krebs 1986), while the ability of humans to rely on alternative resources such as fish, small game and plants mitigated the adverse consequences of overexploitation of large ungulates (Kay 1997; 1998). The effectiveness of aboriginal hunting techniques allowed hunters to kill prime-age animals in addition to more vulnerable individuals, and to select for female ungulates under certain conditions (Kay 1994a, 1995, 1997, 1998; Stiner 1990).

There is some evidence that aboriginal hunting in Alaska resulted in declines or local extirpation of some species including Dall sheep, muskox, moose and brown bears during the late Holocene (Campbell 1978; Lent 1998; LeResche et al. 1974; Coady 1980; Birkedal 1993). Similarly, unregulated hunting by aboriginal people appears to have been an important factor in the decline or disappearance of muskoxen in parts of Canada (Stefansson 1921; Hone 1934; Harington 1961; Tener 1965; Gunn et al. 1984; Will 1984). Lent (1998) suggests that hunting could have led to the disappearance of muskoxen in Alaska even though they were not a staple resource for most indigenous people. Three Athabascan elders who contributed to our study suggest that hunting played a role in the disappearance of wood bison in Alaska, citing the relative ease with which bison were hunted and killed, periodically high human populations, and the need to hunt intensively during periods when food was scarce. Other historical accounts from northeastern Alaska and the Yukon Flats also suggest that human populations sometimes affected the abundance of big game in the region. The *Gwich'in* stories of Johnny and Sarah Frank of Venetie (Mishler, 1995: 11-13) describe how in the early days "...there were hardly any big game animals because there were too many people." These accounts refer to fluctuations in human populations in this region, and describe how "...when big game animals were scarce, lots of people froze to death." References to famines are common in the legendary history of the Chandalar and Tranjik *Gwich'in* (McKennan 1965; Nelson 1973).

The value of bison as a resource and their vulnerability to the hunting implements and techniques of early humans (West 1982) may have resulted in hunting being an additive mortality factor which, in combination with predation and other sources of mortality, affected the population dynamics of bison. The apparent scarcity of moose in much of Alaska during the late Holocene (Yesner 1989; LeResche et al. 1974; Coady 1980) may have accentuated the importance of bison as a big game resource, as would the seasonal or annual scarcity of caribou in many areas (Burch 1972). The behavioral characteristics of female and juvenile bison, which occur almost exclusively in herds (Komers et al. 1993), would increase their vulnerability (Lande 1988) and the likelihood that more than one animal would be taken in an encounter with human hunters (Fisher and Roll 1997). The possibility that changes in hunting technology improved the effectiveness of hunters should also be considered. The development of archery or other improvements in technology during the late Holocene (Blitz 1988) may have altered the relationship between hunter effort and return.

Habitat Availability

Guthrie (1982:314) indicates that in contrast to some other species represented in Pleistocene megafauna such as horses (*Equus sp.*) or proboscideans, bison are able to cope with relatively deep snow. Some earlier studies suggest that windswept areas adjacent to mountainous terrain were the most important habitat for Holocene bison in Alaska, based on the premise that bison are not well adapted to substantial snow accumulation (cf. Ager 1974; Holmes and Bacon 1982). However, the distribution of wood bison remains (Figure 3) and geographic referents in oral accounts indicate the presence of late Holocene bison in low elevation habitat. Current evidence suggests that such areas were an important habitat for wood bison in Alaska and Yukon. Despite indications that meadow habitat declined during the late Holocene, recent studies show that the Yukon Flats and other parts of interior Alaska continue to offer suitable habitat for bison herds at or above the level currently estimated to represent a minimum viable population (Newmark 1987; Berger et al. 1995; Dubois and Stephenson 1998; Whitman and Stephenson 1998; ADF&G unpublished data; Wood Bison Recovery Team 1998). The quantity and quality of potential bison habitat in Alaska is substantially greater than in Yukon (Reynolds et al. 1982b; M. Hoefs, pers. comm.), and rivals or exceeds that found in wood bison range in other parts of Canada (Berger et al. 1995; Gates 1992, H. Reynolds, pers. comm.).

Extant wood bison populations occur primarily in low elevation habitat where sub-hygric meadow systems are intermixed with boreal forest (Larter and Gates 1991), further suggesting that Holocene bison were not restricted to windswept habitats. Bison are adapted to temperatures of -48 F or lower (Fuller 1962; Peters and Slen 1964; Christopherson et al. 1978), and prosper in areas where snow depths average 24-30 inches or more (Larter and Gates 1991; Carbyn et al. 1993). Meagher (1973, 1976) concluded that bison can tolerate deeper snow than other northern ungulates, based on the fact that plains bison have prospered in Yellowstone National Park despite a harsh winter climate, foraging in snow depths exceeding 50 inches. However, snow depth and hardness influence bison foraging behavior, as demonstrated by Meagher (1976), who found that snow depth had a major influence on habitat selection in Yellowstone Park. Wind-packed or ice crusted snow also presents difficulties for bison, and snow density was the principle factor influencing the selection of feeding sites by wood bison in the Slave River lowlands (Reynolds and Peden 1987).

There are a number of references to historic changes in vegetation in interior Alaska and other northern regions, with a trend toward expansion of shrubs and forests during the last few centuries (Vanstone 1974; McKennan 1965); a variety of biotic and abiotic factors appear to be involved (cf. Jacoby, and D'Arrigo. 1989; Campbell et al. 1994; Zimov et al. 1995; Chowns et al. 1998; Overpeck et al. 1997). Several elders from communities on the Yukon Flats also referred to oral traditions indicating the area was formerly more open, with fewer trees and shrubs and more extensive meadow systems (O'Brien 1997:37).

At present the Yukon Flats is characterized by a diverse mosaic of mixed spruce-poplar and spruce hardwood forests, spruce muskeg, and successional and climax stands of willow and alder. Both wet and dry meadows are common, comprising approximately 12 percent of the total area. Despite the apparent reduction in the amount of available meadow habitat, the Yukon Flats and parts of the Tanana Valley continue to offer suitable habitat for viable bison populations (Berger et al. 1995; ADF&G unpublished data).

Bison are eclectic, with an ability to exist in a variety of climatic and habitat regimes ranging from subarctic boreal and eastern deciduous forest, to relatively arid conditions in the southwestern United States (van Zyll de Jong 1986, Dary 1989). In recent decades wood bison have been reintroduced in the southern Yukon, and plains bison have been introduced in the Delta, Farewell, Copper River and Chitina River areas in Alaska. Except for the Copper and Chitina River herds, which settled in poor quality habitat (Miquelle 1985), these herds increased at high rates and continue to show high productivity (DuBois and Stephenson 1998; Whitman and Stephenson 1998).

It is clear that suitable bison habitat is still available in parts of the range once inhabited by wood bison, a fact that opposes the idea that changes in habitat caused their decline and disappearance. Further, the nature of recent environmental changes that are reported to have occurred in interior Alaska indicate that wood bison declined and disappeared at a time when the extent of meadow habitat was greater than at present. Lent (1998) observed that the successful reintroduction of muskoxen to northern Alaska demonstrates that suitable habitat for this once extirpated species continues to exist. However, changes in the distribution of suitable habitat, in combination with hunting and other mortality factors, probably contributed indirectly to the disappearance of wood bison in Alaska.

Rather than being more widely distributed over the landscape, as was the case during the Pleistocene, Holocene bison habitat became more limited through time (Guthrie 1990), although it was not eliminated as has sometimes been suggested (Guthrie 1982). The transition from relatively treeless steppe to increasingly forested terrain during the late Pleistocene and Holocene would have gradually created a mosaic in which bison habitat was largely confined to low-lying areas with upland or riparian meadow habitat. This would have included the Yukon and Tanana lowlands in Alaska, and/or windswept habitats adjacent to large mountain valleys (Holmes and Bacon 1982; Guthrie 1990). This is the general pattern of distribution suggested by the occurrence of skeletal remains of Holocene bison (Table 1; Fig. 3). However, more widespread sedimentation and greater human activity at low elevations would probably increase bone preservation and enhance the likelihood of discovery, making it difficult to assess the degree to which uplands might have been used by late Holocene bison.

Habitat for late Holocene bison populations most likely occurred in discrete patches that, while capable of supporting viable populations, were nonetheless separated geographically. Once a subpopulation was extirpated, geographical constraints and stochastic events affecting founding populations would limit the likelihood of recolonization. This is consistent with the behavior of existing northern bison populations, with distribution being substantially limited by geographic barriers such as mountains or extensive stands of unbroken forest (Gates and

Larter 1990; Larter and Gates 1991). The seasonal and annual movements of extant northern bison populations are largely constrained by the distribution of habitat (Larter and Gates 1990). The distribution of late Holocene bison was probably limited in comparison to moose, caribou and sheep, which persist through the present.

Lent (1998) concludes that small numbers of muskoxen persisted in the eastern Brooks Range, after their extirpation in northwestern Alaska in the mid 1800s, in part because of the relatively low human population density in the mountainous terrain in the area. Historic accounts, including those presented here, indicate that resource availability and other attributes caused areas adjacent to large rivers to be a year-round focus of human habitation during the late Holocene (cf. Schneider 1986; Luke 1998). This pattern of late Holocene human land use would have placed wood bison populations in juxtaposition with human populations, making them accessible to human hunters on a year-round basis, and limiting the occurrence of refugia (Taylor 1984) where bison were not subject to human exploitation. This situation contrasts with that on the Great Plains, where migration reduced the vulnerability of plains bison to predation (Kay 1994a; 1996).

Influence of Climate

Limitations in the current understanding of the late Holocene climatic regime make it difficult to draw precise inferences regarding the influence of climate on the geographic distribution of wood bison in the northwestern part of their original range. Nevertheless, it is difficult to account for the disappearance of wood bison from Alaska and northwestern Canada on the basis of environmental or climatic changes alone, unless there were catastrophic weather events on a regional scale, or broad, uniform and sustained changes that led to the elimination of suitable habitat. There is currently no evidence for either of these phenomena.

The boreal forest is the product of frequent and often pronounced climatic fluctuations since its establishment following the most recent glacial period (Juday et al. 1998). During the last glacial period the arid climate in lowland central Alaska, central Yukon and eastern Russia precluded the formation of ice sheets, and cool and arid conditions prevented the expansion of forests. Following the end of full glacial conditions and a warming trend beginning about 14,000 years ago, forests expanded northward (Pielou 1991). In addition to climatic and vegetational changes that have occurred on a scale of millennia, decadal and century long climatic changes altered plant species abundance and distribution (Bartlein 1988).

Based on studies of Greenland ice cores, tree-rings and other paleoecological proxies, Holocene climate was characterized by significant diversity, with environmental change occurring on an increasingly regional basis. Continuous paleoclimatic records from the GISP2 ice core show variability on an annual to millennial scale, indicating that Holocene climate was significantly more complex than glacial ice age climate (Greenland Ice Core Project 1993; O'Brien et al. 1995; Overpeck et al. 1997; Ice Core Working Group 1998). The Holocene climate was also comparatively stable over variable periods of time (Guthrie 1990). Some oral accounts (see above) mention periods of deeper snow or extreme cold in the past (see also Fleener 1998), although these are difficult to reconcile with Little Ice Age (LIA) models at this point.

There is general agreement that glaciers and arctic sea ice expanded globally at various times during the Little Ice Age (LIA), which is generally viewed as occurring from the 13th through the 19th centuries (Grove 1988). Overpeck et al. (1997) review evidence for climatic and environmental change in the Arctic during the last four centuries, based on a variety of paleoenvironmental indicators. Their study reveals large-scale variability over the past 400 years, but suggests that declines in summer insolation and other factors led to successively cooler summers, culminating in the Little Ice Age. The onset of the LIA occurred about A.D. 1400 following several centuries of warming during the so-called Medieval Warm Period (MWP). Cold, dry conditions and increased atmospheric circulation generally characterized the LIA. Similar conditions appear to have prevailed at various times during the Holocene (Kreutz et al. 1997). Although patterns related to the LIA were global in nature, the period is not defined by uniform cooling on a regional scale. There is no unambiguous and uniform LIA signal recorded in tree-rings (Jacoby and D'Arrigo 1989; D'Arrigo and Jacoby 1992). Instead, the period includes both warm and cold anomalies that varied geographically and may have persisted into the 20th century.

Increases in the severity of winter storms and extent of sea ice during the LIA were accompanied by changes in agricultural production which contrasted with milder conditions during the Medieval Warm Period (Lamb 1995). The GISP2 paleo-atmospheric record indicates the LIA had the most abrupt onset of any of the rapid Holocene climate change events (O'Brien et al. 1995), and involved the most dramatic change in atmospheric circulation and surface conditions in the last 4000 years, with similar timing and magnitude in both polar hemispheres. Summer temperatures from A.D. 1597 to 1730 in both hemispheres were the lowest in the last 500 years (Kreutz et al. 1997). However, the LIA was one of the shortest cold intervals during the last 110,000 years, and was substantially shorter than other major climatic changes during the Holocene. No single factor appears to be responsible for bipolar changes during the LIA. Moreover, there is no indication of a general increase in volcanic activity during the last 600 years (Kreutz et al. 1997), although a steady warming trend from 1840 to the mid 1940s may have been precipitated, in part, by frequent, sulfur-producing volcanic eruptions (Overpeck et al. 1997).

Studies of past climates based on tree-rings reveal a pattern of temperature variation in interior and northern Alaska from the 1500s to the present. This record is in general agreement with overall trends at high latitudes in the Northern Hemisphere. Temperatures in interior Alaska were distinctly cooler during the LIA than during the 20th century, and reconstruction from tree-rings shows the effects were most pronounced in the early part of the record (from at least the 1500s to 1700). This was followed by a partial warming in the early 1700s, an abrupt return to colder conditions from about 1800 to 1840, and a steady warming from 1840 to the mid 1940s (Jacoby and D'Arrigo 1989; D'Arrigo and Jacoby 1992; Overpeck et al. 1997). This pattern is apparent in tree-ring data obtained at several sites in northern Alaska and northwestern Canada (Overpeck et al. 1997).

Warmer winters may have been correlated with heavier snowfall, rendering bison more vulnerable to human hunters and other predators, or with windy or icing conditions that redistribute and harden snow and thus interfere with the foraging ability of bison. Russell (1898) reports that exceptionally deep or ice-crusted snow during a single winter (apparently the winter of 1870) decimated wood bison populations in the Slave River area. Similarly, Harington and Cinq-Mars (1995) suggest that sudden winter thaws may have affected the survival of saiga antelope (*Saiga tatarica*) and other megafauna during the late Pleistocene in northwestern North America.

Murray (1910) notes that moose were fairly common in the Fort Yukon area in 1847-48, while oral narratives (McKennan 1965; this study) indicate they were scarce in earlier times. Murray's observations pertain to a period immediately following the most extended period of increased weather severity in the LIA, and suggest moose had increased in numbers and distribution following the period of absence or extreme scarcity described in historical accounts. It is unlikely that bison populations would have been decimated at the same time that moose populations expanded, considering that bison are relatively well adapted to deep snow and cold (Peters and Slen 1964; Christopherson et al. 1978, 1979; Larter and Gates 1991; Carbyn et al. 1993; Meagher 1973, 1976). Similarly, Lent (1998) concludes there is no clear evidence that climatic change during the late Holocene caused the extinction of Alaskan muskoxen, noting that larger fluctuations in climate had occurred previously (Haworth 1989). Kay (1997: 145-146) concludes that the LIA or other climatic or weather events do not account for the general scarcity of moose prior to the very late Holocene in much of North America.

Bison persisted in Alaska and northwestern Canada for at least several hundred thousand years despite climatic fluctuations that were frequent, dramatic and abrupt in comparison to the relatively benign climatic regime during the Holocene (Guthrie 1990). This suggests that the temporally limited and modest climatic changes during the LIA are an unlikely cause for the demise of wood bison. The LIA did not result in the extirpation of other large ungulates in the region including moose, caribou or Dall sheep. These species can be strongly affected by severe winter weather and snow conditions (Fancy and White 1985, 1987; Heimer et al. 1994; Adams and Dale 1998), with snow depth being particularly important in affecting the population dynamics of moose (Kelsall 1969; Peterson and Allen 1974; Gasaway et al. 1992).

Wood Bison Biogeography during the Late Holocene

The results of our study support earlier indications that bison persisted during much of the Holocene in Alaska (Holmes and Bacon 1982; Guthrie 1990), and that substantial populations persisted in some areas longer than was previously thought. The available skeletal remains of wood bison, associated radiocarbon dates, and oral and written narratives establish that: 1) bison were widely distributed on the late Holocene landscape, and that wood bison populations persisted in northern and eastern Alaska and adjacent Canada until their extirpation during the last few hundred years; 2) bison were a resource of some importance for Athabascan and other Native peoples during the late Holocene; 3) bison populations were high enough to persist over a period of several thousand years and to support significant human use; and 4) low elevation meadow systems, windswept areas at higher elevation, and some areas that are presently arctic tundra supported late Holocene bison populations.

Based on written accounts by Europeans who traveled in northwestern Canada during the late 18th and early 19th centuries (Gates et al. 1992), as well as oral accounts and skeletal remains, van Zyll de Jong (1986) indicated that the so-called "historic" range for wood bison included northern Alberta, the southwestern NT, and parts of northern British Columbia and Saskatchewan (Figure 3). Skinner and Kaisen (1947) had earlier suggested that wood bison were found in the eastern Rocky Mountains as far south as New Mexico. However, subsequent craniometric analyses indicate the southern limit of wood bison distribution was some distance north of the US-Canada border, and that specimens from areas west of the Great Plains in the United States represent *B.b. bison* (van Zyll de Jong 1986).

The recent distribution of bison in southern Canada, where wood and plains bison populations might have overlapped, was particularly germane to van Zyll de Jong's study of the taxonomic status of recent bison. Based on data from three subfossils in Alaska and four from Yukon and northern British Columbia, van Zyll de Jong (1986) further suggested that the "prehistoric" range of wood bison extended "...north and west into the Yukon and Alaska." In consultation with van Zyll de Jong, Canada's National Recovery Team for Wood Bison subsequently revised the boundaries for the original range of wood bison to include additional parts of northern Alaska and Canada (C.C. Gates, pers comm. 1998), based on additional subfossil evidence (Harington 1990). This region included interior, arctic and western Alaska and part of northwestern arctic Canada, encompassing parts of the north that were also classified as original range by Skinner and Kaisen (1947).

Based on existing historical documentation, we describe three general categories relative to Holocene bison distribution in northern North America (Figure 3). The region that appears to have been inhabited by small horned *B. bison* (*sensu* Wilson 1978; 1996) at some time during the Holocene is shown. This region includes areas previously referred to as "prehistoric" and "historic" range for wood bison (van Zyll de Jong 1986), as well as an additional part of the arctic coastal region in Canada where recent paleontological data reveals their presence during the early Holocene. The known distribution of wood bison during the late Holocene (i.e. the last 5000 years) is also shown. This region is based on information obtained from skeletal remains as well as oral and written historic documentation. It reflects knowledge of distribution in the southern portion of the late Holocene range of wood bison, based on previous analyses of paleontological, oral and written information (van Zyll de Jong 1986; Gates et al. 1992), as well as similar information now available regarding areas further north. Finally, Figure 3 shows the region where oral and written accounts describe the presence of wood bison within the last few hundred years. The occurrence of additional dated and undated specimens east and west of the area where there is currently abundant evidence for the recent occurrence of wood bison suggests that future studies may reveal a wider late Holocene distribution than is presently known. Small-horned bison similar to wood bison also occurred in northern Eurasia during the Holocene (Flerov 1979; van Zyll de Jong 1986, 1993; Lazarev et al. 1998).

Zooarcheological, oral and written historical documentation demonstrate that much of Alaska and adjacent Canada is part of the original range of wood bison (Figure 3), and that this area is part of the "historic" range for the subspecies. However, we caution that in the absence of objective and biologically meaningful criteria, the implied dichotomy between "history" and "prehistory" obscures the continuum of time, introducing an ethnocentric bias (Lyman 1996) and confusing the study of the past (Tudge 1997:17). For example, if written records were selected as the sole criteria for "historic" status, the potential depth of "history" would range from thousands of years in parts of the Old World to less than 200 years in interior Alaska. The oral history of indigenous people as well as zooarchaeological and paleontological data would be excluded from the "historic" record. Oral, written, zooarchaeological and paleontological documentation are complimentary and supplementary sources of historical information and should be used in combination (Lyman 1996; 1998).

The available information supports the conclusion that geographical isolation and hunting are factors that acted in combination and led to the extirpation of wood bison. The discontinuous nature of late Holocene habitat probably played an important, albeit indirect, role while hunting is the most likely proximate factor that reduced numbers and prevented the recovery of subpopulations or recolonization of suitable habitat. Bison have recently prospered in suitable habitat in Yukon and other parts of northwestern Canada, and in Alaska, and we now know that additional suitable habitat exists in Alaska. The recent expansion of wood bison populations demonstrates that earlier declines, and the extirpation of bison in various regions, were not caused solely by changes in habitat.

There appear to be similarities in the timing and causes for the extirpation of wood bison and muskoxen in Alaska. Lent (1998) concludes that muskoxen were rare, and nearly extinct, in northern Alaska prior to the arrival of Europeans and the availability of firearms. A similar pattern characterizes the extirpation of wood bison in the northwestern portion of their original range. Some of the last records of isolated groups or individuals of both wood bison and muskoxen in Alaska, and wood bison in Yukon, involve the taking of groups or individuals with firearms in the late 1800s or early 1900s. Hunting by humans appears to have been an important factor in the disappearance of both species (Lent 1998). Although populations of both animals had declined or, in some areas, disappeared before firearms were widely available, continued hunting of remnant populations with firearms is an additional factor that may have ultimately precluded repopulation of their original habitats.

A combination of factors has had various effects on the status of bison in North America and parts of the Old World. Over millennia the "Great Bison Belt" that extended across Eurasia into Beringia and southward across North America, diminished in size (Guthrie 1980, 1990: 51). However, the genus *Bison* was still widely distributed on North America's Great Plains, in eastern woodlands, and in northwestern Canada and Alaska during the late Holocene (Soper 1941; Dary 1989). It appears that bison herds numbering in the millions were unique to the Holocene Great Plains, with its huge expanse of contiguous grasslands (Guthrie 1980, 1982, 1990). Populations in other habitats did not fare as well as bison on the Great Plains. However, even on the plains and prairies, bison abundance and distribution were affected by human populations, particularly after the availability of horses altered patterns of mobility and changed procurement strategies, and horses became sufficiently abundant and widespread to compete with bison for forage (Flores 1996; Fisher and Roll 1997). A number of studies describe the significant role of predation by humans in the dynamics of late Holocene plains bison populations (Roe 1951; Guthrie 1980; Speth 1983; Flores, 1991, 1996; Belue 1996; Dobak 1996; Fisher and Roll 1997; Haynes 1997; Morgan 1997; Martin and Szuter 1999; Isenberg 2000).

The relationships between bison, human populations and other environmental factors have been diverse, with no single defining pattern. Nevertheless, it is clear that during the last millennia bison populations were dramatically reduced in Eurasia and much of North America in areas where the amount and distribution of suitable late Holocene habitat were more limited than on the Great Plains. The ecological history of bison suggests that bison and humans coexisted in Alaska and adjacent regions for a period exceeding 10,500 years, although the nature and extent of local and/or regional fluctuations in bison numbers are unknown. Wood bison were not extirpated in part of their original range in northern Canada, although they rapidly approached extinction following over-hunting during the 19th century. European bison also declined during the Holocene, with habitat reduction and overhunting being key factors causing their near extinction in the early 1900s, when a remnant population of less than 100 wisent (*B.b. bonasus*) persisted only in the forests of eastern Europe (Gstalter and Lazier 1996). Bison persisted in northern Eurasia into the middle or late Holocene but apparently disappeared earlier than in Alaska or parts of adjacent Canada (van Zyll de Jong 1993; Rusanov 1975; Flerov 1979; Archipov 1989; Lazarev et al. 1998). Plains bison persisted in a large region in North America despite being hunted extensively before the introduction of firearms. The status of contemporary wood bison, plains bison and wisent has improved as a result of conservation efforts including reintroductions into suitable habitat, elimination of livestock diseases, and regulation of hunting (Dary 1989; Gates et al. 1992; Gstalter and Lazier 1996).

Notes

[1] The *Gwich'in* are familiar with muskoxen (McKennan 1965) based on their historic presence some distance north and northeast of the Yukon Flats. Lent (1998) notes an absence of physical evidence for the occurrence of Holocene muskoxen in northeast Alaska, and cites oral accounts indicating that muskoxen were scarce in the eastern Brooks Range during the late 1800s, and absent farther south. In contrast, the remains of bison are common in interior Alaska, including in those areas specifically referred to in oral accounts as formerly inhabited by bison. Lent cites a report by Reed (1946) that describes an account by Henry Rapelle of Fairbanks who, in 1895, observed the head and skin of a young bull muskox that had recently been killed by an Indian on the Kandik River. While this appears to be the only record of a muskox occurring in low elevation habitat south of the Brooks Range during the period in question, archaeological data demonstrate the presence and intermittent human use of Holocene muskoxen in northwest Alaska and adjacent Canada (Gerlach and Hall 1988; Lent 1998).

[2] Studies of contemporary bison show the density of primary hair at the skin surface is the highest among bovids, and is substantially greater than in artiodactyls such as moose, elk (*Cervus canadensis*) or deer (*Odocoileus sp.*), contributing to the scarcity of ticks on bison (Peters and Slen 1964; Mooring and Samuel 1998) and making them a relatively poor host for other ectoparasites (B. Samuel, pers. comm.). However, there have been no studies of the effect of mosquitoes on bison.

[3] These *Gwich'in* terms and phrases were transcribed by Kathy Sikorski, Alaska Native Language Center, University of Alaska Fairbanks, from an audio recording of Rev. Salmon provided by T. O'Brien.

[4] Chief Christian was a prominent leader in the Venetie area during the early 1900s and provided substantial information to early ethnographers (see McKennan 1965). He was William Salmon's cousin (T. O'Brien, pers. comm.).

[5] Anecdotal information from commercial bison ranchers also indicates that bison hides are non-allergenic for humans, possibly because the hair is not oily, and the high density of hair discourages insects (Mike Fogel, pers. comm.).

[6] Ms. Titus' said her grandfather, Charlie Johnson, recounted stories from his father, which held that the disappearance of bison from Alaska was related to a decision by Native elders. Shamans decided that when a certain group of Native people left Alaska for the south, they would take the "big animal" (bison) with them to live separately from the "small animals," including moose and caribou, which remained in Alaska. She later explained they had taken bison with them in a spiritual sense by taking a small amount of bison hide, hair, and meat with them in a basket. The decision that some people would move south was prompted by the fact that there were too many people, extremely cold winters, and the desire to find a warmer place. She was told that people would one day say the buffalo died out, but that actually they followed Athabascan people out of Alaska and would one day return.

[7] Recent studies show that North American bison sometimes harbor a limited number of ectoparasites, including lice and mites (Reynolds et al. 1982a). The ectoparasites found specifically on wood bison are not well known, although a species of chewing louse (*Damalinia* sp.) probably occurs. This species would probably not be apparent to, or affect, humans (W. Samuel, pers. comm.).

[8] As noted in our discussion, bison travel and forage in snow exceeding 50 inches in depth.

[9] A date of 470±90 BP was previously reported for this specimen, based on an assay conducted soon after the specimen was found in 1969 (McDonald 1981). The more recent date is based on current AMS techniques. The skull was also radiometrically dated to 170±70 BP (D. Hood, Beta Analytic, Inc., pers. commun.) The suggestion that this specimen might represent a domestic cow (Dixon 1993: 33) is incorrect. The largely intact skull with horn sheaths has been identified by Guthrie, and by McDonald (1981) as representing *Bison*.

[10] Alaska Polar Regions Dept., Elmer E. Rasmussen Library, Robert A. McKennan Collection, Series 2, Box 3, Folder 16.

[11]These linguistic issues were discussed with linguists Richard Mueller of Fairbanks and James Kari and Lillian Garnett, Alaska Native Language Center, University of Alaska Fairbanks *(pers.* comm.). Mueller translates *ch'itthay* to mean "it's flesh" or "it's meat," while *ik* refers to "shirt." Kari translates *ch'itthay* as a nondescript term referring to a game animal and the phrase *ch'itthay ik* as "game coat." Mueller suggests that perhaps the informant was referring to *ch'itthay dighan* which he translates as "humped game." As described above, Rev. Salmon applied a similar term to bison.

[12] We reviewed this information with T.E. Taylor in January 1999. He obtained many of the local place names in northern Alaska, as reflected on USGS topographic maps, through discussions with local residents in the Yukon, Tanana and Kuskokwim drainages during the 1950s. His notes are on file at USGS, National Mapping Division, Anchorage, AK.

ACKNOWLEDGEMENTS

We are grateful for the hospitality and encouragement of people on the Yukon Flats and in other Alaskan communities, who assisted in this study by providing information, ideas, and bison specimens. In particular we acknowledge Annie James, Earl Erick, Daniel Flitt, Steven Henry, Elliot Johnson Sr., Peter John, Kias Peter, Sr., Richard James, Rev. David Salmon, Mary Sam, Moses Sam, Virginia Titus, Joe Herbert, Fred Thomas, Ann Fisher, William Joseph, Howard Luke, Hyacinthe Andre, Gabe Andre and the late Moses Cruikshank, Sarah Malcolm, Steven Peter, Julia Tritt, Myra Roberts and Isaac Tritt, Sr., for sharing their historical knowledge of bison and Athabascan history. We are indebted to Art James Jr. of Fort Yukon for first bringing to our attention the existence of oral history relating to bison. Paul Williams Sr. of Beaver, Sam Roberts of Fort Yukon and Bill Goebel of Eagle provided valuable assistance in translating or facilitating interviews, and Craig Fleener of Fort Yukon contributed information from interviews with two elders. Virginia Titus of Fairbanks provided additional details based on her conversations with Julia Tritt and Elliot Johnson, and Elliot Johnson III. shared additional details based on earlier conversations with his grandfather. We also thank Tom O'Brien for his generous assistance in providing additional information based on his extensive collaboration with Rev. David Salmon, and William Schneider for his assistance in reviewing information provided by Moses Cruikshank. The manuscript benefited from critical reviews by Ernest E. Burch, Jr., Jack Brink, F. Stuart Chapin, C. Cormack Gates, Charles E. Holmes, David R. Klein, Ray Le Blanc, Paul Matheus, Phyllis Morrow, Tom O'Brien, Polly Wheeler, and several anonymous reviewers. We also acknowledge the assistance of personnel from Beta Analytic, Inc. who provided advice regarding the interpretation of radiometric data. Manfred Hoefs and Rick Farnell (Yukon Renewable Resources), Paul Matheus (Quaternary Center, University of Alaska Fairbanks), and Cydny Martin (National Park Service) generously contributed unpublished radiometric data for Holocene bison. Roland Gangloff, University of Alaska Museum, provided access to bison specimens. Tom Chowns contributed historical information from N.T. We are grateful for the assistance regarding place names provided by T.E. Taylor, and by James Kari, Lillian Garnett, Richard Mueller, Tom O'Brien, and Kathy Sikorski in clarifying linguistic issues. Ray LeBlanc provided unpublished information about his on-going study of the journals of Archdeacon Robert McDonald. We also thank Troy Ellsworth, Craig Fleener, Tom Seaton, Kalin Kellie and Beth Lenart for assistance in preparing illustrations and Gail Harington for assistance in compiling radiocarbon data from northwestern Canada.

This study was supported by the Alaska Department of Fish and Game, Alaska Federal Aid in Wildlife Restoration, the University of Alaska Fairbanks, a 1993 Geist Fund grant from the University of Alaska Museum, and US Fish and Wildlife Service Regional Archaeologist Chuck Diters, who provided logistic support for the 1993 archeological survey in the vicinity of Buffalo Mountain. We thank Knut Kielland, Steve Ebbert and Ron Standlee for their assistance in the archaeological survey.

REFERENCES CITED

Adams, L.G and B.W. Dale
1998 Reproductive Performance of Female Alaskan Caribou. *Journal of Wildlife Management* 62(4):1184-1195.

Ager, T.
1974 *Late Quaternary Environmental History of the Tanana Valley, Alaska.* Institute of Polar Studies Report No. 54. Ohio State University, Columbus.

Archipov, N.D.
1989. Drevnye Kul'tury Iakutii. zdatel,stvo,Yakutsk.

Armbruster, P. and R. Lande.
1993 A Population Viability Analysis for African Elephant (*Loxodonta africana*): How Big Should Reserves Be? *Conservation Biology* 7:602-610.

Balikci, A.
1963a *Vunta Kutchin Social Change: A Study of the People of Old Crow, Yukon Territory.* Northern Co-ordination and Research Centre, Department of Northern Affairs and National Resources. Ottawa.

1963b Family Organization of the Vunta Kutchin. *Arctic Anthropology* 1(2):62-69.

Bancroft, H.H.
1886 *History of the Pacific States.* The History Co. Publishers, San Fransisco.

Bartlein, P.J.
1988 Late-Tertiary and Quaternary Paleoenvironments. In *Vegetation History*, edited by B. Huntley and T. Webb III, pp. 113-152. Kluwer Academic Publishers, Boston.

Beaudry. M.C., ed.
1994 *Documentary Archaeology.* Cambridge University Press, Cambridge.

Beck, M.
1996 On Discerning the Cause of Late Pleistocene Megafaunal Extinctions. *Paleobiology* 22:91-103.

Belue, T.F.
1996 *The Long Hunt: Death of the Buffalo East of the Mississippi.* Stackpole Books, Mechanicsburg, Pennsylvania.

Berger, M., R.O. Stephenson, P. Karczmarczyk, and C.C. Gates
1995 *Habitat Inventory of the Yukon Flats as Potential Wood Bison Range.* Alaska Department of Fish and Game, Fairbanks.

Binford, L.R.
1978 *Nunamiut Ethnoarchaeology.* Academic Press, New York.

1981 *Bones; Ancient Men and Modern Myths.* Academic Press, London.

1983 *In Pursuit of the Past.* Thames and Hudson, London.

Birkedal, T.
1993 Ancient Hunters in the Alaska Wilderness: Human Predators and Their Role and Effect on Wildlife Populations and the Implications for Resource Management. In *Partners in Stewardship; Proceedings of the Seventh Conference on Research and Resource Management in Parks and on Public Lands*, edited by W.E. Brown and S.D. Veirs, pp. 228-234. The George Wright Society, Hancock.

Blitz, J.H.
1988 Adoption of the Bow in Prehistoric North America. *North American Archaeologist* 9:123-145.

Bork, A.M., C.M. Strobeck, F.C. Yeh, R.J. Hudson and R.K. Salmon
1991 Genetic Relationship of Wood Bison and Plains Bison Based on Restriction Fragment Length Polymorphisms. *Canadian Journal of Zoology* 69:43-48.

Broughton, J.M.
1997 Widening Diet Breadth, Declining Foraging Efficiency, and Prehistoric Harvest Pressure: Icthyofaunal Evidence from the Emeryville Shellmound, California. *Antiquity* 71:845-862.

Brown, J.H.
1978 The Theory of Insular Biogeography and the Distribution of Boreal Mammals and Birds. In *Intermountain Biogeography: A Symposium*, edited by K.T. Harper and J.L. Reveal, pp. 209-228. Great Basin Naturalist Memoirs 2.

1986 Two Decades of Interaction Between the MacArthur-Wilson Model and the Complexities of Mammalian Distributions. *Biological Journal of the Linnaean Society* 28:231-51.

Burch, E.S., Jr.
1972 The Caribou/Wild Reindeer as a Human Resource. *American Antiquity* 37(3):339-368.

1975 *Eskimo Kinsmen: Changing Family Relationships in Northwestern Alaska*. West Publishing Co., St. Paul.

1991 From Skeptic to Believer: The Making of an Oral Historian. *Alaska History* 6(1):1-16.

Burch, E.S. Jr. and C.W. Mishler
1995 The Di'haii Gwich'in: Mystery People of Northern Alaska. *Arctic Anthropology* 32(1):147-172.

Burke, A. and J. Cinq-Mars
1998 Paleoecological Reconstruction and Taphonomy of Equus lambei from the Bluefish Caves, Yukon Territory, Canada. *Arctic* 51(2):105-115.

Campbell, C., I.D. Campbell, C.B Blyth, and J.H. Mcandrews
1994 Bison Extirpation May Have Caused Aspen Expansion in Western Canada. *Ecography* 17(4): 360-362.

Campbell, J.M.
1978 Aboriginal Human Overkill of Game Populations: Examples from Interior Alaska. In *Archaeological Essays in Honor of Irving B Rouse*, edited by R.C. Dunnell and E.S. Hall, pp. 179-208. Mouton Publishers, New York.

Campisi, J.
1993 Final report on the Neets'aii Gwich'in Tribe of Alaska. United States District Court, Fairbanks, Alaska. Plaintiffs Exhibit 171 No. F86-075. 58 pp. plus appendices.

Carbyn, L.N., S.M. Oosenburg, and D.W. Anions
1993 *Wolves, Bison, and the Dynamics Related to the Peace-Athabasca Delta in Canada's Wood Buffalo National Park*. University of Alberta. Circumpolar Research Series 4.

Charlie, D. and D.W. Clark
1993 Łutthi Män and Tachän Män Hudé Hudä In Frenchman and Tatchun Lakes: Long People. Edited by R.M. Gotthardt, pp.1-26. Little Salmon-Carmacks First Nation, Carmacks, Yukon.

Chowns, T., C. Gates, and F. Lepine
1998 Large Scale Free Burning to Improve Wood Bison Habitat in Northern Canada. In *International Symposium on Bison Ecology and Management in North America*, edited by L. Irby and J. Knight, pp. 205-213. Montana State University, Bozeman.

Christopherson, R.J., R.J. Hudson and R.J. Richmond
1978 Comparative Winter Bioenergetics of American Bison, Yak, Scottish Highland and Hereford Calves. *Acta Theriologica* 23(2):49-54.

Christopherson, R.J., R.J. Hudson and M.K. Christopherson
1979 Seasonal Energy Expenditures and Thermoregulatory Responses of Bison and Cattle. *Canadian Journal of Animal Science* 59:611-617.

Cinq-Mars, J.
1974 Preliminary Archaeological Study, Mackenzie Corridor. (Second Report) *Northern Pipelines Taskforce on Northern Oil Development, Environment-Social Committee.* Report 74-11. Ottawa.

1991 Appendix I: NOGAP AMS Dates. In *NOGAP Archaeology Project: An Integrated Archaeological Research and Management Approach*, edited by J. Cinq-Mars and J.L. Pilon, pp. 149-159. Canadian Archaeological Association Occasional Paper No. 1.

Cinq-Mars, J., C.R. Harington, D.E. Nelson and R. MacNeish
1991 Engigstciak Revisited: A Note on Early Holocene AMS Dates from the "Buffalo Pit". In *NOGAP Archaeology Project: An Integrated Archaeological Research and Management Approach* edited by J. Cinq-Mars and J.L. Pilon, pp. 33-44. Canadian Archaeological Association Occasional Paper No. 1.

Clarke, C.H.D.
1945 *Biological Reconnaissance of the Lands Adjacent to the Alaska Highway in Northern British Columbia and the Yukon Territory*. Unpublished report in the National Archives of Canada, RG85 Collection, vol. 1191, file 400-2, parts 1-2.

Clark, D.W. and R.M. Gotthardt
1997 *Archaeological Settlement Survey at the Frenchman-Tatchun Lakes Near Carmacks, Yukon, Canada.* Unpublished report on file with Archaeological Survey of Canada, Hull, Quebec, and Yukon Heritage Branch, Whitehorse.

Coady, J.W.
1980 History of Moose in Northern Alaska and Adjacent Regions. *Canadian Field Naturalist* 94:61-68.

Cruikshank, J.
1981 Legend and Landscape: Convergence of Oral History and Scientific Traditions in the Yukon. *Arctic Anthropology* 17(2): 67-93.

Cruikshank, M.
1986 *The Life I've Been Living*. University of Alaska Press.

Dall, W. H.
1898 Travels on the Yukon and in the Yukon Territory in 1866-1868. In *The Yukon Territory, the Narrative of W.H. Dall, Leader of the Expedition to Alaska in 1866-1868: The Narrative of an Exploration Made in 1887 in the Yukon District by George M. Dawson*. Extracts from the Report of an Exploration Made in 1896-1897 by William Ogilvie, pp. 1-242. Downey, London.

D'Arrigo, R.D. and G. Jacoby
1992 Secular Trends in Northern High Latitude Temperature Reconstructions Based on Tree Rings. *Climatic Change* 25:163-177.

Dary, D.A.
1989 *The Buffalo Book: The Full Saga of the American Animal*. Swallow Press/Ohio University Press.

Dixon, E.J.
1993 *Quest for the First Americans*. University of New Mexico Press, Albuquerque.

Dixon, E.J., Jr., R.M. Thorson, and D.C. Plaskett
1985 Cave Deposits, Porcupine River, Alaska. *National Geographic Research Reports* 20, 1979, edited by Winfield Swanson, pp. 129-154. National Geographic Society, Washington, D.C.

Dobak, W.
1996 Killing the Canadian Buffalo. *Western Historical Quarterly* 27(1):33-52.

Donahue, P.F.
1974 Appendix D: Old Chief Faunal Analysis (1973). In *Preliminary Archaeological Study, Mackenzie Corridor*, by Jacques Cinq-Mars. (Second Report) Northern Pipelines Taskforce on Northern Oil Development, Environment-Social Committee Report 74-11, Ottawa.

DuBois, S.D. and R.O. Stephenson.
1998 Alaska's Delta Bison Herd: Managing Free-ranging Bison in an Area with Diverse Land Uses. In *International Symposium on Bison Ecology and Management in North America*, edited by L. Irby and J. Knight, pp. 211-213. Montana State University, Bozeman.

Dundes, A.
1965 *A Study of Folklore*. Prentice-Hall, New Jersey.

Dyke, A.S. and V.K. Prest
1987 Paleogeography of Northern North America 18,000-5,000 Years Ago. Geological Survey of Canada Map 1703A.

Fancy, S.G. and R.G. White.
1985 Energy Expenditures by Caribou While Cratering in Snow. *Journal of Wildlife Management* 49:987-993.

1987 Energy Expenditures for Locomotion by Barren Ground Caribou. *Canadian Journal of Zoology* 65:122-128.

Feit, H.A.
1987 North American Native Hunting and Management of Moose Populations. *Swedish Wildlife Research Supplement* 1:25-42.

Fisher, J.W. Jr. and T.E. Roll
1997 Ecological Relationships Between Bison and Native Americans During Late Prehistory and the Early Historic Period. In *International Symposium on Bison Ecology and Management in North America*, edited by L. Irby and J. Knight, pp. 283-302. Montana State University, Bozeman.

Fleener, C.
1998 *A Study of Recent Environmental Changes in the Yukon Flats, Alaska, Based on Oral History and Climate Records*. Senior Thesis. School of Agriculture and Land Resources Management. University of Alaska Fairbanks.

Flerov, K.K.
1979. Systematics and evolution. In *European bison. Morphology, systematics, evolution, ecology*. Edited by V.E. Sokolov, pp. 9-127. Nauka Publishers, Moscow.

Flores, D.
1991 Bison Ecology and Bison Diplomacy. *Journal of American History* 78:465-485.

1996 The Great Contraction: Bison and Indians in Northern Plains Environmental History. In *Legacy: New Perspectives on the Battle of the Little Bighorn*, edited by C.E. Rankin, pp. 3-22. Montana Historical Society Press, Helena.

Franklin, J.
1970 *Narrative of a Journey to the Shores of the Polar Sea in the Years 1819, 20, 21, and 22*. 1824 Reprint. Charles E. Tuttle, Rutland.

1971 *Narrative of a Second Expedition to the Shores of the Polar Sea in the Years 1825, 1826, and 1827*. 1828 Reprint. M.G. Hurtig, Edmonton.

Frison, G.C.
1991 *Prehistoric Hunters of the High Plains*. Academic Press, San Diego.

Frison, G.C. (ed.)
1996 *The Mill Iron Site*. University of New Mexico Press, Albuquerque.

Fuller, W.A.
1962 *The Biology and Management of the Wood Bison of Wood Buffalo National Park*. Canadian Wildlife Service Wildlife Management Bulletin Series 1(16).

Garraghan, G.J.
1940 *A Guide to Historical Method*, edited by Jean Delanglez. Greenwood Publishers, Connecticut.

Gasaway, W.C., R.D. Boertje, D.V. Grangaard, D.G. Kelleyhouse, R.O. Stephenson and D.G. Larsen.
1992 The Role of Predation in Limiting Moose at Low Densities in Alaska and Yukon and Implications for Conservation. *Wildlife Monographs* 120.

Gates, C.C.
1992 Cursory Evaluation of the Habitat Potential of the Yukon River Flats, Alaska, for a Reintroduction of Wood Bison. Unpublished Report. 7 pp.

Gates, C.C., T. Chowns, and H.W. Reynolds
1992 Wood Buffalo at the Crossroads. In *Buffalo*, edited by J. Foster, B. Harrison, and I.S. MacLaren, pp. 139-165. Alberta Nature and Culture Series, University of Alberta Press, Edmonton.

Gates, C.C. and N.C. Larter
1990 Growth and Dispersal of an Erupting Large Herbivore Population in Northern Canada: The Mackenzie Wood Bison (*Bison bison athabascae*). *Arctic* 43:231-238.

Geist, V.
1991 Phantom Subspecies: The Wood Bison *Bison bison "athabascae"*, Rhoads 1897, is not a Valid Taxon, but an Ecotype. *Arctic* 44:283-300.

Gerlach, S.C. and E.S. Hall
1988 The Later Prehistory of Northern Alaska: The View from Tukuto Lake. In *The Late Prehistoric Development of Alaska's Native People*, edited by R.D. Shaw, R.K. Harritt, and D.E. Dumond, pp. 107-137. Aurora, Alaska Anthropological Association Monograph Series, No. 4.

Gordon, B.C., and H. Savage
1973 Appendix C: Research Report. MjTp-1 The Whirl Lake Site, Northwest District of Mackenzie, Northwest Territories, Canada. In *Preliminary Archaeological Study, Mackenzie Corridor*, edited by Jacques Cinq-Mars, pp. c1-C43. (IAND Publication QS-1506-000-EE-Xa1) Northern Pipelines Taskforce on Northern Oil Development, Environment-Social Committee Report 73-10, Ottawa.

1974 Whirl Lake: A Stratified Indian Site Near the Mackenzie Delta. *Arctic* 27(3): 175-188.

Grayson, D.K.
1991 The Biogeographic History of Mammals in the Great Basin: Observations on the Last 20,000 years. *Journal of Mammalogy* 68:359-75.

Greenland Ice Core Project
1993 Climate Instability During the Last Interglacial Period Recorded in the GRIP Ice Core. *Nature* 364:203-207.

Greer, S.S.
1986 *Kusawa Lake Archaeology: Management, Research and Interpretation. Final Report of the 1985 Kusawa Lake Archaeology Project.* MS on file with Archaeological Survey of Canada, Canadian Museum of Civilization, Hull.

In prep. Pre-Contact Land Use Patterns in the Carcross-Tagish Area: Archaeological and Oral History Perspectives. Manuscript in Possession of the Author.

Grove, J.M.
1988 *The Little Ice Age*. Methuen and Company, London.

Gstalter, A. and P. Lazier
1996 *Le Bison d'Europe*. Traces/E&C éditions. Paris.

Gunn, A., R. Decker and T.W. Barry
1984 Possible Causes and Consequences of an Expanding Muskox Population, Queen Maud Gulf Area, Northwest Territories. In *Proceedings of the First International Muskox Symposium*, edited by D.R. Klein, R.G. White and S. Keller, pp. 41-46. Biological Papers of the University of Alaska, Special Report 4.

Guthrie, R.D.
1966 Bison Horn Cores: Character Choice and Systematics. *Journal of Paleontology* 40:328-340.

1968 Paleoecology of the Large-Mammal Community in Interior Alaska During the Late Pleistocene. *American Midland Naturalist* 79:346-363.

1980 Bison and Man in North America. *Canadian Journal of Anthropology* 1:55-73.

1982 Mammals of the Mammoth Steppe as Paleoenvironmental Indicators. In *Paleoecology of Beringia*, edited D. M. Hopkins, J.V. Matthews, Jr., C.E. Schweger, and S.B. Young, pp. 307-329. Academic Press, New York.

1985 Woolly Arguments Against the Mammoth Steppe: A New Look at the Palynological Data. *Quarterly Review of Archaeology* 6:9-16.

1990 *Frozen Fauna of the Mammoth Steppe: The Story of Blue Babe*. The University of Chicago Press, Chicago.

1995 Mammalian Evolution in Response to the Pleistocene-Holocene Transition and the Break-up of the Mammoth Steppe: Two Case Studies. *Acta Zool. Cracov*. 38(1): 139-154.

Hardisty, W. L.
1872 "The Loucheux Indians." *Annual Report of the Smithsonian Institution for 1866*. Pp. 311-30. Government Printing Office, Washington, D.C.

Harington, C.R.
1961 *History, Distribution and Ecology of the Muskoxen*. Unpublished M.Sc. thesis. McGill University.

1977 Pleistocene Mammals of the Yukon Territory. Unpublished PhD. dissertation, University of Alberta, Edmonton.

1978 Quaternary Vertebrate Faunas of Canada and Alaska and Their Suggested Chronological Sequence. *Syllogeus* 15:1-105.

1980a Radiocarbon Dates on Some Quaternary Mammals and Artifacts from Northern North America. *Arctic* 33:815-832.

1980b Pleistocene Mammals from Lost Chicken Creek, Alaska. *Canadian Journal of Earth Sciences* 17:168-198.
1984 Mammoths, Bison and Time in North America. In *Quaternary Dating Methods*, edited by W.C. Mahaney, pp. 299-309 Elsevier Science Publications, Amsterdam.

1989 Pleistocene Vertebrate Localities in the Yukon. In *Late Cenozoic History of the Interior Basins of Alaska and the Yukon*, edited by L.D. Carter, T.D. Hamilton and J.P. Galloway. U.S. Geological Survey Circular 1026: 93-98.

1990 Arctic Bison. *BIOME* (Canadian Museum of Nature) 10(2):4.

Harrington C.R. and J. Cinq-Mars
1995 Radiocarbon Dates on Saiga Antelope (*Saiga tatarica*) Fossils from Yukon and Northwest Territories. *Arctic* 48(1):1-7.

Haworth, L.
1989 *Holocene Glacial Chronologies of the Brooks Range, Alaska and Their Relationship to Climate Change.* Unpublished Ph.D. dissertation. State University of New York at Buffalo.

Haynes, T.
1997 Bison Hunting in the Yellowstone River Drainage, 1800-1884. In *International Symposium on Bison Ecology and Management in North America*, edited L. Irby and J. Knight, pp. 303-311. Montana State University, Bozeman.

Heimer, W.E., F.J. Mauer, and S.W. Watson Keller
1994 The Effects of Physical Geography on Dall Sheep Habitat Quality and Home Range Size. *Proceedings of the Biennial Symposium*, North American Wild Sheep and Goat Council 9:144-148.

Hollin, J.T. and D.N. Shilling.
1981 Late Wisconsin-Weichsalen Mountain Glaciers and Small Ice Caps. In *The Last Great Ice Sheets*, edited by G.H. Denton and T.J. Hughes, pp. 179-206. John Wiley & Sons, New York.

Holmes, C.E.
1979 *Report of Archaeological Reconnaissance: Withdrawal Areas, Fort Greeley, Alaska.* Report to the Alaska District, Corps of Engineers (DACA85-79-M-0001).

1996 Broken Mammoth. In *American Beginnings: The Prehistory and Paleoecology of Beringia*, edited by F. H. West, pp. 312-318. The University of Chicago Press, Chicago.

1998 *Archaeological Testing and Evaluation of the Gerstle River Quarry, East-Central Alaska, 1996.* Office of History and Archaeology Report Number 65, Division of Parks and Outdoor Recreation, Alaska Department of Natural Resources.

Holmes, C.E., and G. Bacon
1982 *Holocene Bison in Central Alaska: A Possible Explanation for Technological Conservatism.* Paper Presented at the 9th Annual Meeting of the Alaska Anthropological Association, 2-3 Apr 1982, Fairbanks.

Hone, E.
1934 *The Present Status of Muskox in Arctic North America and Greenland.* American Committee for International Wildlife Protection Special Publication No. 5.

Ice Core Working Group
1998 *Ice Core Contributions to Global Change Research; Past Successes and Future Directions.* National Ice Core Laboratory, Science Management Office, University of New Hampshire.

Isenberg, A.C.
2000 *The Destruction of the Bison: An Environmental History, 1750-1920.* Cambridge University Press, Cambridge.

Jacoby, G. and R.D. D'Arrigo
1989 Reconstructed Northern Hemisphere Annual Temperature Since 1671 Based on High Latitude Tree-ring Data from North America. *Climatic Change* 14:39-59.

Janetski, J.C.
1997 Fremont Hunting and Resource Intensification in the Eastern Great Basin. *Journal of Archaeological Science* 24:1075-1088.

Jeckell, G.A.
1933 Letter and Article to M.T. Stephens. In Yukon Archives Collection, YRG-1, series 3, file 12-13A, part 1, Sept. 8, 1933.

Jennings, J.D.
1968 *Prehistory of North America.* McGraw-Hill, New York.

Joly, D.O., F.A. Leighton and F. Messier
1998 Tuberculosis and Brucellosis Infection of Bison in Wood Buffalo National Park, Canada: Preliminary Results. In *International Symposium on Bison Ecology and Management in North America*, edited L. Irby and J. Knight, pp. 23-31. Montana State University, Bozeman.

Jones, S.
1872 *The Kutchin Tribes. Notes on the Tinneh or Chipewyan Indians of British and Russian America.* Annual Report of the Smithsonian Institute for the Year 1866:320-327, Washington.

Juday, G.P., R.A. Ott, D.W. Valentine, and V.A. Barber.
1998 *Forests, Climate Stress, Insects and Fire. In Implications of Global Change in Alaska and the Bering Sea Region.* Proceedings of a Workshop, 1997:23-49. Center for Global Change and Arctic System Research, University of Alaska Fairbanks.

Kay, C.E.
1994a Aboriginal Overkill: The Role of Native Americans in Structuring Western Ecosystems. *Human Nature* 5:359-398.

1994b The Impact of Native Ungulates and Beaver on Riparian Communities in the Intermountain West. *Natural Resources and Environmental Issues* 1:23-44.

1995 Aboriginal Overkill and Native Burning. Implications for Modern Ecosystem Management. *Western Journal of Applied Forestry* 10:121-126.

1996 Ecosystems Then and Now: A Historical-Ecological Approach to Ecosystem Management. In *Proceedings for the Fourth Prairie Conservation and Endangered Species Workshop*, edited by W.D. Williams and J.F. Dormaar, pp 79-87. Natural History Occasional Paper 23, Provincial Museum of Alberta.

1997 Aboriginal Overkill and the Biogeography of Moose in Western North America. *Alces* 33:141-164.

1998 Are Ecosystems Structured from the Top-down or Bottom-up: A New Look at an Old Debate. *Wildlife Society Bulletin* 26(3):484-498.

Kelsall, J.P.
1969 Structural Adaptations of Moose and Deer for Snow. *Journal of Mammalogy* 50:302-310.

Kirkby, W.
1865 *A Journey to the Youcan, Russian America*. Annual Report of the Smithsonian Institution for the Year 1864, pp. 416-420. Washington.

Komers, P. E., F Messier, and C.C. Gates.
1992 Search or Relax: The Case of Bachelor Wood Bison. *Behavioral Ecology and Sociobiology*. 31(3):195-203.

Komers, P.E., F. Messier and C.C. Gates.
1993 Group Structure in Wood Bison: Nutritional and Reproductive Determinants. *Canadian Journal of Zoology* 71:1367-1371.

Komers, P.E., F. Messier, P.F. Flood, and C.C. Gates
1994 Reproductive Behavior of Male Wood Bison in Relation to Progesterone Level in Females. *Journal of Mammalogy* 75(3):757-765.

Krech, S., III
1976 The Eastern Kutchin and the Fur Trade, 1800-1860. *Ethnohistory* 23(3):213-235.

1978 On the Aboriginal Population of the Kutchin. *Arctic Anthropology* 15(1):89-104.

1979 The Nakotcho Kutchin: A Tenth Aboriginal Kutchin Band? *Journal of Anthropological Research* 35(1):109-121.

1981 "Throwing Bad Medicine": Sorcery, Disease, and the Fur Trade Among the Kutchin and Other Northern Athabaskans. In *Indians, Animals, and the Fur Trade*, edited by Shepard Krech III, pp. 73-108. The University of Georgia Press, Athens.

1987 The Early Fur Trade in the Northwestern Subarctic: The Kutchin and the Trade in Beads. In *Le Castor Fait Tout*, edited by Bruce G. Trigger, Toby Morant, and Louise Dechene, pp. 236-77. Lake St. Louis Historical Society, Montreal.

Kreutz, K.J., P.A. Mayewski, L.D. Meeker, M.S. Twickler, S.I. Whitlow, and I.I. Pittalwala.
1997 Bipolar Changes in Atmospheric Circulation During the Little Ice Age. *Science*. 277(29):1294-1296.

Lamb, H.H.
1995 *Climate History and the Modern World*. Routledge.

Lande, R.
1988 Genetics and Demography in Biological Conservation. *Science* 241:1455-1460.

Larsen, H.
1968 Trail Creek, Final Report on the Excavation of Two Caves at Seward Peninsula, Alaska. *Acta Arctica*. 15:7-79.

Larter, N.C. and C.C. Gates
1990 Home Ranges of Wood Bison in an Expanding Population. *Journal of Mammalogy* 71(4):604-607.

1991 Diet and Habitat Selection of Wood Bison in Relation to Seasonal Changes in Forage Quantity and Quality. *Canadian Journal of Zoology* 69:2677-2685.

1994 Home-range Size of Wood Bison: Effects of Age, Sex, and Forage Availability. *Journal of Mammology* 75(1):142-149.

Lazarev P.A., Boeskorov G.G., Tomskaya A.I.
1998 *Mlekopitauschie Antropogena Yakutii*. Edited by Yu.V. Labutin. Sibirskoe otdelenie Rossiiskoi Akademii Nauk. Yakutskii Nauchnyi Centr, Yakutsk.

Le Blanc, R.J.
1984 *The Rat Indian Site and the Late Prehistoric Period in the Interior Northern Yukon*. National Museum of Man Mercury Series, Archaeological Survey of Canada, Paper No. 120. Ottawa.

1988 *Archaeological Research in the Mackenzie Delta Region, N.W.T., Archaeological permit 87-617*. Report on file, Archaeological Survey of Canada, Canadian Museum of Civilization. Hull, Quebec.

1991 New Data Relating to the Prehistory of the Mackenzie Delta Region of the NOGAP Study Area. In *NOGAP Archaeology Project: An Integrated Archaeological Research and Management Approach*, edited by J. Cinq-Mars and J.L. Pilon, pp. 65-76. Canadian Archaeological Association Occasional Paper 1.

Leechman, D.
1954 *The Vanta Kutchin*. National Museum of Canada Bulletin 130. Anthropological Series 33. Department of Northern Affairs and National Resources, Ottawa.

Lent, P.
1998 Alaska's Indigenous Muskoxen: A History. *Rangifer* 18:133-144.

LeResche, R.E., R.H. Bishop and J. Coady
1974 Distribution and Habitat of Moose in Alaska. Le *Naturaliste Canadian* 101:143-178.
Libby, D., and L. Medlock

1979 *An Analysis of Lithic Debris from Old John Lake in Northeastern Alaska*. Ms. on file, Bureau of Indian Affairs, ANCSA Division, Anchorage.

Lotenberg, G.
1996 *History of Wood Bison in the Yukon: A Reevaluation Based on Traditional Knowledge and Written Records*. Report Submitted to Yukon Renewable Resources Department, Whitehorse.

Loy, T. and E.J. Dixon
1998 Blood Residues on Fluted Points from Eastern Beringia. *American Antiquity* 63(1):21-46.

Luke, H.
1998 My Own Trail. Edited by J. Steinbright Jackson. Alaska Native Knowledge Network. Fairbanks, Alaska.

Lyman, R.L.
1994 *Vertebrate Taphonomy*. University of Cambridge Press, Cambridge.

1996 Applied Zooarchaeology: The Relevance of Faunal Analysis to Wildlife Management. *World Archaeology* 28(1):110-125.

1998 *White Goats, White Lies: The Abuse of Science in Olympic National Park*. The University of Utah Press, Salt Lake City.

MacArthur, R.H. and E.O. Wilson
1967 *The Theory of Island Biogeography*. Princeton University Press, Princeton.

Mackenzie, A.
1801 *Voyages from Montreal, on the River St. Lawrence, through the Continent of North America, to the Frozen and Pacific Oceans; in the Years 1789 and 1793, with a Preliminary Account of the Rise, Progress, and Present State of the Fur Trade of the Country*. T. Cadell and W. Davies, London. (Reprinted: University Microfilms, Ann Arbor, Michigan, 1966).

MacNeish, R.S.
1964 *Investigations in Southwest Yukon: Archaeological Investigations, Comparisons and Speculations*. Papers of the Robert S. Peabody Foundation for Archaeology 6(2):xiii, 199-488. Phillips Academy, Andover.

Martin, P. S. and R. G. Klein, editors.
1984 *Quaternary Extinctions: A Prehistoric Revolution*. University of Arizona Press, Tucson.

Martin, P. and H. Wright
1989 *Pleistocene Extinctions: The Search for a Cause*. Yale University Press, New Haven.

Martin, P.S. and C.R. Szuter
1999 War Zones and Game Sinks in Lewis and Clark's West. *Conservation Biology* 13(1): 36-45.

Matheus, P.E., W. Payton, and M. Kunz
1999 *Isotope Ecology of Late Quaternary Mammals in Eastern Beringia*. Paper Presented at the 32nd Annual meeting of the Canadian Archaeological Association, Whitehorse, Yukon. April 28-May 2.

Matthews. J.V., Jr.
1982 East Beringia During Late Wisconsin Time: A Review of the Biotic Evidence. *In Paleoecology of Beringia*, edited by D.M. Hopkins, J.V. Matthews, Jr., C.E. Schweger, and S.B. Young, pp 127-150. Academic Press, New York.

McClellan, C.
1975 *My Old People Say: An Ethnographic Survey of Southern Yukon Territory*. Canada, National Museum of Man, Publications in Ethnology 6. Ottawa.

1981 Inland Tlingit. In *Handbook of North American Indians. Vol. 6. Subarctic*, edited by W.C. Sturtevant, pp. 469-480. Smithsonian Institution, Washington.

McDonald, J. N.
1981 *North American Bison, Their Classification and Evolution*. University of California Press, Berkely.

McGhee, R.
1996 *Ancient People of the Arctic*. University of British Columbia Press, Vancouver.

McKennan, R.A.
1965 *The Chandalar Kutchin*. Arctic Institute of North America Technical Paper 17, Montreal.

Meagher, M.
1973 *The Bison of Yellowstone National Park*. National Park Service Scientific Monograph. No. 1. USGPO. Washington, D.C.

1976 Winter Weather as a Population-Regulating Influence on Free-Ranging Bison in Yellowstone National Park. In *Research in the Parks. Trans. National Park Centennial Symposium*, pp. 29-39. American Association for the Advancement of Science. Dec. 28-29, 1971. Ser. No. 1. USGPO, Wash. D.C.

Melton, D.A., N.C. Larter, C.C. Gates and J.A. Virgl
1989 The Influence of Rut and Environmental Factors on the Behaviour of Wood Bison. *Acta Theriologica* 34(12):179-193.

Mills. R.O.
1993 *Geist Fund Narrative and Financial Report. Kutchin Athabaskan Protohistoric Bison Procurement.* Geist Fund Narrative and Financial Report on file at University of Alaska Museum.

Miquelle, D.G.
1985 *Food Habits and Range Conditions of Bison and Sympatric Ungulates on the Upper Chitina River, Wrangell-St. Elias National Park and Preserve.* National Park Service, Alaska Region Research/Resources Management Report AR-8.

Mishler, C., editor
1995 *Neerihiinjik: We Traveled from Place to Place, the Gwich'in Stories of Johnny and Sarah Frank.* Alaska Native Language Center, University of Alaska Fairbanks, Fairbanks.

Mooring, M.S. and W.M. Samuel.
1998 Tick Defense Strategies in Bison: The Role of Grooming and Hair Coat. Behaviour 69:1-26.

Morgan, R.G.
1997 The Destruction of the Northern Bison Herds. In *International Symposium on Bison Ecology and Management in North America*, edited L. Irby and J. Knight, pp. 312-325. Montana State University, Bozeman.

Morlan, R. E.
1972a *NbVk-1: An Historic Fishing Camp in Old Crow Flats, Northern Yukon.* National Museum of Man, Mercury Series, Archaeological Survey of Canada Paper 5, Ottawa.

1972b *The Cadzow Lake Site (MjVi-1): A Multicomponent Historic Kutchin Camp.* National Museum of Man, Mercury Series, Archaeological Survey of Canada Paper 3, Ottawa.

1973 *The Later Prehistory of the Middle Porcupine Drainage, Northern Yukon Territory.* National Museum of Man, Mercury Series, Archaeological Survey of Canada Paper 11, Ottawa.

1999 Canadian Archaeological Radiocarbon Database. www.canadianarchaeology.com/radiocarbon/card/card.htm

Morrison, D.A.
1997 *Caribou Hunters in the Western Arctic: Zooarchaeology of the Rita-Claire and Bison Skull Sites.* Canadian Museum of Civilization, Mercury Series, Archaeological Survey of Canada Paper 157, Ottawa.

Murray, A.H.
1910 *Journal of the Yukon, 1847-1848, by Alexander Hunter Murray*, edited by Lawrence J. Burpee. Publication of the Canadian Archives- No. 4. Government Printing Bureau, Ottawa.

National Oceanic and Atmospheric Administration.
1986 *Climatological Data Annual Summary, 1986.* Vol. 72. No. 13. National Climatic Data Center, Asheville.

Nelson, R.K.
1973 *Hunters of the Northern Forest: Designs for Survival Among the Alaskan Kutchin.* University of Chicago Press, Chicago.

Newmark, W.D.
1987 A Land-bridge Perspective on Mammalian Extinctions in Western North American Parks. *Nature* 325:430-432.

O'Brien, S.R., P.A. Mayewski, L.D. Meeker, D.A. Meese, M.S. Twickler and S.I. Whitlow
1995 Complexity of Holocene Climate as Reconstructed from a Greenland Ice Core. *Science* 270:1962-1964.

O'Brien, T.
1997 *Athabaskan Implements from the Skin House Days as Related by Reverend David Salmon.* Unpublished M.A. thesis, University of Alaska Fairbanks.

n.d. *The Life Story of David Salmon.* Unpublished manuscript in possession of the author.

Orth, D.J.
1971 *Dictionary of Alaska Place Names.* Geological Survey Professional Paper 567. U.S. Government Printing Office.

Osgood, C.
1936 *Contributions to the Ethnography of the Kutchin.* Yale University Publications in Anthropology 14, Yale University Press, New Haven.

Overpeck, J., K. Hughen, D, Hardy, R. Bradley, R. Case, M. Douglas, B. Finney, K, Gajewski, G. Jacoby, A. Jennings, S. Lamoureux, A. Lasca, G. MacDonald, J. Moore, M, Retelle, S. Smith, A. Wolfe, and G. Zielinski
1997 Arctic Environmental Change of the Last Four Centuries. *Science* 278:251-1258.

Peden, D.G. and G.J. Kraay.
1979 Comparison of Blood Characteristics in Plains Bison, Wood Bison, and Their Hybrids. *Canadian Journal of Zoology* 59(9):1778-1784.

Peek, J.M., D.G. Miquelle, and R.G. Wright
1987 Are Bison Exotic in the Wrangell-St. Elias National Park and Preserve? *Environmental Management* 11(2): 149-153.

Peters, H.F. and S.B. Slen
1964 Hair Coat Characteristics of Bison, Domestic x Bison Hybrids, Cattalo, and Certain Domestic Breeds of Beef Cattle. *Canadian Journal of Animal Science* 44:48-57.

Peterson, R.O. and D.L. Allen
1974 Snow Conditions as a Parameter in Moose-Wolf Relationships. *Le Naturaliste Canadian* 101:481-492.

Petitot, E.
1970 The Amerindians of the Canadian North-West in the 19th Century, as seen by Emile Petitot. *Volume 2, The Loucheux Indians*, edited by Donat Savoie. Northern Science Research Group, Department of Indian Affairs and Northern Development, Ottawa.

Péwé, T.L.
1975a *Quaternary Geology of Alaska*. US Geological Survey Professional Paper 835: 1-145.

1975b *Quaternary Stratigraphy and Nomenclature in Unglaciated Central Alaska*. US Geological Survey Professional Paper 862:1-32.

Pielou, E.C.
1991 *After the Ice Age*. University of Chicago Press. Chicago.

Powers, R., R.D. Guthrie, and J.F. Hoffecker
1983 *Dry Creek: Archaeology and Paleoecology of a Pleistocene Alaskan Hunting Camp*. Report Submitted to the National Park Service, Contract CX-9000-7-0047.

Reed, I.M.
1946 *Appendix C to a letter to The Hon. Julius A. Krug, Secretary of the U.S. Department of Interior, November 16, 1946*. Stefansson Collection, Dartmouth College Library.

Reppening, C.A., D.M. Hopkins, and M. Rubin
1964 Tundra Rodents in a Late Pleistocene Fauna from the Tofty Placer District, Central Alaska. *Arctic* 17(3):177-197.

Reynolds, H.W., and C.C. Gates
1991 Managing Wood Bison: A Once-Endangered Species. In *Wildlife Production: Conservation and Sustainable Development*, edited by L.A. Renecker and R.J. Hudson, pp. 363-371. AFES Miscellancous Publication 1-6. University of Alaska Fairbanks.

Reynolds, H.W., R.D. Glaholt and A.W.L. Hawley
1982a Bison. In *Wild Mammals of North America: Biology, Management, Economics*, edited by J.A. Chapman and G.A. Fledhamer, pp. 979-1007. Johns Hopkins University Press, Baltimore.

Reynolds, H.W., R.M. Hansen and D.G. Peden
1978 Diets of the Slave River Lowlands bison herd, Northwest Territories, Canada. *Journal of Wildlife Management* 42:581-590.

Reynolds, H.W. and A.W.L. Hawley, eds.
1987 *Bison Ecology in Relation to Agricultural Development in the Slave River Lowlands, NWT*. Canadian Wildlife Service Occassional Paper No. 63. Occasional Paper No. 63.

Reynolds, H.W., J.R. McGillis and R.D. Glaholt
1982b *Range Assessment of the Nisling River Valley, Yukon Territory as Habitat for Wood Bison*. Unpublished Canadian Wildlife Service Report, Edmonton.

Reynolds, H.W. and D.G. Peden
1987 Vegetation, Bison Diets, and Snow Cover. In *Bison Ecology in Relation to Agricultural Development in the Slave River Lowlands, N.W.T.* edited by H.W. Reynolds and A.W.L. Hawley, pp. 39-44. Canadian Wildlife Service Occasional Paper No. 63.

Reynolds, P.E.
1998 Dynamics and Range Expansion of a Restablished Muskox Population. *Journal of Wildlife Management* 62(2):732-744.

Richardson, Sir J.
1851 *Arctic Searching Expedition: A Journal of a Boat-voyage Through Rupert's Land and the Arctic Sea, in Search of the Ships Under Command of Sir John Franklin with an Appendix on the Physical Geography of North America*. Longman, Brown, Green, and Longmans, London.

Roe, F.G.
1951 *The North American Buffalo: A Critical Study of the Species in its Wild State*. University of Toronto Press, Toronto.

Roseneau, D., P. Stern, and C Warbelow
1975 *The Kutchin Caribou Fences of Northeastern Alaska and the Northern Yukon*. Arctic Gas Biological Report Series, Volume 32. Canadian Arctic Gas Study, Limited and Alaskan Arctic Gas Study Company.

Rusanov, B.S.
1975 Iskapaemye bizony Iakutii. Iakutskoe Knizhnoe Izdatel'stvo, Yakutsk.

Russell, F.
1898 *Explorations in the Far North*. State University of Iowa Press.

Schneider, W.S.
1986. *On the Back Slough. In Interior Alaska: A Journey Through Time*, edited by J.S. Aigner, R.D. Guthrie, M.L. Guthrie, R.K. Nelson, W.S. Schneider, and R.M. Thorson, pp. 147-194. Alaska Geographic Society, Anchorage.

Sidney, A., K. Smith and R. Dawson.
1977 *My Stories are My Wealth: As Told to Julie Cruikshank*. Council for Yukon Indians. Willow Printers, Whitehorse, Yukon Territory.

Simms, S.R.
1984 *Aboriginal Great Basin Foraging Strategies: An Evolutionary Analysis*. Unpublished PhD. Dissertation. University of Utah, Salt Lake City.

Skinner, M.F., and O.C. Kaisen
1947 The Fossil Bison of Alaska and Preliminary Revision of the Genus. *Bulletin of the American Museum of Natural History* 89:126-256.

Slobodin, R.
1962 *Band Organization of the Peel River Kutchin*. National Museum of Canada Bulletin 179. Anthropological Series 55.

Department of Northern Affairs and National Resources, Ottawa.

1981 Kutchin. In *Handbook of North American Indians, Volume 6 Subarctic*, edited by June Helm, pp. 514-532. Smithsonian Institute, Washington.

Smith, E.A.
1983 Anthropological Applications of Optimal Foraging Theory: A Critical Review. *Current Anthropology* 24:625-651.

Smith, E.A. and B. Winterhalder, eds.
1992 *Evolutionary Ecology and Human Behavior*. Aldine de Gruyter, New York.

Soper, J.D.
1941 *History, Range and Home Life of the Northern Bison*. Ecological Monographs 11: 347-412.

Speth, J.D.
1983 *Bison Kills and Bone Counts*. University of Chicago Press, Chicago.

Stefansson, V.
1921 *The Friendly Arctic*. Macmillan and Co., New York.

Stephens, D.W. and J.R. Krebs
1986 *Foraging Theory*. Princeton University Press, Princeton.

Stevenson, M.G.
1996 Indigenous Knowledge and Environmental Assessment. *Arctic* 49(3):278-291.

Stiner, M.C.
1990 The Use of Mortality Patterns in Archaeological Studies of Human Predatory Adaptations. *Journal of Anthropological Archaeology* 9:205-251.

Strobeck, C., R.O. Polziehn and R. Beech
1993 Genetic Relationship Between Wood and Plains Bison Assayed Using Mitochondrial DNA Sequence. In *Proceedings, North American Public Bison Herds Symposium*, July 27-29, edited by R.E. Walker, pp. 209-221. Custer State Park, Lacrosse.

Taylor, R.J.
1984 *Predation*. Chapman and Hall, New York.

Tener, J.S.
1965 *Muskoxen in Canada: A Biological and Taxonomic Review*. Queen's Printer, Ottawa.

Tudge, C.
1997 *The Time Before History: 5 Million Years of Human Impact*. Touchstone, New York.

Vansina, J.
1986 *Oral Tradition as History*. The University of Wisconsin Press. Madison.

Vanstone, J.W.
1974 *Athabascan Adaptations: Hunters and Fisherman of the Subarctic Forests*. Chicago; Aldine.
van Zyll de Jong, C.G.

1986 *A Systematic Study of Recent Bison, with Particular Consideration of the Wood Bison. (Bison Athabascae Rhoads* 1898). Publications in Natural Sciences 6:69. National Museums of Canada, Ottawa

1993 Origin and Geographic Variation of Recent North American Bison. *Alberta* 3:21-35.

van Zyll de Jong, C.G., C.C. Gates, H. Reynolds, and W. Olson
1995 Phenotypic Variation in Remnant Populations of North American Bison. *Journal of Mammalogy* 76:391-405.

Vereschagin, N.K. and G.F. Baryshnikov
1982 Paleoecology of the Mammoth Fauna in the Eurasian Arctic. In *Paleoecology of Beringia*, edited by D.M. Hopkins, J.V. Matthews, Jr., C.E. Schweger, and S.B. Young, pp. 267-280. Academic Press, New York.

Ward, P.D.
1997 *The Call of Distant Mammoths: Why the Ice Age Mammals Disappeared*. Copernicus, Springer-Verlag New York.

Wenzel, G.W.
1999 Traditional Ecological Knowledge and Inuit: Reflections on TEK Research and Ethics. *Arctic* 52:113-124.

West, F.H.
1963 *The Netsi Kutchin: An Essay in Human Ecology*. Unpublished PhD dissertation, Department of Anthropology, Louisiana State University, Baton Rouge.

1982 *The Archaeology of Beringia*. Columbia University Press, New York.

Whitman, J.S. and R.O. Stephenson.
1998 History and Management of the Farewell Bison Herd, Alaska. In *International Symposium on Bison Ecology and Management in North America*. Edited by L. Irby and J. Knight, pp. 267-270. Montana State University, Bozeman.

Whitney-Smith, E.
1998 Late Pleistocene Extinctions by Second Order Overkill, or Why the Bison Made It! In *International Symposium on Bison Ecology and Management*, edited by. L. Irby and J. Knight, pp. 337-345. Montana State University, Bozeman, Montana.

Whymper, F.
1868 A Journey from Norton Sound, Bering Sea, to Fort Yukon (Junction of Porcupine and Yukon Rivers). *Journal of the Royal Geographical Society* 38:219-237.

Will, R.T.
1984 *Muskox Procurement and Use on Banks Island by Nineteenth Century Copper Inuit*. Biological Papers of the University of Alaska. Special Report No. 4:153-161.

Wilson, A.C., R.L. Cann, S.M. Carr, M. George, U.B. Gyllensten, K.M. Helm-Gychowski, R.G. Higuchi, S.R. Palumbi, E.M. Prager, R.D. Sage and M. Stoneking.
1985 Mitochondrial DNA and Two Perspectives on Evolutionary Genetics. *Biological Journal of the Linnaean Society* 26: 375-400.

Wilson, G.A. and C. Strobeck.
1999 *Genetic Variation Within and Relatedness Among Wood and Plains Bison Populations*. Genome 42:483-496.

Wilson, M.C.
1978 Late Pleistocene and Holocene Bison on the Northern Great Plains. Abstracts of the Fifth Biennial Meeting of the American Quaternary Association, Edmonton, Alberta, September 2-4, p. 97.

1996 Late Quaternary Vertebrates and the Opening of the Ice-Free Corridor, with Special Reference to the Genus Bison. Quaternary International 32:97-105.

Winterhalder, B.
1981a Optimal Foraging Strategies and Hunter-Gatherer Research in Anthropology: Theory and Models. In *Hunter-Gatherer Foraging Strategies*, edited by B. Winterhalder and E.A. Smith, pp. 13-35. University of Chicago Press, Chicago.

1981b Foraging Strategies in the Boreal Forest: An Analysis of Cree Hunting and Gathering. In *Hunter-Gatherer Foraging Strategies*, edited by B. Winterhalder and E.A. Smith pp. 66-98. University of Chicago Press, Chicago.

Wood Bison Recovery Team
1998 *Draft National Recovery Plan for the Wood Bison*. Report Prepared for the Committee for the Recovery of Nationally Endangered Wildlife. Ottawa, Canada.

Workman,. W.B.
1978 *Prehistory of the Aishikik Kluane Area, Southwest Yukon Territory*. National Museum of Man Mercury Series, Archaeological Survey of Canada, Paper No. 74. Ottawa.

Yesner, D.R.
1989 Moose Hunters of the Boreal Forest? A Reexamination of Subsistence Patterns in the Western Subarctic. *Arctic* 42(2):97-108.

1994 Seasonality and Resource "Stress" Among Hunter-Gatherers: Archaeological Signatures. In *Key Issues in Hunter-Gatherer Research,* edited by E. S. Burch, Jr. and L. J. Ellana, pp. 151-167. Berg, New York.

Zamora, L.E.
1983 *An Analysis of Bison Erythrocyte Antigens and Blood Proteins*. Unpublished M.Sc. thesis. Texas A&M University, College Station.

Zimov. S.A., V.I. Chuprynin, A.P. Oreshko, F.S. Chapin III, J.F. Reynolds, and M.C. Chapin
1995 Steppe-Tundra Transition: A Herbivore- Driven Biome Shift at the End of the Pleistocene. *The American Naturalist*. 146(5): 765-794.

THE NORTH POINT WET SITE AND THE SUBSISTENCE IMPORTANCE OF PACIFIC COD ON THE NORTHERN NORTHWEST COAST

Peter M. Bowers
Northern Land Use Research, Inc., Fairbanks, Alaska

Madonna L. Moss
Department of Anthropology, University of Oregon

INTRODUCTION

Although the role of salmon continues to attract considerable attention among archaeologists studying the prehistory of the southern coastal regions of Alaska and especially the Northwest Coast (see Cannon, this volume), the role of Pacific cod has received comparatively little notice. The seasonal importance of Pacific cod in the subsistence of prehistoric residents of southeast Alaska was proposed in the analysis of the Hidden Falls faunal assemblage (Moss 1989a), and cod are abundant in other Alaskan and Northwest Coast assemblages (Ackerman *et al.* 1985; Croes 1995; Souders 1997). Cod may have been heavily exploited at times when other resources, especially salmon, were limited in abundance due to seasonal shortfalls, difficulties with storage, or periods of climatically-induced deterioration of key salmon spawning and migration areas. Cod may also have been chosen simply because of a desire for fresh food as opposed to stored food, or because of personal preferences. The current southern limit of this species suggests that Pacific cod was not a significant subsistence factor south of the Columbia River, although this species is adapted to a fairly narrow sea temperature range. The distribution and use of cod by people has no doubt changed over the course of prehistory.

In this paper, we describe the North Point site (49-SUM-25), and provide an analysis of the faunal assemblage. The assemblage is unique in its depositional setting within the intertidal zone, and in its clear association with both late Holocene microblade and ground slate industries on the northern Northwest Coast. The North Point archeofauna is dominated by Pacific cod. This is not unique, however, and the significance of this species, especially as a pivotal resource during late winter-early spring and possibly as a key resource during periods of Neoglacial cooling, has yet to be adequately evaluated across the North Pacific region.

SITE BACKGROUND

In 1994-95, survey and excavations were carried out in the vicinity of Port Houghton and Cape Fanshaw, southeast Alaska (Figure 1). A more detailed review of the environmental setting and survey in the Port Houghton area is provided in Bowers *et al.* (1995a, 1995b, 1996), and in the Environmental Impact Statement for the proposed Port Houghton/Cape Fanshaw timber sale (Parametrix 1998)[1].

As a result of this research, Component I of the North Point site was discovered and test excavated. This intertidal/wet site (Figure 2) was discovered in 1994 as part of a survey of a proposed log transfer facility located about 400 meters to the south of a previously-known historic site (Sealaska 1975; Stanford and Lightfoot 1981). Component II of the North Point Site consists of a historic moss-covered cabin ruin in the uplands, historic artifacts in the intertidal zone and in upland test pits, a stone-lined cache pit, and culturally modified trees, all dating to the late 19th or early 20th centuries (Bowers *et al.* 1996). The historic component is not discussed further in this paper.

During two seasons of fieldwork, twenty-five 50 x 50 cm test pits were excavated: six in the vegetated uplands, and nineteen in the intertidal zone (Figure 2). Seven 1 x 1 m test pits were placed in the mid-to-lower intertidal zone, and one in the uplands. Dozens of soil probes were also placed in the intertidal zone and uplands. The total area of the site tested is 14.25 m^2, representing less than 1% of the estimated 50 x 80 m site area comprising Component I.

The North Point site is about 35 km northeast of the north end of Kupreanof Island, 25 km east of Admiralty Island, and about 64 km north of Petersburg, Alaska. The study area lies on the Alaskan mainland, at the juncture of Stephens Passage and Frederick Sound (Figure 1). 49-SUM-25 is located within a small cove on the south shore of Port Houghton at the tip of a broad peninsula forming the narrow entrance into the North Arm of Port Houghton (Figures 1 and 2). Offshore, a discontinuous line of partly submerged bedrock lies at a right angle to the narrow beach, offering a natural breakwater and partial protection to the small beach from winds and tides (Figures 1 and 2). Less than 50 m inland, the land rises steeply and becomes more densely vegetated. Other than a minor ephemeral drainage flowing out of the forest from the south, there is no stream or other freshwater source in the site vicinity today. The area surrounding the North Point site is comprised of saltwater shorelines, islands, forested mountains, river valleys, a salt chuck, and estuaries. Elevations range from sea level to summits reaching over 1,524 m in the eastern portion of the study area.

A variety of maritime and terrestrial natural resources exist in the area today. Five species of salmon are seasonally abundant, as are other anadromous fish and Pacific herring, halibut, flounder, sablefish, Pacific cod and other bottom fish. Sea mammals such as harbor seals and Steller sea lion exist in the area, as well as occasional humpback whales. Marine invertebrates are found in Sandborn Canal, and at Steamboat Bay near the entrance to Port Houghton. Seasonal use of the Port Houghton Salt Chuck, Sandborn Canal, and the North Arm of Farragut Bay by waterfowl for nesting and molting areas could have provided important waterfowl hunting and egg gathering opportunities for subsistence users of the area. Mountain goats are found in the higher terrain at the head of Port Houghton, Sandborn Canal, and Farragut Bay. In addition, Sitka black-tailed deer, black bears and brown bears, wolf and moose, as well as a variety of small mammals are reported in the region today (Parametrix 1998). The North Point site lies within the Northwest Coast coniferous rainforest, dominated by western hemlock, Sitka spruce, Alaska yellow cedar, and mountain hemlock. At the head of the small pocket beach at 49-SUM-25 are found lush stands of sedge; the upper intertidal zone grades directly into first an alder zone, and then a hemlock-spruce forest. Vegetation in the intertidal zone is dominated by *Fucus* sp. while the dominant offshore macroalgae are brown kelps (Parametrix 1998).

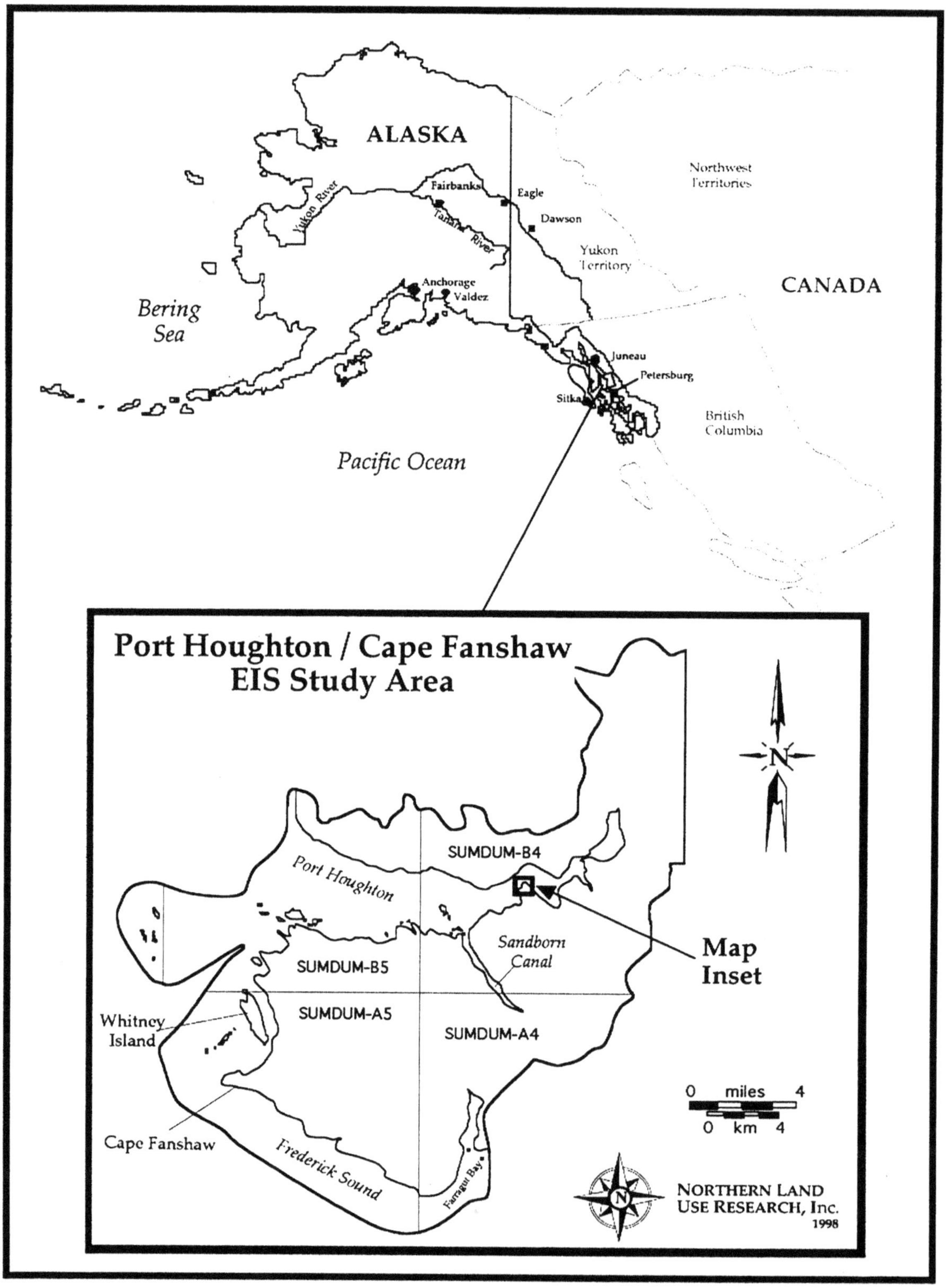

Figure 1. Port Houghton/Cape Fanshaw study areas.

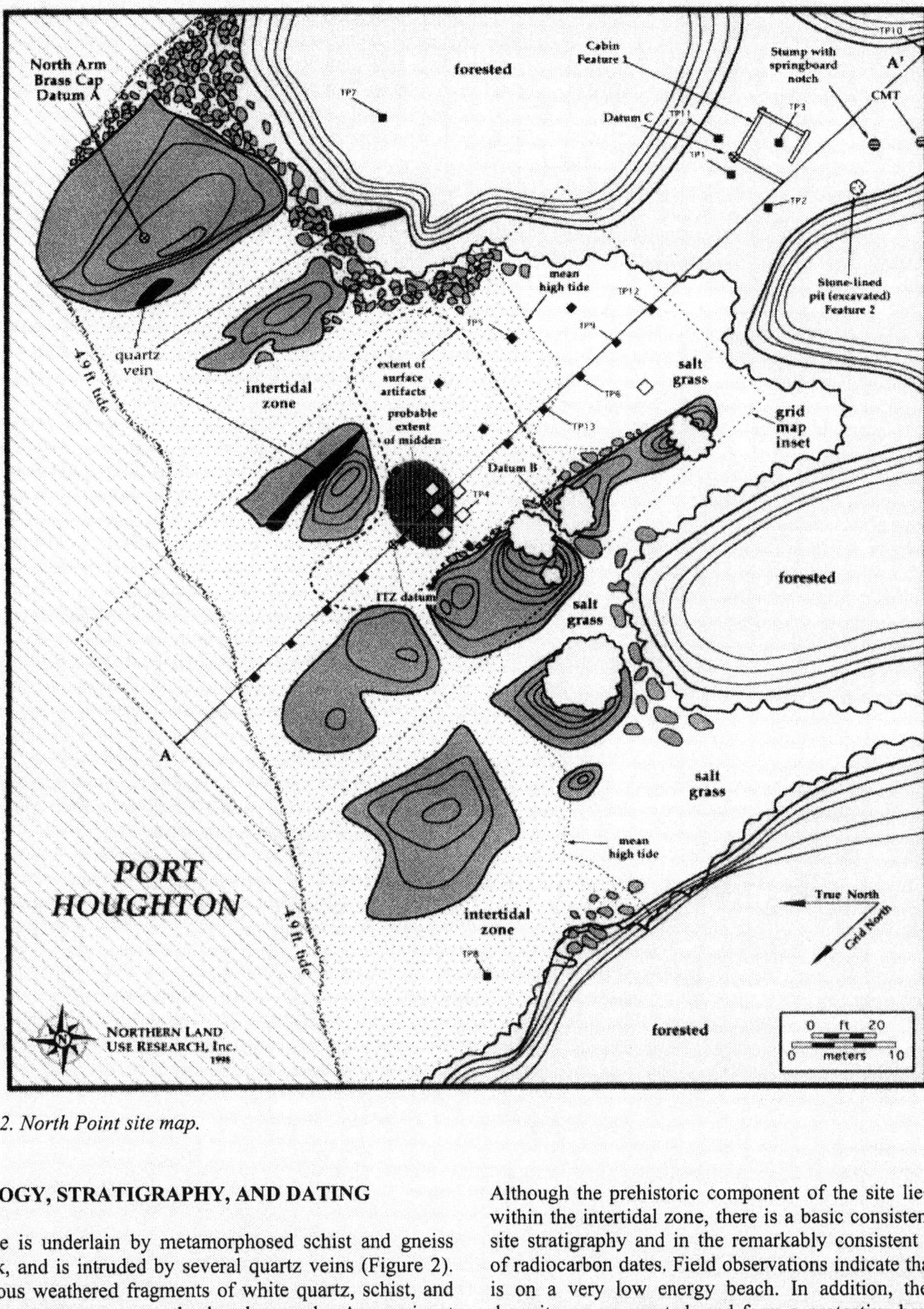

Figure 2. North Point site map.

GEOLOGY, STRATIGRAPHY, AND DATING

The site is underlain by metamorphosed schist and gneiss bedrock, and is intruded by several quartz veins (Figure 2). Numerous weathered fragments of white quartz, schist, and gneiss are present among the beach gravels. A prominent north-south trending thrust fault, the Megalignment, lies several kilometers to the east of the site, and may have been responsible for some 5 m of submergence of the cultural midden since it was deposited (Brew and Ford 1978; Bowers *et al.* 1996; Bowers and Mason 1998).

Although the prehistoric component of the site lies entirely within the intertidal zone, there is a basic consistency in the site stratigraphy and in the remarkably consistent sequence of radiocarbon dates. Field observations indicate that the site is on a very low energy beach. In addition, the midden deposits are compacted, and form a protective barrier from wave and tidal activity for lower lying deposits. No compelling evidence suggests that the artifacts and faunal remains were redeposited from upslope of their present locus: (1) there is no obvious upland source or mechanism for movement of the cultural materials, such as a cut bank or steep slope, (2) the artifacts all retain sharp edges (even diminutive microblades are razor sharp), suggesting little downslope transport, and (3) there is no evidence for the sorting of archaeo-clasts, as would be the case if downslope movement was prevalent.

The site stratigraphy reveals five major sedimentary units (see Bowers *et al.* 1996). The basal Unit 5 consists of a culturally-sterile blue-green silty clay (Figures 3 and 4). Across most of the site's intertidal zone, another culturally-sterile unit, a brown silt and fine to medium sand unit, Unit 4, overlies the Unit 5 silt clay, suggesting a dramatically different sediment source from Unit 5. Unit 3 consists of dark clay silt with abundant wood fragments; this unit contains as many as three subdivisions. Unit 2, a sandy silt, encloses a shell midden and >80% of the buried archaeological deposits from the site. Overlying Unit 2 sandy silt is a coarse sand unit, Unit 1. The surface of the beach, especially below the limit of the salt grass (Figure 3), is littered with subangular and rounded cobbles of several lithologies, as well as rock fragments derived from *in situ* weathering. Moving upslope from the intertidal zone into the forest, the terrain slopes gradually to the locus of the historic cabin (Component II), which occurs on a break in the slope at about the 7-8 m asl elevation (Figures 2 and 3). Test pits and a radiocarbon date of 1030+/-50 ^{14}C years B.P. in the vicinity of the cabin revealed no beach deposits to suggest an ancient raised marine terrace. The stratigraphy of the upland test pits (test pit 10; Figures 2 and 3) is completely distinct from those in the intertidal zone, revealing none of the sedimentary units found in the beach excavations.

A total of eight radiocarbon dates were obtained from the site (Table 1). The dates, taken together with geologic and stratigraphic observations, suggest a conservative age estimate for the prehistoric occupation of between 2090 and 2780 B.P., with the most likely estimation of the intertidal midden and the site's true age being 2505-2760 B.P. (Cal. BC 555 to 810). Given our observations of the relatively thin midden and the character of the artifact assemblage, we do not believe that the prehistoric occupation was particularly long-lived, and most likely is not more than about 250 years long (Bowers *et al.* 1996).

A number of natural, cultural, and taphonomic processes formed and transformed the site. Prior to about 4000 B.P., a silty-clay (Unit 5) was deposited. This may be a regional deposit, given similar deposits observed elsewhere in the area. Unit 5 is interpreted as an early Neoglacial deposit, carried as a suspension load by turbid waters originating from expanding glaciers at the head of Port Houghton (Bowers *et al.* 1996).

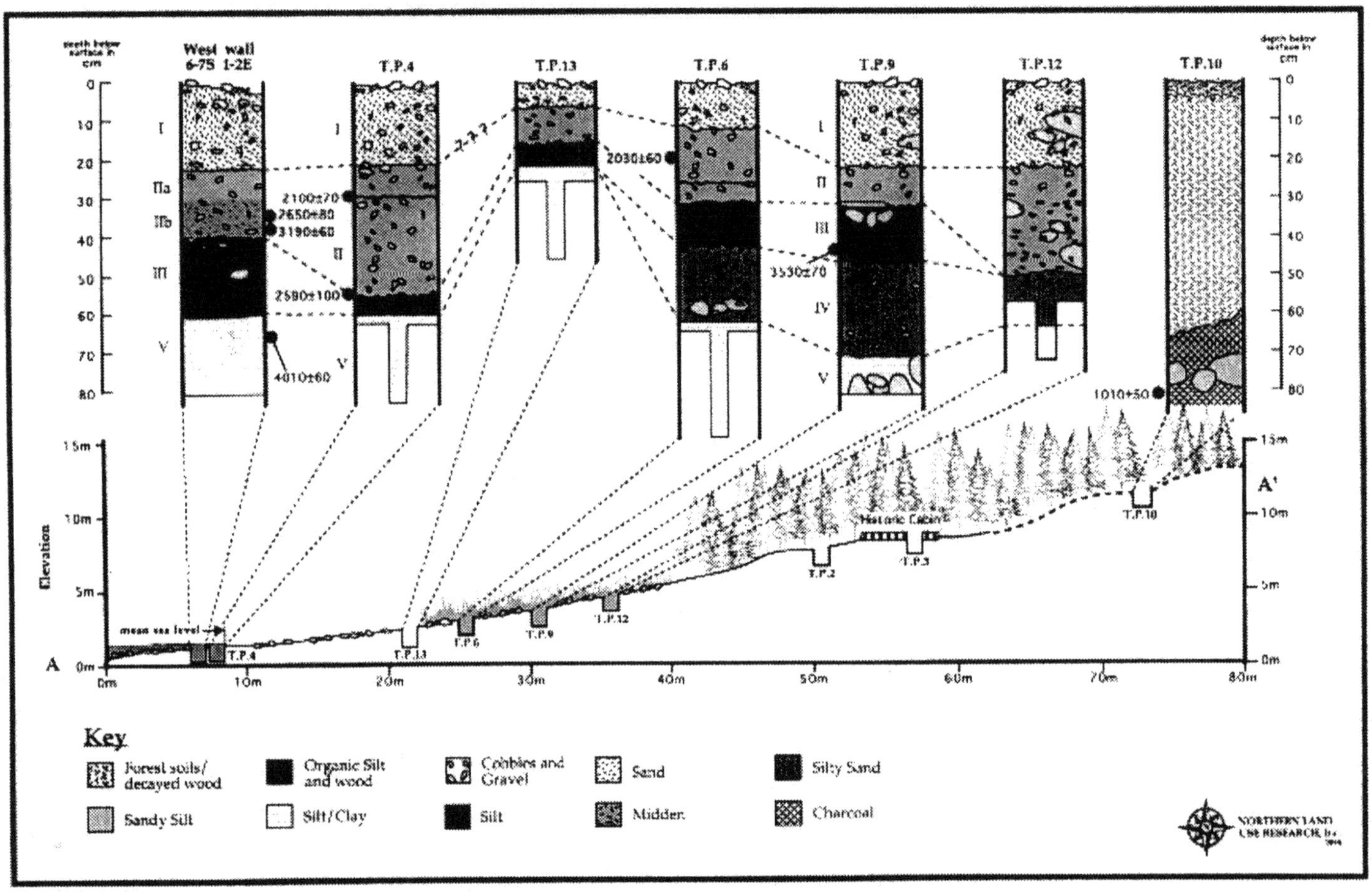

Figure 3. North Point Site, correlation of stratigraphic units and beach profile.

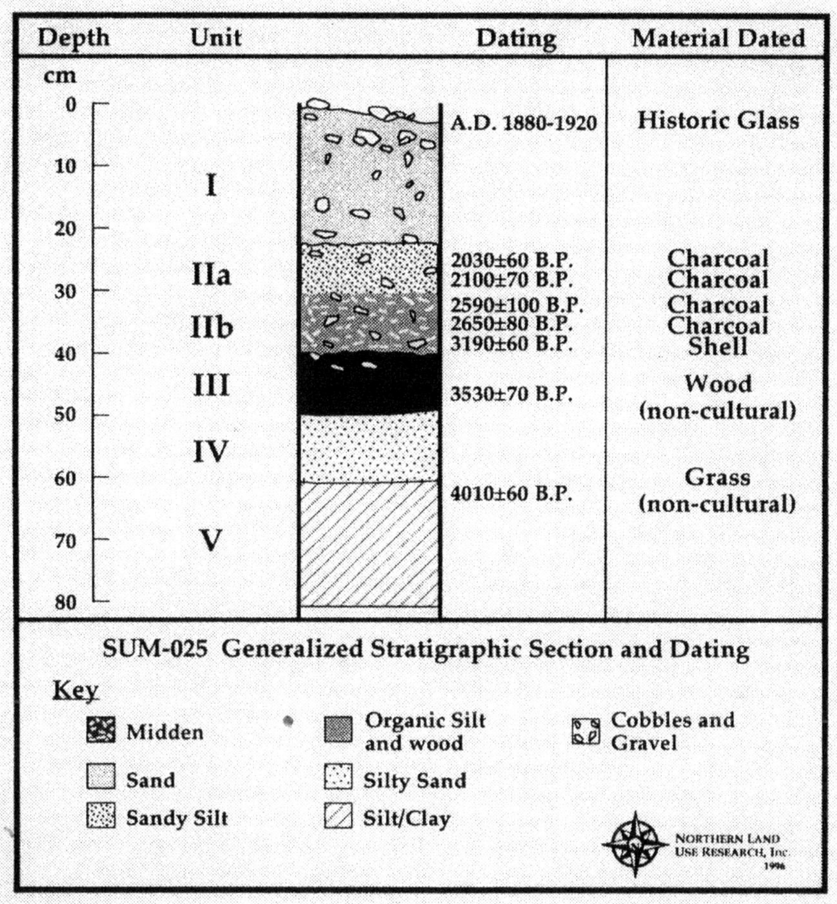

Figure 4. Generalized stratigraphic section and dating.

FIELD#	LAB # BETA-	MEASURED ^{14}C AGE[1]	^{13}C ^{14}C[2]	CONVENTIONAL ^{14}C AGE[3]	CALIBRATED AGE[4]	INTERCEPT AGE[5]	MATERIAL	PROVENIENCE[6]	SIGNIFICANCE
025-5	75460	2110 ± 70	-25.7	2100 ± 70	BC 360 TO 280 BC 250 TO AD 65	Cal BC 100	charcoal	T.p. #4, 30cm	younger than or equivalent to age of artifacts in Unit 2
025-26	75461	2090 ± 60	-28.7	2030 ± 60	BC 180 to AD 100	Cal BC 20	charcoal	T.p. #6, 15-25 cm	younger than or equivalent to age of artifacts in Unit 2
025-20	76049	3620 ± 70	-30.5	3530 ± 70	BC 2025 to 1675	Cal BC 1880	wood	T.p. #9 40-44 cm	older than artifacts; dates base of wood layer in Unit 3; non-cultural
025-32	76050	1030 ± 50	-26.7	1010 ± 50	AD 970 to 1165	Cal AD 1020	charcoal	T.p. #10 70-90 cm	uplands test pit; non -cultural
025-34	76998	2600 ± 100	-25.6	2590 ± 100	BC 825 to 760 BC 670 to 550	Cal BC 795	charcoal	T.p. #4 50-60 cm	dates cultural occupation in Unit 2
NLUR-95-1	83499	4070 ± 60	-28.7	4010 ± 60	BC 2850 to 2820 BC 2630 to 2350	Cal BC 2490	grass	4-5S/2-3W 60-70 cm from clay sample	dates sterile clay, lower limiting date for site
NP-95-433	85165	2780 ± 60	-0.4	3190 ± 60	BC 770-375	Cal BC 555	shell	6-7S/1-2E 30-35 cm	dates midden deposit located in 1995 testing; shell date slightly older than paired charcoal date
NP-95-318	85166	2690 ± 80	-28.0	2650 ± 80	BC 930 to 760 BC 670 to 550	Cal BC 810	charcoal	6-7S/1-2E 30-40 cm NE quad	dates midden deposit located in 1995 testing; confirms shell date on same level

Table 1. Radiocarbon dates from the North Point Site. Calibration based on Stuiver and Reime (1993).

[1]In years Before Present (B.P.). [2]Ratio in 0/00. [3]Conventional date in years B.P. [4]2 sigma range: 95% probability. [5]Intercept of ^{14}C age with calibration curve. [6]Depth below surface in cm.

The overlying Units 4 and 3 appear to be related. Both units are organic-rich, especially Unit 3, which contains large amounts of wood dated to 3530±70 ^{14}C years B.P. Unit 3 sediments were probably incorporated into the site as redeposited forest debris, on top of Unit 4 beach sediments. This material was then buried under a coarser and less organic-rich sediment (Unit 2), which includes the midden identified in 1995.

The processes forming Unit 2 apparently lasted for some 700 years, between about 2090 and 2780 ^{14}C B.P. It is within this period, between about 2505 and 2760 B.P., that the Component I artifactual and faunal materials were deposited within the shell midden. The prehistoric human occupation of the site appears to have taken place during a major period of climatic cooling. Although the Neoglacial period is poorly represented in the archaeological record of southeast Alaska, climatic cooling and concomitant glacier expansion probably had profound effects on human populations. Neoglacial conditions likely affected the distribution and seasonal availability of at least some subsistence species, and anadromous stream productivity may have been disrupted. Near the margins of glaciers, inhabitable landforms may have been lost, and the suitability of other areas for settlement may have been reduced (Streveler *et al.* 1995).

The location of a buried midden in the lower intertidal zone is consistent with discard patterns recorded ethnographically for Tlingit settlements. Broken artifacts along with shell, wood, and food refuse were conveniently tossed down the gently sloping beach into the water, with the expectation that the tides and waves would remove them from the main habitation areas. Observations made near the village of Angoon in 1949-1950 may provide a useful analog to the North Point midden deposits:

> ...all garbage and refuse and discarded artifacts are thrown onto the beach, where they are eventually washed away or buried by silt and gravel. Whether this practice may have originated because of notions that fish and sea mammal remains must be returned to the sea, we cannot say, but it evidently represents a deep-seated habit of the Tlingit. This certainly distinguishes them from the Eskimo, whose past and present houses are built on or in the accumulated middens of generations, and whose house cleaning is apt to consist of little more than throwing refuse out the door (de Laguna 1960:34).

The deposition of the overlying Unit 1 materials was probably fairly rapid. The sources of Unit 1 sediments are believed to be *in situ* weathering of bedrock and locally-available unconsolidated sediments. The surface of the site represents a lag deposit, where smaller clasts have been removed by tidal action, leaving a pavement of boulders, cobbles, and coarse sand. While erosional activities in the cove appear to be very limited, some vertical movement of the underlying Unit 2 sediments has probably occurred through faunalturbation by burrowing clams, which may represent the mechanism that moved some artifacts up to the surface (Bowers *et al.* 1996).

ARTIFACT AND FAUNAL ASSEMBLAGE DISTRIBUTION

As shown in the plan view of the site (Figure 2), cultural materials are divided according to intertidal and upland settings. All of the prehistoric materials are located within the intertidal zone; only historic artifacts and features have been found in the uplands. A few historic artifacts were also recovered from the upper beach deposits.

The vertical distribution of the North Point artifacts and fauna were examined through the use of log-likelihood statistical analysis (Gould, in Bowers *et al.* 1996). The analytical units chosen for study were the main five midden-bearing excavation units (EUs); others were excluded from the distribution analysis primarily because artifacts and faunal frequencies were simply too low. Our primary concern was whether or not deposits could be differentiated by natural or cultural strata, especially: (1) whether or not quartz microblade technology and ground slate co-occur at the site and (2) whether or not lithic and faunal composition vary with depth (Gould, in Bowers *et al.* 1996).

Using arbitrary levels corresponding to above-midden, in-midden, and below-midden for each of the main midden-bearing EUs, it was shown that proportions of slate and quartz items do not significantly change among the three strata, and proportions of mammal and fish remains also remain consistent throughout. It is important to reiterate that 80% of the artifacts and 85% of the faunal assemblage lie within the core midden zone (Gould, in Bowers *et al.* 1996).

The North Point data indicate a stable pattern of use over the limited period of time reflected in the midden-area deposits. Throughout the time period ca. 2505-2760 ^{14}C B.P., we see a consistent mix of mammal and fish bones, as well as a consistent distribution of slate and quartz. It appears that whatever made North Point an attractive site remained constant (Gould, in Bowers *et al.* 1996). Based upon spatial analysis of the site, combined with our geologic interpretations, Component I is thus considered as a single prehistoric horizon. Although we recognize that several occupations may have occurred during the few years or few hundreds of years of occupation possibly represented at the site, we have not been able to distinguish individual events or seasons of human occupation.

ARTIFACTS

Artifacts from Component I are briefly described below; for greater detail including artifact illustrations, see Bowers *et al.* (1996). A total of 868 prehistoric artifacts were collected from the site; of these, 83 prehistoric artifacts were collected from the surface, and 785 from the subsurface. No hearths, burials, structures, or other discernible archaeological features were observed.

Artifacts that appear to relate predominately to *procurement* include: unilaterally barbed projectile points, bone points with line holes, wooden and bone harpoon points, slate points, and cordage (two strand twist line and knots presumably used as leaders on fish hooks; cf. Croes 1993,

1995, 1997). Artifacts probably used for *processing* include: obsidian[2] and quartz microblades, quartz microliths and a possible wooden haft for microliths (cf. Flenniken 1981; Croes 1995), ground slate knives, quartz and slate scrapers, slate awl, fire cracked rock, modified flakes, and charcoal. Likely *manufacturing* artifacts include: quartz crystal microblade cores and core fragments, quartz and obsidian microblades discarded during manufacture, lithic debitage, abraders, wood chips, wooden wedges with root grommets, bone point preforms and fragments, cobble manuports, and one obsidian biface. Included in the subclass of *unknown* or *ornamental* artifacts are quartz crystals and spherical stones.

The results of investigations at 49-SUM-25 are significant for several reasons (Bowers *et al.* 1996). The microblades are at least 2,200 years more recent than others reported on the Northern Northwest Coast. Although there are a number of wet sites in southeast Alaska (Betts 1998; Moss and Erlandson 1998), North Point is the only one in which residential occupational debris has been found in a wet context. Other southeast Alaskan wet sites are fishing weirs and traps of various types. In addition, the North Point artifact assemblage is remarkably similar to the contemporaneous Locarno Beach archaeological culture centered some 800 to 1,000 km to the south (Matson 1989; Croes 1995; cf. Matson and Coupland 1995:154-182). Finally, the co-occurrence of quartz microblade technology and ground slate tools is unique among sites on the Northern Northwest Coast (Croes 1995; cf. Matson and Coupland 1995:154-182).

THE NORTH POINT FAUNAL ASSEMBLAGE

Invertebrate Fauna

The shellfish are listed in Table 2. Only a species list is presented here as invertebrates were not systematically collected or analyzed. Many of the identified taxa are common in intertidal habitats found in the Port Houghton area today. At least five are found in low energy, protected rocky shoreline microenvironments like North Point (Foster 1991; Ricketts *et al.* 1985). However, three identified species (*Clinocardium nuttalli, Mya sp., and Saxidomus giganteus*) are found in sand/mud flat bays and estuaries.

Common Name	Scientific Name
Baltic macoma	*Macoma balthica*
stained macoma	*Macoma inquinata*
softshell	*Mya sp.*
mussel	*Mytilus sp.*
Pacific littleneck	*Protothaca staminea*
Washington clam, butter clam	*Saxidomus giganteus*
fat gaper	*Tresus capax*
Nuttall cockle	*Clinocardium nuttalli*
Arctic hiatella	*Hiatella arctica*
limpet	*Lottia pelta*
chiton/black Katy	*Katharina tunicata*
periwinkle	*Littorina sitkana*
rock snail	*Nucella lamellosa*
rock snail	*Nucella lima*

Table 2. Shellfish identified from the North Point Site.

Vertebrate Fauna

The vertebrate faunal assemblage consists of 3,096 identified specimens[3]. The distribution of faunal specimens (NISP) by cultural strata across all excavation units (EUs) is presented in Table 3. The skeletal elements recovered for each taxon are summarized in Table 4, and the distribution of faunal remains (NISP) by the five midden 1 x 1m EUs used in our analysis is presented in Table 5.[4] Considering its comparatively small size, the North Point faunal assemblage is quite diverse, with 28 taxa identified, 25 to the genus level. Fish are the most abundant class of fauna, comprising 96% of the specimens by NISP. Mammals make up approximately 2.5% of the assemblage, with birds comprising a mere 0.8%. Pacific cod dominate the assemblage, and other fish, in order of abundance, include rockfish, salmon, greenling, halibut, herring, walleye pollock, flatfish, sculpin, lingcod, and requiem shark. Considering the extremely small size of the mammal and bird assemblages (76 NISP and 23 NISP, respectively), these are surprisingly diverse. The nine land mammal taxa represented include porcupine, mountain goat, mink, canid, wolverine, beaver, Sitka black-tailed deer, snowshoe hare, and river (land) otter. Seals are the only sea mammal identified, although they are represented by just four bones, and comprise only about 5% of the mammalian assemblage. The seven identified bird taxa include scoter, common murre, pelagic cormorant, loon, western grebe, scaup, and gull.

Nearly all the faunal remains were recovered from the central site area containing the intertidal shell midden (Figure 2). Preservation of specimens is remarkably good, especially considering the site context in which bones were inundated with salt water twice a day. Unit excavations extended to variable depths, ranging to a maximum of 70+ cm below the surface. For analysis, these 10 cm levels were grouped into the following named cultural stratigraphic layers or levels: "above midden", "midden", "below midden", and "non-midden."

Faunal remains were recovered from each of these levels, and while the proportions of fish and mammals remained the same throughout (Bowers *et al.* 1996:31), the quantities of bone preserved in the midden level were substantially larger than those found above and below. Approximately 81% of the top ten taxa were found in the midden level, followed by 10% below the midden, and 9% above the midden (Table 3). Because the "above midden" layer was usually comprised of three 10 cm deep levels, whereas the "below midden" layer usually contained only one (but in a single case two) levels, it appears that volumetrically, faunal recovery of below midden levels was actually greater than that of above midden levels.

This suggests that conditions for bone preservation below the midden were more favorable than those above the midden. Unit 3, onto which the shell midden was deposited, is an organic silt containing considerable amounts of woody debris. This substrate provided a geochemical and physical environment conducive to bone preservation. The younger assemblage of faunal remains recovered from above the midden has not fared as well within a matrix of sandy silt. As proposed by Bowers *et al.* (1996:22), Unit 3 represents the remains of forest litter or woody debris that washed into the relatively low energy intertidal zone at the site with the tides. As far as we know, an intertidal depositional context such as this, and its effect on faunal preservation, has not been previously documented on the Northwest Coast.

The faunal assemblage is assigned to a single site component, dated between 2505 and 2760 ^{14}C years B.P. The vertical variation in the assemblage described above evidently relates more to depositional factors rather than to any temporal cultural change. With respect to horizontal patterning, Unit 4-5S/2-3W produced the greatest quantity of faunal remains (32%), followed by Unit 6-7S/1-2E (27%) and Unit 5-6S/0-1W (26%). Thus, 85% of the faunal assemblage comes from the three units that lie entirely within the core shell midden area (Bowers *et al.* 1996).

ANIMALS AND HABITATS EXPLOITED

Mammals and Birds

Before focusing on the most abundant faunal class, the fish, we shall consider the numerically limited, but unusually diverse mammal and bird assemblages. Although the small numbers preclude evaluation of their relative subsistence or economic significance to the prehistoric residents of North Point, it is worth noting that land mammals outnumber sea mammals, with just a few seal bones identified. This is somewhat unusual for archaeological sites located in coastal settings across the Northwest Coast region, although it may reflect the site's mainland vs. island setting (Ackerman 1988). North Point site residents were targeting a wide range of terrestrial habitats. Mountain goat habitat occurs in high-elevation areas both to the north and south of Port Houghton; today, several discrete populations of goats reside in the area (Parametrix 1998:3-24). Forested shoreline areas adjacent to the site provide suitable habitat for deer, land otter, mink, porcupine, and wolverine. Beaver were probably procured near the lakes draining into the Salt Chuck at the head of Port Houghton. The distribution of snowshoe hare in the site vicinity is not known, and this species is an unusual occurrence in southeast Alaskan coastal archaeological assemblages.

All the birds present in the North Point assemblage are resident shorebirds; none are migratory waterfowl. Freshwater lakes in the site vicinity would comprise suitable habitat for nesting loons and grebes, and all the rest of these aquatic birds (scaup, scoter, gull, cormorant, common murre) are found in the waters adjacent to the site year-round.

Fish

Of 2,965 total fish remains, 2,089 or 70% are Pacific cod. Vertebrae are the most common element present, but head bones are also fairly common. Based on size and shape, the vast majority of the vertebrae are *Gadus macrocephalus* (Pacific cod), but *Theragra chalcogramma* (walleye pollock) may also be present. The next most common category is unidentified fish (N=550), making up nearly 19% of the fish

TAXON	ABOVE MIDDEN	IN MIDDEN	BELOW MIDDEN	NON-MIDDEN	TOTAL
BIRDS					
Gavia immer	1				1
Uria aalge		2		1	3
Larus sp.	1				1
Phalacrocorax pelagicus		3			3
Aythya marila or A. affinis		1			1
Melanitta sp.	1	4		2	7
Aechmophorus occidentalis	1				1
unknown bird	2	4		1	7
Bird Total	6	14		4	24
FISH					
Gadus macrocephalus	170	1568	214	137	2089
Pleuronectidae	1	2		1	4
Hexagrammos sp.		8		1	9
Hippoglossus stenolepis	1	5			6
Clupea harengus pallasi		5	1		6
Hexagrammos sp.	1				1
Carcharhiniadae		1			1
Sebastes sp.	20	100	19	16	155
Salmonidae	17	92	6	20	135
Cottidae	1	3			4
Theragra chalcogramma		5			5
unknown fish	31	454	36	29	550
Fish Total	242	2243	276	204	2965
MAMMAL					
Castor canadensis				1	1
Canis sp.		1		1	2
Odocoileus h.s.				1	1
Phoca vitulina	1	3			4
Lepus americanus				1	1
Mustela vison		2		1	3
Oreamnos americanus	2	2		1	5
Erethizon dorsatum	1	5	2		8
Lutra canadensis		1			1
Gulo gulo		1	1		2
Sea mammal	1	3			4
unknown mammal	6	33	5	5	49
Mammal Total	11	51	8	11	81
Unknown	4	2	19	1	26
TOTAL	263	2310	303	220	3096

Table 3. Faunal remains (NISP) by cultural strata across excavation units, North Point Site.

BIRDS	
Gavia immer	1 illium
Uria aalge	1 carpometacarpus, 1 scapula, 1 vertebra
Larus sp.	1 radius
Phalacrocorax pelagicus	3 vertebrae
Aythya marila or *A. affinis*	1 tarsometatarsus
Melanitta sp.	1 carpometatarsus, 1 humerus, 2 radii, 1 scapula, 2 ulnae
Aechmophorus occidentalis	1 phalange
unknown bird	1 humerus, 6 unknown bones
FISH	
Gadus macrocephalus	34 angular/articulars, 20 basioccipitals, 18 ceratohyals, 5 cleithra, 27 dentaries, 2 epibranchials, 5 epihyals, 2 exoccipitals, 5 frontals, 11 hyomandibulars, 2 interopercles, 1 mandible, 17 maxillae, 1 mesethmoid, 1 metapterygoid, 2 opisthotics, 2 otoliths, 6 palatines, 12 parasphenoids, 1 parietal, 3 pharyngeal plates, 3 pharyngobranchials, 7 posttemporals, 26 premaxillae, 8 preopercles, 2 prootics, 2 pterotics, 26 quadrates, 3 sphenotics, 4 supracleithra, 1 symplectic, 1819 vertebrae, 11 vomers
Pleuronectidae	4 vertebrae
Hexagrammos sp.	9 vertebrae
Hippoglossus stenolepis	1 interopercle, 5 vertebrae
Clupea harengus pallasii	1 cleithrum, 1 dentary, 1 hyomandibular, 2 interopercles, 1 vertebra
Ophiodon elongatus	1 vertebra
Carcharhiniadae	1 tooth
Sebastes sp.	2 ceratohyals, 1 dentary, 1 epihyal, 1 maxilla, 2 pharyngeal plates, 1 pharyngobranchial, 2 posttemporals, 2 quadrates, 143 vertebrae
Salmonidae	1 maxilla, 134 vertebrae
Cottidae	1 quadrate, 3 vertebrae
Theragra chalcogramma	1 angular/articular, 1 epihyal, 3 parasphenoids
unknown fish	547 unknown bones, 3 vertebrae
MAMMAL	
Castor canadensis	1 acetabulum
Canis sp.	1 basioccipital, 1 radius
Odocoileus hemionus sitkensis	1 rib
Phoca vitulina	1 femur, 1 occipital, 1 phalange, 1 unknown bone
Lepus americanus	1 rib
Mustela vison	1 mandible, 1 metapodial, 1 sacrum
Oreamnos americanus	1 astragulus, 1 horn core, 1 molar, 2 phalanges
Erethizon dorsatum	1 claw, 3 incisors, 1 phalange, 1 scapula, 2 teeth
Lutra canadensis	1 tibia
Gulo gulo	1 femur, 1 metatarsal
unknown mammal	1 phalange, 8 ribs, 1 sternabra, 1 tibia, 1 tooth, 30 unknown bones, 7 vertebrae

Table 4. Skeletal elements identified for each taxon from the North Point Site.

remains as measured by NISP. Most of these unknown fish remains are small and light bones: vertebral spines, fin rays, and other skeletal elements too small and fragmentary to identify. Rockfish make up 5% of the total fish NISP followed by salmon at 4.5%. The other fish, greenling, halibut, herring, walleye pollock, flatfish, sculpin, lingcod, and shark are present in quantities less than 1% as measured by NISP. The herring, as well as small sculpins and flatfish, are undoubtedly underrepresented in the assemblage, due to the use of 1/4 inch screens during recovery.[5] Shark is represented by a single tooth that may be more meaningfully interpreted as a curated artifact rather than as the remains of systematic subsistence activity.

The skeletal element representation of the three most common fish is presented in Table 6. In addition to listing the NISP of each element, MNE calculations are provided. For elements that occur singly, like the parasphenoid, the MNE and NISP are the same, in the case of Pacific cod, this number is 12. For paired elements like the dentary, the NISP is halved to derive the MNE, in other words, a minimum of 14 (MNE) cod is necessary to account for 27 cod dentaries in the assemblage. Since cod average 52 vertebrae per individual (Hart 1980), the 1,819 cod vertebrae in the assemblage represent a minimum of 35 individual cod. The number of vertebrae in each fish species varies considerably; the rockfish have an average of 26 vertebrae per individual while salmon have 64 vertebrae (Hart 1980).

Although the sample is quite small, rockfish show a marked discrepancy between MNE for vertebrae (6 MNE) vs. that for cranial elements (1 MNE). Because rockfish cranial bones are dense and tend to preserve relatively intact, the archaeological findings suggests that rockfish head bones were not deposited with the vertebrae for some reason. For salmon, only one cranial bone (one maxilla) is present while the MNE for vertebrae is three. While the absence or low percentages of salmon cranial bones in Northwest Coast fish assemblages have been commonly interpreted as evidence for stored salmon (see Matson and Coupland 1995:166-169 for historical background and a defense of this interpretive practice), the relatively small number of salmon bones found at North Point may also suggest that salmon were not primary economic species targeted by the North Point residents during the season or the period of occupation.

For cod, the most frequent element that occurs singly is the basioccipital (20 MNE). The most frequent paired element is the angular/articular (17 MNE). As mentioned above, the 35 MNE for vertebrae suggests that cod heads are somewhat underrepresented in the assemblage compared to vertebrae. This may be due to cultural factors, i.e., people were probably treating cod heads differently than the post-cranial carcass at some stage of butchery, cooking, processing, or consumption. The implications of this pattern will be discussed more fully below after a number of other issues are discussed.

Taxon	Unit 5-3	Unit 6-1	Unit 6-7	Unit 7-2	Unit 9-1	Total
BIRD						
Gavia immer				1		1
Uria aalge			1	2		3
Larus sp.		1				1
Phalacrocorax pelagicus		3				3
Aythya marila or *A. affinis*	1					1
Melanitta sp.	1	1	2	2	1	7
Aechmophorus occidentalis				1		1
unknown bird	3	1		2		6
Bird Subtotal	5	6	3	8	1	23
FISH						
Gadus macrocephalus	713	484	136	577	178	2088
Pleuronectidae	1	2	1			4
Hexagrammos sp.		1	1	7		9
Hippoglossus stenolepis	4	2				6
Clupea harengus pallasii	1	5				6
Hexagrammos sp.					1	1
Carcharhiniadae		1				1
Sebastes sp.	9	30	16	76	24	155
Salmonidae	24	29	20	30	32	135
Cottidae		2		1	1	4
Theragra chalcogramma		1		4		5
unknown fish	188	216	29	105	12	550
Fish Subtotal	940	773	203	800	248	2964
MAMMAL						
Castor canadensis			1			1
Canis sp.	1					1
Odocoileus hemionus sitkensis			1			1
Phoca vitulina	3			1		4
Lepus americanus			1			1
Mustela vison	2		1			3
Oreamnos americanus	2		1	1	1	5
Erethizon dorsatum	3	2		2	1	8
Lutra canadensis	1					1
Gulo gulo	1	1				2
Sea mammal	3			1		4
unknown mammal	27	5	5	7	5	49
Mammal Subtotal	40	8	10	11	7	76
Unknown	1	16	2	1	4	24
TOTAL	986	803	218	820	260	3087

Table 5. Faunal Remains (NISP) by Excavation Unit from the North Point Site.

DISCUSSION

Possible Effects of Neoglacial Cooling

Although the most likely explanation for the relative abundance of cod and paucity of salmon remains from North Point is due to seasonality (see below), this difference may also reflect the effects of Neoglacial conditions which appear to be widespread during the third millennium B.P. (Calkin 1988; Mann and Hamilton 1995). Although forest vegetation similar to today's was established by 4000 B.P. in the northern Northwest Coast (Peteet 1986; Cwynar 1990), and sea level was more or less stabilized by 4500 B.P. (Bowers and Mason 1998), a wetter and cooler climate is indicated by pollen records after ca. 3300 B.P. (Heusser 1985; Heusser *et al.* 1985)[6].

Holocene glacier expansions in southern Alaska were widespread by 3000 B.P. (Calkin 1988:179,180). Although no glacial studies have been made in Port Houghton, three distinct glacial expansions were recorded from the Fairweather area, one of which was between 3500-2500 B.P. (Calkin 1988:461). In the Glacier Bay region, Neoglacial advances have been dated from 4200 B.P. to 2200 B.P. (Goldthwait 1966, in Mann 1986:257). Miller (1975) reports Neoglacial advances beginning 3500-2500 B.P. in the Juneau area. Work in the Taku Glacier area near Juneau suggests that clays dated to about 2600 B.P. relate to lacustrine sedimentation in a glacier-dammed Neoglacial lake (Motyka and Beget 1996). Other studies of the Juneau icefield indicate that climatically-induced expansion of glaciers occurred at ca. 3200 B.P., 2800-2600 B.P., and 2200 B.P. (Rothisberger 1986, in Motyka and Beget 1996). In sum, the geologic record from southeast Alaska suggests a general synchroneity between the basal deposits from 49-SUM-25 and Neoglacial evidence from adjacent areas. Minor recessions occurred in most of Alaska about 2000 B.P. (Calkin 1988).

What were the environmental effects of this regional cooling on human subsistence activities? From an ethnographic perspective, Streveler, *et al.* (1995) observe that the environmental changes of Little Ice Age glacier expansion included "cooler summers, lowered snowlines, and increased storminess." They suggest that spawning lakes of red salmon (*Oncorhynchus nerka*) would have been made unproductive due to ice, outwash, or saltwater:

> ...people found shores more difficult to use than today and less productive in food resources. Ice episodically excluded people from Glacier Bay, Yakutat Bay, and Icy Bay. When European explorers arrived in the late 18th Century, Tlingit people were expanding northward into territories occupied by the Eyak Indians and Chugach Eskimos. This expansion may have been a response to resource limitations imposed by changes in sea level, in climate, and in glacier extent (Streveler *et al.* 1995).

The low frequencies of salmon bones from 49-SUM-25 may be explained in part by data and inferences from Glacier Bay, summarized by Mann and Hamilton (1995:461):

> Salmonids do not colonize cold and turbid streams heading in glaciers (Milner and Bailey 1989). Even after the disappearance of glacial ice, rapid downcutting in recently-deglaciated stream systems discourage large salmon runs (Benda *et al.*, 1992).

Weir and trap technologies allowing for the capture of large numbers of salmon were in use elsewhere in southeast Alaska at the time of the deposition of the North Point midden (Moss *et al.* 1990; Moss and Erlandson 1998). During the Neoglacial, however, the Port Houghton area may not have been a particularly productive habitat for salmon. Whether we are looking at climatic influence on spatial and/or temporal variation in salmon productivity is unknown at this time. It is interesting to note, however, that the distribution of radiocarbon dated fish weirs and traps located in southeast Alaska shows a decline in the frequency of radiocarbon dates between 3000 and 2500 B.P. (Moss and Erlandson 1998:189, Figure 2). Possibly, climate had adverse effects on salmon populations during the 3000-2500 B.P. period in at least some areas of southeast Alaska. However, further study is necessary to examine this possibility.

Seasonality

The presence of large quantities of Pacific cod in the North Point assemblage strongly suggests a late winter-early spring occupation of the site. This interpretation is supported by biological information about Pacific cod behavior and by the information available from Tlingit ethnography. Both data sources will be briefly reviewed here. Although we know that longliners fish the marine waters adjacent to the Port Houghton area for Pacific cod and halibut today, not much is documented about patterns of spatial and seasonal abundance of Pacific cod in southeast Alaska. Information from contiguous waters along British Columbia and in the Bering Sea is therefore useful.

Pacific cod (Figure 5) has been the most important trawl-caught bottom fish in British Columbia (Hart 1980), and is typically taken at depths between 100-120 m (Ketchen 1961:518). While such depths are accessible to modern trawlers, pre-contact fisherfolk probably did not regularly access them unless perhaps they were targeting large Pacific halibut. Following spawning, and starting in late winter and through spring, however, schools of Pacific cod move into shallower waters, between 55-65 m and in some years asshallow as 31-37 m (Ketchen 1961). Pacific cod are present in these shallower waters in some areas of British Columbia until summer.[7] They return to deeper waters starting in mid-summer. The cod of northern British Columbia are adapted to slightly colder temperatures than those in southern Brtitsh Columbia, with fish in the north found in 6-7°C waters. Ketchen (1961:555) also points out

Element	Cod *Gadus m.*	Rockfish *Sebastes* sp.	Salmonidae
angular/articular	34		
basioccipital	20		
ceratohyal	18	2	
cleithrum	5		
dentary	27	1	
epibranchial	2		
epihyal	5	1	
exoccipital	2		
frontal	5		
hyomandibular	11		
interopercle	2		
mandible	1		
maxilla	17	1	1
mesethmoid	1		
metapterygoid	1		
opisthotic	2		
otolith	2		
palatine	6		
parasphenoid	12		
parietal	1		
pharyngeal plate	3	2	
pharyngobranchial	3	1	
posttemporal	7	2	
premaxilla	26		
preopercle	8		
prootic	2		
pterotic	2		
quadrate	26	2	
sphenotic	3		
supracleithrum	4		
symplectic	1		
vertebra	1819	143	134
vomer	11		

Table 6. Skeletal element representation of most common fish at 49-SUM-025.

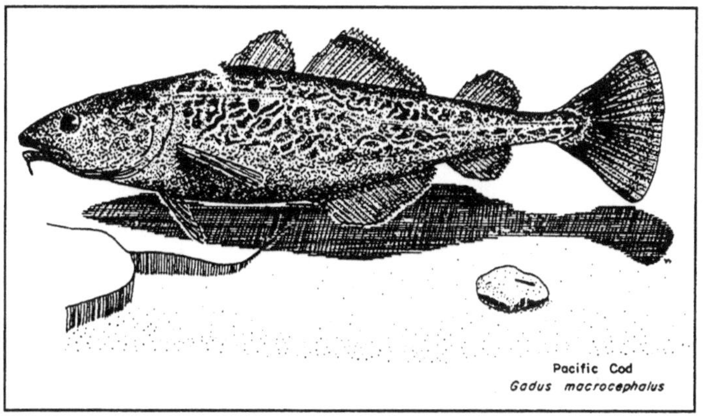

Figure 5. Pacific cod.

that cod congregate and move in shoals of varying density, that within a locality cod are adapted to a fairly narrow temperature range, and that when confronted with an intrusion of water warmer or colder than their preferred range, the fish can disappear quite suddenly. British Columbia cod have a 10 year cycle of alternating high and low biomass (Stepanenko 1995), therefore success in codfishing may have varied quite dramatically over the duration of occupation at an archaeological site like North Point.

The Bering Sea region is perhaps the most productive area for Pacific cod in the North Pacific. In 1884, commercial fishing began in the southeastern Bering Sea and along the southern shores of the Alaska Peninsula (Ketchen 1961:515-516). The hook and line cod fishery peaked in 1916 (3,893,000 fish; 44,000,000 lb.), but the fishery declined to a low level by 1933, where it remained until the early 1940s. New technology, particularly large trawls, is responsible for the increase in fish taken after 1945. The seasonal movements of Pacific cod in the Bering Sea differ from those along the Northwest Coast as they are determined by the presence of winter pack ice. In late September, the fish move offshore the Bering Sea shelf to avoid the intense cooling of inshore waters where pack ice forms (Shimada and Kimura 1994:814). In winter, they converge in large spawning masses, most notably between Unalaska and Unimak islands, southwest of the Pribilof Islands, and near the Shumagin Islands (Shimada and Kimura 1994:808). Following spawning, they move inshore when the coastal shelf warms in the summer.

All of this suggests that the seasonal movements of Pacific cod are closely dependent on seawater temperature regimes. While we can estimate the seasons of occupation at North Point as having occurred sometime within late winter through spring, it is hard to specify particular months, especially if we consider the possibility that the period of occupation at North Point may have coincided with an episode of Neoglacial cooling. Even localized events, like the advance of a nearby valley glacier, or particularly heavy seasonal runoff (cold or warm), could potentially affect the movements and local abundance of Pacific cod.

Ethnographic information on the Tlingit indicates that codfishing was an early spring activity. As Oberg (1973:73) wrote, by February, "stored food had become impalatable" and fresh fish were highly desired. According to Oberg (1973:66), the Tlingit fished for cod in March when schools of Pacific cod migrate to shallower waters. In 1879, Tlingit fishing near Angoon used bark lines with wooden iron-pointed fishhooks and considered 30-40 fish a good day's catch (Cobb 1916:27). Henry Katasse from Petersburg remembers his father coming ashore with a boat "loaded with gray cod" in March or April (Newton and Moss 1984:12).

The abundance of Pacific cod in the North Point faunal assemblage strongly suggests occupation during late winter to spring. Rockfish, the second most abundant taxon, are resident bottom-dwellers that could also have been caught at the same time as Pacific cod. The North Point salmon may represent a few early salmon caught incidental to cod fishing, or perhaps they do represent a small number of stored fish. The North Point site appears to have served as a late winter - early spring base camp, from which a relatively small number of people fished the nearshore waters, collected local shellfish, and went hunting in the uplands and along the forested shoreline in Port Houghton - Salt Chuck vicinity.

Ethnographic and Historical Information on Pacific Cod Use

In general, the primary ethnographic sources on the Tlingit do not provide much information on the use of Pacific cod (e.g., de Laguna 1972:402; Oberg 1973:73-74). Emmons 1991:121) reported that cod, along with several other species, "...were taken with a simple compound V-shaped hook, made with a wooden shank and a barb of sharpened bone or wood. The line was of twisted spruce root, and the sinkers...were unworked stones, generally oblong or of a convenient shape for the attachment of the line." Long lines of kelp were generally attached to these hooks (de Laguna in Emmons 1991:121; Stewart 1977). Referring to Pacific cod and a number of fish other than salmon, Emmons (1991:148-149) states that they were eaten fresh, they were not cured and "formed an unimportant part of the diet as long as fresh or dried salmon or halibut were to be had." This statement is contradicted, however, by some modern Tlingit elders who value Pacific cod quite highly. In the early 1980s, Henry Katasse from Petersburg described Pacific cod as "the most delectable smoke dried fish to have, by way of a change" (Newton and Moss 1984:12). Katasse described the process of cutting off the cod heads and opening the fish lengthwise to prepare them for layering between hemlock branches in a shallow pit to act as a poultice to remove any worms. After three or four days, the fish could be smoked, and later, the smoked fish could be soaked and cooked into stews and soups with seal fat and potatoes.

During the last quarter of the 19th century, Pacific cod in Alaska were commercially exploited for a variety of purposes. Cod roe was considered an excellent food product (Cobb 1916:62). Fishermen saved cod livers, placed them in large casks, and as the fish rotted, the oil would come to the surface. This cod liver oil was a very popular industrial lubricant. Fresh cod tongues were considered a great delicacy; after being thoroughly cleaned, then dried with a clean cloth, they were rolled in bread or cracker crumbs, and fried like oysters (Cobb 1916:70). Cobb (1916:69) explains that the cod's tongue is attached to the lower jaw. In describing other features of the anatomy and food value of cod heads, Cobb (1916:63) wrote,

> As nearly all the nutriment is in the lower half of the head, a small band saw could be installed and the upper half of the head, which is bony and contains little nutriment, cut off and thrown away, and only the lower part, which contains the fleshy cheeks and the succulent tongue, saved. When glue and fertilizer plants are established at the stations, as will doubtless be done in the near future, the

upper part of the head, which is rich in glue, could be used for this purpose.

The entire process of cod butchery used at the turn of the century aboard commercial vessels has been described:

> As soon as enough fish have accumulated on the deck the dress gang begins its work. The "throater" seizes the fish by the head in the left hand, places the back on the edge of a table or tub, and by means of a short knife with pointed end makes a cut each side of the throat just behind the gills (the front of the throat has previously been cut by the fisherman in order to bleed the fish) and another slit is made from the belly to the vent. The "header" then receives the fish, and grasping the head and body, backward pressure is made across the edge of the table or tub, resulting in breaking off the head at the first vertebra. He then opens the belly with the left hand and tears out the viscera. It is then passed on to the "splitter," the most important member of the gang, who places the back of the fish against a cleat on a board and by means of a short, heavy knife, rounded at the end, and with the blade slightly curved flatwise, continues the split down the belly to near the end of the tail, care being taken to keep near the backbone. At about three-fifths of the distance from the neck to the tail the backbone is cut across, and is loosened so that he can catch the end in his fingers. Grasping this with his left hand he cuts under it toward the head of the fish and separates the upper part of the backbone from the fish. In this operation the knife blade is kept close to the backbone to prevent loss of flesh, and a good splitter will drive the knife no deeper than is absolutely necessary, as otherwise the thick flesh at the back would be almost cut in two, thus spoiling the fish for middles.... The fish are then passed to the "black skinner," who, with an old glove or piece of bagging, rubs off the nape skins or membrane covering the napes, also any blood spots, and then drops the fish into a tub of salt water. Here the fish are soused around until thoroughly clean by the lesser members of the gang, who are called "idlers," when they are removed and passed through a chute into the hold, where the "salters" receive them (Cobb 1916:56-57).

Although this description is taken from an industrial setting, it can be compared with descriptions of indigenous butchery and processing practices. Together, such descriptions help to interpret patterns of Pacific cod skeletal representation at the site. Such information is useful in evaluating how North Point residents made specific use of cod.

Because detailed descriptions of cod processing and cooking are rare in Tlingit ethnography and oral history, we quote Henry Katasse's remembrances of processing fish caught near the entrance of Saginaw Bay on Kuiu Island:

> ...one day my father came ashore with a row boat loaded with gray cod. My grandfolks went down with sharp knives and proceeded to cut the heads off and open the cod lengthwise, alongside the backbone from the back until the knife reached the viscera and gutted and washed them. We, as children put them in containers and hauled them up; it wasn't very heavy but there were two of us that hauled it above the tidewater line. We dug an indentation about six inches deep. Others were bringing hemlock branches, and lining this shallow area with hemlock branches. We hauled the fish and laid them close together on top of the branches, flesh down toward the branches. When the area was filled, another layer of branches was thrown on top of the other and another layer of cod went on top, etc. until the cod were all gone. The final stages were to cover the last layer of fish with a thick layer of branches so it will get some air. We would cover it and leave it for four days and three nights. In the meantime, preparation was being made in the smokehouse. Others were cutting wood and everyone was busy on assigned details. Before they are hung in the rack in the smokehouse, sticks are put through the flesh crosswise, to keep it from bending or swerving. After it is dried, they cut a hole near the tail and hang them upside down in the smokehouse. The reason for going to all the trouble of getting hemlock boughs was to get rid of the worms inside the flesh of the cod. Contrary to what people say, it's a known fact that gray cod, the Pacific cod, do have worms. You must remember that this is the month of March and April. The branches acted as a poultice in getting the worms out which were not visible at the time. We were told to shake them off before taking them to the smokehouse. The way to prepare for cooking is to soak the smoked cod until it becomes soft enough to cook with boiled, fresh seal fat and boiled potatoes. The smoked dried cod, is then served, and fit for a king's table.... (Newton and Moss 1984:12).

We will return to this account in the discussion below, but first, it is wise to consider other ways of processing and preparing cod for consumption.

Franz Boas (1921) compiled a number of Kwakiutl[8] accounts documenting different ways of preparing cod. What these accounts suggest, in combination with those of the commercial fishing industry, is that the Pacific cod yielded a variety of edible products with which some of us might be unfamiliar today. For example, one account describes the Kwakiutl "housewife" first cutting off the cod's pectoral fins in one piece with the gills (Boas 1921:379-382). Then she cut across the bone in the neck and pulled out the guts. Although she disposed of the intestines, she gathered the stomachs in a basket, and subsequently boiled them in a kettle. These cod stomachs were then served to four to six friends of the man who caught the fish. Perhaps the most common cooking method was boiled codfish, in which the fish was served in a milky fish broth, and this dish was popular with both Kwakiutl and Tlingit. Another recipe favored by the Kwakiutl was tainted codfish:

> I first talked about the fresh boiled codfish. Now I will talk about what the Indians like best, - tainted boiled codfish. When the codfish has been kept for a very long time in the corner of the house, and when it begins to be tainted, the woman takes the tainted codfish and puts it into water that is in a large dish. As soon as the body is soaked she puts it on the fire and turns it over and over; and when the body is warm, she rubs off the scales (Boas 1921:386).

Boas goes on to describe the beating of the fish so the bones would come out easily when it was boiled. The post-cranial bones are removed while boiling, but the head was not taken out of the kettle. After they ate the fish with spoons, the man took its head. Boas (1921:388) explains, "[t]he first thing to be eaten are the eyes; and when he finishes them, he breaks the head and eats the fat of the skull; and when this is finished, he takes a spoon and eats the meat and the liquid...." Tainted codfish heads were also a specialty, "[w]hen the head of the codfish is really tainted and has been kept for a long time in the corner of the house, the woman takes her kettle and puts it down at the place where the heads are" (Boas 1921:388). Then the heads are boiled for a long time, removed from the kettle, and the eating sequence is described as same as above, "...first the eyes, and after that the fat of the skull; and they suck bones and throw them into the fire" (Boas 1921:390). Cod could also be roasted within roasting tongs (Boas 1921:390-391), although whether this was done with fresh, dried, or smoked cod is unspecified.

Use of Pacific Cod at North Point

Ethnographic and historical information can be used to help interpret the pattern of skeletal representation found at the North Point site. In the Tlingit account, Mr. Katasse describes his grandfolks meeting his father's row boat and be-heading the fish, presumably near where the boat has been hauled up. Mr. Katasse and his siblings then haul the gutted post-cranial carcasses, presumably still containing vertebrae, "above the tidewater line" where they dug a 6-inch deep "indentation." In this "shallow area" they place alternating layers of hemlock branches and cod, flesh-side down. It is worth considering a possible linkage between this practice and the context for the fish remains at North Point, where the midden is deposited on a substrate containing forest litter and woody debris. While we would not expect the bones of cod to have remained in this intermediate processing area, and certainly other faunal remains and cultural debris are found in the midden as well, conceivably these could have been disposed in the "shallow indentation" some time after it had been used to prepare cod. The presence of cod head bones would appear to argue against this possibility, since in Mr. Katasse's account, the cod heads are apparently disposed of lower in the intertidal zone. However, we should not necessarily expect to find an exact archaeological signature of a single ethnographic account. It is interesting to note that the cod cranial MNE in the assemblage is somewhat lower than the vertebrae MNE, suggesting that cod heads were treated differently than trunks.

Although the Katasse account illustrates the disposal of cod heads, other ethnographic and historical data presented above indicate dietary use of cod heads. The food value of Pacific cod is/was not limited to the post-cranial white flesh of the fish; we have seen that the stomach, liver, eggs, eyes, tongue, and meat and fat within the head were also eaten. Mr. Katasse's account probably dates to the early part of the 20th century, presumably at a time in Tlingit history when during spring, people were not as anxious for fresh food (due to the availability of non-local food) as their ancestors might have been in pre-contact times.

We suggest that the reduced number of heads as compared to trunks in the North Point assemblage may result from a number of not mutually exclusive possibilities, none of which can be demonstrated definitively at the moment. First, perhaps, some, but not all cod were decapitated prior to processing, with their heads deposited outside the site area. Second, boiling of whole fish may result in differential destruction of cranial elements vs. vertebrae, depending on when bones were removed from a cooking vessel, and conditioned by the density of the bones themselves. In one of the Boas accounts cited above, the vertebrae are removed prior to serving, while the head bones are handled and potentially damaged during consumption. Third, codfish heads could have been removed from the site to be prepared and eaten at another location in a manner similar to the tainted cod heads described in the Boas account. Perhaps the plausibility of these interpretations can be re-assessed at some time in the future, in the event that we archaeologists turn our attention to studying the variable effects of different processing sequences on bone condition, as well as cod bone density, as Butler and Chatters (1994), have done for salmon.

Other Pacific Cod-dominated Faunal Assemblages

Pacific cod is an important constituent of a small number of Northwest Coast and Alaskan faunal assemblages (Table 7). It is the most abundant fish represented in the Component III assemblage at Hidden Falls, dated to 3000 to 1300 B.P. (Moss 1989a). Pacific cod bones are also prominent in the faunal assemblage at the 8200 year old Chuck Lake site (Ackerman *et al.* 1985; Ackerman 1988, 1992, 1996), the oldest shell midden in southeast Alaska, although the faunal assemblage from this site has not been described in detail. Pacific cod also are abundant in the assemblage recently-recovered from the Russian fort at Sitka (Diane Hanson, pers. comm., 1998). Possibly each of these southeast Alaskan assemblages document use during the Pacific codfishing season of late winter through spring.

Although the Hoko River fish assemblage contains only 15% gadid family (Croes 1995:71, as measured by NISP of identified fish), Croes (1997:605) believes that the particularly numerous self-barbed bentwood fishhooks with knob leader attachment found at the site were used to catch

ARCHAEOLOGICAL SITE	PACIFIC COD (NISP)	PACIFIC COD AS %AGE OF FISH (NISP)	RATIO OF CRANIAL MNE TO VERTEBRAE MNE	REFERENCE
North Point	2089	70%	20:35	this paper
Hidden Falls III	4789	50%	73:84	Moss 1989a
Hoko River	586*	15%	N/A	Croes 1995:71
Nash Harbor	2171	70%	33:4	Souders 1997

Table 7. Selected data on Pacific cod from a few Northwest Coast and Alaskan faunal assemblages. *This number includes Pacific cod and other gadids.

Pacific cod, as opposed to the composite fishhook style. This interpretation is based on fishing experiments in the Seattle Aquarium using modern replicas of the hooks, and is based in part on the observation that cod are aggressive strikers capable of breaking traditional composite hooks (Croes 1995:100-104). A few examples of the same hook type have been found at the Locarno Beach-aged Musquean Northeast site, Marpole phase Water Hazard site, Biederbost, Little Qualicum , and Axeti. Croes states that this hook type is not found in the historic period of the central coast, but is considered the main cod-fishing hook on the north coast (citing Stewart 1977).[9]

Particularly interesting is a comparison of the three sites for which data on skeletal representation are available. We have already discussed the North Point cod which display a somewhat reduced number of cranial elements compared to vertebrae (cranial to post-cranial MNE = 20:35). This case is not that dissimilar from the situation at Hidden Falls III, although cranial and post-cranial MNE are somewhat closer (73:84 MNE). However, these two assemblages are in stark contrast to the Nash Harbor faunal assemblage recently analyzed by Paul Souders (1997).[10] The Nash Harbor faunal assemblage is similar in size to that of North Point, and is also dominated by Pacific cod. At Nash Harbor, though, the elements of cod heads are considerably better represented than cod vertebrae. The frequency of the most common cranial element, the basioccipital, results in an MNE of 33, compared to a vertebrae MNE of four. As Souders (1997:61) states, "[t]he predominance of cod heads, as opposed to cod vertebrae, suggest that the Nash Harbor people not only took cod in great numbers, but butchered them in a regular way that left cod heads in one place and cod vertebrae someplace else, or destroyed them altogether." The Nash Harbor example would appear to represent a fundamentally different way of cod processing than that represented at Hidden Falls and North Point. Whether this can be linked to seasonal, functional, or ethnic differences are topics worth further exploration in the context of Cup'it and Yup'ik ethnography.

While such an investigation is beyond the scope of this paper, this example serves to illustrate some of the range of variability possible in skeletal representation of Pacific cod in archaeological assemblages.

SUMMARY

The North Point site is surprisingly complex. Several lines of evidence indicate that the site served as a season-specific focal point for diverse collecting and processing activities. North Point was an excellent staging area from which to exploit high elevation alpine areas, the forested shoreline, the intertidal zone and nearshore waters. Subsistence activities included fishing in deep cool water, shoreline procurement of shellfish and waterfowl, a very minor amount of sea mammal hunting, and upland hunting of both small and large mammals.

The well-preserved North Point faunal assemblage is unique in its depositional setting within the intertidal zone and its clear association with both late Holocene microblade and ground slate industries on the northern Northwest Coast. The remarkable diversity of the North Point fauna, especially the high percentages of cod, challenge generalized views of subsistence on the Northwest Coast that sometimes over-emphasize salmon as the primary food resource, a view that has been described elsewhere as an "affliction" known as "salmonopia" (Monks 1987:119). While we do not wish to downplay the critical importance of salmon, especially stored salmon, as the pivotal resource which fueled the development of prehistoric Northwest Coast societies, we suggest that, for at least certain periods and/or seasons during the prehistoric past, people pursued diverse and wide-ranging subsistence activities that were not dominated by salmon procurement and storage. As illustrated by data from North Point and elsewhere, the Pacific cod in particular appears to have been a favored food resource by some people owing to a variety of possible reasons such as seasonal availability, deterioration of salmon spawning areas during Neoglaciation, a desire for fresh food as opposed to stored supplies, or simply because of dietary preferences. We hope this paper can inspire others to seriously examine the subsistence role of the humble Pacific cod.

ACKNOWLEDGMENTS

Field research for this project was undertaken by archaeologists from Northern Land Use Research, Inc., a Fairbanks-based cultural resource management firm, under subcontract to the USDA Forest Service, Tongass National Forest. The purpose of this work was to provide inventory and assessment required under the National Historic Preservation Act for a large tract of land proposed for timber harvest. The study is part of a larger Environmental Impact Statement, contracted by the Forest Service by Parametrix, Inc.

The authors would like to express their appreciation to the editors for their patience in awaiting the completion of this paper, and to Dale Guthrie, who has provided inspiration to several generations of northern archaeologists. We are grateful to the staff of Northern Land Use Research who contributed to this research, especially Catherine Williams and Ron Navarro. Russell Gould kindly provided database support and statistical analyses. Bob Betts contributed significantly to the early stages of the North Point site excavation and analysis; Owen Mason added to our understanding of the regional coastal geomorphology. Robin Mills and Mary Ann Sweeney provided some of the preliminary faunal identifications, and Nora Foster identified the shellfish from the site. Elizabeth Sturm prepared the line drawing of Pacific cod shown in Figure 5. Virginia Butler, Ruth Greenspan, and Barbara Roth are to be thanked for providing access to comparative collections. We would like to thank Kristi Madda for bibliographic assistance in our search to understand the distribution and seasonality of Pacific cod. Moss would also like to thank her former student, Paul Souders, for his forbearance of her cod obsession. We thank also Karen Iwamoto, Mark McCallum and Gerry Clark of the USDA Forest Service and Pam Gunther of Parametrix, Inc. NLUR field crew members include: Bob Betts, Catherine Williams, Owen Mason, Mark Diab, Jim Simon, Alex Pontti, Cheryl Weber, and Brian Davis (whose eagle eye first spotted a microblade core on the beach at the North Point site). Finally, we thank reviewers Aubrey Cannon, Owen Mason, Maribeth Murray, and Robin Mills for their constructive criticism that helped improve this paper. Obviously, any remaining flaws are our own.

NOTES

[1] See also Bowers and Betts (1995a, 1995b), Bowers and Moss (1996), and Bowers and Mason (1998).

[2] Obsidian from this site has recently been analyzed and traced to the Mt. Edziza obsidian source, interior British Columbia (R. Hughes, pers. comm. 2000).

[3] Initial sorting of the North Point faunal remains was accomplished by Northern Land Use Research staff in Fairbanks under the direction of P. Bowers. In September, 1995, part of the assemblage was sent to M. Moss at the University of Oregon. In December, 1995, the rest of the assemblage was sent to Moss for identification. The identifications were performed by Moss using the comparative collection of the Department of Anthropology, University of Oregon. Three other comparative collections were used to supplement those of the University of Oregon: collections at Oregon State University Department of Anthropology in Corvallis, Heritage Research Associates' collections in Eugene, and collections at Portland State University. Cannon (1987) also proved to be an invaluable reference aid in the identification process.

[4] Specimens used in this analysis (n=3087) were taken from the five 1 x 1m EUs in the main midden, shown in the first five columns of Table 5. Nine bones were excluded from this analysis; all of these were found in isolated test pits in the upper intertidal zone, and not assignable to a position in, above, or below the lower intertidal zone midden.

[5] From a bulk sample of shell midden from Unit 6-7S/1-2E (southwest quad), 30-35 cm, a 100ml volume of material was retained in the 2mm mesh screen. From this, 358 bone fragments were sorted, 319 of which were too fragmentary to identify. However, 36 vertebrae and three cranial elements of herring were identified. These results raise the possibility that herring remains are abundant in the shell midden, and that recovery methods have underestimated the importance of this species. Among the Tlingit, herring were of particular importance during at least two seasons, in the spring when they school in large numbers during spawning, and in the fall when they are prized for their high oil content (Moss 1989b).

[6] Data and inferences regarding the extent of Neoglacial cooling are ambiguous, but may indicate regional temperatures on the order of 1 to 4 degrees C lower than today (Calkin 1988; Mann and Hamilton 1995).

[7] Because there has been so little archaeological attention to Pacific cod, it is worth reiterating some background information about the spatial concentrations of cod for the benefit of other investigators. Of British Columbia waters and to the south, Ketchen (1961:517) states, "[t]he most productive fishing areas lie in northern Hecate Strait including four main fishing grounds: "White Rocks" or Banks Island; the Warrior Butterworth "edge" which is confluent with the "Two Peaks" ground at the northern end of the Strait and the Horseshoe-Cumshewa grounds in the lower central area. Farther to the south, in Queen Charlotte Sound, Pacific cod are taken principally along the southeastern edge of Goose Island Bank and along the northern edge of Cape Scott Bank. Off the lower west coast of Vancouver Island, cod are encountered on a number of restricted grounds (known locally as the "Firing Range," "Cabbage Patch" and "Forty-mile" or Big Bank) within the general confines of La Perouse Bank and also on or near Swiftsure Bank at the entrance to Juan de Fuca Strait. ...off the coast of the United States, concentrations are found mainly between Cape Flattery and Destruction Island, Washington. Minor quantities are encountered off the estuary of the Columbia River and the coast of Oregon."

[8] The preferred name for the Kwakiutl today is Kwakwaka'wakw. We defer to Boas's usage here for the ease of those who want to pursue the early 20th century anthropological literature on this First Nation.

[9] Croes' (1997) interpretation is not, however, consistent with Stewart's information. Stewart (1977:40) illustrates a Haida fishhook used to catch black cod (*Anoplopoma fimbria*) not Pacific cod. Each has its own habitat and divergent patterns of geographic and seasonal abundance. Croes (1997:606) states that "numerous examples of these bentwood hooks were collected by early explorers and ethnographers" and that he recorded 98 examples of these in the Royal British Columbia Museum. However, the provenience of the specimens and documentation as to which fish they were used for is not provided. The only fishhooks illustrated by Stewart (1977) that she identified as having been used for "cod" are the composite trolling hooks shown on p. 41. In addition, on p. 57, she illustrates the use of a "cod" lure and related gear.

[10] The Nash Harbor site is located on Nunivak Island, located in the Bering Sea within the territory of the Cup'it Eskimo (Griffin 1999).

REFERENCES CITED

Ackerman, R.E.
1988 Early Subsistence Patterns in Southeast Alaska. In *Diet and Subsistence: Current Archaeological Perspectives.* edited by B.V. Kennedy and G.M. LeMoine. Proceedings of the 19th Annual Conference, Chacmool, Archaeological Association of the University of Calgary. pp. 175-189.

1992 Earliest Stone Industries on the North Pacific Coast of North America. *Arctic Anthropology* 29:18-27.

1996 Early Maritime Cultures of the Northern Northwest Coast. In *Early Human Occupation in British Columbia*, edited by R.C. Carlson and L.D. Bona. University of British Columbia Press. pp. 123-132.

Ackerman, R.E., K.C. Reid, J.D. Gallison, M.E. Roe
1985 Archaeology of Heceta Island: A Survey of 16 Timber Harvest Units in the Tongass National Forest, Southeast Alaska. *Center of Northwest Anthropology Project Report No. 3.* Washington State University. Pullman.

Benda, L., T.J. Beechie, R.C. Wissmar, and A. Johnson
1992 Morphology and Evolution of Salmonid Habitats in a Recently Deglaciated River Basin, Washington State, USA. *Canadian Journal of Fisheries and Aquatic Science.* 49:1246-1256.

Betts, R.
1998 The Montana Creek Fishtrap I Archaeological Investigations in Southeast Alaska. In *Hidden Dimensions: The Cultural Significance of Wetland Archaeology*, edited by K. Bernick, University of British Columbia Press, Vancouver. pp. 239-251.

Boas, F.
1921 *Ethnology of the Kwakiutl.* 35th Annual Report of the Bureau of American Ethnology, 1913-1914, part I. Smithsonian Institution Press. U.S. Government Printing Office, Washington D.C.

Bowers, P.M. and R.C. Betts
1995a *Late Holocene Microblades on the Northern Northwest Coast: Preliminary Report on an Intertidal Site at Port Houghton, Alaska.* Paper presented at the 22nd Annual Meeting of the Alaska Anthropological Association, Anchorage.

1995b *Late Holocene Microblades on the Northern Northwest Coast: Preliminary Report on an Intertidal Site at Port Houghton, Alaska.* Paper presented at the International Conference: Hidden Dimensions: The Cultural Significance of Wetland Archaeology. University of British Columbia, Vancouver, B.C.

Bowers, P., R. Betts, and C. Williams
1995a *Port Houghton/Cape Fanshaw Environmental Impact Statement Cultural Resources Inventory Report.* Northern Land Use Research, Inc., Fairbanks. Report prepared for the USDA Forest Service, Petersburg.

1995b *Port Houghton/Cape Fanshaw Environmental Impact Statement Cultural Resources Effects Report.* Northern Land Use Research, Inc., Fairbanks. Report prepared for the USDA Forest Service, Petersburg.

Bowers, P.M., and O.K. Mason
1998 *Intertidal Archaeology on the Northern Northwest Coast: A Late Holocene Microblade Site From Southeast Alaska.* Paper presented at the 63rd Annual Meeting of the Society for American Archaeology, Symposium on Prehistoric Underwater Archaeology, Seattle.

Bowers, P.M. and M.L. Moss
1996 *Faunal Diversity on the Northern Northwest Coast: A View from the North Point Site, a Late Holocene Wet Site at Port Houghton, Alaska.* Paper presented at the 23rd Annual Meeting of the Alaska Anthropological Association, Fairbanks.

Bowers, P.M., C.M. Williams, R.C. Betts, O.K. Mason, R.T. Gould, and M.L. Moss
1996 *The North Point Site: Archaeological Investigations of a Prehistoric Wet Site at Port Houghton, Alaska.* Report prepared for USDA Forest Service, Chatham Area, Sitka.

Brew, D.A. and A.B. Ford
1978 Megalineament in Southeastern Alaska Marks Southwest Edge of Coast Range Batholithic Complex. *Canadian Journal of Earth Science* 15:1763-1772.

Butler, V.L. and J.C. Chatters
1994 The Role of Bone Density in Structuring Prehistoric Salmon Bone Assemblages. *Journal of Archaeological Science* 21:413-424.

Calkin, P.E.
1988 Holocene Glaciation of Alaska, and Adjoining Yukon Territory, Canada. *Quaternary Science Reviews.* 7:159-184.

Cannon, D.Y.
1987 *Marine Fish Osteology: a Manual for Archaeologists.* Department of Archaeology Publication No. 18, Simon Fraser University, Burnaby, B.C.

Cobb, J.N.
1916 *Pacific Cod Fisheries.* Appendix IV to the Report of the U.S. Commissioner of Fisheries for 1915, Bureau of Fisheries Document No. 830. Department of Commerce, Bureau of Fisheries, U.S. Government Printing Office, Washington, D.C.

Croes, D.R.
1993 Prehistoric Hoko River Cordage: A New Line on Northwest Coast Prehistory. In: *A Spirit of Inquiry, Essays for Ted Wright*, edited by John Coles, Valerie Fenwick, and Gillian Hutchinson. WARP Occasional Paper 7, Exeter Department of History and Archaeology, University of Exeter. pp. 32-36.

1995 *The Hoko River Archaeological Site Complex.* Washington State University Press, Pullman.

1997 The North-Central Cultural Dichotomy of the Northwest Coast of North America, its Evolution as suggested by Wet-Site Basketry and Wooden Fish-hooks. *Antiquity* 71(273):594-615.

Cwynar, L.C.
1990 A Late Quaternary Vegetation History From Lily Lake, Chilkat Peninsula, Southeast Alaska. *Canadian Journal of Botany* 68:1106-1112.

Emmons, G.T.
1991 The Tlingit Indians. In *Anthropological Papers of the American Museum of Natural History, vol. 70.* edited by Frederica de Laguna. University of Washington Press. Seattle and London.

Flenniken, J.J.
1981 Replicative Systems Analysis: A Model Applied to the Vein Quartz Artifacts from the Hoko River Site. *Washington State University, Laboratory of Anthropology. Reports of Investigations No. 59*, Pullman.

Foster, N.R.
1991 *Intertidal Bivalves: A Guide to the Common Marine Bivalves of Alaska*. University of Alaska Press, Fairbanks.

Griffin, D.
1999 *Portrait of Nash Harbor: Prehistory, History and Lifeways of an Alaskan Community.* Ph.D. dissertation, University of Oregon, Eugene.

Goldthwait, R.G.
1966 Glacial History. In *Soil Development and Ecological Succession in a Deglaciated Area of Muir Inlet, Southeast Alaska*, edited by A. Mirsky. Institute of Polar Studies Report 20. Ohio State University, Columbus. pp.1-17.

Hart, J.L.
1980 Pacific Fishes of Canada. *Fisheries Research Board of Canada Bulletin 180,* Ottawa. (see also first edition, 1973).

Heusser, C.J.
1985 Quaternary Pollen Records from the Pacific Northwest Coast: Aleutians to the Oregon-California Border. In *Pollen Records of Late Quaternary North American Sediments*, edited by V.M. Bryant and R.G. Halloway. American Association of Stratigraphic Palynologists Foundation, Dallas. pp. 353-387.

Heusser, C.J., Heusser, L.E., and D.M. Peteet
1985 Late Quaternary Climatic Change On the American North Pacific Coast. *Nature* 315:485-487.

Ketchen, K.S.
1961 Observations on the Ecology of the Pacific Cod (*Gadus macrocephalus*) in Canadian Waters. *Journal of the Fisheries Research Board of Canada* 18(4):513-558.

de Laguna, F.
1960 The Story of a Tlingit Community: A Problem in the Relationship Between Archaeological, Ethnological, and Historical Methods. *Smithsonian Institution Bureau of Ethnology Bulletin 172.* U.S. Government Printing Office, Washington D.C.

1972 Under Mount Saint Elias: The History and Culture of the Yakutat Tlingit. *Smithsonian Contributions to Anthropology, vol. 7(1).* U.S. Government Printing Office, Washington D.C.

Mann, D.H.
1986 Wisconsin and Holocene Glaciation in Southeast Alaska. In *Glaciation in Alaska: The Geological Record,* edited by T.D. Hamilton, K.M. Reed, and R.M. Thorson. Alaska Geological Society, Anchorage. pp. 237-265.

Mann, D.H., and T.D. Hamilton
1995 Late Pleistocene and Holocene Paleoenvironments of the North Pacific Coast. *Quaternary Science Reviews* 14:449-471.

Matson, R.G.
1989 The Locarno Beach Phase and the Origins of the Northwest Coast Pattern. In *Development of Hunting-Fishing-Gathering Maritime Societies Along the West Coast of North America,* edited by B. Onat. Washington State University Press, Pullman.

Matson, R.G. and G. Coupland
1995 *The Prehistory of the Northwest Coast.* Academic Press, San Diego.

Miller, R.D.
1975 Surficial Geology Map, Juneau Urban Area and Vicinity. *U.S. Geological Survey Miscellaneous Investigation Series* Map I-885.

Milner, A.M., and R.G. Bailey
1989 Salmonid Colonization of New Streams in Glacier Bay National Park, Alaska, *Aquaculture and Fisheries Management* 20:179-192.

Monks, G.G.
1987 Prey as Bait: The Deep Bay Example. *Canadian Journal of Archaeology* 11:119-142.

Moss, M.L.
1989a Analysis of the Vertebrate Assemblage. In *The Hidden Falls Site, Baranof Island, Alaska*, edited by S.D. Davis, Aurora V. Alaska Anthropological Association Monograph Series. Department of Anthropology, SUNY Brockport. Brockport, NY. pp. 93-129.

1989b Archaeology and Cultural Ecology of the Prehistoric Angoon Tlingit. Ph.D. Dissertation. University of California, Santa Barbara. University Microfilms, Ann Arbor.

Moss, M.L., and J. M. Erlandson
1998 A Comparative Chronology of Northwest Coast Fishing Features. In *Hidden Dimensions: The Cultural Significance of Wetland Archaeology*, edited by K. Bernick, University of British Columbia Press, Vancouver. pp. 180-198.

Moss, M.L., J.M. Erlandson, and R. Stuckenrath
1990 Wood Stake Weirs and Salmon Fishing on the Northwest Coast: Evidence from Southeast Alaska. *Canadian Journal of Archaeology* 14:143-157.

Motyka, R.J., and J.E. Beget
1996 Taku Glacier, Southeast Alaska, USA: Late Holocene History of a Tide-water Glacier. *Arctic and Alpine Research* 28(1):42-51.

Newton, R.G., and M.L. Moss
1984 *The Subsistence Lifeway of the Tlingit People: Excerpts of Oral Interviews*. USDA Forest Service. Alaska Region Report 179, Juneau.

Oberg, K.
1973 *The Social Economy of the Tlingit Indians*. University of Washington Press, Seattle.

Parametrix, Inc.
1998 *Port Houghton/Cape Fanshaw Timber Sale Project: Draft Environmental Impact Statement*. USDA Forest Service, Tongass National Forest, Juneau.

Peteet, D.M.
1986 Vegetational History of the Malaspina Glacier District. *Quaternary Research* 25:100-120.

Ricketts, E.F., J. Calvin, and J.W. Hedgpeth
1985 *Between Pacific Tides*. 5th edition, revised by D.W. Phillips. Stanford University Press, Stanford..

Sealaska Corporation
1975 *Native Cemetery and Historic Sites of Southeast Alaska*. Sealaska Corporation, Juneau.

Shimada, A.M. and D.K. Kimura
1994 Seasonal Movements of Pacific Cod, *Gadus macrocephalus*, in the Eastern Bering Sea and Adjacent Waters Based on Tag-Recapture Data. *Fishery Bulletin* 92(4):800-816.

Souders, P.
1997 Ellikarrmiut Economy: Animal Resource Use at Nash Harbor (49-XNI-003), Nunivak Island, Alaska. M.S. paper, Department of Anthropology, University of Oregon, Eugene.

Stanford, M. and R. Lightfoot
1981 Archaeological Reconnaissance Report of Port Houghton, Windham Bay. Ms. on file at Chatham Area Office, Tongass National Forest, Sitka.

Stepanenko, M.A.
1995 Distribution, Behavior and Abundance of Pacific Cod, *Gadus macrocephalus*, in the Bering Sea. *Journal of Ichthyology* 35(4)18-27.

Stewart, H.
1977 *Indian Fishing: Early Methods on the Northwest Coast.* University of Washington Press, Seattle.

Streveler, G., D. Mann, and A. Crowell
1995 Biological Impacts of Late Holocene Environmental Change, Northern Southeast Alaska. *Program and Abstracts, 46th Arctic Division Science Conference, American Association for the Advancement of Science*, Fairbanks. p. 70.

Stuiver, M., and P.J. Reimer
1993 Extended ^{14}C Database and Revised CALIB Radiocarbon Calibration Program, *Radiocarbon* 35(1):215-230.

WAS SALMON IMPORTANT IN NORTHWEST COAST PREHISTORY?

Aubrey Cannon
Department of Anthropology, McMaster University

The timing and relative importance of a storage-based, salmon-fishing economy has become a key issue in recent discussions of Northwest Coast prehistory (Ames 1994; Croes 1992; Matson 1992; Moss and Erlandson 1995). Intensive salmon fishing, long recognized as a critical element of ethnographically-recorded subsistence economies, is commonly seen as the basis for the dense population, complex social organization, and elaborate artistic specialization characteristic of Northwest Coast cultures; the timing of its development is therefore key to understanding the role of economic production in the region's social and political history. Although opinions vary widely, two major schools of thought have emerged in recent years. One ascribes a major role to salmon fishing in the early subsistence economies of the region (ca. 6000-2000 B.P.), but views subsequent social and cultural developments as more or less independent longer-term historical processes (Cannon 1991; 1998, Carlson 1983, 1996, 1998). The other sees a direct causal linkage among relatively late intensification of salmon production and storage (ca 3500-3000 B.P.), population growth, resource ownership, and social differentiation (Coupland 1998; Croes and Hackenberger 1988; Matson 1992).

Although debate on this issue once turned on mere supposition as to whether early economies were more maritime or terrestrial oriented, zooarchaeological data have provided a basis for much more detailed and sophisticated arguments. At the heart of the debate are two conflicting common-sense notions of what the economic prehistory of the Northwest Coast should be like. One considers the obvious advantages of a salmon-based economy, noting that the resource is abundant and seasonally predictable, suitable for mass harvesting and processing for storage, nutritious, and available in late summer and fall just prior to the winter season when most alternative foods are in relatively short supply. From this perspective, salmon is the resource of first choice for any subsistence economy in the region. The other, equally common-sense notion appears to take the principle of least effort as its basis, and views salmon fishing as analogous to other intensive subsistence activities such as seed gathering or horticulture. From this perspective, intensification and storage are unwarranted until such time as population or social demands force the required increase in labor investment.

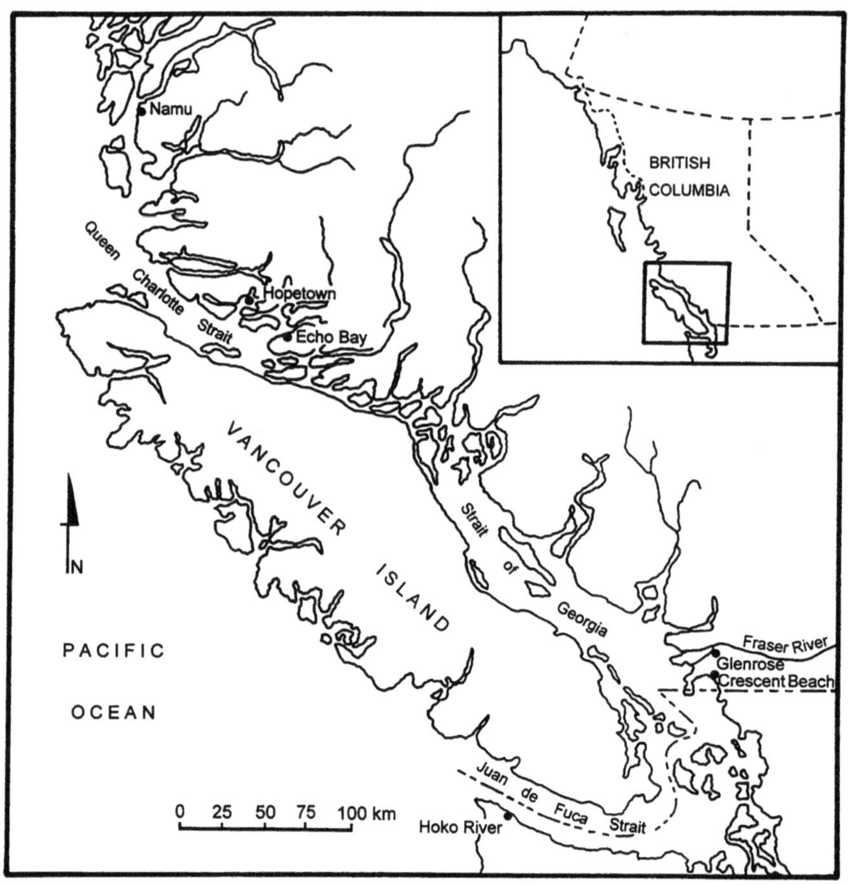

Figure 1. Location of sites mentioned in the text.

The philosophical differences that underlie these positions may be irreconcilable, but the archaeological evidence should provide unequivocal support for one position or the other. The limited evidence available thus far, however, has been used to lend support to both sides of the debate, and it is now time to review the nature of the evidence, and the forms and appropriate standards of evidence required to resolve whether early prehistoric economies were based largely or in part on intensive salmon production. A review of evidence in support of models of later intensification, and clear evidence of an early fishery at Namu, British Columbia (Figure 1) suggest some standards for evaluating current arguments. The inadequacy of current data to resolve the overall debate, however, also indicates the need to develop more effective research strategies in zooarchaeology to extract better, and more reliable information from the available evidence.

ARGUMENTS FOR LATE INTENSIFICATION

Beginning in the late 1980s, archaeologists brought a much-needed critical light to bear on the largely unquestioned assumption that salmon was the only important resource in Northwest Coast prehistory. Researchers questioned whether its value had been overemphasized based on contemporary impressions of desirability (Croes 1988), and the term "salmonopeia" was coined to describe the archaeological neglect of most other foods (Monks 1987:119). Salmon's importance was questioned largely because excavations at various sites had revealed remains of other resources that were far more numerous (e.g. see Bowers and Moss, this volume). Unfortunately, in some cases the interpretation of these data went well beyond the issue of salmon's ubiquitous importance to become the basis for elaborate evolutionary and historical models developed to explain the economic prehistory of the entire Northwest Coast.

Three models derived wholly or in part from zooarchaeological evidence currently purport to demonstrate and explain the relatively late development of intensive salmon fisheries. One scenario, developed by Donald Mitchell (1988) based on the results of his excavations at sites in the Queen Charlotte Strait region on the south-central coast of British Columbia, claims only to explain the southern expansion of an already-developed fishery. The claim is based on observed contrasts in the artifact and faunal assemblages between an early period characterized as the Obsidian Culture Type, dating to 5000-2500 B.P., and a later pattern termed the Queen Charlotte Strait Culture Type, dating to 1650-200 B.P.. These contrasts, including a dramatically increased emphasis on salmon fishing and sea-mammal hunting in the later period are attributed to the migration of Wakashan-speaking peoples from areas to the north. Although not an explanation for the inception of intensive salmon fishing, this evidence does indicate a clear contrast in subsistence economies over time that lends support to the notion that salmon may not have been particularly important in early prehistory. It also lends weight to the argument that an emphasis on salmon is not necessarily the inevitable result of any superior qualities such as abundance, nutrition, or predictability.

The only evidence that casts doubt on Mitchell's interpretation is the nature of the resources exploited during the early period. The faunal remains show a heavy emphasis on species such as deer and ratfish, which are nutritionally inferior and relatively more difficult to harvest in quantity (Cannon 1995a). Such marginal resources are unlikely to have been the focus of subsistence simply as a matter of choice. The only alternative explanation is that for some reason salmon was less readily available at these sites during the early period, though this conclusion is not supported by any evidence other than the propensity for local productivity to vary over time (Cannon 1996a). The patterns observed at the Queen Charlotte Strait sites are interesting examples of temporal variability in regional subsistence economies, but they do not necessarily point to any general long-term trends.

In contrast, models developed initially by Croes and Hackenberger (1988) and later elaborated by Croes (1992) and Matson (1992) do purport to explain the gradual evolutionary transition toward more intensive storage-based, fishing economies. Based on contrasting faunal assemblages at two sites on the Hoko River on Washington's Olympic Peninsula, Croes developed an elaborate model that outlined a gradual process of increasing reliance on storage. In this model, the processing and air drying of summer-caught flatfish to supplement fresh winter supplies of deer and shellfish, allowed for population growth and resource depletion, which eventually culminated in the need to invest the time, labor, and new technologies necessary to intensively harvest, process, and store salmon. The model accounted for faunal assemblages that showed a predominance of flatfish and deer remains at one site dated to 3000-2200 B.P. and the predominance of salmon at another nearby site dated to 900-100 B.P..

Matson's (1992) excavations at the Crescent Beach site produced further evidence that he interpreted as confirming many of Croes' suggestions. This includes an abundance of flatfish remains in earlier deposits and a predominance of salmon in later deposits in the area where he excavated. Matson also interpreted the appearance of quartz microliths as a technological innovation that enabled more efficient processing of salmon for storage. Slate knives are considered an improvement on this initial processing technology. Matson also claimed that variable representation of salmon skeletal elements over time was the result of a shift from the consumption of fresh fish, indicated by an abundance of cranial elements, toward more intensive processing for storage, indicated by the almost complete absence of cranial elements, which are assumed to have been discarded. Although based on a small excavated area of a single site, the points of correspondence between Matson's data and Croes' model provide the basis for a substantial argument.

These models are all supported by empirical data, and fit closely with a pervasive evolutionary perspective, but they also incorporate and depend on a number of unwarranted assumptions that taken together raise serious doubts about their validity. Each assumes that the evidence from a limited number of sites can be taken as representative of broader regional trends. They also assume that a capacity for increased production and storage is the result of a gradual long-term process. In addition, they incorporate a number of particular assumptions about the difficulties and necessary technologies needed for storage, the nutritional viability of resource alternatives, and the taphonomic processes responsible for particular patterns of skeletal representation. When examined in turn, these assumptions appear individually unfounded, and collectively suggest that the case for relatively late intensification of salmon fishing and storage is not as strong as it may seem.

EVALUATION OF THE ARGUMENTS

Scale and Variability

Basic statistical reasoning provides ample argument against any conclusion based on a small sample drawn from a variable population, and general models of Northwest Coast subsistence are based on very few sites. In some cases, patterns evident at one or two sites have been used to characterize trends for the entire region. This would be a questionable practice for any comparably-sized region, but is particularly unwarranted in a region where local micro-environments vary widely, and where ethnographic accounts clearly indicate seasonal and locational variability in subsistence activities.

Although the degree of seasonal mobility evident in the period following European contact may not extend to earlier periods (Ford 1989), seasonal variability in resources and site usage is clearly evident in the archaeological record. The restricted seasonal availability of salmon and herring suggests that different patterns of site usage should be considered as a probable reason for any significant variability in their relative abundance. Unfortunately, though Croes and Hackenberger (1988:21-22) acknowledge that the Hoko River sites probably represent different seasonal occupations, this more likely explanation is subsequently ignored in favor of a more elaborate model of long-term economic transition. Although the sites represent different time periods, their contrasting locations and seasonally-distinct occupations provide the best explanation for their exhibited contrasts in subsistence activity. Matson's interpretation of the Crescent Beach fauna, though based on different periods of occupation at a single site, also fails to consider the more likely possibility that the assemblages simply represent different seasonal activities. An inferred focus on flatfish fishing in the spring (Matson 1992:412) would suggest this was the season of the early-period occupation. Shell analysis also indicates different seasons of shellfish collection between the early and later components (Rankin 1993).

In addition to local seasonal patterns, site usage and resource availability are also subject to longer-term variability. Salmon productivity is particularly subject to the effects of local and regional environmental conditions, and the potential for dramatic change in average abundance at particular site locations is well-documented (Cannon 1996a). The archaeological data from Namu show a clear pattern of growth and decline in production that appears to be closely related to the environmentally-governed productivity of the local fishery (Cannon 1991). The increases in production observed at Crescent Beach and sites in the Queen Charlotte Strait region could equally be due to environmental conditions that enhanced local productivity. Although direct investigation of this alternative might be difficult, the effects of local environmental change should be considered a likely explanation unless a wider regional pattern of economic transition becomes evident.

Any interpretation of regional subsistence patterns and trends requires a large sample of sites to control for local, seasonal, and temporal variability. By this measure, the regional models developed by Croes and Matson are an over-interpretation of very limited site data, which may have value as working hypotheses, but which require vigorous evaluation against a much larger sample of sites (see Hanson, this volume).

Foresight and Technology

The empirical data in support of long-term gradual economic intensification may be limited, but the notion is supported by the pervasive assumption that storage-based economies represent a conceptual and technological advance that requires time to develop. The assumption is implicit in statements citing the need to develop the appropriate technologies for efficient processing and storage (Matson 1992:383, Schalk 1977:231-232). The association of microliths with inferred evidence of storage at Hoko River and Crescent Beach is cited as evidence of this essential technological development (Croes 1992, Matson 1992). The relative ease of air-drying flatfish is also described as a reason why this would likely be the first step toward development of salmon storage (Croes 1992:345).

The efficiency of microliths and later slate knives has never been evaluated, but indirect evidence strongly suggests that the availability of processing technology was never a key to the advent of salmon storage. Evidence for mass-harvesting technology in the form of a fish weir at the Glenrose Cannery site dated to 4500 B.P. (Eldridge and Acheson 1992), suggests ample ability to process the number of fish caught, and the consistent abundance of salmon at Namu, dating from as early as 6000 B.P. (Cannon 1996b), also suggests that the necessary processing technology and storage techniques were well developed prior to that time. The techniques of drying and smoking for storage have been described as basically simple (Ames 1994:217), and would not have required a long period of development. Although processing technologies may have marginally increased the efficiency of later fisheries, no evidence suggests they were critical to the initial development of storage.

Direct evidence of the capacity to construct storage boxes and the existence of rectangular surface dwellings that could also have served as storage facilities have been cited as evidence for initial development of a storage-based economy at around 3500-3000 B.P. (Ames 1994:217-219). These forms of evidence, however, may not be appropriate criteria for identifying the initiation of storage on the Northwest Coast, where facilities typically consisted of perishable baskets, boxes, and roof rafters (Suttles 1968:63-64). The capacity for storage could have been enhanced by the advent of cedar woodworking technology ca. 5000 B.P. (Cannon 1998; Hebda and Mathewes 1984), but there is little reason to believe that the potential for storage changed at all after that time.

Given the readily available capacity for mass harvest, processing, and storage, there is little justification for any model that assumes the need for gradual development in skills, or a "push" toward storage as the result of resource depletion and population growth. The need to provide for winter subsistence, combined with the timely seasonal availability of salmon leads to the more likely conclusion that storage was an early development. The foresight to utilize abundant resources to offset seasonal shortfall is considered such a defining human characteristic that it is used to draw a distinction between modern humans and Neandertals in Europe (Binford 1981). The development of intensive salmon fishing and storage is also postulated to have occurred in eastern Siberia as early as the late Upper Palaeolithic, around 12,000-9000 years ago (Tabarev 1996, Vasil=evskii 1998). Denying this potential to the early inhabitants of the Northwest Coast is therefore indefensible.

Alternative Resource Strategies

The ability to sustain a regional subsistence economy without a general reliance on salmon is a key assumption for models of later intensification. Certainly a vast array of other foods were important, and the range of plant, fish, and mammal foods available and used is well documented ethnographically and archaeologically. Zooarchaeological evidence also shows that salmon was not the single most important resource at many specific site locations (e.g. Bowers and Moss this volume). Although this variety of subsistence options has never been seriously questioned, the range of alternatives capable of sustaining winter subsistence is less certain. Croes (1992) and Croes and Hackenberger (1988), for example, postulate an early-period winter economy based on shellfish collection and deer hunting, supplemented by stores of dried flatfish, all of which he considers less costly, but equally viable alternatives to stored salmon.

The problem with this model, however, is that it seeks only to balance relative amounts of winter food between the early and later periods. It demonstrates that available alternatives could supply the amount of calories required for winter subsistence, particularly before the onset of regional population growth, but it does not take account of the relative nutritional qualities of salmon and other foods. The main difference between salmon on the one hand, and deer, shellfish, and dried flatfish on the other is their relative fat content, and fat is a critical component in the diets of hunter-gatherers, particularly in northern regions where carbohydrates are often in short supply (Speth and Spielmann 1983). Carbohydrate sources on the Northwest Coast are restricted to a range of plant foods that are mostly available in the spring and summer. None of these plants occur in great local abundance, and though berries and some roots were dried and stored for winter use, fat in the form of fish oil and rendered sea-mammal fat was critically important in the local diet (Suttles 1968:63).

Deer can be an adequate source of animal fat in late summer and fall, but its value in this regard diminishes rapidly over the course of the winter, though it remains a good source of dietary protein. Shellfish are another good source of lean protein, but they are also particularly lacking in fat. The average fat content of clams and mussels is no more than two percent (Watt and Merrill 1963). The leanness of flatfish meat is the specific property that makes it easy to dry and then store for long periods of time (Croes and Hackenberger 1988). Flatfish, like a number of other fish species, contain most of their oil in the liver rather than the muscle tissue. The flesh of halibut and most other flatfish is less than one percent oil (Bailey 1952; Watt and Merrill 1963). Salmon belongs to the group of fish that contain most of their oil within the muscle tissue, and the average oil content of the various salmon species ranges between five and thirteen percent, and as high as twenty percent. This is the property that makes them a particularly important food source in winter when other fat- and carbohydrate-rich resources are in shortest supply. If for no other reason, salmon would have to be considered the overwhelmingly preferred alternative to deer, shellfish, and dried flatfish as a basis for winter subsistence.

Taphonomy

Although it has only been used as a supporting line of evidence to bolster an existing argument for later transition to salmon storage, the representation of skeletal elements has been cited widely in ongoing debates and analyses of faunal remains (Ames 1994; Butler and Chatters 1994; Croes and Hackenberger 1988). The premise, as set forward by Matson (1992) based on data from the Crescent Beach site, is that expedient use of fresh-caught salmon for food results in the immediate disposal of the entire skeleton. The intensive processing of fish for storage, in contrast, is said to result in the initial disposal of heads, and therefore cranial elements, at the site of processing, and later disposal of vertebrae at the site of consumption. The higher representation of cranial elements in earlier deposits at Crescent Beach is the basis for the conclusion that salmon storage was a relatively late development.

Although the argument is clear and logical, it also contains a number of unwarranted assumptions and ignores many equally-viable explanations for the same pattern. Salmon heads were processed and smoked for storage in some ethnographically-described cases (Duff 1952:66; Stewart 1977:137), and this would result in very different patterns of preservation and representation. The presence of dogs or other scavengers could also affect the relative survival of skeletal elements. Arguably, the surfeit of fish heads potentially made available by mass harvest and processing (assuming they were discarded) would result in more rather than fewer preserved cranial elements if dogs were the main agents of bone destruction. By the same measure, smaller numbers of fish caught for fresh consumption would result in more complete consumption by dogs of any portions that happened to be discarded. Variation in the organization and use of space at a site over time could also affect the relative distribution of skeletal elements regardless of how fish were actually used. Efforts to control for the taphonomic processes responsible for salmon bone preservation and recovery in order to evaluate the effects of processing and storage patterns (e.g. Butler and Chatters 1994) would therefore seem unwarranted unless a consistent temporal pattern is demonstrated at a much larger number of sites. In the meantime, cranial representation appears to be very weak supporting evidence for any major temporal trend toward later intensification of salmon production.

Summary

The argument that intensive salmon fishing and storage was a relatively late development in Northwest Coast prehistory is not strongly indicated by any of the available archaeological evidence. Empirical zooarchaeological support for this position is present at only a small number of sites, and this sample is insufficient to allow for any extrapolation to the regional scale. Specialized technologies and skills are not required to process and store large quantities of salmon, and conclusive early evidence of the capacity to mass-harvest this seasonally-restricted resource indicates a capacity for storage from an equally-early date. Alternatives to winter supplies of stored salmon, such as deer, dried flatfish, and shellfish, though possibly adequate as sources of protein and calories, lack the critical fat needed to maintain a viable winter diet. Finally, the evidence of differential cranial representation is simply subject to too many potential influences to support a fundamental shift in salmon fishing and processing. Although zooarchaeological data can be assembled in support of the view that intensive production and storage occurred at a relatively late date, the simpler argument that salmon could be stored and was used to sustain winter subsistence from a very early period requires only a single well-documented example. At present, the evidence from Namu makes a compelling case, and is more than sufficient to undermine regional models that would go so far as to deny an early capacity for storage.

THE ROLE OF SALMON AT NAMU BRITISH COLUMBIA

The fish and other faunal remains recovered in excavations by Roy L. Carlson of Simon Fraser University in 1977 and 1978 provide a clear record of the local subsistence economy during the period 6060-480 B.P.. Of the almost 100,000 fish remains recovered, the overwhelming majority are salmon, and this is a pattern that is consistent throughout the period for which faunal remains are available (Figure 2) (Cannon 1991). Recovery methods certainly resulted in the under-representation of herring, which, based on the number of whole vertebrae recovered in auger samples, may have been up to as fifty times as abundant as salmon in some periods (Cannon 1996b) and at least thirty times as abundant overall, but the undeniable subsistence value of herring does not detract from salmon's early and consistent contribution to the local subsistence economy. Although the presence of large quantities of salmon bones has been described as only weak evidence for storage (Lyman 1991:287), the fact that anadromous runs are available for a limited time each year implies that large quantities represent more than the amount that could be consumed fresh. This fact, together with the timing of runs just before the period of relative winter food scarcity suggests that large quantities of salmon almost certainly represent a pattern of harvest and processing for storage.

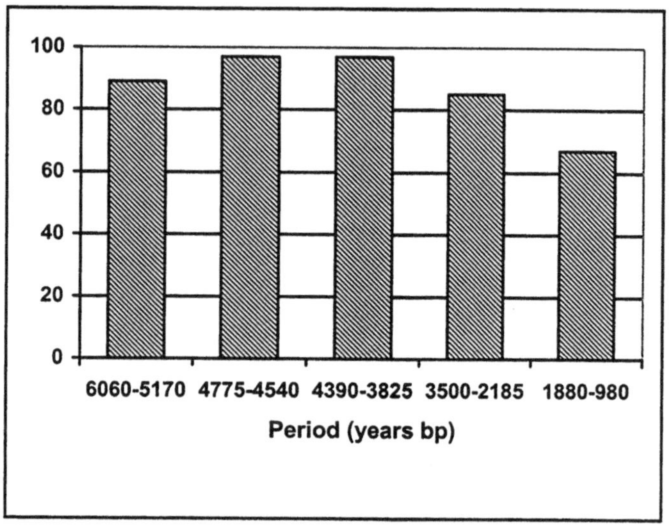

Figure 2. Abundance of salmon remains from Namu, British Columbia, as a percentage of all identified fish by period (dates in radiocarbon years B.P.).

The response to variable productivity in the Namu fishery also points to salmon's critical subsistence role. A decline in production after 3500 B.P., which has been attributed to local environmental change (Cannon 1991), is correlated with increasing reliance on more marginal resources including ratfish and deer (Cannon 1995a). The implication is that the early population had the capacity to harvest and store salmon, and preferred this subsistence option, but were later forced to resort to marginal alternatives to compensate for a shortfall in the local fishery. This conclusion is further supported by the results of recent isotopic analysis of dog bones representing the periods before, during, and after the peak in salmon production (Cannon *et al.* 1999). These show that marine resources contributed 80 to 100 percent of the protein portion of the dogs' diet, which is identical to the results of an analysis of human bone samples from Namu (Chisholm *et al.* 1983). The Namu dog bone samples, however, show a tendency for the marine contribution to vary directly with the level of salmon production. The mean contribution of marine protein to the diet of dogs is the least and its variability the greatest in the period after 3500 B.P., when the salmon fishery appears to have been in decline. These results are consistent with a pattern in which salmon production was increasingly subject to periodic shortfall. The clear implication is that the extent to which terrestrial resources (primarily deer) were utilized is directly related to the availability of salmon. Analysis of variable fracture patterns in deer phalanges also indicates a significant increase after 3500 B.P. in bones exhibiting medial-lateral fractures (Zita 1997), which is a pattern identified by Binford (1978:148) as a characteristic result of human efforts to break open phalanges to obtain marrow. These additional data confirm patterns indicated by the relative abundance of fish remains, and lend support to the conclusion that salmon consumption varied according to local availability rather than long-term evolutionary trends.

Although no definitive evidence of storage in the form of special technologies or facilities is available from Namu, such evidence is unlikely to be present at any early Northwest Coast site. Direct zooarchaeological evidence, however, is more than sufficient to show a long-term pattern of intensive harvest and storage. The multiple lines of evidence that support this interpretation include large quantities of this seasonally-restricted resource and evidence of the periodic need to resort to more marginal alternatives, which is indicated by faunal remains and isotopic confirmation. Even the commonly-cited evidence of low cranial element representation supports the early presence of storage at Namu. The representation of salmon cranial elements, as noted above, can be affected by many different factors and therefore may not have any necessary relationship to the presence or absence of storage, but cranial elements are actually less abundant at Namu from as early as 6000 B.P. (Cannon 1991:18) than they are during any period at Crescent Beach.

Namu of course is just one site, and in itself is not representative of a general regional pattern, but this one example is sufficient to negate the argument that the region as a whole is characterized by a gradual, long-term evolutionary trend toward intensive salmon fishing. The issue of regional subsistence cannot be resolved one site at a time, but the zooarchaeological investigations at Namu indicate the type of alternative research strategy that may be needed to develop a more comprehensive understanding of Northwest Coast subsistence.

AN ALTERNATIVE STRATEGY

The one clear indication from this review is that local data and subsistence economies are too highly variable to allow for the direct inference of general economic trends. The deposition, taphonomy, and recovery effects that determine the quantities and types of bone obtained from coastal shell midden sites demand either extraordinary investment in area and volume excavation (Lyman 1991) or the use of independent, but mutually supportive evidence to draw robust conclusions. Patterns in the Namu fauna that suggest early intensity but variable productivity in the local salmon fishery are supported by the relative quantity of salmon bones, the negatively-correlated abundance of more marginal resources, the isotopic content of dog bones, and shell species data that show a correlated increase in the abundance of silt- and sand-dwelling species of clams (Conover 1978), suggesting increasing siltation of the Namu River estuary, which could have had a deleterious effect on spawning runs. None of these lines of evidence is directly related to the others except in relation to the productivity of salmon. Indicators of seasonality, the only other likely cause of major variability in local resource use, remain constant over time and show a consistent pattern of multi-seasonal occupation.

Mutually-supportive evidence for the later intensification of fisheries at Crescent Beach and Hoko River is much less closely related to the single issue of subsistence. Direct relationships linking technology, cranial element representation, and the intensity of salmon fishing have yet to be established. In the meantime, the simpler argument that observed contrasts in salmon abundance are due to seasonal differences in site use over time or between locations should prevail over any effort to build a regional model on such a limited data base.

The small number of site samples and the lack of an integrated regional approach to subsistence studies constitute the greatest impediment to achieving better understanding of salmon's overall importance in prehistory. The difficulty and prohibitive costs of shell midden excavation, and the extremely limited number of test excavations undertaken to date, suggest that comprehensive regional models, if at all applicable, are a distant prospect. An alternative approach, recently applied in the area around Namu, involves the recovery of small samples of faunal material from a relatively large number of sites through a combination of coring and bucket-auger sampling of midden matrix (Cannon 1995b). Preliminary testing at Namu showed this was an effective method for monitoring trends in the herring and salmon fisheries. The density of whole elements and identifiable fragments of salmon vertebrae in the auger samples (Figure 3) showed the same overall trend for salmon as was evident in the analysis of fauna recovered from the large-scale excavations undertaken in 1977 and 1978 (Figure 2).

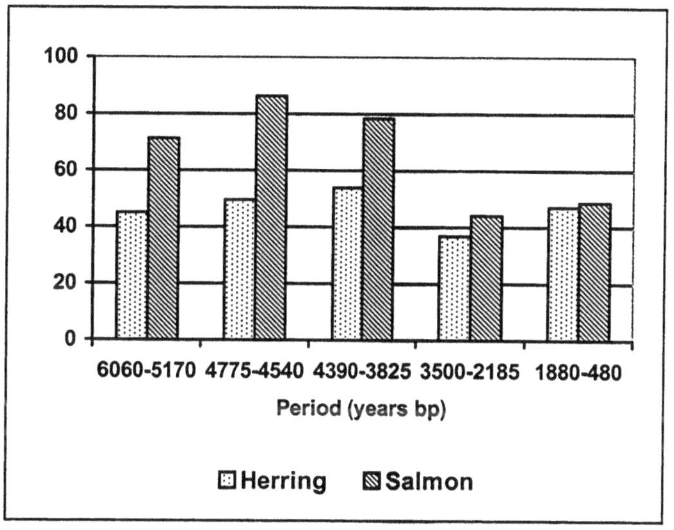

Figure 3. Density of salmon and herring remains in bucket-auger samples from Namu by period (number of specimans per liter of matrix finer than 2 mm).

The intensity of the herring fishery remained more or less consistent throughout the last 6000 years of the Namu occupation. The enhanced recovery of vertebrate faunal remains from the matrix samples, which were washed through 2 mm mesh screen and later examined under a microscope, showed that the overall density of herring vertebrae was twenty-six times greater than that of whole salmon vertebrae. The early spring herring fishery was clearly as important as the fall salmon fishery to the overall subsistence economy.

The preliminary testing at Namu showed that bucket-auger sampling was a valid alternative to full-scale excavation as a means of monitoring the intensity of the most common and most important fisheries at the site. On this basis, an expanded study was undertaken in 1996 and 1997 to obtain further samples from other sites in the Namu area. When combined with coring to control for stratigraphic context and to enable radiocarbon dating, bucket-auger sampling provided an effective means of establishing the presence and relative importance of salmon fisheries between locations and over time. The analysis and interpretation of the results of site testing in the Namu vicinity are still underway, but the completed analysis of the vertebrate fauna recovered from bucket-auger samples obtained at two nearby village sites illustrates the value of this approach, and clearly shows the type of inter-site variability in the timing and intensity of salmon fisheries that might be expected even within a limited local area.

Figure 4 shows the location of two of the village sites recently tested in the Namu vicinity. The Kisameet site (ElSx-3), some 12 km to the north, was test excavated by researchers from the University of Colorado in 1969 and 1970 (Hester and Nelson 1978), though no analysis of the fish remains was apparently undertaken and none was reported. The other village site (EkSx-12) is located near the mouth of the Koeye River, 9 km to the south. The Kisameet site shows evidence of relatively recent habitation (Luebbers 1978:17) and the Koeye village site exhibits the remains of five rectangular house platforms on the surface.

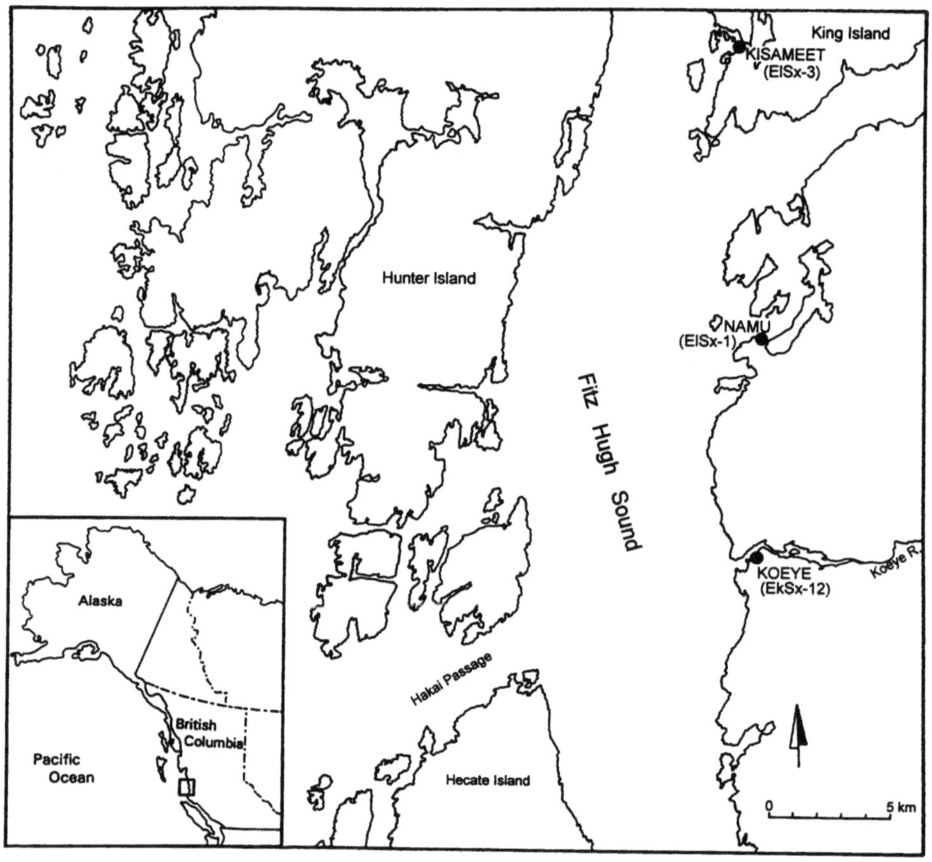

Figure 4. Location of Kisameet and Koeye vollages in relation to Namu.

SITE	TIME RANGE	# OF SAMPLES	TOTAL VOL	DENSITY/ TOT. VOL	< 2 MM VOL	DENSITY/<2 MM VOL
Namu (ElSx-1)	4775-3825 B.P.	69	22.9	65.5	18.4	81.3
Namu (ElSx-1)	3500-480 B.P.	98	31.4	37.6	25.1	47.1
Kisameet (ElSx-3)	2290 B.P. - contact	110	30.7	50.7	22.9	67.9
Koeye R. (EkSx-12)	2030-230 B.P.	88	26.2	61.1	23.0	69.4

Table 1. Density of salmon elements and identified vertebra fragments (number of identified specimens per liter of matrix) in bucket-auger samples obtained at Namu, and at the Kisameet and Koeye Village Sites (matrix volumes in liters, dates in radiocarbon years B.P.).

Table 1 shows the density of salmon elements and vertebra fragments in the auger samples from these sites compared to the Namu samples associated with the periods encompassing the peak and subsequent decline in the local salmon fishery.

The figures in Table 1 show the densities for the total volume of recovered matrix and for the volume of matrix finer than 2 mm. The latter figure is probably a better basis for comparison between sites and time periods since it controls somewhat for variability in the rate of shell deposition. The presence of large quantities of whole and coarsely-broken shell disproportionately increases the matrix volume, and consequently decreases the apparent density of fish remains. As the total and less-than-2 mm volumes show, the volumes of coarse and whole shell vary considerably between sites and between periods at Namu. In either case, the results of the auger sample analysis are consistent. The initiation of settlement at the two nearby village sites at around 2400-2300 B.P. dates to the period of decline in the Namu salmon fishery, but the density of salmon at both sites is considerably higher at this time, and approaches, though does not quite reach, the earlier peak levels at Namu.

These results show that the productivity of salmon fisheries not only varies over time at particular site locations, but can also vary between even nearby site locations at a given point in time. Similar studies based on limited auger samples could quickly and inexpensively establish the focus and intensity of local fisheries in areas throughout the Northwest Coast. On this basis it should be possible to assess the degree of site-specific variability in salmon fisheries and accumulate data from sufficient numbers of sites to establish whether any long-term regional developments in the intensity of salmon fishing have actually taken place.

Zooarchaeology has come to assume a prominent place in current interpretations of Northwest Coast prehistory, and this has resulted in increasing demands for more and better quality data. It has also pushed researchers to extract more information from the evidence they have available. The need now is to move beyond bone counts as part of routine post-excavation analysis, and develop new research strategies specifically designed to address zooarchaeological research questions. In the interpretation of previously-excavated faunal collections, more attention should be directed toward seasonal variability in site use as the cause of variation in faunal abundance. The effects of areal sampling and recovery methods also have to be considered. Greater consideration can be given to the presence and relative abundance of marginal resource indicators such as ratfish and dogfish (Cannon 1995a, 1999), which might indicate local longer-term decline in or inaccessibility to preferred fisheries, such as salmon. Techniques such as the isotopic analysis of dog bones (Cannon *et al.* 1999) and analysis of fracture patterns in deer phalanges (Zita 1997) can also be used to develop independent lines of evidence to show the extent of variability in fisheries and the periodic need to utilize marginal alternatives.

In planning new research, greater attention obviously has to be paid to recovery methods and sampling strategies to maximize the interpretative value of fish remains recovered in excavation projects. Based on my experience in bucket-auger sampling and on earlier work utilizing core and column sampling (Casteel 1970, 1976), complete, laboratory-controlled analysis of even limited samples of matrix, especially if obtained from a variety of sampling locations, should be considered paramount to the analysis of field-collected vertebrate faunal samples, regardless of the screen size or water-screening methods used. This may be the best means of establishing the intensity of local salmon and herring fisheries. Small fragments of salmon bone are readily identified, as are the remains of herring and a variety of other smaller species that are likely to be found in bulk matrix samples. The remains of less-commonly-occurring fish recovered in full-scale excavation, however, will also still provide important information about local subsistence economies. In some cases the abundance of fish species in excavation samples will be critical for identifying the focus of local fisheries and for showing the importance of species such as cod relative to salmon. This is the type of information that would be impossible to establish based only on the abundance of the more readily identifiable fragments of salmon bone found in matrix samples.

More important than improved recovery and analysis of fish remains from individual site locations is the pressing need to expand research strategies beyond the individual site to encompass a variety of site locations within specific regions. This is the only way to gain a better appreciation of the extent of regional, seasonal, and longer-term temporal variability in the focus and intensity of local fisheries. Regional research programs would also provide a way to escape the current tendency to see each and every site as more or less representative of longer-term trends that are sometimes thought to characterize the entire Northwest Coast area. An assessment of the timing and relative intensity of salmon fisheries in particular is needed from a very large number of sites before any greater trend, which may or may not exist, will become apparent. Site testing on this scale will be impossible to achieve anytime soon based on the application of traditional excavation methods, but programs of limited coring and auger sampling of midden sites may provide an alternative means of assessing variability in at least the salmon and herring fisheries. Problems in controlling for variable deposition, sampling, preservation, and quantification will undoubtedly arise in interpretations based on this type of fine-scaled recovery and analysis, as they have in all other zooarchaeological analyses, but a regional program of limited testing will at least shift the scale of recovery and interpretation to a level that is more appropriate for addressing the questions of whether, when, and how salmon was important in Northwest Coast prehistory.

ACKNOWLEDGMENTS

I would like to thank David Yesner and Craig Gerlach for the invitation to participate in the symposium "Zooarchaeology in the North" at the 23rd Annual Meeting of the Alaska Anthropological Association. A travel grant from the Arts Research Board, McMaster University enabled me to attend to present an earlier version of this paper. The bucket-auger sampling undertaken at Namu in 1994 was sponsored by a grant from the Arts Research Board, McMaster University. Site testing in the Namu vicinity in 1996 and 1997 was undertaken with the financial support of a research grant from the Social Sciences and Humanities Research Council of Canada. I also want to acknowledge the helpful comments and suggestions made by Madonna Moss and an anonymous reviewer.

REFERENCES

Ames, K.M.
1994 The Northwest Coast: Complex Hunter-Gatherers, Ecology, and Social Evolution. *Annual Review of Anthropology* 23:209-229.

Bailey, B.E.
1952 *Marine Oils with Particular Reference to those of Canada.* Bulletin of the Fisheries Research Board of Canada No. 89. Ottawa.

Binford, L.R.
1978 *Nunamiut Ethnoarchaeology*. New York: Academic Press.

1981 *Bones: Ancient Men and Modern Myths*. New York: Academic Press.

Butler, V.L. and J.C. Chatters
1994 The Role of Bone Density in Structuring Prehistoric Salmon Bone Assemblages. *Journal of Archaeological Science* 21:413-424.

Cannon, A.
1991 *The Economic Prehistory of Namu: Patterns in Vertebrate Fauna*. Simon Fraser University, Publications in Archaeology No. 19. Burnaby, British Columbia.

1995a The Ratfish and Marine Resource Deficiencies on the Northwest Coast. *Canadian Journal of Archaeology* 19:49-60.

1995b Bucket-Auger Sampling at Namu, British Columbia: Small-Scale Analysis and the Prospects for Regional Interpretation. Paper presented at the 28th Annual Meeting of the Canadian Archaeological Association, Kelowna, British Columbia.

1996a Scales of Variability in Northwest Salmon Fishing. In *Prehistoric Hunter-Gatherer Fishing Strategies*, edited by M.G. Plew, pp. 25-40. Boise: Boise State University Monograph Series.

1996b The Early Namu Archaeofauna. In *Early Human Occupation in British Columbia*, edited by R.L. Carlson and L. Dalla Bona, pp. 103-111. Vancouver: University of British Columbia Press.

1998 Contingency and Agency in the Growth of Northwest Coast Maritime Economies. *Arctic Anthropology* 35:57-67.

1999 Faunal Remains as Economic Indicators on the Northwest Coast. In *Animal Bones, Human Societies*, edited by P. Rowley-Conwy. Oxford: Oxbow. In Press.

Cannon, A., H.P. Schwarcz, and M. Knyf
1999 Marine-based Subsistence Trends and the Stable Isotope Analysis of Dog Bones from Namu, British Columbia. *Journal of Archaeological Science* 26:399-407.

Carlson, R.L.
1983 Prehistory of the Northwest Coast. In *Indian Art Traditions of the Northwest Coast*, edited by R.L. Carlson, pp. 13-32. Burnaby: Archaeology Press, Simon Fraser University.

1996 The Later Prehistory of British Columbia. In *Early Human Occupation in British Columbia*, edited by R.L. Carlson and L. Dalla Bona, pp. 215-226. Vancouver: University of British Columbia Press.

1998 Coastal British Columbia in the Light of North Pacific Maritime Adaptations. *Arctic Anthropology* 35:23-35.

Casteel, R.W.
1970 Core and Column Sampling. *American Antiquity* 35:465-467.

1976 Fish Remains from Glenrose. In *The Glenrose Cannery Site*, edited by R.G. Matson, pp. 82-87. National Museum of Man, Mercury Series, Archaeological Survey of Canada Paper No. 52. Ottawa.

Chisholm, B.S., D. E. Nelson, and H.P. Schwarcz
1983 Marine and Terrestrial Protein in Prehistoric Diets on the British Columbia Coast. *Current Anthropology* 24:396-398.

Conover, K.
1978 Matrix Analyses. In *Studies in Bella Bella Prehistory*, edited by J.J. Hester and S.M. Nelson, pp. 67-99. Simon Fraser University, Publications in Archaeology No. 5. Burnaby, British Columbia.

Coupland, G.
1998 Maritime Adaptation and Evolution of the Developed Northwest Coast Pattern on the Central Northwest Coast. *Arctic Anthropology* 35:36-56.

Croes, D.R.
1988 Review of *British Columbia Prehistory*, by K. Fladmark. *B.C. Studies* 77:73-75.

1992 Exploring Prehistoric Subsistence Change on the Northwest Coast. In *Long-term Subsistence Change in Prehistoric North America*, Research in Economic Anthropology, Supplement 6, edited by D.R. Croes, R.A. Hawkins, and B.L. Isaac, pp. 337-366. Greenwich, Connecticut: JAI Press.

Croes, D.R. and S. Hackenberger
1988 Hoko River Archaeological Complex: Modeling Prehistoric Northwest Coast Economic Evolution. In

Prehistoric Economies of the Pacific Northwest Coast, Research in Economic Anthropology, Supplement 3, edited by B.L. Isaac, pp. 19-85. Greenwich, Connecticut: JAI Press.

Duff, W.
1952 *The Upper Stalo Indians of the Fraser Valley, British Columbia*. British Columbia Provincial Museum Memoir No. 1. Victoria.

Eldridge, M. and S. Acheson
1992 The Antiquity of Fish Weirs on the Southern Coast: A Response to Moss, Erlandson and Stuckenrath. *Canadian Journal of Archaeology* 16:112-116.

Ford, P.J.
1989 Archaeological and Ethnographic Correlates of Seasonality: Problems and Solutions on the Northwest Coast. *Canadian Journal of Archaeology* 13:133-150.

Hebda, R.J. and R.W. Mathewes
1984 Holocene History of Cedar and Native Indian Cultures of the North American Pacific Coast. *Science* 225:711-713.

Hester, J.J. and S.M. Nelson (eds.)
1978 *Studies in Bella Bella_Prehistory*. Simon Fraser University, Publications in Archaeology No. 5. Burnaby, British Columbia.

Luebbers, R.
1978 Excavations: Stratigraphy and Artifacts. *In Studies in Bella Bella Prehistory*, edited by J.J. Hester and S.M. Nelson, pp. 11-66. Simon Fraser University, Publications in Archaeology No. 5. Burnaby, British Columbia.

Lyman, R.L.
1991 *Prehistory of the Oregon Coast: The Effects of Excavation Strategies and Assemblage Size on Archaeological Inquiry*. San Diego: Academic Press.

Matson, R.G.
1992 The Evolution of Northwest Coast Subsistence. In *Long-term Subsistence Change in Prehistoric North America*, Research in Economic Anthropology, Supplement 6, edited by D.R. Croes, R.A. Hawkins, and B.L. Isaac, pp. 367-428. Greenwich, Connecticut: JAI Press.

Mitchell, D.
1988 Changing Patterns of Resource Use in the Prehistory of Queen Charlotte Strait, British Columbia. In *Prehistoric Economies of the Pacific Northwest Coast*, Research in Economic Anthropology, Supplement 3, edited by B.L. Isaac, pp. 245-290. Greenwich, Connecticut: JAI Press.

Monks, G.G.
1987 Prey as Bait: The Deep Bay Example. *Canadian Journal of Archaeology* 11:119-142.

Moss, M.L. and J.M. Erlandson
1995 Reflections on North American Pacific Coast Prehistory. *Journal of World Prehistory* 9:1-45.

Rankin, L.
1993 The Role of Shellfish at the Crescent Beach Site: Interpreting Prehistoric Economy. Paper presented at the 26th Annual Meeting of the Canadian Archaeological Association, Montreal.

Schalk, R.F.
1977 The Structure of an Anadromous Fish Resource. *In For Theory Building in Archaeology*, edited by L.R. Binford, pp. 207-249. New York: Academic Press.

Speth, J.D. and K.A. Spielmann
1983 Energy Source, Protein Metabolism, and Hunter-Gatherer Subsistence Strategies. *Journal of Anthropological Archaeology* 2:1-31.

Suttles, W.
1968 Coping with Abundance: Subsistence on the Northwest Coast. In *Man the Hunter*, edited by R.B. Lee and I. DeVore, pp. 56-68. Chicago: Aldine.

Stewart, H.
1977 *Indian Fishing: Early Methods on the Northwest Coast*. Vancouver: J.J. Douglas.

Tabarev, A.V.
1996 Comment on The Big-Game Focus: Reinterpreting the Archaeological Record of Cantabrian Upper Paleolithic Economy, by M. Kornfeld. *Current Anthropology* 37:648-649.

Vasilevskii, R.S.
1998 Fishing as an Early Adaptation on the Pacific Coast of Northeast Asia. *Arctic Anthropology* 35:281-295.

Watt, B.K. and A.L. Merrill
1963 *Composition of Foods: Raw, Processed, Prepared*. U.S. Department of Agriculture Handbook 8. Washington, D.C.

Zita, P.
1997 Hard Times on the Northwest Coast: Deer Phalanx Marrow Extraction at Namu, British Columbia. In *Drawing Our Own Conclusions: Proceedings from the 1997 McMaster Anthropology Society Students Research Forum*, edited by A. Dolphin and D. Strauss, pp. 62-71. Hamilton, Ontario: McMaster Anthropology Society.

ANIMAL PROCUREMENT AND SEASONALITY: A STUDY OF FAUNAL REMAINS FROM A PREHISTORIC SOD HOUSE RUIN IN NORTHWEST ALASKA

Rebekah DeAngelo
Laboratory for Circumpolar Studies, Brown University
and Paragon Archaeological Consultants, Denver

THE SITE AT SAPUN CREEK (XBM-103)

Archaeological excavations at Sapun Creek (XBM-103) in 1994 and 1995 uncovered a late prehistoric semi-subterranean sod house and an associated midden. XBM-103 is located approximately 180 miles upriver from Kotzebue, on a four meter terrace overlooking the floodplain of Sapun Creek, which flows from the Baird Mountains into the Noatak (Figure 1). The site consists of six visible house depressions, only one of which has been excavated. This feature is five meters in diameter, with a one by three meter entrance tunnel facing northwest. The tunnel is placed at a 45 degree angle to the current terrace edge (Figure 2). Lithic artifacts, antler arrowhead bases, and bird blunts stylistically similar to those from Kangiguksuk, on the lower Noatak River (Hall 1971) suggest an occupation during the 16th or 17th centuries A.D. Faunal preservation at Sapun Creek was excellent. Approximately 6,000 bone fragments were collected from the floor and entrance tunnel of the house, and an additional 22,000 bone fragments were collected from the midden. This paper reports on the analysis of 9,000 of these bone fragments.

CONTEXTUAL INFORMATION

The total area excavated encompassed a five by six meter unit around the house depression, and a two by three meter unit incorporating the entrance tunnel. Both the house structure and associated overlying midden were uncovered. The horizontal and vertical provenience of all objects was recorded using a total station transit. Hence, detailed information about the placement of each bone and artifact is available for future research. Due to field conditions, dry screening was not possible and wet screening was difficult. The latter was reserved for the house floor matrix. In addition, several bulk samples were collected from the house floor and midden.

The excavated house and associated midden are separated from the other houses on the site by three meters. The midden is roughly circular in shape, with the densest concentration of cultural material forming a ring around the perimeter of the house. Faunal remains are comparatively scarce within the house. This pattern is consistent with the discard of refuse out the smoke hole and the entrance tunnel, supporting the idea that the midden is associated with the occupation of the house. If the midden had formed post-abandonment, one would expect to find a higher concentration of bones in the center of the collapsed house. However, to mitigate any possible contamination of the assemblage by later occupants, 2,000 (7.1 %) bones from the top of the deposit were excluded from the analysis.

There is some indication that the house was occupied more than once. There are multiple floors in the entrance tunnel and one rebuilding episode visible in the western portion of the house.

THE FAUNAL SAMPLE

The midden contained approximately 22,000 bones, 7,000 of which (32%) are identified. The floor and the tunnel of the house contained approximately 6,000 bones, 2,000 of which (33%) have been identified. The total number of identified specimens (NISP) provides a general idea of the composition of identified sample.

Bone preservation is excellent. The probability of bone survival is high as is indicated by the presence of fragile elements such as bird crania, fish bone and the porous trabecular bone of caribou. Heavy attrition by carnivores is unlikely as no evidence for the presence of dogs/wolves was observed.

Various sampling strategies were used for different portions of this analysis. To establish seasonality using caribou, all bones from the site that fit the desired criteria were used. These were juvenile caribou bones containing erupting molars (MNI=34). Seasonality was also investigated using bird and fish bones. All identified bones were used to examine patterns of procurement and processing.

SEASON OF OCCUPATION

Alaskan archeological and ethnographic records show that winter is the typical time for occupation of semi-subterranean sod houses (Cantwell 1889; Giddings 1952; Gubser 1965; Burch 1984; Anderson 1984;). In this case, biological data on bird migrations, fish runs, and rates of juvenile caribou tooth eruption provide the basis for establishing a season of occupation at Sapun Creek in conjuction with ethnographic information.

Birds

Migratory species are obviously limited in the length of time that they stay in the region. Thus, the presence of migratory birds in a faunal assemblage may be used as a seasonal marker. However, this method is not without problems because the birds could have been cached or transported for later use. To minimize problems associated with long term use of bird carcasses and bones, those elements fashioned into tools were not included in seasonality determinations. The two migratory bird species found at the site are the tundra swan (MNI=1) and gull (MNI=1).

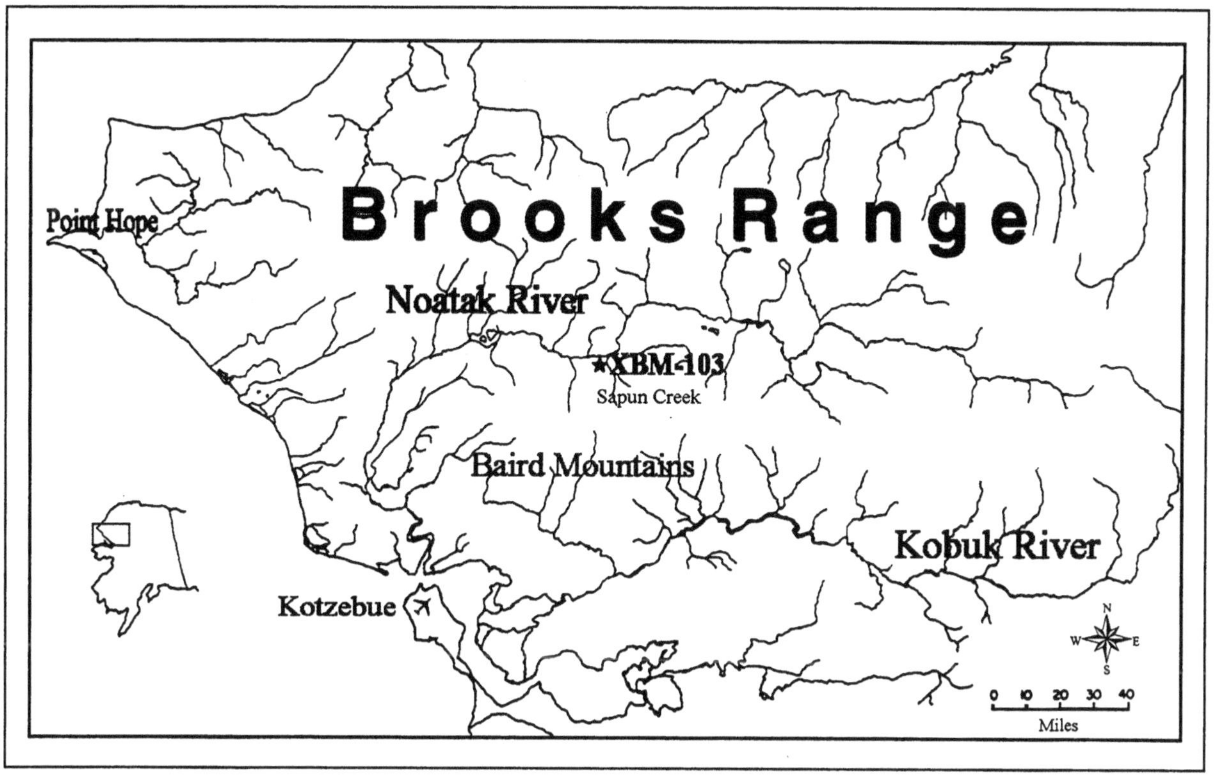

Figure 1. Northwest Alaska.

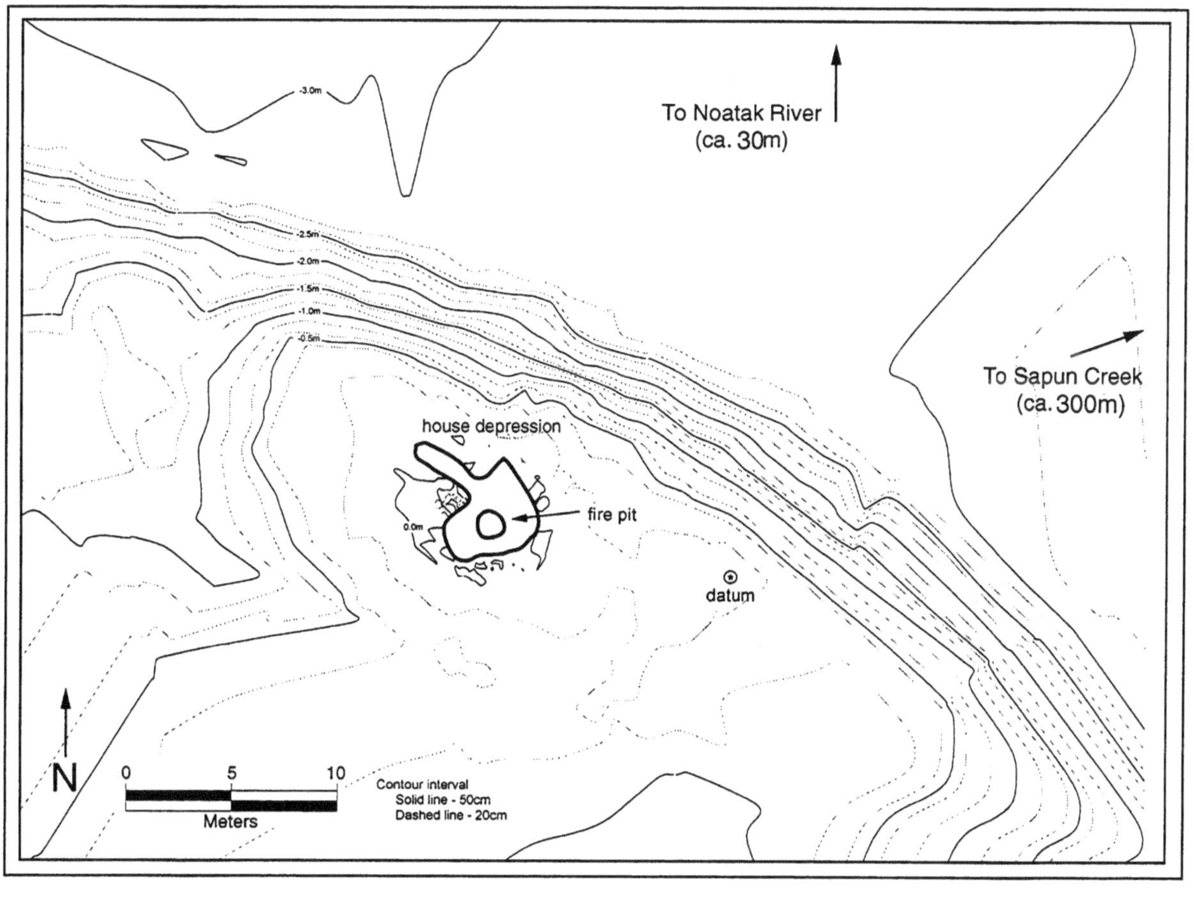

Figure 2. Site map (adapted from Anderson and Anderson 1997).

SPECIES	PLACE	ARRIVAL	DEPARTURE	CITATION
Tundra Swan	Anaktuvuk	May 15	No data	Irving, 1960
	Colville River	June 6	September	Kessel and Cade 1958
	Kotzebue	early May	October	Uhl and Uhl 1977
	Kotzebue		October	Armstrong 1990
Gull	Kotzebue	Spring	Fall	Armstrong 1990

Table 1. Presence of tundra swans and gulls in Northwest Alaska.

Tundra swan (*Cygnus columbianus*) and gull (*Larus* sp.) suggest a spring through fall occupation (Table 1). However, ethnographic information from Wainwright, suggests narrowing the period of harvest as there they "are actively hunted only during the fall, from late September through October (Nelson, 1969:164)."

Ptarmigan (*Lagopus lagopus*) are year round residents in the region. However at Anaktuvuk Pass their numbers fluctuate during the year. "The number of nesting willow ptarmigan is very small in comparison with the great numbers which migrate northward through the valley in late winter" (Irving 1960:248). Hunting techniques differ through the year and people use snares in the winter and nets in the spring and fall (Hall 1971:49). However, the ethnographic information suggests they are "easiest to catch in late winter/early spring (Spiess 1979:166)." This, plus their increased availability suggests that most ptarmigan at XBM-103 were caught in the late winter and early spring.

Fish

Scales and the jaw of a grayling (*Thymallus arcticus*), the jaw of a salmon (*Oncorhynchus sp.)*, and the jaw of a burbot (*Lota lota*) are present. Grayling and burbot can be caught in the summer with a net or line, and in the winter through the ice (Nelson 1983:75). Salmon make seasonal runs up the Noatak River and its' presence indicates some occupation in late summer/early fall (August/September). Ethnographic descriptions of fishing support a pattern of fall harvesting.

> In late August the Noatak people began their homeward trips upriver from Kotzebue. Fish camps are established for the seining of salmon, the primary early fall activity. ...When river ice has formed, fishing for trout, white fish, or grayling with hooks and nets, becomes more popular (Foote 1959:32).

This account is from the middle of the 20th century. However, if the pattern of seasonal availability of salmon was similar in the 1500's, it is likely that fishing in late August also occurred when Sapun Creek was occupied.

Caribou

The caribou in the area of Sapun Creek belong to the Western Arctic Herd, which today numbers about 200,000 animals. This herd migrates south in the fall, passing the Noatak River in August and the Kobuk River in September (Lent 1966:481). Mating occurs in October and the winter is spent concentrated in areas near the Selawik River. These areas are south or east of Sapun Creek, but winter habitation in 1961 and 1962 occurred in the Baird Mountains along the lower Noatak River (Lent 1966:509). In the spring caribou migrate north into the Brooks Range, again passing Sapun Creek and the Noatak River, when they calve in late May and early June.

One of the most valued parts is the hide and seasonal variation in caribou hide condition is an important consideration for hunters. In the late summer, warble flies lay eggs in the hair of the caribou, the larvae then hatch and burrow into the skin. In spring, when the larvae mature they exit the caribou by boring large holes in the hide (Gubser 1965:315). Consequently hides taken in February or March, are worthless for fashioning clothes (Gubser 1965:300). In the fall, however, damage is minimal, and it is at this time that the animals are fattening in preparation for the rutting season. In the fall before the rut (mid-August to late October) the males are in their prime, while the females are at their prime from October until February, as are non-reproducing animals. During the rut, the male caribou cease to eat and become very thin. In the winter both male and female animals become thinner, and less desirable as a food source.

The season of occupation at XBM-103 is inferred from the pattern of molar eruption in juvenile caribou. Because it is difficult to obtain a season of death for adult caribou, except through tooth cross-sections (Spiess 1979:70; Gerlach 1989:199), only juvenile caribou are considered here. The seasonality data are actually reporting the season in which the juvenile caribou were harvested and not necessarily the season of site occupation. Antler size and shape are not used as an indicator of season of death because of the complicated relationship between sex and nutrition in antler growth and shedding (Spiess 1979:101), as well as the possibility that antler was collected after it was shed.

Caribou are born in late May and early June. At birth their jaws contain three deciduous premolars. At three months, permanent molars begin to erupt successively behind those temporary teeth. At about 20 months, after the molars are in place, permanent premolars begin to replace the original, deciduous ones. The sequence of molar and premolar eruption is used to determine the season of death within two to twelve months depending on the tooth. Establishing seasonality from erupting molars is less accurate as caribou mature. There is substantial variability in the timing of the eruption of the third molar (M3) and using this tooth the time of death can only be estimated at less than one year (Spiess 1979:70). The eruption patterns of Ml and M2 are more useful and the high number of erupting Ml is the strongest indicator of seasonality at this site.

Caribou Molars

The 30 juvenile caribou jaws in this sample divide into three groups: 1) Ml erupting (N=ll MNI=8); 2) M2 erupting (N=ll, MNI=8), and 3) M3 erupting (N=17, MNI=14). Note the two

EVENT	AGE (in months.)	SEASONALITY
M1 erupting	3-5	Sept. – Nov
DM high relief wear	1-7	July – Jan
M1 in occulsion	3-13	Sept. – July
M2 absent		
M2 erupting	10-15	April – Sept.
DM moderate relief	10-15	April – Sept.
M2 in, M3 absent	12-22	July – April
M3 erupting	15-29	Any time of year

Table 2. Molar eruption in juvenile caribou and season of death (after Spiess 1979:79). Spiess used mid May because of his interest in exploring reindeer populations in southern France. I use late May/early June because that is the calving time for more northerly populations like the Western Arctic Herd (Gerlach 1989:222).

FN	YEAR	SIDE	BONE	HEIGHT OF DPM3	HEIGHT OF M1 FROM BONE	POSSIBLE COMPANIONS: OTHER INFO	TOTAL MNI
M1 ERUPTION							
3003592	1995	left	mandible	8 mm	0 mm	None; M1 not erupted yet	1
7776669	1995	right	mandible	9 mm	4 mm	2996466; in lot bag from roof	2
2996466	1995	left	maxillary	8 mm	4 mm	7776669	2
9840	1994	right	mandible	8.5 mm	5 mm	7866, or 8588, and/or 4802	3
4802	1994	left	maxillary	8.5 mm	5 mm	9840	3
5755	1994	left	mandible	9 mm	5.5 mm	None	4
7866	1994	left	mandible	9 mm	5.5 mm	9840	5
8588	1994	left	mandible	8.5 mm	6 mm	9840	6
937890	1995	right	maxillary	9 mm	7 mm	None	7
7-44	1994	right	maxillary	8 mm	9 mm	10610	8
10610	1994	left	maxillary	7 mm	9.5mm	7-44	8
M1 IN PLACE, M2 ERUPTION							
FN	YEAR	SIDE	BONE	HIGH OF DPM3	HEIGHT OF M2 FROM BONE	POSSIBLE COMPANIONS: OTHER INFO	TOTAL MNI
9456	1994	left	mandible	8 mm	0 mm	6570711? Bone not broken through.	9
2997504	1995	left	mandible	7 mm	0 mm	6570853?	10
6570853	1995	right	mandible	7 mm	0 mm	2997504; in lot bag	10
9379803	1995	left	mandible	8 mm	0 mm	None	11
9379811	1995	right	mandible	8 mm	0 mm	None	12
6570711	1995	left	maxillary	6 mm	0 mm	9456? Bone not broken through.	12
18005	1995	left	mandible	8 mm	0.1 mm	None	13
3001440	1995	right	mandible	6 mm	0.5 mm	None	14
8684	1994	left	maxillary	7 mm	4 mm	None	15
5338	1994	right	mandible	6 mm	6.5 mm	7811	16
7881	1994	left	mandible	6.5 mm	6.5 mm	5388	16
M1, M2 IN PLACE, M3 ERUPTION							
FN	YEAR	SIDE	BONE	HIGH OF DPM3	HEIGHT OF M3 FROM BONE	POSSIBLE COMPANIONS: OTHER INFO	TOTAL MNI
-	-	left	mandible	-	0 mm	None. Bone not broken through.	17
16039	1995	right	mandible	5 mm	0.5 mm	None. Bone broken slightly.	18
3001138	1995	left	mandible	5 mm	1 mm	15654	19
15654	1995	right	mandible	5 mm	1 mm	3001138	19
3003319	1995	right	mandible	6 mm	1 mm?	None. Bone broken slightly, M3 absent.	20
17910	1995	left	mandible	NA	2 mm	17889, only M2, M3 present.	21
17889	1995	right	mandible	4 mm	2 mm	17910	21
2434540	1995	left	mandible	NA	3 mm	3001128, or 6570853, but no indicators	22
3001128	1995	right	mandible	5 mm	3 mm	2434540? No indicators.	23
6570853	1995	right	mandible	NA	3 mm	2434540? No indicators.	23
4031	1994	left	mandible	NA	6 mm	None. M3 third crest not visible	24
6571700	1995	left	mandible	5 mm	7.5 mm	None. M3 third crest visible	25
17991	1995	left	mandible	NA	8 mm	None. M3 third crest not visible	26
2434962	1995	left	mandible	NA	8 mm	None. M3 third crest visible	27
4777	1994	right	mandible	6 mm	8 mm	None. M3 third crest not visible	28
2996928	1995	left	mandible	NA	11 mm	None. M3 third crest not visible	29
5815	1994	left	mandible	NA	11 mm	None. M3 third crest visible	30

Table 3. Molar eruption in juvenile caribou from XBM-103. FN = field identification number. Year = year excavated. Side = body side. Height of DPM3 = in the mandible this is the height of the outside of the middle crest of deciduous premolar three; in the maxilla it is the height of the caudal crest of deciduous premola three. Height of M1 (or M2 or M3) from bone = height of the outside of that tooth. Possible companions = the bones (if any) that might represent the same individual. Criterion for this category are not limited to the characteristics listed here, but concern the length of the diastema, placement of features.

phases of M2 eruption (Table 3). There are eight (MNI=6) cases in which M2 has not broken through the bone, or where only a slight portion is showing. There are no other M2s erupting except for three (MNI=2) cases in which M2 has erupted more than 4mm. The lack of M2s in later stages of eruption indicate animals close to 11 months old, and suggests a spring kill.

Caribou Premolars

In juvenile caribou, premolars erupt in the last stages of jaw development. There is a window of eruption time similar to that of M2. A more accurate assessment of caribou age at death can be inferred only from the eruption of M1. Indomestic reindeer, premolar eruption is reported to occur at twenty months (Zhigunov 1968:15). However, considering the care and good nutrition that domesticated reindeer receive, this eruption is probably earlier than eruption in wild caribou. Spiess (1979:79) records the eruption of premolars in caribou at 22 months.

Either of these models are applicable to the record at Sapun Creek, as all the premolars were in the first stages of eruption, probably reflecting spring kills. There are only a small number of juveniles in which premolars were erupting (MNI=4). Therefore, these do not play a large part in the assessment of seasonality.

Figure 3 illustrates the combined information from molar and premolar eruption of juvenile caribou at XBM-103. There is a peak in September and October and a second small one in March. The peak in September and October reflects the large number of jaws in which M1 is erupting. The second peak in March reflects the eruption of permanent premolars.

The presence of swan and gull in the spring coincide with the spring occupation as indicated by the eruption of caribou teeth. However, the presence of both bird species in the fall supports a stronger correlation with the season as indicated by the juvenile caribou. This is not to say that the occupation of the house spanned only one season. In fact, the evidence from juvenile caribou molar eruption supports an occupation from fall through spring. However, since migrating birds, salmon, and caribou are available from August through October, this is the narrowest period for occupation.

ANIMALS OTHER THAN CARIBOU

The seasonal movements of species are important factors that determining the length and season of human occupation at a given location. For example, as caribou migrated out of the range in the summer, the Nuataagmiut traveled to the coast to hunt sea mammals and to trade (Burch 1984: 307). Hunters ho rely on caribou during the fall and spring migrations must find other resources when caribou are no longer present.

Optimal foraging theory states that at times when the most efficient source of protein and nutrients is unavailable, a predator will widen its prey selection to attain the needed food. The same may be said of humans: "when caribou are not plentiful, a larger resource base is needed (Spiess 1978: 173)." Humans however, also use animals for purposes other than consumption. For instance, waterproof bags are made from bird skins (Ray 1984:289). Aside from the fish and birds discussed above, the sample from XBM-103 also contained ptarmigan, ringed seal, hare, and red fox. Below I examine the range of animal species in the assemblage and explore resource exploitation patterns.

Ptarmigan

The willow ptarmigan (*Lagopus lagopus*) is by far the most common bird species in the sample (MNI=40). It is a year round dweller in the area, and also migrates locally. The assemblage at Sapun Creek contains about 400 identifiable bone fragments or whole bones (i.e., proximal or distal humerus, etc.), and about 44 fragments that are probably ptarmigan. The high numbers of ptarmigan at the site indicate that there was either a steady reliance on these animals throughout the duration of the house occupation or a heightened dependence during a few seasons or events. If caribou hunting was poor, it may have been necessary to harvest more ptarmigan. It is possible that a majority of the ptarmigan bones were consumed when a low in caribou occurred. "During the winter months of a poor migration, the caribou-hunting tribes eked out their rations with hares and ptarmigan caught in snares or nets" (Ray 1984:289). At XBM-103, reliance on these birds may have fluctuated with caribou availability.

> Although the snaring and hunting of ptarmigan and rabbits can be extremely important to Noatak village, the net meat income from these sources was of secondary importance in 1959-1960 (Foote and Cooke 1960:9).

At Sapun Creek, ptarmigan are also of secondary importance. In the field, the lower and upper portions of the deposit appeared to contain similar proportions of ptarmigan bones, suggesting that ptarmigan use was probably constant through time. Further research is necessary to determine whether the reliance on ptarmigan fluctuated seasonally, or with episodes of low caribou availability, or if consumption was constant but at a low level. Fluctuations in the quantity of secondary resources like ptarmigan, provide clues to changes in resource availability, and food preferences.

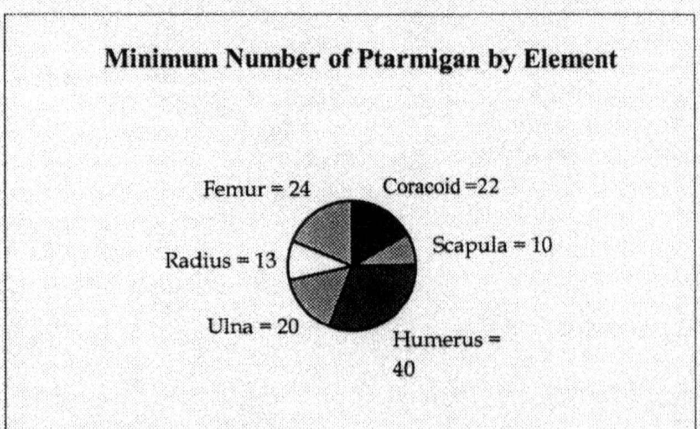

Figure 4: Minimum number of ptarmigan by element.

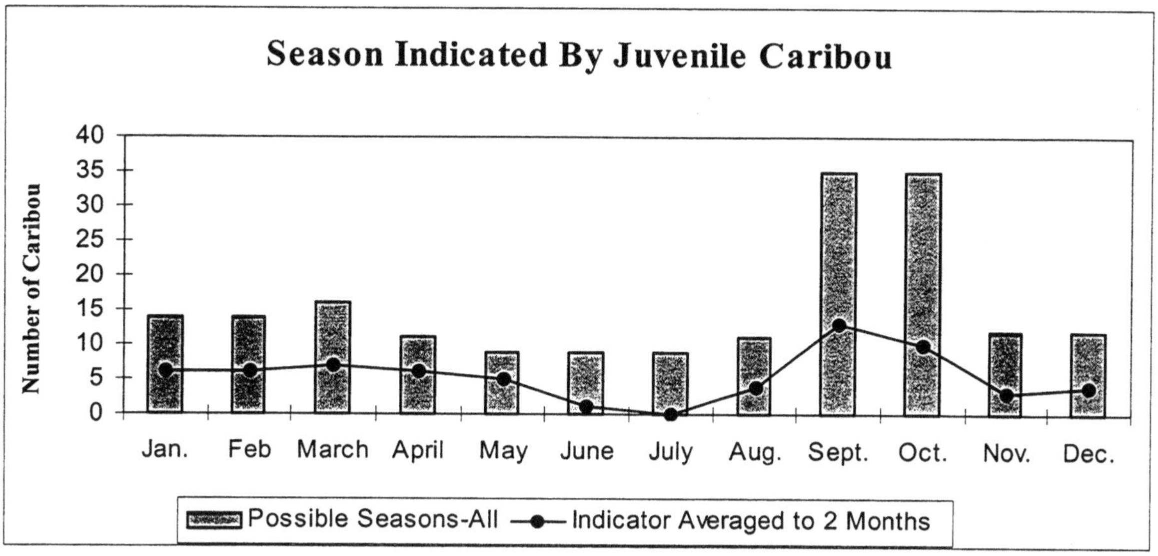

Figure 3. Distribution of the seasons indicated by the molar and premolar eruption in juvenile caribou, with and without averaging the possible age of individuals. Note the difference between the graph in which the season indicated was reduced to a two month period for every tooth eruption, versus the block representation of the entire duration of possible season indicate of permanent premolars and the averaging of M3 eruptions. The seven cases in which M3 was in late stages of eruption were averaged between September and October, however the most probable time for this eruption was in October considering the height of 11 mm above bone for specimen number 5815

Seal

Although XBM-103 is approximately 180 river miles from the ocean, the ulna, tibia, metacarpal and phalanx of a ringed seal (*Phoca hispida*) were found. The uses for seal are many. Their fat is used as fuel in lamps, and is a staple in the diet of coastal Inupiat. Their meat is an ideal source of protein and calories and their skins are used as storage containers, (or poke). "The usual poke was made from a seal skin turned inside out with any openings sewn or tied shut" (Jones 1983:125). These pokes are ideal for the storage of greens, oil, fish or berries. While the phalanx in the sample could have come from a decomposed poke in which the extreme digits were not removed, this does not satisfactorily explain the presence of the tibia and the ulna.

The seal bones may indicate a movement of the site inhabitants from the coast, or suggest trade with people who have traveled to the coast. The entire animal may not have been transported up the Noatak, but at least portions of it were. Even if just the skin was transported it may have included the flippers. "Once I obtained a green [freshly killed] bearded seal hide from Point Barrow for rawhide line, which still contained the two front flippers" (Gubser 1965: 82). Although in this instance transportation was quick because of the use of airplanes, in the past the transporting food could have been accomplished if stored in oil or cooled in the water.

Indications of group contact

Two groups, the Napaaqtugmiut and the Nuataagmiut, occupied the Noatak River drainage in the late 1800s. Both spent their summers on the coast at Sisualik, near present-day Kotzebue, hunting seals and other coastal animals. At this time ten different groups in the Kotzebue basin were dependent upon one another for trade which was most concentrated in the summer gatherings at Sisualik (Burch 1984:306). Social relations among groups changed over time and the treatment of neighboring groups fluctuated within the context of a complex system of appropriate behavior (Burch 1984:306).

> Inter district travel in summer for subsistence purposes was extensive, and it was a key element in the institutionalized pattern of inter-societal relations. Almost the entire population of Nuataagmiut passed right through the heart of Napaaqtugmiut country (Burch 1984: 307).

XBM-103 is located in an area historically occupied by the Nuataagmiut. The seal bones from the site are important because they suggest that the pattern of contact, as outlined by Burch (1980, 1984) for the historic period may have operated similarly in the past. The presence of a non-local food source indicates a network, or travel route to the coast, and contact and toleration of others from different groups. These seal elements suggest that the practice of 180 mile movements, and the use of trade routes was long-standing in the region.

CARIBOU PROCUREMENT AND PROCESSING

The overwhelming majority by number and volume (approximately 7,000), of identified bones are caribou, indicating that the occupants of XBM-103 preferentially procured and consumed this animal. Caribou skeletal element representation can be used to suggest possible strategies of carcass transport and use (Binford 1981;

Landals 1990; Bunn 1993; and Emerson 1993). For example, element frequencies have been used to model kill site dynamics. Assemblage structure is a result of the season of kill, the distance to the kill site, the intended use of each bone (i.e., for grease and protein), and the situation regarding the time of consumption (i.e., was the meat cached or eaten fresh) (Binford 1978). If particular elements are selected for a property only attainable through bone destruction, (ie. grease), then these elements will be under-represented in an assemblage. Thus, the use factor I emphasize is the destruction of bones – primarily juvenile unfused bones (Gerlach 1989:226). Below I present the caribou element data from XBM- 103 and propose models that may account for the assemblage structure.

There are approximately 3,300 caribou elements, 2,747 long bone fragments, 589 rib fragments, 180 porous bone fragments, about 90 cranial fragments and many sternal elements. Caribou abundance is expressed as the minimum number of elements (MNE) necessary to produce the number annd type of each element present. The minimum number of individuals in any sample is the number of individuals necessary to account for all of the identified bones (Klein and Cruz-Uribe 1984:26; Gerlach 1989:188). In most cases, the minimum number of caribou elements (MNE) is a direct translation of the most frequent portion of bone (i.e., distal right humerus) plus the number of complete bones of that type. I use MNI for each element because the final tallies were reached using data sorted by right, left, proximal and distal fragments (Figure 5). This is similar to the procedure for the computation of the MNI. Age data are not factored

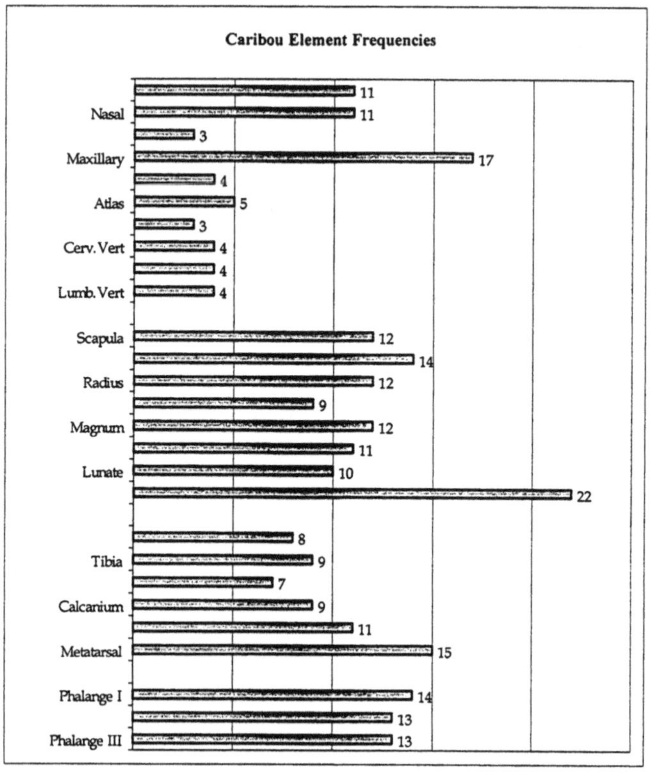

Figure 5: Caribou element frequencies. A compilation of elements reduced to a single caribou unit (i.e., the number of antler bases reported here is the total number found divided by two).

into this calculation, though doing so would likely increase the numbers reported. There is a conspicuous lack of vertebral and some cranial elements in the sample, while postcranial elements are abundant and there are a large number of phalanges. There are also real differences between the number of fore and hind quarters and in most cases the right elements are more abundant than left (Figure 6). There are 43 distal metacarpals that, although they cannot be sided, provide a MNI of 22. The numbers of bones given here are most certainly minimums.

Transportation

Carcass transport is an important factor in the formation of a faunal assemblage. However, since many other factors also contribute to assemblage formation, it is often difficult to unabiguously identify transport-related patterns.

Through ethnographic study of the Nunamiut of Northwest Alaska, Binford (1978) recognized patterns in the treatment of combinations of caribou elements that were a result of butchering. The composition of 'meat packs' transported to dwellings depended on the season of kill and the distance to be traveled, and/or the potential uses of the meat. The most frequent elements left behind were the hooves (phalanges) and possibly the distal portions of legs (metatarsals and metacarpals), along with the skull and other low value elements.

At XBM-103, there is an abundance of both distal leg bones and cranium. Low vertebral counts suggest that those elements may not have been brought to the house from the kill site. The cranial elements are primarily mandibles (MNI=42), maxillary bones (MNI=14), nasal bones

The most often recovered caribou element at XBM- 103 is the metacarpal. Elements reduced to a single caribou unit.(MNI=12), and antler cores (MNI=11). The assemblage is characterized by an abundance of cranial elements, and it is unlikely that only some elements of the head were left at the kill site. There is however, the possibility that some vertebrae such as the atlas and axis are underrepresented because they were abandoned at the kill site.

It is difficult to isolate kill conditions such as season and distance of the kill site in a midden. The assemblage resulting from the transportation of animals from various kill sites under different conditions can at best produce a generalized picture of procurement. Elucidating the conditions surrounding carcass recovery is difficult because of the uncertain treatment those elements receive after transport to the dwelling. Below I examine such treatment, which includes the smashing of bones for marrow and grease, as well as the manufacturing of tools from bone elements. These processes, along with the preferential transport of material may have structured the assemblage seen here.

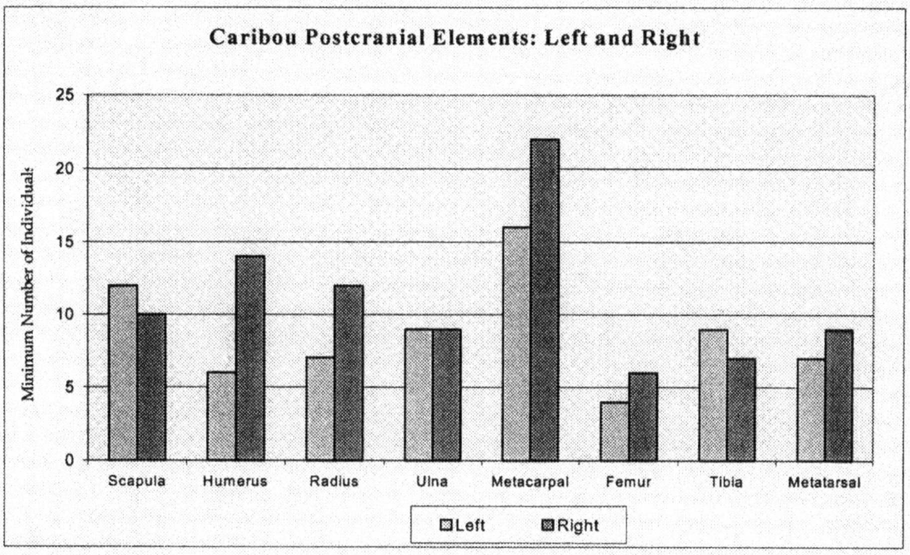

Figure 6. Caribou postcranial elements, the greater of proximal or distal, recovered from the right and left side.

Nutrition

An adult caribou can supply between 31 (young female) and 47 (male) kg of meat, while a yearling caribou will supply about 25 kg of meat. These values differ from season to season. If most kills were in the fall, the parameters from Spiess (1979) to figure food values are useful. In this model 5% of cow and young bull weight is fat and 55% is meat. These fall killed caribou would have had skins in good condition and a high percentage of fat.

Aside from desired meat, marrow is commonly removed from the inside of caribou long bones and eaten. Caribou bones also provide grease when boiled. Where there is a concern with fat availability, long bones will not only be cracked for their marrow (as usual) but also smashed for bone grease (Spiess 1979:173). The production of this grease, however, is not always confined to use in response to the pressures associated with hunger. Bone grease was not only used in times of scarcity, but also by hunting parties (Anderson, personal communication). Leechman (1951), in his experience with people in the Yukon, also noted that bone grease was used "much as we use butter, that it could be eaten alone when it was frozen hard, … and that it could be used in cooking."

All the long bones in the assemblage show signs of reduction for the extraction of marrow, with most long bone fragments ranging between 3 and 20 cm in length. There are also isolated deposits of much smaller fragments between 0.5 and 2 cm, further evidence of the reduction of bones for or due to boiling. Bone cavity size, bone density and volume can be used to rank bones for marrow and grease ultility (cf. Binford 1978). Alternatively, rank can be established by substracting the weight of bone from the weight of bone and associated soft tissue (Metcalfe and Jones 1988). The two utility ranks, GUI (Binford 1978) and FUI (Metcalf and Jones 1988) are comparable. Both these methods employ many characteristics of the element to derive the ranking (i.e., quantities of meat, marrow, and grease).

The element frequencies at XBM-103 do not correspond to the pattern that these indices predict, so it appears that another process is at work. Perhaps elements were selected for only one of the variables factored into the utility indices. Alternatively, elements may have been selected according to the model's predictions and some post harvesting activity has obscured patterns (e.g. destruction of elements at the processing stage, or a redistribution of the elements during disposal).

An important factor in the rendering of grease from bone is the percentage of oleic acid in different bones. Each bone contains a quantity of yellow or white grease; white grease contains a higher percent of oleic acid, and is found in higher proportions in the distal portions of long bones. The percentage increases as one moves farther down the leg. The hind legs contain more of this high quality grease than the front legs. Since oleic acid has a significantly lower melting point (14 degrees centigrade) than yellow grease it provides incentive for extraction (Gerlach 1989:310).

The percentage of oleic acid in bones was one factor used by Binford (1978:24) in his calculation of the GUI. He determined the 'grease value' of elements: second phalange and distal metacarpal (equal at 78), distal metatarsal and first phalange (equal at 76), third phalange (74), proximal metacarpal (73), astragalus (72), proximal metatarsal, distal tibia, and tarsals (71), next come the carpals (68) and distal radio-cubitus (67), proximal tibia (54), distal femur (53), proximal radio-cubitus (48), and proximal femur (44).

If the rendering of grease was important the highest rate of fracturing should occur among elements that contain the highest percentage of this grease. The extraction of grease would skew the assemblage to over represent the forequarters and proximal portions of elements. At XBM-103 forequarters are more abundant than hind, and distal portions of elements are more abundant than proximal ends. The distal portions of legs are better represented than the proximal elements (Figure 7).

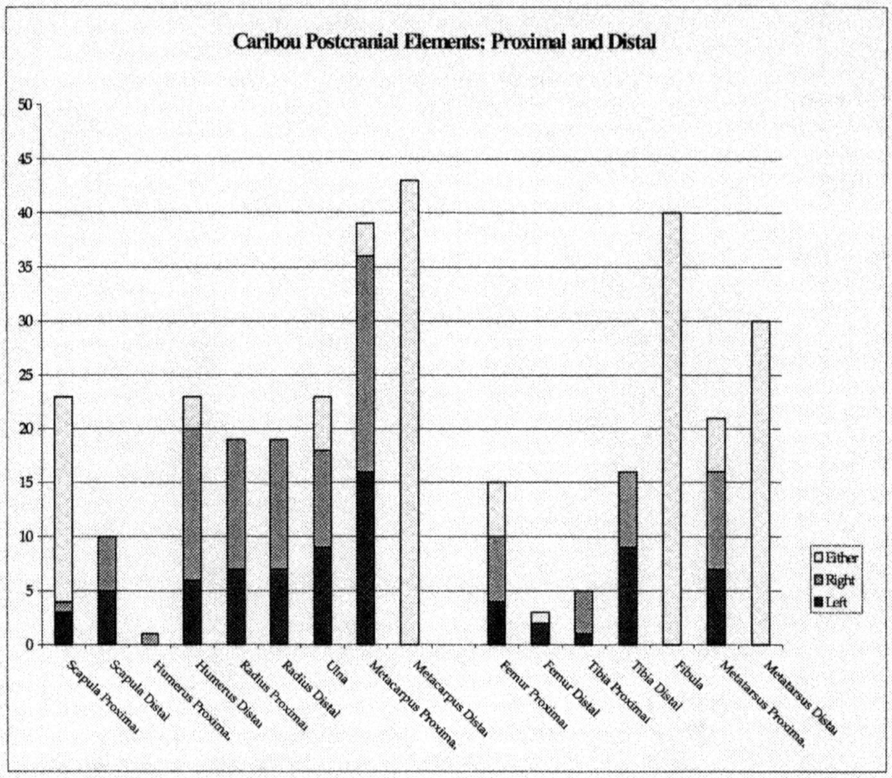

Figure 7. Caribou postcranial elements, proximal and distal.

Phalanges 1 (phl) and 2 (ph2) at XBM- 103 are commonly (about 50%) broken. Phalanges 3 (ph3) are complete, possibly because they are harder to break and there is no marrow cavity to speak of. The broken ph1 and ph2 indicate that these elements were selected for the extraction of marrow and/or used in the production of grease. But as only a little over half are broken, it suggests that they were not always the preferred element.

The percentage of oleic acid in bones provides a reasonable explanation for some of the trends seen here. The poor representation of femora and tibiae may reflect a preference for the reduction of hind-quarters for white grease extraction. However, the pattern may also result from greater breakage of proximal long bones, (femur), than of metacarpals.

Bone Fragmentation

Though the fragmentation of bones for grease production may account for the low numbers of hind-quarters at XBM-103, it does not account for high representation of the distal portions of legs (i.e., phalanges and metacarpals). It is possible that the distal portions of legs were not desired for grease extraction, or that these elements are more resistant to such processes because of bone density or other factors of durability. The process of bone fragmentation for the extraction of grease should result in the destruction of articular ends of long bones that are less dense. This is particularly true if the long bones are juvenile with unfused epiphyses. The articular ends of juvenile long bones are more porous and less dense than those of adult bones. Note the late date of epiphyseal closure for proximal humerii, distal femurs, and proximal tibii (Table 4). These three portions are markedly underrepresented at Sapun Creek with MNI of one, two, and four respectively. Compare this to the early complete fusion of distal humerii, proximal radii and distal tibii. These articular ends have MNI of 14, 12 and 9 respectively.

In some cases, as with the humerii, disparity in representation of the proximal and distal portions may be due to the different ages at which epiphyseal fusion occurs. The proximal end of the humerus fuses at 35 months, while the distal end begins fusion between 11 and 14 months of age. At XBM-103 there are many more distal than proximal units. This is also the case with tibii and femurs. These are also the three most under represented elements in the assemblage.

Fragmentation accounts for significant differences in the number of bone units that survive and suggests that a large proportion of the animals harvested at Sapun Creek were juvenile, as their unfused bones were most susceptible to fragmentation. Further examination of age data will help determine the extent to which bone fusion was a factor. The caribou mandible data indicates that many individuals in the assemblage were juvenile (MN=34).

Manufacture

Caribou are not just important because of the food that they supply. Their hides are an essential raw material. The skins from the legs of caribou are used for many items including bags to carry sewing items and for the manufacture of boots

(Gubser 1965:83-84). If there was a preference for lower leg hide, and if the hooves were not severed prior to transport, the number of phalanges at the site would be elevated as compared to other bone elements.

One striking example of caribou bone element use at XBM-103 is an unfused portion of a proximal ph1 or 2; it is hollowed and shaped for use as a bird blunt. Only one such modified phalange was recovered, but its resemblance to ethnographically known bird blunts is striking.

	DEER (*Odocoileus*)		CARIBOU (*Rangifer*)
Skeletal Element	Epiphyses Open	Epiphyses Closed	Fusion State at 26-27 Months
Proximal Radius	0 – 8	14 - 15	Epiphysis fused/fusing. Epiphyseal line visible.
Distal Radius	0 – 29	34 - 60	Epiphysis not yet fusing.
Proximal Ulna	0 – 29	34 – 60	No data.
Distal Ulna	0 – 29	34 – 60	Epiphysis not yet fusing.
Distal Humerus	0 – 11	14 – 17	Epiphysis fused. No epiphyseal line visible.
Proximal Humerus	0 – 35	35 – 60	Epiphysis not yet fusing.
Phalanges (I and II)	0 – 12	14 – 15	Epiphysis fused.
Calcanium	0 – 29	60 – 65	No data.
Metatarsus	0 – 29	29 – 35	Epiphysis fusing. Synostosis beginning (Data for distal metatarsal only).
Metacarpus	0 – 27	29 – 35	Epiphysis fusing. Synostosis beginning (Data for distal metacarpal only).
Proximal Femur	0 – 29	29 – 34	Epiphysis fusing. Synostosis beginning.
Distal Femur	0 – 34	35 – 52	Epiphysis unfused.
Proximal Tibia	0 – 29	34 – 60	Epiphysis unfused.
Distal Tibia	0 – 15	15 – 19	Epiphysis fused/fusing. Epiphyseal line visible.
Distal Fibula	No Data	14 – 15	Epiphysis fused/fusing. Epiphyseal line visible.
Proximal Fibula	No Data	34 – 60	Epiphysis not yet fusing.

Table 4. Epiphyseal closure sequence(in months) in deer (Odocoileus) and caribou in good nutritional condition. Data summarized from Todd and Todd (1938), Lewell and Cowan (1963) and Spiess (1979). Taken from Gerlach (1989:228).

DISCUSSION

In the XBM-103 faunal assemblage there are more forequarters than hind, perhaps indicating that the hindquarters were preferentially broken for the extraction of white grease and oleic acid. Fragmentation of bone is common, as all of the caribou long bones are broken. The assemblage consists of a large number of juvenile caribou, and taphonomic factors affecting that age class are important.

Differential bone density is the most plausible explanation for the limited number of proximal and distal portions of elements although sharing or consumption in the field may have also occurred. Patterned, recurring activities such as thses might account for the large number of missing left humerii.

The large number of phalanges, either whole or broken in half, may be due to their relatively small size and lack of marrow. The larger leg bones have a greater chance of being broken beyond recognition during the processes of marrow extraction and grease production. Caribou phalanges may be present or because they were used in the manufacture of tools, or simply because they were not detached from hides prior to transport to the house. Only a little over half of the phalanges are broken and seems unlikely that they are present because of a deliberate selection for marrow or grease, though this no doubt occurred some of the time. It is more likely that they were brought back to the house because of their fortuitous association with hides.

The assemblage from XBM-103 results from multiple decisions and actions related to transportation, nutrition needs, and bone processing. The assemblage holds clues to the treatment of caribou during both procurement and processing phases of use.

CONCLUSIONS

I have tried to present some trends in the subsistence practices and the treatment of animals by the people who occupied the Sapun Creek site (XBM-103) in Northwest Alsaka. Using faunal data from one house my aim is to create a picture of food harvesting. The evidence indicates that the Inupiat Eskimos of the middle Noatak River were engaged in productive fall and spring hunts for caribou and that caribou were their primary prey species. A large number of the caribou harvested were juvenile. Ptarmigan was also hunted while seals were obtained either through trade with people on the coast or through travel to the coast for sealing.

In accordance with models where killed animals are transported short distances, cranial elements, metatarsals and metacarpals are abundant at this site. The assemblage contains more fore quarters than hind, perhaps because of a preference for skeletal elements containing an abundance of oleic acid.

There are more left than right elements and this pattern is not referable to models of procurement or processing. Sharing may explain disparity between right and left elements, although this is difficult to assess.

Further work with age classes and spatial data will help decipher the effect of processing and the other events that comprise subsistence. Activities from kill through preparation are a web, each strand of which needs be considered in our efforts to develop a working knowledge of assemblage formation and of the actions of long ago peoples.

ACKNOWLEDGMENTS

Funding for the excavations at XBM-103 was provided by the National Park Service, Northwest Alaska Areas, and the National Science Foundation. I would most like to thank Douglas Anderson, the project's Principle Investigator. Daniel Odess and David Gregg were great help in preparing previous versions of this manuscript. This paper was adapted from an undergraduate thesis submitted to Brown University in May of 1996. Thanks are due to many, specifically Richard Gould and Bob Gal, for all their help during this project.

REFERENCES

Anderson, Douglas
1984 Prehistory of Northern Alaska. In *Handbook of North American Indians*. Edited by David Damas, pp. 80-93. Arctic Vol. 5. Smithsonian Institution, Washington DC.

Anderson, Douglas and Wanni Anderson
1997 *Human Subistence Practices in Response to Environmental Fluctuations in Northwest Alaska over the Last Millennium*. National Science Foundation Report.

Armstrong, Robert
1990 *Guide to the Birds of Alaska*. Alaska Northwest Books, Seattle.

Binford, Lewis
1978 *Nunamuit Ethnoarchaeology*. Academic Press, Inc., New York.

1981 *Bones Ancient Men and Modern Myths*. Academic Press, Inc., New York.

Bunn, Henry T
1993 Bone Assemblages at Base Camps: A Further Consideration of Carcass Transport and Bone Destruction by the Hadza. In *From Bones to Behavior: Ethnoarchaeological and Experimental Contributions to the Interpretation of Faunal Remains*, edited by Jean Hudson, pp. 156-168. Occasional Paper No. 21. Center for Archaeological Investigations, Southern Illinois University.

Burch, Ernest S.
1980 Traditional Eskimo Societies in Northwest Alaska. In *Alaska Native Culture and History*. Edited by H. Kotani and W. Workman, pp. 253-304. *Senri Ethnological Studies* No. 4. National Museum of Ethnology, Osaka, Japan.

1984 Kotzebue Sound Eskimo. In *Handbook of North American Indian*. Edited by David Damas, pp. 303-319. Arctic Vol. 5. Smithsonian Institution, Washington DC.

Cantwell, J.C.
1889 A Narrative account of the exploration of Kowak River, Alaska. In *Report of the Cruise of the Revenue Marine Steamer Corwin in the Arctic Ocean in the Year 1884*, pp. 21-52. US Revenue Cutter Service, Washington DC.

Emerson, Alice, M.
1993 The Role of Body Part Utility in Small-scale Hunting Under Two Strategies of Carcass Recovery. In *From Bones to Behavior: Ethnoarchaeological and Experimental Contributions to the Interpretation of Faunal Remain*. Edited by Jean Hudson, pp. 138-155. Occasional Paper No. 21. Center for Archaeological Investigations, Southern Illinois University.

Foote, Don Charles
1959 *The Economic Base and Seasonal Activities of Some Northwest Alaskan Villages: A Preliminary Study*. Report for Bioenvironmental Studies of Project Chariot, United States Atomic Energy Commission.

Foote, Don C., and Alan Cooke
1960 *The Eskimo Hunter at Noatak, Alaska: Winter 1960*. Report for Bioenvironmental Studies of Project Chariot, United States Atomic Energy Commission.

Gerlach, S Craig
1989 *Models of Caribou Exploitation, Butchery, and Processing at the Croxton Site, Tukuto Lake Alaska*. Ph.D. Dissertation, Brown University.

Giddings, J.L.
1952 *The Arctic Woodland Culture of the Kobuk River*. The University Museum. University of Pennsylvania, Philadelphia.

Gubser, Nicholas
1965 *The Nunamuit Eskimos, Hunters of Caribou*. Yale University Press, New Haven.

Hall, Edward
1971 Kangiguksuk. *Arctic Anthropology*. 8:1-98

Irving, Laurence
1960 Birds of Anaktuvuk Pass, Kobuk and Old Crow: A Study in Arctic Adaptation. *United States National Museum Bulletin* 217. Washington D.C.

Jones, Anore
1983 *Nauriat Niginaqtaut: Plants That We Eat*. Maniilaq Association, Kotzebue, Alaska.

Kessel, Brian and Tom Cade
1958 Birds of the Colville River, Northern Alaska. *Biological Papers of the University of Alaska*, Number 2. The Intelligence Printing Company, Pennsylvania.

Klein, Richard, G and Kathryn Cruz-Uribe
1984 *The Analysis of Animal Bones from Archeological Sites*. The University of Chicago Press, Chicago.

Landals, Alison
1990 The Maple Leaf site: Implications of the analysis of small scale Bison kills. In *Hunters of the Recent Past*. Edited by Leslie Davis and Brian Reeves, pp. 122-147. Unwin Hyman Ltd., Boston.

Leechman, Douglas
1951 Bone Grease. *American Antiquity* 16.

Lent, Peter, C.
1966 The Caribou of Northwestern Alaska. In *Environment of the Cape Thompson Area.* Edited by Norman Wilimovsky, pp. 481-518. United Stated Atomic Energy Commission.

Metcalfe, Duncan, and Kevin T. Jones
1988 Reconsideration of Animal Body-part Utility Indices. *American Antiquity.* 53(3):486-504. Washington, DC.

Nagy, Muriel
1990 Caribou Exploitation at the Trail River Site, Northern Yukon. *Occasional Papers in Archaeology* Number 2. Government of the Yukon, Heritage Branch.

Nelson, Richard K.
1969 *Hunters of the Northern Ice.* University of Chicago Press, Chicago.

1983 *Make Prayers to the Raven.* University of Chicago Press, Chicago.

Ray, Dorothy Jean
1984 Bering Strait Eskimo. In, *Handbook of North American Indians.* Edited by David Damas, pp. 285-302. Arctic Vol. 5. Smithsonian Institution. Washington.

Spiess, Arthur E.
1979 *Reindeer and Caribou Hunters. An archaeological Study.* Academic Press, New York.

Uhl, William R., and Carrie Uhl
1977 Tagiumsinaaqmitt: Ocean Beach Dwellers of the Cape Krusenstern Area, Subsistence Patterns. *Occasional Paper, Anthropology and Historic aPreservation, Cooperative Studies Unit,* no. 14. University of Alaska, Fairbanks.

Zhigunov, P.S.
1968 *Reindeer Husbandry.* US Department of The Interior, Bureau Of Land Management. Juneau, Alaska.

TYRANNY IN THE ARCHAEOLOGICAL RECORD OF SPECIALIZED HUNTERS

Jack L. Hofman
Department of Anthropology, University of Kansas

Lawrence C. Todd
Department of Anthropology, Colorado State University

In the past decade there has been an increasing tendency for archaeologists to view hunting, and particularly specialized hunting of single species of large mammals, to be a rare and unusual or non-typical form of human economic adaptation (e.g., Meltzer 1993a; Kornfeld 1996). This perspective perhaps began in earnest with Lee's (1968) paper from the "Man the Hunter" conference in which he summarized the general economic orientations of a broad range of known hunter-gatherer groups. His review indicated that the vast majority of historically documented hunter-gatherer groups, except those in the highest latitudes, derived the bulk of their diet from plant foods, fish, small animals, or collected resources rather than from hunting large animals. It should be noted that he excluded the Plains tribes from his sample. Recent interpretations of Great Plains Paleoindian economies have relied increasingly upon a localized "broad spectrum" foraging component with less emphasis on specialized megafauna hunting (Johnson 1977, 1987; Greiser 1985; Meltzer and Smith 1986; Grayson 1988; Thurmond 1990; Bamforth 1991; Meltzer 1993a, 1993b). These perspectives are generally supported by use of ethnographically based examples combined with nutritional arguments indicating that human tolerance of animal protein is limited.

In light of these developments our concerns in this discussion are with (1) the role of bison hunting in Folsom economy, (2) an assessment of whether Paleoindian bison hunting represented an economic specialization or an embedded activity in a generalized foraging pattern, (3) evaluation of the Lipscomb site evidence as a case for discussion, and (4) a review of some attempts to integrate archaeological interpretations of Paleoindian hunting with the ecology of late Pleistocene/early Holocene bison.

With an increased awareness of the vagaries of discovery, preservation, and recovery of small scale and delicate economic remains from archaeological sites, the potential over representation of large faunal remains has been given increasing consideration (e.g., Struever 1968; Watson 1976; Johnson 1977; Binford 1981; Lyman 1984; Schiffer 1987). Also, systematic sampling and regional studies have helped document site variability and the place of kill sites within Paleoindian economies (e.g., Wendorf and Hester 1962, 1975; Judge 1973; Hester and Grady 1977; Meltzer 1986; Amick 1994). Increased awareness of background fauna has shown that not all materials recovered from archaeological sites resulted from human action (e.g., Binford 1977, 1981; Brain 1981; Stanford 1983; Speth 1991a; Lyman 1994). The impact of background fauna, carnivore and large rodent activities, and the role of human scavenging (e.g., Binford 1981, 1983, 1984), all served to simultaneously erode the position of "man as hunter," even in pre-agricultural settings.

With increased awareness given to the importance of overall nutrition in hunter-gatherer diets, the limited tolerance of humans for protein-rich or protein-dominated diets, and the critical need for lipids and carbohydrates, the role of hunting as a dominant economic orientation or life way was further questioned (e.g., Speth and Spielmann 1983; Speth 1983, 1987, 1989, 1991b). It became evident, and is now accepted, that many if not most hunter-gatherer groups relied heavily, and by necessity, on a variety of non-protein resources from plants and animals. It also is recognized that a certain percentage of protein (usually less than 25-35% overall) in the diet is essential to good health and maintenance of an adequate metabolic rate, and that protein deficiencies are a common problem among agriculturalists with carbohydrate-rich diets in the New World (e.g., Larsen 1983; Cohen and Armelagos 1984).

A partial counter-weight to this overall trend de-emphasizing the role of hunting, is provided by developments in evolutionary ecology and an increased awareness and use of optimal foraging theory in hunter-gatherer studies (including archaeological, ethnographic and ethnoarchaeological research; Winterhalder and Smith 1981; Pianka 1983; Foley 1987; Smith and Winterhalder 1992). One of the limitations of Lee's (1968) original argument against the significance of hunting and for the importance of foraging was his lack of consideration of processing time and its impact on the energetic and technological returns from hunting versus foraging (e.g., O'Connell and Hawkes 1981). Hunting large animals can be more risky or more prone to failure than plant collecting, but generally provides a greater return for the overall investment when processing time is considered.

Humans as omnivores have the *potential* in most environments to be generalists, or to have a diverse and "broad spectrum" diet. For technologically advanced humans, then, it is commonly assumed that they did or will have a complex dietary pattern dependent upon their technology and the resources available within their economic territories (Kelly 1995).

Plains Paleoindians, however, have traditionally been characterized as specialized hunters (Willey 1966; West 1983; Wedel 1986; Frison 1991), and bison were a key economic species in the Plains region from the late Pleistocene until historic times (Ewers 1955; Davis and Wilson 1978; Frison 1987, 1991; Bement and Buehler 1997). The common occurrence in the Plains area of bison and mammoth bone beds which contain distinctive projectile points, has provided substance for the position that these Paleoindians were hunters of megafauna (West 1983; Bonnichsen et al. 1987; Kelly and Todd 1988; Frison 1991). The emphasis on bison hunting by some historic Plains groups was also seen as support for this position. Recent discussions about site recognition, preservation, sampling, recovery, and reporting have, however, provided a basis for critical reassessment of the role of megafauna in Paleoindian economy (Greiser 1985; Fawcett 1987; Meltzer and Smith 1986; Grayson 1988; Kornfeld 1988, 1996; Bamforth 1991; Davis and Greiser 1992; Meltzer 1993a, 1993b;). It is frequently argued that sites with preserved mammoth or bison bone are more easily recognized than sites without these remains and so are studied out of proportion to their significance or relative importance (Fawcett 1987; Meltzer 1993a;). However, the great majority of identified, studied, and reported archaeological sites in the Plains region do not have bone beds (the many series of CRM contract reports from throughout the Plains show this very well), and one reason that bonebed sites have received special attention is because of the special kinds of information which can be derived from their study (Wheat 1972; Davis and Wilson 1978; Frison and Todd 1987; Davis and Reeves 1990; Frison 1991).

WERE FOLSOM PEOPLE ECONOMIC SPECIALISTS?

This question or problem actually includes several issues. If bison were plentiful in the Plains region during the terminal Pleistocene and earliest Holocene, could people have lived primarily by specializing their economic pursuits toward this single resource? How do we define specialized as opposed to broad-spectrum or generalized economies (Pianka 1983:266-270)? Is specialization a measure of the amount of time spent in a particular pursuit, the gearing up time and specialized technology required to enable a given activity, the relative contribution of a particular prey species in relation to others, or the percentage of economic and technological contributions a particular activity makes to the overall economy? Obviously, specialization can be any or all of these. Ultimately, an economy can be seen as specialized based on the variety and intensity of species used (diet breadth and evenness), only on a comparative basis when contrasted with other economies, whether these are between distinct groups or different seasonal activities of a single group.

Among fully modern humans, to suggest that a particular group was economically specialized (occupying a narrow economic niche) should not be taken to indicate that they were technologically, intellectually, or socially incapable of effective deviation from a specialized tract. (An economy focused on corn horticulture for example.) Rather, specialization can in some situations provide a selective advantage. As discussed by Pianka (I 983:266-267),

> Thus specialists are often relatively rare, whereas generalists are more abundant. Rare organisms may, however, frequently occur in clumps so that their local density need not necessarily be low. If specialization involves becoming less abundant, why have organisms become specialized at all? Since generalized organisms can usually exploit more food types, occupy more habitats, and build up larger populations, they might be expected by their very numbers to out reproduce slightly more specialized competing members of their own population and thereby swamp the population gene pool. The answer to this apparent dilemma lies in the old adage that a jack-of-all-trades is a master of none. More specialized individuals are more efficient on their own ground than are generalists. (Pianka 1983:266-267).

Obviously, the ability to compete as specialists is dependent upon population pressure and competition for resources (Keeley 1988). The Folsom economy provides a significant test case for the applicability of foraging theory and derivative models to human cases (e.g., Winterhalder 1986; Belovsky 1988). Early New World hunter-gatherers had the technological capability to select from a wide range of economic options. The significance of bison hunting in relationship to other economic pursuits remains the key issue. The question is not whether bison were hunted and important to Folsom economy (but see Fawcett 1987), but how significant were bison in relation to other plant and animal resources on a seasonal, yearly, and long term basis? How are economic decisions reflected in technology and in the archaeological record?

Economic variation should not simply be modeled on a hypothetical binary or uni-dimensional scale from "specialized" to "generalized," as such categorization is inadequate to capture the yearly, seasonal, regional, situational, and idiosyncratic variation we can expect in hunter-gatherer behaviors. If a group behaves as economic specialists in one season, there is no requirement that they must do so in other seasons or situations.

People can not live on meat alone, as has been eloquently discussed by Speth (1983, 1987, 1989, 1990; Speth and Spielmann 1983). People can, however, live for an extended period on a combination of meat and other animal products including fat, marrow, blood, and rumen content (Wheat 1972; Todd 1991). Extended periods of high protein intake, more than about 25-35% of the total diet, can be deleterious (Speth 1989). Bison, however, provide enough fat (especially marrow and bone grease), carbohydrates (from rumen), fluids, vitamins (especially from uncooked meat), and technological materials (notably hides) to enable people to live just on this resource for an extended period (Wheat 1972; Todd 1991;). One problem is that an economy focused in large part on bison requires a specialized technology and high mobility at least during some seasons (e.g., Ewers 1955). This has a potentially dramatic impact on technology (Osborn 1999).

Many mid-latitude hunter-gatherer groups documented in the ethnographic and ethnoarchaeological records have relatively broad-spectrum economic orientations (Lee 1968; Oswalt 1976; Kelly 1983, 1995; Gamble 1986), with Plains bison hunters being notable exceptions (Kelly 1983). It is often suggested or implied that hunter-gatherers in general were not specialists, but had broad spectrum and highly flexible economies generally adapted to local conditions (e.g.,

Beardsley et al. 1956; McMillan and Klippel 1981; Greiser 1985; Leonard 1989:499; Thurmond 1990). While this was a common Holocene pattern, it provides a weak basis for a more generalized argument concerning early New World occupants during the late Pleistocene (e.g., West 1983; Price and Brown 1985; Todd 1987a, 1991; Kelly and Todd 1988;). The Holocene and historic periods can provide only a limited range of scenarios and should not be used to delimit explanation of early Paleoindian behavior in the New World (Kelly and Todd 1988; Hofman 1994a; Kelly 1996).

THE ARCHAEOLOGICAL RECORD AT LIPSCOMB

While developing such models of Paleoindian behavior may provide an intractable (and therefore interesting) problem, we are optimistic that it does not present us with an unsolvable case. In seeking answers to questions about the early archaeology of the Americas we have been pursuing several superficially disparate lines of research. Studies of site specific formational histories (; Todd 1987a, 1987b, 1991; Todd et al. 1992, 1997; M.G. Hill 1994; M.E. Hill 1996), regional organizational properties of lithic technology (Hofman 1991, 1992, 1995a; Ingbar 1992, 1994; Amick 1994, 1996), and consideration of general behavioral models (Kelly and Todd 1988; Hofman 1994a,) have all played a role in our ongoing research program. In this discussion we review one case study of research with a single archaeological site, Lipscomb, and then refocus the discussion on questions of the resource structure available to Plains Paleoindian groups.

Lipscomb is a Folsom age bison bonebed in the northwestern corner of the Texas panhandle (Figure 1). The site was first investigated in 1939 (Schultz 1943), with additional work in 1946, and between 1988 and 1993 (Hofman and Todd 1990; Todd et al. 1990, 1992; Hofman et al. 1991). A summary of current interpretations concerning the Lipscomb site provides a basis for discussing "specialization" in the economy of Folsom hunter-gatherers on the Southern Plains.

The minimum number of bison represented at Lipscomb is now at 56 (based on counts of left astragali). The mature animals are represented by 75% females and 25% males. The season of death, based on study of tooth eruption and wear, was late summer or early fall (approximately .2 to .4 tenths of a year after the birth pulse). The consistency of dental wear patterns indicates that a single event or very closely spaced events are represented. In addition to the consistent pattern from tooth wear, other lines of evidence support the interpretation that Lipscomb may represent a single kill event. These include the patterned distribution of carcasses (Figure 2), lack of stratigraphic evidence for multiple levels, and the lack of evidence for trampling or disturbance of the lowermost bones. The Lipscomb bonebed is relatively thin (less than 20 cm) and shallowly buried within 60 cm of the surface. The skeletal element representation appears to reflect several factors including bone volume densities, subareal weathering ost depositional weathering or crushing, and collection biases (Todd et al. 1992:146). Bones partially exposed by modern erosion and those nearest the surface were most fragmentary, whereas the bones at the base of this concentration were intact and articulated (Figure 3).

A bison kill rather than a natural death and scavenging event is indicated by the numerous impact- damaged projectile points found in the bone bed; a minimum of 30 projectile points are represented. A number of tools, presumably utilized in skinning and butchering were also found (13 tools and 20 resharpening flakes are documented). The butchering and processing was not comprehensive or intensive. Many carcasses remained articulated, suggesting a "gourmet" style selective butchery (Todd 1987a, 1987c; Todd et al. 1992).

The abundance of artifacts lost or abandoned at the site and the less than complete utilization of the carcasses are believed to reflect a combination of factors including season of the kill, site setting, and surface conditions (Hofman 1999). The warm season kill would have limited the length of time available between the kill-butchery event and putrification of carcasses. This, and the sandy substrate in the bonebed area, may have contributed to the loss of numerous stone tools that remained fully functional. The kill apparently occurred in the upper end of an arroyo or gully that constrained the movement of the bison. Access to some carcasses for butchering after the kill may also have been restricted, especially at the northern end of the bonebed.

A series of radiocarbon dates are available from the Lipscomb site, but only one is of Folsom age and suggests that the kill may have occurred about 10,800 years ago (Hofman 1995b). This period is generally interpreted to have been equable with a somewhat less seasonal and highly productive environment as compared to the Holocene period in the area. Stratigraphic, paleoecological, and paleobotanical evidence support this position (Leonhardy 1966; Wendorf and Hester 1975; Graham 1987; Johnson 1987; Wyckoff and Albert 1984; Holliday 1997;). Playa lakes still commonly held water on the High Plains and species diversity was greater than during the following millennia. The potential diet breadth of Folsom people was arguably greater than for Holocene hunter-gatherers in the region.

Lipscomb was not a permanent habitation site, and given the season it is unlikely that people camped by this kill for an extended period of time. As a result we can appropriately ask, where and how did the Lipscomb people spend the remainder of their economic cycle? This problem has been expressed by Meltzer (1993b:111), "Paleoindians, ... if they weren't overkilling megafauna, what were they doing?" And Bamforth's (1991:360) question, "...what happened when [Paleoindian] people were not hunting communally," remains a central issue in Paleoindian studies.

We need to learn what parts of the landscape were used by specific Folsom groups (mobility and land use), what seasons the different areas were occupied, and what combination of resources were used on a seasonal and long term basis. The Folsom technological system on the Southern Plains has been widely discussed, and included a bifacial core and flake-blank based system which was stylized, highly flexible, and economizing (Tunnell 1977; Boldurian 1990, 1991; Hofman et al. 1990; Hofman 1992; Amick 1994, 1999; Collins 1999). Southern Plains Folsom lithic assemblages, and that from Lipscomb in particular,

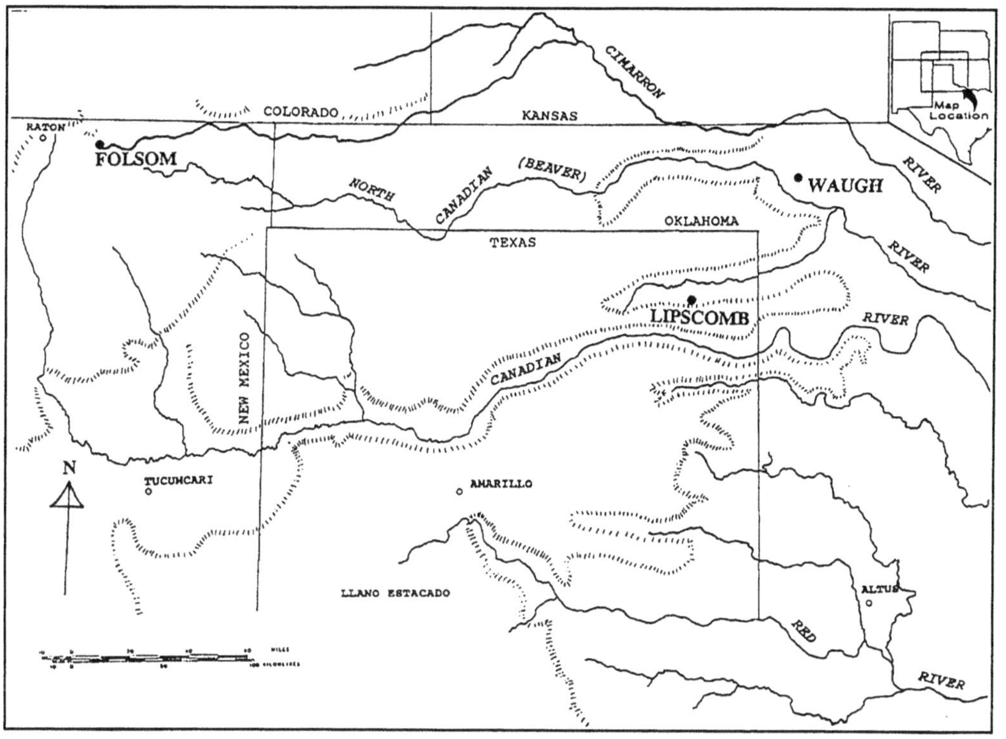

Figure 1. Location of the Lipscomb site.

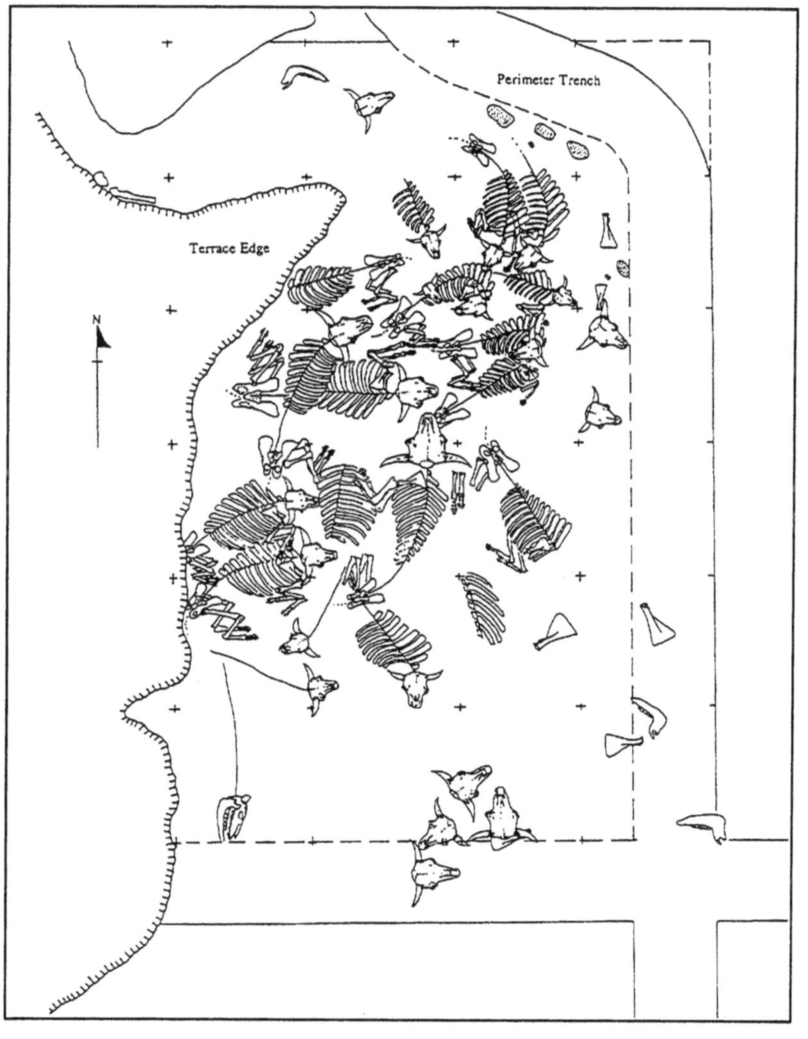

Figure 2. Patterned distribution of carcasses at Lipscomb in 1939 (from Hofman 1995b).

provide some clues to the multi-faceted problem of long-term land use practices .

Two primary stone types are represented by the Lipscomb artifacts. The majority are made from high quality central Texas "Edwards" cherts which are available in a variety of gravel, residual, and bedrock sources located 350 km and more to the southeast of Lipscomb. Edwards cherts are the predominant stones used in most Southern Plains Folsom assemblages (Hester 1972; Hofman 1991a, 1993, 1995a). At Lipscomb, many of the Edwards chert projectile points had been previously damaged and resharpened prior to their use and loss. This suggests that the bison hunters at Lipscomb had been involved in hunting during the summer prior to their arrival at Lipscomb.

The second stone type represented is Alibates "flint," with primary source areas along the Canadian River in the central Texas Panhandle and in Canadian River gravels to the south and southwest of Lipscomb about 100 to 125 km. Folsom tools at Lipscomb made from Alibates are larger than Edwards tools and Alibates points had not been resharpened. This suggests limited attrition of the Alibates component of the tool assemblage prior to the actions at Lipscomb. Procurement of lithics from the Alibates source may have been embedded into the overall hunting activities being conducted on the High Plains as a means of refurbishing depleted tool-stone reserves. For this discussion we are assuming direct procurement of primary toolstone, rather than trade (e.g., Hayden 1982) because of assumed low population, high mobility, and the resulting unpredictable or potentially unreliable nature of exchange as a source for critical materials (Meltzer 1989; 1993a). The Lipscomb evidence indicates that this particular Folsom group had moved onto the Southern High Plains after collecting Edwards chert and probably a variety of other resources, from central Texas. At least two distinct sources of Edwards chert are represented in the Lipscomb sample based on ultraviolet light response (cf. Hofman et al. 1991). During the summer this group had moved north up the Southern Plains, perhaps along the High Plains escarpment, making it to the vicinity of the Alibates source by late summer. The dominant pattern of Edwards chert in Folsom assemblages on the Southern High Plains indicates that this was a common and recurrent pattern. Current evidence indicates that the group did not spend an extended period at Lipscomb after the kill. The very limited occurrence of Edwards chert artifacts in Folsom assemblages to the north and northwest of the Texas Panhandle (Hofman 1991a, 1994b, 1995a) suggests that the Lipscomb Folsom people may have moved into the western Oklahoma area where Edwards chert and Alibates Folsom artifacts are common (Hofman 1993). Folsom bison kills are known in western Oklahoma (Bement 1997, 1999; Hofman and Carter 1991; Hill and Hofman 1997), but a variety of other resources may also have attracted Folsom people to the area including wood, good water, sheltered settings, and diverse flora and fauna (Hofman 1988, 1990, 1991b, 1994a).

The limited availability of quality lithic sources in western Oklahoma and on most of the Southern High Plains may have served in part to tether Folsom people to the central Texas region. Folsom groups appear to have left the Edwards chert source area with enough lithics to last through an extended hunting period (and distance). Lithic materials from the Edwards source area were perhaps supplemented on a situational, logistical, and on an as needed basis. Lithics and bones are the most visible evidence from Lipscomb, but the primary resources gained from the Lipscomb event were almost certainly meat, fat, organs, blood, hides, sinew, skin containers, bones for tool manufacture, and rumen content. These products could not only sustain a hunter-gatherer group for an extended period, but many of these products were critical to the technology and to long term survival. Such critical resources as hides, containers, meat, and fat are assumed to have been transported from Lipscomb when the group moved from the site. These products would have been key to sustaining the group for a significant portion of the year. Evidence from the Hell Gap period Casper site suggests that even when large quantities of meat have been transported from Paleoindian bison kills, it may have been in the form of relatively bone free portions such as the hump or meat stripped from the limb bones (Todd et al. 1997). Occupation sites associated with products derived from muscle-stripping processing activities are not expected to contain abundant faunal evidence of meat consumption. Evidence from kill sites suggests that the lack of large number of bison bones at other locations cannot be taken as a measure of the relative contribution of bison products to the diet. This problem is at the heart of the "tyranny" in the visible archaeological record of specialized hunters when they are not at kill sites. The archaeological record of Folsom bison hunters can be expected to have very little "hard evidence" of bison even when a considerable portion of the resources being used are bison-derived.

It is likely that bison hunting success was highly variable as to frequency and size of kills during any particular year or season. As a result, the variety of alternative economic strategies and economic mixes employed by Folsom people, as with most hunter-gatherers were probably numerous. However, bison hunting is not something that prehistoric hunters could simply do on a whim without adequate preparation or lacking sophisticated technology (Frison 1974, 1987, 1991). Preparation, practice, training, equipment, knowledge of bison behavior, and group cooperation are all critical to successful kills, especially to large ones such as Lipscomb. Even during the periods that Folsom people were not hunting bison, they were probably using bison resources (from previous hunts) and making preparations for future hunts. If we consider the full range of species documented for Folsom age sites (Wilmsen and Roberts 1978; Forester 1985; Walker 1982; Davis and Greiser 1992; Amick 1994), it is no longer reasonable to argue or assume that Folsom people relied solely upon bison or on hunting. To consider bison hunting as an incidental embedded backup strategy would, however, also be inappropriate.

The evidence from Lipscomb suggests that bison hunting was not embedded into lithic resource procurement, seeking shelter, wood procurement, vegetal food collecting, or antelope, deer, or rabbit hunting. Each of these and similar activities may, however, likely have been embedded in an

(1986:320), Brown (1986:320) following Meltzer and Smith (1986) suggests that,

> "...the big-game associations with the Paleoindians do not necessarily imply that they specialized in large game hunting. There is no reason to believe that hunting this megafauna really implies an early Amerindian specialization. The hunting of these large animals was probably opportunistic and embedded in other strategies."

We disagree with this position and suggest that Folsom people were bison hunting specialists for some seasons of most years, and that the tyranny of the archaeological record is such that the use of many critical bison resources such as hides, sinew, meat, fat, and containers are simply not recovered at most sites where these products were important. If lithics, wood resources, protected settings, or biotic diversity were the driving forces in the mobility, land use, and economy of Folsom people at Lipscomb and other sites in the western Plains region, they would never have left central Texas for the southern High Plains.

OTHER RECORDS: THE BISON

So far, we have argued that although there is no reason to make the *a priori* assumption that Paleoindians were either generalized foragers or specialized hunters, a preponderance of the archaeological evidence provides greater support for some form of specialization. However, over the last decade it has become increasingly evident that the archaeological record of late Pleistocene/early Holocene human populations in the Americas is difficult to incorporate into either extreme of simple categorical, analogical classificatory schemes. Although, as outlined above, there are multiple lines of archaeological evidence that implicate the importance of hunting to Paleoindian adaptations, the specifics of big game hunting are still open to question. Regardless of these specifics, there is abundant evidence that bison hunting was at times a component of the Paleoindian behavioral repertoire. One of the difficulties in evaluating models of Paleoindian behaviors is the inherent potential for arguments to become tautological. When observations on the archaeological record have been used to develop models of Paleoindian behavior, exclusive use of that same data set to evaluate the models is inappropriate.

Since the nature of this human/bison interaction is at the heart of the question of Paleoindian hunting specialization, we see a need not only for a greater concern with reevaluating the archaeological record, as just described for the Lipscomb site, but also to spend commensurate effort examining the prey ecology. It is to this "other record" that we now turn in order to expand our potential to effectively evaluate different models of Paleoindian behavior. In shifting focus to the prey rather than on the hunters, there are a number of important questions about which we still need finer-grained data. Several of these include: what factors may have conditioned late Pleistocene and early Holocene bison abundance and distribution? To what degree are behavioral observations on recent bison (Geist 1991; Berger and Cunningham 1994) relevant to their ancestral forms (in terms of foraging ecology, herd dynamics, migration patterns) and how can we arrive at proxy measures of these activities for prehistoric bison populations? While comprehensive answers to these questions are not currently available, we want to close our discussion with an emphasis on preliminary attempts to address these questions as an illustration of the importance of a paleoecological, community approach to investigating questions of Paleoindian adaptations.

Prior to the late Pleistocene, North American bison lived with numerous potential predators (Byers 1997), and in some cases evidence of predator action is clear cut (Guthrie 1990; Carbyn et al. 1993;). However, there is little indication that predation was a major factor in the determination of bison population characteristics. From a bison's-eye-view, the North American Plains may have been a community with two trophic levels - the grasses and herbivores (bison and other large grazing mammals). This is not to imply that other community interactions were not relevant. Predation may have played a central role molding the behaviors and biology of pronghorn (Byers 1977) and other herbivores, which may have in turn influenced community structure, grazing conditions, and hence bison ecology. Plains bison were not immune to predation, but to offer a preliminary model we see the most significant bison predation in terms of multi-animal mass kills, which were a common feature of the Plains only after 11,000 B.P. These mass kills would have had very different potential impacts on populations than occasional, single animal culling as may have been enacted by predators such as wolves.

Clearly such community level interactions need additional research, but at present we are simply outlining some aspects of our on-going research program to integrate studies of stone tools and archaeological patterns (the human record) with the study of Plains paleoecology. The need for Paleoindian archaeologists to develop a better understanding of bison behavioral ecology has been advocated for a number of years, especially by George Frison (e.g., 1974, 1978, 1991). Guthrie's research (1984, 1990) has provided an appropriate model illustrating the benefits of conducting zooarchaeological and paleoecological investigations in concert. Here, we discuss a research framework that begins by modeling an artificially simplistic single herbivore, grassland community in which predation is limited. Such a model provides an important point of departure or frame of reference for further modeling. In discussing the influences that intensive large-scale human hunting may have had on Plains ecology, we need to consider the nature of community structure both with and without human hunting. This is an issue of predator-prey interactions that can initially be modeled at a general level. One of the long standing debates in ecological studies (Powers 1992) is between adherents of "top-down" perspectives in which the predation and consumption by the species in the higher trophic levels are emphasized as being primary factors in conditioning population dynamics of species in subordinate levels, and "bottom-up" proponents who see the primary controls as being from the nutrient and energy fluxes from lower trophic levels. One model of two trophic level community interaction (Hairston et al. 1960), which emphasizes that both bottom-up and top-down controls need to be considered, suggests that "...as each trophic level is added, the controlling process for the plant trophic level shifts alternately from bottom-up (odd number of trophic levels) to top-down (even number of trophic levels). Moreover, moving vertically up from the bottom trophic level to those above, there is again an alternation between top-down and bottom-up control" (Begon et al. 1996:834). We are using this model not as an explanatory tool, but as a framework for discussing a variety of studies to bring the dynamic interrelationships between Plains bison and human populations into clearer focus.

Figure 3 illustrates the basic structure of this model. In the unlikely situation that a single trophic level grassland community existed without herbivores, the population dynamic of the plants would be exclusively under the control of nutrient availability and energy fluxes. Ecosystem processes dominate and community interactions are driven by competition for available energy. Addition of herbivores to the setting is modeled as changing in the factors controlling plant biomass from being exclusively biogeochemical to being dominated by herbivore consumption. Vegetative population dynamics are less influenced by competition as herbivore consumption increases. In such communities the population characteristics of the plants are subject to strong top-down, predator-conditioned influences, while the herbivores are subject to bottom-up influences (competition for forage as more important than avoidance of predators). In other words, in such systems the grazers may have more influence on grass population characteristics than factors of nutrient availability. In terms of North American Plains bison-plant interactions, this is the situation that we would expect prior to the appearance of human predators or at very low levels of human predation - large herbivores would have had a major effect on the structure of the grassland communities (Owen-Smith 1987; Mack and Thompson 1982). Guthrie's theory of late Pleistocene extinctions (1984) emphasizes this bottom-up condition, and has provided a number of key conceptual components for several models of Paleoindian hunting (e.g., Kelly and Todd 1988). Guthrie argued that climate change, and particularly the structure of 'seasonality,' had a dramatic influence on the structure of vegetative communities and hence on the megafaunal herbivores.

With the introduction of predators (or as in late Pleistocene North America, novel predators as the known/habituated predators were diminishing in numbers) and the appearance of a three level system (Figure 3), the interactive arrows are again changed with vegetative productivity coming under greater bottom-up influence (as was the case in the one level system) while herbivores are being subjected to greater top-down influence (predation becomes more important than competition). Although human hunting has been a common theme in Pleistocene overkill arguments (Martin 1967), novel predation need not be modeled exclusively in terms of the direct, devastating, lethal effects on the prey (Guthrie 1990). Our concern here is with the possible effects of novel predation on one of the species that not only survived the terminal Pleistocene but also flourished during the Holocene - *Bison bison*. If increased, or novel, predation at the end of the Pleistocene influenced Plains food web dynamics, we should expect to see evidence of top-down effects on bison. In thinking about the interactions of early Paleoindian hunters and their prey, consideration should be given to not only the animals that are killed, but also the influence which a new predator and increased risk of predation could have on prey species' foraging behaviors, growth, and fecundity (Schmitz et al. 1997; McPeek and Peckarsky 1998). Casual, embedded hunting (the "no they weren't" side of early Paleoindian specialization discussions) is expected to influence prey populations in different ways than a more regular, focal predation pattern (the "yes they were" group). One of the ways that such focal, top-down control might be expressed in the paleontological record is in terms of prey (bison) body-size as a response to changing predator avoidance strategies and foraging behaviors.

Measurements of bison upper forelimb bones from a number of Plains sites are summarized in Table 1. These data are subdivided into several chronological groupings, an early Paleoindian group (Chron 1: Folsom sites), a mid-Paleoindian group (Chron 2: Agate Basin and Hell Gap sites), a late Paleoindian group (Chron 3: Cody Complex sites), an Archaic group (Chron 4), and a Late Prehistoric/modern group (Chron 5). Figure 4, which plots averages of one distal humerus measurement by chronological group, indicates that in terms of postcranial size reduction, nearly 50% of the change in Holocene bison body size took place early, with a very rapid change between bison associated with Folsom assemblages (Chron 1) to those associated with later Paleoindian materials (Chron 3). Folsom-age bison were 18- 20% larger than modern bison, while Cody-age bison were only 9-10% larger than modern forms. This pattern in postcranial size reduction contrasts with the more gradual pattern of horn-core diminution, for which a more rapid rate of change takes place during the mid-Holocene as documented by Wilson (1978) for many of the same sites. Continued study of rates of mosaic morphological changes of Holocene bison is provocative and has the potential to yield significant data on more general issues of evolutionary change.

Can we argue that the dramatic postcranial bison size reduction on the Plains is clear-cut evidence of focal hunting by early Paleoindians? Although predation (Schmitz et al. 1997) by a novel predator would combine top-down potentially significant influences of lethal direct effects (hunting), lethal indirect effects (increased likelihood of starvation or winter kill from foraging disruption), and *nonlethal indirect effects* (changing habitat selection or herd composition) into a system with impact on the bison, we cannot uncouple these from the bottom-up effects of the rapidly changing Late Pleistocene/early Holocene vegetation and competing herbivore communities (Guthrie 1984). In order to make full use of the bison record to assess the nature of late Pleistocene/early Holocene hunting, research designed to provide information on the relative contributions of climate, diet, competition, and predation to changing bison behavioral and morphological responses is needed.

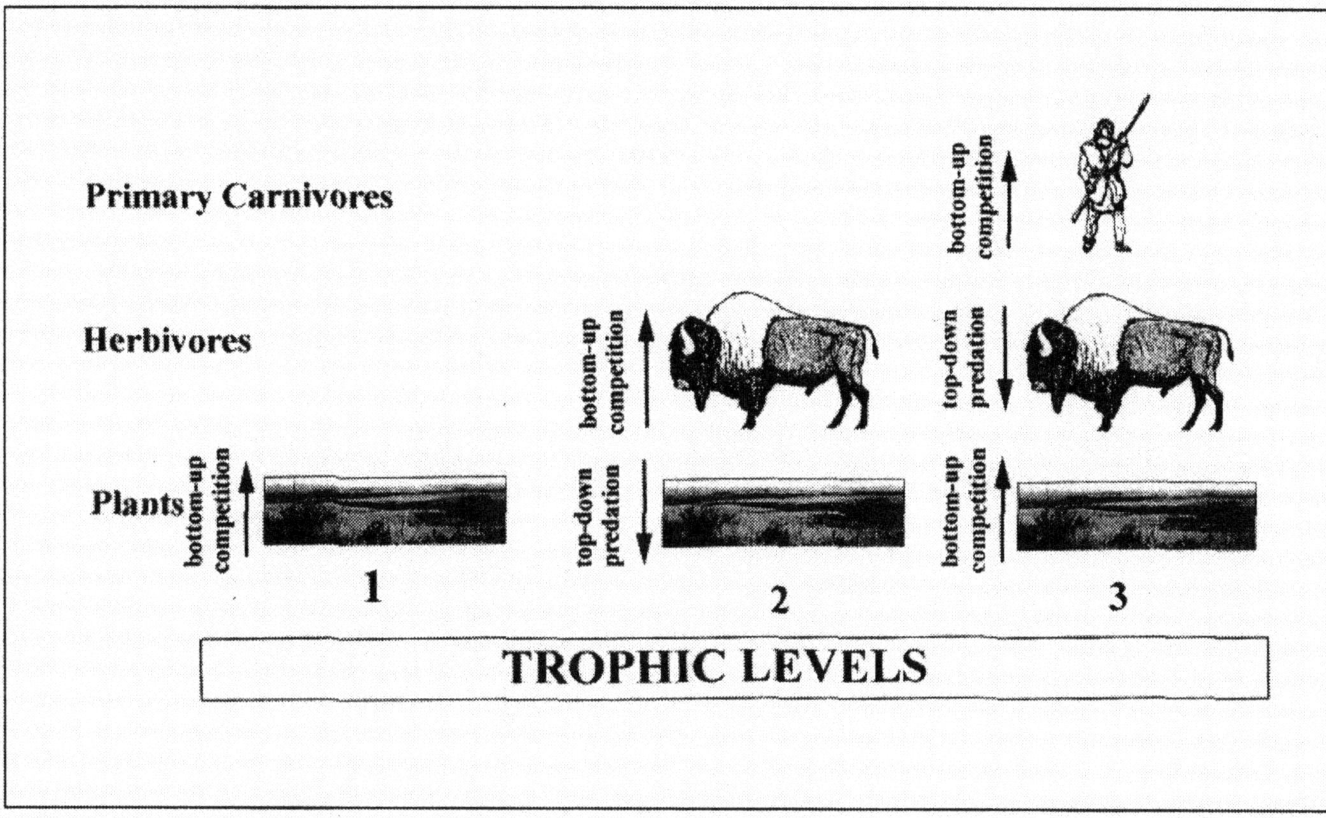

Figure 3. Model of North American Plains that emphasizes the interaction between grasslands, bison, and humans (adapted from Begon et al., 1996, Figure 22.2; bison illustration modified from Geist 1991, Figure 11).

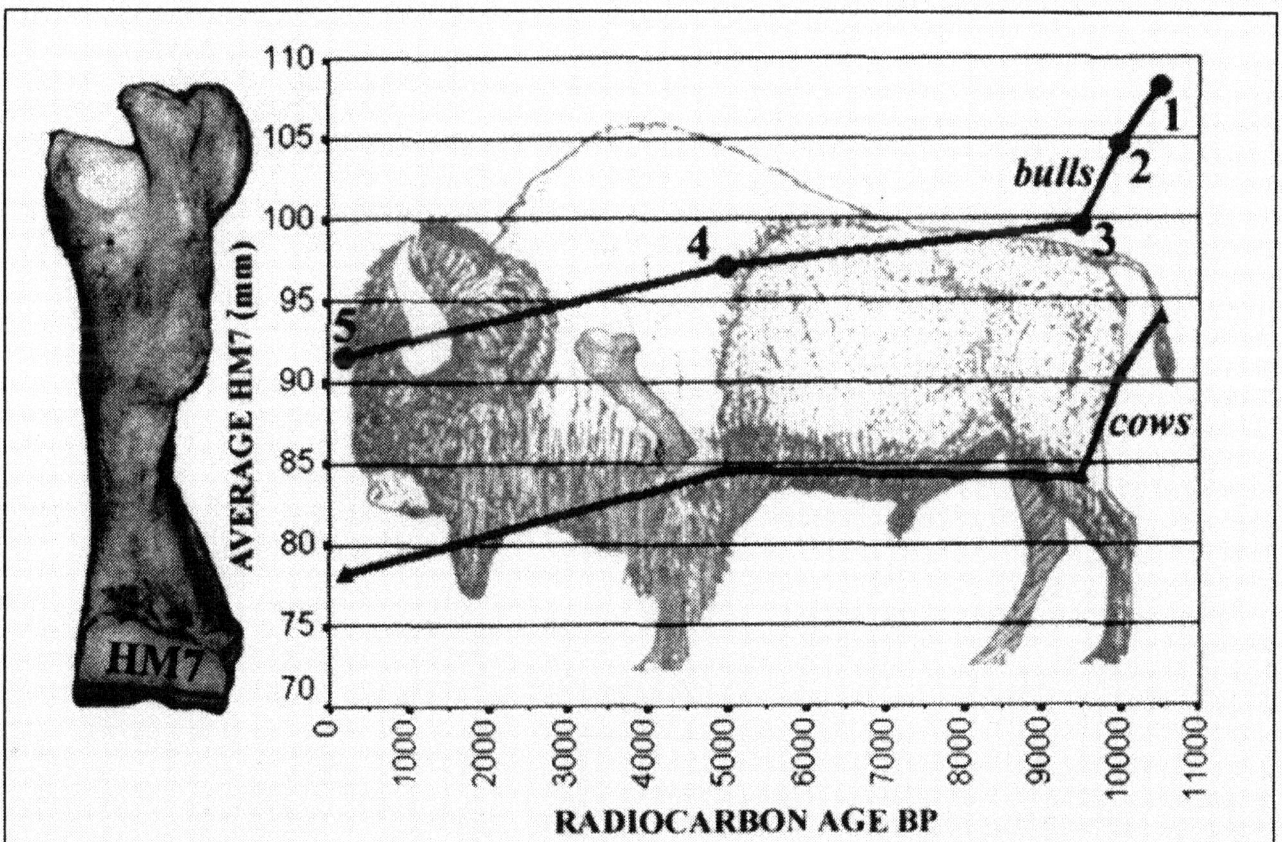

Figure 4. Postcranial size reduction in North American Plains Bison as indicated by measurement of distal humeri.

SITE	COWS No	Mean	Max	Min	SD	BULLS No	Mean	Max	Min	SD	Reference	Chron
HM7 (distal articular breadth)												
Agate Basin Main Folsom	4	95.8	97.3	94.8	1.2	2	113	116.1	109.8	4.4	MG.Hill 1994 (Table 3.12)	1
Folsom	9	95.4	100.9	91.5	3.4	8	106.8	113.3	103.7	2.9	Todd and Hofman	1
Lipscomb	14	93.0	99.4	89.1	2.9	7	107.9	113.1	104.2	3.6	Todd et al. 1992	1
Lindenmeier	2	91.7	93.5	89.9	2.5	8	105.1	108.3	101.0	2.4	unpublished data (M.G. Hill)	1
12 Mile Creek	0	-	-	-	-	4	106.0	110.0	105.0	2.0	M.G. Hill 1994 (Table 4)	1
Jones-Miller	179	89.8	98.1	81.2	3.1	15	105.3	109.3	101.5	2.5	Todd 1987c	2
Casper	56	90.2	98.5	84.2	3.0	6	105.6	107.7	103.5	1.7	unpublished data (Todd & M.G. Hill	2
Frazier	3	91.2	95.6	87.4	3.4	2.	103.0	103.0	1.03.0	0.0	unpublished data (Todd & Hofman)	2
Finley	14	84.4	87.0	78.0	2.6	30	98.4	106.0	90.0	4.3	Haspel and Frison 1987	3
Horner II	13	83.2	87.0	80.0	2.7	10	99.8	103.0	97.0	2.0	Todd 1987c	3
Frasca	15	86.6	93.0	83.2	2,9	17	101.3	109.8	96.4	3.4	Todd 1987c	3
Lamb Spring	3	82.4	85.2	78.5	3.5	10	100.6	106.9	92.1	5.0	Todd 1987c	3
Hudson-Meng	44	83.7	91.2	71.6	3.6	4	97.9	105.0	93.6	5.1	Todd and Rapson 1991	3
Hawken	11	85.1	90.7	80.5	3.3	13	97.1	102.4	91.3	3.1	this report	4
Scoggin	5	84.2	89.5	74.8	5.9	1	97.2	-	-	-	this report	4
River Bend	28	76.8	82.2	64.8	3.7	15	92.3	99.5	87.0	2.9	McKee 1988 (Table 5.1)	5
Vore	62	79.3	84.6	74.4	2.5	67	91.9	99.9	86.2	3.0	this report	5
Glenrock	25	79.0	82.9	75.3	1.9	51	89.8	98.7	83.4	3.1	this report	5
Bugas-Holding	2	78.6	78.7	78.5	0.1	8	93.0	95.7	90.2	1.7	unpublished data (Todd & Rapson)	5
Roberts	5	79.0	84.8	74.6	3.3	1	90.0	-	-	-	this report	5
modern	15	76.3	81.0	72.0	2.8	10	92.8	95.9	89.0	2.3	Todd 1987c	5
HM1 1 (distal depth)												
Agate Basin MF	4	108.0	106.1	111.1	2.3	2	127.9	130.5	125.3	3.7	Hill 1994 (Table 3.12)	1
Folsom	9	106.6	109.9	100.6	3.0	7	119.0	121.0	115.0	2.1	Todd and Hoffman	1
Lipscomb	14	103.8	107.0	95.7	3.3	5	120.4	123.6	115.3	3.3	Todd et al. 1992	1
Lindenmeier	3	101.1	105.1	97.8	2.5	7	121.0	125.8	115.4	3.5	unpublished data (M.G. Hill)	1
12 Mile Creek	0	-	-	-	-	3	116.0	121.0	108.0	7.0	M.E. Hill 1996 (Table 4)	1
Jones-Miller	91	101.0	109.1	93.8	3.6	4	119.2	121.2	117.2	1.9	Todd 1987a	2
Casper	53	100.5	107.9	92.5	3.4	4	116.4	118.8	111.1	3.6	unpublished data (Todd & M.G. Hill)	2
Frazier	2	99.5	104.3	94.7	4.8	1	116.8	-	-	-	Todd and Hofman	2
Finley	10	95.4	103.0	87.0	4.3	23	110.9	123.0	108.0	4.9	Haspel and Frison	3
Horner II	14	94.9	100.0	85.0	3.9	8	112.6	118.0	1.5.0	3.9	Todd 1987c	3
Frasca	7	98.5	99.7	97.5	0.9	10	115.3	124.0	110.8	3.8	Todd 1987c	3
Lamb Spring	1	94.7	-	-	-	2	116.7	123.6	109.8	9.8	Todd 1897c	3
Hudson-Meng	22	96.1	100.4	86.3	3.2	2	105.0	105.0	105.0	0.0	Todd and Rapson 1991	3
Hawken	10	95.9	103.8	90.6	3.9	10	108.7	105.0	114.6	3.1	this report	4
Scoggin	0	-	-	-	-	1	104.8	-	-	-	this report	4
River Bend	29	87.4	100.0	78.5	4.4	14	104.5	108.0	100.2	2.6	McKee 1988 (Table 5.1)	5
Vore	59	90.5	97.5	84.8	2.7	60	104.1	112.3	97.8	3.5	this report	5
Glenrock	21	90.5	94.4	86.3	2.1	39	101.8	110.2	96.1	3.4	this report	5
Bugas-Holding	1	89.2	-	-	-	7	105.4	110.2	101.4	3.0	unpublished data (Todd & Rapson)	5
Roberts	5	91.7	96.6	87.4	3.7	1	108.5	-	-	-	this report	5
modern	15	88.4	93.0	85.0	2.3	10	105.4	110.5	101.7	2.8	Todd 1987	5
HM14 (distal articular depth)												
Agate Basin MF	4	49.2	50.4	47.8	1.4	2	53.9	54.9	53.0	1.4	M.G. Hill 1994 (Table 3.12)	1
Folsom	10	48.4	53.4	45.5	2.3	5	53.1	54.4	507.0	1.4	Todd and Hofman	1
Lipscomb	20	46.0	49.8	43.2	1.6	8	52.8	55.6	50.4	1.9	Todd et al. 1997	1
Lindenmeier	3	42.7	44.7	41.3	1.8	8	52.6	53.9	50.7	1.1	unpublished data (M.G. Hill)	1
12 Mile Creek	0	-	-	-	-	0	-	-	-	-	M.E. Hill 1996 (Table 4)	1
Jones-Miller	177	44.6	50.5	39.7	2.1	15	50.5	53.5	47.9	1.8	Todd 1987c	2
Casper	53	44.4	48.5	40.4	1.9	6	52.2	54.5	50.1	1.6	unpublished data (Todd & M.G. Hill)	2
Frazier	3	46.9	53.4	43.4	4.6	2	51.7	52.9	50.5	1.7	Todd and Hofman	2
Finley	13	42.8	44.0	39.0	1.7	31	47.8	52.0	44.0	1.9	Hespel and Frison	3
Horner II	13	42.1	45.0	39.0	1.8	11	48.8	52.0	46.0	1.7	Todd 1987c	3
Frasca	15	42.4	45.7	34.3	2.8	16	49.7	46.4	53.1	1.9	Todd 1987c	3
Lamb Spring	3	41.8	43.3	39.2	2.3	13	48.1	52.9	44.9	2.3	Todd 1987c	3
Hudson-Meng	42	42.7	38.3	47.5	2.0	4	47.6	48.6	45.5	2.0	Todd and Rapson 1991	3
Hawken	11	41.7	44.3	38.9	2.1	10	46.6	44.0	34.6	2.9	this report	4
Scoggin	5	39.7	43.7	32.5	4.1	0	-	--	-	-	this report	4
River Bend	33	38.4	42.0	35.2	1.9	15	44.8	47.7	42.2	1.7	McKee 1988 (Table 5)	5
Vore	62	39.3	42.9	36.6	1.6	67	44.2	47.9	40.9	1.8	this report	5
Glenrock	25	38.9	42.1	36.5	1.4	52	43.3	46.6	39.8	1.7	this report	5
Bugas-Holding	2	41.1	41.3	40.8	0.3	7	44.7	43.7	45.5	0.7	unpublished data (Todd & Rapson)	5
Roberts	5	39.9	42.4	36.6	2.2	1	46.5	-	-	-	this report	5
modern	15	37.8	40.0	36.0	1.2	10	44.9	42.2	48.8	2.1	Todd 1987c	5

Table 1. Summary of Distal Humerus Measurements (mm) of Bison from North American Plains late Pleistocene-Holocene Bonebeds (for descriptions of measurements see Todd 1987b).

It is along these lines that we see some of the most promising developments in recent paleoecological investigations. Over the last five years there has been a rapid growth and refinement in a number of isotopic and molecular techniques to investigate past communities. Data on changing atmospheric patterns on the Plains (Amundson et al. 1997), rapidly changing vegetation structure (Fredlund and Tieszen 1997) and seasonal temperatures (Elias 1996), and a variety of advances in the methods for sampling and interpreting stable isotope evidences of past herbivore diets (Tieszen 1994; Wang and Cerling 1994; Brooks 1995; R.M. Larson 1995; Bryant et al, 1996; Fricke and O'Neil 1996; Kohn et al. 1996; MacFadden and Cerling 1996; Gadbury 1997; Gadbury et al. 1998; Jahren et al. 1998; Kelly et al. 1998) are all beginning to provide a better foundation on which to build models for interpreting bison-human-vegetation-climate interactions. Investigations of ancient DNA (Chambers 1998) hold promise for providing information on late Pleistocene-Holocene bison population structure. All of these approaches should provide information enabling development of more refined models concerning the paleoecological settings to which bison, grasses, and Paleoindians were adapting during the last 11,000 years.

We consider these techniques to be extremely valuable additions to the archaeological and zooarchaeological records which have customarily been used as the primary sources of information for interpreting early Paleoindian hunting patterns. We have argued that the archaeological and zoological evidence from Lipscomb and other sites supports ideas that early Paleoindians were at least seasonally focusing their activities on bison hunting. In order to more fully evaluate these, and alternative, suggestions of Paleoindian behaviors, we see the need for a greater emphasis on the incorporation of provocative data from multiple paleoecological data sets. As noted by Futuyma and Moreno (1988:207), discussion of "...specialization and generalization must be defined with reference to particular axes (e.g., temperature, range of food particle sizes)." One of the legacies of Dale Guthrie's research has been the growing awareness among Plains archaeologists that bison paleoecology is one such essential axis of Folsom niche breadth models. Folsom stone tools are highly distinctive and are often seen as representing one end of a spectrum of specialized lithic production. However, evaluation of ideas about the relative specialization of Folsom people in terms of subsistence resource selection requires a better understanding of not only the Folsom lithic technology and land use, but also of their key prey. While we do not have a definitive answer to the degree of hunting specialization of the Folsom hunters at Lipscomb, we still think it is too early to quit asking the somewhat unfashionable question: "*but what if they really were* big game hunters?" We still see this not only as a viable, but also reasonable alternative that needs to be incorporated when modeling Paleoindian behaviors.

ACKNOWLEDGMENTS

A version of this paper was presented at the Society for American Archaeology Meeting in 1995 and we wish to thank Britt Bousman and Mike Collins for including us and for making that session possible. Thanks to M. E. Hill for measurements of the Lindenmeier site bison humeri. Special thanks to Craig Gerlach for his willingness to include this version of the paper in the current volume.

REFERENCES CITED

Amick, D.S.
1994 Folsom Diet Breadth and Land Use in the American Southwest. Unpublished Ph.D. dissertation, Department of Anthropology, University of New Mexico, Albuquerque.

1996 Regional Patterns of Folsom Mobility and Land Use in the American Southwest. *World Archaeology* 27:411-426.

Amick, D.S. (editor)
1999 *Folsom Lithic Technology: Explorations in Structure and Variation*. International Monographs in Prehistory, Archaeological Series 12.

Amundson, R., 0. Chadwick, C. Kendall, Y. Wang, and M.J. DeNiro
1997 Isotopic Evidence for Shifts in Atmospheric Circulation Patterns during the Late Quaternary in North-North America. *Geology* 24(l):23-26.

Bamforth, D.
1991 Flintknapping Skill, Communal Hunting, and Paleoindian Projectile Point Typology. *Plains Anthropologist* 36:309-322.

Beardsley, R.K., P. Holder, A. Krieger, B. Meggers, J. Rinaldo, and P. Kutsche
1956 Functional and Evolutionary implications of Community Patterning. In *Seminars in Archaeology l955*. Society for American Archaeology Memoir ll:129-157.

Begon, M., J.L. Harper, and C.R. Townsend
1996 *Ecology: Individuals, Populations and Communities* (3rd edition). Blackwell Scientific, Cambridge, MA.

Belovlsky, G.
1988 An Optimal Foraging-Based Model of Hunter-Gatherer Population Dynamics. *Journal of Anthropological Archaeology* 7:329-372.

Berger, J and C. Cunningham
1994 *Bison: Mating and Conservation in Small Populations*. Columbia University Press, New York.

Bement, L.C.
1997 The Cooper Site: A stratified Folsom Bison Kill in Oklahoma. In, *Southern Plains Bison Procurement and Utilization from Paleoindian to Historic*. Memoir 29, Plains Anthropologist, edited by L.C. Bement and K.J. Buehler, pp. 85-99.

1999 *Bison Hunting at Cooper Site, Where Lightning Bolts Drew Thundering Herds*. University of Oklahoma Press. Norman.

Bement, L.C. and K.J. Buehler (editors)
1997 *Southern Plains Bison Procurement and Utilization from Paleoindian to Historic*. Memoir 29, Plains Anthropologist.

Binford, L.R.
1977 Introduction. In, *For Theory Building in Archaeology*, edited by L. R. Binford, pp. 1-10. Academic Press, New York.

1981 *Bones: Ancient Men and Modern Myths*. Academic Press, New York.

1983 *In Pursuit of the Past: Decoding the Archaeological Record.* Thames and Hudson, London.

1984 *Faunal Remains from Klasies River Mouth.* Academic Press, Orlando.

Boldurian, A.
1990 *Lithic Technology at the Mitchell Locality of Blackwater Draw: A Stratified Folsom Site in Eastern New Mexico.* Memoir 24, Plains Anthropologist.

1991 Folsom Mobility and Organization of Lithic Technology: A View from Blackwater Draw, New Mexico. *Plains Anthropologist* 39(150):445-464.

Bonnichsen, R., D. J. Stanford, and J.L. Fastook
1987 Environmental Change and Developmental History of Human Adaptation Patterns: The Paleoindian Case. In *North America and Adjacent Oceans During the Last Deglaciation: The Geology of North America, vol. K-3,* edited by W.F. Ruddiman and H.E. Wright, Jr., pp. 403424. Geological Society of America, Boulder.

Brain, C.K.
1981 *The Hunters or the Hunted? An Introduction to African Cave Taphonomy.* University of Chicago Press, Chicago.

Brooks, M.
1995 *Paleoclimate of the Late Pleistocene and Holocene: Stable Carbon Isotope Evidence of a Moist-Hot Altithermal Period from Prehistoric Bison Bone Collagen.* Unpublished M.A. Thesis, Department of Anthropology, University of Wyoming, Laramie.

Brown, J. .
1986 Discussion. In, *Foraging, Collecting, and Harvesting: Archaic Period Subsistence and Settlement in the Eastern Woodlands,* edited by S. Neusius, Center for Archaeological Investigations, Southern Illinois University, Carbondale.

Bryant, J.D., P.H. Forelich, W.J. Showers, and B.J. Genna
1996 A Tale of Two Quarries: Biologic and Taphonomic Signatures in the Oxygen Isotope Composition of Tooth Enamel Phosphate from Modern and Miocene Equids. *Palaios* 11(4):397-408.

Byers, John A.
1997 *American Pronghorn: Social Adaptations and the Ghosts of Predators Past.* University of Chicago Press. Chicago.

Carbyn, L.N., S.M. Oosenbrug, and D.W. Anions
1993 *Wolves, Bison and the Dynamics Related to the Peace-Athabasca Delta in Canada's Wood Buffalo National Park.* Circumpolar Research Series No. 4, Canadian Circumpolar Institute, University of Alberta.

Chambers, K.E.
1998 *Genetic Diversity in Pre- and Post-Bottleneck Populations of North American Bison bison: Ancient DNA Analysis of Archaeological Bison Skeletal Remains,* Unpublished Ph.D. Dissertation, Department of Anthropology, The Pennsylvania State University, University Park.

Cohen, M.N. and G. Armelagos (editors)
1984 *Paleopathology and the Origins of Agriculture.* Academic Press, New York.

Collins, M.B.
1999 Clovis and Folsom Lithic Technology on and Near the Southern Plains: Similar Ends, Different Means. In, *Folsom Lithic Technology: Explorations in Structure and Technology,* edited by D. S. Amick, pp. 12-38. International Monographs in Prehistory, Archaeological Series 12. Ann Arbor.

Davis, L.B. and S.T. Greiser
1992 Indian Creek Paleoindians: Early Occupation of the Elkhorn Mountains' Southeast Flank, West-Central Montana. In *Ice Age Hunters of the Rockies,* edited by D.J. Stanford and J.S. Day, pp. 225-283. University Press of Colorado, Niwot.

Davis, L. B., and B. O. K. Reeves (editors)
1990 *Hunters of the Recent Past.* Unwin Hyman, London.

Davis, L.B. and M.B. Wilson (editors)
1978 *Bison Procurement and Utilization: A Symposium.* Plains Anthropologist Memoir 14.

Elias, S.A.
1996 Late Pleistocene and Holocene Seasonal Temperatures Reconstructed from Fossil Beetle Assemblages in the Rocky Mountains. *Quaternary Research* 46:311-318.

Ewers, J.
1955 The Horse in Blackfoot Indian Culture. *Bureau of American Ethnology Bulletin* 159.

Fawcett, W.
1987 *Communal Hunts, Human Aggregation, Social Variation, and Climatic Change: Bison Utilization by Prehistoric Inhabitants of the Great Plains.* Unpublished Ph.D. dissertation, Department of Anthropology, University of Massachusetts, Amherst.

Foley, R.
1987 *Another Unique Species: Patterns in Human Evolutionary Ecology.* Longman Scientific. New York.

Forester, R. E.
1985 Horn Shelter Number 2: The North End, A Preliminary Report. *Central Texas Archeologist* 10:21-36.

Fredlund, G.G. and L.L. Tieszen
1997 Phytolith and Carbon Isotope Evidence for Late Quaternary Vegetation and Climate Change in the Southern Black Hills, South Dakota. *Quaternary Research* 47:206-217.

Fricke, H.C., and J.R. O'Neil
1996 Inter- and intra-tooth Variation in the Oxygen Isotope Composition of Mammalian Tooth Enamel Phosphate: Implications for Paleoclimatolgical and Paleobiological Research *Palaeogeography, Palaeoclimatology, Palaeoecology 126:91-99.*

Frison, G. C.
1974 *The Casper Site: A Hell Gap Bison Kill on the High Plains* Academic Press, New York.

1978 Animal Population Studies and Cultural inference. In, *Bison Procurement and Utilization: A Symposium*, edited by L.B. Davis and M. Wilson, pp. 44-52. Plains Anthropologist Memoir 14.

1987 Prehistoric, Plains-Mountain, Large-Mammal, Communal Hunting Strategies. In, *The Evolution of human Hunting,* edited by H.H Nitecki and D.V. Nitecki, pp. 225-226. Plenum Press, New York.

1991 *Prehistoric Hunters of the High Plains* (2nd edition), Academic Press, San Diego.

Frison, G.C. and L.C. Todd (editors)
1987 *The Horner Site, The Type Site of the Cody Cultural Complex*. Academic Press, Orlando.

Futuyma, D. J. and G. Moreno
1988 The Evolution of Ecological Specialization. *Annual Review of Ecology and Systematics* 19:207-233.

Gadbury, C. A.
1997 *Paleoecological Interpretations of Isotope and Trace Element Variation in Bison Tooth enamel* Unpublished M.A. thesis, Department of Anthropology, Colorado State University, Fort Collins.

Gadbury, C.A., L.C. Todd, A.H. Jahren, R. Amundson
1998 Variation in Carbon and Oxygen Isotope Analysis of Early Holocene Bison Tooth Enamel Bulk Samples and Subsamples. Manuscript submitted for review.

Gamble, C.
1986 *The Palaeolithic Settlement of Europe.* Cambridge University Press, Cambridge.

Geist, V.
1991 Phantom Subspecies: The Wood Bison, *Bison bison "athabascae"* Rhoads 1897 is Not a Valid Taxon, but an Ecotype. *Arctic* 44(4):283-300.

Graham, R. W.
1987 Late Quaternary Mammalian Faunas and Paleoenvironments of the Southwestern Plains of the United States. In *Late Quaternary Mammalian Biogeography and Environments of the Great Plains and Prairies,* edited by R-W. Graham, H.A. Semken, Jr., and M.A. Graham, pp. 24-86. Illinois State Museum Scientific Papers 22. Springfield.

Grayson, D.K.
1988 Perspectives on the Archaeology of the First Americans. In, *Americas Before Columbus: Ice-Age Origins,* edited by R. C. Carlisle, pp. 107-123. Ethnology Monograph 12. University of Pittsburgh, Department of Anthropology.

Greiser, S.T.
1985 *Predictive Models of Hunter-Gatherer Subsistence and Settlement Strategies on the High Plains.* Plains Anthropologist Memoir 20.

Guthrie, R.D.
1984 Mosaics, Allelochemics, and Nutrients: An Ecological Theory of Late Pleistocene Megafaunal Extinctions. In *Quaternary Extinctions: A Prehistoric Revolution,* edited by P.S. Martin and R.G. Klein, pp. 259-298. University of Arizona Press, Tucson.

1990 *Frozen Fauna of the Mammoth Steppe: The Story of Blue Babe.* University of Chicago Press, Chicago.

Hairston, N.G., F.E. Smith, and L.B. Slobodkin
1960 Community Structure, Population Control, and Competition *American Naturalist 94:421-424.*

Hayden, B.
1982 Interaction Parameters and the Demise of Paleo-Indian Craftsmanship. *Plains Anthropologist* 27(96):109-124.

Hester, J. J.
1972 *Blackwater Locality No. 1: A Stratified Early Man Site In Eastern New Mexico.* Fort Burgwin Research Center Publication no. 9. Ranchos de Taos, New Mexico.

Hester, J.J. and J. Grady
1977 Paleoindian Social Patterns on the Llano Estacado. In Paleoindian Life Ways, edited by Eileen Johnson. *The Museum Journal* 17:78-96.

Hill, M.E., Jr.
1996 Paleoindian Bison Remains from the 12 Mile Creek Site in Western Kansas. *Plains Anthropologist* 41(158):359-372.

Hill, M E., Jr. and J.L. Hofman
1997 The Waugh Site: A Folsom-Age Bison Bonebed in Northwestern Oklahoma. In, *Southern Plains Bison Procurement and Utilization from Paleoindian to Historic,* edited by L. C. Bement and K. J. Buehler, pp. 63-83. Plains Anthropologist Memoir 29.

Hill, M.G.
1994 *Subsistence Strategies by Folsom Hunters at Agate Basin, Wyoming: A Taphonomic Analysis of the Bison and Pronghorn Assemblages.* Unpublished M.A. Thesis, Department of Anthropology, University of Wyoming, Laramie.

Hofman, J. L.
1988 Folsom Research in Western Oklahoma: Rethinking the Folsom Occupation of the Southern Plains. *Transactions of the 23rd Regional Archaeological Symposium for*

Southeastern New Mexico and Western Texas, pp. 6-30. Southwest Federation of Archaeological Societies. Amarillo.

1990 Salt Creek, Recent Evidence from the Eastern Folsom Margin in Central Oklahoma. *Plains Anthropologist* 35:367-374.

1991a Folsom Land Use: Projectile Point Variability as a Key to Mobility. In, *Raw Material Economies Among Prehistoric Hunter-Gatherers*, edited by A. Montet-White and S. R. Holen, pp. 335-355. University of Kansas Publications in Anthropology 19. Lawrence.

1991b Sites, Localities, and Isolates: A Regional Perspective on Southern Plains Folsom Evidence. Paper presented at the Society for American Archaeology Meeting, New Orleans. To appear in, *Folsom Archaeology*, edited by M.A. Jodry and D.J. Stanford. In press.

1992 Recognition and Interpretation of Folsom Technological Variability on the Southern Plains. In Ice-Age *Hunters of the Rockies*, edited by D.J. Stanford and J.S. Day, pp. 193-224. University Press of Colorado, Niwot.

1993 An Initial Survey of the Folsom Complex in Oklahoma. *Bulletin of the Oklahoma Anthropological Society* 41:71-105.

1994a Paleoindian Aggregations on the Great Plains. *Journal of Anthropological Archaeology* 13(4):341-370.

1994b Kansas Folsom Evidence. *Kansas Anthropologist* 15(2):31-43.

1995a Unbounded Hunters: Folsom Bison Hunting on the Southern Plains Circa 10,500 BP, The Lithic Evidence. Paper presented in Toulouse, France. To appear in, *Bison Subsistence from Palaeolithic to Paleoindian Times*, edited by J-P. Brugal, J. Jaubert, and F. David. CNRS.

1995b Dating Folsom Occupations on the Southern Plains. *Journal of Field Archaeology* 22(4):421-437.

1999 Folsom Fragments, Site Types, and Assemblage Formation. In, *Folsom Lithic Technology: Explorations in Structure and Variation*, edited by D. S. Amick, pp. 122-143. International Monographs in Prehistory, Ann Arbor.

Hofman, J.L. and B.J. Carter
1991 The Waugh Site: A Folsom-Bison Association in Northwestern Oklahoma. In, *A Prehistory of the Plains Border Region*, Guidebook, 9th Annual Meeting, South-Central Friends of the Pleistocene, edited by B. J. Carter and P.A. Ward, pp. 24-37. Oklahoma State University, Agronomy Department, Stillwater.

Hofman, J.L., D.S. Amick, and R.O. Rose
1990 Shifting Sands: A Folsom-Midland Assemblage from a Campsite in Western Texas. *Plains Anthropologist* 35:221-253.

Hofman, J.L., L.C. Todd, C.B. Schultz, and W. Hendy
1991 The Lipscomb Bison Quarry: Continuing Investigation at a Folsom Site on the Southern Plains. *Bulletin of the Texas Archaeological Society* 60:149-189. (For 1989)

Hofman, J.L. and L.C. Todd
1990 The Lipscomb Bison Quarry: 50 Years of Research. *Transactions of the 25th Regional Archaeological Symposium for Southeastern New Mexico and Western Texas*, pp. 43-58. Southwest Federation of Archaeological Societies. Midland.

Holliday, V.T.
1997 *Paleoindian Geoarchaeology of the Southern High Plains*. University of Texas Press. Austin.

Jahren, A.H., L.C. Todd, and R.G. Amundson
1998 Stable Isotope Dietary Analysis of Bison Bone Samples from the Hudson-Meng Bonebed: Effects of Paleotopography. *Journal of Archaeological Science*, in press.

Johnson, E. (editor)
1977 *Paleo-Indian Lifeways. The Museum Journal*, 17, Lubbock.

1987 *Lubbock Lake: Late Quaternary Studies on the Southern High Plains. Texas A&M Press*, College Station.

Judge, J.W.
1973 *Paleoindian Occupation of the Central Rio Grande Valley* in *New Mexico*. University of New Mexico Press, Albuquerque.

Keeley, L.
1988 Hunter-Gatherer Economic Complexity and "Population Pressure": A Cross-Cultural Analysis. *Journal of Anthropological Archaeology* 7:373-411.

Kelly, E. F., S. W. Blecker, C. M. Yonker, C. G. Olson, E. E. Wohl, and L. C. Todd
1998 Stable Isotope Composition of Soil Organic Matter and Phytoliths as Paleoenvironmental Indicators. *Geoderma 82:59-81.*

Kelly, R. L.
1983 Hunter-gatherer Mobility Strategies. *Journal of Anthropological Research* 39:277-306.

1995 *The Foraging Spectrum: Diversity in Hunter-Gatherer Lifeways*. Smithsonian Institution Press, Washington., D.C.

1996 Ethnographic Analogy and Migration to the Western Hemisphere. In *Prehistoric Mogoloid Dispersals,* edited by T. Akazawa and E.J.E. Szathmary, Oxford University Press, Oxford.

Kelly, R.L. and L.C. Todd
1988 Coming into the Country: Early Paleoindian Hunting and Mobility. *American Antiquity* 53:231-244.

Kohn, P.L., M.J. Schoeninger, and J.W. Valley
1996 Herbivore Tooth Oxygen Isotope Compositions: Effects of Diet and Physiology. *Geochimica et Cosmochimea Acta* 60(20):3889-3896.

Kornfeld, M.
1988 The Rocky Folsom Site: A Small Folsom Assemblage from the Northwestern Plains. *North American Archaeologist* 9(3):197-222.

1996 The Big-Game Focus. *Current Anthropology* 37(4):629-657.

Larson, C.S.
1983 Behavioral Implications of Temporal Change in Cariogenesis. *Journal of Archaeological Science 10:* 1-8.

Leonard, R.D.
1989 Resource Specialization, Population Growth, and Agricultural Production in the American Southwest. *American Antiquity* 54:491-503.

Leonhardy, F.C. (editor)
1966 *Domebo: A Paleo-Indian Kill in the Prairie-Plains.* Contributions of the Museum of the Great Plains No. 1, Lawton, Oklahoma.

Lee, R. B.
1968 What Hunters Do for a Living, or, How to Make Out on Scarce Resources. In *Man the Hunter,* edited by R.B. Lee and I. DeVore, pp.30-55. Aldine, Chicago.

Lyman, R. L.
1984 Bone Density and Differential Survivorship of Fossil Classes. *Journal of Anthropological Archaeology* 3:259-299.

1994 *Vertebrate Taphonomy.* Cambridge University Press, Cambridge.

MacFadden, B.J. and T.E. Cerling
1996 Mammalian Herbivore Communities, Ancient Feeding Ecology, and Carbon Isotopes: A 25 Million-Year Sequence from the Neogene of Florida. *Journal of Vertebrate Paleontology* 107:103-115.

Mack, R.N., and J.N. Thompson
1982 Evolution in Steppe with Few Large, Hooved Mammals. *The American Naturalist 119(6):757-*773.

Martin, P.S.
1967 Pleistocene Overkill. In *Pleistocene Extinctions: The Search for a Cause,* edited by P.S. Martin and H.E. Wright pp. 75-120. Yale University Press, New Haven.

McKee, D. F.
1988 *A Faunal analysis of the River Bend Site (48NA202): Evidence of Protohistoric Subsistence in the Northwest Plains.* Unpublished MA Thesis, Dept. of Anthropology, University of Wyoming, Laramie.

McMillan, R.B., and W.E. Klippel
1981 Post-Glacial Environmental Change and Hunting Gathering Societies of the Southern Prairie Peninsula. *Journal of Archaeological Science 8:215-245,*

McPeek, M.A., and B.L. Peckarsky
1998 Life Histories and the Strengths of Species Interactions: Combining Mortality, Growth, and Fecundity Effects. *Ecology* 79(3):867-879.

Meltzer, D.J.
1986 The Clovis Paleoindian Occupation of Texas: Results of the Texas Fluted Point Survey. *Bulletin of the Texas Archaeological Society* 57:27-68.1989

Was Stone Exchanged Among Eastern North American Paleoindians? *In Eastern Paleoindian Lithic Resource Use,* edited by C.J. Ellis and J. Lothrop, pp. 11-39. Westview Press, Boulder.

1993a *In Search of the First Americans.* Smithsonian Library. St. Remy Press, Montreal.

1993b Is there a Clovis Adaptation? In, *From Kostenki to Clovis: Upper PaleolithicBPaleo-Indian Adaptations*, edited by O. Sofer and N.D. Praslov, pp. 293-310. Plenum Press, New York.

Meltzer, D.J. and B.D. Smith
1986 Paleo-Indian and Early Archaic Subsistence Strategies in Eastern North America. In *Foraging, Collecting, and Harvesting: Archaic Period Subsistence and Settlement in the Eastern Woodlands,* edited by S. Neusius, pp. 1-30. Center for Archaeological Investigations, Southern Illinois University, Carbondale.

O'Connell, J. and K. Hawkes
1981 Alyawara Plant Use and Optimal Foraging Theory. In, *Hunter-Gatherer Foraging Strategies*, edited by B. Winterhalder and E.A. Smith, pp. 99-125. University of Chicago Press, Chicago.

Osborn, A.J.
1999 From Global Models to Regional Patterns: Possible Determinants of Folsom Hunting Weapon Design, Diversity, and Complexity. In, *Folsom Lithic Technology, Explorations in Structure and Variation*, edited by D.S. Amick, pp. 188-213. International Monographs in Prehistory, Archaeological Series 12. Ann Arbor.

Oswalt, W.
1976 *An Anthropological Analysis of Food Getting Technology*. Wiley and Sons. New York.

Owen-Smith, N.
1987 Pleistocene Extinctions: The Pivotal Role of Megaherbivores. *Paleobiology* 13(3):351-362.

Pianka, E.R.
1983 *Evolutionary Ecology*. Harper and Row, New York.
Powers, M.E.

1992 Top-Down and Bottom-Up Forces in Food Webs: Do Plants Have Primacy? *Ecology* 73(3):733-746.

Price, T.D. and J.A. Brown
1985 Aspects of Hunter-gatherer Complexity. In *Prehistoric Hunter-gatherers: 7he Emergence of Cultural Complexity*, edited by T.D. Price and J.A. Brown, pp. 3-30. Academic Press, Orlando.

Schiffer, M.B.
1987 *Formation Processes of the Archaeological Record*. University of New Mexico Press, Albuquerque.

Schmitz, O.J., A.P. Bekerman, and K.M. O'Brien
1997 Behaviorally Mediated Trophic Cascades: Effects of Predation Risk on Food Web Interactions. *Ecology* 78(5): 1388-1399.

Schultz, C.B.
1943 Some Artifact Sites of Early Man in the Great Plains and Adjacent Areas. *American Antiquity* 3:242-249.

Smith, E.A. and B. Winterhalder (editors)
1992 *Evolutionary Ecology and Human Behavior*. Aldine De Gruyter, New York.

Speth, J.D.
1983 *Bison Kills and Bone Counts: Decision Making by Ancient Hunters*. University of Chicago Press, Chicago.

1987 Early Hominid Subsistence Strategies in Seasonal Habitats. *Journal of Archaeological Science* 14:13-29.

1989 Early Hominid Hunting and Scavenging: The Role of Meat as an Energy Source. *Journal of Human Evolution* 18: 329-343.

1990 Seasonality, Resource Stress, and Food Sharing in So-Called "Egalitarian" Foraging Societies. *Journal of Anthropological Archaeology* 9:148-188.

1991a Taphonomy and early Hominid Behavior: Problems in Distinguishing Cultural and non-Cultural Agents. In *Human Predators and Prey Mortality*, edited by M. Stiner, pp. 31-40. Westview Press, Boulder.

1991b Some Unexplored Aspects of Mutualistic Plains-Pueblo Food Exchange. In *Farmers, Hunters, and Colonists: Interaction Between the Southwest and the Southern Plains*, edited by K.A. Spielmann, pp. 18-35. University of Arizona Press. Tucson.

Speth, J.D., and K.A. Spielmann
1983 Energy Source, Protein Metabolism, and Hunter-Gatherer Subsistence Strategies. *Journal of Anthropological Archaeology* 2:1-3 1.

Stanford, D.J.
1983 Pre-Clovis Occupation South of the Ice Sheets. In, *Early Man in the New World*, edited by Richard Shutler, Jr., pp. 65-72. Sage Publications, Beverly Hills.

Struever, S.
1968 Flotation Techniques for the Recovery of Small Scale Archaeological Remains. *American Antiquity* 33(3):353-362.

Tieszen, L.L.
1994 Stable Isotopes on the Plains: Vegetation Analyses and Diet Determinations. In *Skeletal Biology in the Great Plains*, edited by D.W. Owsley and R.L. Jantz, pp. 261-282. Smithsonian Institution Press, Washington, D.C.

Thurmond, J.P.
1990 *Late Paleoindian Utilization of the Dempsey Divide on the Southern Plains*. Plains Anthropologist Memoir 25.

Todd, L.C.
1987a Analysis of Kill-Butchery Bonebeds and Interpretation of Paleoindian Hunting. In, *The Evolution of Human Hunting*, edited by H.H. Nitecki and D.V. Nitecki, pp. 225-226. Plenum Press, New York.

1987b Appendix 1: Bison Bone Measurements. In, *The Horner Site: The Type Site of the Cody Cultural Complex*, edited by G.C. Frison and L.C. Todd, pp. 371-403. Academic Press, Orlando.

1987c Taphonomy of the Horner II Bone Bed. In *The Homer Site: The Type Site of the Cody Cultural Complex*, edited by G.C. Frison and L.C. Todd, pp. 107-198. Academic Press, Orlando.

1991 Seasonality Studies and Paleoindian Subsistence Strategies. In, *Human Predators and Prey Mortality*, edited by M.C. Stiner, pp. 217-238. Westview Press, Boulder.

Todd, L.C., M.G. Hill, D.J. Rapson, and G.C. Frison
1997 Cutmarks, Impacts, and Carnivores at the Casper Site Bison Bonebed. In *Proceedings of the 1993 Bone Modification Conference Hot Springs, South Dakota*, edited by L.A. Hannus, L. Rossum, and R.P. Winham, pp. 136-157. Occasional Publication No 1, Archaeology Laboratory, Augusta College, Sioux Falls.

Todd, L.C., J.L. Hofman, and C.B. Schultz
1990 Seasonality of the Scottsbluff and Lipscomb Bison Bonebeds: Implications for Modeling Paleoindian Subsistence. *American Antiquity* 55(4):813-827.

1992 Faunal Analysis and Paleoindian Studies: A Reexamination of the Lipscomb Bison Bonebed. *Plains Anthropologist* 37(139):137-165.

Tunnell, C.D.
1977 Fluted Projectile Point Production as Revealed by Lithic Specimens from the Adair-Steadman Site in Northwest Texas. In Paleoindian Life ways, edited by E. Johnson. *The Museum Journal* 17:140-168.

Walker, D.N.
1982 Early Holocene Vertebrate Fauna. In, *The Agate Basin Site, A Record of Paleoindian Occupation of the Northwestern High Plains*, edited by G. C. Frison and D. J. Stanford, pp. 274-308. Academic Press, New York.

Watson, P.J.
1976 In Pursuit of Prehistoric Subsistence: A Comparative Account of Some Contemporary Flotation Techniques. *Midcontinental Journal of Archaeology* l(l):77-100.

Wang, Y., and T.E. Cerling
1994 A Model of Fossil Tooth and Bone Diagenesis: implications for Paleodiet Reconstruction from Stable Isotopes. *Palaeogeogrphy, Palaeoclimatology, Palaeoecology* 107:281-289.

Wedel, W.R.
1986 *Central Plains Prehistory: Holocene Environments and Culture Change in the Republican River Basin.* University of Nebraska Press, Nebraska.

Wendorf, F. and J.J. Hester
1962 Early Man's Utilization of the Great Plains Environment. *American Antiquity* 28:158-171.

1975 *Late Pleistocene Environments of the Southern High Plains*. Publications of the Fort Burgwin Research Center, no. 9. Taos.

West, F.H.
1983 The Antiquity of Man in America. *In Late Quaternary Environments of the United States, Volume 1: The Late Pleistocene,* edited by S.C. Porter, pp. 364-382. University of Minnesota Press, Minneapolis.

Wheat, J.B.
1972 *The Olsen-Chubbuck Site: A Paleo-Indian Bison Kill.* Society for American Archaeology, Memoir 26.

Willey, G. R.
1966 *An Introduction to American Archaeology: North and Middle America.* Prentice Hall, Englewood Cliffs.

Wilmsen, E.N., and F H.H. Roberts, Jr.
1978 *Lindenmeier, 1934-1974: Concluding Report on Investigations.* Smithsonian Contributions to Anthropology, no. 24. Smithsonian Institution, Washington, D.C.

Wilson, M.C.
1978 Archaeological Kill Site Populations and the Holocene Evolution of the Genus *Bison.* In *Bison Procurement and Utilization: A Symposium*, edited by L.B. Davis and M. Wilson, pp. 9-22. Plains Anthropologist Memoir 14.

Winterhalder, B.
1986 Diet Choice, Risk, and Food Sharing in a Stochastic Environment *Journal of Anthropological Archaeology* 5:369-392.

Winterhalder, B and E.A. Smith (editors)
1981 *Hunter-Gatherer Foraging Strategies.* University of Chicago Press, Chicago.

Wyckoff, D.G., and L. Albert
1984 Oklahoma Environments: Past and Present. In *Prehistory of Oklahoma*, edited by R.E. Bell, pp.1-43. Academic Press, Orlando.

SEAL PUPS AND THEIR MOMS: A PREFERRED PALAEOESKIMO HUNTING STRATEGY IN IVUJIVIK (NUNAVIK, EASTERN ARCTIC).

Murielle Nagy
GÉTIC, Université Laval, Quebec City, Canada

INTRODUCTION

Ivujivik, located at the northwest extremity of the Canadian province of Quebec, was a strategic place in the peopling of Arctic Quebec. The first human populations that inhabited Hudson Bay, the Hudson Strait, and possibly the Labrador coast, may have passed through the Ivujivik Peninsula (Figure 1), although Inuit oral histories from Ivujivik and Salluit mention that people first came into the area from the Hudson Strait (Graburn 1972:54). The region is rich in animals and a polynya at Ivujivik allows the exploitation of marine mammals year round (Roy 1971a:122). Ivujivik also contains lithic sources as evidenced by a near-by quartz quarry (Taylor 1960:2), and two steatite quarries southeast and west, and one on Digges Island (Roy 1971a:267). Archaeological surveys and excavations in the Ivujivik region have shown that people were there during the Pre-Dorset, Dorset, and Thule periods (Figure 2). The Ivujivik region has the second highest density of place names along the Nunavik coastline, reflecting the intensive use and knowledge of the area by the Ivujivimiut today and in the recent past (Müller-Wille 1990:41).

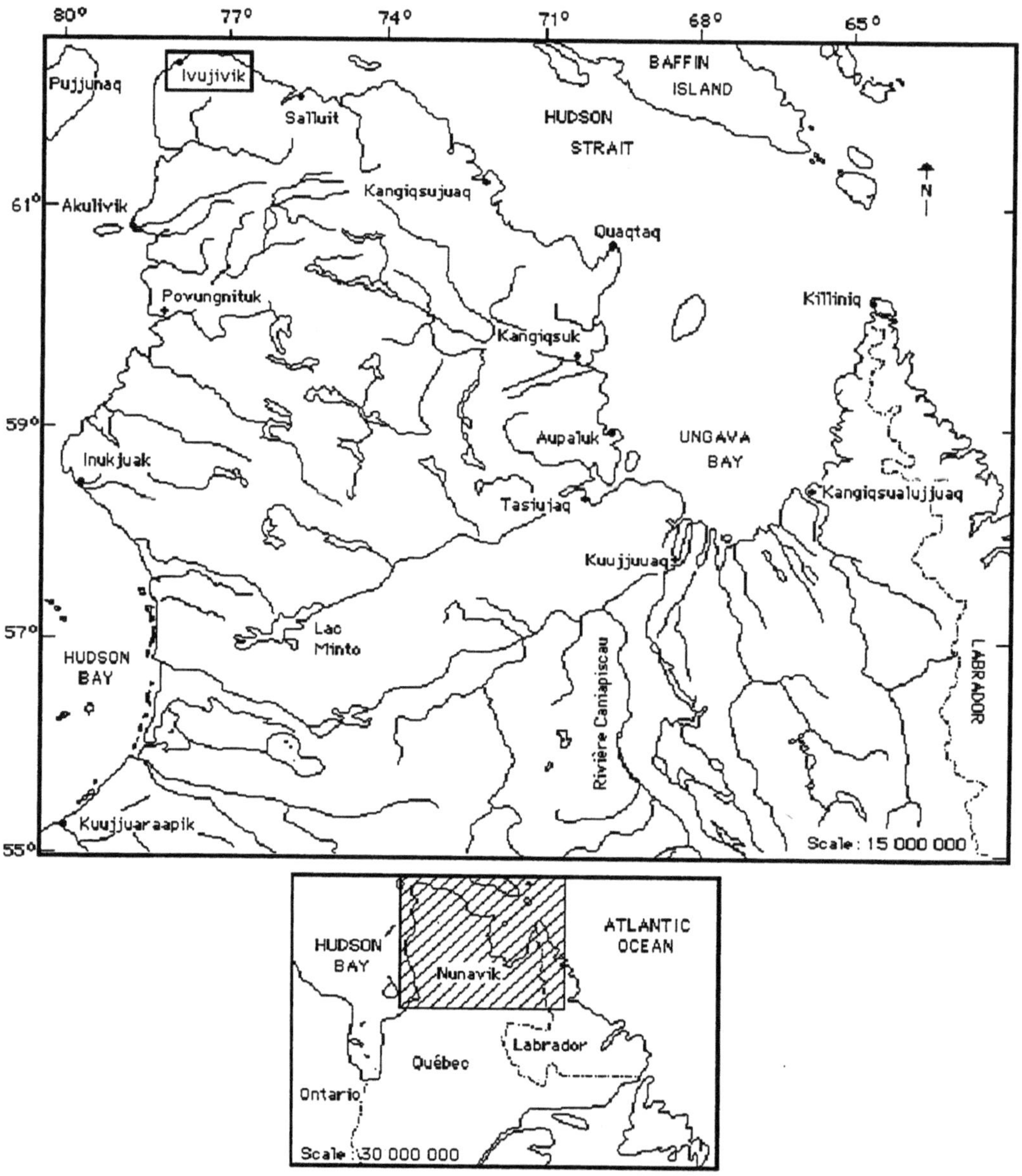

Figure 1. Location of Ivujivik (Nunavik, Northern Quebec).

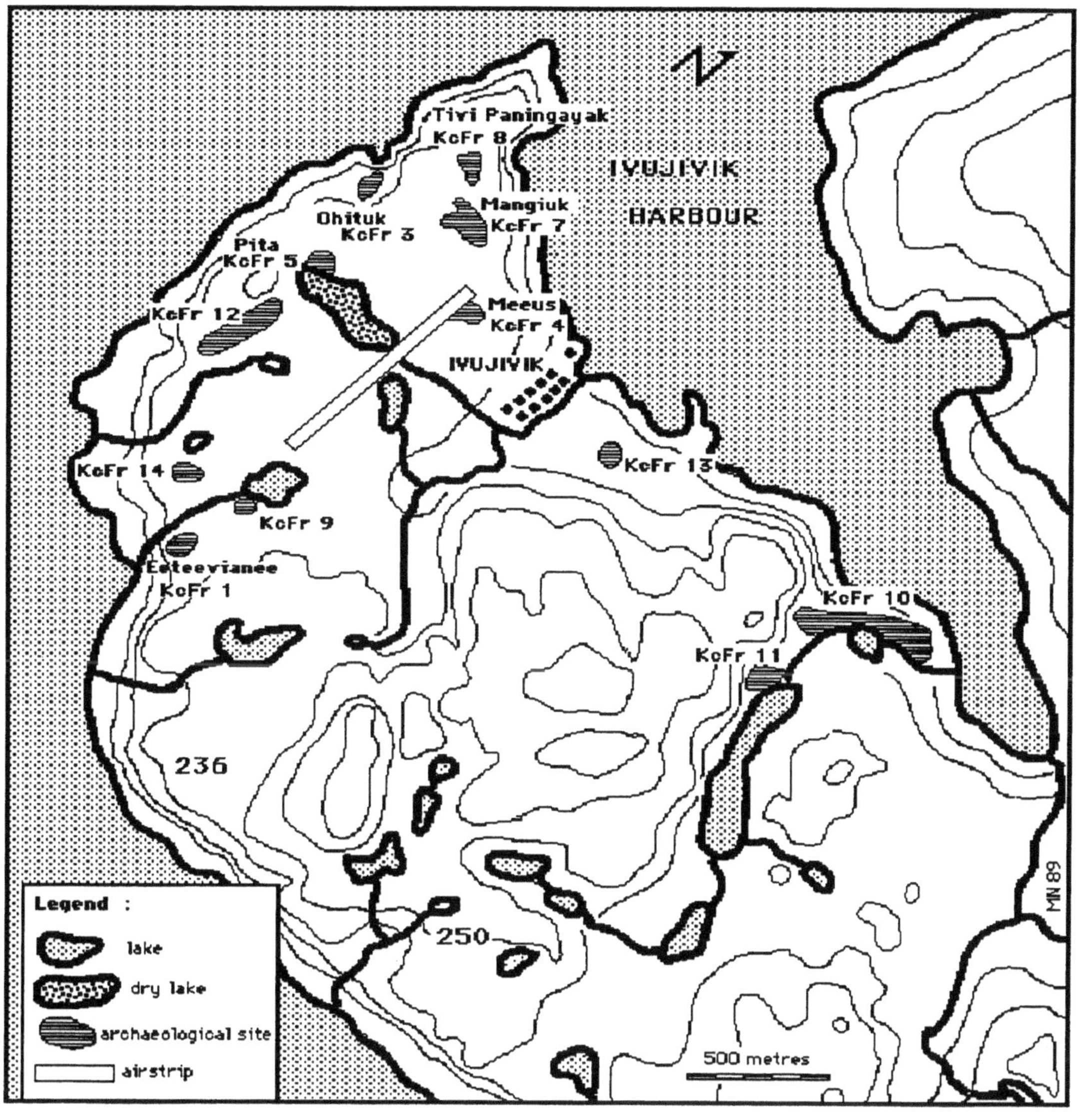

Figure 2. Archaeological sites in Ivujivik.

Ivujivik is also important in the history of archaeological theories about palaeoeskimo culture change in the eastern Arctic. Archaeologist William E. Taylor Jr. first worked in Ivujivik before excavating the Arnapik and the Tyara sites. Analyses of artifacts from the latter sites, as well as those from Ivujivik, led Taylor to conclude that there was cultural continuity between Pre-Dorset and Dorset palaeoeskimos in the eastern Canadian Arctic, and an *in situ* development of Dorset culture (Taylor 1968).

This paper presents faunal data from three sites in Ivujivik excavated between 1988 and 1990. Two sites date to the Pre-Dorset/Dorset transition (800-500 BC), and the third site is a Dorset site (500 BC-AD 500). The faunal assemblages are used compare the subsistence patterns of people who occupied the Ivujivik region during the Pre-Dorset/Dorset transition and the Dorset periods.

ANALYTICAL METHODS

The quantitative units used in this study are NISP, MNI, MNE and MAU. Many discussions have risen from the use and misuse of these units (e.g., Grayson 1984; Lyman 1994; Ringrose 1993) although a review of these is beyond the scope of this paper. Thus, only a brief definition of each quantitative unit is presented. NISP is the number of identified specimens per taxon. MNI is the minimum number of individuals per taxon. In this study, the age and side of bones are considered in the calculation of the MNI. MNE is the minimum number of skeletal elements of a taxon. It is calculated with attention to complete and incomplete elements, as well as age categories, but without consideration of side for similar elements that could be paired. Comparison of NISP and MNE illustrates the degree of bone fragmentation (Lyman 1994:102). MAU is the minimum number of animal units necessary to account for the skeletal parts of a taxon. To derive MAU, the MNE is calculated and divided by the number of times the given skeletal part is present in a specific taxon. MAU accounts for the degree of fragmentation and also for the fact that various skeletal parts are differentially represented in a complete skeleton.

%MAU values are also used to present the data. These are normed values calculated by equating the most frequent skeletal part to 100 and scaling all the other parts to that value. Separate MNI and MAU values are not calculated for the bones recovered from individual dwelling structures as the elements found in caches, middens and between structures also reflect animals consumed at the site. Limiting calculations of MNI and MAU to bones discarded within tents or houses would certainly give smaller values than those calculated for whole site assemblages.

Although a meat utility index for phocid seals has been published (Lyman et al. 1992), it is not used to interpret the Ivujivik data since its authors conclude that there are few significant correlations between the indices and the skeletal part frequencies from archaeological sites. Meat utility indices are used to predict and explain which parts of animals are left at the kill site and which are brought back to camp. The Ivujivik sites discussed in this study were all located near sea level at the time of their occupation, and it is reasonable to assume that most seal carcasses were complete when brought back to the sites. Ethnoarchaeological data collected in the eastern Arctic support this assumption (see Lyman et al. 1992:544). Furthermore, attempts by LeBlanc (1994) to apply the Lyman et al. (1992) meat utility index to seal remains from a palaeoeskimo site in the western Canadian Arctic proved equally unsuccessful. LeBlanc (1994:107) concludes that some of the patterns observed in the seal bone frequencies were due to post-depositional attrition rather than to specific transport decisions. More recently, Diab (1998) presented a ringed seal meat utility index which he compared with preference ranks and bone density, and then with archaeofauna. His results support previous research, demonstrating that the absence of taphonomic information on factors such as weathering and carnivore destruction limits the interpretive strength of utility data (Diab 1998:1).

The faunal remains in this study were analyzed using the bone collection at the Zooarchaeology Laboratory of the Anthropology Department at the University of Alberta. Ringed seal and harbour seal are difficult to distinguish osteologically, although the former species is slightly smaller than the latter (Banfield 1981). However, since there is sexual dimorphism in both species (King 1983), the bones of a male ringed seal could easily be confused with those of a female harbour seal. For these reasons, all bones belonging to *Phoca* species are identified as "small seal."

The bulla, mandible and maxilla (Amorosi 1992) best differentiate the ringed seal and harbor seal. Only two mandibles (out of an MNE of 158), one maxilla (out of an MNE of 38), and one bulla (out of an MNE of 165) in all three assemblages belong to harbour seal. It is therefore reasonable to assume that most of the small seal bones are from ringed seal. All the other mandibles (MNE=27) and bullae (MNE=48) identifiable to species belong to ringed seal. In the case of skeletal elements from large seals that are not identifiable to species, contemporary data on seal species of the Ivujivik region (Roy 1971a, 1971b) suggests that these are bearded seals.

The seal bones are divided into five age categories: 1) foetal/newborn; 2) juvenile; 3) subadult; 4) adult; and 5) subadult or adult. The first category comprises foetal and 1-2 month old individuals. The juveniles are individuals from 3-12 months old. Most limb bones identified as juvenile were very small and belonged almost certainly to 3-5 month old seals as indicated by one of the seal specimens of that age in the comparative collection. The subadults are yearlings and sexually immature individuals up to 6 and 7 years old, whose bones were not yet fused. The adults are sexually mature individuals, that is, older than 7 years, with completely fused bones. The last category (subadult/adult) is used to classify bone fragments where epiphyses are absent and which may belong to either subadult or adult seals.

BRIEF DESCRIPTIONS OF THE SITES AND FAUNAL ASSEMBLAGES

Pita Site (KcFr-5)

Three tent rings and two houses with small sod walls, and spaces between these structures, were excavated at the Pita site. The stylistic diversity of artifacts at the Pita site as a whole may indicate that the habitation structures were occupied at different times from the Pre-Dorset to the Dorset period. However, within individual structures there is also high stylistic diversity. This suggests that the site was occupied during the Pre-Dorset/Dorset transition. This inference is corroborated by the AMS dating of a sample of willow charcoal collected from the hearth of one of the structures (see Nagy 1997) to 2580±60 BP uncalibrated (Beta-51804). Calibrated with the intercept method at 1 sigma (Stuiver and Becker 1986), the date is 2755-2611 years BP (or 806-662 years calibrated BC). This date corresponds to an occupation during the early part of Pre-Dorset/Dorset transition.

The majority of the 15,572 bones recovered came from the midden area and from some of the caches. Of these, 84.5% are identifiable to taxon and the majority are sea mammals (Table 1). The most common taxon is small seal (96.20 %NISP and 82.97 %MNI). A total of 12,659 small seal bones were excavated at the Pita site. The total MNI is 190 individuals and the highest MNI (83) is for juvenile individuals. All skeletal elements are represented. The degree of fragmentation, as indicated by the difference between the NISP and the MNE values, is highest for the ribs, cranium, scapula, innominate, and the limb bones (Figure 3). The %MAU demonstrates that the humerus is the best represented element. The most underrepresented bones are the carpals, baculum and sternum.

TAXA	PITA SITE				OHITUK SITE				TIVI PANINGAYAK SITE			
	NISP	%NISP	MNI	%MNI	NISP	%NISP	MNI	%MNI	NISP	%NISP	MNI	%MNI
BIRDS												
Anas sp.	10	0.08	5	2.18	1	0.07	1	2.38				
Black guillemot	1	0.01	1	0.44								
Canada goose	5	0.04	2	0.87	3	0.22	1	2.38				
Common snipe	2	0.02	2	0.87								
Dovekie	2	0.02	1	0.44								
Eider/scoter	21	0.16	6	2.62	17	0.79	3	7.14				
Ptarmigan	16	0.12	4	1.75								
Snow goose	6	0.05	2	0.87								
Thick-billed murre	22	0.17	2	0.87	11	0.79	3	7.14	3	0.3	1	2.6
Subtotal	85	0.65	25	10.92	32	2.31	9	21.47	3	0.3	1	2.6
LAND MAMMALS												
Arctic fox	36	0.27	3	1.31	1	0.07	1	2.38	1	0.1	1	2.6
Arctic/red fox	4	0.03	–	–	2	0.14	-	-	1	0.1	-	-
Arctic hare	1	0.01	1	0.44								
Dog/wolf	1	0.01	–	–								
Wolf	2	0.02	1	0.44								
Caribou	80	0.61	2	0.87	56	4.04	2	4.76	1	0.1	1	2.6
Polar bear	13	0.10	1	0.44	4	0.29	1	2.38	1	0.1	1	2.6
Subtotal	137	1.04	8	3.49	63	4.55	4	9.52	4	0.4	3	7.9
SEA MAMMALS												
Beluga	2	0.02	1	0.44	1	0.07	1	2.38	2	0.2	1	2.6
Beluga/narwhal	15	0.11	–	–	4	0.29	-	-	7	0.7	-	-
Large whale	10	0.08	1	0.44	1	0.07	1	2.38	6	0.6	1	2.6
Bearded seal	32	0.24	3	1.44	2	0.14	1	2.38	9	0.6	1	2.6
Large seal	54	0.41	–	–	1	0.07	-	-	3	0.3	-	-
Large seal/walrus	2	0.02	–	–								
Small seal (phoca sp.)	12659	96.20	190	82.97	1271	91.70	25	59.52	1025	95.6	30	78.9
Walrus	163	1.24	1	0.44	11	0.79	1	2.38	13	1.2	1	2.6
Subtotal	12937	98.31	196	85.59	1291	93.15	29	69.05	1065	99.3	34	89.5
Total identifiable	13159	100.00	229	100.00	1386	100.00	42	100.00	1072	100.0	38	100.0
UNIDENTIFIABLE	N	%			N	%			N	%		
Unid. class	13	0.54			12	3.02			2	1.1		
Unid. mollusc	72	2.98			6	1.51			-	-		
Unid. bird	98	4.06			25	6.28			1	0.5		
Unid. mammal	1019	42.23			301	75.63			106	57.9		
Unid. land mammal	70	2.90			19	4.77			1	0.5		
Unid. sea mammal	1141	47.29			36	9.05			73	39.9		
Subtotal	2413	100.00			399	100.00			183	100.0		
Grand total	15572				1785				1255			

Table 1. List of taxa, Pita, Ohituk and Tivi Paningayak sites.

Ohituk Site (KcFr-3A)

Identification of individual habitation structures at the Ohituk site is difficult due to the seemingly random distribution of small boulders at the site. Nevertheless, the spatial distribution of artifacts and faunal remains suggest the presence of a tent structure. It is difficult to determine if all the artifacts found at the Ohituk site were deposited during a single occupation, and two radiocarbon dates were obtained to help resolve the issue. The first sample is from three fragments of willow charcoal associated with the remains of a hearth feature. It dates to 1450±130 BP uncalibrated (Beta-62290). Calibrated with the intercept method at 1 sigma (Stuiver and Becker 1986), it dates between 1580-1320 years BP (or 435 to 675 years calibrated AD). Such a recent date places the occupation of the site well within the Dorset period, but is rejected as it does not corroborate an earlier affiliation based on artifact typology

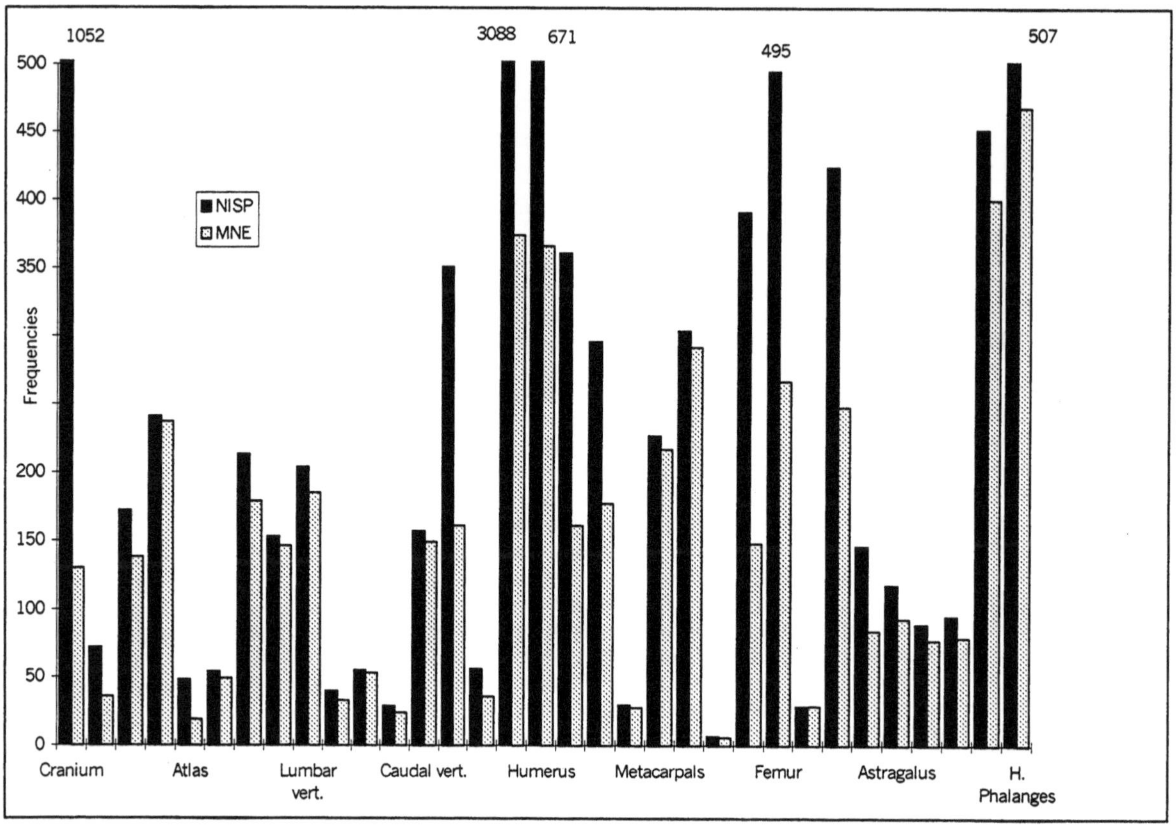

Figure 3. Pita site (KcFr-5): NISP and MNE of small seal skeletal parts (all ages)caches containing foetal/newborn and juvenile seals.

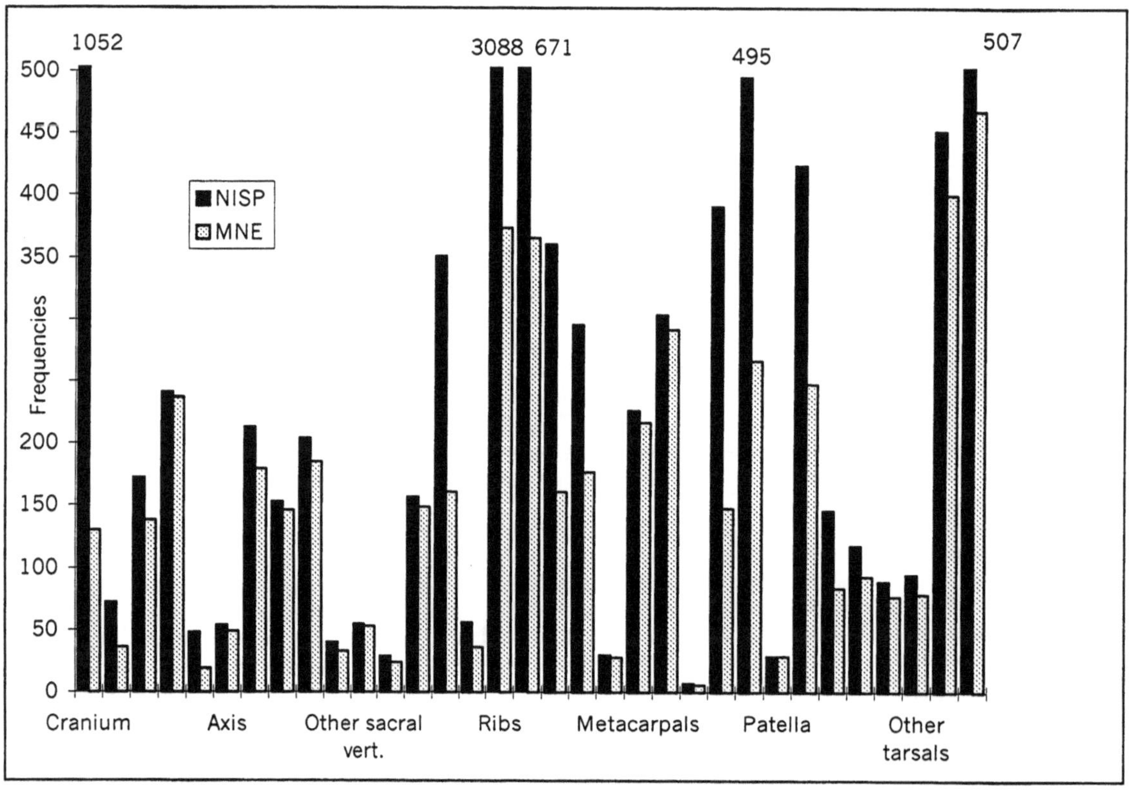

Figure 4. Ohituk site (KcFr-3A): NISP and MNE of small seal skeletal parts (all ages)caches containing foetal/newborn and juvenile seals.

(see Nagy 1997). A second sample from caribou bones dates to 2520±80 BP (AECV-1795C). Calibrated using the intercept method at 1 sigma (Stuiver and Becker 1986), it dates between 2747-2466 years BP (or 798-517 years calibrated BC). This date places the occupation of the Ohituk site during the Pre-Dorset/Dorset transition.

A total of 1,785 bones were recovered at the Ohituk site. Of these, 77.6% are identifiable to taxon and the majority are sea mammals (Table 1). The most common taxon was small seal (91.70 %NISP and 59.52 %MNI). Differences between NISP and MNE values show minimal fragmentation for those identified elements.

A total of 1,271 bones of small seals was identified in the faunal assemblage from the Ohituk site. The total MNI is 26 individuals and the highest MNI (13) is for juvenile individuals. With the exception of the baculum, all skeletal elements are represented. Bone fragmentation is highest for the ribs, cranium, scapula, innominate, and the limb bones (Figure 4). The %MAU shows that the best represented bones are the humerus and the femur. The most underrepresented bones are the baculum and sternum.

Tivi Paningayak Site (KcFr-8A)

Two tent rings, one house with small sod walls, and one semi-subterranean house were excavated at the Tivi Paningayak site. No radiocarbon dates were obtained, but the style of the lithic artifacts links it to the middle Dorset period (i.e.; 500 BC-AD 500).

Of the 1,255 bones recovered, 85.4% are identifiable to taxon, and the majority are sea mammals (Table 1). The most common taxon is small seal (95.6 %NISP and 78.9 %MNI). The site yielded a total of 1,025 bones from small seals. The total MNI is 30 individuals and the highest MNI (14) is for the subadults. All skeletal elements are represented, with the exception of the maxillae and baculum. Bone fragmentation is highest for the ribs, innominate and limbs, especially the humerus (Figure 5). The %MAU indicates that the best represented bone is the humerus. Underrepresented bones include carpals and patella, while the baculum and maxillae are totally absent.

SITE SEASONALITY

Pita Site (KcFr-5)

Ten ringed seal canines from the habitation were cross sectioned to identify the season of death. Eight canines provided good cross-sections. The method used to "read" the dentine layers was the same as that used by McLaren (1958) and Smith (1973) for ringed seals. All canines came from seals that died between April and June. This evidence, and the presence of bones from foetal/newborn small seals, migratory birds, and beluga whale, indicates that people occupied the Pita site at least during the spring/early summer. The inhabitants may have stayed longer as suggested by the remains of two houses with small sod walls, and the presence of food remains. These caches suggest storage of spring and summer killed animals for later consumption, possibly the fall/winter. Multiseasonal use of Pre-Dorset/Dorset transitional sites has been documented for Groswater sites in Ungava (Gendron 1990) and Newfoundland (Renouf 1994).

Ohituk Site (KcFr-3A)

As indicated by the presence of foetal/newborn and juvenile seals, juvenile caribou, and migratory birds, the Ohituk site was occupied during the spring and part of summer. The relatively thin cultural deposit and the absence of caches suggest that the occupation period was short.

Tivi Paningayak Site (KcFr-8A)

The presence of few foetal/newborn and juvenile small seals at the site indicates that people were present during the spring and part of the summer, or that they consumed seals hunted during those seasons. The high proportion of subadult seals suggests people stayed at the site until at least June when subadults are available on the coast (Smith 1973:21). However, the absence of migratory birds, the low diversity of bird species exploited, and the presence of a subterranean house at the site, also point to a cold season occupation, at least for one structure. People who occupied that house might have lived on cached sea mammals hunted during the summer as suggested by the presence of caches all around the site. With the exception of the midden area, the low density of bones in and around the habitation structures indicates that the site was occupied for a short period.

BONE DISTRIBUTION AT THE IVUJIVIK SITES

Comparison of the %MAU of the skeletal elements from all Ivujivik sites shows that the transitional sites (KcFr-5 and KcFr-3A) and the Dorset site (KcFr-8A) have similar patterns of %MAU, but that the early and late transitional sites are most alike (Figure 6). The upper limbs and the innominate are the best represented. The high representation of humerus and femur is certainly due to the high bone density of these elements; they preserve better than most other elements (Lyman 1995:56). The most under-represented elements are the baculum and the flippers, although the hind flippers are somewhat better represented than the front flippers.

The near absence of bacula might be due the exclusive exploitation of juvenile seals and their mothers. When hunting subadult seals, there are as many chances to catch females as males. If most hunting is done in open water where hunters can look at their prey (in contrast to breathing hole hunting), a distinction between males and females may be made. In the spring male seals have a "very strong, offensive, musky odour" (Banfield 1981:373) and according to Inuit informants, a very strong taste which makes people favour female seals over males. Rutting male ringed seals have blackened face (Ryg et al. 1990:470) and are easily spotted and avoided by Inuit hunters. Palaeoeskimo hunters may have had similar preferences, or if they did hunt male seals, they may have detached bacula prior to transporting the seals to camp. Another possibility is that bacula were used to make tools such as flint knapping punches. However,

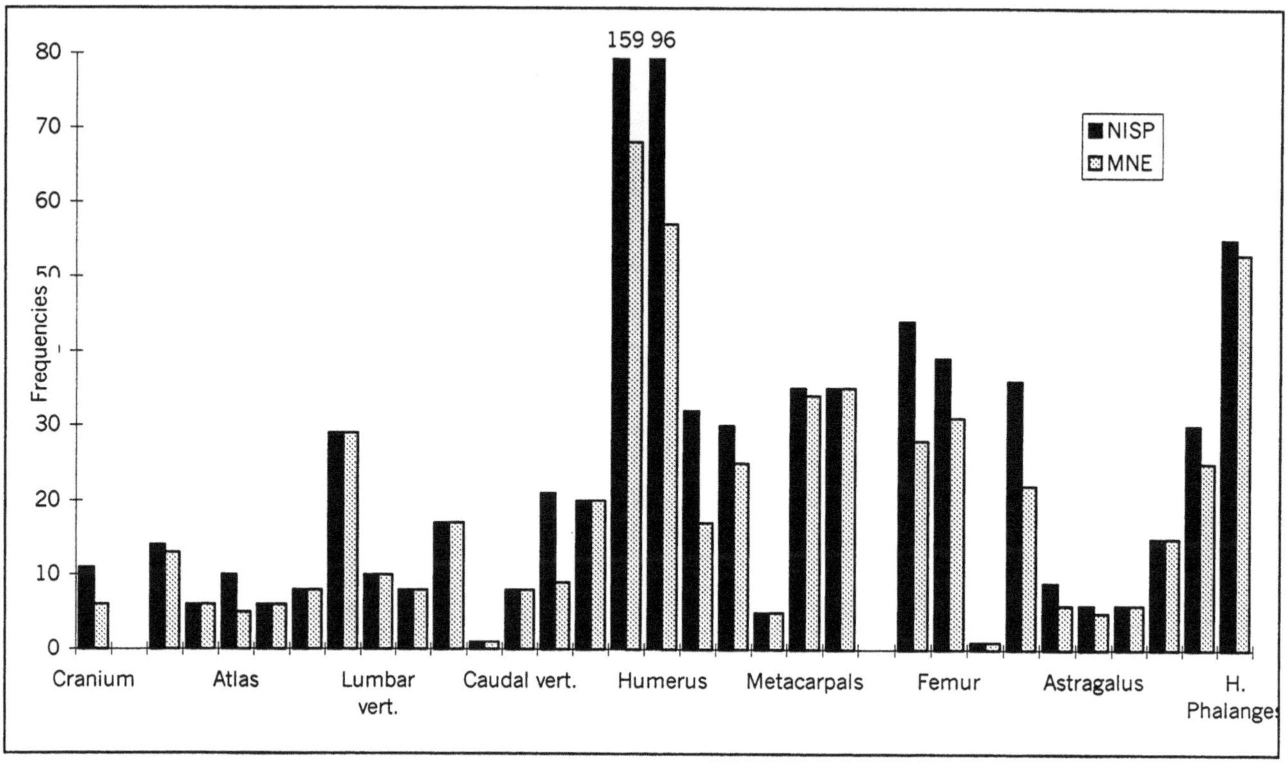

Figure 5. Tivi Paningayak (KcFr-8A): NISP and MNE of small seal skeletal parts (all ages).

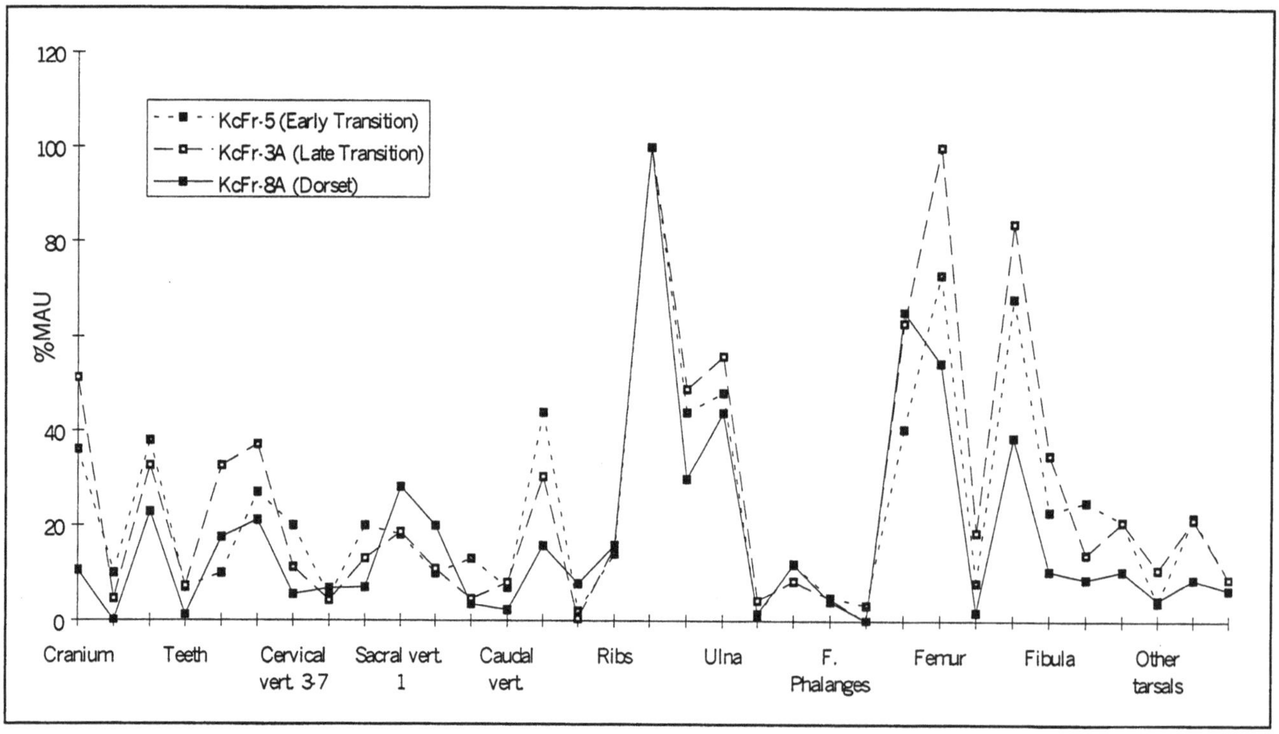

Figure 6. %MAU of small seal skeletal parts (all ages) from Ivujivik sites.

only one tool made from a baculum was recovered from the Ivujivik sites. A low incidence of bacula among harp seal remains was also noted by Murray (1992) in the faunal assemblage from a Dorset dwelling in Newfoundland, but in that case it is attributed to differential availability of male and female seals.

The low %MAU for flippers may suggest that they were cut off before the carcasses were transported.Alternatively they were left attached, and were later chewed to unidentifiability by carnivores, rodents or even people. Flippers are highly valued as food item and for bone games (e.g., Jenness 1922; Van de Velde 1956), and it is unlikely that they would have been discarded at the kill site. Another possibility is that flippers were consumed away from the site perhaps as snack food. Stenton (1983) reports similar low distributions of flippers from a Thule site on Baffin Island. In contrast, high %MAU for flippers was calculated for small seal bones from a Dorset site in Newfoundland (Murray 1992:92) and from a transitional site from the western Canadian Arctic (LeBlanc 1994:103).

Aside from the low representation of elements from the flippers and the lack of bacula, the overall pattern of %MAU suggests that whole animals were brought back to the sites during the transition and the Dorset periods. At the Dorset site, most skeletal elements from small seals were recovered, but the %MAU for the crania and maxillae is very low and might indicate a different treatment of seal heads compared with previous periods of occupation. The seal heads might have been discarded elsewhere near the site or thrown back into the sea in the same way that beluga heads are disposed of by the Inuit from Ivujivik (Roy 1971a:153). McGhee (1981:31) also reports a lack of cranial elements from the Dorset occupation in Port Refuge (High Arctic Canada). In contrast, data from a Dorset site in Newfoundland showed that %MAU of small seal heads were the highest (Murray 1992:92).

Finally, comparison of %MAU for all age groups shows generally similar results in skeletal representation with the exception of the foetal/newborn category where low MNE are certainly responsible for the drastic differences in %MAU.

PALAEOESKIMO SUBSISTENCE PATTERNS IN IVUJIVIK

In order to estimate whether the subsistence patterns of the people who occupied the Ivujivik sites changed through time, the number of species was tabulated for each major class (i.e., birds, land mammals and sea mammals). The species diversity of birds decreases between the early transition to the Dorset periods, while sea mammal species diversity remains the same (Figure 7).

Land mammal species diversity decreases from the early to late transition, but increases slightly during the Dorset period. These trends should be treated with caution as they may also reflect the season of occupation. For example, if a site was occupied during winter, migratory birds would not be available, unless they had been cached during spring/summer for winter consumption. Furthermore, if a site was occupied for longer periods or during repeated visits, then the opportunities to catch a wider variety of species would be increased.

As expected at coastal sites, hunting focused on sea mammals (Figure 8). Marine mammal exploitation fluctuates over time, decreasing in the late transitional period with a subsequent increase in the Dorset period. This corresponds to an increase in the exploitation of other species in the late transition and a decrease in the exploitation of those same species in the Dorset period.

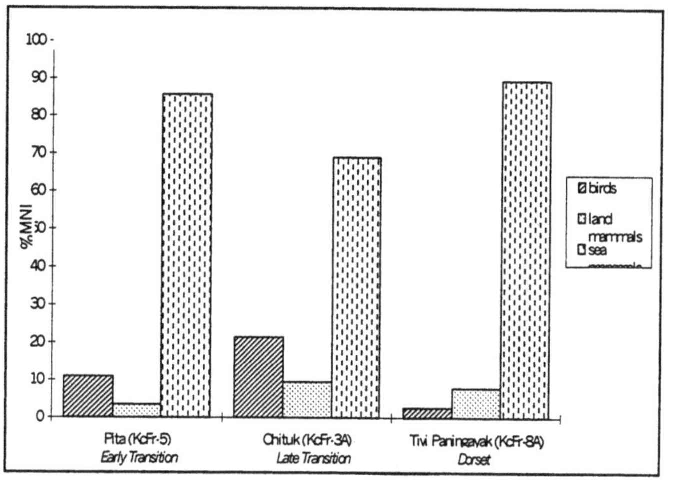

Figure 7. Number of species per class from Ivujivik sites.

Comparisons of %MNI of small seals by age categories through time indicate similar patterns in the hunting of juveniles during early and late transition, while subadults were more heavily exploited during the Dorset period (Table 2, Figure 9). This pattern is likely linked to the season of exploitation. At sites occupied during the spring/early summer, people favoured the hunting of immature seals. The very low count of bacula suggests that few males were hunted. This plus the age data suggest females and their pups (foetal/newborn and juvenile) were the target group. Young seals and their lactating mothers are reported to be easy prey at least until June when most pups have been weaned (Smith 1973:21; Stirling and McEwan 1975). Ringed seals are born between mid-March and mid-April and stay in their dens during the first two months of their life (Banfield 1981:374). The identification of newborn seals in the faunal assemblages indicates that the hunters took the seals from their dens where they could have been easily clubbed (see Lyman 1992, 1995). Incidentally, no complete crania of small seals were recovered at the Pita site, which reinforces the interpretation that they were clubbed on the head. Foetal/newborn seals may also have been scavenged since those born too early in the spring risk death by freezing (McLaren 1958:63).

As for the seal mothers, the task of hunting them might have been trickier since they are quick to return to the sea, and the ice is dangerous for the hunters during the spring. The spearheads recovered from the Ivujivik sites were likely used to hunt seals. At the beginning of the century, Inuit from northern Quebec were reported to hunt sun basking seals on the spring ice with spears (Low 1906:153).

Young seals are desirable because of their high fat content. After birth, ringed seal pups feed on their mother's milk for the next two months (McLaren 1958:57) and rapidly increase in weight and blubber with an average daily mass gain of 350 to 480g and a weaning weight of 20±2.9kg after 39 days of suckling (Lydersen et al. 1992:365, 367). While by April lactating females have lost 25% of their fat (Ryg et al. 1990:472), the pups at this time are approximately 45% fat by wet weight (Stirling and McEwan 1975). The meat of a

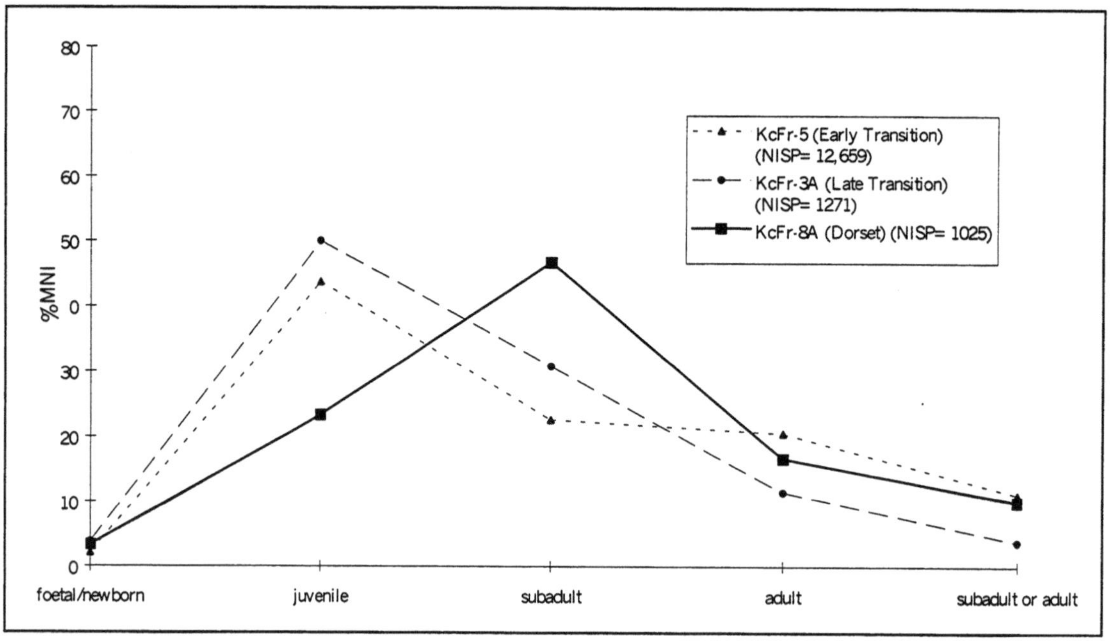

Figure 8. %MNI of birds, land mammals and sea mammals from Ivujivik sites.

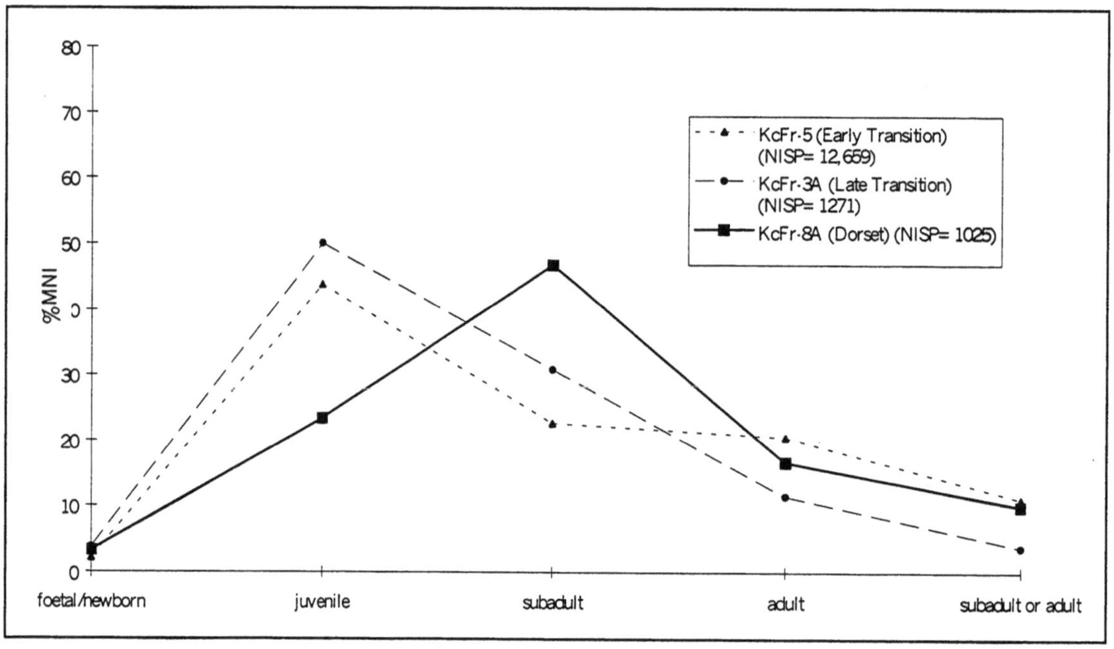

Figure 9. %MNI of small seals from Ivujivik sites per age categories.

SITE	NISP	SMALL SEAL %NISP	SMALL SEAL %MNI	HIGHEST SMALL SEAL %MNI per age group
Pita (KcFr-5)	15572	96.20	82.97	43.70
early Pre-Dorset/Dorset transition		(N=12659)	(N=190)	juvenile (N=83)
Ohituk (KcFr-3A)	1785	91.70	59.52	50.00
late Pre-Dorset/Dorset transition		(N=1271)	(N=25)	juvenile (N=13)
Tivi Paningayak (KcFr-8A)	1255	81.70	78.90	46.70
Dorset		(N=1025)	(N=30)	subadult (N=14)

Table 2. Comparisons of small seal %NISP, %MNI, and highest %MNI per age group.

lactating female with the blubber of a pup might have provided a return equivalent to hunting a female seal with normal blubber content.

Although the selective hunting of immature seals may seem maladaptive (i.e., leading to population extinction) on the part of the hunters, Krupnik (1993) suggests that it is the adult population that hunters should avoid since even if the young individuals are killed, the sexually mature will be able to reproduce again. Furthermore, Stirling and Øritsland (1995:2608) conclude that "the greater that the proportion of young-of-the-year and subadults taken is, the higher the assemblages may indicate a summer occupation. Subadults could also have been hunted during cold weather seasons when they are also present along the Ivujivik peninsula. However, the identification of caches at both the early transition and the Dorset sites suggest that a surplus of food. hunting activities since people had accumulated and cached food all through the summer. Similar subsistence strategies have been reported at Dorset and Thule sites (e.g., Morrison 1983; Murray 1992; Savelle 1987). The presence of caches at the early transitional site from Ivujivik suggests that collector strategies (see Binford 1980) were not exclusive to the Dorset and Thule adaptations.

UNDERSTANDING THE PRE-DORSET/DORSET TRANSITION

Archaeological evidence at Pre-Dorset sites from the Ivujivik area, and also from other sites in the Arctic, suggests that the Pre-Dorset people were still pioneers in the Arctic (e.g., Grønnow 1994:205). They travelled in small family units, changed camps frequently and exploited resources as they encountered them. In other words, Pre-Dorset people were in the process of exploring and understanding vast areas of the Arctic. For hundreds of years they were rather conservative in their manufacture of lithic tools and almost exclusively used chert as their major raw material (Maxwell 1985). They used the land as foragers (Binford 1980) who stayed for short periods of time at any given site and exploited resources in an opportunistic manner. There are a few exceptions to this pattern, particularly in Greenland where some groups of Pre-Dorset people were able to specialize in the hunting of caribou and/or musk-ox (e.g., Knuth 1967; Grønnow 1994). However, in general the land use pattern of the Pre-Dorset people was similar to that described for Paleoindians who entered North America (see Kelly and Todd 1988). This does not mean that people were maladapted, inflexible or incapable of coping with the different environments of the Arctic. Nevertheless, they were still in the process of acquiring knowledge concerning available resources in the areas they visited.

Around 2800 years BP, the descendants of the Pre-Dorset people started to use the Ivujivik area more intensively. After occupying the territory in a highly nomadic and rather exploratory manner during the Pre-Dorset period, people were now coming back year after year to the same hunting grounds. This was a new adaptation based on the exploitation of the territory according to the availability of animals at specific seasons. On the Ivujivik Peninsula, people began mainly to hunt seals during the spring/early sustainable kill can be" and Law (1979) estimates that up to 65% of the young-of-the-year can be harvested annually from a stable population of ringed seals. Finally, a large population of ringed seals would have prevented the overhunting of young individuals. Indeed, in the late 1960s the estimated number of ringed seals for the southeast Baffin Island region was about 166,000 (Smith 1973:31) and they must have been more numerous before the introduction of the gun.

Subadult ringed seals gradually move near the coast in June (Smith 1973:21) and their remains in these archaeological summer and accumulated food surpluses for later use. This strategy is associated with collectors (see Binford 1980) who had a seasonal round of sites to exploit. People also started to use a wider range of lithic sources and made changes in their use of technology (Nagy 1997).

The reasons for the shift from a foraging to a collecting pattern are difficult to identify, especially when looking at sites from a single area. Traditionally most cultural changes in the Arctic are explained with reference to environmental conditions: the climate become colder or warmer, this affects animal populations and ultimately human behaviour. In an article on transitional and Dorset sites from Phillip's Garden in Newfoundland, Renouf (1994) reviews climate studies and links the period between 3000-2200 BP (i.e., the transition) to unstable cold conditions which means unpredictable sea mammal resources. According to Renouf, the people who lived during the transition had to generalize their subsistence base and to live in small groups "so that moves to new resource locations can be made quickly in response to changing circumstances" (Renouf 1994:207)., Renouf (1994) equates the beginning of the Dorset period with warmer temperatures at around 2200 BP. At that time people made more intensive use of sea mammal resources, stayed longer at sites and group size was larger (Renouf 1994).

It should be noted that although the transitional sites from Port au Choix are small with little cultural deposit, as expected from a short term occupation, the inhabitants exploited seals almost exclusively (Renouf 1994:201). It appears to me that the faunal data from these sites indicate the behaviour of foragers who were "serial specialists" (Binford 1980), that is foragers who focused their hunting on specific resources at least during some seasons, rather than generalizing their subsistence base at all times. In fact, the people who occupied Phillip's Garden during the transition might have been in the process of shifting from a foraging to a collecting strategy, as were the people who occupied the early transitional site (KcFr-5) in Ivujivik.

This said, if the unpredictability of sea mammal resources was the norm during the transition through the whole Eastern Arctic, and if the transitional sites from Phillip's Garden represent the remains of generalists (see Renouf 1993), then the people who occupied the Ivujivik sites responded differently to that situation. Their strategy was to specialize in the hunting of seals, at least in the spring/early summer and to accumulate food surpluses, which implies returning to the same site. A similar pattern is associated with the Dorset occupation. Thus, in the case of the Ivujivik sites, the change in adaptation is not between the transition and the Dorset period but between the Pre-Dorset and the transition (and Dorset) period.

Although the climate change hypothesis is attractive, it does not explain why cultural responses came at different rates in the eastern Arctic. Furthermore it does not take into consideration the fact that people would have needed sufficient knowledge of the land in order to change their adaptive strategy. Another explanation accounting for the different land use patterns observed in the palaeoeskimo periods is related to people's knowledge of the environment. It is reasonable to assume that over the generations, people reached a point where familiarity with the land and the exploitation of specific sites at different seasons was preferred to high mobility over vast areas. This proposition is speculative as it cannot be known with any certainty how long it takes for small groups of people to accumulate sufficient information to exploit their surrounding environment in a more efficient manner.

It is nevertheless possible that around 3000 to 2800 BP, people who exploited the Ivujivik area had accumulated enough traditional knowledge to infer the location of the best hunting grounds in relation to the annual cycle of animals. Knowledge of the land would also include that of safe travelling routes. Thus during the transition, people became more specialized in their exploitation of sea mammals, particularly small seals in the spring/early summer. Furthermore, they accumulated surpluses of food for later use. New lithic sources were procured from nearby quarries and probably from trading networks. People living in the Ivujivik area must have included the exploitation of islands and inland resources in their seasonal round. It is also likely that a sense of territoriality developed during the Pre-Dorset/Dorset transition, since at that time people would have identified themselves with the area they exploited through the year.

CONCLUSION

Three faunal assemblages were analyzed in order to compare the subsistence patterns of people who occupied the Ivujivik region during the palaeoeskimo periods. As expected from coastal sites, faunal remains indicate that hunting focused on sea mammals. The variety of species from all the taxa exploited decreased from the early Pre-Dorset/Dorset transition to the Dorset period, suggesting that specialized seal hunting became more important through time. The analyses and comparisons of faunal assemblages indicate that people hunted mainly juvenile (i.e., about 3-5 month old) seals and their mothers during the Pre-Dorset/Dorset transition while more subadults were exploited during the Dorset period. Pups and their mothers were easy prey to catch during the spring/early summer, while subadults were available only in June and later during winter. Although female seals have low fat content while feeding their pups, the latter rapidly acquire an enormous amount of fat. It was proposed that the hunting of pups with their mothers is roughly equivalent to the hunting of adult seals in prime shape. The presence of caches at the early transitional site attests that in Ivujivik the accumulation of food surpluses for later use had started shortly after the Pre-Dorset period. The results also indicate a change from a forager mode of subsistence to that of more logistically oriented collector mode during the Pre-Dorset/Dorset transition.

ACKNOWLEDGMENTS

I want to thank Jack Brink for sending me recent literature on ringed seal body fat. The comments of David Morrison and two anonymous reviewers (including editorial help from one of the latter) were greatly appreciated. Financial support for this research was provided by the Avataq Cultural Institute, the Ministère des Transports du Québec, the Boreal Institute of Alberta, the Northern Science Training Grant, the Ministère des Affaires Culturelles du Québec, the Northern Science Training Grant, the University of Alberta, the Fonds pour la Formation de Chercheurs et l'Aide à la Recherche and the Conseil de la Recherche en Sciences Humaines du Canada. Dennis Carmel (U of Alberta) did the seal teeth thin sections. Anne M. Rick (ZAP, Canadian Museum of Nature) and Jim Woollett (CUNY) identified problematic faunal specimens.

REFERENCES

Amorosi, T.
1992 Non-metric Trait Distinctions for the North Atlantic-Eastern Arctic Phocidae. Bioarchaeological Laboratory, Department of Anthropology, Hunter College, New York. Unpublished.

Arundale, W.H.
1981 Radiocarbon Dating in Eastern Arctic archaeology: A flexible approach. *American Antiquity* 46:244-271.

Banfield, A.W.F.
1981 *The Mammals of Canada*. National Museum of Natural Sciences, National Museums of Canada, University of Toronto Press, Toronto.

Binford, L.R.
1980 Willow Smoke and Dog's tails: Hunter-gatherer Settlement Systems and Archaeological Site Formation. *American Antiquity* 45:4-20.

Diab, M.C.
1998 Economic Utility of the Ringed Seal (Phoca hispida): Implications for Arctic Archaeology. *Journal of Archaeological Science* 25:1-26.

Gendron, D.
1990 The JgEj-3 site: A Groswater Occupation in Northwestern Ungava Bay. Paper presented at the 23rd Annual meeting of the Canadian Archaeological Association, Whitehorse.

Graburn, N.
1972 *Eskimos of Northern Canada*. 2 vols. Human Relation Area files. New Haven.

Grayson, D.K.
1984 *Quantitative Zooarchaeology: Topics in the analysis of Archaeological faunas*. Academic Press, Orlando.

Grønnow, B.
1994 Qeqertasussuk - The Archaeology of a Frozen Saqqaq Site in Disko Bugt, West Greenland. In *Threads of Arctic Prehistory: Papers in Honour of William E. Taylor, Jr.* , edited by David Morrison and Jean-Luc Pilon, pp. 197-238. Canadian Museum of Civilization, Mercury series, Archaeological Survey of Canada, Paper 149. Hull.

Jenness, D.
1922 Life of the Copper Eskimos. *Report of the Canadian Arctic Expedition, 1913-1918*. Vol. 12(B), Ottawa.

King, J.E.
1983 *Seals of the World*. Oxford University Press, London.

Kelly, R. and L.C. Todd
1988 Coming into the Country: Early Paleoindian Hunting and Mobility. *American Antiquity* 53:231-244.

Knuth, E.
1967 *Archaeology of the Musk Ox Way*. Contributions du centre d'études Arctiques et Finno-Scandinaves no. 5. École pratique des Hautes Études, Sorbonne. Paris.

Krupnik, I.
1993 *Arctic adaptations. Native Whalers and Reindeer Herders of Northern Eurasia*. Darmouth College, University Press of New England, Hanover.

Law, R.
1979 Harvest Optimization in Populations with Age Distributions. *American Naturalist* 114:250-259.

Le Blanc, R.J.
1994 *The Crane site and the Palaeoeskimo Period in the Western Canadian Arctic*. Mercury Series, Archaeological Survey of Canada, Paper 148. Canadian Museum of Civilization, Hull.

Low, A.P.
1906 *The Cruise of the Neptune. Report on the Dominion Government Expedition to Hudson Bay and the Arctic Islands on board the D.G.S. Neptune, 1903-1904*. Government Printing Bureau. Ottawa.

Lydersen, C., M.O. Hammill and M.S. Ryg
1992 Water Flux and Mass Gain During Lactation in Free-Living Ringed Seal (*Phoca hispida*) Pups. *Journal of Zoology, London* 228:361-369.

Lyman, R.L., J.M. Savelle, and P. Whitridge
1992 Derivation and Application of a Meat Utility Index for Phocid seals. *Journal of Archaeological Science* 19:7-22.

Lyman, R.L.
1992 Prehistoric Seal and Sea-lion Butchering on the Southern Northwest Coast. *American Antiquity* 57(2): 246-261.

1994 Quantitative Units and Terminology in Zooarchaeology. *American Antiquity* 59(1):36-71.

1995 On the Evolution of Marine Mammal Hunting on the West Coast of North America. *Journal of Anthropological Archaeology* 14:45-77.

McGhee, R.
1981 *The Dorset occupations in the vicinity of Port Refuge, High Arctic Canada*, National Museum of Man, Mercury Series, Archaeological Survey Canada, Paper 105. National Museums of Canada, Ottawa.

McLaren, I.A.
1958 The Biology of the Ringed Seal (*Phoca hispida Schreber*) in the Eastern Canadian Arctic, *Fisheries Research Board of Canada*. Bulletin 118, Ottawa.

Maxwell, M.S.
1985 *Prehistory of the Eastern Arctic*. Academic Press, New York.

Morrison, D.
1983 *Thule culture in western Coronation Gulf, N.W.T.* National Museum of Man, Mercury Series, Archaeological Survey Canada, Paper 116. National Museums of Canada, Ottawa.

Müller-Wille, L.
1990 Places and Names in Nunavik. *Tumivut* 1:40-44

Murray, M.S.
1992 Beyond the Laundry List: The Analysis of Faunal Remains from a Dorset dwelling at Phillip's Garden (EeBi 1), Port au Choix, Newfoundland. Unpublished Master's thesis, Department of Anthropology, Memorial University of Newfoundland, St. John's.

Nagy, M.
1997 Palaeoeskimo Cultural Transition: A Case Study from Ivujivik, Eastern Arctic. Unpublished Ph.D. dissertation, Department of Anthropology, University of Alberta, Edmonton.

Renouf, M.A.P.
1994 Two Transitional Sites at Port au Choix. In *Threads of Arctic prehistory: Papers in Honour of William E. Taylor, Jr.* edited by David Morrison and Jean-Luc Pilon, pp: 165-196, Mercury series, Archaeological Survey of Canada, Paper 149. Canadian Museum of Civilization, Hull.

Ringrose, T.J.
1993 Bone Counts and Statistics: A Critique. *Journal of Archaeological Science* 20:121-157.

Roy, C.
1971a Le Village Esquimau d'Ivujivik et de sa Région. Quelques Aspects Géographiques. Direction Générale du Nouveau-Québec. 2 vols. Ministère des Richesses Naturelles du Québec, Québec.

1971b La Chasse des Mammifères Marins chez les Ivujivimmiut. *Cahiers de Géographie.* 36:509-521.

Ryg, M., T.G. Smith and N.A. Øritsland
1990 Seasonal Changes in Mass and Body Composition of Ringed Seals (*Phoca hispida*) on Svalbard. *Canadian Journal of Zoology* 68:470-475.

Savelle, J.M.
1987 *Collectors and Foragers. Subsistence-Settlement System Change in the Central Canadian Arctic, A.D. 1000-1960.* BAR International Series no. 358, Oxford.

Smith, T.G.
1973 Population Dynamics of the Ringed Seal in the Canadian Eastern Arctic. *Fisheries Research Board of Canada,* Bulletin 173.

Stenton, D.R.
1983 An Analysis of Faunal Remains from the Peale Point Site (KkDo-1), Baffin Island, NWT. Unpublished Master's thesis. Department of Anthropology, Trent University, Peterborough.

Stirling, I. and E.H. McEwan
1975 The Caloric Calue of Whole Ringed Seals (*Phoca hispida*) in Relation to Polar Bear (*Ursus maritimus*) Ecology and Hunting Behavior. *Canadian Journal of Zoology* 53:1021-1027.

Stirling, I. and N.A. Øritsland
1995 Relationships Between Estimates of Ringed Seal (*Phoca hispida*) and Polar Bear (*Ursus maritimus*) Populations in the Canadian Arctic. *Canadian Journal of Fisheries and Aquatic Science* 52:2594-2612

Stuiver, M. and B. Becker
1986 High Precision Decadal Calibration of the Radiocarbon Time Scale A.D. 1950-2500 B.C. *Radiocarbon* 28 (2B):863-910.

Taylor, W.E. Jr
1960 Archaeological Work, Ivugivik and Mansel Island. *Arctic Circular* 13(1):1-4.

1968 *The Arnapik and Tyara sites: An Archaeological Study of Dorset Culture Origins.* Memoirs of the Society for American Archaeology, Paper no. 22.

Van der Velde, F.
1956 Les Règles du Partage des Phoques pris par la Chasse aux Aglus. *Anthropologica* 3:5-15.

PROTOHISTORIC FAUNA FROM THE KITLIK RIVER SITE ON THE SEWARD PENINSULA, ALASKA

Becky M. Saleeby
National Park Service, Anchorage

Angela L. Demma
Minnesota Historical Society, St. Paul

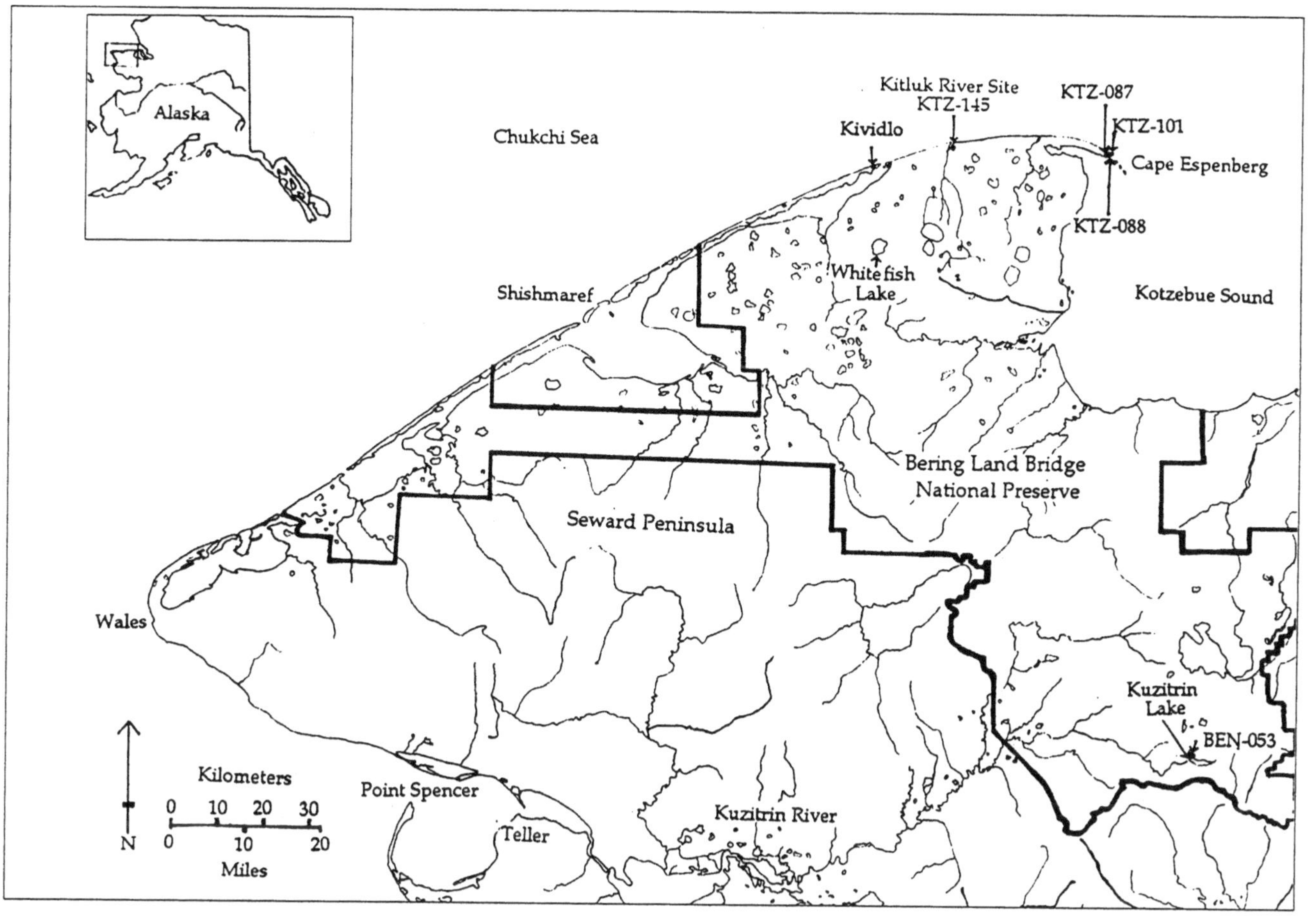

Figure 1. Map of the project area on the northern Seward Peninsula.

The Kitluk River site (KTZ-145) is located on a coastal dune formation at the mouth of the Kitluk River in northwest Alaska. It lies some 40km west of Cape Espenberg at the northern tip of the Seward Peninsula, within Bering Land Bridge National Preserve (Figure 1). When first recorded in 1986, this village site, also known as Killak village, was rapidly eroding, to the point where it was eventually undercut by a river channel. Excavation of what remained of the well-preserved house and midden features was undertaken in 1993. These features date to ca. AD 1800, when direct trade with Europeans was just beginning in the area. Most of the cultural material was recovered from a 40-m long area along the precipice of the dune. The excavation of the Kitluk River site produced a faunal assemblage of just over 10,700 specimens. Data from this assemblage provides information on protohistoric subsistence on the northern Seward Peninsula. Of particular interest in the Kitlik River assemblage is the relative abundance of caribou bones, especially as compared to the frequencies of caribou bones from other Cape Espenberg sites which are located some 30-40 km to the east.

ENVIRONMENTAL AND CULTURAL SETTING

The Kitluk River flows northeast for 24 km from the North Devil Mountain maar lake to the Chukchi Sea. As it crosses the coastal plain, the river channel passes extensive thaw lake formations and then widens near its mouth, where it is bordered by low sedge wetlands. On the east side of the river mouth is a localized coastal dune formation, about 5 m high, which has eroded on the seaward side, forming a "knife-backed" profile. The Kitluk River site lies on this eroding dune face, fronted by a 200-m wide stretch of sandy beach (Schaaf 1988:II:403).

Characteristic natural features of the northwest coastline of the Seward Peninsula are a series of salt water lagoons, fringed by a discontinuous barrier bar of sand and river deposits stretching from Lopp Lagoon to Kividlo, about 15 km southwest of the Kitluk River. To the east lies the Espenberg River; still further east, at the tip of the peninsula, lies Cape Espenberg, formed by a parallel series of prograding beach ridges. Numerous prehistoric sites are located on these relict ridges (Schaaf 1988 II; Mason 1990; Harritt 1994).

In 1986 the Kitluk River site appeared to consist of five house features and at least two unidentified features (Schaaf 1988II:403). Weathering and slumpage continued at the site over the next seven years so that by 1993 only six features were discernable. They included midden (F1); the eroded remains of a house storm shed (F3); one house with a storm shed and two passageways (F4); a connecting house (F5) with adjoining kitchen (F6); and an organic lens (F7) (Figure 2). Aeolian sand deposits more than a meter in thickness encased the collapsed roof and walls of houses constructed of driftwood poles and hewn planks. A blackened, oil-soaked sand veneer lay atop the plank-lined floors in some features. Since no cultural material lay beneath the wooden floor surfaces, it appears that there was only one occupation of the site. The style of the Kitluk River house is typical of late prehistoric houses located along the coast from Espenberg southwest to the Ikpik Lagoon (Schaaf 1996:44).

CHRONOLOGY

The site has not been radiocarbon dated so artifacts are the primary source of information about chronology. The artifact assemblage consists of worked bone, antler, wood artifacts, a variety of precontact ceramics, a few lithic and ivory implements, and trade items which include wound blue beads, a piece of copper, two small iron fragments, and two carved white stone pipe bowls (Schaaf 1995:73, 173). The people of the northern Seward Peninsula did not join the international trade network until quite late, probably around 1789 when the Chukchi established a large trading market on the Anyui tributary of the Kolyma River in Siberia, about 800 miles (1300 km) west of the Bering Strait. One primary node in the northwest Alaska trade network was located at

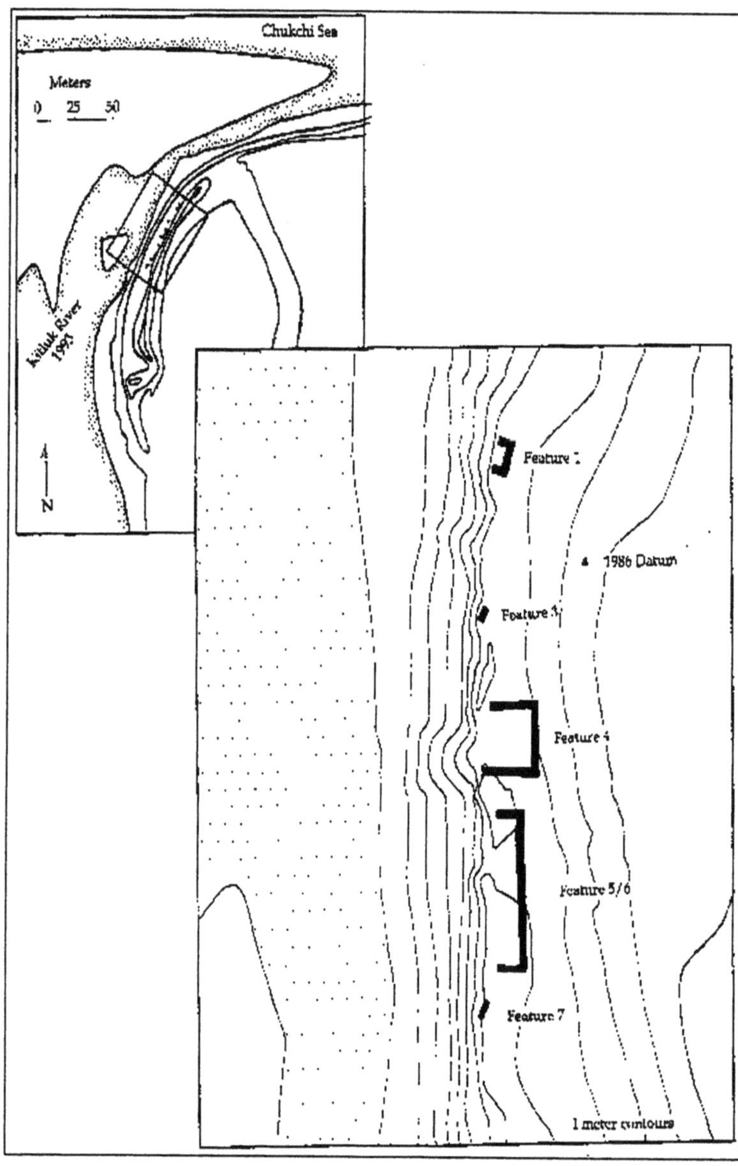

Figure 2. The Kitlik River site (KTZ-145)

Sheshalik (Sisualik) on the northern shore of Kotzebue Sound. Glass beads, iron, metal implements, and tobacco from the Siberian end were exchanged here for Alaskan marten, beaver, fox, muskrat, river otter, lynx, and wolverine. Walrus tusks, seal and walrus hides, and baleen also figured in this trade (Burch 1988b:235). Tobacco, the most highly sought-after commodity after 1810 (Burch 1988b:235), is an indigenous American plant which was transported virtually around the globe eastward from the southern reaches of North America to the Bering Strait. Ray (1975:97-102), recorded an entry into Alaska no earlier than 1750, but it was not until the voyage of Otto von Kotzebue to the northern coast of the Seward Peninsula in early August of 1816 that an explorer actually observed Eskimos using tobacco. Somewhere between the Kitluk (Killak) River and Cape Espenberg, Kotzebue traded knives, looking-glasses, and tobacco with a group of men in umiaks who arrived from the east. Another group from the west, possibly from the Killak, Sinik and/or Qivalauq (Schaaf 1996:56) later joined them. Two pipe bowls, along with the other sparse evidence for trade, bracket the Kitliuk River site occupation around the beginning of the nineteenth century.

When the Kitluk River site was occupied, people were still practicing traditional patterns of settlement and subsistence, unaltered by later events, such as the introduction of reindeer herding or the decline of whale, walrus, and caribou populations (Schaaf 1995:73, 174). The site occupants lived within the territory of the nineteenth century Tapqagmiut (or Tapqarmiut, also referred to as the Shishmaref Society). Burch (1980:293) defines the territory to include the northwest shore of the Seward Peninsula from a point just east of Wales to a point just west of Cape Espenberg. Burch (1980:293) describes their annual subsistence cycle as follows:

> At freshwater freezeup the Shishmaref people were distributed among one large and several medium-sized villages along the outer coast of their territory. During the fall and winter they hunted seals, fished, and occasionally hunted caribou in the mountains to the south. In the spring the people moved right onto the ice to hunt seals, which they continued after breakup from small camps on shore. After the ice left a few families went to the Sisualik fair, but most of them moved across the lagoons to fish, hunt waterfowl, and pursue caribou and small game in the mountains. They continued these activities until the approach of freezeup, when they returned to their fall-winter settlements.

The Pittagmiut (Pitarmiut or Goodhope Bay Society) held the adjoining territory to the east, from Cape Espenberg to Goodhope Bay (Burch 1980:288, 1998:285-304). Ray (1984:286), on the other hand, delineates these nineteenth century boundaries differently by including the entire northeast tip of the Seward Peninsula, from Kividlo around the cape to a point near the mouth of the Goodhope River, in the "Cape Espenberg" tribal territory (Figure 3). The Cape Espenberg people and the Shishmaref people practiced what Ray (1975:104; 1983:175) calls the small-sea-mammal pattern of subsistence based on seals, beluga, fish, and caribou, and other products of the rivers and sea. In the following section on the faunal assemblage from the Kitluk River site, we provide a more detailed look at the range of subsistence resources that the Tapqagmiut (or Cape Espenberg people) relied upon in their seasonal round.

THE FAUNAL ASSEMBLAGE

The 10,708 specimens that comprise the faunal assemblage range from large fragments of whale bone to the most fragile pieces of bird and fish bone. Table 1 presents the NISP (number of identified specimens) for the assemblage by species and by site feature. Although NISP values are used for most quantitative purposes, such as describing and comparing relative frequencies of specimens by species, other quantitative methods, are used to answer specific questions about caribou procurement. The identifiable portion of the assemblage numbers 2885 specimens (27%), while the unidentifiable portion consists of 7823 specimens (73%).

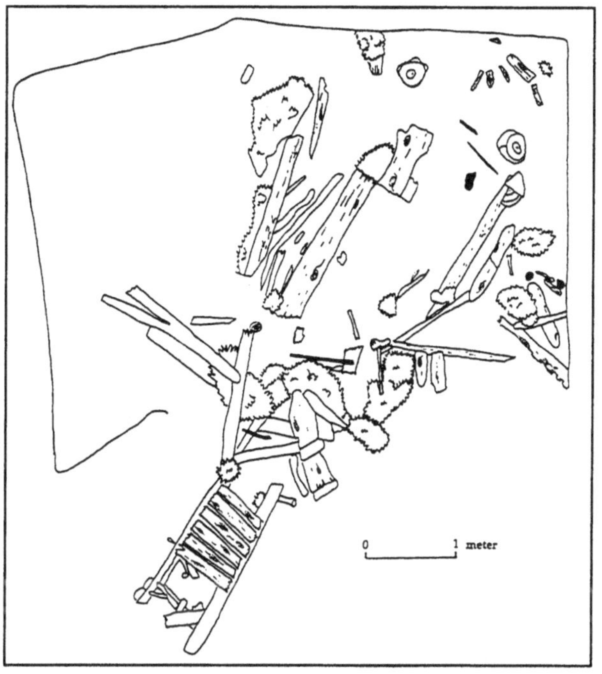

Figure 4. Feature 4, planview.

Over half of the faunal material was recovered from Feature 4 (Figure 4), while most of the remainder was recovered from either Feature 6 (the kitchen of the adjoining house) or from the Feature 1 midden. Although Table 1 only shows the distribution of bones by feature, the data were divided into smaller units within each feature for finer grained analysis. For example, there were a total of nine separate cultural units within Feature 4 (entry tunnel, a south passageway, a storm shed, and others), and these were used to evaluate possible differences in bone disposal patterns throughout the house. Although we compared the presence or absence of various species within these units, we were unable to define any discernable and unambiguous disposal patterns. The bones

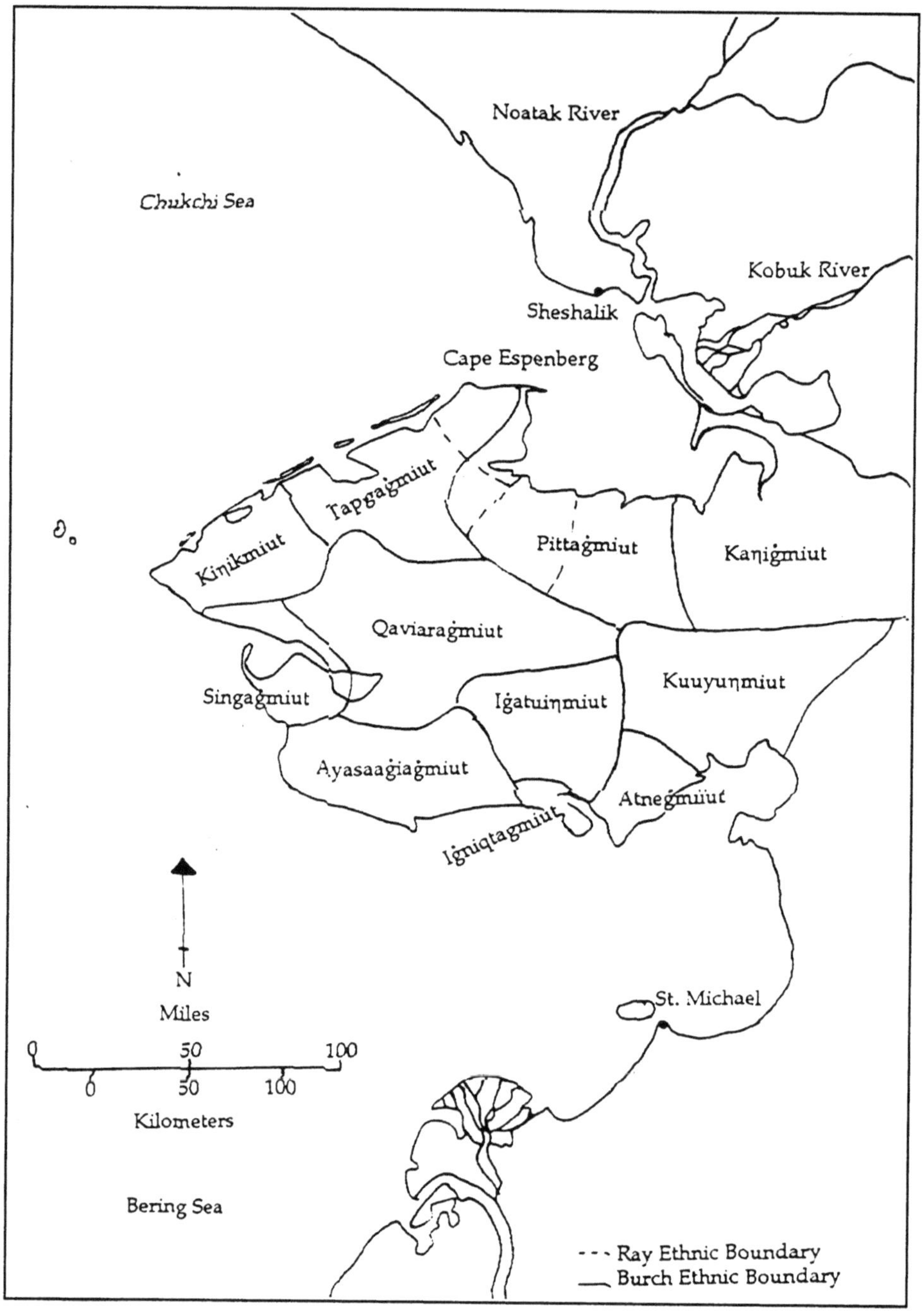

Figure 3. Nineteenth century territorial boundaries as defined by Burch (1980, 1984, 1998) and Ray (1984) (after Schaaf 1996).

appeared to be randomly discarded throughout the house, although we did find slightly different patterns when comparing the distribution of bones within the house features to that from the Feature 1 midden.

In the midden, (N=1500 specimens), over three-quarters of the bones were either burned or calcined. A majority of specimens were still identifiable to species. Table 1 lists these as *Phoca hispida* (ringed seal); *Phoca largha* (spotted or larga seal); *Erignathus barbatus* (bearded seal), *Alopex lagopus* (arctic fox), as well as some birds and fish. All *Rangifer tarandus* (caribou) bones recovered from the midden were unburned. Completely absent were the bones of *Odobenus rosmarus* (walrus), *Phoca fasciata* (ribbon seal), *Canis*, *Vulpes*, and *Lepus* species. Bones of cetaceans (whales) were also absent from the intact section of the midden, but were found nearby in a slump on the dune edge near the midden. In short, the faunal diversity within the house features was not mirrored in the midden assemblage. Comparable results are reported for an historic Eskimo settlement in eastern Greenland, Ikaasap Ittiva, and at KTZ-145, where seal predominated and many of the less common species identified within the house features were totally absent from the midden (Mobjerg *et al.* 1990:246).Sea mammal bones dominate the identifiable portion of the Kitluk assemblage. The small ringed seal, *Phoca hispida*, is the most abundant sea mammal, with 870 specimens recovered. Virtually all skeletal elements of this species, from the cranium to the caudal vertebrae are represented. Gideon K. Barr, Sr., an Inupiat elder whose family lived at Ublasaun, a village only a few kilometers down the coast from the Kitluk River, is the one of our best sources of information about the hunting of ringed seal and many other animals identified at the site. In transcribed interviews Barr states that ..." seal hunting is just perfect in the winter time. That's the reason why you live on the coastline" (Barr 1988:3).

The primary adaptation of the ringed seal is its ability to maintain breathing holes by abrading the ice with claws of its foreflippers (Nowak and Paradiso 1983:1116, 1119). From ethnographic sources we know that hunters in the Bering Strait region traveled on ice, often using dog teams to capture seals, either with harpoons, nets or guns, at their breathing holes or in leads where the ice separated. Ringed seals, not fully mature until the age of six or seven years (Mobjerg *et al.* 1990:247), are seasonally segregated by age. Between November and January, when the ice is solid, older seals are commonly found close to the coast, while immature seals remain offshore except for short periods in the spring and fall when they move to the outer edges of the shorefast ice to haul out (Foote 1965:269; Lentfer 1988:60). In the Kitluk River assemblage, at least 37% of the ringed seal specimens are immature individuals, probably indicating that this small seal was an important resource in the spring and fall as well as in the winter.

The second most numerous taxon in the assemblage is the genus *Phoca*, species (sp.) undetermined, which accounts for 580 specimens. The genus *Phoca* includes the small ringed seal as well as two medium-sized seal species: *Phoca largha* (the spotted seal) and *Phoca fasciata* (the ribbon seal). The category of *Phoca* sp. includes seal bones that could not be identified to one of these three species, but which with some certainty could be attributed to the genus. Many of the bones in the *Phoca* category were immature, lacking some of the diagnostic anatomical features such as the epiphyses that are necessary to make a positive species identification.

Bones specifically identified as *Phoca largha* number 57 specimens. These spotted seals were available to residents of the Cape Espenberg area in the spring, summer, and fall. The Bering-Chukchi population of spotted seals travels northward in large numbers in the spring as they move with the ice front. They are attracted to estuarine areas along the coast when the herring are running. A major summer haul out area is the tip of Cape Espenberg, where 1000 or more animals have been seen at a time (Frost *et al.* 1983:569; 1993:13).

In the fall, spotted seals may be captured in nets that are anchored between offshore ridges as they move south with the ice and feed on smelt in the river mouths (Cross 1988; Eisler 1977:23). The fact that the ribbon seal, much prized for its uniquely patterned fur, is less well represented in the Kitluk River assemblage (numbering only seven) is not surprising as they are found mainly on pack ice in the north-central Bering Sea and seldom occur near shore. Once encountered, however, these elusive animals are easily approached and killed (Lentfer 1988:98-100), usually from the late spring to early fall.

More physically imposing than the *Phoca* species, is the large bearded seal (*Erignathus barbatus)* or *ugruk*. In the Kitluk River assemblage, *ugruk* is represented by 24 specimens, mostly extremities or flipper bones. Large numbers of *ugruk* migrate northward to the Chukchi Sea in the spring and south to the ice front in the fall (Lentfer 1988:79-80). These bearded seals are most vulnerable to hunters when they haul out on the ice close to the beach in the springtime. According to Gideon Barr, spring *ugruk* camps are never in the same place because camps need to be located where ice conditions are right for the large seal to emerge. The Inupiat dried *ugruk* meat, packed it into pokes with seal oil, and cached it in the frozen ground at spring camps. The meat was then brought to winter camp piece by piece, as needed (Barr 1987:tape 1, side 1, p. 9; 1991:tape 3, side 2, pg. 7-10). It appears from the Kitluk River assemblage that *ugruk* camps were probably located at some distance from the village with only select portions of the carcass such as the flippers brought back to the site for consumption.

The NISP for walrus is 43, almost double that for *ugruk*. Walrus is represented by elements from virtually all parts of the skeleton and by both adult and immature elements. Several of the walrus bone fragments were also modified for tool use. This relative abundance of walrus bone is somewhat unexpected as the ethnographic literature indicates that they were an unreliable resource in the area (Koutsky 1981:21). Present-day informants in Shishmaref stated that walrus were not hunted, but their skins and ivory were obtained in exchange for *ugruk* from the people of Wales, at the western tip of the Seward Peninsula (Eisler 1977:19). However, during the spring migration from the Bering Sea northward to the Chukchi Sea, walrus, primarily females and young, could have been taken several miles offshore on the ice if the conditions were particularly favorable. At the time the site was occupied, around AD 1800, the walrus population was presumably still high, and was not yet severely depleted as it was a half-century later when Yankee whalers turned to killing walruses for oil (Lentfer 1988:20-21). A total of 103 whale bone fragments completes the identified sea mammal portion of the faunal assemblage. These fragments, which range from a 40 cm- long rib fragment down to small chunks of cancellous bone no more than 2-3 cm in diameter, were mostly chopped, worked, or chiseled in preparation for making tools. The more intact vertebral centra appear to have been used as "furniture" within the houses. In all likelihood, this bone is from the bowhead, the only whale species taken in the area on a regular basis, at least according to Herbert Anungazuk, a former whaling captain from the village of Wales.

Traditionally Wales has been the focal point of whaling activity on the Seward Peninsula, but oral histories indicate that whales were hunted off Cape Espenberg until the waters became too shallow (Barr 1996). Whales were also

TAXA SEA MAMMALS	F1	F3	F4	F5	F6	TOTAL
Phoca hispida	81	5	450	22	312	870
Phoca larga	2	0	17	1	37	57
Phoca fasciata	0	0	3	0	4	7
P. larga/fasciata	0	0	9	2	5	16
Phoca sp.	102	0	261	11	206	580
Erignathus barbatus	1	0	10	2	11	24
Odobenus rosmarus	0	0	27	10	6	43
Cetacea	6	1	72	7	17	103
Unidentified sea mammal	18	0	60	2	40	120
TOTAL	210	6	909	57	638	1820
LAND MAMMALS	**F1**	**F3**	**F4**	**F5**	**F6**	**TOTAL**
Rangifer tarandus	14	5	149	43	205	416
Alopex lagopus	5	0	158	3	26	192
Vulpes vulpes	0	0	27	1	11	39
Alopex/Vulpes	0	0	31	2	2	35
Lepus othus	0	0	56	1	5	62
Lepus americanus	0	0	1	0	1	2
Lepus sp.	0	0	0	0	3	3
Canis lupus	0	0	2	0	0	2
Canis familiaris	0	0	12	1	2	15
Canis sp.	0	1	2	1	1	5
Ondatra zibethicus	0	0	12	0	2	14
Ovis dalli	0	0	0	0	1	1
Castor canadensis	0	0	1	0	0	1
Spermophilus parryi	0	0	4	0	0	4
Microtus sp.	0	0	4	1	1	6
Lemmus sibiricus	0	0	0	1	0	1
TOTAL	19	6	459	54	260	798
BIRDS	**F1**	**F3**	**F4**	**F5**	**F6**	**TOTAL**
Anas sp.	0	0	2	2	1	5
Larus sp.	0	0	0	1	1	2
Stercorarius sp.	0	0	0	0	1	1
Phasianidae sp.	0	0	0	0	1	1
Aves sp.	7	0	32	58	20	117
TOTAL	7	0	34	60	24	125
FISH	5	0	133	0	4	142
UNIDENTIFIED MAMMALS	**F1**	**F3**	**F4**	**F5**	**F6**	**TOTAL**
Small	2	0	62	4	50	118
Small/Medium	52	0	63	1	36	152
Medium	25	0	229	3	71	327
Medium/Large	776	1	3735	55	1913	6480
Large	1	0	51	2	19	73
TOTAL	1263	1	4376	65	2100	7805
UNIDENTIFIED TAXON	0	0	16	0	2	18
GRAND TOTAL	1504	13	5927	236	3028	10708

Table 1. Number of identified specimens (NISP) from KTZ-145, by feature.

opportunistically hunted when they became stranded in offshore polynyas (Barr 1987, 1988, 1991; Schaaf 1995:62). Active whaling, on the other hand, would have required lengthy trips to either Wales or across Kotzebue Sound to Kivalina (Anungazuk 1995-1996, personal communication). As there is no compelling evidence in the KTZ-145 artifact assemblage to suggest that whales were routinely hunted, we conclude that the bone found at the site was scavenged from the beach when bowhead carcasses washed up on shore.

The land mammal portion of the assemblage is diverse including 13 different species represented by 798 bone specimens. The most abundant terrestrial species is caribou, *Rangifer tarandus* (N=416). Caribou bones, unlike sea mammal bones, are nearly all unburned; and the long bones are in many cases spirally fractured. Both adult and immature caribou are represented in the assemblage. The Seward Peninsula caribou herd, present before 1850, was an overflow population of the Western Arctic herd, which was at its maximum population during the mid-nineteenth century. During this period, the Seward Peninsula caribou

were numerous, but less migratory than the main herd, traveling only short distances during their annual cycle (Burch 1972: 356). An *inuksuk* - kill site referred to by Barr (1987) is located at a lake only 6 km southeast of the Kitluk River site. Ethnographic data indicates that Shishmaref and Cape Espenberg people hunted caribou inland during the fall, along the coast in the summer, and sometimes during the winter (Ray 1975: 90; 1983:175).
The second most abundant land mammal in the Kitluk assemblage is the arctic fox, *Alopex lagopus,* represented by 192 specimens. The bones of the larger red fox, *Vulpes vulpes*, with a total of 39 specimens, are also present. We could not attribute another 35 fox bones to one or the other of these two species, so they were identified as *Alopex/Vulpes*. Fox bones in the Kitluk River assemblage are somewhat problematic as the site was used in recent times for denning purposes, probably by both fox species. Although much of the fox bone is undoubtedly cultural, some of it is intrusive, as is some of the bird bone brought in by foxes. According to Ray (1975:98, 119), the Bering Strait Inupiat did not use fox for food, but did use their pelts to decorate clothing and prized the fur for trade, particularly after the advent of the large trade fair on the Kolyma River in 1789. The best evidence for active involvement in the fur trade from the Kitluk assemblage is an ivory bow drill handle, incised with a series of symbols thought to represent fox pelts. This artifact, recovered from an eroding deposit below Feature 5 is depicted in Figure 5.

Two *Lepus* or hare species occur in the Kitluk assemblage. The bones of the tundra hare, *Lepus othus*, are far more abundant (N=62) than are those of the snowshoe hare, *Lepus americanus* (N=2). The tundra hare is large, often weighing 12 pounds or more, with bones comparable in size to those of an arctic fox. They range across the tundra and hills from December to March, with large concentrations occurring near willows or out on the ice. These large hares, considered primarily a winter starvation food, could be snared in the willows or trapped in holes on the ice. The meat was eaten and the pelts used to make jacket or bedclothes, especially for children. The fur, however was not considered to be very durable (Ray 1975:118).

Other land mammal species all occur in low frequencies. The *Canis* species is represented by two specimens attributed to wolf (*Canis lupus*), 15 tentatively identified as dog (*Canis familiaris*), and five identified to the genus level. There are three other fur-bearers present in the assemblage: muskrat (*Ondatra zibethicus*), ground squirrel (*Spermophilus parryii*), and one possible specimen of beaver (*Castor canadensis*). The most unusual specimen was a single horn core of a Dall sheep (*Ovis dalli*) which may have been a trade item or brought back from a distant hunting trip. Vole (*Microtus* sp.) and lemming (*Lemmus sibiricus*) bones are intrusive to the site.

We have only identified a few of the 125 bird bones and bone fragments from the site, primarily because a comprehensive comparative collection of bird bones was unavailable to us. Ptarmigan or grouse (family Phasianidae), gull (*Larus* sp.), jaeger (*Stercorarius* sp.), and a species of dabbling duck (*Anas* sp.) have been tentatively identified.

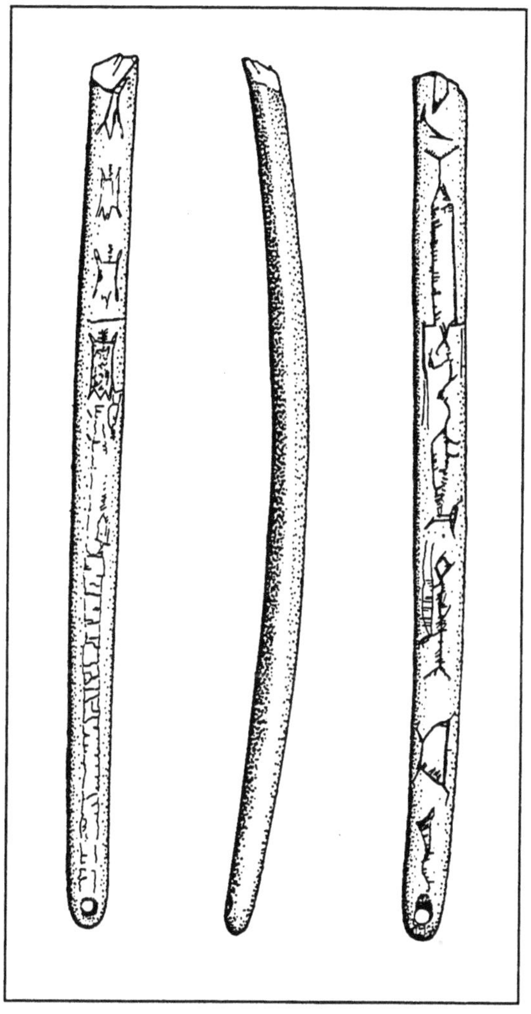

Figure 5. Ivory bow drill handle, incised with symbols representing fox pelts (from KTZ-145) (drawing by Frank Borderick).

Birds and their eggs were important resources to the people of the Kitluk River, particularly during the late spring and summer when migratory waterfowl amassed in large number on the flats between Cape Espenberg and Shishmaref. Ptarmigan were a very important food resource in the winter, and were sometimes a principal part of the diet for weeks at a time (Ray 1975:117).

A total number of 142 fish specimens, identified as sculpins, and tomcod or saffron cod, are also present. These species are taken in winter along the brackish channel edges or through the ice, using lures and hooks (Uhl and Uhl 1977: 21; Eisler 1977:14). One ivory lure was recovered from the site, and an ice scoop and ice pick fragments are represented in the antler and bone tool assemblage (Demma n.d.).

Finally, there are shellfish in the faunal assemblage. Eleven species, both gastropods and bivalves, were identified (Table 2). Generally, shellfish are washed up on shore by storms in the fall and collected around freeze-up. If the harvest was plentiful, they were cooked and stored in seal pokes with seal oil as a preservative (Edgar Ningeulook 1996: *pers. comm.*; Herbert Anungazuk 1996: *pers. comm.*).

The picture that emerges from this faunal inventory is of a village site, strategically located in a rich environment, where food resources were available virtually throughout the year (Table 3). The Kitluk River village was probably occupied principally in fall and winter, intermittently throughout the spring when people were not in *ugruk* camps, and sporadically in the summer. From oral histories and local place names, we can surmise that the summer was spent inland fishing for trout and whitefish in the maar lakes near Devil Mountain at the headwaters of the river, and in collecting plants, such as sourdock (Schaaf 1995: 173; Fair and Ningeulook 1995). This assessment, based on the faunal remains, compares well with Burch's (1980) description of the Tapqagmiut annual subsistence round presented above.

COMPARISONS TO SITES AT CAPE ESPENBERG AND KUZITRIN LAKE

A total of 222 archeological sites were recorded, between 1985-1986, including a large cluster located on or very close to the tip of Cape Espenberg (Schaaf 1988:I and II). In order to evaluate and protect some of these Bering Land Bridge sites, archaeological crews returned to the Preserve during the summers of 1988-1990 for testing and excavation (Harritt 1994). Faunal remains were collected from 11 of the 16 excavated sites. Three of these sites on Cape Espenberg - KTZ-087, KTZ-088, and KTZ-101 (Figure 1), yield substantial faunal assemblages. (Saleeby 1994). KTZ-087, consisting of four features from two houses, dates between 675 years BP (AD 1275) and 508 years BP (AD 1442). KTZ-088, lying only 50 m to the east, has one feature occupied contemporaneously with the village at KTZ-087,and two other younger features, probably inhabited around 300-400 years ago. KTZ-101, lying furthest east on tip of Cape Espenberg, has been radiocarbon dated at 302 years BP, with some sporadic occupation until about AD 1850 (Harritt 1994).

The bones from these three late prehistoric villages indicate that past occupants practiced a subsistence pattern similar to that just described for the Kitluk River site. As at KTZ-145, sea mammals dominate the faunal assemblage and these are primarily ringed seal. The sites were inhabited in the fall and winter, and possibly into the spring (Saleeby 1994:374-375). A striking absence in the sea mammal inventory from the Cape Espenberg sites, however, is the spotted seal, *Phoca largha* One possibility for the lack of these bones might be the selective sampling strategy used; the focus was on obtaining diagnostic artifacts and radiocarbon samples from the sites rather than on the recovery of a broad spectrum of faunal remains (Harritt 1994:65). On the other hand, *Phoca largha* may simply have not been an available. resource to the inhabitants of these sites. Today, Cape Espenberg is a prime location for spotted seals in the summer months, but their migration patterns may have shifted during the last several hundred years during periods of intense storms which occurred in the area beginning around AD 1400 (Mason and Gerlach 1995:109). The environment around the tip of the Cape may have been inhospitable causing the larga seals to migrate past this stretch of coastline.

GASTROPODS	BIVALVES
Neptunia heros	*Mytilus edulis* (blue mussel
Fellina lutea	*Pododesmus macroschisma* (Alaska falsejingle)
Boreoscala groenlandica	*Astarte borealis* (boreal tridonta)
Boccinum angulosum (?)	*Serripes groenlandicus* (Greenland cockle)
	Mactromeris polynyma (Arctic surfclam)
	Siliqua alta (Alaska razor)
	Siliqua patula (Pacific razor)

Table 2. Shellfish identified from the Kitluk River site (KTZ-145).

TAXA	SPRING	SUMMER	FALL	WINTER
Phoca hispida	xxx		xxx	xxx
Phoca largha	xxx	xxx	xxx	
Phoca fasciata	xxx	xxx	xxx	
Erignathus barbatus	xxx		xxx	
Odobenus rosmarus	xxx			
Rangifer tarandus	xxx	xxx	xxx	???
Cetacea (scavenged)		xxx	xxx	
Alopex or *Vulpes*				xxx
Lepus sp.				xxx
Canis lupus				xxx
Ondatra zibethicus	xxx			
Spermophilus parryii	xxx	xxx		
Ovis dalli (trade?)			xxx	xxx
Anas sp.	xxx	xxx		
Phasianidae sp.				xxx
Sculpin, Tomcod			xxx	xxx
Shellfish			xxx	

Table 3. Seasons of most common procurement for resources identified at KTZ-145.

Despite the general similarities in assemblages, there are some notable differences in relative proportions of sea mammals versus land mammals, particularly in comparing. the Kitluk River site with the three Cape Espenberg sites. Sea mammals contribute 63% of the total assemblage at the Kitluk River site, but contribute 83-96% of the total at the Espenberg sites (Table 4). Conversely, the land mammal portion of the assemblage is greater at the Kitluk River site (27% of the total) than at the Espenberg sites (2-12%) remains from the other Cape Espenberg sites (Table 4). At Kitluk, the relatively high frequency of land mammal bones is inflated by the presence of 266 fox bones. These may represent fox pelts, which were valued as items for barter in the protohistoric trade network. Caribou, however, are consistently the greatest contributors to the land mammal portion of the assemblage at each site.

TAXON	KTZ-145	KTZ-087	KTZ-088	KTZ-101
Phoca hispida	870 (30%)	224 (54%)	947 (84%)	273 (64%)
Phoca larga	57 (2%)	0	0	0
Phoca fasciata	7 (<1%)	11 (3%)	8 (<1%)	1 (<1%)
Phoca sp.	596 (21%)	73 (18%)	119 (11%)	49 (12%)
Erignathus barbatus	24 (<1%)	3 (<1%)	10 (<1%)	19 (4%)
Odobenus rosmarus	43 (1%)	42 (10%)	6 (<1%)	10 (2%)
Cetacea	103 (4%)	0*	0*	0*
Unidentified Sea Mammal	120 (4%)	0	0	0
SEA MAMMAL TOTAL	1820 (63%)	353 (85%)	1090 (96%)	354 (83%)
Rangifer tarandus	416 (14%)	26 (6%)	26 (2%)	40 (9%)
Alopex lagopus	192 (7%)	0	2 (<1%)	0
Vulpes vulpes	39 (1%)	0	1 (<1%)	0
Alopex/Vulpes	35 (1%)	0	0	0
All other land mammals	116 (4%)	1 (<1%)	7 (<1%)	13 (3%)
LAND MAMMAL TOTAL	798 (27%)	27 (6%)	36 (2%)	53 (12%)
BIRDS	125 (4%)	32 (8%)	3 (<1%)	17 (4%)
FISH	142 (5%)	1 (<1%)	1 (<1%)	1 (<1%)
TOTAL	2885 (100%)	413 (100%)	1130 (100%)	425 (100%)

*Table 4. Comparison of identifiable portion of faunal assemblages by site and taxon (Values are given in NISP and Percentage of Identifiable Assemblage; data taken from Saleeby 1994).** Large fragments of whale bone were found within site features, but were not included as specimens in the total faunal inventory.

SKELETAL COMPONENT	NUMBER OF BONES
Antler	variable
Cranial	34
Teeth	32
Cervical vertebrae	7
Thoracic vertebrae	14
Lumbar vertebrae	5
Sacrum	1
Caudal vertebrae	9
Ribs	28
Sternum	1
Axial Total	65
Scapula	2
Shoulder Total	2
Humerus	2
Radio-cubitus	2
Forelimb Total	4
Innominate	2
Pelvis Total	2
Femur	2
Tibia/Fibula	2
Patella	2
Hindlimb Total	6
Carpals	14
Calcaneus	2
Astragalus	2
Navicular-cuboid	2
Other tarsals	6
Metacarpals	2
Metatarsals	2
Phalanges (+vestigal)	32
Extremities total	62
TOTAL	207

Table 5. Number of caribou bones by skeletal component .

SKELETAL COMPONENT	KTZ-145	KTZ-101	BEN-053
Antler	7 (--)	4 (--)	2 (--)
Cranial	12 (7%)	1 (2%)	69 (70%)
Teeth	3 (2%)	0 (0%)	70 (75%)
Axial	53 (16%)	15 (15%)	190 (100%)
Shoulder	6 (57%)	3 (100%)	3 (52%)
Forelimb	21 (100%)	3 (50%)	5 (43%)
Pelvis	2 (19%)	0 (0%)	1 (17%)
Hindlimb	9 (29%)	0 (0%)	11 (63%)
Extremity	115 (35%)	14 (15%)	100 (55%)
TOTAL	228	40	451

Table 6. Comparison of caribou MNE and % MAU at three Seward Peninsula sites.

BEN-053 is an extensive late prehistoric village at the west end of Kuzitrin Lake (Figure 1). It dates to ca. 290-500 years BP (Harritt 1994:212). This site is located near the convergence of four territorial boundaries (the Pittagmiut, Qaviaragmiut, Igatuinmiut, and Kuuyunmiut as depicted in Figure 3). People who had a caribou-based subsistencepattern, including fish, seal, and beluga (Ray 1983:175) once occupied the Kuzitrin Lake area. At BEN-053, with the exception of 21 bird bones, caribou constitute virtually the entire identifiable portion (N=512) of the assemblage excavated from four of the site features. The site was probably used primarily as a summer caribou kill and butchery location (Saleeby 1994:363). The results of the faunal analysis from BEN-053 are useful as a comparative model for our goal of attempting to understand the procurement pattern of caribou at the Kitluk River site.

Comparison of Caribou Procurement Strategies

Caribou are represented by virtually all skeletal elements (Table 5). The MNE for each appears in the first column of Table 6 (KTZ-145), along with a %MAU value listed in parentheses. The %MAU values were calculated after Lyman (1994:57). For comparative purposes, the MNE and %MAU values for caribou bone assemblages from KTZ-101, are representative of the Cape Espenberg sites, and BEN-053 are shown in the adjacent columns of Table 6. An unexpected pattern is seen in the comparative %MAU figures depicted above. After analyzing only the MNE figures for KTZ-145 and BEN-053, the interpretation is that caribou were killed in close proximity to the Kitluk River village site, and that most of the carcass was easily transported back to the village for further processing and consumption.

The %MAU values, however, shed a different light on the interpretation. At both of the coastal sites (KTZ-145 and KTZ-101), the %MAU values for skeletal components, except the shoulder and forelimb, are under 50%, while the pattern at BEN-053, the inland kill site, is nearly the opposite. All sites show a low frequency of the innominate bone (pelvis), probably resulting from bone fragmentation. On the basis of these data, we conclude that the relatively high %MAU for the shoulder and forelimb represent differential transport to the coastal sites, while the relatively low %MAU values for these skeletal components reflect selective transport back to the site or simply an indication that caribou were also hunted opportunistically when they migrated close to the coast and to the village.

It is clear that when the people were in residence at coastal villages such as KTZ-145 and KTZ-101, caribou hunting was secondary to the harvesting of marine resources, but it is unclear as to why the relative proportion of caribou in the Kitluk River assemblage is greater than it is in the Cape Espenberg assemblages. One possibility may lie in temporal differences between the occupations. The protohistoric Kitluk River site may be as much as 600 years younger than the prehistoric Cape Espenberg sites. The population of caribou on the Seward Peninsula was at its peak during the mid-nineteenth century, close to the time when the Kitluk River site was occupied. Given the fluctuations in caribou numbers over time (Skoog 1968), it is possible that the population was much lower when KTZ-087, KTZ-088, and KTZ-101 were inhabited, and that caribou were a less reliable resource than in the nineteenth century. A second explanation for disparities among the sea mammal/caribou assemblages may be related to territorial differences in subsistence patterns. Both Burch (1980, 1984, 1988a, 1998) and Ray (1975, 1983, 1984) stress that the annual subsistence round varied from territory to territory depending upon the local availability and quantity of the resources. The Kitluk River site and the three Cape Espenberg sites fall within Ray's delineation of Cape Espenberg territory, so presumably no differences should be expected. According to Burch, however, the border between the Tapqagmiut and the Pittagmiut is drawn between the Kitluk River site and the three Espenberg sites (Figure 3). The question is: how far back in time can we extend these territorial distinctions and the associated variations in settlement and subsistence organization.

In comparing the Kitluk River faunal assemblage to those from the Cape Espenberg sites, it is necessary to refer back to Burch's (1980) brief summaries of the annual cycle of the two groups in question. For the Pittagmiut (Goodhope), he says:

> At freshwater freezeup, the Goodhope Bay people were distributed among medium-sized villages along the larger rivers; Kugruk, Imnatsiaq, and Goodhope. During the winter they subsisted on fish, caribou and small game. In spring they all moved to Cape Espenberg to hunt seals. When the ice left in July a few families went to the Sisualik fair, but most returned to the rivers to fish. Later they moved inland to hunt caribou, returning to the coast in

time to net seals for a week or two before moving to their winter settlements (Burch 1980: 288).

While the Kitluk River Tapqagmiut were in residence at or near their village for much of the time between the fall and early spring, where they hunted seals, fished, and occasionally hunted caribou in the mountains to the south, the Pittagmiut were residing at their primary villages along the major rivers emptying into Goodhope Bay. The Pittagmiut sometimes traveled to Cape Espenberg in winter specifically for seal hunting, but "by mid to late March, virtually all Pittagmiut moved to Cape Espenberg to hunt seals" (Burch 1998: 301). In other words, while residing on Cape Espenberg, the Pittagmiut were focused on one primary subsistence activity, and other pursuits, such as caribou hunting, were of secondary importance or perhaps even incidental.

Although we cannot conclusively answer the question of why there are differing relative percentages of sea mammals vs. caribou at the Kitluk River and Cape Espenberg sites, we may gain further insight by comparing our percentages with those from other sites in the region. The best examples come from Cape Krusenstern to the north on Kotzebue Sound. Giddings and Anderson (1986:56) indicate that a wide variety of resources were important to the people of Cape Krusenstern during late prehistoric times from the 15th to early 19th century. They report relative percentages of fauna excavated from houses dating to this time period as follows: small seals (50%), large seals (10%), caribou (10%), other land game (14%), birds (10%), and miscellaneous (6%).

Despite the late prehistoric dates for the Cape Krusenstern sites, the relative percentages in sea mammal vs. land mammal are more similar to those of the protohistoric Kitluk River site than they are to the Cape Espenberg sites, which also date to late prehistoric times. These data further support the notion that we may be monitoring subtle territorial differences in regional subsistence patterns, rather than change through time. It will take future excavation of faunal material from other protohistoric and prehistoric sites in both the Tapqagmiut and the Pittagmiut territories to fully resolve the questions about time and territory as posed above.

In summary, the excavation of faunal remains from the Kitluk River site provides us with a window through which to view a traditional form of subsistence on the northern Seward Peninsula. The people who once occupied this protohistoric village had a long-standing reliance on a broad spectrum of resources, with a special emphasis on small sea mammals. We propose here that subtle differences in the relative proportion of sea mammals and land mammals, specifically caribou, in the Kitluk River assemblage as compared with prehistoric faunal assemblages from Cape Espenberg may reflect actual territorial differences in the settlement and subsistence patterns of the region.

ACKNOWLEDGEMENTS

We would like to thank Jeanne Schaaf, project director for the Kitluk River site excavation; Cassie Flynn, Megan Partlow, and Martha Olympic Crow for helping with bone identification; Diane Hanson for identifying the fish bones; Nora Foster for identifying the shellfish; Meredith Mengel for database input; Gordon Jarrell for allowing us access to the comparative faunal collections at the University of Alaska Museum; Herbert Anungazuk and the late Edgar Ningeulook for their knowledge of traditional subsistence; Charles Utermohle and Don Callaway for being our statistical consultants for an earlier version of this paper; and two anonymous reviewers for their insightful comments on the shortcomings of our first attempt at presenting these data

REFERENCES CITED

The Alaska Trappers Association
1991 *Alaska Trappers Manual.* Alaska Department of Fish and Game.

Barr, G.K. (Sr.)
1987 Audio tapes and maps: Interview with Jeanne Schaaf, May 1987, transcribed by H.O. Anungazuk, National Park Service, Alaska Regional Office, Anchorage, Alaska.

1988 Audio tapes and maps, "Espenberg area Place Names" transcribed by H.O. Anungazuk, on file at Bureau of Indian Affairs ANCSA Office, Anchorage, Alaska.

1991 Videotape in Inupiaq, translated by Herbert O. Anungazuk. Shared Beringian Heritage Program, on file at National Park Service, Anchorage, Alaska.

1996 Ilaganiq, A Folktale. In *Ublasaun First Light.* U.S. Department of the Interior, National Park Service, Alaska Field Area, Shared Beringian Heritage Program, Anchorage.

Burch, E.S., Jr.
1972 The Caribou - Wild Reindeer As a Human Resource. *American Antiquity* 37:339-368.

1980 Traditional Eskimo Societies in Northwest Alaska. In Alaska Native Culture and History. Y. Kotani and W. Workman, eds. *Senri Ethnological Studies 4.* National Museum of Ethnology, Osaka.

1984 Kotzebue Sound Eskimo. In Arctic. D. Damas, ed. *Handboook of North American Indians Volume 5.* Smithsonian Institution, Washington, D. C.

1988a Toward a Sociology of the Prehistoric Inupiat: Problems and Prospects. In The Late Prehistoric Development of Alaska's Native People. R. D. Shaw, R.K. Harritt, and D.E. Dumond, eds. *Aurora, No. IV.* Alaska Anthropological Association, Anchorage.

1988b War and Trade. In *Crossroads of the Continents.* William W. Fitzhugh and Aron Crowell, editors. Smithsonian Institution Press, Washington, D.C.

1998 *The Inupiaq Eskimo Nations of Northwest Alaska.* University of Alaska Press, Fairbanks.

Cross, B.
1988 Audio tapes, "Espenberg area Place Names." On file at Bureau of Indian Affairs ANCSA Office, Anchorage, Alaska.

Demma, A.L.
n.d. A Macroscopic Study of Manufacture and Use of Protohistoric Antler Tools from the Kitluk River Site, Northern Seward Peninsula, Alaska. Ms. On file, National Park Service, Alaska Support Office, Anchorage.

Eisler, D.
1977 Subsistence Activities of Shishmaref, Brevig Mission, Teller, Wales, and Deering. *Anthropology and Historic Preservation Cooperative Park Studies Unit.* University of Alaska, Fairbanks.

Fair, S.W. and E.N. Ningeulook
1995 *Qamani* Up the Coast, In My Mind, In My Heart. Ms on file, National Park Service, Anchorage, Alaska.

Foote, D.C.
1965 Exploration and Resource Utilization in Northwestern Arctic Alaska Before 1855. Unpublished Ph.D. Dissertation. McGill University, Montreal.

Frost, K.J., L.F. Lowry, and J.J. Burns
1983 Distribution of Marine Mammals in the Coastal Zone of the Eastern Chukchi Sea During Summer and Autumn. *Environmental Assessment of the Alaskan Continental Shelf Final Reports of Principal Investigators* 20:563-650. Outer Continental Shelf Environmental Assessment Program. U.S.D.O.C. National Oceanic and Atmospheric Administration.

Frost, K.J., L.F. Lowry, and G. Carroll
1993 Beluga Whale and Spotted Seal Use of a Coastal Lagoon System in the Northeastern Chukchi Sea. *Arctic* 46:8-16.

Giddings, J.L. and D.D. Anderson
1986 *Beachridge Archeology of Cape Krusenstern.* National Park Service Publications in Archeology 20. U.S. Department of the Interior, Washington, D.C.

Harritt, R.K.
1994 *Eskimo Prehistory on the Seward Peninsula, Alaska.* U.S.D.O.I., National Park Service, Anchorage, Alaska, Resources Report NPS/ARORCR/CRR-93/21.

Koutsky, K.
1981 Early Days of Norton Sound and Bering Strait: An Overview of Historic Sites in the BSNC Region. Volume I: The Shishmaref Area. *Anthropology and Historic Preservation Cooperative Park Studies Unit Occasional Paper No. 29.* University of Alaska, Fairbanks.

Lentfer, J.W.
1988 *Selected Marine Mammals of Alaska: Species Accounts with Research and Management Recommendations.* Marine Mammal Commission, Washington, D. C.

Lyman, R.L.
1994 Quantitative Units and Terminology in Zooarchaeology. *American Antiquity* 59:36-71.

Mason, O.K.
1990 Beach Ridge Geomorphology of Kotzebue Sound: Implications for Paleoclimatology and Archaeology. Unpublished Ph.D. dissertation, University of Alaska Fairbanks.

Mason, O.K. and S.C. Gerlach
1995 Chukchi Hot Spots, Paleo-Polynyas, and Caribou Crashes: Climatic and Ecological Dimensions of North Alaska Prehistory. *Arctic Anthropology* 32:101-130.

Mobjerg, T., A. Robert-Lamblin, and J. Robert-Lamblin
1990 The Settlement at Ikaasap Ittiva, East Greenland. *Acta Archaeologica* 60:229-262.

Murdoch, J.
1892 *Ethnological Results of the Point Barrow Expedition.* Ninth Annual Report of the Bureau of American Ethnology for the Years 1896-1897. Washington, D.C.

Nowak, R.M., and J.L. Paradiso
1983 *Walker's Mammals of the World II* (Fourth Edition). Johns Hopkins University Press, Baltimore.

Phebus, G. Jr.
1972 *Alaskan Eskimo Life in the 1890s as Sketched by Native Artists.* Smithsonian Institution Press, Washington, D.C.

Ray, D.J.
1975 *The Eskimos of Bering Strait, 1650-1898.* University of Washington Press, Seattle, Washington.

1983 *Ethnohistory in the Arctic: The Bering Strait Eskimo.* R.A. Pierce, ed. The Limestone Press, Kingston, Canada.

1984 Bering Strait Eskimo. In Arctic. D. Damas, ed. *Handboook of North American Indians Vol. 5.* W. Sturtevant, gen. ed. Smithsonian Institution, Washington, D. C.

Rearden, J.
1981 *Alaska Mammals.* Alaska Geographic, Volume 8, No. 2. Alaska Geographic, Anchorage.

Saleeby, B.M.
1994 Results of the Faunal Analysis. In *Eskimo Prehistory on the Seward Peninsula, Alaska*, by Roger Harritt, U.S.D.O.I., National Park Service, Anchorage, Alaska, Resources Report NPS/ARORCR/CRR-93/21:317-377.

Schaaf, Jeanne M.
1988 *Bering Land Bridge National Preserve: An Archeological Survey (Vols. I and II)*. National Park Service, Alaska Region, Anchorage. Research/Resources Management Report AR-14.

1995 Late-Prehistoric Inupiaq Societies, Northern Seward Peninsula, Alaska: An Archeological Analysis AD 1500-1800. Unpublished Ph.D. Dissertation. Department of Anthropology, University of Minnesota, Minneapolis.

1996 Before Our Fathers' Time. In *Ublasaun First Light.* U.S. Department of the Interior, National Park Service, Alaska Field Area, Shared Beringian Heritage Program, Anchorage.

Skoog, R.O.
1968 Ecology of the Caribou (*Rangifer tarandus granti*) in Alaska. Unpublished Ph.D. dissertation. University of California, Berkeley.

Uhl, Wi.R. and C.K. Uhl
1977 Tagiumsinaaqmiit - Ocean Beach Dwellers of the Cape Krusenstern Area - Subsistence Patterns. *Anthropology and Historic Preservation Cooperative Park Studies Unit*, University of Alaska, Fairbanks.

HUMAN PREDATORS AND MIGRATORY MEGAFAUNA: THE CASE OF THULE INUIT BOWHEAD WHALING

James M. Savelle
Department of Anthropology, McGill University

Allen P. McCartney
Department of Anthropology, University of Arkansas

Arthur S. Dyke
Terrain Sciences Division, Geological Survey of Canada

Mortality patterns in mammalian archaeofaunas have played an increasingly central role in the investigation of prehistoric subsistence. As noted by Stiner (1991b:1), such patterns "reflect some very basic ecological relationships between humans and their prey," and they have been instrumental in the investigation of such topics as early hominid hunting vs. scavenging (e.g., Klein 1982 1987; Stiner 1990), Upper Paleolithic and Paleoindian mammoth and mastodon procurement techniques (e.g., Saunders 1980; Fisher 1987; Haynes 1985, 1991), prehistoric bison procurement (e.g., Frison 1978; Reher 1974; Guthrie 1980), and early animal domestication (e.g., Jarmon and Wilkinson 1972; see also papers in Stiner 1991a). With few exceptions, however, these studies have concentrated on terrestrial faunas.

In this paper we present mortality patterns found in archaeofaunas of the bowhead whale (*Balaena mysticetus*), the largest animal hunted by any traditional native society (Figure 1). Specifically, we examine local and regional variation in bowhead mortality profiles created by prehistoric Thule Inuit of the Canadian Arctic Archipelago. We demonstrate that estimated bowhead size and relative position along their spring/summer migration round are reflected in the zooarchaeological record (note that we deal with these data from the perspective of whaling 'success rates' in Savelle and McCartney 1999).

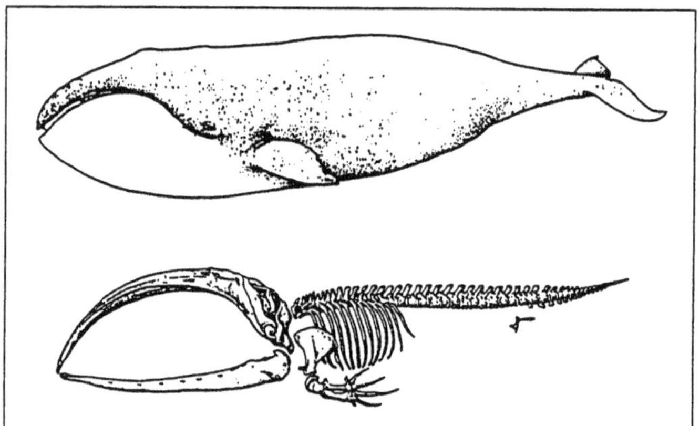

Figure 1. Bowhead whale and skeleton (from McCartney 1980).

This study differs from most megafaunal analyses in two ways. First, it treats the largest of fauna, baleen whales, that were major prey species along various coasts of the Old and New Worlds (Heizer 1941, 1968). The size/weight constraints in hunting and processing baleen whales are common to the various species, such as bowhead, right, humpback, gray, sei, and fin whales. Marine mammalian faunas have been only rarely investigated in the context of mortality patterns represented at archaeological sites. When they have, the focus has been on the more common, smaller pinnipeds, such as harp seals (Cox and Spiess 1980), ringed seals (Morrison 1983; Danielson 1994), various otariids (Lyman 1991), and small odontocete whales (Savelle 1995; Friesen and Arnold 1995; Friesen et al. 1998). Second, we have collected sufficient biometric data for bowheads to permit a fairly fine-grained reconstruction of seasonal predation of young animals along their migration paths. This reconstruction is possible because of the fast growth of immature bowheads.

BACKGROUND TO THE STUDY

Many prehistoric Thule Inuit sites in the Canadian Arctic Archipelago (c. AD 1000-1600) are characterized by the remains of winter semisubterranean dwellings and other structures in which bowhead whale bones figure prominently as construction materials. Larger sites of 25-50 dwellings often contain the remains of over 100 whales and sites associated with whale processing and caching are also common. We have previously demonstrated that Thule Inuit hunted, rather than scavenged, bowheads in the Canadian Arctic Archipelago. Osteological data were collected in 1978 by McCartney, in 1988 and 1990 by McCartney and Savelle, and in 1992, 1993 and 1994 by Savelle, on Somerset, Prince of Wales, Bathurst, Cornwallis, Devon, and Baffin islands (McCartney and Savelle 1993; for other measurement techniques, see Gerlach et al. 1993). In previous papers, we have described Thule Eskimo whale hunting and subsistence (McCartney and Savelle 1985), settlement patterns and social organization (Savelle 1987; Savelle and McCartney 1988, 1990), specialist vs. generalist subsistence patterns (Savelle and McCartney 1988, 1991), the distinction between natural and human-derived bowhead mortality patterns (Savelle and McCartney 1991), and spatial zonation of bowhead whaling (Savelle and McCartney 1994). The present study uses whale mortality data to reconstruct a positive relationship between whale size and position on the annual migration route.

WHALE MIGRATION AND HUMAN PREDATION

Bowhead migration and human hunting activity may be summarized as follows. (1) Bowheads winter below or at the edge of the ice pack throughout the circumpolar North. The Bering Sea stock winters in the western and central Bering

Sea and the Davis Strait stock winters from western Greenland to the Labrador coast (Reeves et al. 1983; Moore and Reeves 1993). (2) In the late spring, these whales begin their long migration to the middle Arctic, following the opening ice until they reach their summer ground by mid- to late summer. (3) Bowheads are channeled along the leads in the pack ice or along the ice floe edges, and hunters have traditionally taken whales as they pass their particular coasts. (4) Bowheads tend to migrate in a linear fashion in one direction along the coast with no significant doubling back. (5) They tend to travel in age-segregated groups that include dam or cow-calf pairs, immatures, and adults. (6) Once the seasonal pack ice clears by mid-summer, bowheads tend to swim far offshore and out of the reach of shore-based hunters. When migrating south in the fall, bowheads are not constrained by pack ice as they are in their spring movement to the north. (7) Calves are born in April-early June, and thus age-based, and with younger animals, size-based, cohorts representing different year classes of animals are recognizable. (8) Because bowhead bones are heavy and cumbersome, it is anticipated that they were mainly used locally and not distributed between many archaeological sites within a region. (9) Bones from the same sized (almost all small) bowheads are found both at beaching/butchering locales and used in the construction of winter houses, and thus the bones that are counted and measured can be expected to reflect the animals killed and butchered by a site's occupants.

Given these conditions, we should see distinctive differences in animal sizes at sites if different age cohorts were hunted, and/or slightly different sized animals at sites in different regions of the migration route, reflecting different ages in months, if the same age cohort was being hunted. In fact, we commonly find the same sized (estimated length about 8.5m) animals at Thule winter sites. However, while no major difference are noted, below we present data to suggest that minor size differences may well express different ages (in weeks or months) of the same age cohort.

CANADIAN ARCTIC ARCHIPELAGO BOWHEAD PATTERNS

Bowhead whales are the largest of arctic animals. Adults attain lengths of 18-20 m and weigh up to 90,000 kg (Nerini et al. 1984; Reeves and Leatherwood 1985; George et al. 1990). They are slow swimming animals that are rich in blubber and baleen and readily float when killed. They are highly ice-adapted, favoring clear ice-edge waters and loose pack of up to 4/10th-5/10th ice cover, and are restricted to ice-bound arctic seas where they follow the seasonally advancing and retreating ice margins. Bowheads migrating into the central and eastern Canadian Arctic Archipelago are part of the Davis Strait stock. Due to intensive commercial whaling up to 1915, this stock has been reduced in size to about 250 (Finley 1990) to 350 (Zeh et al. 1993) animals. As late as 1825, the stock numbered at least 11,000 animals (Mitchell and Reeves 1981). Prior to 19th century commercial whaling, the population would have been much larger, as the estimate for 1825 follows more than a century of commercial whaling by Scottish and American fleets.

Although historic whaling and sighting records offer the most direct evidence of Canadian Arctic bowhead distributions and migration routes (Ross 1979, 1993; Reeves et al. 1983), these observations are few in the Canadian Arctic Archipelago, which is the focus of this study (Figure 2). Most whaling voyages concentrated on hunting in the easily accessible Davis Strait-Baffin Bay area, and it was only after c. 1850 (Lubbock 1937) that intensive efforts were made to harvest whales in the central Arctic Archipelago region. But by the 1880s, bowheads in eastern and central Canadian Arctic waters had become increasingly rare, as reflected in fewer whales taken by the European and American fleets (Ross 1979). Also, apart from the flurry of voyages in search of Sir John Franklin in 1848-1859 (see e.g., Holland 1994), few expeditions penetrated the study area and recorded bowhead distributions. Finally, recent Inuit and Euroamerican settlement in the study area post-dates the whaling era, and thus any 'modern' sightings are associated with remnant bowhead population levels only.

Between 1819 (when the region, then uninhabited, was first explored by Europeans) and 1859, only 15 bowhead sightings (single animals or small groups) are known in the study area from all sources, 18 sightings are known between 1859 and 1913, six between 1914 and 1974, and 19 between 1975 and 1979 (Reeves et al. 1983). The last are particularly indicative of the present population, since this was a period of intensive aerial whale surveys.

It is therefore difficult to reconstruct in any detail probable past intraregional bowhead distributions and migration routes, based on the modern population. However, sea ice distributions are better known. Because bowheads generally follow the floe edge, their migrations should mimic ice clearance and formation. The most detailed observations of seasonal sea ice dynamics in the Canadian Arctic Archipelago are those of Lindsay (1977, 1982) and especially Markham (1981). Figures 2-4 summarize typical ice distribution patterns in monthly intervals from mid-July, when Lancaster Sound becomes accessible to bowheads, to mid-September, when bowheads can reach their maximum range extensions just prior to freeze-up. Ice remains longest in embayments and restricted passages.

On the basis of these sea ice patterns, bowheads enter eastern Lancaster Sound by early summer (June to mid-July; Figure 2). By mid-summer (early to mid-August), they have penetrated Prince Regent Inlet and Barrow Strait (Figure 3). By late summer/early fall (early to mid-September), they have penetrated westward to their furthest extent (Figure 4). Pack ice blocks their passage further west, north, and south. For purposes of comparison below, we examine measurement data at the maximum extent of the summer bowhead range: (1) Prince Regent Inlet, (2) Peel Sound, and (3) McDougall Sound-Queen's Channel. In addition, we treat separately the (4) Lancaster Sound-Barrow Strait corridor that serves as a common entryway for these three areas. Although the timing of annual breakup and freeze-up may vary from year to year or between major climatic periods (Alt et al. 1985; Dyke at al. 1996), the sequence of breakup and freeze-up can be expected to have remained fairly consistent within and between these summer termini.

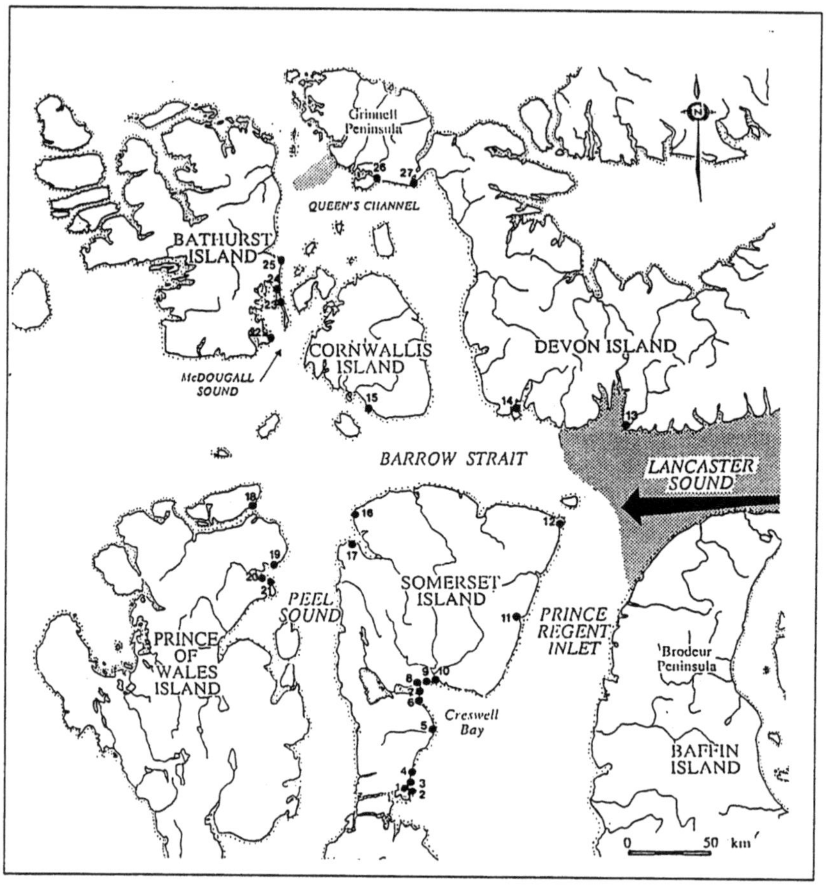

Figure 2. Typical sea ice patterns in the central Canadian Arctic Archipelago in mid-July. Bowhead whale migration routes indicated by arrows. Open water is indicated by shading. Numbers refer to Thule sites listed in Table 1.

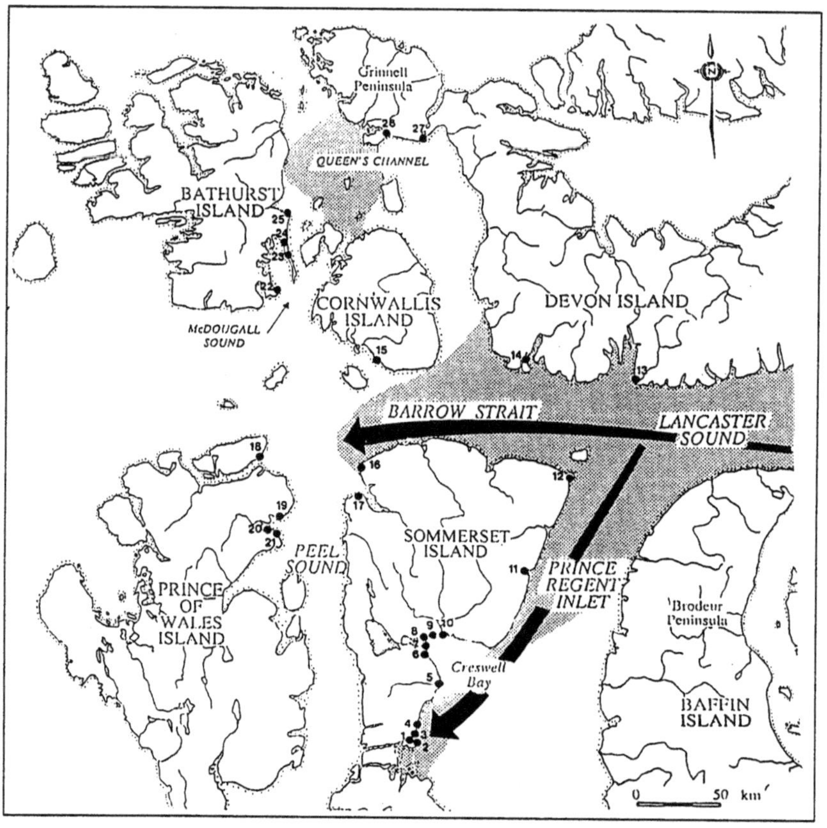

Figure 3. Typical sea ice patterns in the central Canadian Arctic Archipelago in mid-August. Bowhead whale migration routes indicated by arrows. Open water is indicated by shading. Numbers refer to Thule sites in Table 1.

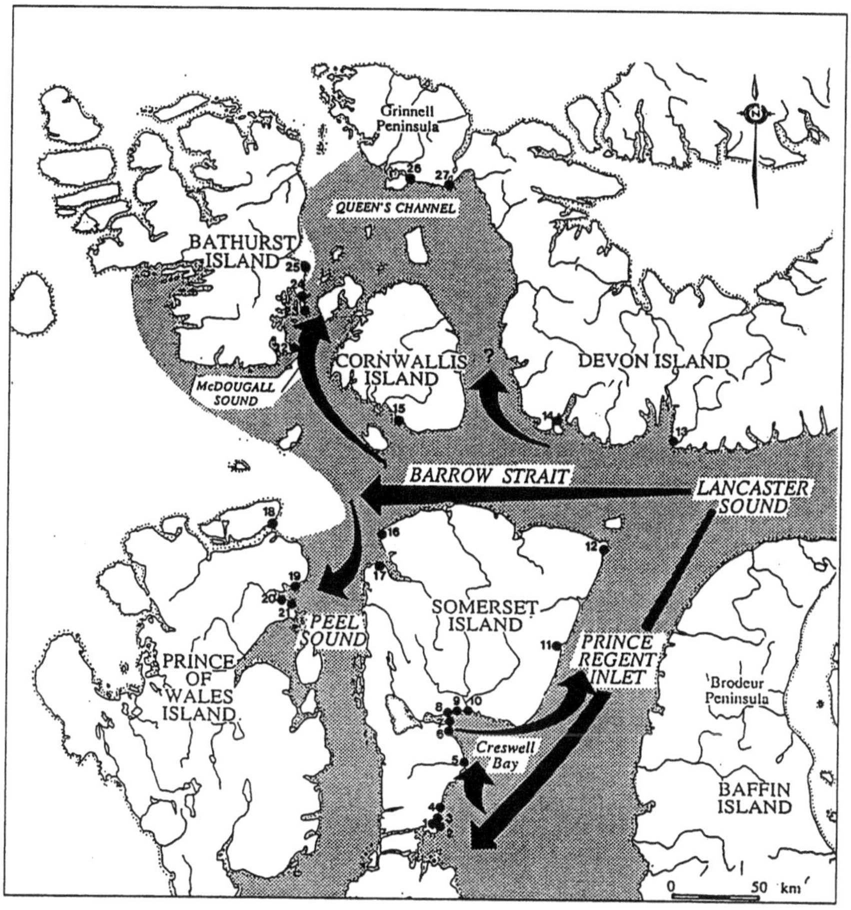

Figure 4. Typical sea ice patterns in the central Canadian Arctic Archipelago in mid-September. Bowhead whale migration routes indicated by arrows. Open water is indicated by shading. Numbers refer to Thule sites listed in Table 1.

BOWHEAD WHALE GROWTH CHARACTERISTICS

Traditional methods for determining age at death, such as those based on tooth eruption and wear and dental annuli development, do not apply to bowheads because they lack teeth. Instead, they feed by filtering krill through baleen. No anatomical method is presently available to determine a bowhead's age, although Schell et al. (1989) and Schell and Saupe (1993) report promise in analysis of baleen layers (see below). Attempts have been made to distinguish growth layers in auditory bulla of balaenid whales (Omura et al. 1969; Mitchell 1984), but no definitive ageing method has resulted. After bowheads reach sexual maturity at 12-14 m in length, they grow slowly to maximum lengths of 18-20 m. Some large bowheads taken in northern Alaska in recent years held stone-bladed harpoon heads that can be dated typologically to 80-100 years old (Philo et al.:1993).

Bowhead biologists typically rely on size, principally length, to estimate age/maturation, using either measurements of harvested animals (e.g., Nerini et al. 1984; Philo et al. 1993) or photogrammetry (Cubbage et al. 1984; Davis et al. 1983; see also compilations and summary in Koski et al. 1993). Age cohorts are assumed to be of approximately the same length. If so, cohorts will form separate peaks in living population length frequencies, similar to crown height frequencies of some terrestrial mammals (e.g., Klein and Cruz-Uribe 1984). At the youngest ages, body shape, coloration, and behavior help discriminate between calves and yearlings (Koski et al. 1993:255). With only bones present at archaeological sites, we must rely on size alone to estimate ages.

Neonates (usually born in a short period between April and early June) measure approximately 3.6-4.5 m in length. Calves grow rapidly, typically attaining lengths of 6.0-7.0 m by early fall, thus growing about 1.5 cm per day (Koski et al. 1993). Some calves are born in June and would be expected to grow less in the subsequent 12 month period than those born earlier in the year. While differing estimates have been suggested for yearling length, the various studies summarized by Koski et al. (1993:259-263) suggest that a range from approximately 7.0 m (late spring) to 9.0 m (late fall). Note that we (Savelle and McCartney 1991, 1994) have previously used 6.0-9.4 m as the length range for yearlings, based on data in Cubbage et al. (1984) and Nerini et al. (1984). However, additional and more recent information presented by Koski et al (1993) suggest that 7.0-9.0 is a much more realistic estimate for yearling length during the summer and fall (i.e. the period during which bowheads would have been present in the study area). Finally, note that some length frequency plots from Alaska (Koski et al. 1993:263) and age estimates based on ^{13}C concentrations of baleen (Schell et al. 1989; Schell and Saupe 1993) suggest that some whales in the 8.7-9.0 m range may be older than yearlings. While of concern to cetacean biologists, this question is not critical to our study. That is, because the greatest frequency of Thule-associated bowhead length estimates is 8.0-8.5 m, well within the yearling length of 7.0-9.0 m noted above, we assume that our measured sample derived primarily from yearlings.

Finally, because of the rapid growth during their first 18 months, and because they are in the inter-island channels only during the summer, the relative time of death during summer occupation (early, middle, late) may be approximated for these younger animals. Older immature animals or young adults could only be aged in a general manner, and older adults could not be aged at all osteometrically.

DETERMINING ORIGINAL BOWHEAD LENGTH

Bowhead bone measurements that might best reflect early growth initially were suggested by Edward D. Mitchell, a cetacean biologist with Fisheries and Oceans Canada. These included 16 cranial, 12 mandibular, five scapular, and five cervical vertebral measurements. The measurements were used to generate a series of multiple step-wise regression models, based on a series of 14 Alaskan bowhead skeletons collected by Floyd E. Durham and housed as the Los Angeles County Museum (McCartney and Savelle 1993). These models, in turn, can be applied to measurements of bowhead bones in order to estimate whale length (McCartney 1978; McCartney and Mitchell 1988).

DATA BASE

Bowheads in three different contexts are available for comparing measurements: archaeological (hunted) animals, stranded Holocene animals, and modern Western Arctic animals.

Archaeological Specimens

Approximately 10,500 bowhead bone elements were recorded by us between 1978-1994 at Canadian Thule sites along the bowhead summer migration routes (Figures 3-5). These bones represent approximately 1000 whales. A total of 441 crania, 798 mandibles, 217 scapulae, and 122 fused cervical vertebrae units were measured, of which 354 crania, 791 mandibles, 213 scapulae and 120 cervical vertebrae units yielded sufficient measurements to determine animal length. As multiple mandible and cranial measurements provide the most precise length estimates, they are the only elements used in the present study. Furthermore, estimates based on crania and mandibles are consistent, showing the same distribution patterns (see Savelle and McCartney 1994:291). Because mandibles are the most abundant elements at these sites in terms of either minimum number of individuals (MNI) or minimal animal units (MAU), the Thule examples discussed below are based on mandibles only (n=791).

Stranded Specimens

Lengths were estimated from cranial and mandible measurements from 515 naturally-beached bowhead whales dating from approximately 10,500 B.P. to recent times and found within and adjacent to the study area (note that this is a considerably larger sample than the 231 animals reported in Savelle and McCartney 1994). These remains are typically embedded within uplifted Holocene beaches, resulting from postglacial isostatic rebound, and were dated by Dyke either directly through C14 methods or indirectly according to local relative sea level curves (see e.g., Dyke and Morris 1990; Dyke et al. 1991, 1996). Most of the remains predate Thule whaling and few, if any, are thought to have been hunted.

Modern Bowhead Populations

No data are available on frequency distributions of whale lengths for the modern Davis Strait bowhead population, but such data are available for the Bering Sea stock. Koski et al. (1988, 1993) provide the most comprehensive data in summaries of length frequencies determined photogrammetrically on 1988 bowheads photographed in the Beaufort Sea between 1981-1986.

ASSESSING THE NATURE AND EXTENT OF THULE WHALING

Hunting vs. Scavenging

Thule Inuit very likely hunted bowheads individually because of their large size and the necessity of using several umiaks to tow the carcass to shore. Because of the evident work requirements, we would expect some preference for smaller whales; they are by far the easiest to kill and process (see e.g., McCartney and Savelle 1985). Historically, North Alaskan peoples, the direct descendants of Thule Inuit in northern Alaska, also preferred small whales (McCartney 1995). On the other hand, if Thule Inuit merely collected bones for architectural purposes from naturally stranded animals, we would expect the size frequency distribution of whale lengths (hereafter referred to as 'profiles') at Thule sites to resemble those of naturally stranded animals. The major differences between the Thule (Figure 5) and the stranded bowhead profiles (Figure 6) clearly argue against the scavenging hypothesis.

Furthermore, in contrast to length distributions of the live population (Figure 8), the vast majority (649=82%) of Thule-derived whales were in the overall yearling size range (7.0-9.0 m). Few were calves or adults, although those measuring between 9.0 and 10.4 m (about 17%) probably represent older subadults. This selection for yearlings, in contrast to calves and adults, can be attributed to both bowhead size and behaviour. Based on data in George et al. (1990), 4.5-5.5 m bowhead calves weigh 1200-2000 kg, a 7.5 m bowhead weighs about 5100 kg, a 9 m animal 12,100 kg, and an 11 m animal 14,750 kg. The sharp decline in animals longer than 9 m in the Thule profile suggests that this length approaches the upper length/weight limit for normal manageability in killing and processing. The absence of calves requires an explanation of a different sort. Calves might initially seem to be the highest valued whales for ease of killing and processing. However, since calves are typically weaned at 9-12 months (Koski et al. 1993:264-265), they would probably be extremely difficult to kill and retrieve, due to their constant proximity to defensive mothers

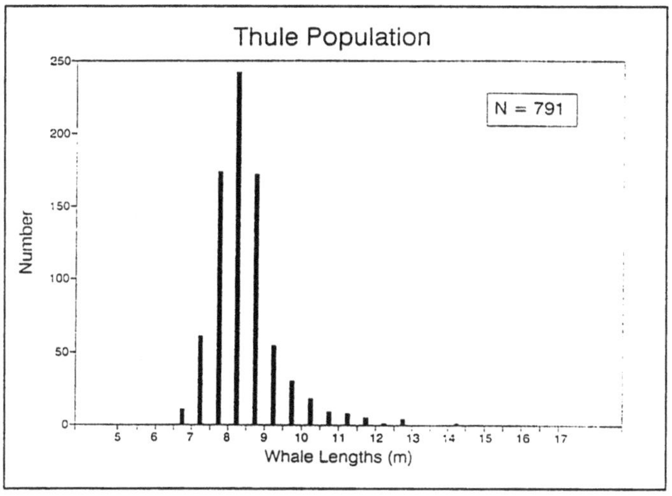

Figure 5. Estimated length distribution of Thule-derived bowheads measured at sites (and site regions) 1-27 in Figures 2-4.

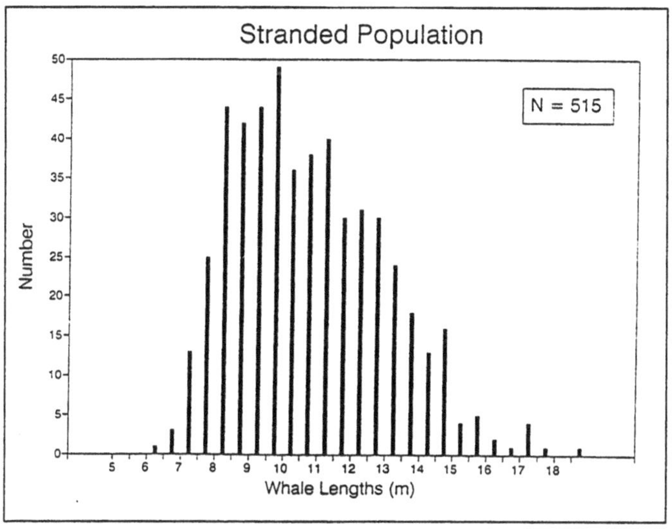

Figure 6. Estimated length distributions of Thule-derived bowheads measured at sites (and site regions) 1-27 in Figures 2-4.

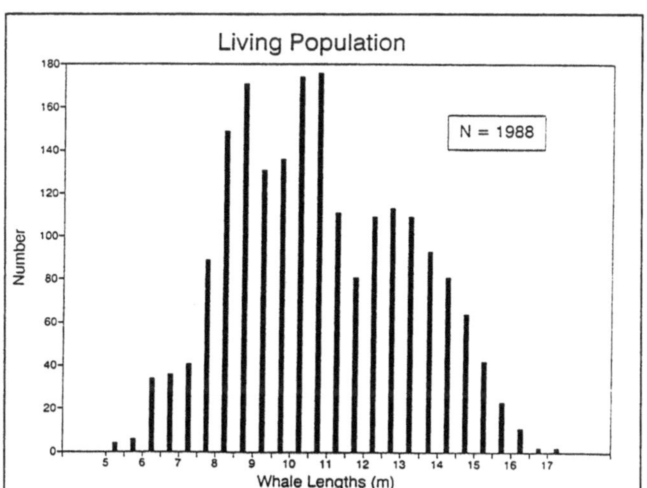

Figure 7. Estimated lengths distributions of live bowheads measured photogrammetrically in the Chukchi and Beaufort seas (1981-86; estimated figures from Koski et al. 1988).

(see e.g. Anungazuk 1995:342). Alternatively, mother-calf pairs may be particularly wary, and thus be much more difficult to approach. Thus, returning yearlings during the summer migration to the Canadian Arctic Archipelago would be the ideal 'targets' due to their relatively small size, aggregation in cohort sets separate from adults (Wursig and Clark 1993:173), and, possibly, general inexperience.

One might argue that the yearling selection expressed in the Thule whale population profile is due to the fact that primarily yearling carcasses were deposited on active beaches through natural taphonomic processes. Comparison of the Thule profile with the stranded profile (Figures 5 and 6), however, indicates that this is not the case. In fact, the beached profile strongly resembles the living population profile (Figure 7), suggesting that random or catastrophic events, such as killer whale attacks or ice entrapment (Philo et al. 1993), are probably responsible for the majority of deaths of naturally deposited animals within the study area. In addition, typically it is only the crania that are exposed at these naturally stranded sites; for example, of 767 individual stranded bowheads recorded during 1990-1994, single or mandible pairs of only 27 were exposed. Taken together, we consider these data to be conclusive evidence for active whale hunting by Thule Inuit.

REGIONAL VARIATION IN THULE BOWHEAD LENGTH DISTRIBUTIONS

Previously, we have subdivided Thule whale profiles according to core and peripheral (Savelle and McCartney 1991) and core, intermediate, and peripheral (Savelle and McCartney 1994) whaling zones, based on the probable overall summer abundance of bowheads during Thule times. Comparison of profiles within each of these zones indicated that, while there was an overall selection for yearlings, the proportion of larger animals increased from core to periphery.

Here, we examine variations in mortality profiles within and between the central Canadian Arctic Archipelago channels shown in Figures 3-5, and relate these specifically to local ice breakup patterns (for regional ice discussions, see Dunbar and Greenaway 1956:415ff). In particular, we focus on variation in *yearling* mortality patterns within each corridor, the analysis of which follows below. The neighbouring residential sites that form clusters are treated as single units (for example, sites 1-4 in the Prince Regent Inlet range), since the arrival time during the migration would be similar. In addition, associated processing sites (see e.g., Savelle 1997, n.d.) are included within site totals. This, together with the aggregation of samples from adjacent sites, results in larger samples, an important consideration when testing for significant differences between samples. Specifically, in the following analyses we follow Drennan's (1996:109, 164) suggestion regarding sample size in the context of the central limit theorem; that is, minimum sample sizes for statistical comparison between means of normally-distributed populations should be at least 30, although we still comment upon any apparent general trends. Mortality data for the sites or site clusters indicated in the respective corridors are presented in Table 1.

Prince Regent Inlet

Prince Regent Inlet presents the best illustration of the effect of local ice conditions on the timing of bowhead harvesting. The sequence of ice breakup (Figures 2-4) is such that bowheads are accessible just offshore first at Hazard Inlet at the southeastern tip of Somerset Island, because of an open water 'wedge' that follows the shore-fast ice northward. As this shore-fast ice melts or otherwise dissipates, the open water 'wedge' widens from south to north, eventually clearing Creswell Bay. Finally, during late summer, the only remaining ice-covered areas are the restricted bays (e.g., Port Leopold and Batty Bay). In order of ice clearing, then, the sequence is Hazard Inlet-Cape Garry-Creswell Bay-Batty Bay/Port Leopold.

While there is considerable variation between sites within individual site clusters, there is nevertheless an overall trend of increasing mean yearling size (Table 1) with later relative position of the site or site clusters along the whale migration route. Specifically, mean yearling sizes are lowest within the Hazard Inlet cluster (sites 1-4) and Cape Garry (site 5), and highest in the Creswell Bay (sites 6-10) and Batty Bay-Port Leopold (sites 11 and 12) clusters. Statistical analyses applied to sites or clusters with sample sizes of larger than 30 support this trend. That is, the Hazard Inlet cluster mean is significantly different (smaller) than either the Creswell Bay cluster mean ($t=-6.435$; $p<.000$) or the Batty Bay-Port Leopold cluster mean ($t=-3.698$; $p<.000$). There is no significant difference, on the other hand, between the means of the Creswell Bay and Batty Bay-Port Leopold clusters ($t=0.172$; $p=.864$).

McDougall Sound-Queen's Channel

Trends in mortality patterns in McDougall Sound-Queen's Channel are consistent with those described above. Breakup in this area generally proceeds from south to north, with enclosed bays clearing last (Figures 3-5). The ice clearance sequence of sites adjacent to open water is, initially, Cape Evans (site 22), followed by Brooman Point and Black Point (sites 23 and 25), and then Port Refuge and Porden Point (sites 26 and 27). The Deblicquy site (site 24), although in the southern part of the region, overlooks a late clearing bay. Increases in mean yearling lengths match this progression (Table 1), increasing from 7.83 m at Cape Evans, to 8.20 m at Brooman Point, 8.38 m at Black Point, and 8.38 m, 8.49 m and 8.65 m at Deblicquy, Port Refuge and Porden Point, respectively. For the purposes of minimum sample size requirements for statistical analyses, we consider the earlier clearing sites (Cape Evans, Brooman Point, Black Point) as one sample, and the later clearing sites (Deblicquy, Port Refuge, Porden Point) as a second sample. The test results indicate there is a significant difference between the means of the two populations, with that of the later clearing sites being greater than the earlier clearing sites ($t=-2.761$; $p=.007$).

Site/Site Cluster	Mean Whale Size (N)	Mean Yearling Size (N)
1) PRINCE REGENT INLET		
Hazard Inlet		
1. Mount Oliver	7.84 (86)	7.78 (74)
2. Ditchburn Point	8.19 (54)	8.11 (51)
3.PaJs 4	8.14 (4)	8.43 (3)
4. PaJs 13	8.25 (30)	8.14(28)
Sites 1-4 cluster	8.03 (174)	7.97 (156)
Cape Garry		
5. Cape Garry	8.17 (30)	8.07 (25)
Cresswell Bay		
6. Idlout Pt. South	8.07 (11)	7.98 (10)
7. Idlout Pt. North	8.26 (5)	8.26 (5)
8. PeJr 1	8.57 (127)	8.27 (102)
9. PeJr 2	8.26 (44)	8.15 (40)
10. PeJq 1	8.64 (114)	8.32 (88)
Sites 5-10 cluster	8.53 (301)	8.25 (245)
Batty Bay/Port Leopold		
11. Batty Bay	8.40 (26)	8.28 (24)
12. Port Leopold	8.38 (33)	8.21 (28)
Sites 11-12 cluster		
2) BARROW STRAIT and LANCASTER SOUND		
13. Fellfoot Pt.	8.72 (6)	8.12 (5)
14. Radstock Bay	8.44 (28)	8.26 (26)
15. Resolute	8.22 (8)	8.10 (7)
3) PEEL SOUND		
16. Aston Bay North	8.12 (35)	8.05 (33)
17. Aston Bay South	8.03 (9)	8.05 (7)
18. Cape Walker	8.19 (7)	8.01 (6)
19-21. Back Bay	8.54 (1)	8.54 (1)
Sites 18-21 cluster	8.23 (8)	8.09 (7)
4. McDOUGALL SOUND and QUEEN'S CHANNEL		
22. Cape Evans	8.74 (8)	7.83 (6)
23. Brooman Pt.	8.70 (33)	8.20 (25)
25. Black Pt.	9.55 (11)	8.38 (7)
Sites 22-23, 25 cluster		
24. Deblicquy	9.14 (52)	8.38 (29)
26. Port Refuge	8.66 (18)	8.49 (15)
27. Pordon Pt.	9.74 (11)	8.65 (4)
Sites 24, 26-27 cluster	9.12 (81)	8.44 (48)

Table 1. Sites indicated in Figures 2-4, and means of whale sizes (metres) for each site and site clusters.

Peel Sound

While admittedly based on much smaller samples, the trends in yearling mortality in the Peel Sound region are generally consistent with ice-clearing patterns. Ice clears first along the northeastern coast of Peel Sound, including northern Aston Bay, then along northern Prince of Wales Island and to the south of Aston Bay, and finally within restricted bays such as Back Bay. There is a generally positive relationship between the relative timing of ice clearing and average yearling size per site or site cluster (Table 1). Average yearling bowhead lengths are lowest at sites on northern Aston Bay (a series of sites represented by >site= 16; 8.05 m) and immediately south of Aston Bay (site 17; 8.05 m), and increase, admittedly only very slightly, in the Cape Walker-Back Bay region (sites 18-21; 8.09 m). However, as noted above, the samples are small, and in fact the increase in the Cape Walker-Back Bay region is dependent upon a single large yearling from Back Bay (see Table 1)

Lancaster Sound-Barrow Strait

As discussed previously, this is the common entryway to the other inlets and sounds. Ice clears first in the east (Fellfoot Point, site 13), later in the west (Resolute, site 15), and finally in the enclosed bays (Radstock Bay, site 14). Note, however, that Fellfoot Point *may* have been accessible to bowheads during much of the winter because of the recurring polynyas and winter-long unconsolidated ice in that area (see e.g. Smith and Rigby 1981; Richard et al. 1998). Overall, mean yearling size is consistent with the clearing pattern (Table 1). That is, bowhead means are lower at Fellfoot Point (8.12 m) and Resolute (8.10 m) and higher at Radstock Bay (8.26 m). However, as with the situation for Peel Sound, samples sizes are not large enough to confirm these apparent differences.

DISCUSSION

The mortality data presented here support the interpretation of a causal relationship between certain details of mortality profiles and local summer bowhead migratory patterns.

Selective Hunting

Comparison of archaeological faunal assemblage mortality profiles to both the living population and the attritional (natural death) population is common in distinguishing hunting from scavenging (Klein 1982). While the approach employed here follows the same rationale and procedures as those of previous studies (see e.g., the reviews in Stiner 1990, 1991a), it also introduces animal size as a prime factor in explaining mortality profile variability. Selection of animals of a specific age/size is typically associated with prime-dominated hunting, wherein animals in their prime (i.e., young adults) are preferred because they yield proportionately the greatest amount of meat, fat, and other products. Only rarely has selection of juveniles been documented. Where it has been, as in juvenile caribou selection by Nunamiut Eskimos (Binford 1978), it occurred as an exception within the broader context of an overall adult prey preference (Stiner 1990:331).

As discussed earlier, we interpret the preference for small yearlings as a product of available technology. Specifically, larger yearlings (animals about 9.0 m in length) approached the maximum weight, 12,000 kg (approximately double the weight of a large adult male African elephant; Haynes 1991:14), that could be killed and processed using traditional techniques. This interpretation accords with the suggestion that traditional Alaskan Eskimo whalers selected almost exclusively smaller subadult bowheads (see summary in McCartney 1995); it is only with the introduction of 19th century Euroamerican whaling technology (bomb darts, shoulder guns, and block and tackle; VanStone 1958; Bockstoce 1986) that larger and full grown whales were hunted regularly.

Thus, while size selection for the largest terrestrial megafauna hunted (mammoths and mastodons) does not appear to have been constrained by an upper size limit (see e.g., Saunders 1980; Haynes 1991), large baleen whale hunters, at least, appear to have had this restriction. This size restriction is based on the maximum carcass size that could be towed ashore and butchered (given local winds, currents, available boats, and butchery participants) prior to putrefaction, which starts within 24-48 hrs of death (see McCartney 1995:98-99). Furthermore, we suspect that size selection of other baleen whales and the larger sirenians by traditional hunting societies may have been similarly restricted.

Mortality Profiles and Bowhead Migration

An advantage of using biometrical data, especially from large and fast-growing animals, is that it is possible to identify when, during the migration cycle, the animals were harvested. The fact that bowheads were available to Thule Inuit for only 3-4 months a year suggests that separate size modes should appear in a profile of hunted animals, as represented by the various site/site cluster whale bone assemblages. In this regard, the present analysis has demonstrated not only that Thule Inuit were harvesting whales at various points along the migration ranges, but, in addition, that the timing of the harvest at any one locality corresponded closely with the arrival of the whales. That is, we are dealing with a clearly defined intercept strategy common to collectors preying on migrating animals (cf., Binford 1978, 1980). Such availability and hunting correspondence appears to be the most reasonable explanation for the progressive increase in length of harvested yearlings along the migration route of the whales.

An implication of this interpretation of intercept strategy is that it would appear that structural bone elements from most, if not all, of the harvested bowheads were incorporated into structures in villages that were immediately adjacent to the hunting and processing areas. This is not to suggest that villages did not engage in cooperative hunting, but only that most structural whale bones, if not meat, were destined for the nearest village(s). The weight of bowhead bones alone would dissuade transport for any great distance from their processing locations.

IMPLICATIONS FOR THE INVESTIGATION OF HUMAN-MEGAFAUNA RELATIONSHIPS

Much has been written on human-megafaunal relationships, particularly how these relationships may have influenced settlement patterns and group mobility, including the expansion of humans into the New World and the extinction of many species of megafauna (see e.g., Martin 1967; Martin and Klein 1984; Saunders 1980; Kelly and Todd 1988; Haynes 1991; Yesner 1995). These relationships, however, have typically been evaluated from zooarchaeological assemblages that are relatively small in size and are derived from a few, widely dispersed sites. For example, less than 70 mammoths and mastodons are represented at the 16 New World Clovis sites listed by Haynes (1991:286-287) which contain their remains. Not surprisingly, there is considerable disagreement concerning the role of hunting vs. scavenging, harvesting techniques and size selection, requisite logistical organization, as well as the basic question of whether particular megafauna were locally or regionally migratory (see e.g., Soffer 1985:193). We feel that the present study, based as it in on large regional (and even site) sample sizes and relatively sensitive age-at-death estimates, is instructive in three respects for modeling human-megafauna relationships.

First, the results of our analysis suggest that much megafauna procurement involved a highly organized intercept strategy that required a considerable amount of positioning at predictable animal encounter locations, the success of which relied on knowledge of animal behavior. The implications of this strategy are presumably applicable to many other human-megafauna contexts: (a) intercept scheduling of hunting when animals first appear, and (b) relatively permanent, stable residential bases.

Second, the study has demonstrated that size selection may be more dependent upon mother-calf bonds and carcass manageability than on overall size per se. That is, while larger megafauna in general may be desirable from a cost-effectiveness standpoint, a limit is reached at which relative cost-effectiveness is reduced and finally lost as individual animal size increases. Therefore, the optimal animal preference is actually for the smallest individuals within the largest species. For bowheads, the upper limit appears to have been reached at approximately 10,000 kg or 8.5-9.0 m in length. Calves, on the other hand, remain with their mother during their first summer, and apparently were very rarely available to hunters. This may have been due to aggressive defense and/or a heightened state of vigilance on the part of the mother.

Finally, mortality profiles have traditionally been derived from absolute animal ages or relative degree of maturity (juvenile, prime adult, old adult; see e.g., Stiner 1990). The biometric method employed here, while initially developed out of necessity, provides relatively precise age/size-at-death estimates that allows us to 'track' bowheads into and along the various migration corridors. This method should be equally applicable to other fast-growing, young migrating megafauna, allowing researchers to 'track' sequential interceptions and hunting localities.

While we fully recognize that there are considerable differences between terrestrial and marine megafauna and habitats, we nevertheless believe that patterns similar to those found at prehistoric Thule Eskimo whaling sites may be relatively common in other megafauna hunting contexts.

SUMMARY

This paper demonstrates that osteometric data enables us to reconstruct the human predation patterns that were imposed on bowhead whales by Thule Inuit hunters in the Canadian Arctic Archipelago. With relatively large sample sizes from many coastal locales, it is possible to confirm the expected trend of hunting yearling animals that grow perceptibly in length during a short summer season. Starting with disarticulated bowhead bones found at Thule sites, regression analyses permit the estimation of original animal lengths and these lengths may be compared between hunting locales along the annual bowhead migration routes. This approach of studying fast-growing, young animals as represented in archaeofaunas may have applicability to other regions where migrating megafauna were hunted in the past.

ACKNOWLEDGMENTS

We gratefully acknowledge the sponsoring institutions, field assistants, and principal colleagues who assisted the studies summarized here. The 1978 fieldwork was sponsored by the Northern Environmental Protection Branch, Department of Indian and Northern Affairs. In addition to McCartney, the field crew consisted of Allen Clarke, Michelle McLaughlin, and David Sudlovenick. Field advisors and facilitators were Bryan Kemper and Ed Mitchell. The 1988 fieldwork was conducted under the sponsorship of the Social Sciences and Humanities Research Council of Canada; in addition to Savelle and McCartney, the field crew consisted of Karen Digby, Elisa Hart, and Terry Manik. The 1990, 1992, 1993 and 1994 fieldwork was funded by the Geological Survey of Canada, the National Sciences and Engineering Research Council of Canada, and the Social Sciences and Humanities Research Council of Canada; in addition to Savelle (1990, 1992, 1993, 1994) and McCartney (1990), field crew members were Ruth Edelstein, Colin Grier, Elisa Hart, Krista Kieswetter, Katherine Patton, and Matthew Sturgess. Sandra Parker, Susan Houston, Brian Schnarch, Matthew Sturgess and Robert Rosenswig have, variously, assisted in the data analysis since 1978, and Amanda Crandall assisted in the production of Figures 2-4.

The Polar Continental Shelf Project, Department of Energy, Mines and Resources (Canada), assisted us with logistical support and fixed-wing and helicopter service during all of the fieldwork described here; we especially wish to thank the staff of the Resolute PCSP base for their invaluable assistance.

Finally, the many valuable comments by Craig Gerlach and two anonymous referees prompted us to rethink several critical issues.

REFERENCES CITED

Alt, B.T., Koerner, R.M., Fisher, D.A. and J.C. Bourgeois
1985 Arctic Climate During the Franklin Era, as Deduced from Ice Cores in The Franklin Era in Canadian Arctic History, 1845-1859, pp. 69-92. *National Museum of Man Mercury Series, Archaeological Survey of Canada Paper* 131.

Binford, L.R.
1978 *Nunamiut Ethnoarchaeology*. Academic Press, Inc., New York.

1980 Willow Smoke and Dogs' Tails: Hunter-Gatherer Settlement Systems and Archaeological Site Formation. *American Antiquity* 45:4-20.

Bockstoce, J.R.
1986 *Whales, Ice, and Men: The History of Whaling in the Western Arctic*. University of Washington Press, Seattle.

Cox, S.L. and Spiess, A.
1980 Dorset Settlement and Subsistence in Northern Labrador. *Arctic* 33:659-669.

Cubbage, J.C., Calambokidis, J. and D.J. Rugh
1984 *Bowhead Whale Length Measured Through Stereophotogrammetry*. Final Report from Cascadia Research Collective to National Marine Mammalogy Laboratory.

Danielson, R.A.
1994 *Ringed Seal Mortality Patterns as an Aid in the Determination of Thule Eskimo Subsistence Strategies*. Unpublished M.A. thesis, Department of Anthropology, McGill University, Montreal.

Davis, R.A., Koski, W.R., and Miller, G.W.
1983 *Preliminary Assessment of the Length-Frequency Distribution and Gross Annual Reproductive Rate of the Western Arctic Bowhead Whale as Determined with Low-Level Aerial Photography, with Comments on Life History*. Report from LGL Ltd. to National Marine Mammalogy Laboratory.

Drenan, R.D.
1996 *Statistics for Archaeologists: A Commonsense Approach*. Plenum Press, New York.

Dunbar, M. and Greenaway, K.R.
1956 *Arctic Canada from the Air*. Canada Defence Research Board, Ottawa.

Dyke, A.S., and T.F. Morris
1990 Postglacial History of the Bowhead Whale and of Driftwood Penetration; Implications for Paleoclimate, Central Canadian Arctic. *Geological Survey of Canada Paper* 89-24.

Dyke, A.S., Morris, T.F. and D.E.C. Green
1991 Postglacial Tectonic and Sea Level History of the Central Canadian Arctic. *Geological Survey of Canada Bulletin* 397.

Dyke, A.S., Hooper, J. and J.M. Savelle
1996 A History of Sea Ice in the Canadian Arctic Archipelago Based on Postglacial Remains of the Bowhead Whale (*Balaena mysticetus*). *Arctic* 49(3):235-255.

Finley, K.J.
1990 Isabella Bay, Baffin Island: An Important Historical and Present-Day Concentration Area for the Endangered Bowhead Whale (*Balaena mysticetus*) of the Eastern Canadian Arctic. *Arctic* 43:137-152.

Fisher, D.C.
1987 Mastodont Procurement by Paleo-Indians of the Great Lakes Region: Hunting or Scavenging? In *The Evolution of Human Hunting*, edited by M.H. Nitecki and D.V. Nitecki, pp. 309-421. Plenum Press, New York.

Friesen, T.M. and Arnold, C.D.
1995 Prehistoric Beluga Hunting at Gupuk, Mackenzie Delta, Northwest Territories, Canada in *Native Whaling in the Western Arctic and Subarctic: Hunting the Largest Mammals*, edited by A.P. McCartney, pp.109-125. Canadian Circumpolar Institute, Edmonton.

Friesen, T.M., Savelle, J.M., and Smith, T.G.
1998 Mandibular Growth Layers as a Means of Determining the Age at Death of Beluga Whales (*Delphinapterus leucas*) in Zooarchaeological Assemblages. *Journal of Archaeological Science* 25:697-706.

Frison, G.C.
1978 *Prehistoric Hunters of the High Plains*. Academic Press, New York.

George, J.C., Philo, L.M. and G.M. Carroll
1990 Observations on Weights of Subsistence Harvested Bowhead Whales in Fifth Conference on the Biology of the Bowhead Whale, *Balaena mysticetus*: Extended Abstracts and Panel Discussion, edited by T.F. Albert, pp.89-93. North Slope Borough, Barrow.

Gerlach, C, J.C. George and R. Suydam
1993 Bowhead Whale (*Balaena mysticetus*) Length Estimations Based on Scapula Measurements. *Arctic* 46:55-59.

Guthrie, R.D. 1980 Bison and Man in North America. *Canadian Journal of Anthropology* 1:55-73.

Haynes, G.
1985 Age Profiles in Elephant and Mammoth Bone Assemblages. *Quaternary Research* 24:333-345.

1991 *Mammoths, Mastodons, and Elephants*. Cambridge University Press, Cambridge.

Heizer, R.F.
1941 Aboriginal Whaling in the Old and New Worlds. Unpublished Ph.D. dissertation, Department of Anthropology, University of California, Los Angeles.

1968 A Bibliography of Aboriginal Whaling. *Journal of the Society of Bibliographers for Natural History* 4:344-362.

Holland, C.
1994 *Arctic Exploration and Development c 500 B.C. to 1915: An Encyclopedia*. Garland Publishing, New York.

Jarmon, M.R. and P.F. Wilkinson
1972 Criteria of Animal Domestication in *Papers in Economic Prehistory*, edited by E.S. Higgs, pp. 83-96. Cambridge University Press, Cambridge.

Kelly, R.L. and L.C. Todd
1988 Coming into the Country: Early Paleoindian Hunting and Mobility. *American Antiquity* 53:231-244.

Klein, R.K.
1982 Age (Mortality) Profiles as a Means of Distinguishing Hunted Species from Scavenged Ones in Stone Age Archaeological Sites. *Paleobiology* 8:151-158.

1987 Reconstructing How Early People Exploited Animals: Problems and Prospects in *The Evolution of Human Hunting*, edited by M.H. Nitecki and D.V. Nitecki, pp. 11-45. Plenum Press, New York.

Klein, R.K. and K. Cruz-Uribe
1984 *The Analysis of Animal Bones from Archaeological Sites*. University of Chicago Press, Chicago.

Koski, W.R., Miller, G., and R. Davis
1988 The Potential Effects of Tanker Traffic on the Bowhead Whale in the Beaufort Sea. *Indian and Northern Affairs Canada, Environmental Studies* 58.

Koski, W.R., Davis, R.A., Miller, G.W., and D.E. Withrow
1993 Reproduction in *The Bowhead Whale*, edited by Burns, J.J., Montague, J.M. and C.J. Cowles, pp.239-274. Special Publication No.2, Society for Marine Mammalogy.

Lindsay, D.G.
1977 *Sea Ice Atlas of Arctic Canada 1969-1974*. Department of Energy, Mines and Resources, Canada.

1982 *Sea Ice Atlas of Arctic Canada 1975-1978*. Department of Energy, Mines and Resources, Canada.

Lubbock, B.
1937 *The Arctic Whalers*. Brown, Son and Ferguson, Glasgow.

Lyman, R.L.
1991 *Prehistory of the Oregon Coast*. Academic Press, San Diego.

Markham, W.E.
1981 *Ice Atlas: Canadian Arctic Waterways*. Supply and Services Canada, Ottawa.

Martin, P.S.
1967 Prehistoric Overkill in *Pleistocene Extinctions: The Search for a Cause*, edited by P.S. Martin and H.E. Wright, pp.75-120. Yale University Press, New Haven.

Martin, P.S. and R.G. Klein
1984 *Quaternary Extinctions: A Prehistoric Revolution*. University of Arizona Press, Tucson.

McCartney, A.P.
1978 Study of Archaeological Whale Bones for the Reconstruction of Canadian Arctic Bowhead Whale Stocks and Whale Use by Prehistoric Inuit. Preliminary report prepared for Northern Environmental Protection Branch, Department of Indian and Northern Affairs, Ottawa.

1980 Study of Archaeological Whale Bones for the Reconstruction of Canadian Arctic Bowhead Whale Stocks and Whale Use by Prehistoric Inuit. Final report prepared for the Northern Environmental Protection Branch, Department of Indian and Northern Affairs, Ottawa.

1995 Whale Size Selection by Precontact Hunters of the North American Western Arctic and Subarctic in *Hunting the Largest Animals: Native Whaling in the Western Arctic and Subarctic*, edited by A.P. McCartney, pp. 83-108. Canadian Circumpolar Institute, Studies in Whaling No.3.

McCartney, A.P. and E.D. Mitchell
1988 Thule Eskimo Bowhead Whale Selection on Somerset Island, Arctic Canada. Paper presented at the 53rd Annual Meeting of the Society for American Archaeology, Phoenix.

McCartney, A.P. and Savelle, J.M.
1985 Thule Eskimo Whaling in the Central Canadian Arctic. *Arctic Anthropology* 22(2):37-58.

1993 Bowhead Whale Bones and Thule Eskimo Subsistence-Settlement Patterns in the Central Canadian Arctic. *Polar Record* 29(168): 1-12.

Mitchell, E.D.
1984 Ecology of North Atlantic Boreal and Arctic Monodontid and Mysticete Whales in *Arctic Whaling*, edited by s'Jacob, H.K., Snoeijing, K. and R. Vaughan, pp.65-78. Arctic Centre, University of Groningen, Groningen.

Mitchell, E.D. and R.R. Reeves
1981 Catch History and Cumulative Catch Estimates of Initial Population Size of Cetaceans in the eastern Canadian Arctic. *Report of the International Whaling Commission* 31:645-682.

Moore, S.E. and Reeves, R.R.
1993 Distribution and Movement in *The Bowhead Whale*, edited by Burns, J.J., Montague, J.M. and C.J. Cowles, pp. 313-386. Special Publication No.2, Society for Marine Mammalogy.

Morrison, D.
1983 Thule Culture in Western Coronation Gulf. *National Museum of Man Mercury Series, Archaeological Survey of Canada* No. 116.

Nerini, M.K., Braham, H.W., Marquette, W.M., and D.J. Rugh.
1984 Life History of the Bowhead Whale, *Balaena mysticetus* (Mammalia: Cetacea). *Journal of Zoology* 204:443-468.

Omura, H., Ohsumi, S., Nemoto, T., Nasu, K. and Kasuya, T.
1969 Black Right Whales in the North Pacific. Tokyo, *Scientific Report of the Whales Research Institute* 21:1-78.

Philo, L.M., Shotts, E.B., and J.C. George
1993 Morbidity and Mortality in *The Bowhead Whale*, edited by Burns, J.J., Montague, J.M. and C.J. Cowles, pp.275-312. Special Publication No.2, Society for Marine Mammalogy.

Reeves, R.R. and S. Leatherwood
1985 Bowhead Whale, *Balaena mysticetus* Linnaeus, 1958 in Ridgway, S.H. and R.J. Harrison (editors). *Handbook of Marine Mammals: the Sirenians and Baleen Whales* pp. 305-344. Academic Press, London.

Reeves, R.R., Mitchell, E.D., Mansfield, A. and M. McLaughlin
1983 Distribution and Migration of the Bowhead Whale, *Balaena mysticetus*, in the Eastern North American Arctic. *Arctic* 36:5-64.

Reher, C.A.
1974 Population Study of the Casper Site Bison in *The Casper Site: A Hell Gap Bison Kill on the High Plains*, edited by G.C. Frison, pp. 113-124. Academic Press, New York.

Richard, P.R., Orr, J.R., Dietz, R. and Dueck, L.
1998 Sightings of Belugas and Other Marine Mammals in the North Water, late March 1993. *Arctic* 51:1-4.

Ross, W.G.
1979 The Annual Catch of Greenland (Bowhead) Whales in Waters North of Canada 1919-1915: A Preliminary Compilation. *Arctic* 32:91-121.

1993 Commercial Whaling in the North Atlantic Sector in *The Bowhead Whale*, edited by Burns, J.J., Montague, J.M. and C.J. Cowles, pp. 511-591.Special Publication No.2, Society for Marine Mammalogy.

Saunders, J.J.
1980 A Model for Man-Mammoth Relationships in Late Pleistocene North America. *Canadian Journal of Anthropology* 1:87-98.

Savelle, J.M.
1987 Collectors and Foragers: Subsistence-Settlement System Change in the Central Canadian Arctic, A.D. 1000-1960. *British Archaeological Reports (Int. Series)* 358.

1995 An Ethnoarchaeological Investigation of Inuit Beluga Whale and Narwhal Harvesting in *Native Whaling in the Western Arctic and Subarctic: Hunting the Largest Mammals*, edited by A.P. McCartney, pp.127-148. Canadian Circumpolar Institute, Edmonton.

n.d. Information Systems and Thule Eskimo Whaling in *Animal Bones, Human Societies*, edited by P. Rowley-Conwy. Oxbow Books, London. In press.

Savelle, J.M. and McCartney, A.P.
1988 Geographical and Temporal Variation in Thule Eskimo Subsistence Economies: A Model. *Research in Economic Anthropology* 10:21-72.

1990 Prehistoric Thule Eskimo Whaling in the Central Canadian Arctic: Current Knowledge and Future Research Directions in *Canada's Missing Dimension: Science and History in the Canadian Arctic Islands*, edited by C.R. Harington, pp.695-723. Canadian Museum of Nature.

1991 Thule Eskimo Bowhead Whale Procurement and Selection in *Human Predators and Prey Mortality*. Edited by M. Stiner pp.201-216. Westview Press, Boulder.

1994 Thule Inuit Bowhead Whaling: A Biometrical Analysis in Threads of Arctic Prehistory: Papers in Honour of William E. Taylor, Jr., edited by D. Morrison and J.-L. Pilon, pp.281-310. *Canadian Museum of Civilization, Mercury Series, Archaeological Survey of Canada Paper* No. 149.

1999 Thule Eskimo Bowhead Whale Interception Strategies. *World Archaeology* 30:437-451.

Schell, D.M. and S.M. Saupe
1993 Feeding and Growth as Indicated by Stable Isotopes in *The Bowhead Whale*, edited by Burns, J.J., Montague, J.M. and C.J. Cowles, pp.491-509. Special Publication No.2, Society for Marine Mammalogy.

Schell, D.M., Saupe, S.M. and N. Haubenstock
1989 Bowhead Whales (*Balaena mysticetus*) Growth and feeding as Estimated by 13C Techniques. *Marine Biology* 103:433-443.

Smith, M. and Rigby, B.
1981 Distribution of Polynyas in the Canadian Arctic. In *Polynyas in the Canadian Arctic* edited by I. Stirling and H. Cleator). Canadian Wildlife Service, Occasional Paper 45:7-28.

Soffer, O.
1985 *The Upper Paleolithic of the Central Russian Plain*. Academic Press, Orlando.

Stiner, M.C.
1990 The Use of Mortality Studies in Archaeological Studies of Hominid Predatory Patterns. *Journal of Anthropological Archaeology* 9(4):305-351.

1991a (editor) *Human Predators and Prey Mortality*. Westview Press, Boulder.

1991b Introduction: Actualistic and Archaeological Studies of Prey Mortality in *Human Predators and Prey Mortality*, edited by M.C. Stiner, pp.1-14. Westview Press, Boulder.

Todd, L.C., Hofman, J.L. and C.B. Schultz
1990 Seasonality of the Scottsbluff and Lipscomb Bison Bonebeds: Implications for Modeling Paleoindian Subsistence. *American Antiquity* 55:813-827.

VanStone, J.W.
1958 Commercial Whaling in the Arctic Ocean. *Pacific Northwest Quarterly* 41:1-10.

Wursig, B. and C. Clark
1993 Behavior in *The Bowhead Whale*, edited by Burns, J.J., Montague, J.J. and C.J. Cowles, pp.157-199. Society for Marine Mammalogy, Special Publication No. 2.

Yesner, D.R.
1995 Whales, Mammoths and Other Big Beasts: Assessing Their Roles in Prehistoric Economies in *Hunting the Largest Animals: Native Whaling in the Western Arctic and Subarctic*, edited by A.P. McCartney, pp.149-164. Canadian Circumpolar Institute, Studies in Whaling No.3.

Zeh, J.E., Clark, C.W., George, J.C., Withrow, D., Carroll, G.M. and W.R. Koski
1993 Current Population Size and Dynamics in *The Bowhead Whale*, edited by Burns, J.J., Montague, J.J. and C.J. Cowles, pp.409-489. Society for Marine Mammalogy, Special Publication No. 2.

CARCASS UTILITY INDICES AND BISON BONES FROM THE WARDELL KILL AND BUTCHERING SITES

Jack W. Brink
Archaeological Survey, Provincial Museum of Alberta

Following T. E. White (1952, 1953a, 1953b, 1954, 1955), Binford (1978) designed a series of utility indices that are now popular in the interpretation of skeletal abundance at archaeological sites. These indices consist of models of economic anatomy; that is, an assessment of the value or utility of different carcass portions according to the quantity and/or quality of edible food associated with specific skeletal parts. The theoretical underpinning of Binford's anatomical model is that aboriginal hunters utilize large game carcasses in a rational and maximizing fashion (Binford 1978:19, 44, 453-455). In other words, carcasses are used primarily as a source of food (as opposed to other potential uses), and in such a manner so as to obtain the maximum amount of either bulk or nutritious food. In practice, the models are implemented by juxtaposing frequencies of archaeological skeletal specimens with a measure, or series of measures, of element utility in an attempt to explain the relative abundance of skeletal material.

Binford's provocative ideas spawned a generation of research examining skeletal element representation from the perspective of nutritional utility. Although both his methods and theory have been criticized (Lyman 1984, 1985; Jones and Metcalfe 1988; Metcalfe and Jones 1988; Grayson 1989; Brink 1997), his strategy has still been widely employed. Many have applied Binford's own models of economic anatomy, based on caribou and sheep, to their faunal assemblages (Binford 1981; Thomas and Meyer 1983; Marshall 1988; Speth 1983; Landals 1990). Others have formulated new models based on the anatomy of other animals and typically use these data in the analysis of faunal samples (Will 1985; Blumenschine and Caro 1986; Brink and Dawe 1989; Borrero 1990; Emerson 1990; Lyman 1992; Lyman *et al.*, 1992; Savelle and Friesen 1996; Savelle *et al.*, 1996; Brink 1997; Diab 1998; Outram and Rowley-Conwy 1998). In the majority of these studies a method of analyzing the relative abundance of skeletal material has been developed that mimics Binford's original model. However, there are a series of issues about the application of this model that need to be questioned.

Here data from a previously excavated bison kill site are used to address three issues relevant to the use of anatomical indices in the analysis of skeletal representation. First, I question the nearly universal use of total food utility indices to interpret skeletal abundance. Instead, I argue that selective consideration of specific food indices may be more powerful explanatory tools when examining faunal patterns for very large game animals such as bison. Second, I argue that it is often inappropriate to employ all skeletal elements in the application of a specific utility index. The use of specific subsets of the skeleton compared with specific nutritional measures may be more instructive. Third, I suggest that, for very large game animals, foods directly linked with the skeletal frame (such as muscle attached to individual bones) are more useful in charting the fate of individual elements than will those resources that are only weekly associated with specific bones (like internal fats lacking skeletal attachments).

These three arguments will assist in the interpretation of a bison skeletal assemblage from the High Plains of western North America. The Wardell bison kill and processing site is chosen for analysis because it contains two separate but related perspectives on the treatment of skeletal elements by Plains hunters: a kill area and a nearby processing site. Viewing the transformation of a kill assemblage to a processing assemblage affords a rare opportunity to document the processes by which Aboriginal butchers made decisions about bison skeletal elements. The paper concludes with a consideration of the concept of maximization and how deviations from this idealized model may be equally informative.

ON THE APPLICATION OF UTILITY INDICES

My first concern regarding trends in the use of indices of economic anatomy is that skeletal frequencies are typically compared with some measure of total available food (Binford 1981; Speth 1983; Emerson 1990; Landals 1990; Le Blanc 1994; Morlan 1994; Koch 1995; Hjermstad 1996). A few authors have compared only selected parts of total nutrition with skeletal frequency (Hjermstad 1996; Morrison 1997). Ideally, the formulation of a general or total utility index accounts for all available nutrition including all muscle and fat tissue. In applying indices based on total available food to archaeological samples the logic appears to be that people will select and reject carcass portions according to the complete package of all food associated with skeletal elements. Indices based on total food value may be appropriate for the analysis of differential skeletal treatment of smaller game; animals that can be transported nearly intact by one or a few people, such as caribou and sheep as used in Binford's original indices. Larger game, like bison, require more complex decision making processes associated with butchering, transport and processing carcass portions (Bunn *et al.*, 1988; O'Connell *et al.*, 1989, 1990; Bunn 1993).

This is dramatically apparent for bison. As the largest land mammal on the continent, there are almost no small, light-weight parts to a bison. Even the lower legs weigh many kilograms and would not likely be "schlepped" along with other parts of the carcass simply because it requires more effort to remove them than it does to haul them (Perkins and Daly 1968). The tail was often left with the hide and the hyoid with the tongue (Frison 1970; Wheat 1972) but otherwise schlepping was irrelevant for bison. The great size and weight of bison would undoubtedly have resulted in careful consideration of the specific value of each carcass portion when deciding on skeletal retention and abandonment. This likely consisted of an evaluation of one or a few food sources known to be strongly linked with certain skeletal elements, such as marrow and grease, rather than consideration of the total food value associated with any particular carcass portion. I support Jochim (1976) and Speth (1983, 1990; Speth and Spielmann 1983) in arguing that fat was the dominant resource of interest to hunters and gatherers, as fat is the prime source of energy in game carcasses.

Below I discuss the Wardell bison kill site in some detail. For the moment I draw upon data from the kill site to illustrate the point about comparing total food products to skeletal frequencies. Figure 1 plots the frequency of major skeletal elements recovered from the Wardell kill site against the average total food value (AVGTP) of each element and element portion as determined by Emerson (1990) in her detailed formulation of utility indices for bison. Emerson's total food value considers muscle tissue weight and all major fat stores including those associated with bone and muscle. The two layers of the Wardell kill have been combined in this analysis (see discussion below). Bone frequency as used here is deduced as best as is possible from Frison (1973), converting frequency counts of diagnostic element portions into MAU (minimal animal units). Minimal animal units are the count of minimum elements divided by the frequency of that element in the skeleton (see Todd 1987). For the six major leg bones (femur, tibia, metatarsal, humerus, radius, metacarpal) I have used the highest count of either the proximal or distal end. This is based on the premise that the whole bone, rather than a portion, was of interest and hence the highest count best reflects the presence of the element. Frison's presentation of data for the pelvis does not lend itself to calculation of an MAU value, but since Emerson (1990) combines the sacrum and pelvis I was able to use Frison's unambiguous counts for the sacrum. From Figure 1 it is apparent that there is little relationship between total food products and differential skeletal abundance at the Wardell kill (r=0.05, P>.05). I argue that limiting this analysis to select portions of total nutrition may be more useful in explaining skeletal abundance.

My second concern with the use of utility indices is a corollary to the issue addressed above; that most utility index applications compare bones from the entire animal to some measure of nutrition (Speth 1983; Thomas and Meyer 1983; Will 1985; Borrero 1990; Landals 1990; Morlan 1994; Savelle *et al.*, 1996; Savelle and Friesen 1996; Diab 1998; Koch 1995). A few authors, however, have presented interesting variations using only select groups of bones (Emerson 1990; Hjermstad 1996; Morrison 1997). Typically, all major skeletal elements are plotted against a measure of nutrition. This approach ignores the fact that bones from different portions of the carcass tend to be associated with certain food resources and not with others. Upper leg bones are typically heavy with muscle tissue, but this meat is relatively lean. In contrast, parts of the backbone and rib cage are associated with moderate amounts of meat but this tissue is usually rich in fat. Subcutaneous fat is draped over the backbone, pelvis and part of the ribs, but is virtually absent in legs of large game. Conversely, the vast majority of marrow and grease are associated with the legs and are scarce in the axial skeleton. Hard fats of the thoracic and pelvic cavities are not directly associated with any skeletal elements. This variability makes it unlikely that *all* skeletal elements will be sensitive to any single index of nutrition, regardless of whether it is based on total food (as in Figure 1), or on a limited subset of available food from the carcass (such as just muscle tissue).

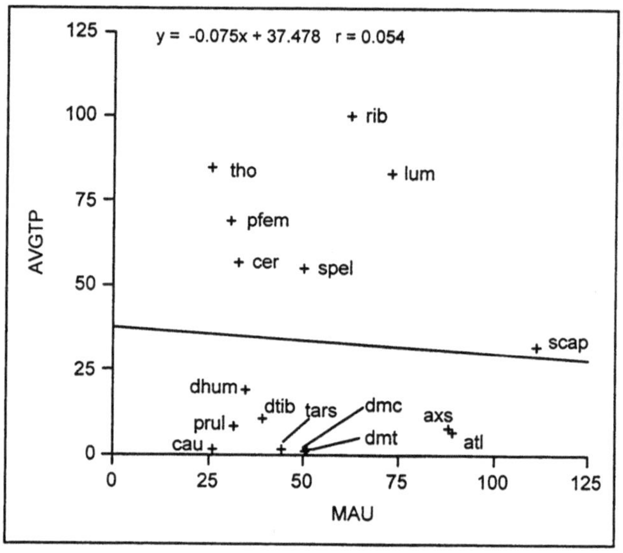

Figure 1. Relationship between total food products (AVGTP) for bison skeletal elements and bone frequency at the Wardell kill site (with top and bottom levels of the kill combined). Wardell data (Frison 1973); bison data (Emerson 1990). Abbreviations: atl=atlas; axs=axis; cer=cervical; tho=thoracic; lum=lumbar; pel=sacrum+pelvis; cau=caudal; scap=scapula; p=proximal; d=distal; hum=humerus; rad=radius+ulna; mc=metacarpal; fem=femur; tib=tibia; tars=tarsals; mt=metatarsal.

Even restricting analysis to logical subsets of the skeleton still requires comparison with relevant measures of nutrition. For example, Figure 2 plots only the frequency of proximal and distal ends of leg bones from the Wardell bison kill against Emerson's (1990) averaged total products. Given the above argument, this subset of the skeleton might be expected to correlate more strongly with a utility index. Such is not the case. The total available food model used in Figure 2 incorporates all of the available nutrition for each element, including muscle, all relevant carcass fats, marrow and grease. The low correlation coefficient in this relationship (r=0.484, P>0.05) indicates that the total food value of the articular ends accounts poorly for the relative abundance of these elements at the site.

A prime reason for these relatively weak correlations is the position of both portions of the femur in the relationship (Figure 2). The proximal and distal femur are extreme outliers, having vastly higher total products values than all other leg bone portions. This is due to the great weight of muscle tissue associated with the femur and not with other leg bones in bison. The femur accounts for two-thirds of the muscle in the legs (Emerson 1990). The high value of a particular element for one specific food resource highlights the complexities of combining numerous food resources into a single index and using this to evaluate skeletal abundance - be this all bones or only logical subsets of the skeleton. At times, for certain size animals and for certain subsets of the skeleton, it may be appropriate to employ models based on total food value. At other times applications of economic anatomy need to be limited to select elements and to restricted measures of food.

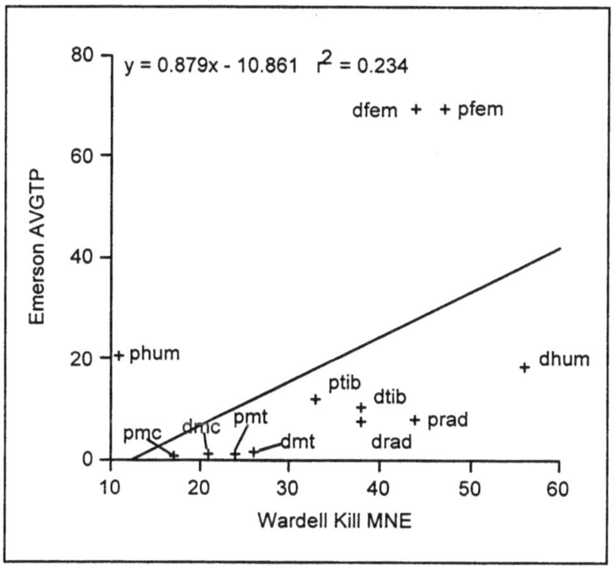

Figure 2. Relationship between frequency of leg bone portions at the Wardell kill site and Emerson's (1990) measure of average total products (AVGTP) for all food value associated with each bone.

My final contention about trends in the use of utility indices concerns how food sources are "linked" with individual skeletal elements. All utility indices are formulated by associating individual skeletal elements with the various edible products surrounding them. Yet exactly how these carcass products "associate" with specific bones is highly variable, ranging from foods contained within the bones themselves to foods only remotely related to any skeletal element. Many foods linked with specific bones in large game may simply be removed during butchering with little attention directed to the bone itself. For example, subcutaneous fat covers the hump and back of healthy large ungulates in temperate and northern regions (Anderson *et al.* 1972; Torbit *et al.* 1985; Adamczewski *et al.* 1987a), and is a soft, delicate fat that was highly valued (Binford 1978). In the standard formation of anatomical indices some portion of subcutaneous fat weight would be assigned to the thoracic, lumbar and sacral vertebrae, and to the pelvis and the ribs. In reality this fat could be cut away after removal of the hide without retaining or modifying any of the bones. The same is true of the hard fats of the body cavity. Fats of the kidney, the lining of the stomach and attached to other organs (Dauphiné 1975; Finger *et al.* 1981; Johns *et al.*, 1984) have little association with any skeletal element and yet must be linked with specific bones in the formation of utility indices. An opposite situation occurs with the recovery of intramuscular fat. This fat is found within muscle tissue, it cannot be cut away, and requires retention of the associated muscle if the fat is to be recovered (Link *et al.* 1970; Berg and Butterfield 1976; Leat and Cox 1980;). Bone marrow and especially grease represent extreme examples of an absolutely direct link between skeletal material and the desired food (Brink 1997).

In bison-sized animals the sheer weight of bones is such that at least part of the decision regarding retention versus abandonment of each element must be based on the perceived value of the bone itself, rather than on the total value of all products associated with each element. While this may be less a concern with smaller game, in very large prey the association between bone units and rankings based on nutritional indices must be rigorously considered in order to assess the degree to which bone portions are actually linked with a desired food product. In particular, bones must be looked at from the perspective of what foods they may actually contain in the structure of the bone itself.

Large, heavy bones that have little nutrition contained within their structure may not be transported from a kill just because they are surrounded by a substantial weight of muscle or fat (Bunn *et al.*, 1988; Monahan 1998; O'Connell *et al.*, 1989, 1990; O'Connell and Marshall 1989). Muscle and fat tissue can be stripped away to save the energetic drain of schlepping bones that are of little further use. In contrast, bones of any size and weight that have a substantial amount of food contained within their structure may be expected to be retained from a kill regardless of the amount of other foods that surround them. Marshall and Pilgrim (1991) and Brink (1997) have argued that charting the fate of individual bones from very large game may be more profitably viewed through consideration of foods directly contained within the bones themselves rather than through a total package of food, some of which have little association with each element.

In the following analysis I examine leg bones from the Wardell kill and camp from the perspective of their direct association with the two fats contained within the bone structure; marrow and grease. I argue that correlation between subsets of the skeleton and specific foods known to be strongly associated with individual elements provides an important new understanding of the butchering and processing events at the Wardell site.

The Wardell Bison Bone Assemblage

The Wardell bison kill and butchering site is located in the upper reaches of the Green River Basin in western Wyoming (Figure 3). Wardell is a Late Prehistoric communal kill where a wooden corral was used to kill several hundred bison over a period of about 500 years (Frison 1973). Wardell is one of the rare sites where both a kill and processing (camp) component of a single site have been excavated and where bone counts are reported. The presence of two related samples permits scrutiny of the continuum of human decision making from the point of initial kill to the subsequent rendering of final products from the bison carcasses. Knowing that one assemblage (the processing site) must be the consequence of an intentional selection from the universe of another assemblage (kill site), permits

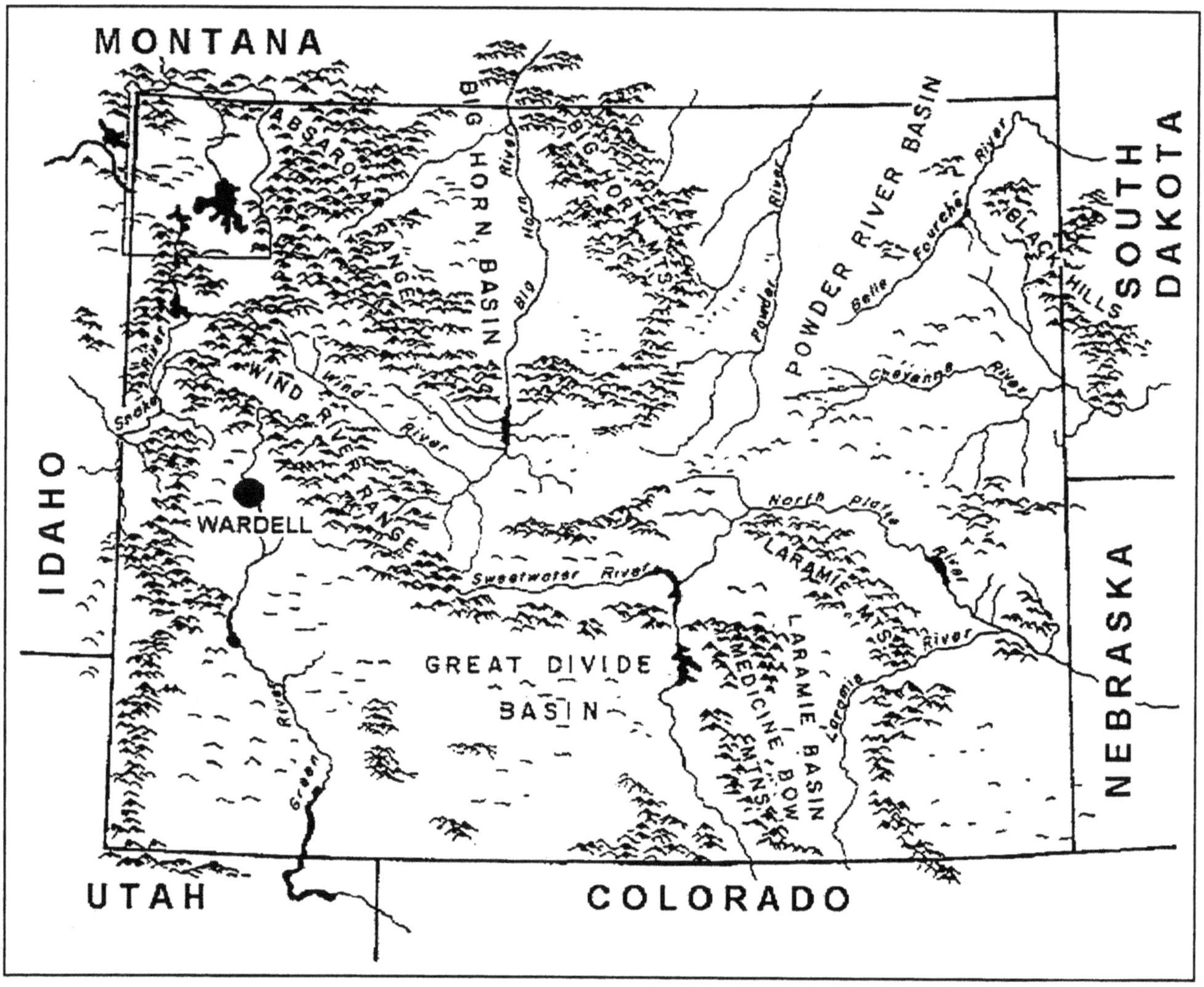

Figure 3. Location of the Wardell site in west central Wyoming.

assessment of the criteria that led to formation of both bison bone assemblages. Determining whether or not rational decisions were made with bison carcass portions at the Wardell site is facilitated by the ability to view the skeletal sample from two related perspectives rather than the more common single perspective. Here I will examine only leg bones from the Wardell site. While all bones are certain to be instructive, leg bones fit with arguments made above because they are a specific subset of the skeleton and have directly associated food resources, especially marrow and bone grease.

The Wardell site was excavated and reported on by Frison (1973) who provides faunal counts for both the kill and processing components (Table 1). The kill site assemblage was recovered and reported on in two distinct stratigraphic layers. I have combined these two layers to form a larger sample size. While this risks masking potential differences between the two kill events, Figure 4 suggests that, for my interests, the archaeological assemblages are similar enough (r=690, P<0.05) to warrant combining the two layers. I recognize that counts used in some of the following correlations are lower than desired for statistical significance, but I employ them to illustrate general trends.

	MNE Kill	MNE Camp
Complete humerus	14	0
Proximal humerus	11	0
Distal humerus	56	14
Complete radius	20	0
Proximal radius	44	4
Distal radius	38	11
Complete metacarpal	80	0
Proximal metacarpal	17	30
Distal metacarpal	21	27
Complete femur	14	0
Proximal femur	47	1
Distal femur	44	2
Complete tibia	41	0
Proximal tibia	33	1
Distal tibia	38	14
Complete metatarsal	76	0
Proximal metatarsal	24	18
Distal metatarsal	26	20

Table 1. Frequency of bison bone elements from the kill and processing portions of the Wardell site. Note: Data from Frison (1973), with "Top" and "Bottom" layers of the kill combined.

Skeletal element counts provided by Frison (1973) are presumed to be equivalent to what is now commonly called the minimum number of elements, or MNE (see Todd 1987). Use of MNE is considered an acceptable measure of abundance in this study because all bones under consideration occur with the same frequency in the living animal (two of each element and each proximal and distal portion). Overall, the number of bones recovered from the camp area is low relative to the kill area. Frison (1973) attributes this to a number of possibilities, including relatively poor preservation in the camp area, not having excavated in areas where certain processing activities may have taken place, and loss of bone through crushing and pounding. Numbers at both components of the site are, however, sufficient to illustrate trends in the data.

Since the camp assemblage originated from the kill there should be a patterned relationship between the two. For leg bones, the Wardell kill consists of both complete elements and the proximal and distal portions of these bones. The camp area lacks complete bones and consists only of proximal and distal leg elements.

Figure 5 compares the complete leg bones from the kill versus those that were recovered in broken condition. A strong, inverse and significant relationship (r=0.894, $P<0.05$) is apparent. Bones most commonly recovered in complete condition are those least often broken, and vice versa. A proximal to distal trend is apparent, with the metacarpal and metatarsal most often recovered complete, the femur and humerus least often, and the tibia and radius are intermediate. I believe that these results can be interpreted in light of food resources associated with the leg bones, as discussed below.

Since only broken bones were recovered from the camp no similar comparison can be made for this area of the Wardell site, but it is possible to compare the broken bones of the camp with those recovered from the kill (Figure 6). In general there is an inverse relationship between the camp and kill assemblages, with bones commonly recovered in broken condition at the camp tending to be less often broken at the kill and vice versa. The relationship, however, is not significant (r=0.394, $P>0.05$) and is thus of little predictive value. At first glance this relationship seems unusual in that it is presumed that the camp assemblage must have derived from the kill site, thus a stronger correlation between the two assemblages should be apparent.

Like mirror images, subtraction of one from the other should have predictable and corresponding faunal patterns at each location. Of course, this may have once been the case and the action of subsequent cultural or natural taphonomic processes could have obliterated the expected reflective pattern. This analysis, however, has not yet accounted for all leg bones from the Wardell site.

The lack of a strong relationship between the broken bones from the kill and those from the camp ignores the complete bones from the kill. Complete bones can also be viewed as additional specimens of proximal and distal ends of bones at the kill - ones that for whatever reason were never broken.

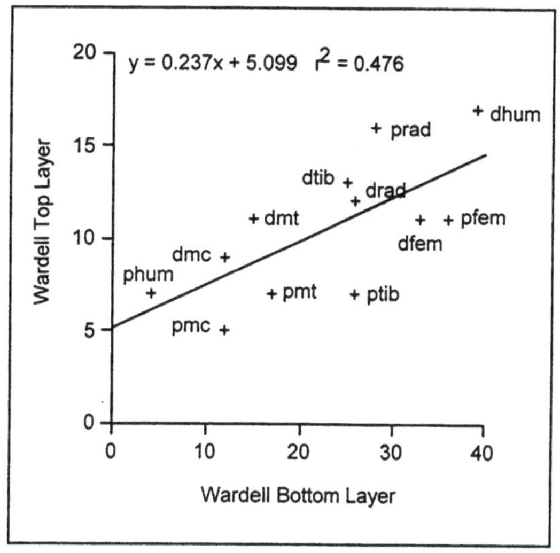

Figure 4. Scatterplot of twelve portions of bison leg bones from the top and bottom layers of the Wardell kill site. Data from Frison (1973).

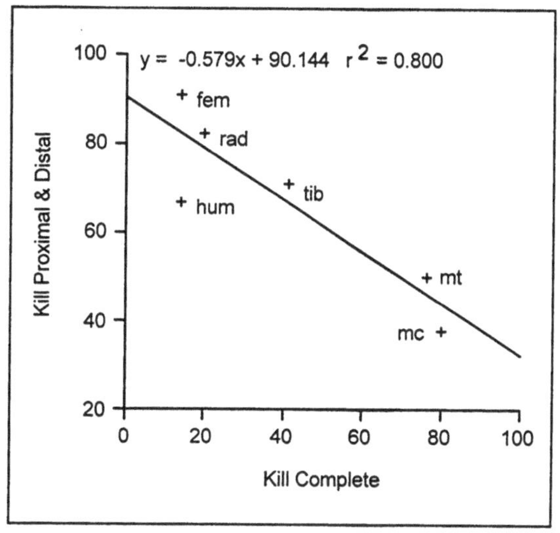

Figure 5. Relationship between complete and broken leg bones from the Wardell kill site.

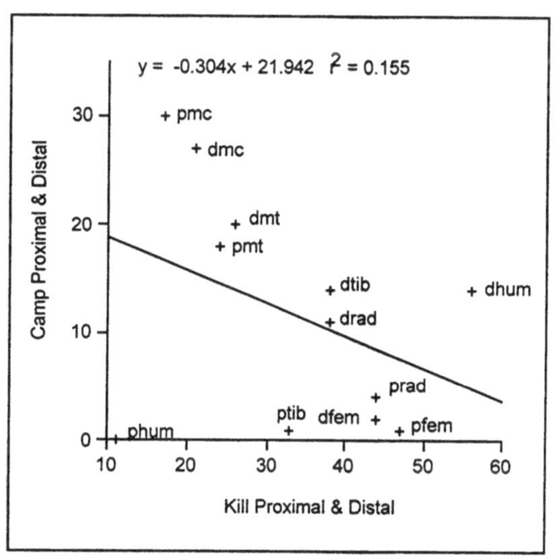

Figure 6. Relationship between proximal and distal limb bone elements recovered from the Wardell kill site versus the camp area.

Figure 7 compares the camp leg bone assemblage with the kill assemblage, with complete bones counted as additional specimens of proximal and distal ends. A moderately strong, significant (r=0.817, P<0.01) and positive relationship between the two samples is seen. There is a general proximal to distal trend in Figure 7, with metapodials the most common elements at each locality of the Wardell site, and the ends of the femora and humeri the least common (proximal humeri are especially rare at both the kill and camp).

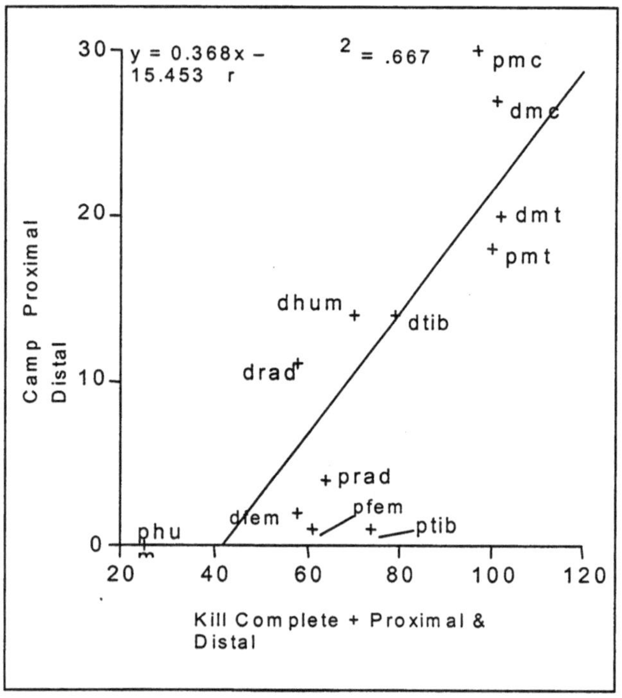

Figure 7. Relationship between proximal and distal limb bone elements recovered from the Wardell camp versus those from the kill with complete bones included in total count.

The positive trend to Figure 7, where bones common at one part of the Wardell site tend to also be common at the other part, is contrary to expectations. Subtraction of the camp assemblage from the kill site should produce an inverse relationship. I suggest, however, that the strong negative relationship between the complete versus the broken leg bones from the kill (Figure 5) and the positive relationship between the broken leg bones from the camp versus the total count of proximal and distal ends from the kill (Figure 7) are as expected if the hunters used the skeletal assemblages in a manner which maximized the return of specific food sources. Maximizing the recovery of fats associated with the bones themselves, marrow and bone grease, best explains the faunal patterns at the two localities of the Wardell site. Before pursuing this argument it is necessary to discuss how these two fats are quantified.

BONE MARROW AND BONE GREASE

Bone marrow and grease represent two forms of fat that, in addition to carcass fats, are important parts of the total fat composition of large mammals. The distinction between these two bone fats has not been documented in the animal science literature and is of interest primarily to archaeologists. Both refer to fatty acids contained within bones. Bone marrow generally refers to the fatty acids found inside the medullary cavities of the bones of most animals. Bone grease is the term archaeologists use to refer to the fats contained within the structure of the bone itself, in particular cancellous tissue. Chemically, the two fats are likely indistinguishable. Marrow is typically recovered as a cohesive mass after exposing the marrow cavity. Grease is mixed in with bone tissue and can only be adequately recovered by crushing the bones and boiling in water to render the grease from the bone chips (Leechman 1951; Vehik 1977; Binford 1978).

The dynamic nature of the fatty acids contained within the marrow cavity and cancellous tissue of bone is as complex and as variable as all body fats. While grease has been little studied, a wealth of literature documents the composition, abundance, variability and purpose of bone marrow (Ranson 1965; Ratcliffe 1980; Suttie 1983; Davis *et al.,* 1987; Marquez & Coblentz 1987; Cederlund *et al.,* 1989;). General reviews of grease and marrow are provided by Binford (1978), Emerson (1990) and myself (Brink 1997). Of interest is the fact that both sources of fat represent potentially important sources of nutrition. This is especially so in bison because of the large size of the bones of the animal. It is also true because the fats contained within marrow, and presumably grease, represent the richest source of fat available from the carcass. That is, the true chemical fat content of marrow (as opposed to non-fat components such as ash, protein, water) is substantially higher than it is in carcass fat tissue. In healthy adults the chemical fat in marrow is frequently over 90%, although this is seen to vary with the seasons (Anderson *et al.* 1972; Adamczewski *et al.* 1987a; Adamczewski *et al.* 1987b; Franzmann and Arneson 1976; Ballard and Whitman 1987; Davis *et al* 1987; Marquez and Coblentez 1987). In contrast, carcass fat tissue is composed, on average, of about 70% chemical fat in healthy adults, and considerably less in animals under stress (Berg and Butterfield 1976; Jones *et al.* 1981).

In addition, the specific fatty acids that comprise marrow and grease tend to be dominated by the more unsaturated fatty acids, especially oleic acid (Binford 1978:23-4). Ruminants are known for being generally high in saturated fatty acids and low in unsaturated fatty acids relative to other large mammals (Emery 1969:240). If Binford is correct in asserting that the soft, unsaturated fatty acids were more desired by hunters, then repositories containing predominantly these fatty acids may have been highly prized despite the fact that the absolute weight of fat is minor compared to that of other carcass fats. Certainly there is evidence in the historic literature that marrow was a favourite item of the carcass among recent bison hunters (Wheat 1972).

Finally, the two bone fats are the most stable of all carcass fats and will be present in animals that are otherwise fat depleted. This may be of profound significance in hunting and gathering cultures at times of great nutritional stress. Marrow and presumably grease are the last fats to be catabolized when the animal is stressed, such as at the end of a long winter in temperate and northern regions where winter feed is poor and cold temperatures cause depletion of other, more accessible, fat stores (Harris 1945; Dauphiné 1976; Franzmann 1985; Cederlund *et al.*, 1989). Accordingly, dependable bone fats may have been critically important to hunting cultures at times when game animals, and people themselves, were otherwise in poor condition (Speth 1990). At such times the value of bone fat (otherwise a relatively minor component of total carcass fat) increases dramatically in response to the decline of other carcass fats (Speth and Spielmann 1983) and may have been targeted for preferential selection. Thus, the relative importance of bone fats to aboriginal hunters should not be measured simply as a weight or percentage of total carcass fat in peak season, as this blurs differences that only are manifest upon close inspection of large herbivore bioenergetic principles (Hudson and White 1985). To my knowledge, only Emerson (1990) and I (Brink and Dawe 1989; Brink 1990, 1997) have quantified the marrow and grease content of bison bones. Emerson has done this for nearly all bones in the body, and I for just the leg bones where the majority of both marrow and grease are located. Insofar as possible these data are merged into a combined view of the potential nutritional content of individual elements. While this risks combining data from bison that may have differed in some ways and which were studied independently by different researchers, it has the advantage of producing more reliable inferences because it is based on larger sample size. In studies of carcass composition large sample size is the single most important variable. Single animals may vary for any number of unpredictable reasons, including sickness, genetic differences or unique environmental stress, but distinct biological trends will emerge given sufficient sample size. Given that Emerson and I both sampled small numbers of bison I have elected to combine our results into single measures of marrow and grease.

Bone Marrow

Emerson quantified the marrow weight from bones of four animals; two of each sex, and two killed each during the spring and fall seasons. I recorded marrow weight from five bison, three females and two males. Emerson recorded marrow weight for most elements of the carcass; however, only values for major leg bones are reported here. Emerson recorded separate marrow weights for the radius and ulna while I combined the two elements. For comparability Emerson's values for the ulna have been added to the radius. Table 2 provides mean values for weight and percentage of marrow weight for Emerson's bison, those I studied, and the two samples combined.

Comparing the two studies, Emerson consistently recorded greater weight of marrow than did I. This is likely due to the fact that the animals used by Emerson were, on average, older and thus had larger marrow cavities. In both studies the femur, tibia and humerus are the dominant marrow bones of the legs, with each bone contributing between 21% to 28% of total marrow. Emerson had the greatest percentage of marrow associated with the tibia followed by the femur and humerus. I identified the humerus as the top marrow producer, followed by the femur and tibia. Using Emerson's more complete data, the six major bones of the legs account for nearly 97% of all marrow in the appendicular skeleton. Because these six elements were studied by both Emerson and myself, the lower portion of Table 2 provides a calculation of the combined mean percentage of marrow weight. When the data for both studies are combined, the rank order of the top three bones is neither exactly as Emerson or I had determined. The tibia ranks first, the humerus second and the femur third. Both researchers found the metatarsal to be associated with a slightly greater percentage of total marrow than the metacarpal. The radius is intermediate between the high ranking elements and lower ranking metapodials. Although Emerson and I recorded noticeably different weights of marrow, Figure 8 shows how when expressed as a percentage for the six major bones our results are extremely similar (r=0.960, P<0.01).

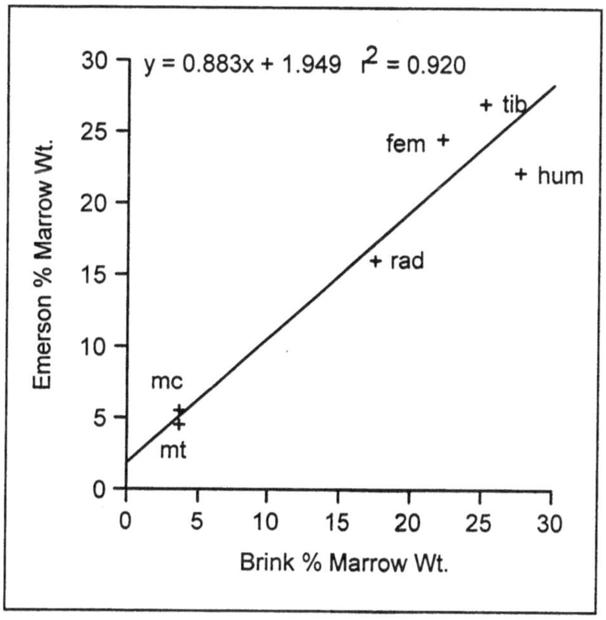

Figure 8. Comparison between percentage of marrow in six bison leg bones as recorded by Brink and Dawe (1989), Brink (1990), and by Emerson (1990).

Bone Grease

Bones may also have been preferentially selected or rejected by hunters as a result of their potential to yield what has commonly been termed bone grease (Leechman 1951; Vehik 1977; Binford 1978). Bones must be boiled in order to render the grease, and the process is vastly more successful if the bones are first broken into many small pieces (Leechman 1951; Vehik 1977; Brink 1997). Bison are believed to have quite greasy bones relative to other large terrestrial game. In historic times the large marrow bones of the legs were sometimes burned as a source of heat on cold nights (Maximilian 1905, II:196).

Emerson (1990:184-5) collected bone grease data from the same four bison as with the marrow analysis. This was done by smashing bones and boiling them. Fat was collected from the water surface, weighed and then oven dried. From the dried samples the chemical fat was extracted with a solvent, and is reported as weight of chemical fat for each skeletal portion. I recorded bone grease from three of the five animals that were studied for marrow. Grease was recorded in two ways: by boiling leg elements, and by crushing element portions and extracting the grease with chemical solvent. Details are given in Brink and Dawe (1989) and Brink (1990, 1997). Only the data obtained through boiling are reported here as these are directly comparable with Emerson's data. Note that I was unable to sample distal tibiae; the combined values in Table 3 are based solely on Emerson's results.

In Table 3 I again combine the bone grease values for Emerson's four bison into a mean weight and mean percentage of grease fat weight. This obscures differences between the animals of different sex and age. For example, Emerson (1990:378) found that the two females produced proportionally more bone grease than the two males, despite the size advantage of the bones of a four year old male. This being the case, bones of the female bison might well be preferred over those of males for bone grease rendering. However, my present concern is the ranking of skeletal elements in terms of their relative production of bone grease.

	EMERSON'S MARROW DATA			BRINK'S MARROW DATA		
	Mean Wt. (g)	SD	Mean %	Mean Wt. (g)	SD	Mean %
Humerus	124.98	35.77	22.25	111.62	22.58	27.70
Radius-ulna	89.65	11.86	15.96	70.28	17.9	17.44
Metacarpal	25.58	3.6	4.55	15.26	1.4	3.79
Femur	138.08	37.63	24.58	89.32	13.39	22.17
Tibia	152.15	24.03	27.09	101.42	12.84	25.17
Metatarsal	31.25	5.66	5.56	15.00	3.65	3.72
			100.00			100.00

COMBINED EMERSON & BRINK MARROW DATA			
	Mean Wt. (g)	SD	Mean %
Humerus	117.56	28.00	24.83
Radius-ulna	78.89	17.81	16.66
Metacarpal	19.84	5.95	4.19
Femur	110.99	35.79	23.44
Tibia	123.97	31.84	26.18
Metatarsal	22.22	9.59	4.69
			100.00

Table 2. Mean weight and mean percentage of bone marrow weight for bone elements from bison.
Note: Data from Brink and Dawe (1989), Brink (1990), Emerson (1990:338).

	BRINK'S GREASE DATA			EMERSON'S GREASE DATA		
	Mean Wt. (g)	SD	Mean %	Mean Wt. (g)	SD	Mean %
Proximal humerus	187.33	56.4	20.2	64.63	33.27	17.15
Distal humerus	82.11	0	8.854	24.75	11.25	6.567
Proximal radius	126.28	17.7	13.62	18.88	10.77	5.009
Distal radius	42.45	33.7	4.578	37.1	16.29	9.843
Proximal metacarpal	15	5.52	1.618	3.4	1.85	0.902
Distal metacarpal	12.2	0.71	1.316	16.43	4.9	4.359
Proximal femur	101.33	44.9	10.93	45.33	21.74	12.03
Distal femur	193.17	43.6	20.83	86.45	44.68	22.94
Proximal tibia	136.33	69.8	14.7	39.43	26.81	10.46
Distal tibia	—	—	—	16.98	3.91	4.505
Proximal metatarsal	17.3	0	1.866	3.08	1.77	0.817
Distal metatarsal	13.85	2.33	1.494	20.45	4.96	5.426
	927.35		100.0	376.91		100.0

COMBINED EMERSON & BRINK GREASE DATA		
	Mean Wt. (g)	Mean %
Proximal humerus	125.98	19.07
Distal humerus	53.43	8.09
Proximal radius	72.58	10.99
Distal radius	39.78	6.02
Proximal metacarpal	9.20	1.39
Distal metacarpal	14.31	2.17
Proximal femur	73.33	11.10
Distal femur	139.81	21.16
Proximal tibia	87.88	13.30
Distal tibia	16.98	2.57
Proximal metatarsal	10.19	1.54
Distal metatarsal	17.15	2.60
		100.0

Table 3. Bone grease studies of bison leg bones. Note: Data from Brink (1990, 1997) and Emerson (1990:390).

It should be possible to establish predictions for sequential butchering based on consistent patterns of relative abundance of nutritional materials. The bottom portion of Table 3 provides a combined view of the percentage of grease associated with each limb bone portion. The greater weight of grease for the bison that I studied probably results from the fact that these bones were thoroughly smashed with the aid of a hydraulic press, while Emerson's were only moderately smashed with a hammer. Emerson extracted and weighed only true chemical fat from the boiled residue, while my results are for the total weight of the rendered fat (which includes non-fat materials).

In the case of marrow recovery, the logical unit of analysis is the entire bone or paired broken proximal and distal ends. Single bones must be snapped in half to recover the marrow, and neither articular end is more diagnostic of this activity than the other. Grease rendering depends on intense smashing of bones to facilitate recovery of fats from within bone tissue. Evidence for this activity will be negative. Bones missing from the assemblage may have been processed for grease (or are missing for other reasons), while bones present have certainly not been subjected to grease rendering. Practically, at most sites the frequency of proximal and distal ends of leg bones will be the unit of analysis for arguments relating to grease rendering.

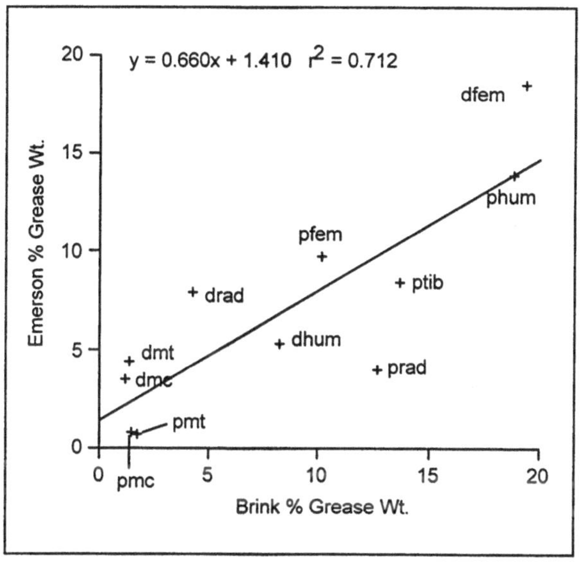

Figure 9. Comparison between percentage of bone grease in twelve major leg bone portions as recorded by Brink and Dawe (1989), and by Brink (1990, 1997), and Emerson (1990).

Grease rendering differs greatly from marrow recovery in that the former is messy and imprecise while the latter is relatively exact. Marrow plugs can be completely removed from the shafts and the quantity of marrow calculated. In grease rendering, bones were crushed with great force and, if warm, grease would fly over the work area. Grease recovery from water is incomplete, as fat adheres to walls of the pots or hides and to the utensils. Emerson (1990:186-7) mentions bones still being greasy after many hours of boiling followed by weeks of soaking in de-greasing agents. Given the imprecision of the endeavor, I am encouraged by the general compatibility of my results with those of Emerson (r=0.844, P<0.01, Figure 9).

The archaeological signature of bone marrow recovery will be the presence of separated proximal and distal ends of leg bones, as mid-shaft breakage of long bones would be the most practical method of marrow recovery. It is recognized, however, that subsequent taphonomic processes, including human modification, could alter the configuration of the assemblage of articular ends. Complete bones are evidence of the decision to ignore bone marrow. While complete bones may initially seem the preferable form of evidence (of negative marrow recovery) because subsequent taphonomic destruction of that particular bone can be ruled out, this does not eliminate the possibility that other complete bones may be missing from the assemblage due to other taphonomic processes. Thus, counts of both separated articular ends and complete leg bones may be skewed from the original faunal population for different reasons. This is unavoidable, and below I examine both sets of data from the Wardell site from this perspective.

BONE MARROW AND BONE FREQUENCY AT THE WARDELL SITE

Figure 10 compares the count of complete leg bones from the Wardell kill with the mean percentage of marrow as deduced from the combined data of Emerson and Brink (Table 2). The relation is strong, significant and negative (r=0.857, P<0.05). Bones associated with the greatest percentage of marrow are least often found complete and those with the least amount of marrow are most often complete. This is as would be expected if the aboriginal hunters were exploiting primarily those bones that offered the greatest return of nutritious (high fat content) marrow. Exploitation was not complete, as the marrow content of a large number of bones was ignored including a surprising number of tibiae. Why a significant number of prime marrow-producing tibiae were rejected at the Wardell kill is not known.

Figure 10 illustrates two clusters: one composed of two metapodials, the other of four larger leg bones. This clustering reflects the great difference in the count of the two groups of complete bones at the Wardell kill, especially the high numbers of metapodials. Clustering also reflects the considerable disparity between marrow content of the six bones. Although data do not conform entirely to a model based on maximization of marrow, comparison of complete elements and percentage of marrow are in keeping with a rational exploitation of marrow at the kill.

Alternatively, Figure 11 compares the count of proximal and distal ends from the Wardell kill with the percentage of marrow. Proximal and distal ends in Table 1 are combined to yield a total MNE for each articular end from the kill. Relationships in all following graphs are identical regardless whether MNE or MAU [minimal animal unit, see Binford (1978) and Todd (1987)] is used because all leg bones occur with the same frequency in the carcass. The relationship in Figure 11 is moderately strong and positive (r=0.775,

$P>0.05$) although not significant with this small sample size. This pattern is again in keeping with the prediction that bones associated with the greatest amount of marrow will be preferentially broken, while those low in marrow will more often be ignored. Two clusters are again apparent, the metapodials vs. the four larger elements, indicating a wide disparity in counts of the two groups. The results of both of the above analyses are generally, though not completely, consistent with expectations based on a strategy of maximum marrow fat recovery. Bones are broken, or are abandoned in complete condition, in approximate agreement with their relative value as sources of bone marrow.

Turning to the Wardell camp and processing site, it will be recalled that there are no complete leg bones in this assemblage. Figure 12 plots the count of leg bone articular ends against the percentage of marrow. The relationship is strong (r=0.890, $P<0.05$) negative and significant.

Bones associated with the greatest amount of marrow are less well represented at the camp, while the most abundant bones have the least marrow. This relationship parallels that of the complete bones from the kill (Figure 10), but is the opposite of the plot of the proximal and distal leg bone portions from the kill (Figure 11). Thus, the frequency of leg bone portions at the Wardell camp appears to be contrary to the expectations of a marrow maximizing model which predicts a preponderance of bones linked with greater marrow content.

It is logical to expect that bones common at the kill area were seldom transported to the camp (for example, 91 broken femora from the kill vs. 3 from the camp), while bones relatively rare in the kill would have been more often carried to the camp (38 metacarpals in the kill vs. 57 in the camp). Further, it is unreasonable to expect that the process of bison butchering, as reflected in skeletal element representation, will be the same at both components of the Wardell site. That is, if bones were broken for marrow at the kill then it is unlikely that this activity would continue apace at the camp portion of the site. Bones may have been moved from the kill to the camp for the purpose of continuing to extract marrow, but, given that bones were likely moved for other reasons as well, it is unreasonable to expect the transported assemblage to fully resemble patterns observed at the kill site.

Presumably, the camp assemblage formed through subtraction of bones from the kill for a number of reasons in addition to marrow recovery. These may have included the desire to convert certain bones into tools, or to use them for ceremonial or spiritual purposes (Frison 1991). At a massive kill site I suspect that non-food uses of bones would make a minimal impact on the overall structure of the assemblage. Bones were likely transported to the camp for reasons that include the presence of additional food sources directly associated with element portions, including bone grease. The meaning of Figure 12, showing the opposite tendency observed in Figure 11, and suggesting a non-maximizing marrow recovery strategy, is addressed below.

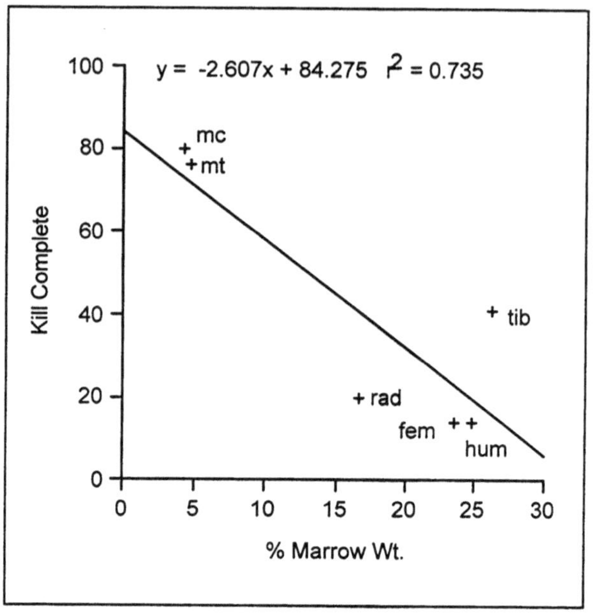

Figure 10. Relationship between frequency of complete bison limb bones at the Wardell kill and percentage of marrow using combined Emerson and Brink data.

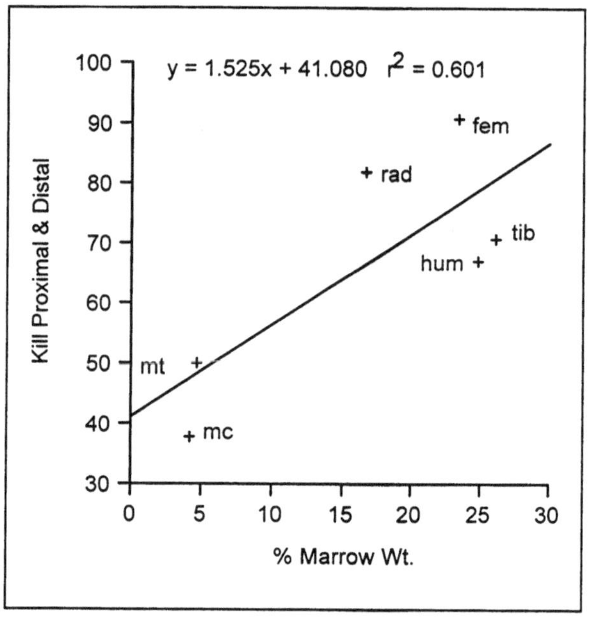

Figure 11. Relationship between frequency of proximal and distal bison limb bones at the Wardell kill and percentage of marrow using combined Emerson and Brink data.

BONE GREASE AND BONE FREQUENCY AT THE WARDELL SITE

The archaeological signature of bone grease rendering differs dramatically from that of marrow extraction in that recovery of bone grease requires the virtual destruction of the element portion. Thus, selection of a bone for grease rendering will be marked by the absence of this element in the archaeological record. Common elements at a site are those rejected for bone grease. This has important consequences for interpretation of the archaeological record.

Archaeologists must begin to think of absence as an indication of preference, and to further realize that the extraordinarily destructive signature of grease rendering will "over-print" the hallmarks of other bone processing activities. For example, if certain bones were first stripped of muscle and fat tissue, then broken for marrow recovery and finally thoroughly smashed for grease rendering, only the latter activity will remain in the archaeological record. Evidence of initial meat stripping and marrow recovery will have been obliterated or severely altered. I believe that this continuum of subtraction of food resources from skeletal elements is at least partially responsible for the structure of the faunal assemblage at the Wardell processing site.

The units of analysis for grease rendering are the proximal and distal ends of the six major leg bones, not whole bones. Some bison leg bone articular ends are characterized by widely different grease content at opposite ends of the same bone (Table 3). The proximal and distal ends of the humerus, femur and tibia exhibit this trait, while the metapodials show little variability in the articular ends. Furthermore, shafts of leg bones have extremely low amounts of bone grease (Brink 1997). Accordingly, it is unreasonable to expect entire leg bones to be pounded for grease rendering. Selection was likely made of those articular ends that offered the greatest return of grease.

Figure 13 compares the proximal and distal elements from the Wardell camp against the percentage of bone grease as provided by combining my data with Emerson's (Table 3). The relationship is relatively strong, significant (r=0.852, P<0.01) and negative, suggesting the use of a grease maximizing strategy at the Wardell camp. Bones high in grease content, such as the proximal humerus, distal femur and proximal tibia, are very rare at the camp. Conversely, low grease-bearing bones, such as all the metapodials and the distal tibia, are relatively common. This interpretation must be tempered by the knowledge that other processes may mimic the pattern seen in Figure 13.

Grease-rich bones may also be singled out for preferential consumption or transport by carnivores or other scavengers (Marean *et al.* 1992; Brink 1997). Grease-rich bones are characterized by lower density (Emerson 1990; Brink 1997), and lower density bones are more susceptible to natural decay (Lyman 1984). Thus, a death assemblage at a bison kill site left relatively untouched by human hunters, may through time come to resemble an assemblage that has been intentionally culled of the most useful grease-bearing bones. I cannot rule out any of these possibilities, although the presence of a large number of bone processing features at the Wardell camp site (Frison 1973) suggests that some reduction of bone took place on site. And while both humans and scavengers may be responsible for the final configuration of the Wardell fauna, it is noteworthy that Lupo (1995) argues that bones processed by humans through cooking and heating are subsequently less attractive to carnivores.

A dichotomy is apparent. The frequency of bison leg bone portions from the Wardell camp site exhibits a relationship with the percentage of marrow weight that seems opposite of a fat maximizing strategy (Figure 12), but these same bones exhibit a negative relationship with the percentage of grease weight that is in agreement with a maximizing model (Figure 13). Given that the camp assemblage must have derived from the kill, what were the decision-making processes that led to the selection of fauna from the kill area for eventual transport and processing, and what were the criteria behind those decisions?

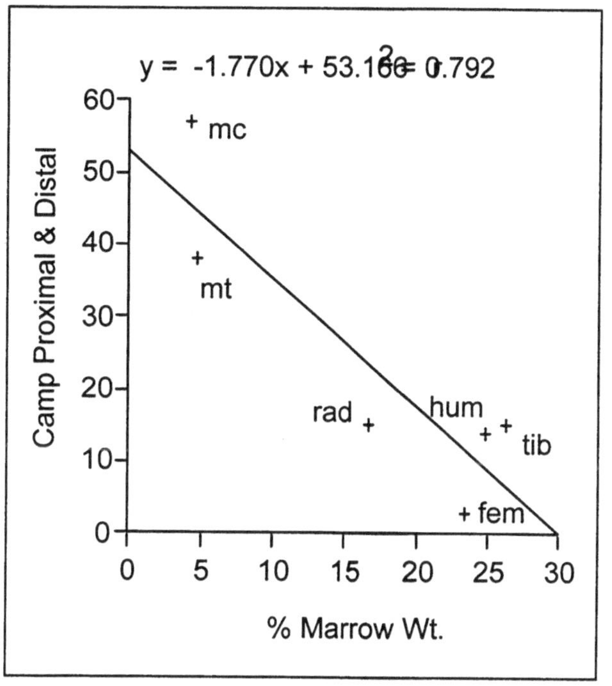

Figure 12. Relationship between frequency of proximal and distal leg bones from the Wardell camp site and percentage of marrow.

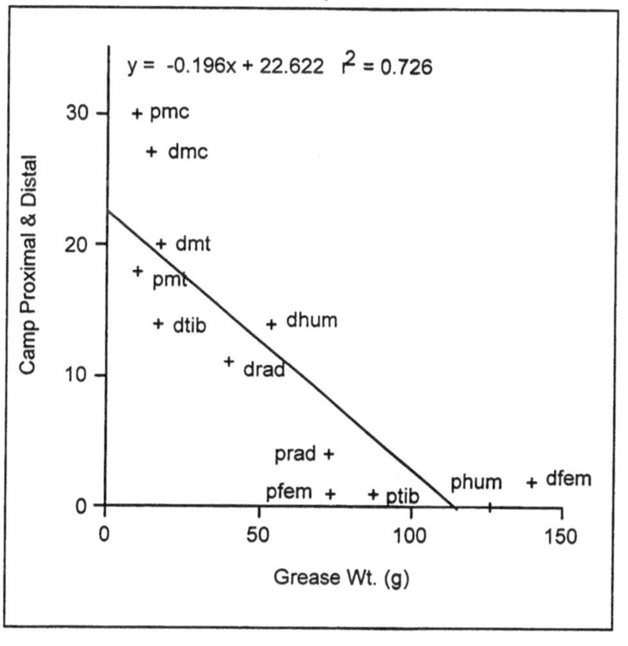

Figure 13. Relationship between frequency of proximal and distal leg bone portions at the Wardell camp site and percentage of bone grease.

DISCUSSION

A reasonable interpretation of the data presented so far is as follows: many (though not all) bison leg bones were broken for marrow at the kill site. This was done in a manner that capitalized on the bones that yielded the greatest amount of marrow. The effects of this are seen in two separate analyses: through the preferential rejection of complete bones that have the least amount of marrow (Figure 10), and through preferential breakage of bones with the greatest amount of marrow (Figure 11). The fact that broken leg bone ends from the camp area do not correspond with expectations of a marrow index (Figure 12), but are in agreement with a grease maximizing strategy (Figure 13), suggests that the bones brought to the camp were selected from among the broken bones at the kill. That is, once bones were broken for marrow at the kill the separated articular ends were selectively culled from those element portions offering the greatest return of grease. These were transported to the camp. Once grease rendering took place in the camp, the final signature of the leg bone assemblage is one of a negative relationship with grease content owing to the preferential destruction (hence removal from the archaeological record) of high grease-bearing portions.

Although this scenario is in agreement with data presented so far, I don't believe this is how the Wardell camp assemblage formed. First, if limb bones were first broken for marrow and then culled for differential grease content, the kill site assemblage should be characterized by highly skewed counts of opposing articular ends, especially in cases where an individual bone is characterized by highly divergent grease content. This is not the case (Table 1). With one exception, the count of proximal and distal portions of individual bones at the Wardell kill site is very similar. For example, there are 44 proximal and 38 distal radii, there are 24 proximal metatarsals and 26 distal, there are 47 proximal femora and 44 distal, and so forth. The counts for the tibia are particularly instructive. The proximal end of this bone is the third ranked limb element for grease production, while the distal end is ranked eighth. Had the Wardell kill been culled for productive grease elements after marrow recovery then there should be an abundance of distal tibiae and a shortage of proximals. Yet the proximal end is represented by 33 specimens and the distal end by 38 bones. Only the humerus shows a wide divergence between frequency of articular ends (11 proximal vs. 56 distal). This discrepancy, common to many bison kill sites, is discussed below.

Thus, while there is a considerable difference in the frequency of different leg bones in the kill assemblage (e.g., 91 femora vs. 38 metacarpals), with the exception of the humerus, each bone is represented by about the same number of proximal and distal ends. This suggests that, except for the proximal humerus, there is no evidence for the preferential removal of separated articular ends of individual limb bones from the kill area. This contrasts with the counts of articular ends for individual elements from the camp area where numbers of proximal and distal ends are quite divergent. As expected, this is especially true of the bones with the greatest difference in the relative grease value of opposing ends (e.g., proximal humeri =0, distal =14; proximal tibiae =1, distal =14). Conversely, bones that have proximal and distal ends of nearly equal grease value are represented by fairly equal numbers at the camp site. This is true for both high and low value elements: both ends of the femur are high ranking grease elements and both are nearly absent from the camp; both ends of the metacarpal are low ranking and both are present in high, and nearly equal, numbers.

The second problem with the interpretation that limb bones from the kill were first broken for marrow and then culled for bones high in grease content to transport to the camp is that there should be a pattern of broken leg bone ends left at the kill that also shares a strong relationship with the grease index. That is, preferential removal of element portions from the kill, for the expressed purpose of grease rendering, leaves behind as much an opposite signature of the elements rejected for grease rendering as it does of the items selected.

Figure 14 plots the relationship between the broken leg bone portions recovered from the kill with the percentage of grease. If the above interpretation is correct, this figure should illustrate a negative relationship with the grease index, not because of preferential destruction of grease-rich bones (as seems to have happened at the camp), but because of the preferential transport of grease-rich bones to the camp. As seen in Figure 14, there is little relationship (r=0.207, P>0.05) between the frequency of broken leg bones at the kill and a measure of grease content. This suggests no culling of high grease-yield element portions from among bones already subjected to marrow removal. Given that the scenario of preferential marrow recovery at the Wardell kill followed by selective culling and transport of grease-rich bones to the camp is not substantiated by the faunal evidence, the question remains how did the camp assemblage originate from the kill population? Answering this question requires reconsideration of the full leg bone assemblage from the kill, not just the broken elements. Complete bones recovered from the Wardell kill have been shown to be a sensitive index of the selective procurement of marrow in that they occur in inverse proportion to their marrow content (Figure 10). Complete bones are viewed as bones not chosen for marrow recovery. In addition, complete bones are comprised of proximal and distal ends that were also not selected for bone grease rendering. Thus, the counts of separated proximal and distal leg bone ends from the Wardell kill assemblage represent elements broken for marrow but not chosen for grease rendering, while the complete bones represent elements rejected for both marrow and grease. In the following analysis the counts of complete bones are added to the separated articular ends (Table 1) to yield a total count of limb bone ends. For example, the 11 complete humeri represent an additional 11 proximal and distal ends that are added to the previous count of articular ends to yield totals of 25 and 70 for opposing ends of the humerus.

When the total count of articular ends at the kill is adjusted to include those specimens represented by complete elements, correlation with percentage of grease is in agreement with a maximizing strategy (Figure 15). Previously (Figure 14), there was no correspondence between the grease index and broken bones from the kill.

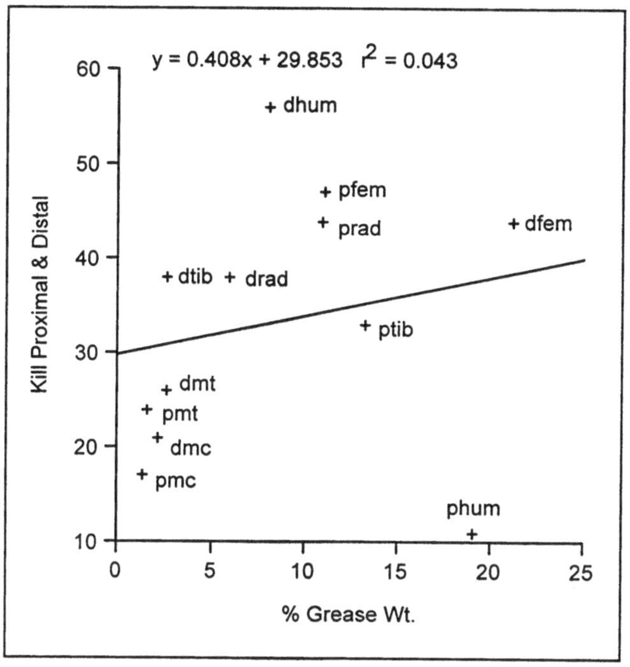

Figure 14. Relationship between frequency of proximal and distal leg bone portions at the Wardell kill and percentage of bone grease.

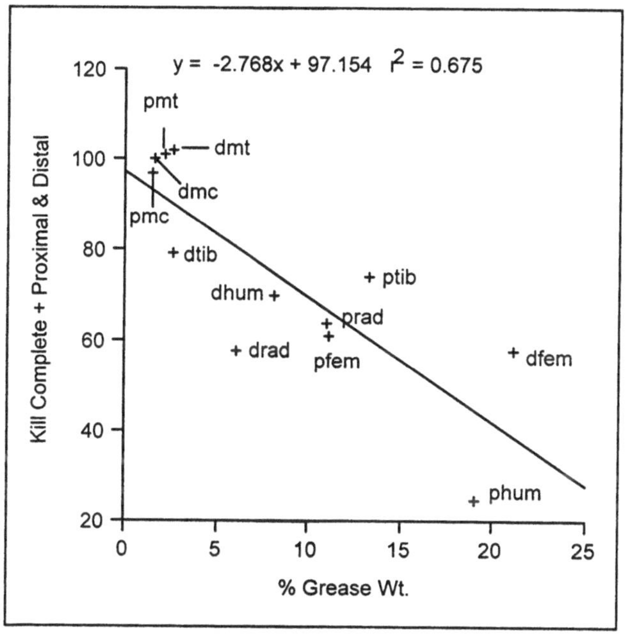

Figure 15. Relationship between the frequency of bison leg bone portions including complete bones from the Wardell kill and the percentage of grease.

When complete elements are included the relationship becomes inverse, significant and moderately strong (r=0.822, P<0.01). Remembering that a grease maximizing strategy will be inverse due to the destruction of the desired elements, Figure 15 indicates that articular ends of bones associated with the greatest amount of grease are rare at the kill while those low in grease are most abundant. This relationship now resembles that from the camp (Figure 13) where leg bone portions were negatively correlated with grease.

That the addition of complete leg bones to the count of articular ends recovered from the Wardell kill results in a strong, negative relationship with the grease index suggests an explanation for how the camp assemblage was formed. Hypothetically, imagine that butchering at the kill site included a considerable amount of breaking of bones for marrow recovery. Favouritism was directed towards bones associated with the greatest amount of marrow, as indicated by preferential breakage of bones with the largest marrow cavities (Figure 11), and the preferential discard of complete bones associated with the lowest amount of marrow (Figure 10). The broken portions of the bones from which marrow was recovered at the kill were subsequently ignored and not removed to the camp area. This is reflected in the similarity in count of proximal and distal ends (excluding the proximal humerus), and by the lack of any significant relationship between grease content and the frequency of articular ends at the kill (Figure 14). Instead, the limb bones transported to the camp area must have originated as complete limb bones transported from the kill.

In a maximizing strategy, removal of complete elements from the kill to the camp makes economic sense. Complete bones offer both fat resources, marrow and grease, while bones already processed for marrow offer only grease. Complete bones may have also had muscle and fat tissue still attached, to be cut away later in the camp. If broken bones were transported to the camp, muscle and fat tissue would have been largely removed at the kill, as necessary to break the bones in the first place. If the limb bone faunal assemblage at the Wardell camp was formed primarily by the transport of complete elements from the kill, then it would be predicted that the elements chosen for transport should be those offering the greatest return of both marrow and grease. Demonstrating this argument is complicated by the fact that the residual assemblage at any true processing site is, by definition, a ravaged collection of the bones deemed to be least useful (by both humans and scavengers) for food purposes. Evidence of selection and transport of choice marrow bones from the kill to the camp, for example, will have been largely obliterated by subsequent grease rendering. However, it is possible to investigate the preferential transport of bison skeletal elements away from the kill to the camp (or elsewhere) through an examination of what elements are missing from the kill relative to the entire Wardell assemblage.

Table 4 provides the count of MNE for leg bones from the entire site including the kill and the camp. All counts include the calculation of complete elements as an occurrence of both a proximal and a distal end. For the total site it can be seen that the greatest count of articular ends is for the distal metacarpal (n=128). Thus, it can be assumed that a minimum of 128 proximal and distal ends (MNE) of all leg bones should have once been present at the Wardell kill site. This value is used to obtain the percentage of each element present at the kill and the number and percentage missing from the kill assemblage. For example, the 25 proximal

	Total Site MNE/MAU	% of Expected Kill Assemblage	# MNE/MAU Missing from Kill	% MNE/MAU Missing from Kill
Proximal humerus	25	19.53	103	80.47
Distal humerus	84	54.69	58	45.31
Total humerus	54.5	37.25	80	62.75
Proximal radius	68	50.00	64	50.00
Distal radius	69	45.31	70	54.69
Total radius	68.5	47.84	66.5	52.16
Proximal metacarpal	127	75.78	31	24.22
Distal metacarpal	128	78.91	27	21.09
Total metacarpal	127.5	77.65	28.5	22.35
Proximal femur	62	47.66	67	52.34
Distal femur	60	45.31	70	54.69
Total femur	61	46.67	68	53.33
Proximal tibia	75	57.81	54	42.19
Distal tibia	93	61.72	49	38.28
Total tibia	84	60.00	51	40.00
Proximal metatarsal	118	78.13	28	21.88
Distal metatarsal	122	79.69	26	20.31
Total metatarsal	120	79.22	26.5	20.78

Table 4. Count of limb bones from Wardell kill site. Note: Data from Frison (1973).

humeri recovered from the kill represent about 20% of the number expected based on the total count of 128 distal metacarpals, which means that about 103, or 80%, of the expected proximal humeri are missing from the kill assemblage. Element portions are MNE, whole elements are MAU. Because marrow values can only be assigned to whole elements (values for proximal and distal ends would simply be equal halves of the total), the same calculations are then conducted for the combined counts of articular ends divided by two (MAU values). The highest MAU value (127.5) is for the metacarpal.

The bone most commonly absent from the expected totals of the kill site is the proximal humerus, with 80% of these bone portions missing (Table 4). This is well in excess of any other bone, as the next ranked elements are the distal femur, distal radius, proximal femur and proximal radius—all with between 50% to 55% of element portions missing. The proximal and distal tibia are absent 38% to 42% of the time, respectively. The figures for the tibia and the radius in Table 4 are interesting in that the percentage which they are absent from the kill is approximately reversed from their ranking as providers of marrow (although the two bones are about equal in rank for grease if articular ends are combined). Why radii are relatively abundant while the tibiae are scarce is unknown. The ends of the metapodials are consistently missing from the kill between 20% to 24% of the time.

As noted above, it may be most appropriate to compare the percentage of bones missing from the Wardell kill with a combined nutritional measure of marrow and grease. Accordingly, I have calculated the percentage of both marrow and grease weight associated with each of the six major limb bones. This was done by combining the mean weight of grease associated with the proximal and distal portions of each major leg bone and adding this to the mean weight of marrow for each element (Tables 2 and 3). These combined grease and marrow weights were then converted to a percentage. Figure 16 plots the combined marrow and grease index against the percentage of bones missing from the kill for the six major leg bones. The relationship is linear, positive, strong and significant (r=0.916, P<0.05). Bones associated with the greatest percentage of marrow and grease are more often missing from the kill, and vice versa. Three distinct pairs are apparent: the humerus and femur as the most often missing and most nutritious, the tibia and radius as intermediate, and the metapodials widely separated from the other four bones as seldom missing and of lowest nutritional value.

Overall, the data presented above indicate that the combined percentage of marrow and grease weight account for a great deal of the variability in the relative abundance of leg bones at the Wardell kill. This is not to say that only marrow and grease content can account for differential skeletal abundance, nor that all missing elements were taken to the camp, as it is equally possible that some were removed entirely from the general site area. Indeed, this is strongly suggested by the great difference between the counts of certain bones missing from the kill and the number recovered

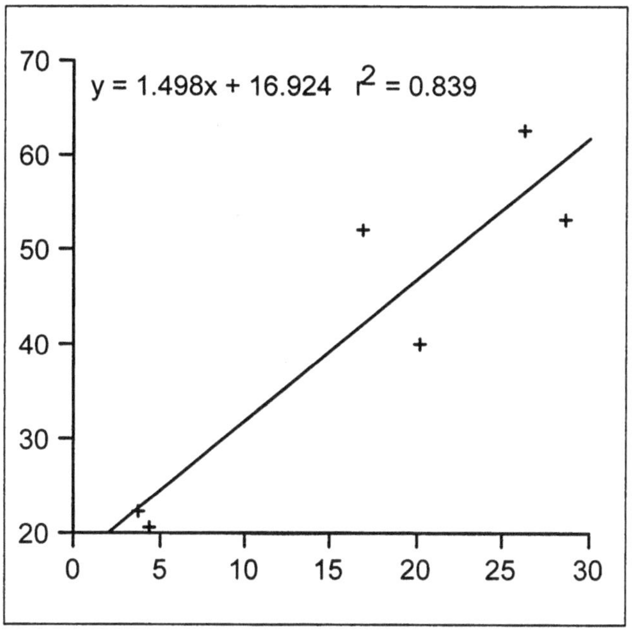

Figure 16. Relationship between the percentage of elements missing from the Wardell kill and the combined percentage of bone marrow and grease.

from the camp. For example, some 103 proximal humeri are missing from the kill while none were recovered from the camp. With this being a prime grease-bearing element, many were probably destroyed through crushing for grease rendering. However, processing at the Wardell camp would have to have been exhaustive in order to account for the absence of this element portion. In the bone grease studies reported above (Table 3) the proximal humerus ranked second behind the distal femur in weight of bone grease. A separate measure of bone grease using solvent extraction (Brink and Dawe 1989; Brink 1997) ranked the proximal humerus as number one, being composed of 40% fat by weight. The proximal humerus also has the lowest density value of all limb bones in bison (Kreutzer 1992). These data may account for the scarcity of the proximal humerus in the all Wardell assemblages.

CONCLUSION

It must be stressed that a variety of other taphonomic factors may produce faunal patterns similar to those postulated to have been caused by human action at the Wardell site. Indeed, patterns of carnivore selection of limb bones are very similar to those illustrated for both components of the Wardell site (Binford *et al.,* 1988; Blumenschine 1988; Marean and Spencer 1991; Lam 1992; Marean *et al.,* 1992). A strong correlation between the grease content of bison leg bone portions and the selective consumption by carnivores has been noted (Brink 1997).

Of the issues raised at the outset of this paper, first I question interpretations of skeletal frequency of very large game animals such as bison with models of nutritional value based on all food resources available in the carcass. As documented above (Figure 1), Emerson's total products model of bison anatomy, which includes all muscle and fat resources, accounts poorly for the frequency of most bones recovered from the Wardell kill, and poorly (Figure 2) for the restricted analysis of only the leg bones. Historical records document that bison were butchered in highly selective ways, reflecting extreme differential distribution of desired foods (Wheat 1972; Frison 1991). It seems inappropriate to ignore this information by standardizing carcass differences with a total products model.

Second, I argue that it is inappropriate to juxtapose utility indices with all bones from the skeleton. Again, the documentary record provides compelling evidence for the differential treatment of bison carcass portions in the butchering and processing stages (Frison 1970, 1973, 1974, 1991; Wheat 1972). Combining vertebrae, ribs, leg bones, pelves and cranial parts into a graph of any measure of food utility, regardless of whether it is based on total products or only a specific food such as grease, may ignore an abundance of information about aboriginal butchering patterns, and can lead to extraordinary interpretations of perplexing graphic displays of element frequency (Binford 1978, 1981; see Ringrose 1993). Instead, it seems prudent to employ anatomical information on carcass composition to assist in isolating those carcass parts that can reasonably associated with specific food resources.

Third, I suggest that interpretation of skeletal frequencies from archaeological sites, when based on anatomical utility indices, employ the nutritional resources directly relevant to particular carcass portions. For leg bones, I have argued that marrow and grease content may be the most appropriate models by which to measure preferential human selection. As stated at the outset, some nutritional resources are only weekly associated with the skeleton and may be obtained with little or no direct manipulation of skeletal elements. Greater emphasis on consideration of those food resources most directly associated with the skeleton, such as marrow and grease, may hold greater explanatory power in the interpretation of skeletal frequencies than has been previously acknowledged. This is believed to be especially true for very large game animals where muscle and fat tissue may be removed from the skeleton rather than transported with the bones intact (Marshall and Pilgrim 1991). In such cases, decision making regarding transport and processing of the bones may be based as much on the food resources directly contained within - or closely linked to - the bones themselves. For other parts of the skeleton, such as ribs and vertebrae, grease and marrow content is low compared to the legs and the relative abundance of other foods will likely be the relevant predictors of skeletal treatment.

A few final thoughts on the concept of maximization are in order. While the basic assumption that hunters and gatherers use large game carcasses in a manner so as to maximize the return of nutritious food can be supported, it is also important to recognize that the real world will be far more complex than that which can be modeled in a simple ranking of animal parts by food value. It may be fair to assume, as Binford (1978) does, that most carcass parts are used for food in a rational manner based on their relative ability to sustain life. But deviations from predicted utility models occur (see Woollett 1991) and they must be viewed as just as informative about human behaviour as are actions that conform to the model.

The documentary record for the North American Plains offers a wealth of evidence that bison carcass parts, including bones, served purposes other than sheer nutrition. Choice pieces of bison were frequently offered to the spirits in the aftermath of a successful bison kill or at ceremonies (Fletcher and La Flesche 1911:273; Catlin 1926:144; Glover 1962:259; Mandelbaum 1979:205). Tails were sometimes taken as trophies (McDermott 1940:193; Audubon 1960, II:104). Skulls were often associated with the spiritual world and treated in special, non-food related, ways (Maximilian 1905, II:375, 389; Farnham 1906:268; Densmore 1918:444; Lowie 1922; Audubon 1960, II:12). Following special feasts bison bones were sometimes burned so as to prevent dogs from feeding on them and thereby bringing insult to the honoured guests (Schoolcraft 1851,V:171). Bison bones were also burned as a source of heat (Spry 1968:235,255; Mandelbaum 1979:93).

It is unlikely that the number of bones involved in non-food behaviour was large enough to alter the fundamental faunal patterns at communal bison kill sites. But at small sites or single kills these practices could have important consequences on interpretation of skeletal treatment. Regardless, the great range of behaviour associated with the treatment of game animals needs to be recognized because not to do so relegates aboriginal hunters to mechanistic maximizers of game animals, which they were not.

In the present study there are a number of examples where the Wardell site occupants did not conform to predicted maximizing behaviour. For example, the excavated faunal samples contained many complete leg bones, a clear sign that rich marrow fat was ignored. Hundreds more complete leg bones must remain unexcavated. It is true that when the rejected bones are scaled according to marrow content there is a consistent trend to use the most productive bones and discard those least useful. But the abandonment of hundreds of complete leg bones must temper the acceptance of a maximizing butchering strategy. Further, there is significant deviation from perfect observance of nutritional rankings. Most notably, the tibia is abandoned more often in complete condition than is predicted by the amount of marrow associated with this element. The tibia is virtually tied with the humerus and femur as a top marrow producer, and has about 10% more marrow than the radius, yet it was recovered complete twice as often as the radius. This may simply reflect the vagaries of excavation bias at the Wardell kill, or it may represent a true pattern of differential human selection for reasons not currently understood. Either way, variations from ideal utility curves need to be addressed. This approach, rather than rigid adherence to expected relationships, holds the promise of exposing the nuances of human behaviour that probably made life interesting.

ACKNOWLEDGEMENTS

My thanks to Craig Gerlach for encouraging me to submit this paper to the Guthrie volume, and to Craig and Maribeth Murray for assistance with the manuscript. The comments of two anonymous reviewers improved the paper considerably, as did comments by Matthew G. Hill. Thanks to Danny Walker for providing me with the map of the Wardell site. I am grateful to Clifford Hickey and Raymond Le Blanc for comments on a larger manuscript that contained parts of the present paper. Thanks also to Bob Hudson of the University of Alberta and Cormack Gates of the University of Calgary for their insights into the anatomy and bioenergetics of bison. Finally, I acknowledge my debt to the writings of Dale Guthrie. His expansive thinking on large herbivore dynamics, ecological relationships and human predation has stimulated and inspired me.

REFERENCES CITED

Adamczewski, J.Z., C.C. Gates, and R.J. Hudson
1987a Fat Distribution and Indices of Carcass Composition in Coats Island Caribou (*Rangifer tarandus groenlandicus*). *Canadian Journal of Zoology* 65:368-373.

Adamczewski, J.Z., C.C. Gates, R.J. Hudson, and M.A. Price
1987b Seasonal Changes in Body Composition of Mature Female Caribou and Calves (*Rangifer tarandus groenlandicus*) on an Arctic Island with Limited Winter Resources. *Canadian Journal of Zoology* 65:1149-1157.

Anderson, A.E., D.E. Medin, and D.C. Bowden
1972 Indices of Carcass Fat in a Colorado Mule Deer Population. *Journal of Wildlife Management* 36:579-594.

Audubon, M.R.
1960 *Audubon and His Journals*. 2 vols. Dover Publications, New York.

Ballard, W.B., and J.S. Whitman
1987 Marrow Fat Dynamics in Moose Calves. *Journal of Wildlife Management* 51:66-69.

Berg, R.T., and R.M. Butterfield
1976 *New Concepts of Cattle Growth*. John Wiley and Sons, New York.

Binford, L.R.
1978 *Nunamiut Ethnoarchaeology*. Academic Press, New York.

1981 *Bones, Ancient Men and Modern Myths*. Academic Press, New York.

Binford, L.R., M.G.L. Mills, and N.M. Stone
1988 Hyena Scavenging Behavior and its Implications for the Interpretation of Faunal Assemblages from FLK 22 (the Zinj floor) at Olduvai Gorge. *Journal of Anthropological Archaeology* 7:99-135.

Blumenschine, R.J.
1988 An Experimental Model of the Timing of Hominid and Carnivore Influence on Archaeological Bone Assemblages. *Journal of Archaeological Science* 15:483-502.

Blumenschine, R.J., and T.M. Caro
1986 Unit Flesh Weights of Some East African Bovids. *African Journal f Ecology* 24:273-286.

Borrero, L.A.
1990 Fuego-Patagonian Bone Assemblages and the Problem of Communal Guanaco Hunting. In *Hunters of the Recent Past*, edited by L. B. Davis and B. O. K. Reeves, pp. 373-399. Unwin Hyman, London.

Brink, J.W.
1990 Unpublished notes on marrow and bone grease content of bison limb bones. Ms. on file Provincial Museum of Alberta, Edmonton.

1997 Fat Content in Leg Bones of *Bison bison*, and Applications to Archaeology. *Journal of Archaeological Science* 24:259-274.

Brink, J., and B.Dawe
1989 *Final Report of the 1985 and 1986 Field Season at Head-Smashed-In Buffalo Jump, Alberta*. Manuscript Series No. 16. Archaeological Survey of Alberta, Edmonton.

Bunn, H.T.
1993 Bone Assemblages at Base Camps: A Further Consideration of Carcass Transport and Bone Destruction by the Hadza. In *From Bones to Behavior*, edited by J. Hudson, pp. 156-168. Center for Archaeological Investigations at Southern Illinois University, Carbondale.

Bunn, H.T., L.E. Bartram, and E.M. Kroll
1988 Variability in Bone Assemblage Formation from Hadza Hunting, Scavenging, and Carcass Processing. *Journal of Anthropological Archaeology* 7:412-457.

Catlin, G.
1926 *North American Indians, Being Letters and Notes on their Manners, Customs, and Conditions*. 2 vols. John Grant, Edinburgh.

Cederlund, G.N., R.J. Bergström, and K. Danell
1989 Seasonal Variation in Mandible Marrow Fat in Moose. *Journal of Wildlife Management* 53:587-592.

Davis, J.L., P. Valkenburg, and D.J. Reed
1987 Correlation and Depletion Patterns of Marrow Fat in Caribou Bones. *Journal of Wildlife Management* 51:365-371.

Dauphiné, J.R.
1975 Kidney Weight Fluctuations Affecting the Kidney Fat Index in Caribou. *Journal of Wildlife Management* 39:379-386.

1976 *Biology of the Kaminuriak Population of Barren-Ground Caribou*. Canadian Wildlife Service Report, Series No. 38. Queen's Printer, Ottawa.

Densmore, F.
1918 *Teton Sioux Music*. Bureau of American Ethnology, Bulletin 61. Smithsonian Institution, Washington.

Diab, M.
1998 Economic Utility of Ringed Seal (*Phoca hispida*): Implications for Arctic Archaeology. *Journal of Archaeological Science* 21:1-26.

Emerson, A.E.
1990 *Archaeological Implications of Variability in the Economic Anatomy of Bison bison*. Unpublished Ph.D dissertation, Department of Anthropology, Washington State University, Pullman.

Emery, R.S.
1969 Lipids and Adipose Tissue. In *Animal Growth and Nutrition*, edited by E. S. E. Hafez and I. A. Dyer, pp. 236-255. Lea and Febiger, Philadelphia.

Farnham, T.J.
1906 *Travels in the Great Western Prairies, the Anahuac and the Rocky Mountains, and in the Oregon Territory*. Early Western Travels: 1748-1846, edited by R. G. Thwaites, vol. 28. Arthur H. Clark Co., Cleveland.

Finger, S.E., I.L. Brisbin Jr., M.H. Smith, and D.F. Urbston
1981 Kidney Fat as a Predictor of Body Composition in White-Tailed Deer. *Journal of Wildlife Management* 45:964-968.

Fletcher, A. C., and F. La Flesche
1911 *The Omaha Tribe*. 27th Annual Report of the Bureau of American Ethnology, pp. 19-672. Smithsonian Institution, Washington.

Franzmann, A.W.
1985 Assessment of Nutritional Status. In *Bioenergetics of Wild Herbivores*, edited by R. J. Hudson and R. G. White, pp. 239-259. CRC Press, Boca Raton.

Franzmann, A.W., and P.D. Arneson
1976 Marrow Fat in Alaskan Moose Femurs in Relation to Mortality Factors. *Journal of Wildlife Management* 40:336-339.

Frison, G.C.
1970 *The Glenrock Buffalo Jump, 48CO304*. Plains Anthropologist Memoir No. 7.

1973 *The Wardell Buffalo Trap, 48 SU 301: Communal Procurement in the Upper Green River Basin, Wyoming*. Anthropological Papers No. 48. Museum of Anthropology, University of Michigan.

1974 The Archaeology of the Casper Site. In *The Casper Site, A Hell Gap Bison Kill on the High Plains*, edited by G. C. Frison, pp. 1-122. Academic Press, New York.

1991 *Prehistoric Hunters of the High Plains*. Second Edition. Academic Press, San Diego.

Glover, R.
1962 *David Thompson's Narrative 1784-1812*. The Champlain Society, Toronto.
Grayson, D.K.

1989 Bone Transport, Bone Destruction, and Reverse Utility Curves. *Journal of Archaeological Science* 16:643-652.

Harris, D.
1945 Symptoms of Malnutrition in Deer. *Journal of Wildlife Management* 9:319-322.

Hjermstad, B.E.
1996 *The Fitzgerald Site: A Besant Pound and Processing Area on the Northern Plains*. Unpublished M.A. thesis, Department of Anthropology, University of Saskatchewan, Saskatoon.

Hudson, R.J., and R.G. White (editors)
1985 *Bioenergetics of Wild Herbivores*. CRC Press, Boca Raton.

Jochim, M.A.
1976 *Hunter-Gatherer Subsistence and Settlement: a Predictive Model.*. Academic Press, New York.

Johns, P.E., M.H. Smith, and R.K. Chesser
1984 Annual Cycles of the Kidney Fat Index in a Southeastern White-Tailed Deer Herd. *Journal of Wildlife Management* 48:969-973

Jones, S.D. M., M.A. Price, and R.T. Berg
1981 Accumulation of Lipids in Rib Cuts from Bull and Heifer Carcasses of Two Breeds. *Canadian Journal of Animal Science* 61:23-26.

Jones, K.T., and D. Metcalfe
1988 Bare Bones Archaeology: Bone Marrow Indices and Efficiency. *Journal of Archaeological Science* 15:415-423.

Koch, A.
1995 The MacIntosh Fauna: Late Prehistoric Exploitation of Lake and Prairie Habitats in the Nebraska Sand Hills. *Plains Anthropologist* 40(151):39-60.

Kreutzer, L.A.
1992 Bison and Deer Bone Mineral Densities: Comparisons and Implications for the Interpretation of Archaeological Faunas. *Journal of Archaeological Science* 19:271-294.

Landals, A.J.
1990 The Maple Leaf site: Implications of the Analysis of Small Scale Bison Kills. In *Hunters of the Recent Past*, edited by L.B. Davis and B.O.K. Reeves, pp. 373-399. Unwin Hyman, London.

Lam, Y.M.
1992 Variability in the Behaviour of Spotted Hyaenas as Taphonomic Agents. *Journal of Archaeological Science* 19:389-406.

Le Blanc, R.J.
1994 *The Crane Site and the Paleoeskimo Period in the Western Canadian Arctic*. Mercury Series, Archaeological Survey of Canada, Paper 148. Canadian Museum of Civilization, Ottawa.

Leat, W.M.F., and R.W. Cox
1980 Fundamental Aspects of Adipose Tissue Growth. In *Growth in Animals*, edited by T. L. J. Lawrence, pp. 137-174. Butterworths, London.

Leechman, D.
1951 Bone Grease. *American Antiquity* 16:355-356.

Link, B.A., R.W. Bray, R.G. Cassens, and R.G. Kauffman
1970 Lipid Deposition in Bovine Skeletal Muscle During Growth. *Journal of Animal Science* 30:6-9.

Lowie, R.H.
1922 *Religion of the Crow*. Anthropological Papers Vol. 25, Pt. 2, pp. 309-451. American Museum of Natural History, New York.

Lupo, K.D.
1995 Hadza Bone Assemblages and Hyena Attrition: An Ethnographic Example of the Influence of Cooking and Mode of Discard on the Intensity of Scavenger Ravaging. *Journal of Anthropological Archaeology* 14:288-314.

Lyman, R.L.
1984 Bone Density and Differential Survivorship of Fossil Classes. *Journal of Anthropological Archaeology* 3:259-299.
1985 Bone Frequencies: Differential Transport, in situ Destruction, and the MGUI. *Journal of Archaeological Science* 12:221-236.

1992 Anatomical Considerations of Utility Curves in Zooarchaeology. *Journal of Archaeological Science* 19:7-22.

Mandelbaum, D.G.
1979 *The Plains Cree*. Canadian Plains Studies, No. 9. Canadian Plains Research Centre, Regina.

Marean, C.W., and L.M. Spencer
1991 Impact of Carnivore Ravaging on Zoological Measures of Element Abundance. *American Antiquity* 56:645-658.

Marean, C.W., L.M. Spencer, R.J. Blumenschine, and S.D. Capaldo
1992 Captive Hyaena Bone Choice and Destruction, the Schlepp Effect and Olduvai Archaeofaunas. *Journal of Archaeological Science* 19:101-121.

Marquez, M., and B.E. Coblentz
1987 Metatarsal and Mandibular Marrow Fat in Black-Tailed Deer. *Journal of Wildlife Management* 51:38-40.

Marshall, F., and T. Pilgram
1991 Meat Versus Within-Bone Nutrients: Another Look at the Meaning of Body Part Representation in Archaeological Sites. *Journal of Archaeological Science* 18:149-163.

Marshall, S.E.
1988 *The Calderwood Buffalo Jump: Bison Procurement in the Porcupine Hills of Southern Alberta*. Unpublished M.A. thesis, Department of Anthropology, Trent University, Peterborough.

Maximillian, Prince of Wied
1905 *Travels in the Interior of North America*. Early Western Travels: 1748-1846, edited by R.G. Thwaites, vols. 22-24. Arthur H. Clark Co., Cleveland.

McDermott, J.F. (editor)
1940 *Tixier's Travels on the Osage Prairies*. University of Oklahoma Press, Norman.

Metcalfe, D., and K.T. Jones
1988 A Reconsideration of Animal Body-Part Utility Indices. *American Antiquity* 53:486-504.

Monahan, C.M.
1998 The Hadza Carcass Transport Debate Revisited and its Archaeological Implications. *Journal of Archaeological Science* 25:405-424.

Morlan, R.E.
1994 Oxbow Bison Procurement as Seen from the Harder Site, Saskatchewan. *Journal of Archaeological Science* 21:757:-77.

Morrison, D.A.
1997 *Caribou Hunters in the Western Arctic*. Mercury Series, Archaeological Survey of Canada, Paper 157. Canadian Museum of Civilization, Ottawa.

O'Connell, J.F., and Marshall, B.
1989 Analysis of Body Part Transport Among the Alyawara of Central Australia. *Journal of Archaeological Science* 16:393-405.

O'Connell, F.J., K. Hawkes, and N. Blurton Jones
1989 Hadza Hunting, Butchering and Bone Transport and their Archaeological Implications. *Journal of Anthropological Research* 44:113-161.

1990 Reanalysis of Large Mammal Body Part Transport among the Hadza. *Journal of Archaeological Science* 17:301-316.

Outram, A., and P. Rowley-Conwy
1998 Meat and Marrow Utility Indices for Horse (*Equus*). *Journal of Archaeological Science* 25:839-849.

Perkins, D., and P. Daly
1968 A Hunter's Village in Neolithic Turkey. *Scientific American* 219(5): 97-106.

Ranson, A.B.
1965 Kidney and Marrow Fat as Indicators of White-Tailed Deer Condition. *Journal of Wildlife Management* 29:397-398.

Ratcliffe, P.R.
1980 Bone Marrow Fat as an Indicator of Condition in Roe Deer. *Acta Theriologica* 25(26):333-340.

Ringrose, T.J.
1993 Bone Counts and Statistics: a Critique. *Journal of Archaeological Science* 20:121-157.

Savelle, J.M., and T.M. Friesen
1996 An Odontocete (Cetacea) Meat Utility Index. *Journal of Archaeological Science* 23:713-721.

Savelle, J.M., T.M. Friesen, and R.L. Lyman
1996 Derivation and Application of an Otariid Utility Index. *Journal of Archaeological Science* 23: 705-712.

Schoolcraft, H.R.
1851 *Historical and Statistical Information Respecting the History, Condition and Prospects of the Indian Tribes of the United States*. 6 vols. Lippincott, Grambo and Co., Philadelphia.

Speth, J.D.
1983 *Bison kills and Bone Counts: Decision Making by Ancient Hunters*. University of Chicago Press, Chicago.

1990 Seasonality, Resource Stress, and Food Sharing in So-Called "Egalitarian" Foraging Societies. *Journal of Anthropological Archaeology* 9:148-188.

Speth, J.D., and K.A. Spielmann
1983 Energy Source, Protein Metabolism, and Hunter-Gather Subsistence Strategies. *Journal of Anthropological Archaeology* 2:1-31.

Spry, I.M. (editor)
1968 *The Papers of the Palliser Expedition 1857-1860*. The Champlain Society, Toronto.

Suttie, J.M.
1983 The Relationship Between Kidney Fat Index and Marrow Fat Percentage as Indicators of Condition in Red Deer Stags (*Cervus elephus*). *Journal of Zoology, London* 201:563-565.

Thomas, D.H., and D. Mayer
1983 Behavioral Faunal Analysis of Selected Horizons. In *The Archaeology of Monitor Valley: 2. Gatecliff Shelter*, edited by D. H. Thomas, pp. 353-391. Anthropological Papers Vol. 59. American Museum of Natural History, New York.

Todd, L.C.
1987 Taphonomy of the Horner II Bone Bed. In *The Horner Site, The Type Site of the Cody Cultural Complex*, edited by G.C. Frison and L.C. Todd, pp. 107-198. Academic Press, Orlando.

Torbit, S.C., L.H. Carpenter, D.M. Swift, and A.W. Alldredge
1985 Differential Loss of Fat and Protein by Mule Deer During Winter. *Journal of Wildlife Management* 49:80-85.

Vehik, S.C.
1977 Bone Fragments and Bone Grease Manufacturing: A Review of their Archaeological Use and Potential. *Plains Anthropologist* 22(77):169-182.

Wheat, J. B.
1972 *The Olsen-Chubbuck Site, A Paleo-Indian Bison Kill.* Memoir No. 26. Society for American Archaeology, Washington, D.C.

White, T.E.
1952 Observation on the Butchering Techniques of Some Aboriginal Peoples: 1. *American Antiquity* 17:337-38.

1953a Observation on the Butchering Techniques of Some Aboriginal Peoples: No. 2. *American Antiquity* 19:160-164.

1953b Method of Calculating the Dietary Percentage of Various Food Animals Utilized by Aboriginal Peoples. *American Antiquity* 18:396-398.

1954 Observation on the Butchering Techniques of Some Aboriginal Peoples: Nos. 3,4,5 and 6. *American Antiquity* 20:254-264.

1955 Observation on the Butchering Techniques of Some Aboriginal Peoples: Nos. 7, 8 and 9. *American Antiquity* 21:170-178.

Will, R.T.
1985 *Nineteenth Century Copper Inuit Subsistence Practices on Banks Island, N.W.T.* Unpublished Ph.D. dissertation, Department of Anthropology, University of Alberta, Edmonton.

Woollett, J.M.
1991 *Cultural Perceptions of Man-Animal Relationships and Carcass Utilization.* Unpublished M.A. thesis, Department of Anthropology, University of Alberta, Edmonton.

A CONSIDERATION OF THE INTER-SPECIFIC APPLICATION OF FOOD UTILITY INDICES, WITH REFERENCE TO FIVE SPECIES IN THE ORDER PINNIPEDIA

T. Max Friesen
Department of Anthropology, University of Toronto

James M. Savelle
Department of Anthropology, McGill University

Mark C. Diab
Department of Anthropology, University of Alaska, Fairbanks

INTRODUCTION

Thirty years ago, Guthrie (1967) discussed differential element preservation in Alaskan palaeontological assemblages, concluding that major factors affecting this phenomenon included element size and element composition. Since then, the study of differential element representation has become a central component of taphonomic studies in both palaeontological and zooarchaeological contexts. More specifically, the interpretation of element frequencies from archaeological sites has been greatly enhanced through the recognition and quantification of a number of biological factors, the two most important being differential bone density (e.g., Behrensmeyer 1975; Brain 1969; Lyman 1984, 1993), which affects the likelihood of bone's survival when subjected to a variety of taphonomic factors, and "utility indices" (Binford 1978), which quantify the amount of meat, fat, and other edible substances attached to particular skeletal elements. To date, utility indices have been calculated for 17 mammalian and one avian species, each of which can be assumed to accurately reflect the anatomy of that species. However, as yet there are no generally accepted guidelines for the application of utility indices calculated for a particular species to element distributions of another, related species (hereafter referred to as the "inter-specific application of utility indices"). This paper represents an attempt to refine our ability to interpret utility indices through a consideration of the degree of similarity between utility indices calculated for different species.

In the past decade, the order Pinnipedia (seals, sea lions, and walruses) has progressed from being completely unknown in terms of utility indices to being perhaps the best-represented mammalian order. Lyman et al. (1992b) published utility indices for harp seal (*Phoca groenlandica*; three specimens) and hooded seal (*Cystophora cristata*; one specimen), Savelle et al. (1996) published a utility index for California sea lion (*Zalophus californianus*; two specimens), and the present report presents utility indices for harbor seal (*Phoca vitulina*; three specimens) and ringed seal (*Phoca hispida*; three specimens). Thus, a total of 12 specimens and five species within this order are available for comparison. Following a description of the development of utility indices for harbor and ringed seals, this relatively extensive data base is used to assess the validity of inter-specific application of utility indices.

DERIVATION OF A UTILITY INDEX FOR HARBOR SEALS

Three harbor seals were processed by two of the authors (T.M.F. and J.M.S.) at the Pacific Biological Station, Canadian Department of Fisheries and Oceans, Nanaimo, British Columbia. In order to ensure comparability of results, the same methods were used to dismember and measure the harbor seals as those employed earlier for harp seals (Lyman et al., 1992b), a hooded seal (Lyman et al., 1992b), and California sea lions (Savelle et al., 1996). These methods were originally developed by Binford (1978), and have since been refined by Metcalfe and Jones (1988) and Lyman et al. (1992b).

The three harbor seals had all drowned in fish nets near Snake Island, coastal British Columbia, in April, 1993. The seals were frozen until they were processed in June, 1993, and were completely thawed before processing. The first harbor seal (Pv-93-1) was a male weighing 30.8 kg, the second (Pv-93-2) a female weighing 25.4 kg, and the third, (Pv-93-3) a female weighing 31.8 kg. Although accurate ages could not be established for the seals at the time of processing, they are assumed to be immature individuals based on the fact that they are approximately half the weight reported for adult harbor seals (average adult male weight = 72.5 kg, average adult female weight = 58 kg; Banfield 1974:370). The seals were provided by Dr. Thomas Smith of the Pacific Biological Station, and during processing the authors were assisted by Thomas Smith, Gregor Beck and George Horonowitsch of the same institution.

Prior to dismemberment, each seal was weighed and a series of standardized measurements were taken, including nose-tail length, axial girth, maximum girth, and blubber thickness (Committee on Marine Mammals, 1967). Scalpels and steel knives were used to dismember the carcass. Initially the sculp, consisting of the hide and attached blubber layer, was removed and weighed as a unit, with the exception of the flippers and tail which are not easily detached from the sculp. The sculp and viscera are not included in meat weight calculations, as they are easily removed from carcasses early in the butchery process, and therefore would have been treated differently from other body parts in subsequent carcass processing and transport. The remainder of the carcass was then dismembered in a manner consistent with the production of a utility index. That is, dismemberment occurred at joints between each category of skeletal element, with meat being left attached to the adjacent skeletal element. As has now become standard practise (e.g., Lyman et al. 1992b; Savelle et al. 1996), we assessed the degree of similarity in the gross weights (total of bone and soft tissues) of the left and right examples of each sided body portion (limb elements and rib sections) for each seal. This allowed the determination of whether our dismemberment was consistent for both sides of each seal. For all three seals, gross weights of left and right body portions are significantly correlated (in all cases, Spearman's rho (r_s)>0.97, P<0.001; see below for a discussion of this statistical technique). Therefore, we averaged left and right weights in the production of all subsequent indices.

Following the calculation of gross weights, each bone or bone group was cleaned manually, boiled, and then weighed. The difference between the gross weight and the bone weight is termed the flesh weight, and it is this weight which forms the basis of the utility index. Exceptions to this method occurred with the calculation of flesh weights for the front and rear flippers. It is difficult to remove the hide from flippers, and in particular to determine precise weights of flipper bone or flesh. This problem was addressed by boiling one front and one rear flipper from specimen Pv-93-3 to determine the proportion of their weights which was composed of bone (carpals/tarsals, metapodials, and phalanges). For the front flipper, bone represented 24.61% (37.7/153.2 g) of gross weight, while for the rear flipper, bone represented 23.72% (104.1/438.8 g) of gross weight. These percentages are then multiplied by the average ratio of bone weight to flesh weight for the flippers of four species of Antarctic seals (Bryden and Felts 1974:592-595). For front flippers, the average meat to bone ratio was 1.49:1, while for rear flippers the ratio was 1.23:1. When the two ratios (bone weight to gross weight; meat weight to bone weight) are multiplied, the meat weight percentage of gross weight (including hide) was determined to be approximately 37% for front flippers, and 29% for hind flippers. These percentages are used to calculate the flipper meat weights for the harbor seals.

Table 1 presents the flesh weight per skeletal part for each harbor seal, as well as an average weight for each part of the three harbor seals. This average weight is then normed on a scale of 0-100 to yield a %MUI (meat utility index) (Binford 1978). The %MUI represents a ranking of the amount of meat which is associated with each class of skeletal element. In zooarchaeological analysis, the relationship between the %MUI and skeletal element frequencies from archaeological sites can be used, in concert with other classes of data, to interpret patterns of bone transport and processing.

A final transformation of the utility index is the production of a modified meat utility index (MMUI). The MMUI is intended to account for the fact that during butchery, carcasses are often divided into anatomical portions rather than individual bones, and that body parts of lower utility will often then be transported as "riders" along with parts of higher utility (Binford 1978:74-75). Accordingly, the index is "modified" by averaging the utility of lower utility parts with adjacent higher utility parts, and assigning the averaged value to the lower part. Finally, the MMUI is normed on a scale of 1-100 to produce a %MMUI.

A number of different methods have been used for the calculation of %MMUIs. In his initial study of Nunamiut processing and transport of caribou body parts, Binford (1978) averaged only the utility values of appendicular skeletal elements, due to his observation that in the Nunamiut cultural context axial elements are not transported as "riders." This procedure was accepted and reinforced by Metcalfe and Jones (1988) in their reanalysis of Binford's methods. However, in a great majority of archaeological contexts, the degree to which different body portions (e.g., appendicular vs. axial, upper limb vs. lower limb) are affected by their proximity to one another during butchery is unknown. Therefore, we believe that *all* adjacent body parts should be averaged to produce the %MMUI, in order to make consistent assumptions about the presence of "riders." Another alternative is simply to use the %MUI, which does not involve any assumptions regarding "riders."

Here, the %MMUI is calculated for all skeletal elements (Table 2). Accordingly, indices for individual skeletal elements are calculated in one of three ways, depending on the relative utility of neighboring body parts. In cases where an element does not directly articulate with another element of higher utility, its value does not change. Where an element articulates with one element which has a higher utility, the average of the two values is assigned to the element of lower utility. Where an element is located between two or more elements of higher utility, it is assigned the average value of the adjacent higher utility elements. In one case (thoracic vertebrae), this resulted in the %MMUI value being calculated as the average of three skeletal elements: cervical vertebrae, lumbar vertebrae, and ribs.

DERIVATION OF A UTILITY INDEX FOR RINGED SEALS

Three ringed seals were obtained from the Department of Wildlife Management of the North Slope Borough, Alaska. The seals were frozen from the time they were harvested by Inupiat hunters in 1991 to that of their butchery and dismemberment in October, 1992 (see Diab 1998 for a more detailed account of the derivation of the ringed seal utility index). Although the heart was removed from all seals prior to the measurements outlined here, the indices are not affected because viscera are not included with any specific element class in this or any other pinniped utility index.

Specimen 1 (RS-B-8-26-91) was a subadult female, aged at 1.5 years based on dental annuli. It weighed 24.6 kg, and contained a relatively thin blubber layer, consistent with its death in August, a period of poor health and, thus, low weight (Frost and Lowry 1981). Specimen 2 (RS-B-1991-A) was a juvenile male, and based on its weight of 11.7 kg and length of 76.5 cm, is considered to be about 2-3 months old at death (Geraci and Lounsbury 1993). Specimen 3 (RS-B-1991-B) was a subadult/adult male, weighing 35.3 kg. Processing procedures followed closely those employed by Lyman et al. (1992b) and outlined above for harbor seals.

ELEMENT	HARBOUR SEAL 1 Flesh wt (gm)	HARBOUR SEAL 1 %MUI	HARBOUR SEAL 2 Flesh wt (gm)	HARBOUR SEAL 2 %MUI	HARBOUR SEAL 3 Flesh wt (gm)	HARBOUR SEAL 3 %MUI	HARBOR SEAL AVERAGE Flesh wt (gm)	HARBOR SEAL AVERAGE %MUI
Head	796	67.5	579	63.8	695	58.6	690	63.7
Cervical	987	83.6	691	76.1	854	72.0	844	78.0
Thoracic	681	57.7	538	59.3	828	69.8	682	63.0
Lumbar	1030	87.3	763	84.0	1186	100.0	993	91.7
Pelvis	782	66.3	669	73.7	636	53.6	696	64.3
Rib	1180	100.0	908	100.0	1160	97.8	1083	100.0
Sternum	136	11.5	149	16.4	205	17.3	163	15.1
Scapula	309	26.2	212	23.3	221	18.6	247	22.8
Humerus	219	18.6	209	23.0	194	16.4	207	19.2
Radius	143	12.1	98	10.8	109	9.2	117	10.8
Front flipper	56	4.7	54	5.9	59	5.0	56	5.2
Femur	85	7.2	79	8.7	106	8.9	90	8.3
Tibia	308	26.1	275	30.3	322	27.2	302	27.9
Rear flipper	117	9.9	113	12.4	124	10.5	118	10.9

Table 1. Flesh weights and % Meat Utility Indices (%MUI) for individual and averaged harbor seals.

ELEMENT	PARTS AVERAGED	HARBOR SEAL 1 Modified wt (gm)	HARBOR SEAL 1 %MMUI	HARBOR SEAL 2 Modified wt (gm)	HARBOR SEAL 2 %MMUI	HARBOR SEAL 3 Modified wt (gm)	HARBOR SEAL 3 %MMUI	HARBOR SEAL AVERAGE Modified wt (gm)	HARBOR SEAL AVERAGE %MMUI
Head	Head, Cerv	891.5	75.6	635	69.9	774.5	65.3	767.0	70.8
Cervical	None	987.0	83.6	691.0	76.1	854.0	72.0	844.0	77.9
Thoracic	Cerv, Lum, Rib	1065.7	90.3	787.3	86.7	1066.7	89.9	973.3	89.9
Lumbar	None	1030.0	87.3	763.0	84.0	1186.0	100.0	993.0	91.7
Pelvis	Lum, Pel	906.0	76.8	716.0	78.9	911.0	76.8	844.5	78.0
Rib	None	1180.0	100.0	908.0	100.0	1160.0	97.8	1083.0	100.0
Sternum	Rib, Ster	685.0	55.8	528.5	58.2	682.5	57.5	623.0	57.5
Scapula	Rib, Scap	744.5	63.1	560.0	61.7	690.5	58.2	665.0	61.4
Humerus	Scap, Hum	264	22.4	210.5	23.2	207.5	17.5	227.0	21.0
Radius-ulna	Hum, Rad	181.0	15.3	153.5	16.9	151.5	12.8	162.0	15.0
Front flipper	Rad, F. flip	99.5	8.4	76.0	8.4	84.0	7.1	86.5	8.0
Femur	Pelv, Tib	545.0	46.2	472.0	52.0	479.0	40.4	499.0	46.1
Tibia	None	308.0	26.1	275	30.3	322.0	27.2	302.0	27.9
Rear flipper	Tibia, R. flip	212.5	18.0	194.0	21.4	223.0	18.8	210.0	19.4

Table 2. Derivation of harbor seal % Modified Meat Utility Indices (%MMUI) from flesh weight.

Viscera and sculp weights, while recorded, are not incorporated into the utility indices for reasons outlined above. Processing was conducted with steel knives, and for Specimen 1, steel ulus (Inuit semi-circular knives) were used. Greatest potential errors occurred with the processing of the skull and associated soft tissues, since each seal had been shot through the head. Right and left gross weights of each sided body portion were compared for all 3 seals, and are positively and significantly correlated (in all cases, r_s>0.89, $P \leq 0.007$).

MUI values are calculated by subtracting bone weight from gross weight, and then standardized as %MUIs based on their percentage of the highest ranking element (Table 3). Flipper meat weights for specimens 1 and 3 are calculated in a manner similar to that for harbor seals; that is, front flipper flesh weights are considered to be 37% of gross weight, and rear flipper flesh weights are considered to be 29% of gross weight. The front flippers of specimen 2 were skinned prior to the measurements described here, and thus flipper weights are based on the average %MUI value of specimens 1 and 3. Finally, the same procedure as that described above for harbor seals was used to calculate ringed seal %MMUIs (Table 4).

COMPARISON OF PINNIPED UTILITY INDICES

In a number of zooarchaeological analyses, utility indices generated on the basis of a particular species have been used to interpret element distributions of a different species. For example, Binford's (1978) original caribou %MGUI has been used to interpret bison (Binford 1978:475; Speth 1983) and gemsbok assemblages (Bartram 1993), a combined harp seal-hooded seal %MMUI has been used to interpret ringed seal element frequencies (Lyman et al. 1992b), and a harbor porpoise %MMUI has been used to interpret beluga whale element frequencies (Savelle and Friesen 1996). In most cases, reservations are expressed regarding the validity of such inter-specific application of utility indices (e.g., Bartram 1993: 126). However, there appears to be no consensus as to how closely related two taxa must be in order to apply utility indices inter-specifically. In this respect, standard taxonomy might be expected to serve as a rough guide to the inter-specific application of utility indices, despite the fact that taxonomic classification is to a degree arbitrary. For instance, it would be useful to determine whether utility indices could be validly applied between species within the same genus, family, order, or other taxonomic category.

ELEMENT	RINGED SEAL 1 Flesh wt (gm)	%MUI	RINGED SEAL 2 Flesh wt (gm)	%MUI	RINGED SEAL 3 Flesh wt (gm)	%MUI	RINGED SEAL AVERAGE Flesh wt (gm)	%MUI
Head	476`	57.0	419	77.6	725	35.3	540	47.2
Cervical	687	82.3	314	58.1	1021	49.7	674	59.0
Thoracic	375	44.9	172	31.9	670	32.6	406	35.5
Lumbar	724	86.7	296	54.8	802	39.0	607	53.1
Pelvis	505	60.5	234	43.3	1274	62.0	671	58.7
Rib	835	100.0	540	100.0	2054	100.0	1143	100.0
Sternum	73	8.7	49	9.1	104	5.1	75	6.6
Scapula	312	37.4	213	39.4	551	26.8	359	31.4
Humerus	323	38.7	125	23.1	271	13.2	240	21.0
Radius	236	28.3	73	13.5	221	10.8	177	15.5
Front flipper	74	8.9	33.5	6.2	72	3.5	60	5.2
Femur	118	14.1	66	12.2	165	8.0	116	10.1
Tibia	371	44.4	169	31.3	488	23.8	343	30.0
Rear flipper	104	12.5	42	7.8	138	6.7	95	8.3

Table 3. Flesh weights and % Meat Utility Indices (%MUI) for individual and averaged ringed seals.

ELEMENT	RINGED SEAL 1 Parts averaged	Modified wt (gm)	%MMUI	RINGED SEAL 2 Parts averaged	Modified wt (gm)	%MMUI
Head	Head, Cerv	581.5	69.6	None	419.0	77.6
Cervical	None	687.0	82.3	Head, Cerv	366.5	67.9
Thoracic	Cerv, Lum, Rib	748.7	89.7	Cerv, Lum, Rib	383.3	71.0
Lumbar	None	724.0	86.7	None	296.0	54.8
Pelvis	Lum, Pelv	614.5	73.6	Lum, Pelv	265.0	49.1
Rib	None	835.0	100.0	None	540.0	100.0
Sternum	Rib, Stern	454.0	54.4	Rib, Stern	294.5	54.5
Scapula	Rib, Hum	579.0	69.3	Rib, Scap	376.5	69.7
Humerus	None	323.0	38.7	Scap, Hum	169.0	31.3
Radius-ulna	Hum, Rad	279.5	33.5	Hum, Rad	99.0	18.3
Front flipper	Rad, F. flip	155.0	18.6	Rad, F. flip	53.3	9.9
Femur	Pelv, Tib	438.0	52.5	Pelv, Tib	201.5	37.3
Tibia	None	371.0	44.4	None	169.0	31.3
Rear flipper	Tib, R. flip	237.5	28.4	Tib, R. flip	105.5	19.5
ELEMENT	RINGED SEAL 3 Parts averaged	Modified wt (gm)	%MMUI	RINGED SEAL AVERAGE Parts averaged	Modified wt (gm)	%MMUI
Head	Head, Cerv	873.0	42.5	Head, Cerv	607.0	53.1
Cervical	None	1021.0	49.7	None	674.0	59.0
Thoracic	Cerv, Lum, Rib	1292.3	62,9	Cerv, Lum, Rib	808.0	70.7
Lumbar	Lum, Pelv	1038.0	50.5	Lum, Pelv	639.0	55.9
Pelvis	None	1274.0	62.0	None	671.0	58.7
Rib	None	2054.0	100.0	None	1143.0	100.0
Sternum	Rib, Stern	1079.0	52.5	Rib, Stern	609.0	53.3
Scapula	Rib, Scap	1302.5	63.4	Rib, Scap	751.0	65.7
Humerus	Scap, Hum	411.0	20.0	Scap, Hum	299.5	26.2
Radius-ulna	Hum, Rad	246.0	12.0	Hum, Rad	208.5	18.2
Front flipper	Rad, F. flip	146.5	7.1	Rad, F. flip	118.5	10.4
Femur	Pelv, Tib	881.0	42.9	Pelv, Tib	507.0	44.4
Tibia	None	488.0	23.8	None	343.0	30.0
Rear flipper	Tib, R. flip	313.0	15.2	Tib, R. flip	219.0	19.2

Table 4. Derivation of ringed seal % Modified Meat Utility Indices (%MMUI) from flesh weights.

Pinniped data accumulated over the past five years allow us to address this issue. As a result of the calculation of meat utility for harbor and ringed seals described here, Pinnipedia can now be considered the best-represented mammalian order in terms of meat utility indices. As such, this order offers the best opportunity for comparison of utility indices within a single species, and between species within the same genus, family and order. Within this single order, 12 specimens representing five species in two families have been processed specifically for the development of utility indices. Family Otariidae is represented by a single species, California sea lion (*Zalophus californianus*). Family Phocidae is represented by four species in two subfamilies; with subfamily Phocinae represented by harbor seal (*Phoca vitulina*), ringed seal (*Phoca hispida*), and harp seal (*Phoca groenlandica*), and subfamily Cystophorinae represented by hooded seal (*Cystophora cristata*). In the remainder of this paper, we assess the degree of correlation in utility indices among these taxa.

Table 5 presents flesh weights and %MUIs for the two California sea lions, three harp seals, and one hooded seal published elsewhere (Lyman et al. 1992b; Savelle et al. 1996), as well as averaged values for each species. These %MUIs are calculated from raw data provided in earlier reports, and in all cases the utility values of flippers are recalculated on the basis of the correction factor outlined earlier in this paper in order to ensure similar derivations of

ELEMENT	CALIFORNIA SEA LION			HARP SEAL				HOODED SEAL
	No. 1	No. 2	Avg.	No. 1	No. 2	No. 3	Avg.	
Head	34.8	25.7	30.7	15.5	21.6	40.0	20.7	69.3
Cervical	100.0	86.6	94.8	34.2	31.1	45.2	34.5	44.2
Thoracic	25.7	17.7	22.0	20.3	24.7	18.8	21.7	44.6
Lumbar	13.3	13.8	13.8	28.4	18.0	63.4	29.1	57.1
Pelvis	21.8	20.2	21.4	37.5	50.1	42.2	42.6	56.7
Rib	96.5	100.0	100.0	100.0	100.0	100.0	100.0	100.0
Sternum	8.7	9.3	9.2	2.4	2.2	4.8	2.6	3.2
Scapula	29.2	27.1	28.6	18.9	22.7	19.3	20.3	16.7
Humerus	19.1	12.0	15.8	9.3	12.2	12.9	10.8	10.2
Radius-ulna	10.8	11.8	11.5	3.8	4.6	6.7	4.4	6.8
Front flipper	3.2	3.5	3.4	1.1	2.0	3.1	1.7	2.1
Femur	3.4	4.3	3.9	3.8	4.3	10.8	4.8	2.3
Tibia	6.2	6.3	6.4	16.0	16.2	20.4	16.6	12.6
Rear flipper	3.0	3.2	3.2	3.7	5.0	6.5	4.5	4.5

Table 5. Percent Meat Utility Indices (%MUI) for California sea lions (Savelle et al 1996), harp seals (Lyman et al 1992b), and a single hooded seal (Lyman et al 1992b).

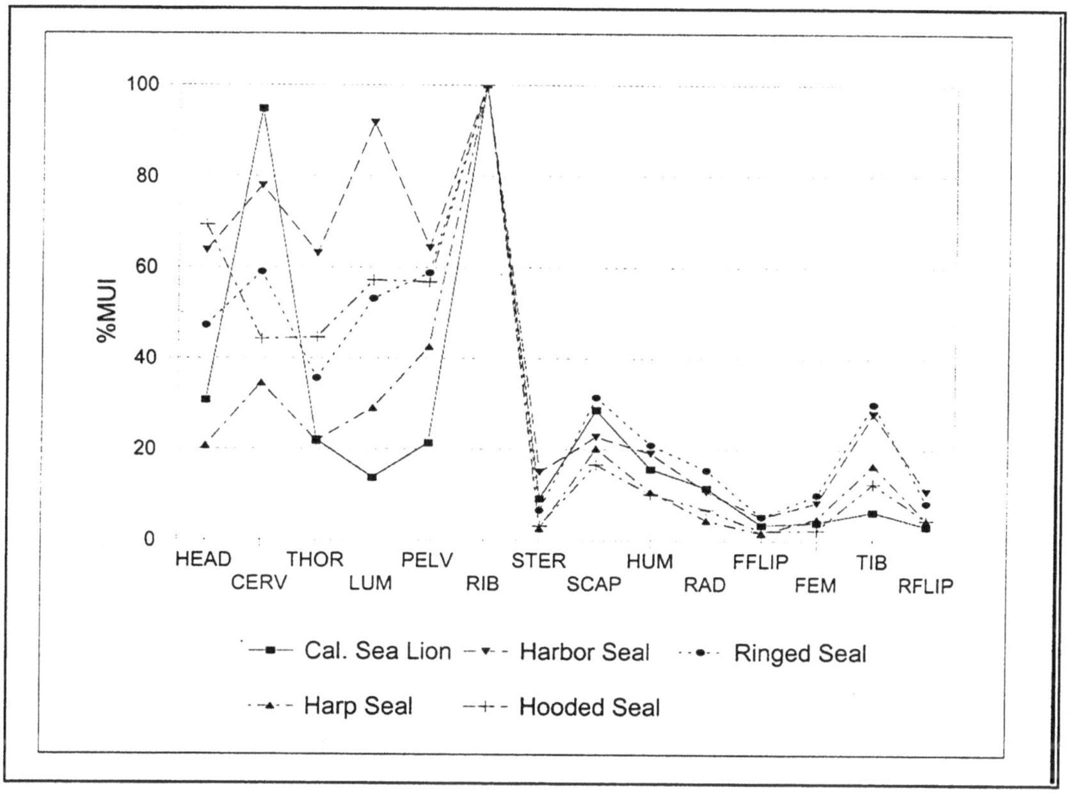

Figure 1. Comparision of %MUI values for five species in the order Pinnipedia.

the indices. Figure 1 graphically compares the five averaged %MUIs for the three previously published species and the harbor and ringed seals described in this paper (in the case of the hooded seal, the %MUI of the single individual is used as an average for this species). Several observations can be made on the basis of these comparisons. First, all five species share several major characteristics in the distribution of flesh. In all cases, the most flesh is associated with the ribs, and more flesh is associated with the axial skeleton than with the appendicular skeleton. Second, all species are similar in the distribution of flesh on the appendicular skeleton, but are more variable in the distribution of flesh on the axial skeleton. Third, the California sea lion stands out as most at variance with the other species. This species contains the highest ranked cervical vertebrae, and the lowest ranked thoracic and lumbar vertebrae, pelvis, and hind limb. These differences are expected, due to the gross anatomical differences which exist between otariids (sea lions and fur seals) and phocids (true seals). At the broadest level, otariids have much more substantial necks, longer forelimbs, and more slender bodies than phocids (e.g., Howell 1929).

The utility data have thus far been presented as interval scale measurements in order to illustrate gross similarities and differences among species. However, due to the small sample sizes of animals processed for each species, and since not all potential variation within each species according to age, sex, or other factors is incorporated into the utility indices, %MUIs should preferably be analysed as ordinal scale measures (Lyman 1994:231-234). Accordingly, in the following comparisons Spearman's rho (r_s), which calculates rank-order correlation and is therefore appropriate for analysis of ordinal scale data, is used to interpret the degree of similarity between the various indices.

Rank-order correlation between all possible pairs of %MUIs for the 12 individual seals (as opposed to the five averaged seal species) indicates that all are significantly and positively correlated (in all cases, $r_s \geq 0.73$, $P \leq 0.003$). Two observations on these coefficients confirm the general expectation that individuals which are closer to one another taxonomically should also exhibit more closely correlated utility indices. First, in all comparisons between different individuals within a single species (for example, comparison of the three harbor seals with each other), the correlation coefficients are relatively high (in all cases, $r_s \geq 0.86$, and $P<0.001$.) Therefore, variability is generally observed to be greater between species than within them. Second, the most divergent species is the California sea lion, which is the only one which differs taxonomically from the others at the family level. This divergence is indicated by the fact that all cases for which $r_s \leq 0.80$ and $P \geq 0.001$ represent correlations of individual sea lions with individuals of other species.

However, it is important to note that in no case are any two individuals significantly different. The same relationship is true of correlations between the averaged data for each of the five species (see Figure 1). In all cases, $r_s \geq 0.77$ and $P \leq 0.001$. Again, the averaged California sea lion index was the most divergent from the other averaged indices, although it is still significantly correlated.

This process was repeated for the %MMUI values (Table 6). The %MMUI values presented here are recalculated from the original data in order to ensure that averaging methods are the same for all indices, and thus in some instances the values are slightly different from earlier published versions. This comparison yielded similar results to the comparison of %MUIs. When %MMUIs are calculated for all possible pairs of the twelve individual pinnipeds, none was significantly different from any other, although correlation coefficients are slightly lower than for %MUIs. In all cases, $r_s \geq 0.64$, and $P \leq 0.014$. The six instances of coefficients less than 0.71 ($P \geq 0.005$), all occur in comparisons between California sea lions and individuals of other species. The averaged %MMUI data indicate the same pattern (Figure 2). In all cases, $r_s \geq 0.74$ and $P \leq 0.002$; and both cases in which $r_s < 0.80$ and $P \geq 0.001$ involved correlation of California sea lions with other seal species.

DISCUSSION

An important remaining issue is whether this pattern of anatomical similarity is restricted to the order Pinnipedia, or if it applies to other mammalian orders as well. This issue can be addressed, at least in part, by considering utility indices developed for the order Artiodactyla, which is the only other mammalian order for which utility indices have been calculated for more than one species. Within the order Artiodactyla, utility indices have been developed for five species on the basis of primary processing data: domestic sheep (*Ovis aries*, Binford 1978), caribou (*Rangifer tarandus*, Binford 1978), domestic llama (*Lama glama*, Mengoni-Gonalons 1991), guanaco (*Lama guanicoe*, Borrero 1990), and bison (*Bison bison*, Emerson 1993). Will's (1985) muskox utility index is not considered here because it was calculated from other species' anatomy. Because these utility indices have been reported and presented with varying degrees of detail, and occasionally with different calculation methods, in this study utility indices developed for domestic llama (Mengoni-Gonalons 1991; reproduced in Lyman 1994), caribou (Binford 1978; Metcalfe and Jones 1988), and bison (Emerson 1993) are compared. Importantly, these three species represent three different families: Camelidae, Cervidae and Bovidae, respectively. The specific utility index to be compared is the %FUI (Food Utility Index; Metcalfe and Jones 1988). This is very similar to the %MMUI outlined above, with the important difference being that averaging of body parts to account for "riders" is performed only for bones of the appendicular skeleton. The %FUI was calculated in the same way for the three species (Table 7).

ELEMENT	CALIFORNIA SEA LION			HARP SEAL				HOODED SEAL
	No. 1	No. 2	Avg.	No. 1	No. 2	No. 3	Avg.	
Head	67.4	56.2	62.8	24.8	26.4	42.6	27.6	69.3
Cervical	100.0	86.6	94.8	34.2	31.1	45.2	34.5	57.0
Thoracic	98.3	93.3	97.4	54.2	65.6	69.5	54.5	78.5
Lumbar	23.8	18.9	21.7	33.0	37.4	63.4	35.8	57.1
Pelvis	21.8	20.2	21.4	37.5	50.1	52.8	42.6	56.9
Rib	96.5	100.0	100.0	100.0	100.0	100.0	100.0	100.0
Sternum	52.6	54.6	54.6	51.2	51.1	52.4	51.3	51.6
Scapula	62.9	63.5	64.3	59.4	61.3	59.6	60.1	58.3
Humerus	24.2	19.5	22.2	14.1	17.4	16.1	15.5	13.5
Radius-ulna	15.0	11.9	13.6	6.6	8.4	9.8	7.6	8.5
F. flipper	7.0	7.6	7.5	2.4	3.3	4.9	3.1	4.5
Femur	14.0	13.2	13.9	26.8	33.2	31.3	29.6	34.6
Tibia	6.2	6.3	6.4	16.0	16.2	20.4	16.6	12.6
R. flipper	4.6	4.7	4.8	9.8	10.6	13.4	10.5	8.5

Table 6. Percent Modified Meat Utility Indices (%MMUI) for California sea lions (Savelle et al 1996), harp seals (Lyman et al 1992b), and a single hooded seal (Lyman et al 1992b).

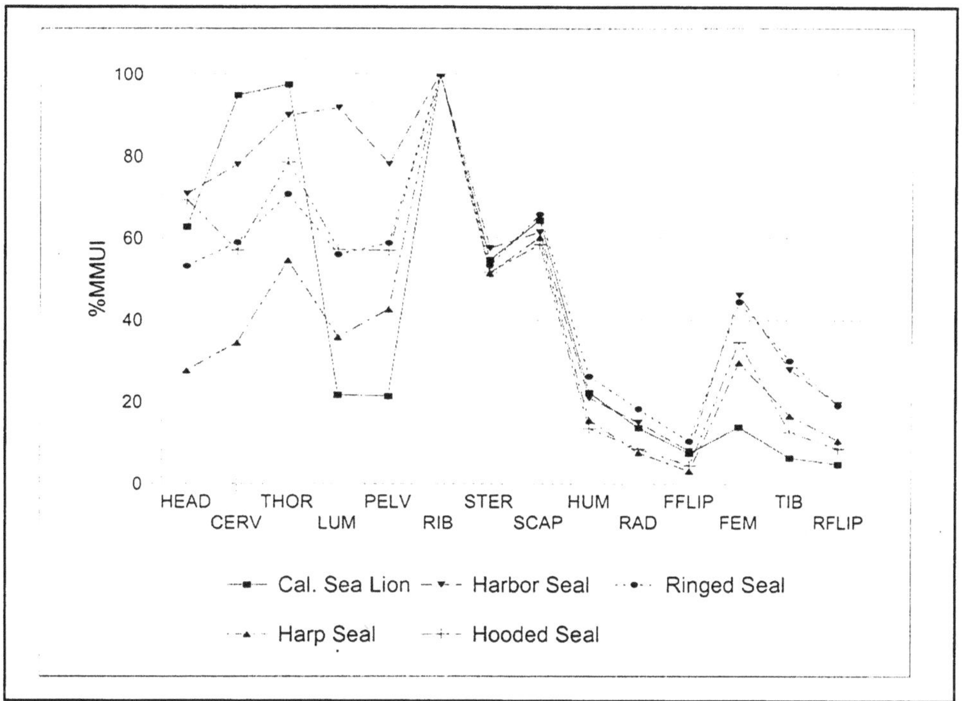

Figure 2. Comparison of %MMUI values for five species in the order Pinnipedia.

ELEMENT	LLAMA (Camelidae)	CARIBOU (Cervidae)	BISON (Bovidae)
Head	24.7	49.3	25.3
Atlas-axis	8.6	10.2	9.1
Cervical 3-7	64.2	37.1	38.6
Thoracic	61.8	47.3	47.4
Lumbar	78.0	33.2	45.1
Pelvis	40.2	49.3	34.7
Rib	100.0	51.6	62.3
Sternum	99.4	66.6	32.4
Scapula	41.7	44.7	27.5
Humerus	36.7	36.8	27.5
Radius-ulna	23.0	25.8	19.2
Metacarpal	6.5	5.2	6.5
Femur	75.9	100.0	100.0
Tibia	43.0	62.8	57.7
Metatarsal	11.5	37.0	16.1

Table 7. Percent Food Utility Index (%FUI) for Llama (Mengoni-Gonalons 1991), Caribou (Binford 1978), and Bison (Emerson 1993).

ELEMENT	HARBOR PORPOISE (Cetacea)	HARP SEAL (Pinnipedia)	CARIBOU (Artiodactyla)
Head	94.3	20.7	49.4
Cervical	4.3	34.5	47.3
Thoracic	49.2	21.7	47.3
Lumbar	100.0	29.1	33.2
Rib	39.7	100.0	51.6
Sternum	2.1	2.6	66.6
Scapula	4.8	20.3	44.7
Front limb	5.8	16.9	50.6

Table 8. Percent Meat Utility Indices (%MUI) for Harbor Porpoise (Savelle and Friesen 1996), Harp Seal (Lyman et al 1992b), and Caribou (Binford 1978).

Figure 3 graphically compares the three artiodactyl %FUIs. Rank-order correlations confirm the pattern observed within the order Pinnipedia, in that all are significantly positively correlated with each other (in all cases, $r_s \geq 0.67$ and $P \leq 0.006$). Thus, within Artiodactyla, utility indices based on any of the three species should be useful for the interpretation of element distributions for species in other artiodactyl families. The fact that within the orders Pinnipedia and Artiodactyla the inter-specific application of utility indices is valid implies that, if suitable caution is applied, the same may be true of other mammalian orders.

This fact, however, raises another question: are any utility indices significantly different from one another, or are they all statistically similar? In order to answer this question, utility indices of species from different mammalian orders must be compared. At present, three orders are available for comparison: Cetacea (whales), Pinnipedia and Artiodactyla. A fourth order, Marsupialia, in which utility indices have been calculated for red kangaroo (O'Connell and Marshall 1989), is inappropriate for comparison because its utility index is created using different body-part divisions.

The orders Cetacea, Pinnipedia, and Artiodactyla exhibit obvious and major anatomical differences relating to habitat and locomotion; they are, respectively, two-limbed marine mammals, four-limbed marine mammals, and four-limbed terrestrial mammals. Rather than compare all available species, the first species within each order for which a utility index was constructed was selected for comparison. Thus, comparison is between harbor porpoise in order Cetacea (Savelle and Friesen 1996), harp seal in order Pinnipedia (Lyman et al. 1992b), and caribou in order Artiodactyla (Binford 1978) (Table 8).

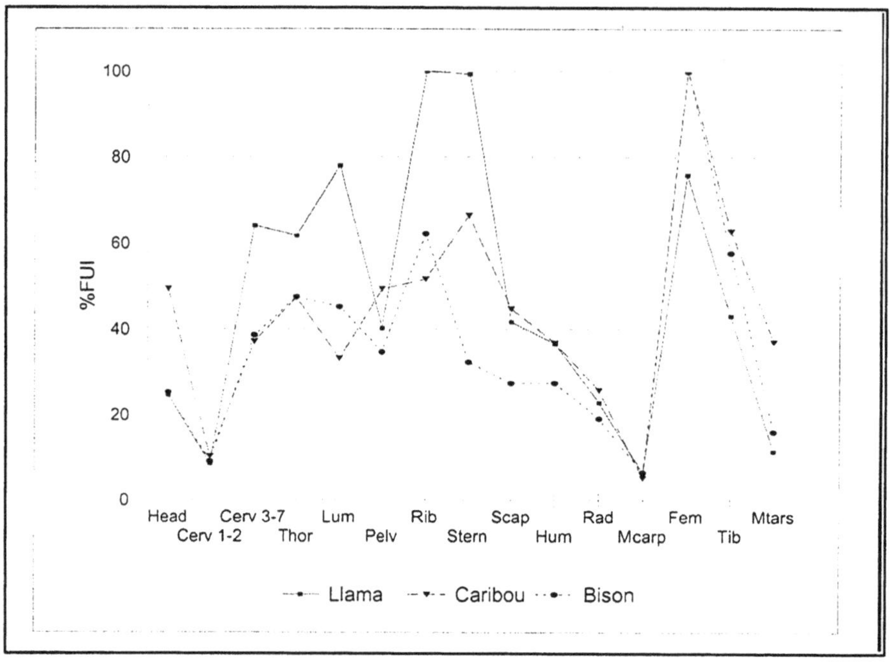

Figure 3. Comparison of %FUI values for three species in the order Artiodactyla.

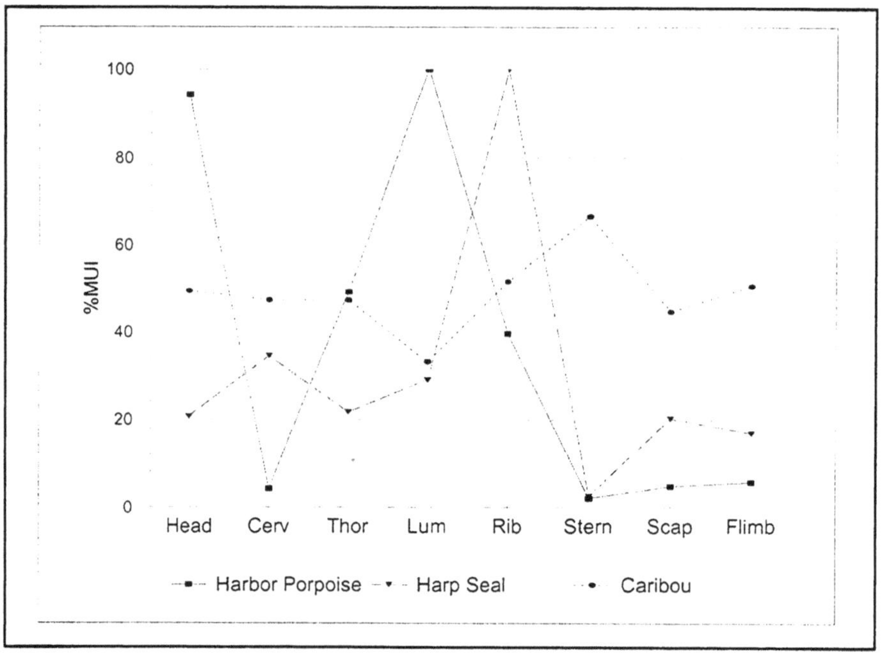

Figure 4. Comparison of %MUI values for harbor porpoise (Cetacea), harp seal (Pinnipedia) and caribou (Artiodactyla).

For this comparison %MUIs are recalculated from original raw data to ensure similarity of index derivation. Therefore, while earlier authors modified the utility values of heads for harbor porpoise (Savelle and Friesen 1996) and caribou (Binford 1978), for present purposes the full flesh weights of heads are used in order to make them directly comparable to the seal data. It should be noted that only eight element classes can be compared, primarily because species in the order Cetacea do not retain hind limbs.

Not unexpectedly, the anatomical patterns of the three orders are markedly different from each other (Figure 4), and none is significantly correlated with any other. Two pairs yielded low negative coefficients (porpoise-caribou: r_s=-0.43, P=0.286; caribou-seal: r_s=-0.29, P=0.490), and the third pair yielded a low positive correlation (porpoise-seal: r_s=0.38, P=0.352). On the basis of these comparisons, it is clear that there is no general mammalian anatomical pattern which can be used as a standardized utility index. Species from different orders are dissimilar enough to make cross-order comparison unjustifiable.

CONCLUSION

Statistical comparisons presented in this paper emphasize the fact that if utility indices are viewed as ordinal measures, it is justifiable to apply a species-specific utility index to a different species within the same order, if a utility index for a more closely related taxon is not available. We believe that the pattern is robust enough that it will be applicable to other measures derived from body part data, such as the ratio of flesh weight to bone weight as advocated by Ringrose (1993). This does not mean that new utility indices should not be developed. Each new index will provide a test of the conclusions outlined here and, when available, it is preferable to use species-specific utility indices. In addition, the production of new utility indices is valuable for the heuristic interpretation of "utility curve" scatter plots comparing utility indices to archaeological element frequencies, as originally performed by Binford (1978) and subsequently by many other zooarchaeologists.

An important corollary issue is the degree to which bone structural density, as opposed to food utility, varies among related taxa. Because structural density is also crucial to zooarchaeological interpretation, analysts must be able to determine whether species-specific density values can be used to interpret element frequencies for different species within the same genus, family, or order.

Appropriate data exist for six taxa within the order Artiodactyla (deer, pronghorn antelope, and domestic sheep (Lyman 1982, 1984), bison (Kreutzer 1992), and guanaco and vicuna (Elkin and Zanchetta 1991)), two within the order Rodentia (woodchuck and yellow-bellied marmot (Lyman et al. 1992a)), and one within the order Pinnipedia (an averaged set of seal values (Chambers 1992, cited in Lyman 1994)). Published statistical comparisons of bone density values, while limited, indicate strong correlation within orders, as between two species of deer (r>0.95, P<0.001; Lyman 1984), between deer and bison (r_s=0.87, P<0.001; Kreutzer 1992), and between two species of marmot (r=0.75, P<0.001; Lyman et al. 1992a). Inter-order comparisons of structural density have been made only between marmot and deer (r=0.55, P<0.001) and marmot and bison (r=0.59, P=0.001), in both cases the correlation is significant but weaker than that within orders (Lyman et al. 1992a). Further discussion of this issue is beyond the scope of the present paper, although additional comparisons should be made between the full range of species in order to determine the degree to which density values can be applied across different taxa.

In conclusion, we believe that our results display explicitly what has previously been implicit in studies of economic anatomy; namely, that if used with caution, utility indices developed for a particular species can be used to interpret element frequencies of related species. If utility indices are viewed as ordinal scale measures, and if they are compared to zooarchaeological data using rank-order correlation, then indices can be used to interpret species in related genera, or even families, within the same order.

ACKNOWLEDGMENTS

The research reported here was made possible through grants from the Social Sciences and Humanities Research Council of Canada. Thomas G. Smith, George Horonowitsch and Gregor Beck, all then of the Pacific Biological Station, Nanaimo, B.C., kindly arranged for the acquisition, and assisted in the processing of the harbor seals. The ringed seals were provided by the Department of Wildlife Management of the North Slope Borough, Alaska, and the hunters of Barrow, Alaska. Francis Fay, Elizabeth Frantz, Craig Gerlach, Donald Hartbauer, Elain Hogard, Robin Mills, and James Nageak helped with their processing and interpretation. R. Lee Lyman has been an important source of advice and encouragement in our studies of utility indices, and provided valuable comments on this paper. We also appreciate the comments of the other reviewers: Craig Gerlach, Lee Kreutzer, and Duncan Metcalfe. Our sincere thanks are extended to all of these individuals and organizations.

REFERENCES CITED

Banfield, A.W.F.
1974 *The Mammals of Canada*. University of Toronto Press, Toronto.

Bartram, L.E.
1993 Perspectives on skeletal part profiles and utility curves from eastern Kalahari ethnoarchaeology. *In From Bones to Behavior*, edited by J. Hudson, pp. 115-137. Center for Archaeological Investigations, Carbondale.

Behrensmeyer, A.K.
1975 The taphonomy and paleoecology of Plio-Pleistocene vertebrate assemblages east of Lake Rudolf, Kenya. *Bulletin of the Museum of Comparative Zoology* 146:473-578.

Binford, L.R.
1978 *Nunamiut Ethnoarchaeology*. Academic Press, New York.

Borrero, L.
1990 Fuego-Patagonian bone assemblages and the problem of communal guanaco hunting. In *Hunters of the Recent Past*, edited by L.B. Davis and B.O.K. Reeves, pp. 373-399. Unwin Hyman, London.

Brain, C.K.
1969 The contribution of Namib Desert Hottentots to an understanding of Australopithecine bone accumulations. *Scientific Papers of the Namib Desert Research Station* 39:13-22.

Bryden, M.M. and W.J.L. Felts
1974 Quantitative anatomical observations on the skeletal and muscular systems of four species of Antarctic seals. *Journal of Anatomy* 118:589-600.

Chambers, A.L.
1992 Seal bone mineral density: its effect on specimen survival in archaeological sites. Unpublished Honors thesis, Department of Anthropology, University of Missouri, Columbia.

Committee on Marine Mammals
1967 Standard measurement of seals. *Journal of Mammalogy* 48:459-462.

Diab, M.C.
1998 Economic utility of the ringed seal (*Phoca hispida*): implications for arctic archaeology. *Journal of Archaeological Science* 25:1-26.

Elkin, D.C. and J.R. Zanchetta.
1991 Densitometria osea de camélidos - aplicaciones arqueológicas. *Actas del X Congreso Nacional de Arqueologia Argentina* 3:195-204. Catamarca.

Emerson, A.M.
1993 The role of body part utility in small-scale hunting under two strategies of carcass recovery. *In From Bones to Behavior*, edited by J. Hudson, pp. 138-155. Center for Archaeological Investigations, Carbondale

Frost, K.J. and L.F. Lowry
1981 Ringed, Baikal and Caspian seals: *Phoca hispida* schreber, 1775; *Phoca sibirica* gmelin, 1778 and *Phoca caspica* gmelin, 1788. In *Handbook of Marine Mammals, Volume 2: Seals*, edited by S.H. Ridgway and R.J. Harrison, pp. 29-53. Academic Press, London.

Geraci, J.R. and V.J. Lounsbury
1993 *Marine Mammals Ashore: A Field Guide for Strandings*. Texas A&M Sea Grant Publication, Galveston.

Guthrie, R.D.
1967 Differential preservation and recovery of Pleistocene large mammal remains in Alaska. *Journal of Paleontology* 41:243-246.

Howell, A.B.
1929 Contribution to the comparative anatomy of eared and earless seals (Genera *Zalophus* and *Phoca*). *Proceedings of the United States National Museum* 73:1-142.

Kreutzer, L.A.
1992 Bison and deer bone mineral densities: comparisons and implications for the interpretation of archaeological faunas. *Journal of Archaeological Science* 19:271-294.

Lyman, R.L.
1982 The taphonomy of vertebrate archaeofaunas: bone density and differential survivorship of fossil classes. Ph.D. dissertation, University of Washington, Seattle. University Microfilms International, Ann Arbor.

1984 Bone density and differential survivorship of fossil classes. *Journal of Anthropological Archaeology* 3:259-299.

1993 Density-mediated attrition of bone assemblages: new insights. In *From Bones to Behavior*, edited by J. Hudson, pp. 324-341. Center for Archaeological Investigations, Carbondale.

1994 *Vertebrate Taphonomy*. Cambridge University Press, Cambridge.

Lyman, R.L., L.E. Houghton and A.L. Chambers.
1992a The effect of structural density on marmot skeletal part representation in archaeological sites. *Journal of Archaeological Science* 19:557-573.

Lyman, R.L., J.M. Savelle and P. Whitridge.
1992b Derivation and application of a food utility index for phocid seals. *Journal of Archaeological Science* 19:531-555.

Mengoni-Gonalons, G.L.
1991 La llama y sus productos primarios. *Arqueologia: Revista de la Sección Prehistoria* 1:179-196.

Metcalfe, D. and K.T. Jones
1988 A reconsideration of animal body part utility indices. *American Antiquity* 53:486-504.

O'Connell, J.F. and B. Marshall
1989 Analysis of kangaroo body part transport among the Alyawara of Central Australia. *Journal of Archaeological Science* 16:393-405.

Ringrose, J.T.
1993 Bone counts and statistics: a critique. *Journal of Archaeological Science* 20:121-157.

Savelle, J.M. and T.M. Friesen
1996 An odontocete (Cetacea) meat utility index. *Journal of Archaeological Science* 23:713-722.

Savelle, J.M., T.M. Friesen and R.L. Lyman
1996 Derivation and application of an otariid utility index. *Journal of Archaeological Science* 23:705-712.

Speth, J.D.
1983 *Bison Kills and Bone Counts: Decision Making by Ancient Hunters*. University of Chicago Press, Chicago.

Will, R.T.
1985 *Nineteenth Century Copper Inuit Subsistence Practices on Banks Island, N.W.T.* Unpublished Ph.D. Thesis, University of Alberta, Edmonton.

THE ROLES OF BONE DENSITY AND CULTURAL BEHAVIOUR IN HEAD-SMASHED-IN BONE BED TAPHONOMY

Brian Kooyman
Department of Archaeology, University of Calgary

SAMPLE CONTEXT

Head-Smashed-In Buffalo Jump is located in the southwestern corner of the province of Alberta, Canada about 150 km south of Calgary (Figure 1). The most prominent feature of the site is the cliff over which the bison were driven. The cliff is about ten to fifteen meters high and at its' base the kill site archaeological deposits are one to about 10 meters deep. Use of the site dates to at least 5500 B.P. The kill deposits are composed largely of butchered buffalo bone. The most common lithic tools recovered are projectile points and lithic debitage is also quite common. Further downslope from the kill deposits is a large flat area that was used for processing and as a camp. Above and behind the cliff, drive lanes stretch out into a 40 square km, bowl-shaped gathering basin.

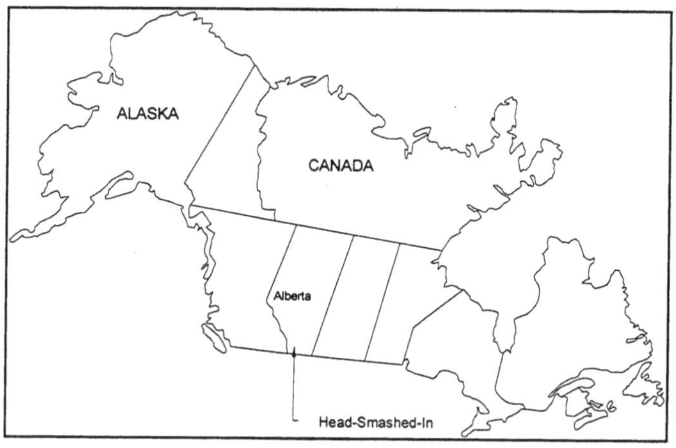

Figure 1. Location of Head-Smashed-Buffalo Jump.

In 1965 research begun by R.G. Forbis and B.O.K. Reeves established the considerable time depth represented at Head-Smashed-In and provided an appreciation of the complexity of the use of the jump (e.g. Reeves 1978). More detailed knowledge of the nature of the processing at the site has been revealed by the work of J. Brink (e.g. Brink *et al.* 1985; see Brink, this volume). My excavations (Kooyman 1990, 1992) were undertaken between 1989 and 1993 in the area known as the North Kill. The final excavation depth was almost four meters below surface. Twelve cultural layers with various subdivisions were defined, from Late Prehistoric Old Women's phase at the top through to Middle Prehistoric Mummy Cave at the bottom. This lowest level provided a radiocarbon date of 4700 B.P. Three-point provenience was taken for all remains recovered *in situ*. All excavation was done by trowel to recover as much *in situ* material as possible. All bones were drawn in place on floor plans. Various observations useful for the taphonomic study were taken on the bones in the field.

Although the quantity of bone recovered was staggering, significant practical problems were encountered because of poor preservation. Many of the bones disintegrated when they were removed from the sediment matrix. As well, many of the *Bison* teeth fell apart in the excavation bags, hence much useful seasonality data was lost. Field recording procedures were devised to address these problems. Cards were printed and filled in once a bone was exposed but before it was removed from the sediment. Bone fragment units, based on a system developed by Brumley (1973, 1995), were recorded, as were observations on fracture type (spiral, etc.), weathering stage, and other data useful for taphonomic studies. All tooth eruption and wear stages were also recorded where appropriate.

In the thicker layers, many of the bones had disintegrated into each other in the sediment, hence the photographic record was very disappointing. Because of this situation, we began doing the floor plan drawings as true artistic renditions of the bone as excavated. The bone layers were so thick in some areas that all the bones from a layer could not be drawn on a single sheet. The floor plans are literally peels of the bone beds, in some cases requiring 10 or 11 separate drawings before all the bones were exposed and removed from a layer.

Before beginning the discussion of the analysis some comments about quantification of the material are appropriate. My standard practice is to use the four basic quantification techniques: Weight; NISP; MNI; and MNE/MAU, together to try to better understand my material. I have based my MNI and MNE/MAU counts on individual landmark totals (Kooyman 1985:123). Hence the tibia values may be based on counts of the proximal caudal shaft nutrient foramen, the mandible values on the caudal cusp of the third molar, the rib values on the head, etc. Each bone has from as few as five or six such landmarks, to as many as 40 to 50. Use of small, discrete, easily identifiable landmarks eliminates many of the problems of interdependence inherent in zooarchaeological quantification. Morlan (1994) has recently developed a similar technique.

TAPHONOMIC ANALYSIS

This paper examines the bone from a single sublayer within one layer of the excavations. Layer 5 dates to Avonlea times and shows five recognizable subdivisions. Five radiocarbon dates from this Layer fall between 1320 and 1550 B.P. (with standard deviations of between 70 and 90 years). Hence the entire Layer comes from a restricted time period and this particular analysis employs material from one of the five subdivisions. Although the actual number of kill events and seasons involved cannot be defined with certainty, in archaeological terms we are looking at a short time period and one would presume relatively uniform general taphonomic conditions.

Sublayer "d" in Layer 5 was unusual in that a portion of the sublayer showed much better bone preservation than did the rest of the sublayer. Using the 1-6 weathering stage scale Todd (Todd *et al.* 1987:64) used based on the earlier work of Behrensmeyer (1978) (1 = unweathered, 6 = falling apart), this better preserved area shows (Table 1) an average weathering stage of only 3.8 compared to the average of 4.5 for the poorly preserved portion this sublayer. This better preservation was apparently due to the fact that a thin film of sediment had washed over this portion of the site. The sediment protected this bone from further subaerial weathering at some stage after the bone was left on the site surface, although obviously some weathering had already occurred prior to burial. This sediment film was very thin, generally only one to two cm thick. Despite its insubstantial nature, this thin layer had a considerable effect on the preservation of the bone. The bone layer in the two areas was clearly continuous and resulted from the same events. The only difference was the overlying thin sediment layer. This sublayer is a sample of the bone left at Head-Smashed-In from a specific series of events and the well and poorly preserved areas are then subsamples of the same sample. In probabilistic terms, then, the two areas may show a few minor differences, but they should have been originally essentially the same. To the extent that they differ, we have a controlled experiment showing the effects of differential preservation. The cultural butchering pattern is also a factor in the actual overall patterns of both samples, of course, and I will examine this aspect as well.

WEATHERING STAGE	WELL PRESERVED		POORLY PRESERVED	
	Number	Percent	Number	Percent
1	0	0	0	0
2	37	7	1	1
3	123	23	15	8
4	267	51	70	40
5	90	17	71	40
6	11	2	20	11
Total	528		177	

Table 1. Bone Weathering Stages, Layer 5d.

The difference in preservation is seen in Figure 2, a graph of the standardized frequency of bones in the well-preserved area (open bars) versus the frequency in the poorly preserved area. With a few exceptions, the bones from the poorly preserved area are underrepresented compared to the well-preserved area. There is no archaeological indication of any activity difference, even minor, between these two portions of this sublayer. The only difference these two samples have experienced is in the amount of post-depositional weathering as a result of the thin sediment film. Despite the clear differences in these two samples there is also an underlying similarity--this similarity is the result of the butchering and discard pattern resulting from this site being a communal kill site. For example, some bone elements have more economic value for food products such as meat and bone marrow (Binford 1978, Emerson 1990), and these are usually removed from kill sites and taken to camp sites. Both sets of data show a high frequency of low value bones such as mandibles, metapodials, phalanges, and tarsals.

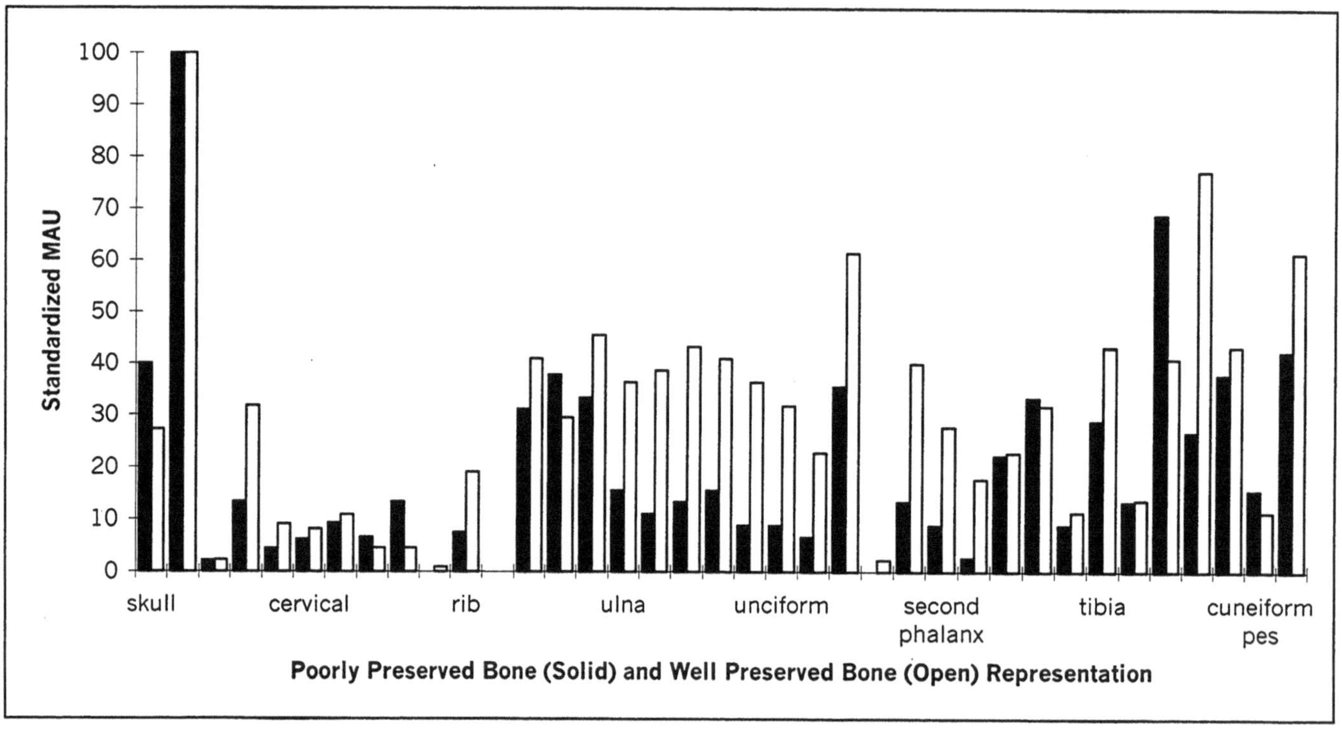

Figure 2. Standardized MAU comparison for well and poorly preserved bone in Layer 5d at Head-Smashed-In.

I have used Kreutzer's (1992) data on *Bison* bone density (volume density) to determine the effects of differential preservation. Since I do my element counts based on anatomical landmarks, I correlated each of Kreutzer's bone density scan sites with specific anatomical landmarks and then calculated an MAU value for each scan site. Figures 3 and 4 show scan site density on the X axis (multiplied by 100) compared to the raw MAU value on the Y axis. Bearing in mind that cultural patterns also influence element representation, the plot for the well preserved area (Figure 4) shows a reasonably good correlation (correlation coefficient 0.4720, p<.001) between increasing density and increasing representation in the remains. Conversely, the plot for the poorly preserved area (Figure 3) shows a noticeably lower though still statistically significant correlation between density and representation (correlation coefficient 0.3567, p<.001). Visual inspection of Figure 3 suggests that all elements are equally poorly represented, with a few exceptions, regardless of bone density. Here bone density has only very generalized predictive value. Instead of bone density proportionally influencing the observed pattern, there seems to be a threshold of some sort which isolates a few bones as highly preserved compared to all the rest.

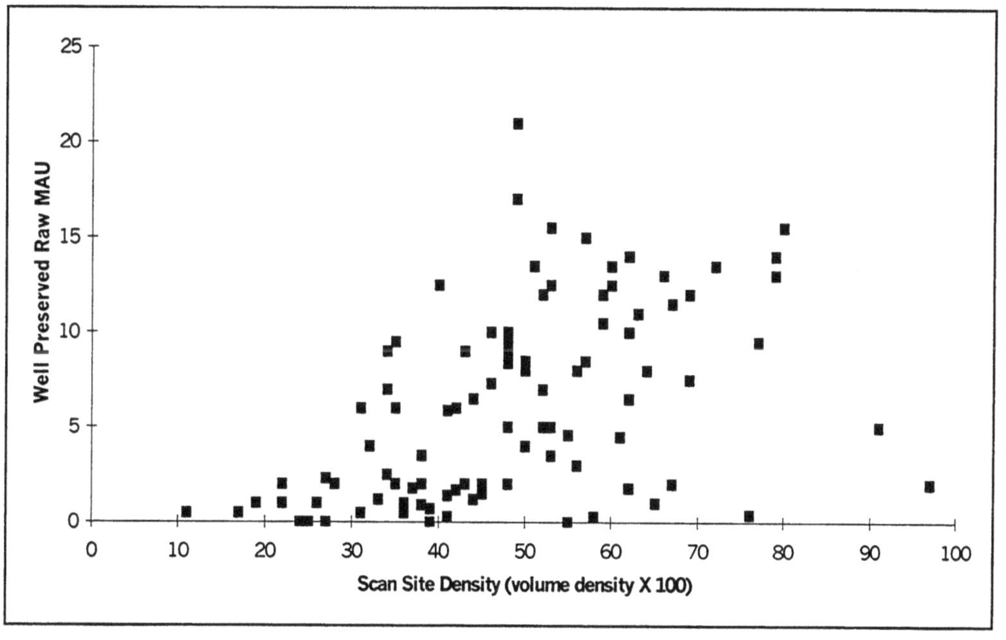

Figure 3. Scan site density (Kreutzer 1992) versus raw MAU for Head-Smashed-In poorly preserved area, Layer 5d

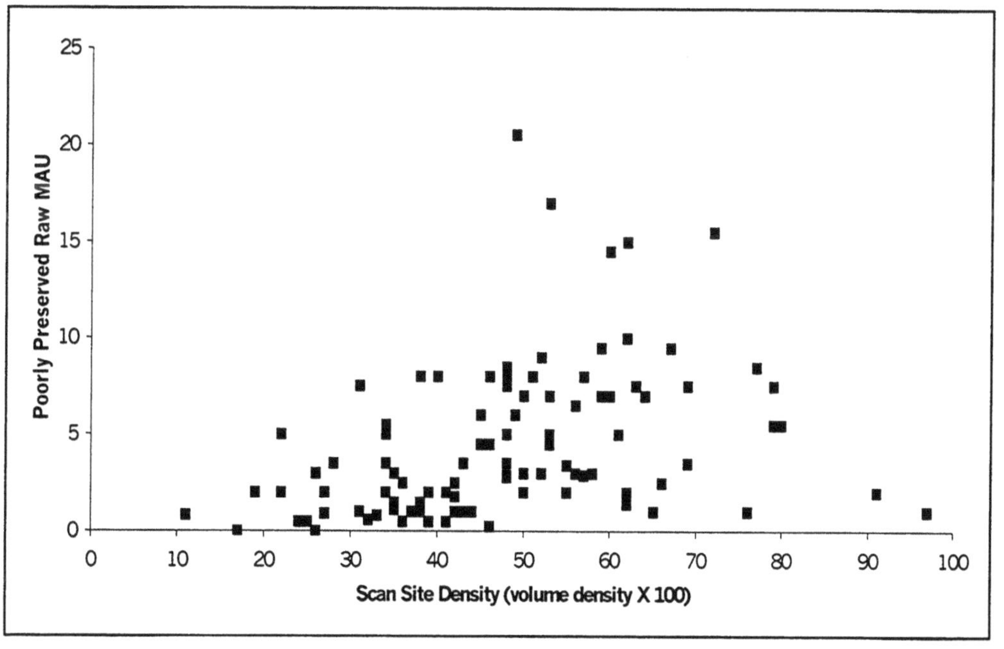

Figure 4. Scan site density versus raw MAU for Head-Smashed-In well-preserved area.

The five high representation values in Figure 3 are all from scan sites on the mandible and astragalus, the two most abundant elements in the sample. The astragalus is a very low value bone (Emerson 1990:624) having little meat, marrow, or grease associated with it, and hence it is usually left at kill sites (e.g. Frison 1970:10, Landals 1990:131, Shortt 1993:71, Zeimens 1982:229). With the tongue removed he mandible has only a relatively small economic utility (for marrow and grease). Since the tongue is easily removed, the mandible is also often one of the most common bones in communal kill sites (e.g. Frison 1970:10, Landals 1990:131, Shortt 1993:71, Todd 1987:135). These elements have survived not primarily by virtue of their inherent durability, but rather because they were disproportionately discarded at the kill site in the first place. A similar overriding influence of the cultural butchering pattern is seen in the two outlying points with a high density but a very low representation. These same two points are also clear outliers on the plot for the well preserved bone area. One is an atlas scan site and the other is an axis scan site; regardless of the high density of these element scan sites, these first two cervical vertebrae are still uncommon in the remains. These two vertebrae have low economic values (Emerson 1990:624) and are often abandoned at kill sites (e.g. Frison 1970:10, Shortt 1993:71, Speth 1983:86, Todd 1987:135). However, they are distinctly uncommon in some kill sites (Landals 1990:131; Zeimens 1982:228).The Head-Smashed-In remains seem to represent a case where the atlas and axis simply were not abandoned at the kill site. They were likely part of a skull butchering unit removed from the kill as a carcass portion. Despite poor preservation, at least some aspects of the original butchering and discard patterns are distinctly visible.

It is relevant to note here that Kreutzer did not obtain any density measurement for the skull and so it is not represented on these graphs. The densest portion of the *Bison* skull is the internal auditory meatus or inner ear. This structure is probably the densest portion of the *Bison* skeleton and I would anticipate that the skull would also be an outlier at the

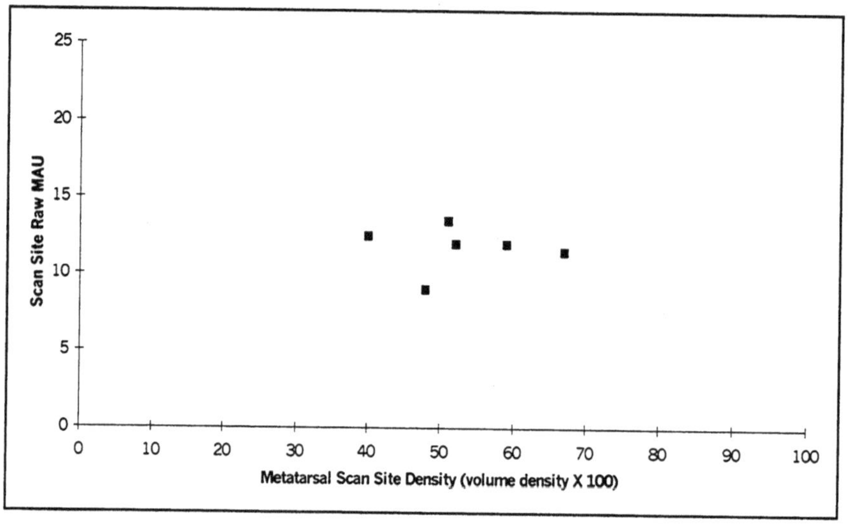

Figure 5. Comparison of metatarsal scan site representation and density.

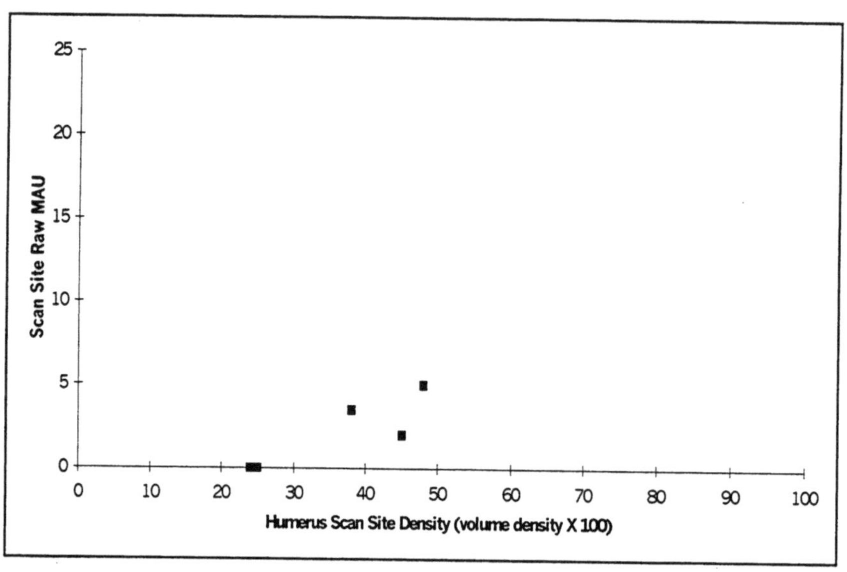

Figure 6. Comparison of humerus scan site representation and density.

top end of the density distribution. Its placement would not be quite so low as the atlas and axis, given its MAU value of five to ten.

Plots of density and representation of selected elements allow further exploration of these trends. A relatively tight cluster of plotted points should result from an element that has scan sites with similar density and that is butchered and treated as a complete bone. This situation can be seen on the metatarsal graph (Figure 5) from the well-preserved bone area. This can also be seen in a comparable humerus plot (Figure 6), although the linear aspect of this plot shows that there is a density-dependent aspect to the bone representation as well. The humerus plot suggests that there is a density threshold below which even relatively well-preserved bone will suffer greater weathering decay. This threshold might be about 30 on these plots, a volume density of .30 in Kreutzer's study. Certainly in the plot of all elements from the well-preserved area (Figure 4) there is a low density elongation or tail to the distribution below a volume density of .30 indicating that the linear relationship between density and representation is lost. The extremely poor preservation in the poorly preserved area obscures this pattern (Figure 3). The metatarsal and humerus plots (Figures 5 and 6) also show the influence of the cultural butchering patterns--the low value metatarsal is three times as common as the higher value humerus, presumably because the relatively high value humerus (Emerson 1990:624) was much more commonly removed from the kill.

The mandible data provide an element plot that does not follow this same pattern (Figure 7, again using data from the well preserved area of Layer 5d). The mandible plot seems to suggest that density has no effect on representation at all, given particularly the six scan sites with almost identical density (ca. 50-60 on the "X" axis) yet as much as a fivefold difference in representation. The three of these six with MAU values of 10 or less (plotted lowest on the "Y" axis) are all from the rostad end of the mandible, in the symphyseal area. These scan sites may be removed in some aspect of butchering, a fact given some support by a mandible fragment recovered in excavation that is missing the symphyseal area and that has a smooth, curved break indicating the break occurred while the bone was still fresh (no cut marks were preserved on any of the mandible fragments, hence it is difficult to use cut mark patterning to address the question). If one takes this explanation as reasonable, the remaining five (high MAU) mandible values can be seen as showing the overwhelming influence of the cultural pattern of discarding the mandibles at the kill. Hence the pattern seen is influenced by butchering strategy, discard patterns, and bone density.

Having begun the discussion of cultural patterning it is appropriate to look at this aspect in more detail since it is the focus of archaeological analysis. For this work I am using Emerson's (1990) data on *Bison* element utility, specifically her averaged total products model which takes riders into consideration in addition to economic utility. I have combined her values for proximal and distal ends of long bones to give values for the complete element. I have used the same value for each of her grouped elements: five for each carpal, ten for each tarsal, three for each phalanx, ten each for each of the skull and mandible. Splitting of these low values did not alter the plots and I had no knowledge of how the group value might be realistically divided between the individual elements.

Looking first at Figure 8, a plot of nutritional value versus frequency in the well preserved area, it is clear that none of the highly valuable elements are present in any quantity. Equally clear is that any abundant elements have very low utility values. Some of the low value elements are not as common as one might anticipate, perhaps bone density being an issue here since even the well preserved bone has been much weathered. Bone density is available only for three of the six elements in the lower left corner of the plot (value 20, MAU frequency 5). However, with a high of .97 for the axis and values of .56 for the lateral malleolus and .32 for thethird phalànx, the range in bone density values is considerable and interpretation is difficult. Density does not seem to be the sole factor involved. Overall the pattern is clearly as expected for a kill site, however: there are few high value elements and many low value elements, with moderate value elements having a generally intermediate representation.

Although this broad pattern is also seen in the poorly preserved bone as shown in Figure 9, the group of under-represented low value elements is even larger and reinforces the suggestion that bone density is a factor. In support of his suggestion, the larger sample of twelve elements (two have same value so cannot be seen separately on the plot) in the corner of this plot includes eleven with density measurements. All but three have densities of .5 or less, suggesting that density is certainly a factor here. Two of the three density values for the well preserved plot (Figure 8) also had values of about .5 or less, so density may indeed be a factor.

The differences between the well and poorly preserved samples, as well as their essential similarity, is clearly shown in Figure 10, a plot of the two samples on the same graph. It is especially clear that the top right half of the graph is totally empty, the area where highly valuable and abundant elements should be graphed. When the standardized MAU values are plotted in anatomical order (Figure 2) there is clearly an anatomical strategy as to which regions of the carcass are abandoned at the kill and which portions are taken away. The vertebral and rib areas are removed from the kill preferentially. The cranial and lower leg (metapodials, carpals, tarsals, probably phalanges) areas are abandoned at the site. The upper front leg is also abandoned with some regularity. This last aspect might relate to the physical condition (poor) of the animals caught, the greater discard of higher value portions when large numbers of animals are caught at once in a communal hunt event or a culturally defined dislike for front leg meat. It might also show that a considerable amount of butchering occurred at the kill itself, with meat being easily stripped from the limb bones, but this is much more difficult for the intricately interlocking vertebral/rib area. In such a scenario, the vertebral/rib area might be preferentially removed to the processing area for the more time-consuming butchering needed, hence few of these elements would be present in the

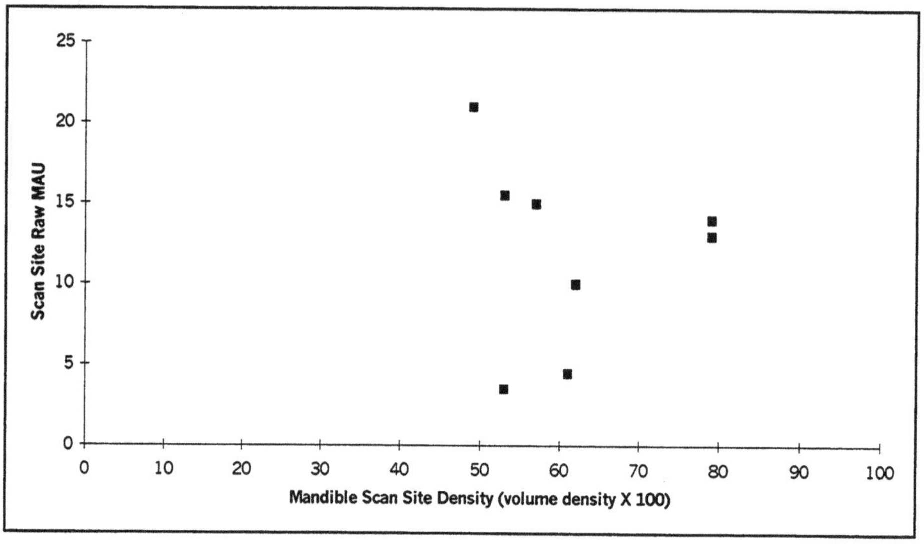

Figure 7. Comparison of mandible scan site representation and density.

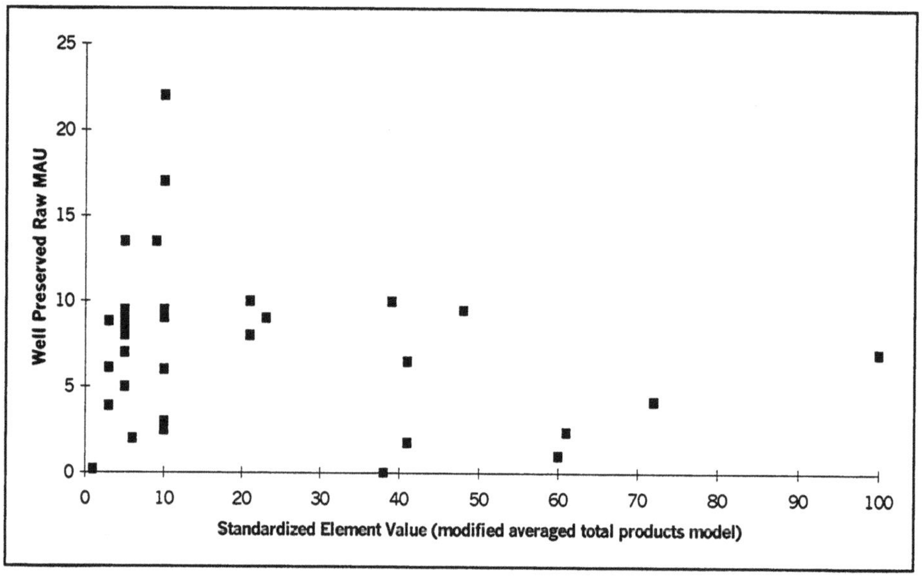

Figure 8. Comparison of element utility (adapted from Emerson 1990) and representation in the well preserved area of Layer 5d.

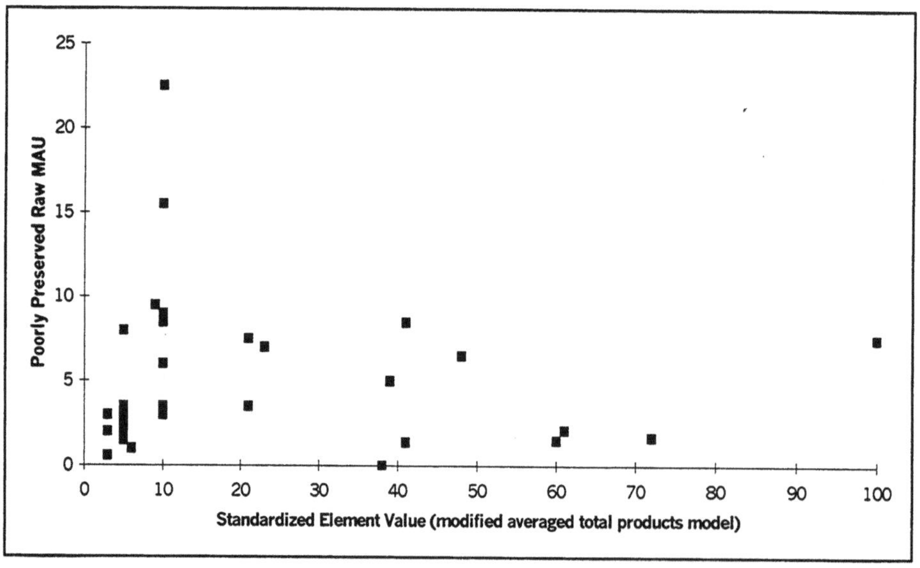

Figure 9. Comparison of element utility and representation in the poorly preserved area of Layer 5d.

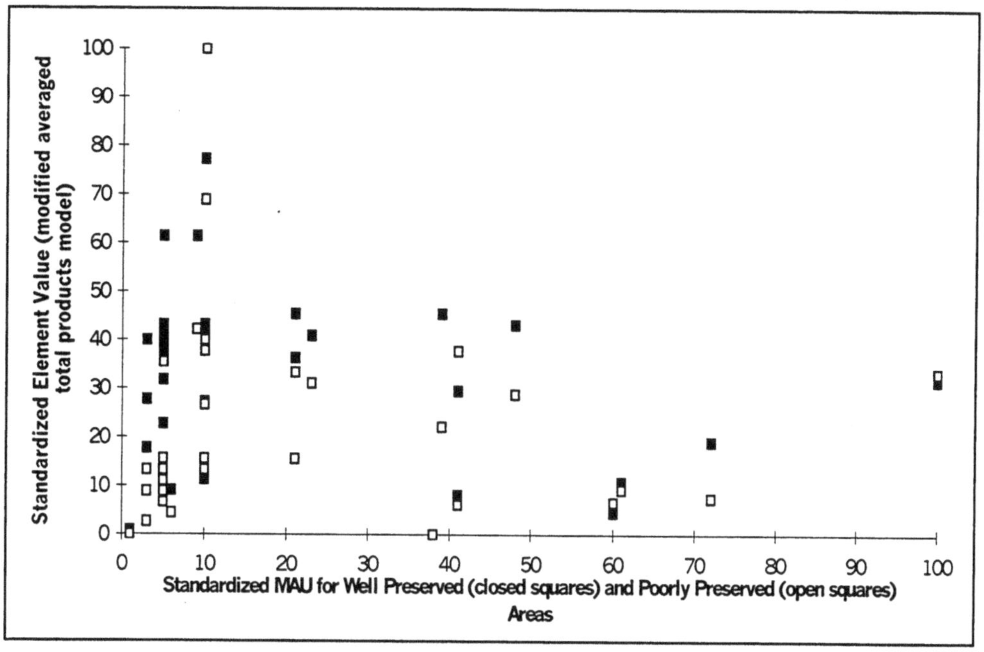

Figure 10. Comparison of utility and frequency of well (closed squares) and poorly (open squares) preserved areas.

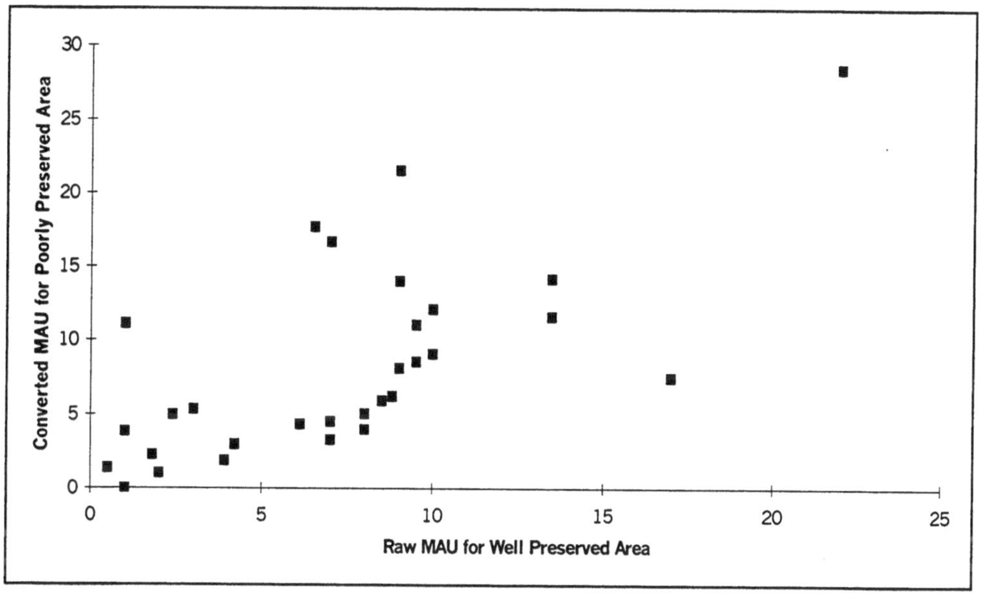

Figure 11. Comparison of converted poorly preserved bone and well preserved bone.

kill deposits. We do have some butchering tools from the kill deposits, although they are uncommon compared to their frequency in the processing area. There are large numbers of small sharpening flakes in the kill deposits, clearly indicating a great deal of cutting tool sharpening in the kill area; such tools would, of course, rarely be abandoned in a kill site so close to a camp area. We also found one storage pit in this sublayer at the site, showing that the processing in the kill was more complex or multifaceted than is often supposed.

Another aspect that I wish to explore is the use of the bone density measurements to "reconstruct" the poorly preserved bone. To attempt this I took the raw MAU values for the poorly preserved bone and multiplied them by the inverse of their density to try to "compensate" for the differential density. This is not a very sophisticated means of assessment, but it can be used to gauge the usefulness of such an exercise. In Figure 11 the "converted" poorly preserved bone MAU is plotted against the well preserved bone MAU - the "unaltered pattern." The patterns are similar and the relationship seen is surprisingly linear (correlation coefficient 0.6645, $p<.001$). This observation suggests that the greater decay of the more weathered bone must be considered before these two subsamples can be used together to look at the original butchering pattern. The question this observation posses is the extent to which even the "well" preserved sample adequately reflects the original event, given that its advanced weathering stage also shows decay. I would argue that given the intact nature of the bone recovered in this "well" preserved area, weathering has not been sufficiently extensive to alter the original pattern. Everything originally there is still, more or less, there, just in a very decayed state. Here too I return to my earlier observation that there was a threshold apparent in comparing the well preserved and poorly preserved areas. Bone samples are probably significantly compromised once the weathering stage reaches 4.5 to 5.

A final interesting aspect of element frequency is that there is consistently a substantial difference between the representation of the skull and mandible, the mandible generally having a MAU value about three or four times greater than that of the skull. The low skull value compared to the mandible suggests that the skull had some value that the mandible did not. This is even more interesting because the skull is also underrepresented in the camp and processing area (e.g. Brink and Dawe 1989:100-123). The horn sheath was of some value as raw material for tools and some skulls may have been used in ceremonies, but the main value of the skull was for the brains which were used in hide tanning. The relative paucity of the skull in all excavations at Head-Smashed-In to date may be because there is an as yet undiscovered hide processing area on the prairie level flats. The site encompassed a large area prehistorically, with only a small fraction of it tested so far.

CONCLUSIONS

Preservation does alter bone frequency patterns to the point where the data are distorted. Bone density measurements do track these changes accurately and can be used to overcome these preservation problems to a considerable extent. Even in badly preserved bone, careful analysis can delineate the main features of cultural butchering patterns. However, many of the details can be lost if bone density is not considered. Attempts to simulate preservation alteration in the laboratory are theoretically useful, but cannot provide a true emulation of change when we do not know all the factors and interactions that are involved. The Layer 5d sample from Head-Smashed-In is an experiment where preservation conditions are real, hence the patterns observed are also real, not simply estimations based upon untested hypotheses. The results of this examination suggest that research into bone density and element utility such as undertaken by Kreutzer and Emerson provide us with very valuable indices that will significantly increase our insight into past human behaviour.

ACKNOWLEDGEMENTS

The Social Sciences and Humanities Research Council of Canada provided major funding for this research. The Alberta Historical Resources Foundation, the Friends of Head-Smashed-In Buffalo Jump Society, the Archaeological Survey of Alberta, and the Department of Archaeology provided additional funding at the University of Calgary. I am much indebted to Joe Crowshoe and Leo Pard for their sharing of Piikáni traditional knowledge. The staff from the Head-Smashed-In Buffalo Jump Interpretive Centre aided my work in numerous ways. Many people worked on the excavation and analysis of this material during the ten years it has taken to date, but in particular I wish to thank Maureen Reeves and Lisa Stephen for their insight and dedication.

REFERENCES CITED

Behrensmeyer, Anna K.
1978 Taphonomic and Ecological Information from Bone Weathering. *Paleobiology* 4:150-162.

Binford, Lewis R.
1978 *Nunamiut Ethnoarchaeology*. Academic Press, Inc., New York.

Brink, Jack, and Bob Dawe
1989 *Final Report of the 1985 and 1986 Field Season at Head-Smashed-In Buffalo Jump Alberta*. Archaeological Survey of Alberta Manuscript Series No. 16. Alberta Culture and Multiculturalism, Edmonton.

Brink, Jack, Milt Wright, Bob Dawe, and Doug Glaum
1985 *Final Report of the 1983 Season at Head-Smashed-In Buffalo Jump Alberta*. Archaeological Survey of Alberta Manuscript Series No. 1. Alberta Culture and Multiculturalism, Edmonton.

Brumley, John H.
1973 Quantitative Methods in the Analysis of Butchered Faunal Remains: A Suggested Approach. *Archaeology in Montana* 14:1-40.

1995 *Prehistoric Settlement and Subsistence in the Plains of Southern Alberta and Northern Montana: The Seasonality Evidence*. Submitted to the Archaeological Survey of Alberta, Provincial Museum of Alberta, Edmonton.

Emerson, Alice M.
1990 *Archaeological Implications of Variability in the Economic Anatomy of Bison bison*. Ph.D. dissertation, Anthropology Department, Washington State University. Ann Arbor: University Microfilms.

Frison, George C.
1970 The Glenrock Buffalo Jump, 48CO304. *Plains Anthropologist Memoir* 7. Plains Anthropological Society, Lincoln.

Kooyman, Brian P.
1985 *Moa and Moa Hunting: An Archaeological Analysis of Big Game Hunting in New Zealand*. Ph.D. dissertation, Anthropology Department, University of Otago.

1992 *Preliminary Report of the 1991 Excavations in the North Kill at Head-Smashed-In Buffalo Jump* (DkPj-1). Submitted to the Archaeological Survey of Alberta, Provincial Museum of Alberta, Edmonton.

1990 *Final Report of the 1989 Excavation in the North Kill at Head-Smashed-In Buffalo Jump* (DkPj-1). Submitted to the Archaeological Survey of Alberta, Provincial Museum of Alberta, Edmonton.

Kreutzer, L. A.
1992 Bison and Deer Bone Mineral Densities: Comparisons and Implications for the Interpretation of Archaeological Faunas. *Journal of Archaeological Science* 19:271-294.

Landals, Alison
1990 The Maple Leaf Site: Implications of the Analysis of Small-scale Bison Kills. *In Hunters of the Recent Past* edited by Leslie B. Davis and Brian O.K. Reeves, pp. 122-151. Unwin Hyman, London.

Morlan, Richard E.
1994 Bison Bone Fragmentation and Survivorship: A Comparative Method. *Journal of Archaeological Science* 21:797-807.

Reeves , Brian O.K.
1978 Head-Smashed-In: 5500 Years of Bison Jumping in the Alberta Plains. In *Bison Procurement and Utilization: A Symposium* edited by Leslie B. Davis and Michael Wilson, pp. 151-174. Plains Anthropologist Memoir 14. Plains Anthropological Society. Lincoln:

Shortt, Mack W.
1993 *The Happy Valley Kill.* M.A. thesis, Archaeology Department, University of Calgary.

Speth, John D.
1983 *Bison Kills and Bone Counts: Decision Making by Ancient Hunters*. Chicago: University of Chicago Press.

Todd, Lawrence C.
1987 Taphonomy of the Horner II Bone Bed. In *The Horner Site: The Type Site of the Cody Complex* edited by G.C. Frison and L.C. Todd, pp. 107-198. Academic Press, Inc., Orlando.

Todd, Lawrence C., Robert V. Witter, and George C. Frison
1987 Excavation and Documentation of the Princeton and Smithsonian Horner Site Assemblages. In *The Horner Site: The Type Site of the Cody Complex* edited by G.C. Frison and L. C. Todd, pp. .39-105. Academic Press, Inc., Orlando.

Zeimans, George M.
1982 Analysis of Postcranial Remains. In *The Agate Basin Site: A Record of the Paleoindian Occupation of the Northwestern High Plains* edited by G.C. Frison and D.J. Stanford, pp. 213-240. Academic Press, Inc., New York.

CARIBOU REMAINS AT KILL SITES AND THE ROLE OF SCAVENGERS IN PRODUCING PATTERNED DISTRIBUTIONS IN BONE ASSEMBLAGES

Audrey J. Magoun and Patrick Valkenburg
Alaska Department of Fish and Game, Fairbanks

INTRODUCTION

In the course of their daily activities, wildlife biologists often have the opportunity to observe the remains of large ungulate carcasses at predator kill sites, and to study remains acted on by scavengers. Remains and records from these sites provide valuable information to zooarchaeologists interested in taphonomic problems. This paper highlights data obtained by reviewing notes from four different wildlife studies carried out in Alaska. In this study we present information on the remains of 53 caribou (*Rangifer tarandus*), and discuss factors that may contribute to the formation and preservation of bone assemblage found at carcass sites. Because our research objectives varied by project, records of caribou remains were not collected in a consistent manner among all study areas. While taphonomy was not initially a focus of any of the studies. we still believe the data presented here will provide zooarchaeologists with some new insights into selected taphonomic processes.

STUDY AREAS AND METHODS

Information was collected from three different study areas. The first was the northern foothills of the eastern Brooks Range, from the Canning River east to Peters-Schrader Lakes (Magoun 1976). The purpose of the study was to document the behavior of scavengers feeding on ungulate carcasses in general, and caribou carcasses in particular. Fieldwork was conducted during the summer when small, scattered groups of caribou resided in the area. Observations were made at 11 caribou carcass sites; remains were usually recorded within 10 days of when the carcasses became available. Most carcasses were established by shooting a single caribou. Nearby, a station for observation of the carcass was established. We did not approach or handle these carcasses. Another researcher who had collected them for a study of parasites and diseases provided two carcasses. These were placed at an observation site by helicopter and had human scent on them, delaying utilization by wolves (*Canis lupus*) but not bears (*Ursus* sp.). The carcasses were observed sequentially, and the carcass sites were separated by at least 2 km. All carcasses were continuously observed until scavengers had removed most of the meat.

The second study area was in the central Alaska Range, in the northern foothills and upper Nenana River drainage (Valkenburg et al. 1996). Two different studies were conducted in this area. The first was a study of wolf predation in March 1989 when 10,700 caribou resided in the area. The second was a study of caribou calf mortality in May 1995 when 4500 caribou were present.

In the first study, radio-collared wolves were tracked daily to caribou kill sites. Other predators and scavengers were sometimes observed near the carcasses. In the second study, radio-collared caribou calves were radio-tracked daily until they were found dead. Predators and scavengers were usually not present when the carcasses were located. Most carcasses were examined and documented for evidence of predators and scavengers (tracks, scats, hair, etc.). Remains of carcasses were generally recorded within 10 days after the caribou died, with many recorded within the first three days after death.

The third study area was in the northern foothills of the western Brooks Range, along the Utukok River drainage (Magoun 1985). This study was an investigation of wolverine (*Gulo gulo*) ecology conducted coincidentally with wolf, grizzly bear (*Ursus arctos*), and caribou studies. About 200,000 caribou migrate through this area in the spring, with smaller numbers moving through in the fall. No caribou were in the area in winter during the study period. Remains of caribou scavenged by wolverines were recorded from February to November. Remains were found by radiotracking wolverines or visiting den sites. Other predators had already fed on most carcasses before the wolverines found them.

RESULTS

The species we documented using caribou carcasses included grizzly bears, wolves, wolverines, coyotes (*Canis latrans*), red foxes (*Vulpes fulva*), ground squirrels (*Spermophilus parryi*), golden eagles (*Aquila chrysaetos*), ravens (*Corvus corax*), gulls (*Larus* sp.), jaegers (*Stercorarius* sp.), and gray jays (*Perisoreus canadensis*). Most of our comments here pertain to wolves, grizzly bears, and wolverines. Birds and smaller mammalian scavengers are capable of removing all of the meat from caribou carcasses, and even from portions of some bones. However, the three largest carnivores, wolves, bears and wolverines, were primarily responsible for utilizing caribou carcasses and/or scattering bones during the periods of observation.

Data on bones found at the carcass sites are presented in two categories: intact and fragmented (Tables 1-5). By intact, we mean the skull was largely in one piece, although the lower jaws may have been detached; the leg bones were largely intact, though ends may have been chewed and hooves removed; large sections of the vertebral column were still present; the rib cage was still mostly intact, though a few ribs may have been missing or tips of the ribs were chewed off; the pelvis was still intact; and meat may or may not have been left on the carcass.

Fragmented bones included pieces of the skull; lower jaws separated and the ends chewed; leg bones cracked and in pieces; vertebral columns disjointed and in small pieces; scapula separated from the rest of the leg and chewed on; pelvis in pieces; bone fragments; and pieces of the hooves.

		INTACT					
PREDATOR	CARCASS NUMBER	LEGS	SKULL	VERT	RIBS	PELVIS	MEAT
Bears	C8	x					
	C49	x					
	C52	x					
	C53	x					
Wolves (no bears)	C2	x					
	C3	x					
	C6	x					x
	C12						x
	C14	x					
	C17	x	x		x		x
	C18	x					
	C19	x	x				x
	C23		x				
	C25	x					
	C27	x					
	C28	x	x				
	C31	x					x
	C33		x				
	C34	x				x	
	C35	x	x			x	
	C36	x					
	C38	x	x			x	
	C40	x		x		x	
	C41	x				x	
	C42	x				x	
	C43	x	x		x	x	x
	C44	x		x	x		

		FRAGMENTED								
PREDATOR	CARCASS NUMBER	LEGS	SKULL	JAW	VERT	RIBS	SCAP	PELV	BONE FRAGS	HOOF PIECES
Bears	C1		x	x						
	C7			x						
	C8									
	C13		x	x						
	C15								x	
	C16									x
	C20								x	
	C24								x	
	C45	x		x		x			x	
	C47	x	x		x	x	x			x
	C48	x		x	x	x	x			
	C49						x			
	C52				x	x				
	C53				x	x				
	C3		x		x		x	x		
Wolves (no bears)	C6		x							
	C11									
	C12		x		x					
	C14		x						x	
	C17				x					
	C18		x							
	C19									
	C21		x							
	C27			x						
	C29			x						
	C30					x				
	C31			x						
	C32			x		x				x
	C34			x						
	C36			x						
	C37			x						
	C39			x		x				
	C41			x			x			
	C42			x						
	C44			x						

Table 1. Remains of caribou carcasses fed on by bears, and by wolves but not bears.

Bears

The presence of bears at the carcass site influenced the degree to which carcasses were utilized, with bears utilizing far more of the carcasses than did wolves over the same time period. We examined 14 carcasses that were visited by bears and 30 carcasses that were not visited by bears (Table 1).

Of carcasses visited by bears, only 11% of the remains were considered intact compared to 63% at carcasses visited by wolves but not by bears. Leg bones were intact at 29% of carcasses fed on by bears and at 67% of carcasses fed on by wolves but not bears. No skulls or pelves were intact at sites where bears had been. Of carcasses fed on by wolves but not bears, 27% had the skull intact and 23% had the pelvis intact. An adult caribou carcass fed on by two wolves over a three-day period in summer had forelegs and hind legs still present, as well as the rib cage and vertebral column. Most of the meat was gone. On the fourth day a bear arrived; when this carcass was checked on the fifth day, a scapula was the only intact bone left at the site. A few small bone fragments were the only remains left at the site a month later.

Number of Wolves

We compared the remains at carcasses that were visited by < four wolves and >11 wolves (Table 2). The number of wolves at a carcass did not appear to affect the type of remains found except for a higher number of intact pelvic girdles at carcasses with <4 wolves present. Over the period of observation, the utilization of carcasses by wolves, regardless of the number in the pack, was similar. This is probably because wolves move on to make other kills when most of the meat has been removed from the carcasses. When prey is scarce, utilization may be higher, especially over long periods of time as wolves are known to revisit kill sites repeatedly over a number of months.

Utilization of Carcasses

Carcasses that were not visited by bears were examined to determine if the number of days of carcass availability affected the utilization pattern of the carcass (Table 3). More leg bones were present and intact after only four days compared to periods > eight days. Rib cages were also more intact. Meat might still be present, but wolves appeared to lose interest in carcasses after about four days. Of course the particular scavengers present at the carcass influence the level of utilization over time. For 10 days one carcass site was visited only by birds (golden eagles, jaegers, gulls, and ravens); all bones were still present at the end of this period except for the tips of the ribs, and the skeleton remained largely intact though most of the meat was gone. Smaller scavengers such as foxes and birds rely on larger carnivores to open adult caribou carcasses (Magoun 1976). When large carnivores are not present, carcasses remain intact for long periods of time.

Size of the Carcass

We compared the difference in remains at sites where wolves had killed or scavenged adult caribou to sites of newborn calf caribou (Table 4). The remains are surprisingly similar despite differences in carcass size. There were more intact pelvic girdles at the adult caribou sites than at calf sites, but many of the other calf bones remained intact. More calf skull fragments remained than adult skull fragments; either the entire adult skull was removed from the site or the skull remained intact. Because wolves sometimes killed more than one calf caribou over a short period of time, some calves were not eaten by the wolves at all; some were eaten by scavengers, including bears, while the wolves cached others by burying them whole in moss or mud.

There were clear differences in remains at sites where bears fed on adult caribou versus calf caribou (Table 5). There were more intact or fractured leg bones at sites where bears fed on adult caribou rather than on calf caribou. Ribs, scapula, or pelvis pieces were also more commonly found at sites where bears fed on adult caribou. Bears consumed many of the calf caribou bones leaving, if anything, only small fragments of bone or pieces of hooves.

SEASON AND AVAILABILITY

Winter

Carcasses that become available in winter generally take longer to disappear than those available in summer. In winter bears are hibernating and meat freezes hard making it difficult for scavengers to dismember the carcass. If prey are readily available, wolves will make fresh kills rather than utilize frozen carcasses. Many of the bones remain at carcass sites for much of the winter, slowly disappearing through the following spring and summer. However, in areas where wolverines are common and food is limited, frozen carcasses will be heavily utilized because wolverines are able to dismember frozen carcasses. Like wolves, if additional carcasses become available, wolverines abandon the older carcasses to feed on fresher kills, periodically returning to the older carcasses through the winter as conditions and need warrant.

Summer

Carcasses, which become available in the summer, usually disappear rapidly if carnivores are common in the area. In particular bears cause carcasses to disappear quickly. Carcasses may remain intact for relatively long periods if caribou availability is high, such as on calving grounds, or along the migration routes of large herds. Furthermore, we have observed breeding wolverines pass up fresh caribou kills during the summer, and even breeding bears sometimes leave carcasses before they are fully utilized.

During the denning period, scavengers may more fully utilize carcasses because they need additional food for their offspring. For example, in summer we observed denning wolves repeatedly return to a carcass to feed, cache meat, and to carry off bones to a den site. Wolverines are believed to cache more carcass parts before and during cub rearing in

INTACT							
PREDATOR	CARCASS NUMBER	LEGS	SKULL	VERT	RIBS	PELV	MEAT
Wolves ≥12	C27	x					
	C38	x	x			x	
	C44	x		x	x		
Wolves ≤4	C28	x	x				
	C31	x					x
	C33		x				
	C34	x				x	
	C35	x	x			x	
	C36	x					
	C40	x		x		x	
	C41	x				x	
	C42	x				x	
	C43	x	x		x	x	x

FRAGMENTED										
PREDATOR	CARCASS NUMBER	LEGS	SKULL	JAW	VERT	RIBS	SCAP	PELV	BONE FRAGS	HOOF PIECES
Wolves ≥12	C27			x						
	C32			x		x				
	C38			x						
	C44			x						
Wolves ≤4	C29			x						
	C30					x				x
	C31			x						
	C34			x						
	C36			x						
	C37			x						
	C39			x		x	x			
	C41			x						
	C42			x						

Table 2. Remains of caribou carcasses fed on by ≤4 wolves and ≥12 wolves.

INTACT							
PREDATOR	CARCASS NUMBER	LEGS	SKULL	VERT	RIBS	PELV	MEAT
Scavengers ≤4 days	C27	x					
	C28	x	x				
	C31	x					x
	C36	x					
	C38	x	x			x	
	C43	x	x		x	x	x
	C44	x		x	x		
Scavengers ≥8 days	C33		x				
	C34	x				x	
	C40	x		x		x	

FRAGMENTED										
PREDATOR	CARCASS NUMBER	LEGS	SKULL	JAW	VERT	RIBS	SCAP	PELV	BONE FRAGS	HOOF PIECES
Scavengers ≤4 days	C27			x						
	C29			x						
	C30					x				x
	C31			x						
	C36			x						
	C38			x						
Scavengers ≥8 days	C32			x		x				
	C34			x						
	C39			x		x	x			

Table 3. Remains of caribou carcasses fed on by scavengers ≤ 4 days and ≥8 days.

		INTACT					
CARIBOU	CARCASS NUMBER	LEGS	SKULL	VERT	RIBS	PELV	MEAT
Calves	C2	x					
	C3	x					
	C6	x			x		
	C11	x				x	
	C12	x					
	C14	x	x				
	C17	x					
	C18	x	x			x	
	C21		x				
	C23	x					
Adults	C27	x					
	C28	x	x				
	C31	x					x
	C33		x				
	C34	x				x	
	C35	x	x			x	
	C36	x					
	C38	x	x			x	
	C40	x		x		x	
	C41	x				x	
	C42	x				x	
	C43	x	x		x	x	x

				FRAGMENTED						
PREDATOR	CARCASS NUMBER	LEGS	SKULL	JAW	VERT	RIBS	SCAP	PELV	BONE FRAGS	HOOF PIECES
Calves	C3		x		x					
	C6		x		x					
	C11		x		x				11	
	C12		x							
	C17		x							
	C19		x						x	
Adults	C27			x						
	C29			x						
	C30					x				x
	C31			x						
	C32			x		x				
	C34			x						
	C36			x						
	C37			x						
	C39			x		x	x			
	C41			x						
	C42			x						

Table 4. Remains of adult and newborn calf caribou carcasses fed on by scavengers other than bears.

		INTACT					
CARIBOU	CARCASS NUMBER	LEGS	SKULL	VERT	RIBS	PELV	MEAT
Calves	C8	x					
Adult	C48	x					
	C49	x					
	C52	x					
	C53	x					

				FRAGMENTED						
	CARCASS NUMBER	LEGS	SKULL	JAW	VERT	RIBS	SCAP	PELV	BONE FRAGS	HOOF PIECES
Calves	C1		x	x						
	C7			x						
	C8								x	
	C13		x	x						x
	C15								x	
	C16								x	
	C20								x	
	C24									x
Adult	C45	x		x		x	x			
	C47	x	x		x	x	x			
	C48			x	x	x	x			
	C49									
	C52				x	x		x		
	C53				x	x	x			

Table 5. Remains of adult and calf caribou carcasses fed on by bears and other scavengers.

late winter and spring (Landa et al. 1997). We also observed wolverines storing caribou bones by burying them in the tundra or by hiding them under remnant snowdrifts during early summer.

Occasionally, a large number of caribou will die within a relatively small area. In 1994 there was a winter die-off on the north slope of Alaska near Cape Thompson. The caribou carcasses disappeared slowly with numerous remains still evident in the area two years later. The sheer number of carcasses probably overwhelmed the capability of scavengers to rapidly dispose of them.

Location and Scavenger Community

The location of a carcass may affect the length of time it lasts and the final disposition of the bones. In areas where bears are numerous, carcasses will generally disappear more rapidly than in areas where bears are scarce. In locations where wolverines are relatively common but where winter food sources are limited, many caribou bones will be fully utilized. For example, in the western Brooks Range study area where wolverines were common and large ungulates were absent in winter, wolverines consumed most of the remains of caribou carcasses over the winter, including most bones that remained from the spring and fall caribou migration through the area (Magoun 1985). They also fed on carcasses during summer. In contrast, wolverines did not visit any carcasses in summer in the eastern Brooks Range during our study. We found old caribou bones lying on the tundra after the winter season more commonly in this area than in the western Brooks Range.

SUMMARY

During four wildlife field studies in Interior and northern Alaska, we recorded the remains of caribou carcasses that had been fed on by predators and scavengers. Most of the information was gathered within 10 days of a carcass becoming available. The factors, which influenced the pattern and rate of disappearance of caribou carcass parts, included the community of carnivores in the area, the number of carcasses that were available, the length of time scavengers fed on the carcasses, the carcass size, and the season in which the carcass became available. The scavengers most efficient at removing bones from carcass sites were bears, wolves, and wolverines. When bears were present, carcasses disappeared more quickly and more completely than when bears were absent. Wolves rarely fed on carcasses for more than four consecutive days, preferring to make new kills rather than to return to carcasses on which little meat remained. Wolverines were capable of utilizing all the bones of a carcass over the winter if food was scarce.

Though taphonomy was not the focus of these field studies, useful taphonomic data was collected. Zooarchaeologists interested in taphonomy could benefit from information gathered by wildlife field biologists, who could provide even better data if co-operative studies were designed before fieldwork was conducted.

REFERENCES

Landa A., O. Strand, J.E. Swenson, and T. Skogland.
1997 Wolverines and Their Prey in Southern Norway. *Canadian Journal of Zoology* 75:1292–1299.

Magoun A.J.
1976 *Summer Scavenging Activity in Northeastern Alaska.* Unpublished MS thesis, University of Alaska, Fairbanks.

1985 *Population Characteristics, Ecology, and Management of Wolverines in Northwestern Alaska.* Unpublished Ph.D. dissertation, University of Alaska, Fairbanks.

Valkenburg P., J.L. Davis, J.M. Ver Hoef, R.D. Boertje, M.E. McNay, R.M. Eagan, D.J. Reed, C.L. Gardner, and R.W. Tobey.
1996. Population Decline in the Delta Caribou Herd with Reference to Other Alaskan Herds. *Rangifer Special Issue* 9:53–62.

RODENT GNAWING AS A TAPHONOMIC AGENT: IMPLICATIONS FOR ARCHAEOLOGY

Meg Thornton
Department of Anthropology, University of Alaska-Fairbanks

Jennifer Fee
Department of Anthropology, University of California-Davis

INTRODUCTION

The role of taphonomic agents in the formation of zooarchaeological assemblages has come under a great deal of scrutiny in the last few decades. Although much research has been directed toward understanding the effects of carnivore gnawing on bone (Binford 1981; Blumenschine and Marean 1993), comparatively little work has been carried out on the pre- and post-depositional destruction of bone resulting from rodent gnawing. Further, the published literature on rodents as taphonomic agents is sometimes contradictory. Here we report observations on rodent gnawing recorded under controlled experimental conditions. Our most notable finding is that rodents may do significantly more damage to the archaeolofaunal record than has previously been realized.

RODENT GNAWING

Typically rodent gnawing on bone is thought to be evidenced by easily identifiable, paired, broad, shallow, flat-bottomed grooves which exhibit minute striae that parallel the principle axis of the scraped surface (see Shipman 1981; Grayson 1988; Fiorillo 1989; Marshall 1989; and Schick et al. 1989, for just a few examples). In most cases gnawing marks result when the upper and lower incisors are drawn towards each other across a bone (Shipman 1981). The striae formed by this action can be seen under low powered magnification (Bonnichsen and Will 1990). Typically, these parallel grooves are perpendicular to the long axes of the bones (Figure 1) (Singer 1956; Hesse and Wapnish 1985).

While most references to rodent gnawing only include descriptions of "clear parallel grooves," other research illustrates that other types of damage can be produced. In addition to parallel grooves, rodents can create tubular shafts by scooping out the ends of long bones (Lyman 1994:195), and they may also chew characteristic "windows" through the long bone shafts (Singer 1956; Bonnichsen and Will 1990).

Investigators differ in opinion over the difficulties of distinguishing rodent gnaw marks from other sources of bone damage. Some (see for example Shipman 1981) argue that rodent gnawing is easily distinguished, by gross inspection, from marks made by other animals or tools while others argue that the damage caused by rodents is difficult to distinguish from that caused by other taphonomic agents. For example, canids can also produce shallow marks with their incisors that might be attributed to rodents (Miller 1975), and the tubular shafts produced by rodent gnawing are similar to those created by carnivore gnawing (Lyman 1994:195). Furthermore it may be difficult or impossible to distinguish between rodent damage and cultural modifications, such as butchery marks (Dixon 1984).

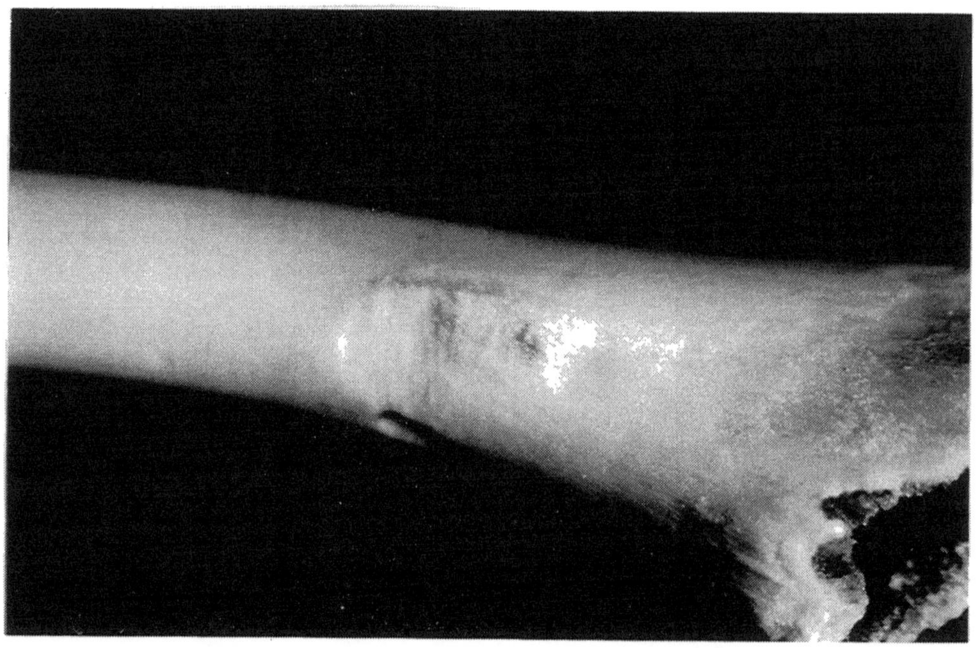

Figure 1. Parallel grooves caused by rodent gnawing.

The identification of rodents as taphonomic agents at work in any given archaeological context is important because rodents can do a significant amount of damage to faunal assemblages. Through gnawing, rodents have been observed to destroy whole sections of bone, and/or to render bones unidentifiable (Brain 1981; Hoffman and Hays 1987). Thick cortical shafts of bone can be reduced to fragile "shells" (Brothwell 1976; Roze 1989), and small and fragile bones may be completely destroyed (Price 1985).

Rodents can also differentially transport or concentrate bones in ways that may significantly affect the representation of different taxa and different skeletal elements in an assemblage. Ethnoarchaeological research among the Aka of the Central African Republic identified the cane rat (*Thryonomys* sp.) as the principle scavenger of bone (Hudson 1993:303). Other rodents such as the packrat (*Neotoma* sp.), North American porcupine (*Erethizon dorsatum*), and African porcupine (*Hystrix africaeaustralis*) are known to remove bones from sites and gather them in their dens (Rainey 1956; Heizer and Brooks 1965; Brain 1981; Finley 1985; Hoffman and Hays 1987; Hockett 1989; Vaughan 1990).

Evidence of rodent gnawing on bones in a faunal assemblage may indicate substantial disturbance. Rodents may alter the spatial relationships among various taxa and elements, and this may distort calculations of relative taxonomic abundance when using the minimum number of individuals (MNI) (Hoffman and Hays 1987). In addition, the transportation of the bones into the dens can lead to the creation of "pseudo-archaeological sites" (Dixon 1984).

Knowledge of rodent behavior is important as it aids our ability to understand and assess the significance of faunal patterning in archaeological samples. Given the potential impact of rodent gnawing on archaeofaunal collections, we decided to document the kinds of damage done by rodents to bone, and to assess the ease, or difficulty, archaeologists might experience in identifying the resulting rodent-gnawed bone.

EXPERIMENTAL SAMPLE

Bones from four taxa, including adult and subadult white-tailed deer (*Odocoileus virginianus*), adult raccoons (*Procyon lotor*), adult Eastern Gray squirrels (*Scuirus carolinensis*), and adult and subadult chickens (*Gallus gallus*) were used. Selection of elements was based on variability in element size and density. Mammals representing three different size classes were selected in order to compare the effects of rodent gnawing on different sized bones. Bird bones were included in order to make cross class comparisons of the damage to bones. For the mammals, mandibles, innominates, lumbar vertebrae, and ribs were included, and from the chicken, the humeri, mandible, pelvis, and ribs were used to generate a sample of one hundred and fifty bones (Table 1).

Each of the animals used in the experiment was macerated and all of the bones were boiled until they were no longer greasy. Brain (1980, 1981) and Andrews (1990) have both observed that rodents clearly prefer weathered and defatted bone over fresh greasy bone. Lyman (1994) has suggested that this is because dry bones are less slippery than fresh bones, and therefore are easier to gnaw. The amount of boiling time varied considerably due to differences in bone sizes.

Animal	Element	Number	Number with parallel striations
Deer	Mandible	10	3
	Vertebrae	10	3
	Rib	7	1
	Humerus	4	
	Pelvis	7	4
Squirrel	Mandible	10	
	Vertebrae	9	
	Rib	7	
	Humerus	9	
	Pelvis	7	
Raccoon	Mandible	10	2
	Vertebrae	2	
	Rib	10	
	Humerus	8	4
	Pelvis	3	
Chicken	Mandible	6	
	Vertebrae	2	
	Rib	10	
	Humerus	12	
	Synsacrum	7	
Total		150	17

Table 1. Sample summary.

Due to the difficulty of observing the behavior of wild rodents, caged lab mice (*Mus musculus*) were used in this experiment. The use of caged animals also provided us with the opportunity to be sure that there were no other taphonomic agents at work. Mice, rather than other rodents, were chosen for this experiment for several reasons. The commensal relationship between humans and mice is a close one, with mice relying on humans for "(1) maintenance of [their] habitat, (2) provision of food, and/or (3) protection from predation" (Tchernov 1991:317). Mice are found in archaeological contexts at numerous sites dating from 10,000 B.C. to the present (Colburn 1985; Tchernov 1991). The morphological similarities between laboratory mice and wild mice (Burt 1976) also played a role in the selection of mice for use in this experiment, as did the ease of obtaining laboratory mice.

Ten different mice were used. Each was kept in an individual cage. Bones were placed in the cage one at a time to better record the amount of damage a single rodent was capable of inflicting on a single bone within specified periods of time.

In a preliminary experiment it was discovered that the largest percentage of damage occurred in the first twenty minutes after the bone was placed in the cage. Little new damage occurred after forty-eight hours. This information was used to establish a schedule for the observation of rodent gnawing activity. Once a bone had been placed in the cage with a mouse, the ensuing twenty minutes of mouse activity were observed closely. During this time notes were taken describing how the mice reacted to the bones and where gnawing took place. At the end of the twenty-minute period, any damage to the bone was recorded on a diagram. The bone was then checked twenty-four and forty-eight hours later. Again, at each of these intervals the damage was drawn on a diagram. After forty-eight hours the bone was removed from the cage. All specimens were examined by gross inspection and with a 10X hand lens to identify the location and appearance of any rodent gnawing.

RESULTS

Three broad categories are used to describe the damage on the bones. The first is "parallel striations," as described above. The second category, "roughening," describes damage of an ambiguous nature while the last category, "missing," denotes the absence of large segments of bone, such as processes.

Most damage to the mammalian mandibles occurred on the mandibular condyles and on the coranoid process. The deer mandibles were scarred by easily identified parallel striations, and of the ten, three showed clear incisor marks. Patches of roughening, mainly on the mandibular condyle and coranoid process occurred on the rest of the mandibles. The squirrel mandible was slightly damaged by roughening on the ascending ramus. The chicken mandible suffered the lightest damage with only slight roughening to the front of the mandible and at the condyle.

The lumbar vertebrae showed gnawing damage concentrated on the transverse and spinous processes. The deer vertebrae showed damage on the spinous and transverse processes. Two vertebrae have clear incisor marks, one on the edge of the transverse process and one on the end of the body. The squirrel vertebrae were missing all of their processes, and a small amount of roughening was present on the body of the vertebrae. Two of the squirrel vertebrae were completely missing. The raccoon vertebrae showed severe damage, including the loss of processes. Most of the vertebral bodies were completely destroyed.

The ribs of all four taxa showed the same kinds of damage. Gnawing occurred on the sternal and vertebral extremities. The squirrel ribs were greatly reduced, as were the raccoon ribs with both the sternal and vertebral ends missing. The chicken ribs showed some roughening, but were not as damaged as the squirrel and raccoon ribs. The deer ribs received most roughening on the sternal extremity and one showed a clear pair of incisor marks.

The humeri were variously damaged depending on the species. Chicken humeri showed almost no gnaw damage, with only a few abrasions on the proximal and distal ends. Similarly, deer humeri showed little damage, with only a few isolated patches of gnawing on the proximal end, typically near the symphysis. The most interesting damage was done to the raccoon humeri, where the proximal symphysis was gnawed so extensively that the bone was nearly bitten in two (Figure 2). This damage occurred within the first twenty-four hours and for many the proximal end was missing by the time of the forty-eight hour removal. On the proximal end of the humeral shaft a few incisor marks were found. For five of the seven squirrel humeri the proximal end was missing.

The raccoon innominates were the most damaged of all the innominates. On several, the ilia were severely damaged; most were reduced approximately fifty percent (Figure 3). There was roughening on the ischium, pubis, the fusion between the ischium and the ilium, and around the acetabulum. The deer innominates were primarily damaged on the proximal end of the ilia, where clear incisor marks were found on four of the bones. There was also some gnawing on the pubic symphysis, however this resulted in relatively little damage, probably due to the size of the bone. The squirrel innominates showed some roughening on the proximal end of the ilia and on the edges of the ischium and pubis. The chicken synsacrum did not sustain much damage, with the ilium showing only a little damage on the distal end.

DISCUSSION

There was very clear patterning in the location of the gnawing modifications. The areas of recorded damage in this study are similar to those recorded in a study of the gnawing habits of the African porcupine (Brain 1981). Typically evidence of rodent gnawing was found on the projections and roughened areas of the bone; rarely was it found on the shafts. This is consistent with other research showing that mice gnaw on the densest bone (Dixon 1984; Lyman 1994). Most of the damage occurred on porous, cancellous bone, although these areas showed no parallel striations.

The patterning observed here is explained by the fact that rodents are limited to gnawing on the parts of bones that are smaller than the area between their incisors. In addition, the mice appear to require at least some sort of roughened patch, such as a symphysis or a process, to catch their teeth on in order to take a bite. We observed that on many occasions the mice would try to chew on various portions of the bone and would be unsuccessful because the bone lacked one of the above characteristics.

We had some concern about how the use of domesticated rodents would affect the outcome of the experiment. While domesticated and wild mice are morphologically similar, they do not necessarily behave similarly. In addition, there were concerns about the differences in the diets of wild and domesticated mice. These cannot be addressed here and require further research. It does seem likely, however, that wild mice would experience the same kinds of functional restraints that the domesticated mice would in regards to gnawing, and that these factors would determine where on the bone they could gnaw.

Wild rodents gnaw on bones for two reasons: to reduce the length of their incisors, and to obtain minerals (Schwartz and Schwartz 1981; Hoffman and Hays 1987; Andrews 1990).

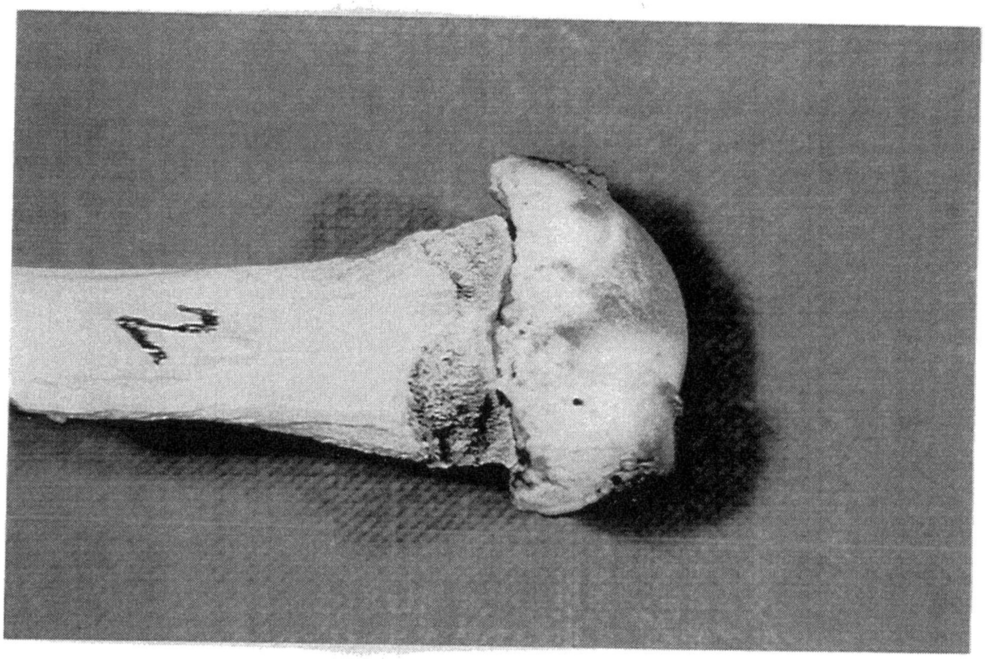

Figure 2. Raccoon proximal humerus gnawed by mice.

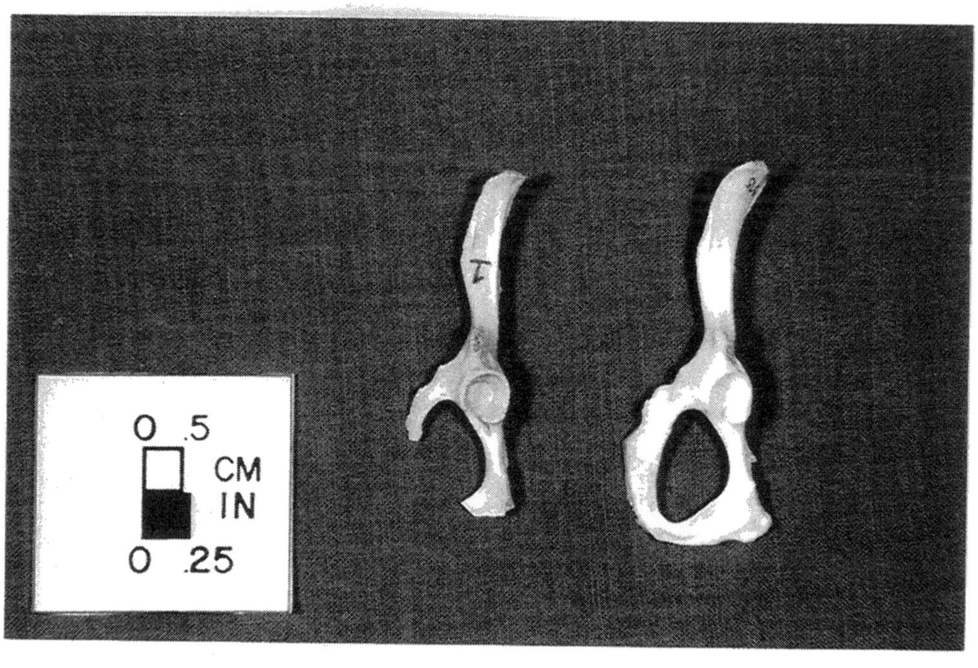

Figure 3. Rodent gnawed raccon innominate (left). Ungnawed innominate (right).

Rodents posses ever-growing incisors that require frequent reduction (Roze 1989), and regular attrition of the incisors is accomplished by gnawing hard surfaces. When rodents gnaw they draw the sides of their lips into the diastema between their incisors and premolars. This seals their mouths so that unwanted fragments are not ingested (Honney 1975; Morrell 1986; Roze 1989). In certain cases, however, the ingestion of gnawed particles is favorable to rodents, who obtain salt (Roze 1989), calcium, and phosphates (Andrews 1990) in this manner. The mice used in this experiment were being fed specially formulated mouse feed and it seems unlikely that they needed to obtain salt, calcium, or phosphates from bone gnawing. Instead they probably gnawed on the bones out of boredom and/or to reduce the length of their incisors.

The concentration of gnawing on the roughened areas or projections of bone has several noteworthy consequences. This focus tends to obliterate many of the characteristics used to identify bone fragments to species. In addition it means that a broken or sawed bone might suffer more extensive damage than a whole bone because it offers a rough surface that is easy for rodents to gnaw. In this way the record may be skewed toward the preservation of whole bones. As a result, numbers of butchery marks and numbers of butchered specimens will be lower than the actual numbers (Grayson 1988).

Perhaps an even more serious problem is the scarcity of the characteristic parallel striations so commonly used as the only indicator of rodent gnawing. In this study a large percentage of the bone damage was in the form of roughened edges or missing pieces, damage that could easily be attributed to other taphonomic processes. These roughened patches were found in conjunction with clear parallel striations on only 11.4% of the grawed sites. (Figure 4). On the remaining 88.7% there are no indicators that the damage was caused by rodents.

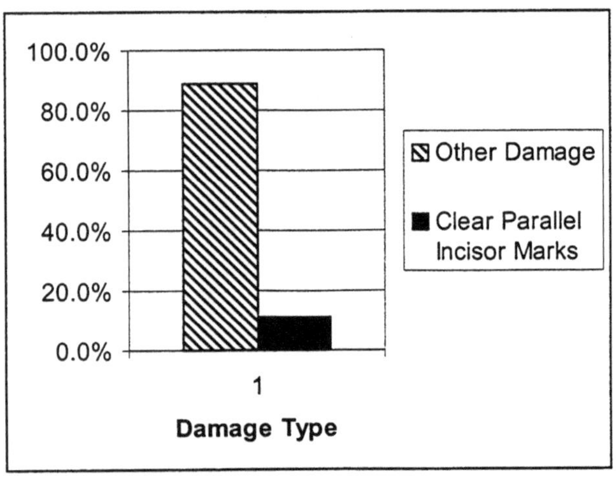

Figure 4. Damage summary.

Further, rodents may act on an assemblage and leave no marks. On the humeri, especially for deer and chicken, we noticed the mice mouthed them and moved them, yet there were no gnawing marks evident on any part of the bone. Similar 'non-damaging' gnawing was also seen in our observations of the other bones. It is possible that bones may be moved significant distances without any evidence for rodent presence.

The difference in bone size may be a factor of gnawing choice. Small rodents like mice may bias samples by deleting small bones, while large rodents bias samples in the opposite way. Research on African porcupines has shown that they have a preference for elements in the range of 1 to 30 grams (Hoffman and Hays 1987).

We had also wanted to determine if there would be a difference in gnawing on mammal and bird bones. In this study there was little difference in the type of damage, but there was a noticeable difference in the amount of damage. Typically the bird bones received much less damage than an equivalent mammal element.

Many aspects of rodent gnawing need further research. As of yet, the extent to which rodents may affect a faunal sample is not clear. The fact that a large percentage of rodent damage goes unnoticed due to the difficulty in identification reinforces our belief that rodents have a profound effect on the archaeological record.

ACKNOWLEDGEMENTS

We would like to thank Fiona Marshall for her help and guidance in structuring and carrying out this experiment. Thanks also to the Washington University Chapter of Sigma Xi for their Grant-in-Aid-of Undergraduate-Research. This research could not have been accomplished without the generous help of Darrell Fee in obtaining animal specimens. We would also like to thank Pat Watson for her help in reading several drafts of this paper, and our reviewers whose comments were of great help.

REFERENCES CITED

Andrews, P.
1990 *Owls, Caves and Fossils*. University of Chicago Press, Chicago.

Binford, L. R.
1981 *Bones: Ancient Men and Modern Myths*. Academic Press, New York.

Blumenschine, R. J. and C. W. Marean
1993 A Carnivore's View of Archaeological Assemblages. In *From Bones to Behavior: Ethnoarchaeological and Experimental Contributions to the Interpretation of Faunal Remains*, edited by Jean Hudson, pp. 273-300. Center for Archaeological Investigations, Occasional Paper No. 21. Board of Trustees, Southern Illinois University.

Bonnichsen, R. and R. T. Will
1990 Cultural Modification of Bone: The Experimental Approach in Faunal Analysis. In *Mammalian Osteology* by

B. Miles Gilbert, pp. 7-30. Missouri Archaeological Society, Columbia MO.

Brain, C.K.
1980 Some Criteria for the Recognition of Bone Collecting Agencies in African Caves. In *Fossils in the Making*, edited by Anna K. Behrensmeyer and Andrew P. Hill, pp. 107-130.

1981 *The Hunters or the Hunted? An Introduction to African Cave Taphonomy*. The University of Chicago Press, Chicago and London.

Brothwell, D.
1976 Further Evidence of Bone Chewing by Ungulates: the Sheep of North Ronaldsay, Orkney. *Journal of Archaeological Science* 3:179-182.

Burt, W. H. and R. P. Grossenheider
1976 *A Field Guide to the Mammals: North America North of Mexico*. Houghton Mifflin Company, Boston.

Colburn, M. L.
1985 Faunal Remains from the Campbell Hollow Archaic Occupations. In *The Campbell Hollow Archaic Occupations: A Study in Intrasite Spatial Structure in the Lower Illinois Valley*, edited by C. Russel Stafford, pp. 108-120. Kampsville Archeological Center, Research Series Volume 4. Center for American Archeology.

Dixon, E. J.
1984 Context and Environment in Taphonomic Analysis: Examples from Alaska's Porcupine River Caves. *Quaternary Research* 22:201-215

Finley, R.B.
1985 *The Wood Rats of Colorado: Distribution and Ecology*. University of Kansas Publications, Museum of Natural History 10(6):213-552.

Fiorillo, A.R.
1989 An Experimental Study of Trampling: Implications for the Fossil Record. In *Bone Modification*, edited by Bonnichsen and Marcella H. Sorg, pp. 61-71. Center for the Study of the First Americans.

Grayson, D. K.
1988 Danger Cave, Last Supper Cave, and Hanging Rock Shelter: The Faunas. *Anthropological Papers of the American Museum of Natural History* 66(1):1-130.

Heizer, R.F. and R.A. Brooks
1965 Lewisville-Ancient campsite or Woodrat Houses? *Southwest Journal of Anthropology* 21:155-165.

Hesse, B. and P. Wapnish
1985 *Animal Bone Archaeology: From Objectives to Analysis*. Taraxacum, Washington, D.C.

Hockett, B. S.
1989 The Concept of "Carrying Range": A Method in Determining the Role Played by Woodrats in Contributing Bones to Archaeological Sites. *Nevada Archaeologist* 7:28-35.

Hoffman, R. and C. Hayes.
1987 The Eastern Wood Rat (*Neotoma floridana*) as a Taphonomic Factor in Archaeological Sites. *Journal of Archaeological Science* 14:325-337.

Honney, P. W.
1975 *Rodents: Their Lives and Habits*. Taplinger, New York.

Hudson, J.
1993 The Impacts of Domestic Dogs on Bone in Forager Camps; or, The Dog-Gone Bones. In *From Bones to Behavior: Ethnoarchaeological and Experimental Contributions to the Interpretation of Faunal Remains,* edited by Jean Hudson, pp. 301-323. Center for Archaeological Investigations, Occasional Paper No. 21. Board of Trustees, Southern Illinois University.

Lyman, R. L.
1994 *Vertebrate Taphonomy*. Cambridge University Press, Great Britain.

Marshall, L. G.
1989 Bone Modification and "The Laws of Burial". In *Bone Modification*, edited by Robson Bonnichsen and Marcella H. Sorg, PP. 7-24. Center for the Study of the First Americans.

Miller, G. J.
1975 A Study of Cuts, Grooves, and Other Marks on Recent and Fossil Bone: II Weathering Cracks, Fractures, Splinters and other Similar Natural Phenomena. In *Lithic Technology: Making and Using Stone Tools*, edited by E.H. Swanson, Jr., pp. 211-228. Mouton Publishers, distributed by Aldine, Chicago.

Morrell, A.
1986 *Beaver Behavior: Architect of Fame and Bane*. Naturegraph, CA.

Price, C. R.
1985 Patterns of Cultural Behavior and Intrasite Distributions of Faunal Remains at the Widow Harris Site. *Historical Archaeology* 19(2): 40-56.

Rainey, D.G.
1956 *Eastern Woodrat, Neotoma floridana: Life History and Ecology*. University of Kansas Publications, Museum of natural History 8(10):535-646.

Roze, U.
1989 *The North American Porcupine*. Smithsonian Institution Press, Washington, D.C., and London.

Schick, K. K., N. Tooth, and E. Daeschler
1989 An Early Paleontological Assemblage as an Archaeological Test Case. In *Bone Modification*, edited by Robson Bonnichsen and Marcella H. Sorg, pp. 121-137. Center for the Study of the First Americans.

Schwartz, C. W. and E. R. Schwartz
1981 *The Wild Mammals of Missouri, Revised Edition.* University of Missouri Press and Missouri Department of Conservation. Columbia and London.

Shipman, P.
1981 Applications of Scanning Electron Microscopy to Taphonomic Problems. In *The Research Potential of Anthropological Museum Collections,* pp. 357-385. Annals of the New York Academy of Science, 376.

Singer, R.
1956 The "Bone Tools" from Hopefield. *American Anthropologist* 58:1127-1134.

Tchernov, E.
1991 Biological Evidence for Human Sedentism in Southwest Asia during the Natufian. In *The Natufian Culture in the Levant*, edited by Ofer Bar-Yosef and Francois R. Valla, pp. 315-340. International Monographs in Prehistory.

Vaughan, T. A.
1990 Ecology of Living Packrats. In *Packrat Middens: The Last 40,000 Years of Biotic Change*, edited by Julio L. Betancourt, Thomas R. Van Devender, and Paul S. Martin, pp. 14-27. University of Arizona Press, Tucson.

www.ingramcontent.com/pod-product-compliance
Lightning Source LLC
LaVergne TN
LVHW071204210926
848482LV00015B/722